Table of Atomic Weights

Element	Symbol	Atomic number	Atomic weight	Element	Symbol	Atomic number	Atomic weight
Actinium	Ac	89	r	Mercury	Hg	80	200.59
Aluminum	Al	13	26.982	Molybdenum	Mo	42	95.94
Americium	Am	95	r	Neodymium	Nd	60	144.24
Antimony	Sb	51	121.76	Neon	Ne	10	20.180
Argon	Ar	18	39.948	Neptunium	Np	93	r
Arsenic	As	33	74.922	Nickel	Ni	28	58.693
Astatine	At	85	r	Niobium	Nb	41	92.906
Barium	Ba	56	137.33	Nitrogen	N	7	14.007
Berkelium	Bk	97	r	Osmium	Os	76	190.23
Beryllium	Be	4	9.0122	Oxygen	O	8	15.999
Bismuth	Bi	83	208.98	Palladium	Pd	46	106.42
Boron	B	5	10.811	Phosphorus	P	15	30.974
Bromine	Br	35	79.904	Platinum	Pt	78	195.08
Cadmium	Cd	48	112.41	Plutonium	Pu	94	r
Calcium	Ca	20	40.078	Polonium	Po	84	r
Californium	Cf	98	r	Potassium	K	19	39.098
Carbon	C	6	12.011	Praseodymium	Pr	59	140.91
Cerium	Ce	58	140.12	Promethium	Pm	61	r
Cesium	Cs	55	132.91	Protactinium	Pa	91	231.04
Chlorine	Cl	17	35.453	Radium	Ra	88	r
Chromium	Cr	24	51.996	Radon	Rn	86	r
Cobalt	Co	27	58.933	Rhenium	Re	75	186.21
Copper	Cu	29	63.546	Rhodium	Rh	45	102.91
Curium	Cm	96	r	Rubidium	Rb	37	85.468
Dysprosium	Dy	66	162.50	Ruthenium	Ru	44	101.07
Einsteinium	Es	99	r	Samarium	Sm	62	150.36
Erbium	Er	68	167.26	Scandium	Sc	21	44.956
Europium	Eu	63	151.96	Selenium	Se	34	78.96
Fermium	Fm	100	r	Silicon	Si	14	28.086
Fluorine	F	9	18.998	Silver	Ag	47	107.87
Francium	Fr	87	r	Sodium	Na	11	22.990
Gadolinium	Gd	64	157.25	Strontium	Sr	38	87.62
Gallium	Ga	31	69.723	Sulfur	S	16	32.066
Germanium	Ge	32	72.61	Tantalum	Ta	73	180.95
Gold	Au	79	196.97	Technetium	Tc	43	r
Hafnium	Hf	72	178.49	Tellurium	Te	52	127.60
Helium	He	2	4.0026	Terbium	Tb	65	158.93
Holmium	Ho	67	164.93	Thallium	Tl	81	204.38
Hydrogen	H	1	1.0079	Thorium	Th	90	232.04
Indium	In	49	114.82	Thulium	Tm	69	168.93
Iodine	I	53	126.90	Tin	Sn	50	118.71
Iridium	Ir	77	192.22	Titanium	Ti	22	47.867
Iron	Fe	26	55.845	Tungsten	W	74	183.84
Krypton	Kr	36	38.80	Uranium	U	92	238.03
Lanthanum	La	57	138.91	Vanadium	V	23	50.942
Lead	Pb	82	207.2	Xenon	Xe	54	131.29
Lithium	Li	3	6.941	Ytterbium	Yb	70	173.04
Lutetium	Lu	71	174.97	Yttrium	Y	39	88.906
Magnesium	Mg	12	24.305	Zinc	Zn	30	65.39
Manganese	Mn	25	54.938	Zirconium	Zr	40	91.224

r = radioactive, no stable isotopes. Atomic weights from *CRC Handbook of Chemistry and Physics*, 81st ed. (2000).

Principles of Environmental Geochemistry

G. Nelson Eby

University of Massachusetts, Lowell

BROOKS/COLE
CENGAGE Learning

Australia • Brazil • Japan • Korea • Mexico • Singapore • Spain • United Kingdom • United States

BROOKS/COLE
CENGAGE Learning™

**Principles of Environmental
Geochemistry**
G. Nelson Eby

Editor: Keith Dodson

Assistant Editor: Carol Ann Benedict

Technology Project Manager: Samuel
Subity

Marketing Manager: Kelley McAllister

Marketing Assistant: Mona Weltmer

Project Manager, Editorial Production:
Tom Novack

Print/Media Buyer: Kristine Waller

Permissions Editor: Karyn Morrison

Production Service: Rockwell Production
Services

Text Designer: John Edeen

Copy Editor: Pamela Rockwell

Illustrator: Precision Graphics

Cover Designer: Roger Knox

Cover Image: Paul Chesley/Getty Images

Compositor: Better Graphics

For product information and technology assistance, contact us at
Cengage Learning Customer & Sales Support, 1-800-354-9706

For permission to use material from this text or product, submit all
requests online at **www.cengage.com/permissions**
Further permissions questions can be emailed to
permissionrequest@cengage.com

Library of Congress Control Number: 2002033358
ISBN-13: 978-0-12-229061-9
ISBN-10: 0-12-229061-5

Brooks/Cole
10 Davis Drive
Belmont, CA 94002-3098
USA

Cengage Learning is a leading provider of customized learning
solutions with office locations around the globe, including Singapore,
the United Kingdom, Australia, Mexico, Brazil, and Japan. Locate your
local office at: **international.cengage.com/region**

Cengage Learning products are represented in Canada by Nelson
Education, Ltd.

To learn more about Brooks/Cole, visit
www.cengage.com/brookscole

Purchase any of our products at your local college store or at our
preferred online store **www.ichapters.com**

Printed in the United States of America
6 7 8 11 10

For Jennifer, Stephanie, and Susan

May a scientific understanding of the environment lead to rational decisions.

Contents

Preface   xi

1 Basic Principles   1

The Atom   1
  The Bohr Atom   1
  Beyond the Bohr Atom—Quantum Mechanics   3
  Ionization and Valences   6
  Spectra and Elemental Analysis   7
  The Nucleus   8
Chemical Bonding   8
Atomic and Molecular Weights   9
Measurement of Concentration   10
Types of Chemical Reactions   12
Determining and Balancing a Chemical
  Equation   13
Gases   15
Structure and Properties of Water   18
Box Models and Geochemical Cycles   20
Questions and Problems   23

2 Equilibrium Thermodynamics and Kinetics   27

The Laws of Thermodynamics   27
  First Law of Thermodynamics   28
  Second Law of Thermodynamics   28
Equilibrium Thermodynamics   29
  Free Energy   30
  Chemical Potential   30
  Activity and Fugacity   30
  The Equilibrium Constant   31
  Henry's Law   32
  Free Energies at Temperatures Other than
    25°C   33
  Le Châtelier's Principle   35
    Changes in Concentration   35
    Changes in Pressure   35
    Changes in Temperature   35
Calculation of Activity Coefficients   36
  Debye–Hückel Model   36
  Truesdell–Jones Model   37
  Pitzer Model   39
  Why Do We Care about Activity-Coefficient
    Models?   39
  Calculation of Activity Coefficients for Uncharged
    Species   39
  Activity of Water   40
Aqueous Complexes   40

Measurement of Disequilbrium   41
Kinetics   42
  Order of Reactions   42
  The Arrhenius Equation   45
  Nucleation   45
  Dissolution and Growth   46
Water-Chemistry Computer Models   47
Case Studies   48
Questions and Problems   51

3 Acid–Base Equilibria   59

Definition of Acids and Bases   59
  Acids   59
  Bases   60
The Dissociation of Water and pH   61
pH of Natural Waters   62
  Strong Acids   62
  Weak Acids   63
    Acetic Acid   63
    Carbonic Acid   63
    Silicic Acid   69
    Carbonic Acid–Carbonate System   70
  Salts of Weak Acids and Strong Bases   71
  Strong Bases   75
Amphoteric Hydroxides   75
Acidity and Alkalinity   75
Acidity and Alkalinity Titrations   77
Buffers   80
  Buffering Index for H_2O 82
  Buffering Index for a Weak Monoprotic Acid   83
  Buffering Index for Weak Polyprotic Acids   84
Mineral–Water Reactions   84
  Buffering Index for the Calcite–Carbonic Acid
    System   85
  Buffering Index for Water–Silicate Mineral
    Systems   86
  Buffering Capacities of Natural Waters   87
  Mineral Reactivity   87
Acid–Base Equilibria Case Studies   88
Questions and Problems   91

4 Oxidation–Reduction Reactions   94

Basic Principles   94
  Electrochemical Cells   94
  Electromotive Series   95
  Electromotive Force   95

Balancing Oxidation–Reduction Equations   97
The Nernst Equation and Eh   98
Oxidation–Reduction Reactions and p*e*   100
Oxidation–Reduction Diagrams   101
Stability of Water   101
Constructing Eh–pH Diagrams   103
The Iron Eh–pH Diagram   103
The Nitrogen Eh–pH Diagram   108
The Role of Microorganisms in
Oxidation–Reduction Reactions   110
Microorganisms and Energy Pathways   111
Examples of Oxidation–Reduction Reactions
Mediated by Microorganisms   112
Oxidation–Reduction Processes in Natural
Systems   114
Redox Buffering   116
Classification of Oxidation–Reduction
Environments   118
Oxidative and Reductive Capacity   118
Oxidation–Reduction Ladders   119
The Redox Interface   121
Questions and Problems   124

5   Carbon Chemistry   129

Basic Structures and Naming of Organic
Compounds   129
Formulas   129
Isomers   130
Hydrocarbons   130
Naming and Types of Hydrocarbon
Compounds   131
Alkanes   131
Alkenes and Alkynes   132
Cyclic Hydrocarbons   133
Aromatic (Aryl) Hydrocarbons   133
*Polycyclic Aromatic Hydrocarbons
(PAH)   133*
Functional Groups   134
Organohalide Compounds   134
Organooxygen Compounds   136
Organonitrogen Compounds   136
Organosulfur Compounds   137
Organophosphorus Compounds   138
Esters   138
Polymers   139
Carbon Compounds in the Environment   139
Natural Sources   139
Anthropogenic Sources   140
Humic Substances   141
Soil and Marine Organic Matter   143
Soil Organic Matter   143
Marine Organic Matter   144
Fossil Fuels   146
Coals   146
Petroleum   147

Natural Carbon Inputs to Surface and Ground
Waters   148
Anthropogenic Carbon Inputs to Surface and
Ground Waters   150
Organic Carbon   150
Pesticides   150
Dioxins, PCBs, and Dibenzofurans (DFs)   152
PAHs   152
NAPLs   153
Geochemical and Biochemical Processes That
Control the Concentration of Organic
Chemicals in Water   156
Sorption/Partition   156
Precipitation   158
Volatilization   158
Oxidation–Reduction   159
Biological Processes   159
Degradation Half-life   159
Questions and Problems   161

6   Isotopes   165

Radioactive Isotopes   165
Basic Principles   165
Radioactive Decay and Growth   167
Measurement of Radioactivity   169
*Radioactive Isotopes Used in Environmental
Studies   169*
Tritium Dating   169
Carbon-14 Dating   171
U-series Disequilibrium Methods of Dating   174
^{230}Th *Dating of Marine Sediments   174*
$^{230}Th/^{232}Th$ *Dating of Marine Sediments   174*
$^{230}Th/^{231}Pa$ *Dating of Marine Sediments   176*
*Activity and Sedimentation-Rate
Relationships   176*
$^{230}Th/^{234}U$ *Dating of Calcium Carbonate   177*
*Why Do We Want to Date Marine Sediments
and Carbonate Minerals?   177*
^{210}Pb *Dating   177*
Radiogenic Isotopic Tracers   178
Rb-Sr System   178
Th-U-Pb Isotopic Systems   181
Stable Isotopes   181
Basic Principles   182
Stable Isotope Fractionation   182
Fractionation Factor   183
The δ (Delta) Notation   184
Oxygen and Hydrogen Isotopes in Water   185
*Factors Affecting the Isotopic Composition
of Water   185*
Climate Change   188
Carbon   189
DIC in Aqueous Systems   189
Sources of Methane   191
DOC in Aqueous Systems   191

Tracers in Food Chains 191
Nitrogen 193
Nitrates in Surface and Ground Waters 193
NH_4^+ *and* NO_3^- *in Rain* 193
Sulfur 194
Mixing 196
Binary Isotopic Mixing 196
Multi-End-Member Mixing of Chemical Species 196
Paleothermometry 197
Questions and Problems 198

7 Environmental Mineralogy 207

Basic Mineralogy 207
Definition of a Mineral 207
Types of Minerals 208
Crystal Chemistry 208
Types of Bonding 208
Coordination Numbers 210
Ionic Substitutions 211
X-ray Crystallography 212
Basic Silicate Structures 214
Clay Minerals 215
Clay Mineral Structures 217
1:1 Clays 217
2:1 Clays 217
Ion-Exchange Properties—Origin 219
Surface Charge 220
Surface Area 220
Ion-Exchange Properties—Determination 221
Batch Method 221
Adsorption Isotherms 221
Column Test Method 223
Zeolites 225
The Asbestos Minerals 227
Crystal Structures of Asbestos Minerals 228
Types of Asbestos and Their Uses 229
Health Effects of Asbestos Exposure 229
Dissolution of Fibers 229
Chemistry of Exposed Fiber Surface 230
Surface Charge 230
Biological Interactions 230
Determining the Effect of Asbestos Exposure 231
Crystalline and Amorphous Silica 233
Dissolution of Silica Minerals 233
Health Effects of Silica Exposure 235
Mineral–Microorganism Interactions 235
Questions and Problems 238

8 The Atmospheric Environment 243

Properties of the Earth's Atmosphere 243
Structure and Composition of the Atmosphere 243
Solar and Terrestrial Radiation 245

Atmospheric Circulation 247
Temperature and Water Vapor Variations 248
Air Pollution 251
Types of Air Pollutants 251
Aerosols 252
Smogs 253
Greenhouse Gases and Climate Change 253
Greenhouse Gases 255
Carbon Dioxide (CO_2) 257
Methane (CH_4) 259
Nitrous Oxide (N_2O) 262
Climate Change and the Geologic Record 262
Ice Cores 262
Sediment Record 263
Ozone 268
Stratospheric Ozone 268
Ozone-Destroying Reactions 271
Kinetic Considerations 272
The Antarctic Ozone Hole 273
Tropospheric Ozone 273
Radon 275
Rainwater Chemistry 278
Ionic Species 278
Gaseous Species 281
National Atmospheric Deposition Program (NADP) 283
pH of Rainwater and Acid Rain 283
Chemistry and Sources of Atmospheric Particulates (Aerosols) 288
Types of Particulates 288
Mineral Dust 288
Sea Salt 288
Sulfates 288
Carbonaceous Particles 289
Removal of Particulates from the Atmosphere 289
Dry Deposition 289
Wet Deposition 289
Particulate Sampling 290
Air Mass Trajectories 290
Source Identification and Apportionment 291
Enrichment Factor 291
Elemental Signatures 292
Molecular Signatures 294
Isotopic Signatures—Pb 296
Isotopic Signatures—C 299
Temporal Variations in Particulate Compositions 300
Questions and Problems 301

9 The Continental Environment 313

The Hydrologic Cycle 313
Weathering 314
Dissolution and Precipitation of Silica 317
Dissolution and Precipitation of Aluminum and

Iron Hydroxides   318
Stability Diagrams   320
Geochemistry of Surface and Ground
    Waters   324
    Variations in River and Groundwater
        Chemistry   324
        *Graphical Representations of Water
            Chemistry   325*
        Hydrofacies   326
    Rivers   328
    Groundwaters   331
    Lakes   334
        Structure and Mixing of Lake Waters   335
        Lake (Box) Models   336
    pH of Surface and Ground Waters   339
        Carbonic Acid System   339
        Water–Mineral Interactions   340
Adsorption–Desorption Processes   341
    Adsorption of Metal Cations   342
    Colloids   343
    Isotherms and Partitioning of Species   344
    Ion Exchange   345
    Adsorption and Contaminant Transport   347
Complex Formation and Chelation   349
Metals   350
    Metal Cycles   350
        Natural   350
        Anthropogenic   352
    Acid Drainage   352
    Heavy Metals   354
        Transition metals, Zn, Cd, and Pb   354
        Selective Extraction   355
        Arsenic and Selenium   356
        Mercury   357
        Metal Variations in Space and Time   358
        *Metal Contamination of Soils and
            Sediments   358*
    Radioactive Isotopes and Radioactive Waste
        Disposal   358
        Nuclear Fuel Cycle   359
        Nuclear Fission   360
        *Radioactive Wastes due to Nuclear
            Fission   360*
        *Mobility and Sequestering of Radioactive
            Isotopes in the Surface Environment   362*
        The Oklo Fossil Fission Reactors   363
Nonmetals   364
    Carbon   364
    Halogens   365
        Fluorine   365
        Cl^- and Br^- 366
    Nitrogen   368
    Phosphorus   369
    Sulfur   371
Questions and Problems   372

10   The Marine Environment   387

Physical Aspects of the Open Ocean   387
    Vertical Structure of the Ocean   388
    Oceanic Circulation   388
        Surface Currents   388
        Upwelling and Downwelling   389
        Thermohaline Circulation   390
    Constancy of Composition and Salinity   392
Chemical Composition of the Open
    Ocean   393
    Residence Time of Seawater Species   393
    The Global Chemical Cycle for Seawater
        Species   394
    Processes Controlling Seawater
        Composition   395
        *Biological Controls on Seawater
            Composition   395*
        Seawater–Sediment Interactions   398
        Seawater–Basalt Interactions   398
        *Summary of Processes Controlling Seawater
            Composition   400*
Seawater Chemistry   400
    Seawater Alkalinity and Dissolved Inorganic
        Carbon (DIC)   401
    Seawater pH   403
    Speciation and Stoichiometric Equilibrium
        Constants   404
    Seawater Buffers and Buffering Capacity   406
    Inorganic and Organic Ligands in
        Seawater   407
        Inorganic Ligands   408
        Organic Ligands   409
    Trace Metals in Marine Colloids   410
    Trace Metals in Seawater   412
        Sources of Trace Metals   412
        *Removal of Trace Metals from the
            Ocean   413*
    Types of Metal Distributions   414
        Conservative-Type Distributions   414
        Nutrient-Type Distributions   415
        Surface-Water Enrichments   415
        Mid-Depth Maxima and Minima   415
    Determination of Scavenging Rates   415
        Sediment Traps   415
        Radioactivity Measurements   416
Geochemistry of Marine Sediments   416
    Distribution of Deep-Sea Sediments   418
        Siliceous Sedimemts   419
        Calcareous Sediments   419
    Authigenesis in Marine Sediments   421
    Age of Seafloor Sediments   421
    Methane Hydrates   422
The Marginal Marine Environment   423
    Estuaries   424

Water Circulation in Semi-Isolated
  Basins   424
  *Evaporation Greater than
    Precipitation   425*
  *Precipitation Greater than
    Evaporation   425*
Flushing Time   426
Water Column Chemistry   428
  Equilbrium Constants   428
  Adsorption–Desorption   428
  Redox Reactions   430
  Suspended Particulate Matter (SPM)   430
Pore Water Chemistry   432
  *Release of Metals to Pore Waters and the
    Water Column   432*
  Transport of Metals in Pore Waters   432
  Removal of Metals from Pore Waters   433
Sediment Chemistry   434
  Sediment Age and Accumulation Rate   434
  Redox Indicators   434

  Metal Indicators   434
  Bioindicators   436
  Organic Contaminants   436
Questions and Problems   437

References   459

**Appendix I. Electronic Structure and Valences of
the Elements   472**

**Appendix II. Standard State (298.15 K, 10^5 Pa)
Thermodynamic Data   474**

**Appendix III. Ionic Radii in Å for Different
Charges and Coordination Numbers   490**

**Appendix IV. Piper Diagram for Water Chemistry
Problems   494**

Answers to Selected Problems   495

Index   497

Credits   511

Preface

Over the past several decades, environmental issues and problems have become an important area of study occupying scientists from a number of disciplines. This textbook is written for geoscience and environmental science students who are interested in the interactions between anthropogenic activities and the natural environment. In addition, the book is suitable for undergraduate chemistry and engineering (civil and chemical) students, and a number of students from these disciplines have profited from a course based on this book. *Environmental Geochemistry* involves an understanding of how natural systems work and the human impact on these systems. While the topical areas are the same as those found in many environmental chemistry courses, the approach is different in that both natural processes and anthropogenic activities are considered. Another fundamental difference is that the historical record is an important component of the geosciences. Thus, you will find that significant attention is given to topics such as paleoclimatology and the historical record of pollution as recorded in sediments. The approach also differs from that of low-temperature (or aqueous) geochemistry courses, which tend to emphasize the functions of the natural system with little emphasis on anthropogenic effects. Hence, this book attempts to bridge the gap between low-temperature geochemistry and environmental chemistry.

The textbook title, *Principles of Environmental Geochemistry*, was not a capricious choice. It is not the intent of the author to provide a broad discussion of environmental issues, but rather to develop the principles that can be applied to understanding and solving environmental problems. Throughout the book, even in the final three synthesis chapters, the emphasis is on processes and the identification and understanding of these processes. There are a number of excellent books and papers, and more appearing almost daily, that deal with the present state of the environment, levels of pollution, estimates of potential impacts, remediation techniques, and policy issues. It is not the intent of the author to cover these topics here.

The book is written for a one-semester course given at the junior-senior or first-year graduate student level, but it can also be used for a full-year course by adding supplemental material. It is anticipated that for any particular course the instructor will not want to cover all the material in the textbook. Thus, the book has a modular form. The first seven chapters develop many, but not all, of the concepts needed to deal with environmental problems. These chapters form the core of the book and, along with selected topics from Chapters 8–10, are intended to comprise an undergraduate course in environmental geochemistry. Chapter 1, "Basic Principles," reviews many of the ideas that would be developed in a university-level chemistry course and that are relevant to the material covered in the textbook. Also included in Chapter 1 is a discussion of steady-state models, which are widely used in the environmental sciences. Chapter 2, "Thermodynamics and Kinetics," Chapter 3, "Acid–Base Equilibria," and Chapter 4, "Oxidation–Reduction," form the core of the aqueous geochemistry portion of the textbook. It is anticipated that all instructors will cover these three chapters, and topics developed in these chapters are revisited in later sections of the book. Chapter 5, "Carbon Chemistry," develops the basic nomenclature for carbon compounds and looks at the interactions between various organic chemicals and the natural environment. This is an area often ignored, or only briefly treated, in aqueous geochemistry textbooks, but a very important topic because of the widespread release of various organic compounds to the environment. Chapter 6, "Isotopes," deals with both radiogenic and stable isotopes. The chapter develops the basic mathematical relationships for the various systems and shows how these systems are used to solve and/or understand environmental problems. This topic is rarely covered in environmental chemistry textbooks, but isotopic systems have come to play an important role in understanding environmental processes. Chapter 7, "Environmental Mineralogy," deals with the role of minerals in the environment. This is an important area as evidenced by a number of recent symposia, short courses, and compendiums. Ion adsorption is dealt with at some length in this chapter because the clay minerals play an important role in this process. This chapter also considers the environmental impact of various minerals. Depending on the emphasis of the course, the instructor may wish to skip some of the material covered in the first seven chapters. Chapter 1 is a review and much of this material can be left to the student. The last section on geochemical cycles should be addressed. Chapters 2, 3, and 4 are essential for the understanding of virtually everything that follows and should be covered. Chapters 5, 6, and 7 do not need to be covered in their entirety. However, the last three chapters do draw on ideas developed in all of the first seven chapters, so the instructor may want to cover selected topics in Chapters 5, 6, and 7. I have tried to structure the text in such a way that this approach is possible.

The last three chapters integrate material developed earlier in the context of a specific environment. Chapter 8, "The Atmospheric Environment," is typically covered in environmental chemistry books, but not in geochemistry textbooks. Atmospheric problems often have an important geological component and many geo- and environmental scientists can expect to deal with these types of problems during their professional careers. This is a long and eclectic chapter, ranging from photochemistry to paleoclimatology. In part, this is because the atmosphere is a topical area that may be least familiar to users of this book. Chapter 9, "The Continental Environment," starts with the processes that control the chemistry of natural waters and then looks at anthropogenic impacts. A number of new ideas are developed in this chapter, including stability diagrams and an extended discussion on ion-exchange processes and adsorption. Chapter 10, "The Marine Environment," is divided into two parts, the first dealing with the open ocean and the second with the generally more highly impacted marginal marine environment. Both sections start with an overview of physical processes, followed by a discussion of seawater chemistry and the processes that control seawater chemistry. As in Chapter 9, basic processes controlling ocean chemistry are developed first followed by more specific environmental issues. The last three chapters, along with selected topics from Chapters 1–7, are intended to form the core of a graduate-level environmental geochemistry course. For this reason, problems in these chapters are largely drawn from the literature and use real data and require analysis of the data to answer the problems.

It is anticipated that the student will have had at least a one-semester course of university-level chemistry, although most ideas are developed from first principles. A fundamental understanding is developed for each topical area, and it is then shown how this understanding is applied to environmental issues. Thus, in a certain sense the text can be regarded as an introductory geochemistry book using environmental examples. Case studies are interspersed throughout the text to show how ideas developed in the text are applied to real-world problems. Problem solving is emphasized and quantitative approaches to problems are embraced. I am a firm believer in the value of homework and the importance of problem solving in gaining an understanding of the material. Problems range from simple "plug-and-chug" to more complex problems that may require students to access the scientific literature. Not all problems have a unique solution. In a number of cases, problems are used to extend ideas developed in the text. The range of problem types gives the instructor flexibility in selecting problems appropriate for the level of her or his course.

I have tried to fully integrate the text, case studies, and problems. On a number of occasions students will need to refer to case studies or earlier chapters in the book in order to fully understand the material being presented or to do a particular problem. This approach is motivated by 30 years of teaching experience, which has made the point that students tend to compartmentalize learning, and is based on the hope that the student will come to appreciate the interrelated nature of the various topics. A good deal of time has been devoted to developing the problems in an attempt to make them interesting, informative, and entertaining. For problems based on the literature, I try to lead the student through the analytical approach used by the original authors in an attempt to provide some insight into real-world problem solving.

A word about computers and problem solving. I have deliberately not used any of the number of computer models that are available for solving water-chemistry problems (although I do reference these programs in the text) or, for that matter, other types of problems. I do, however, strongly encourage the use of spreadsheets, and many of the problems are easily solved using a spreadsheet program. There are two reasons for this approach. First, most scientists have their own favorite set of codes that they use for solving particular types of problems, many of which are available in the public domain but some of which are commercial (and expensive) programs. I do not want to presuppose a preference or availability of any particular code. Second, and most important, I believe that students should understand the problem. Too often, computer codes provide a shortcut to a solution that does not require an understanding of the process. For problems that are computationally demanding and/or for which repetitive calculations are required, spreadsheets are ideal. The student needs to understand the problem in order to write the appropriate equations but can then quickly and easily solve the problem numerically and/or graphically using the spreadsheet program.

A solutions manual is available. The manual comes on a CD-ROM and the text is in pdf format. In addition, the CD-ROM contains all the spreadsheets (Excel) used to solve the various problems in the textbook. Many of these spreadsheets also function as templates and can be used to solve similar problems.

This textbook has been in the "talking stage" since I first started teaching an environmental geochemistry course in the late 1980s. The project finally moved from the talking to the doing stage when I was an Erskine Fellow at the University of Canterbury, New Zealand, in 1998. I thank Steve Weaver, who provided the opportunity for a "quiet time" in which to start "the book." Students in a graduate seminar at the University of Canterbury were the first, albeit somewhat unwitting, test subjects. Since then, a number of my undergraduate classes have served as test subjects, and their comments and corrections were greatly appreciated. What often seems absolutely clear to the instructor is, when seen through the eyes of the student, total gobbledygook. A number of colleagues took on the task of reading various parts of the manuscript. I would like to specifically thank George Chabot, Frank Colby, Susan Eby, Dan Golomb, James Hall, Dan Larsen, and Arnold O'Brien. Dan Larsen critically read the water-chemistry chapters (2–4), which form the core of the book. His comments were most helpful and constructive, although, of course, all remaining errors are my responsibility. The environmental mineralogy chapter was written at a time when a number of books on environmental mineralogy and related topics were scheduled to appear, but as is often the case were delayed or unavailable for a variety of reasons. Various members of the mineralogical community stepped into the gap, helping me acquire copies of the books (in one case, liberating a copy from a warehouse in Budapest) and providing prepublication copies of relevant chapters. I would like to particularly thank Herta Effenberger (European Mineralogical Union), Adrian Lloyd-Lawrence (Mineralogical Society), Robert Martin (Canadian Mineralogist), and Paul Ribbe (Mineralogical Society of America). On behalf of the publisher, Massimo Gasparon, David Ryan, and Johan Varekamp provided useful reviews on parts or all of the textbook. Last, but certainly not least, none of this would have been possible without the support of my wife, Susan, who tolerated being a "book-widow." The project couldn't have been accomplished without her patience and good humor.

G. Nelson Eby

Basic Principles

1

Certain basic principles are fundamental to our understanding of how natural systems work. In this chapter we will consider the structure of the atom and other principles that are necessary in order to describe the chemical and biological processes that can occur in the natural environment. Much of this material would be covered in an introductory university chemistry course. Hence, this chapter will serve as a review. In addition, concepts such as *steady state* and *box models*, widely used in describing interactions in the natural environment, but unlikely to have been covered in a basic chemistry course, will also be developed.

THE ATOM

We can envision two possible models of matter, one in which matter is continually divisible into smaller units (a continuous view of matter) and one in which there is a limit to the size of the units (a discontinuous view of matter). Which is correct was largely a philosophical issue until the early 1800s, when Dalton developed a testable model of matter composed of tiny, indestructible particles (atoms). The structure of atoms is fundamental to our understanding of the physical world.

Atoms were initially viewed as solid, impenetrable units, i.e., the smallest recognizable form of matter. With the discovery of the electron in the late 19th century, the plum-pudding model was proposed by Thomson. In this model the negatively charged electrons were seen to be embedded in a positively charged matrix—like raisins in a plum pudding. In the early 20th century, Rutherford conducted an experiment in which he fired a beam of alpha particles (helium nuclei) at a sheet of gold foil. Many of the alpha particles passed through the gold foil, as expected, but some bounced back at high angles. Rutherford interpreted these results as follows. Most of the atom must consist of empty space; the alpha particles passed through unimpeded. But somewhere in the atom there must be a concentration of mass that is responsible for the deflection of the alpha particles. Because the deflections were relatively rare, he concluded that the mass must be concentrated in a small volume, which he referred to as the nucleus. In this model the nucleus is seen to be positively charged, and it is circled by negatively charged electrons. However, in terms of classical physics the Rutherford atom is unstable because the electrons should quickly spiral into the nucleus; i.e., an atom would have a very short lifetime—clearly a problem because matter had been around for billions of years. This problem was solved by Bohr (the Bohr atom), who proposed that the electrons could only exist in stable orbits at certain discrete distances from the nucleus.

The Bohr Atom

The mathematical description of the Bohr atom is straightforward, and an understanding of the Bohr atom provides an insight into the origin of atomic spectra. We start with a single electron (charge $= -1$) revolving around a nucleus consisting of one proton

1

(charge $= +1$). For a stable atom, the electrostatic attraction between the electron and the proton must equal the centrifugal force. Thus,

$$k\frac{e^2}{r^2} = \frac{mv^2}{r} \qquad (1\text{--}1)$$

Electrostatic force = Centrifugal force

where k is a proportionality constant that depends on the units used for the electrostatic charge and the distance between the charges, e is the charge of the electron, r is the distance between the proton and the electron, m is the mass of the electron, and v is the velocity of the electron in its orbit. Bohr proposed that the electron could exist only at certain discrete distances from the nucleus; i.e., the position of the electron was quantized. The allowable distances were determined by restricting the angular momentum of the electron to multiples of $h/2\pi$, where h is Planck's constant ($h = 6.62607 \times 10^{-34}$ J s). This relationship can be written

$$mvr = \frac{nh}{2\pi} \qquad (1\text{--}2)$$

where n $= 1, 2, 3$, etc. and n is the first, or principal, quantum number. From equation 1–1,

$$mvr = k\frac{e^2}{v} \qquad (1\text{--}3)$$

Combining equations 1–2 and 1–3 gives

$$k\frac{e^2}{v} = \frac{nh}{2\pi} \qquad (1\text{--}4)$$

For any particular allowed orbital radius, the velocity of the electron can be determined by solving equation 1–2 for velocity.

$$v = \frac{nh}{2\pi mr} \qquad (1\text{--}5)$$

Substituting equation 1–5 into equation 1–4 and solving for r gives

$$r = \frac{n^2h^2}{4\pi^2 mke^2} \qquad (1\text{--}6)$$

The energy of the atom is the sum of the kinetic and potential energies. The kinetic energy is due to the revolution of the electron around the nucleus, and the potential energy is due to the electrostatic attraction between the positively charged nucleus and the negatively charged electron.

$$E = \underset{\substack{\text{kinetic}\\\text{energy}}}{\frac{1}{2}mv^2} - \underset{\substack{\text{potential}\\\text{energy}}}{k\frac{e^2}{r}} \qquad (1\text{--}7)$$

Substituting for r (from equation 1–6) gives

$$E = -\frac{2\pi^2 mk^2 e^4}{n^2 h^2} \qquad (1\text{--}8)$$

For the Bohr atom, the energy of each allowed orbital is determined solely by the principal quantum number. In atoms more complicated than the Bohr atom—i.e., atoms of atomic number greater than 1—the picture becomes more complex, as will be described in the next section.

   Before we leave the Bohr atom, let us consider what will happen if an electron moves from one orbit to another. If the radius of the electron orbit was zero, the electron would be at the nucleus, the principal quantum number would be zero, and the atom would have an infinite negative energy. As the electron moves away from the nucleus, the energy becomes less negative. Hence, if an electron moves from a lower to a higher orbit, it must gain energy, and if it moves from a higher to a lower orbit, it must lose energy. When an electron moves from a higher to a lower orbit, the energy is released as a particular wavelength of electromagnetic radiation (EM radiation). Einstein's famous photoelectric experiment (for which he was awarded the Nobel prize) demonstrated that in certain situations EM radiation behaved as a particle (called a *photon*). The relationship between the energy of the photon and the frequency of emitted EM radiation is

$$E = h\nu \tag{1-9}$$

where h is Planck's constant and ν is the frequency of the EM radiation. The velocity of electromagnetic waves is related to frequency (ν) and wavelength (λ) as follows:

$$v = \nu\lambda \tag{1-10}$$

Combining equations 1–9 and 1–10 to find the relationship between energy and wavelength gives

$$E = \frac{hc}{\lambda} \tag{1-11}$$

where c is the velocity of electromagnetic waves in a vacuum ($c = 3.0 \times 10^8$ m s^{-1}).

EXAMPLE 1–1 Calculate the energy released when an electron moves from the third allowed orbit to the second allowed orbit. The mass of the electron (m) = 9.109 × 10^{-31} kg, the charge of the electron (e) = 1.602 × 10^{-19} C, and k = 8.98742 × 10^9 N m^2 C^{-2}. For n = 2,

$$E = -\frac{(2)(\pi)^2(9.109 \times 10^{-31} \text{ kg})(8.98742 \times 10^9 \text{ N m}^2 \text{ C}^{-2})^2(1.602 \times 10^{-19} \text{ C})^4}{(2)^2(6.62607 \times 10^{-34} \text{ J s})^2}$$

$$= -5.4403 \times 10^{-19} \text{ J}$$

For n = 3, $E = -2.4179 \times 10^{-19}$ J. The energy released when the electron moves from orbit 3 to orbit 2 is

$$\Delta E = E_3 - E_2 = -2.4179 \times 10^{-19} \text{ J} - (-5.4403 \times 10^{-19} \text{ J}) = 3.0224 \times 10^{-19} \text{ J}$$

The wavelength of the emitted radiation is

$$\lambda = \frac{hc}{E} = \frac{(6.62607 \times 10^{-34} \text{ J s})(3.0 \times 10^8 \text{ m s}^{-1})}{3.0224 \times 10^{-19} \text{ J}} = 6.578 \times 10^{-7} \text{ m} = 658 \text{ nm}$$

It is known that under the proper conditions of excitation atomic hydrogen will emit EM radiation of particular wavelengths. These so-called lines (as seen in a spectrograph) can be grouped into series, which are named after the person who discovered the particular series. One of these series, called the Balmer series, has a line at λ = 656.3 nm, very close to the wavelength just calculated. The interpretation is that this spectral line is due to the movement of an electron from the third to the second orbit in hydrogen. The set of observed spectral lines is characteristic of atomic hydrogen.   ■

Beyond the Bohr Atom—Quantum Mechanics

Although the Bohr model was successful in predicting spectral observations for hydrogen atoms, it failed to adequately describe the spectral emission of atoms of higher atomic number. The **atomic number** of an atom is *the number of protons (positively charged particles) in the nucleus.* The number of protons in the nucleus determines the element (i.e., 1 proton, hydrogen; 2—helium; 3—lithium; etc.). This problem was addressed by Heisenberg and Schrödinger. It was concluded that the fundamental problem with the Bohr model was that it considered electrons to be particles whose position in space could be determined with certainty. Work by de Broglie had shown that particles can have wave properties. The relationship between the wavelength (λ) of particles, their mass (m), and velocity (v) is expressed by the de Broglie equation

$$\lambda = \frac{h}{mv} \tag{1–12}$$

EXAMPLE 1–2 Calculate the de Broglie wavelength for an electron in the first orbit (n = 1) of the Bohr atom.

From equation 1–6, $r = 5.29 \times 10^{-11}$ m. Substituting into equation 1–5 gives $v = 2.189 \times 10^6$ m s^{-1}. The de Broglie wavelength for the electron is

$$\lambda = \frac{h}{mv} = \frac{6.62607 \times 10^{-34} \text{ J s}}{(9.109 \times 10^{-31} \text{ kg})(2.189 \times 10^6 \text{ m s}^{-1})} = 3.32 \times 10^{-10} \text{ m}$$

This wavelength is similar to that of energetic X rays.   ∎

Schrödinger used the wave properties of the electron to develop the now well-known, if not necessarily well-understood, Schrödinger wave model of the atom. In this model the most probable position of the electron is at a node developed in a standing wave. Multiple nodes are possible, each representing a principal quantum number. In three dimensions, nodes may form along three principal axes, giving rise to three quantum numbers. A fourth quantum number deals with the spin of the electron. These quantum numbers are summarized in Table 1–1. The magnetic quantum number arises when the atom is subjected to a magnetic field. The spin quantum number indicates the direction of spin of an electron in a particular orbital. An orbital cannot contain more than two electrons, and the two electrons must spin in opposite directions (the ***Pauli exclusion principle***, which states that *no two electrons in an atom can have the same quantum numbers*). Four types of wave functions are possible, designated *s*, *p*, *d*, and *f*, which describe the orientation of the electron orbitals in space (Figure 1–1). Only the *s*, *p*, and *d* orbitals are shown in Figure 1–1. The *f* orbitals cannot be pictured in two dimensions. The relationship between the quantum numbers and the electron orbitals is given in Table 1–2.

Table 1–1   Summary of Quantum Numbers

Name	Symbol	Values
Principal	n	$1, 2, 3, \ldots, \infty$
Azimuthal	l	$n-1, n-2, n-3, \ldots, 0$
Magnetic	m	$0, \pm 1, \pm 2, \ldots, \pm(l-1), \pm l$
Spin	s	$\pm \frac{1}{2}$

Table 1–2   Relation Between Quantum Numbers and Electron Orbitals

n	l	m	Number of subshells	Number of orbitals	Designation
1	0	0	1	1	1*s*
2	0	0	2	1	2*s*
	1	$-1, 0, 1$		3	2*p*
3	0	0	3	1	3*s*
	1	$-1, 0, 1$		3	3*p*
	2	$-2, -1, 0, 1, 2$		5	3*d*
4	0	0	4	1	4*s*
	1	$-1, 0, -1$		3	4*p*
	2	$-2, -1, 0, 1, 2$		5	4*d*
	3	$-3, -2, -1, 0, 1, 2, 3$		7	4*f*

Figure 1–1
Shape of various electron orbitals.
From Brownlow (1996).

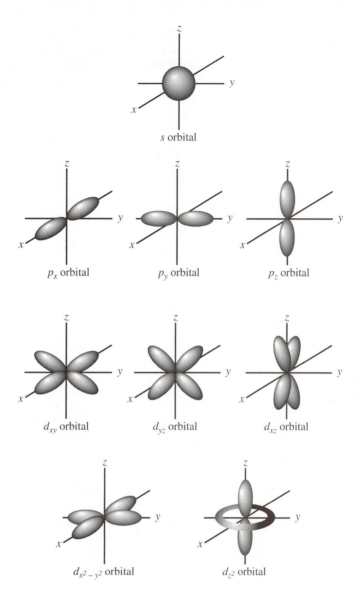

It is important to note that these orbital shapes actually represent probability distributions for the location of the electron. The ***Heisenberg uncertainty principle*** tells us that *it is impossible to simultaneously measure both the position and velocity of a particle*. The very act of measurement will disturb the particle. Hence, the position of a particle, in this case an electron, is only approximately known. Thus, the orbitals shown in Figure 1–1 represent the most probable location of an electron at any particular time. All distances are possible, but some are less probable. What this means is that there are two endpoints for the position of the electron—the nucleus and infinity—both of which have zero probability. At any instant the electron can occupy a position between these two end members, but its most likely position is at a distance defined by the allowable electron orbitals.

By applying the principles of quantum mechanics we can now build the various elements of the periodic table. We do this by adding electrons and protons in equal numbers to the basic hydrogen atom. Remember that the element is determined by the nucleus, i.e., the number of protons in the nucleus. The addition of electrons conforms to the following rules:

1. An electron will enter the available orbital with the lowest energy. The overall energy of the atom is minimized.

2. For each set of orbitals (s, p, d, f) the electrons will first be added singly to each available orbital. After all the orbitals in a set have a single electron, subsequent electrons can enter these orbitals if they have the opposite spin.

3. Atoms attain their maximum stability when the available orbitals are either completely filled, half-filled, or empty.

Because of the electrostatic attractions between the positively charged nucleus and the negatively charged electrons, the filling of the orbitals does not proceed in simple numerical order. For example, the $4s$ orbital is at a lower energy than the $3d$ orbital. Hence, electrons are first added to the $4s$ orbital and subsequent electrons are then added to the $3d$ orbital. This sequence of additions is responsible for the transition series metals in the periodic table. The relative energies of the various orbitals, as a function of atomic number, are shown in Figure 1–2.

The sequence in which the subshells are filled is given in Appendix I. In the periodic table the elements are grouped according to their electronic structure. Row 1 represents the filling of the $1s$ orbital, row 2 the filling of the $2s$ and $2p$ orbitals, and row 3 the filling of the $3s$ and $3p$ orbitals. Row 4 starts with the filling of the $4s$ orbitals followed by the filling of the $3d$ orbitals. This occurs because the $4s$ orbital

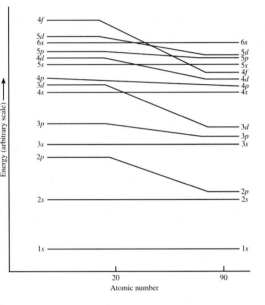

Figure 1–2
Variation of the energy levels for the various subshells as a function of atomic number. From Brownlow (1996).

is at a lower energy than the $3d$ orbitals. The filling of the $3d$ orbitals forms a sequence of 10 elements that behave similarly because all but Cr and Cu have two $4s$ electrons in their outer shell. A similar situation exists for row 5, in which the $5s$ orbital is filled followed by filling of the $4d$ orbitals, leading to another sequence of 10 elements that show similar chemical behaviors. Note that the $4f$ orbitals do not yet contain any electrons. The filling of the $4f$ orbitals occurs in row 6, giving rise to the 14 elements of the lanthanide series. A similar situation exists with respect to the filling of the $5f$ orbitals, which occurs in row 7, giving rise to the 14 elements of the actinide series.

Ionization and Valences

It was noted earlier that atoms are most stable when their orbitals are filled, half-filled, or empty. This observation allows us to understand why different elements show particular valences. **Valence** is *the combining capacity of atoms*. This capacity can be interpreted in several ways, i.e., the number of hydrogen atoms that can combine with a particular atom or the number of single bonds it can form with other atoms. Some authors consider valence to be a pure number, whereas others assign charges to valence. We will use the latter convention. *Atoms that have gained or lost electrons are called **ions**. If the atom has gained electrons and has a net negative charge*, it is an **anion**; and if an *atom has lost electrons and has a net positive charge*, it is a **cation**. The *energy required to remove an electron from an atom and place it at an infinite distance is the **ionization potential***. In any given row of the periodic table, as we move from left to right the ionization potential tends to increase; i.e., it becomes more difficult to remove electrons from the atom. Hence, elements on the left-hand side of the periodic table tend to form cations, and those on the right-hand side tend to form anions.

Consider the element sodium (Na). The 1*s*, 2*s*, and 2*p* orbitals are completely filled. The outer 3*s* orbital contains 1 electron. The most stable configuration for this atom is one in which all the shells (n = 1 and n = 2) are filled. Because there is only a single 3*s* electron, it is readily lost, leading to a cation with a +1 charge. On the other hand, chlorine (Cl) has 5 electrons in the 3*p* orbital. A complete subshell can be formed by adding an electron to the 3*p* orbital, leading to an anion with a charge of −1. Argon (Ar) has a completely filled 3*p* orbital; hence, this element neither gains nor loses electrons. The elements in the eighth column of the periodic table are referred to as *noble gases*, because they do not readily combine with other elements. An inspection of the common valences listed in Appendix I will show that most of them can be explained in terms of completely filled, half-filled, or empty orbitals.

Spectra and Elemental Analysis

In Example 1–1 we calculated the energy that would be emitted, and the corresponding wavelength of EM radiation, during a particular electronic transition in the Bohr atom. In the example the electron moved from a higher to a lower orbit, releasing energy. The result is an emission line. If we consider other possible transitions in the Bohr atom from higher orbits to the same orbit, we get a series of emission lines that form, in the case of Example 1–1, the Balmer series. This series of emission lines is characteristic of the neutral hydrogen atom. For other neutral atoms in the periodic table, transitions between orbitals release different amounts of energy, and the resulting emission spectra are different and characteristic for each element. If the atoms have been ionized—i.e., they have gained or lost electrons resulting in a net negative or net positive charge, respectively—the sequence of emission lines will be slightly shifted because the electrostatic attraction between the nucleus and the electrons will have changed. Thus, emission lines are not only characteristic of particular elements but they can also be used to infer the ionization state of an atom. The *spectrum produced when electrons move from higher orbitals to lower orbitals* is called an **emission spectrum** (Figure 1–3).

Let us suppose that we have a transparent container filled with hydrogen gas. We pass a beam of white light through this container and use a spectroscope to view the beam when it emerges from the other side of the container. We would observe a continuous spectrum (all colors are present) except for certain specific wavelengths for which no light is visible. If we compared the dark bands in the spectrum with the position of the emission lines for neutral hydrogen, we would discover that they are at the same wavelength. We can explain this result as follows (Figure 1–4). The incident beam of white light consists of photons of all possible energies (in the visible region). Those photons that have exactly the

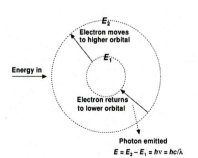

Figure 1–3

An emission spectrum occurs when energy applied to the atom causes an electron to move from a lower orbital to a higher orbital. The electron returns to a lower orbital and emits energy corresponding to the energy difference between the two orbitals $(E_2 - E_1)$. Planck's constant is *h*, ν is frequency, and λ is wavelength.

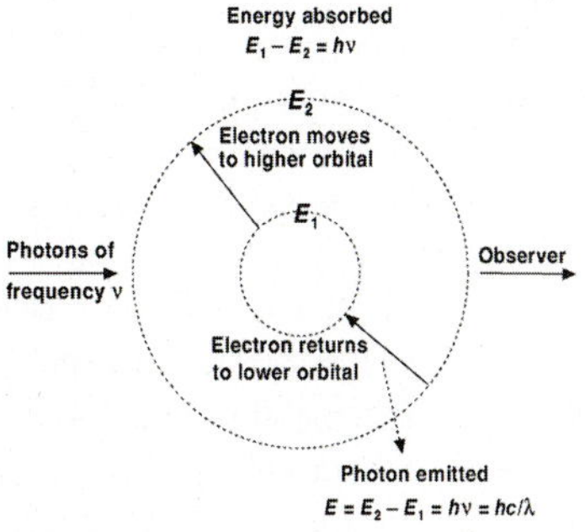

Figure 1–4

An absorption spectrum occurs when photons that have exactly the right energy to move an electron from one orbital to another interact with the atom. When the electron returns to the lower orbital, the emitted photon can travel in any direction. Thus, the observer notices a decrease in the number of photons of this energy (wavelength). Planck's constant is *h*, ν is frequency, and λ is wavelength.

same energy as the energy required to move an electron from one orbital to another in the hydrogen atom will interact with the hydrogen atom and move electrons to higher orbitals. The electrons then return to lower orbitals, emitting energy, but the direction of the emitted photons varies randomly through a 360° solid angle. The probability of an emitted photon traveling in exactly the same direction as the beam of white light and entering the spectroscope is improbably small. Thus, what we observe are *missing wavelengths in the spectrum*, referred to as an ***absorption spectrum***, and the photons that correspond to these energies (wavelengths) have been absorbed by the hydrogen atoms.

Absorption and emission spectra are the basis of a variety of analytical techniques used to determine the elemental composition of natural materials. Examples of analytical techniques that use emission spectra are the electron microprobe (a beam of electrons provides the energy to move the electrons), X-ray fluorescence (X rays provide the energy), ICP-AES (a plasma arc provides the energy), and flame atomic emission spectroscopy (an air-acetylene flame provides the energy). A variety of detectors are used, depending on the energy of the emitted radiation, to measure the rate of photon emission. Absorption spectra are used in atomic absorption spectroscopy (AAS). In this method a hollow cathode tube, the cathode of which consists of a single element or several elements, provides the photon beam. The sample is introduced into a flame or vapor that occupies a position between the source of the photons and the sensor. Changes in the intensity of the beam can be related to the concentration of the element in the flame (or vapor). For example, if the cathode is made of iron, iron atoms in the flame will absorb some of the photons and there will be a decrease in intensity. By relating intensity changes to known concentrations of iron in standards, it is possible to determine the concentration of iron in an unknown. A number of books have been written on these various analytical techniques. The interested student should consult the literature to learn more about spectroscopic analysis.

The Nucleus

The nucleus of an atom consists of two particles, a proton with a charge of $+1$ (the same magnitude but opposite sign from the charge on the electron) and a neutron, which has no charge but is essentially equivalent to the proton in terms of mass. The *number of protons in the nucleus* determines the ***atomic number***, which identifies the element. For example, an atom with 1 proton is hydrogen, an atom with 10 protons is neon. The nucleus can also contain variable numbers of neutrons. *Atoms that have the same number of protons but different numbers of neutrons* are referred to as ***isotopes***. For example, there are three isotopes of hydrogen: common hydrogen, which contains one proton and no neutrons; deuterium, which contains one proton and one neutron; and tritium (a radioactive form of hydrogen), which contains one proton and two neutrons. The *sum of the protons plus neutrons* is the ***mass number*** of the nucleus. By convention, elemental symbols are written with the atomic number at the lower left of the symbol and the mass number at the upper left (Figure 1–5). The atomic number is often omitted when writing isotopic compositions because it is fixed for a particular element. The upper right is reserved for the valence of the atom. Note that the nucleus consists only of positive charges, and, given the small size of the nucleus, strong repulsive forces are developed. The nucleus is held together by what is referred to as the ***strong force***, which acts over a very short range, nominally equal to the diameter of the nucleus. When the size and/or charge of the nucleus becomes too great, the nucleus will break down; i.e., the process known as radioactive decay occurs (see Chapter 6).

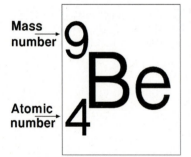

Figure 1–5
Standard format for reporting atomic number and mass number. This isotope of beryllium has four protons and five neutrons.

CHEMICAL BONDING

Two or more atoms may combine to form a compound. The compound is held together by chemical bonds. There are four basic types of bonds: ionic, covalent, metallic, and hydrogen (or van der Waals). ***Ionic bonding*** occurs when *cations and anions combine by electrostatic attraction*. The strength of the bond is a function of the distance between the ions and the total charge. ***Covalent bonding*** occurs when *two or more atoms combine by sharing valence electrons*. In practice, most bonds are not purely ionic or purely covalent, but rather are mixed bonds having characteristics of both types. This topic is more fully discussed in Chapter 7. ***Metallic bonding*** occurs in the case of *pure metals in which electrons are freely shared among all the atoms*. Compounds that are held together by metallic bonding are good conductors of electricity. ***Hydrogen bonding***, of which there are many types, involves *polar molecules (such as water) that have a slight charge bias and form weak electrostatic bonds with other ions*. ***Polar covalent bonds***, which are responsible for the polar characteristics of a molecule, arise *when electrons are unequally shared between atoms*. For example, hydrogen fluoride (HF) forms a polar molecule. The F end of the molecule has a slight net negative charge and the H end of the molecule has a slight net positive charge. The interpretation is that the F atom has a stronger attraction for the shared electron than the hydrogen atom. The important case of the polar water molecule will be described in a later section.

ATOMIC AND MOLECULAR WEIGHTS

The modern atomic mass scale is based on ^{12}C. In this system, ^{12}C is given the mass of exactly 12 atomic mass units (amu). In SI units, 1 amu = 1.66054×10^{-27} kg. All other isotopes are referenced to ^{12}C. This is done using mass spectrometry, a very precise way of measuring mass ratios. For example, mass spectrometric measurements yield a $^{13}C/^{12}C$ ratio = 1.0836129. The mass of ^{13}C is found as follows:

$$\text{Mass } ^{13}C = \text{Mass } ^{12}C \times 1.0836129 = 12 \text{ amu} \times 1.0836129 = 13.0033548 \text{ amu}$$

Masses of other isotopes are found in a similar way. The results of these measurements are tabulated in standard references such as the *CRC Handbook of Chemistry and Physics* or *Chart of the Nuclides*.

In nature, most elements occur in more than one isotopic form. For example, there are four naturally occurring isotopes of iron. The ***atomic weight of an element*** is *the sum of the masses of the individual isotopes weighted according to their abundance*. Using iron as an example, the relative abundances and masses (taken from the *CRC Handbook of Chemistry and Physics*) of the iron isotopes and the calculation of the atomic weight of iron is shown here. The abundance of each iron isotope is multiplied by the mass of each isotope. Note that in doing this multiplication percents are converted to decimals, e.g., 5.81% = 0.0581. The proportionate masses of the individual isotopes are then added to give the atomic weight of iron.

Isotope	% Abundance		Mass isotope (amu)	Mass (amu)
^{54}Fe	5.81	$\times$	53.9396	3.1339
^{56}Fe	91.72	$\times$	55.9349	51.3035
^{57}Fe	2.19	$\times$	56.9354	1.2469
^{58}Fe	0.28	$\times$	57.9333	0.1622
at. wt Fe				55.8465

The ***mole*** (mol) is defined as *the number of carbon atoms in exactly 12 grams of pure ^{12}C*. The ***gram-atomic weight*** of an element is defined as *the atomic weight of the element in grams*. In the case of ^{12}C, the gram-atomic weight is 12 g. Because the atomic weight of all other isotopes are compared to ^{12}C, this means that one gram-atomic weight of any isotope or element contains the same number of atoms. Careful measurements by mass spectrometry have determined the number of atoms in a mole to be 6.022×10^{23} (***Avogadro's number***). From the definition of the mole we can also determine the relationship between amu and grams; i.e., 6.022×10^{23} amu = 1 g and 1 amu = 1.66054×10^{-24} g.

The **gram-molecular weight** is *the weight of a compound in grams*. For example, halite has the formula NaCl. The atomic weight of Na = 22.9898 and the atomic weight of Cl = 35.4527. The gram-molecular weight of NaCl is 58.4425 g. This corresponds to 1 mole of NaCl, which contains 6.022×10^{23} NaCl molecules.

EXAMPLE 1–3 0.02 g of calcium carbonate ($CaCO_3$) is dissolved in 1 liter of water. Calculate the moles of calcium carbonate that have dissolved in the water, the moles of Ca^{2+} ions in solution, and the number of Ca^{2+} ions in solution.

The gram-molecular weight of $CaCO_3$ is calculated as follows:

$$
\begin{array}{rl}
1\ \text{Ca} = & 40.078 \\
1\ \text{C} = & 12.011 \\
3\ \text{O} = & \underline{47.997} \\
& 100.086\ \text{g}
\end{array}
$$

Moles of $CaCO_3$ = 0.02 g/100.086 g mol^{-1} = 1.998×10^{-4} mol. There is 1 mole of Ca in each mole of $CaCO_3$, so Ca^{2+} = 1.998×10^{-4} mol. Number of Ca^{2+} ions in solution = 1.998×10^{-4} mol $\times$ 6.022×10^{23} ions per mol = 1.203×10^{20} ions. ■

EXAMPLE 1–4 A 1-liter solution contains 2×10^{-3} mol of $BaSO_4$. Calculate the mass of $BaSO_4$ dissolved in the water. The gram-molecular weight of $BaSO_4$ = $(137.33) + (32.066) + (4)(15.999)$ = 233.392 g. The grams of $BaSO_4$ in solution = 2×10^{-3} mol $\times$ 233.392 g mol^{-1} = 0.467 g. ■

The **gram-equivalent weight of an ion** is *the molecular or atomic weight divided by the valence*. In the case of an acid or base, it is the number of H^+ or OH^- ions that can be produced when the acid or base is dissolved in water. For example, in $BaSO_4$, barium has a valence of 2 and there are two gram-equivalent weights of Ba^{2+} in each mole of $BaSO_4$. HCl (hydrochloric acid) and H_2SO_4 (sulfuric acid) are both strong acids that will completely dissociate in water. One mole of HCl will release 1 mole of H^+ ions to solution, so there is one gram-equivalent weight of H^+ in this acid. One mole of H_2SO_4 will release 2 moles of H^+ ions to solution; thus, there are two gram-equivalent weights of H^+ in sulfuric acid.

MEASUREMENT OF CONCENTRATION

Concentrations of elements or compounds in the natural environment have been presented in a variety of ways. In principle, the *Système international d'unités* (SI) should be used, but authors routinely violate this system (although some journals now insist that it be used), and in the older scientific literature SI units were often not used. In this section the relationships between the most commonly used units are given.

Concentrations can be reported in terms of absolute mass (Table 1–3), and the SI unit is the kilogram (kg). Concentrations can also be reported on a weight per weight basis, giv-

Table 1–3   Units of Absolute Mass Relative to the Kilogram

Unit	Abbreviation	Kilogram
Kilogram	kg	1
Gram	g	1×10^{-3}
Milligram	mg	1×10^{-6}
Microgram	μg	1×10^{-9}
Nanogram	ng	1×10^{-12}

ing rise to units such as parts per thousand (ppt or ‰; note that ppt is also used for parts per trillion), parts per million (ppm), or parts per billion (ppb). For example, 1 g of As in 1 kg of soil is equivalent to 1 ppt, and 1 mg of As in 1 kg of soil is equivalent to 1 ppm. When analyzing solutions, concentrations are often reported on a weight per volume basis, for example, 1 mg of As in 1 L (liter) of solution. At room temperature, the density of pure water is close to 1000 kg m^{-3}. A liter is equivalent to 1×10^{-3} m^{-3}. Hence, the mass of 1 L of pure water at room temperature is approximately 1 kg. For very dilute solutions, the amount of dissolved material is relatively small and the mass of the total solution is close to 1 kg per L. On a weight per weight basis, 1 mg of As dissolved in 1 L of water would give a concentration of 1 ppm As. The concentration of elements in freshwater are sometimes given in ppm, but note that, strictly speaking, this is not correct because only on rare occasions will the mass of 1 L of solution be exactly 1 kg.

Concentrations in solutions are reported in a variety of ways, including molarity, molality, normality, and mole fraction. ***Molarity*** is the *number of moles of* **solute** *(the substance that is dissolved) per volume of solution in liters.* ***Molality*** is the *number of moles of solute per kg of* **solvent** *(the material in which the solute is dissolved).* ***Normality*** is the *number of equivalents per liter of solution.* Gram-equivalents were defined in the preceding section. Table 1–4 shows the relationship between molarity and normality for some common acids and bases. ***Mole fraction*** is the *ratio of the number of moles of a given component to the total number of moles of solution.* Solutions do not have to be liquids. In fact, many minerals (such as the feldspars and carbonates) form solid solution series.

Table 1–4 *Molarity and Normality for Several Acids and Bases*

Acid or base	Molecular weight	Gram-equivalent weight	Relationship between molarity and normality
HCl	36.5	36.5	1M = 1N
H$_2$SO$_4$	98.0	49.0	1M = 2N
H$_3$PO$_4$	98.0	32.6	1M = 3N
NaOH	40.0	40.0	1M = 1N
Ca(OH)$_2$	74.0	37.0	1M = 2N

EXAMPLE 1–5 An important mineral in many rocks is plagioclase feldspar. The plagioclase feldspars form a solid solution series with the end members pure albite (NaAlSi$_3$O$_8$) and pure anorthite (CaAl$_2$Si$_2$O$_8$). A particular plagioclase contains 5 wt% Ca. Calculate the mole fraction of anorthite in the plagioclase. The molecular weights of Ca, Al, Si, and O are, respectively, 40, 27, 28, and 16.

First we calculate the weight percent of anorthite. This is done by multiplying the molecular weight of anorthite by 5, the weight percent of Ca in the plagioclase, and dividing by the atomic weight of Ca. We divide by the atomic weight of Ca because we want to know the ratio of anorthite molecule to Ca.

$$\text{wt\% CaAl}_2\text{Si}_2\text{O}_8 \text{ (An)} = \frac{(5)(40 + 2 \times 27 + 2 \times 28 + 8 \times 16)}{40} = 34.8$$

We then subtract the weight percent of anorthite from 100% to give the weight percent of albite.

$$\text{wt\% NaAlSi}_3\text{O}_8 \text{ (Ab)} = 100 - 34.8 = 65.2$$

We then determine the relative number of moles of anorthite by dividing the weight percent of anorthite by the molecular weight of the anorthite molecule.

$$\text{Relative number of moles An} = \frac{34.8}{278} = 0.13$$

Similarly, we determine the relative number of moles of albite:

$$\text{Relative number of moles Ab} = \frac{65.2}{262} = 0.25$$

Finally, we determine the mole fraction of anorthite by dividing the relative number of moles of anorthite by the total relative moles of anorthite + albite.

$$\text{Mole fraction An} = \frac{\text{Moles An}}{\text{Moles An + Ab}} = \frac{0.13}{0.13 + 0.25} = 0.34$$ ∎

TYPES OF CHEMICAL REACTIONS

Most chemical reactions are one of the following types: precipitation, acid–base, or oxidation–reduction. During a chemical reaction there is a reorganization of the atoms in one or more substances. Chemical reactions are represented by chemical equations. By convention, the reactants appear on the left-hand side of the equation and the products on the right-hand side of the equation. Consider what happens when methane (natural gas) is combined with oxygen. This reaction can be represented by the following chemical equation:

$$CH_4 + 2O_2 \rightarrow CO_2 + 2H_2O$$

The reactants are methane (CH_4) and oxygen (O_2), and the products are carbon dioxide (CO_2) and water vapor (H_2O). This is the familiar process of combustion, and heat energy is released. Note that there has been a rearrangement of the atoms in the substances. The carbon in the methane has combined with oxygen to form carbon dioxide, and the hydrogen in the methane has combined with oxygen to form water vapor. The various substances in a chemical reaction are referred to as **species**, *chemical entities that can be ions, molecules, solid phases, gases, etc.*

Precipitation reactions occur when two solutions are mixed and a solid, called a **precipitate**, forms. For example, if a silver nitrate solution is added to a sodium chloride solution, a white precipitate is formed. The chemical reaction can be written

$$Ag^+ + NO_3^- + Na^+ + Cl^- \rightarrow AgCl + Na^+ + NO_3^-$$

The silver ions in solution have combined with the chloride ions in solution to form an insoluble silver chloride. As written, this is a **complete ionic equation** because it shows all the individual ions. Inspection of the equation reveals that only Ag^+ and Cl^- participate in the reaction to form AgCl. The other ions do not participate in the reaction and are called **spectator ions**. The reaction is more appropriately written as a **net ionic equation**.

$$Ag^+ + Cl^- \rightarrow AgCl$$

This reaction is the basis of the titration method used to determine the salinity of seawater (see Chapter 10).

Acid–base reactions involve the transfer of protons. By definition, **acids** are *proton donors* and **bases** are *proton acceptors*. We distinguish between two types of acids and bases, strong and weak. Strong acids or bases dissociate completely in water. For example, HCl is a strong acid and when dissolved in water dissociates into H^+ (or H_3O^+) and Cl^- ions. Acetic acid ($HC_2H_3O_2$) is a weak acid and only partly dissociates into H^+ and $C_2H_3O_2^-$ ions when dissolved in water. As an example of an acid–base reaction, consider what happens when solutions of hydrochloric acid (HCl) and potassium hydroxide (KOH), a strong acid and strong base, respectively, are mixed. The complete ionic equation can be written

$$H^+ + Cl^- + K^+ + OH^- \rightarrow H_2O_{(aq)} + K^+ + Cl^-$$

A note about the representation of species in chemical equations: In general, if the species is written as an ion, it is in solution; and if it is written as a neutral compound, it is not in solution. If an uncharged species is in solution, the chemical formula is followed by the subscript (aq). Some textbooks indicate the state of all species by using the subscripts (aq)—aqueous, (s)—solid, (l)—liquid, or (g)—gas.

Note that only the hydrogen and hydroxyl ions are involved in the reaction; the potassium and chlorine ions are spectators. Hence, the net ionic equation is written

$$H^+ + OH^- \rightarrow H_2O_{(aq)}$$

Acid–base reactions will be considered in detail in Chapter 3.

Oxidation–reduction reactions occur when there is a transfer of electrons. As an example of an oxidation–reduction reaction we will consider the oxidation of methane as represented by the following equation:

$$CH_4 + 2O_2 \rightarrow CO_2 + 2H_2O$$

Because the molecules are electrically neutral, the oxidation states of the various atoms in each molecule must sum to zero. Thus, in methane $C = -4$ and $H = +1$, in the oxygen molecule the atoms are electrically neutral, in carbon dioxide $C = +4$ and $O = -2$, and in water $H = +1$ and $O = -2$. During the course of the reaction, the oxidation state of carbon changes from -4 to $+4$ (8 electrons are transferred) and the oxidation state of oxygen changes from 0 to -2. There is 1 carbon atom on the left-hand side of the equation and 4 oxygen atoms; hence, charge balance is maintained—i.e., 8 electrons from carbon = 2 electrons per oxygen $\times$ 4 oxygen atoms. Note that the oxidation state of hydrogen remains unchanged and it does not participate in the oxidation–reduction reaction. Oxidation–reduction reactions will be considered in detail in Chapter 4.

DETERMINING AND BALANCING A CHEMICAL EQUATION

In the previous section we considered different types of chemical reactions. How do we determine which chemical reactions occur and the equations that represent the reactions? While we can use our basic understanding of chemical processes to make some assumptions about possible reactions, ultimately we must do an experiment to determine the chemical equation. This topic is covered in introductory chemistry textbooks and will be briefly considered here.

Let us return to the reaction between silver nitrate and sodium chloride solutions. We know that in these solutions the various species are present in their ionic form. When we mix the two solutions a precipitate is formed. From our experience with chemical reactions we know that silver halide salts are very insoluble; hence, we might conclude that the precipitate is most likely silver chloride. But in order to be certain we must analyze the precipitate. Our analysis reveals that, on a weight basis, the precipitate is 75.3% silver and 24.7% chlorine. Our original guess is correct—the precipitate is an insoluble silver halide. What is the chemical formula of the precipitate? If we assume that we have 100 g of precipitate (we make this assumption in order to simplify the calculation—the actual value doesn't matter), we can calculate the number of moles of silver and chlorine in the compound.

$$\text{Moles of silver} = (100 \text{ g}) \times 0.753/107.9 \text{ g mol}^{-1} = 0.698 \text{ mol}$$

$$\text{Moles of chlorine} = (100 \text{ g}) \times 0.247/35.45 \text{ g mol}^{-1} = 0.697 \text{ mol}$$

The number of moles of silver and chlorine are essentially identical, and the compound has the formula AgCl. Because neither sodium nor nitrate is found in the precipitate, we will make the assumption that these species remain in solution as spectator ions. The net ionic chemical equation for this precipitation reaction is

$$Ag^+ + Cl^- \rightarrow AgCl_{(s)}$$

The procedure we used to determine the formula of the compound can be summarized as follows:

1. Determine the weight percent of each element in the compound.
2. Using 100 g as a basis, determine the weight of each element in the compound.

3. For each element determine the number of moles by dividing the weight of the element by the gram-atomic weight of the element.

4. Using the element with the least number of moles, determine the relative number of moles of each of the other elements. After rounding, this should yield integer values for each of the elements, and these integers represent the formula for the compound.

EXAMPLE 1–6 During a chemical reaction a precipitate forms that has the following composition (in weight percent): Ca = 39.74, P = 18.42, O = 38.07, and F = 3.77. Determine the formula for this compound.

First convert to absolute weight on the basis of 100 g, and then determine the number of moles of each element.

Element	wt%	wt/100 g	at. wt	Moles
Ca	39.74	39.74	40.08	0.992
P	18.42	18.42	30.97	0.595
O	38.07	38.07	16.00	2.379
F	3.77	3.77	19.00	0.198

The element present in the smallest amount is fluorine. Dividing each of the other elements by the number of moles of fluorine yields, per mole of fluorine, 12 moles of oxygen, 3 moles of phosphorus, and 5 moles of calcium. The formula for this compound is $Ca_5P_3O_{12}F$ [or $Ca_5(PO_4)_3F$], the formula for the mineral fluoroapatite. ∎

The formula we calculated in Example 1–6 is the ***empirical formula***, i.e., *the formula that represents the simplest whole-number ratio of the atoms that make up the compound.* This formula, however, may not be the correct formula. For example, if you are familiar with organic compounds, you will know that many organic compounds consist of multiple numbers of all elements found in the compound. We can determine the ***molecular formula*** *(the actual number of each type of atom in the compound)* by comparing the formula weight to the molecular weight of the compound.

EXAMPLE 1–7 The empirical formula of a compound is CH_3 and its molecular weight is 30 g. Calculate the molecular formula for the compound.

The formula weight is 15 g mol^{-1} (1C × 12 g mol^{-1} + 3H × 1 g mol^{-1}). The molecular weight is twice the formula weight (30 g/15 g = 2). Thus, the molecular formula of the compound is C_2H_6. ∎

Based on experiments, we can determine the reactants and products of a chemical reaction. However, simply determining the reactants and products does not completely describe the chemical reaction. We must also balance the number of atoms on each side of the equation. Returning to a previous example involving the combustion of methane, experiments tell us that methane combines with oxygen to produce carbon dioxide and water. The simplest equation that represents this reaction is

$$CH_4 + O_2 \rightarrow CO_2 + H_2O$$

Inspection of this equation reveals that the number of hydrogen and oxygen atoms on each side of the equation is not the same; i.e., the equation is not balanced. To balance the equation, we must add 2 hydrogen atoms to the right side of the equation, giving 2 water molecules. We now have a deficiency of oxygen atoms on the left side of the equation and we must add 2 oxygen atoms, i.e., increase the number of oxygen molecules to 2. The balanced equation is

$$CH_4 + 2O_2 \rightarrow CO_2 + 2H_2O$$

Unfortunately, there are no simple rules for balancing chemical equations. After you have written the equation showing the reactants and products, the balancing of the equation proceeds by inspection. The formulas of the compounds cannot change, so the equation can only be balanced by changing the number of the formula units. It is usually best to start with the most complicated species in the chemical equation and then proceed to the least complicated species. In oxidation reactions, where molecular oxygen and water are involved, these two species are balanced last because they can be added to the reaction in appropriate amounts. A more detailed discussion of the balancing of oxidation–reduction reactions is found in Chapter 4.

EXAMPLE 1–8 In the production of phosphate fertilizer, calcium phosphate rock [$Ca_3(PO_4)_2$] is reacted with sulfuric acid (H_2SO_4) to produce calcium sulfate ($CaSO_4$) and phosphoric acid (H_3PO_4). Sulfuric acid is a strong acid (i.e., it dissociates completely in water), and phosphoric acid is a weak acid (i.e., it only partly dissociates in water). The chemical reaction can be written as follows:

$$Ca_3(PO_4)_2 + 2H^+ + SO_4^{2-} \rightarrow CaSO_4 + H_3PO_{4\,(aq)}$$

To balance this equation, we will start with the most complicated compound, calcium phosphate. Calcium phosphate contains 3 Ca atoms. Ca atoms are found in the product calcium sulfate. In order to balance the calcium, we need 3 calcium sulfate molecules.

$$Ca_3(PO_4)_2 + 2H^+ + SO_4^{2-} \rightarrow 3CaSO_4 + H_3PO_{4\,(aq)}$$

There are now 3 sulfur atoms in the products. We need 3 sulfur atoms in the reactants. Hence, there must be 3 sulfate molecules in the reactants. The equation now becomes

$$Ca_3(PO_4)_2 + 2H^+ + 3SO_4^{2-} \rightarrow 3CaSO_4 + H_3PO_{4\,(aq)}$$

In order to balance the number of phosphorus atoms in the reactants (2), we need 2 phosphorus atoms in the products. The equation becomes

$$Ca_3(PO_4)_2 + 2H^+ + 3SO_4^{2-} \rightarrow 3CaSO_4 + 2H_3PO_{4\,(aq)}$$

We now have 6 hydrogen atoms in the products, but only 2 in the reactants. To balance the hydrogen atoms, we need to add 4 hydrogen atoms to the left-hand side of the equation, which gives

$$Ca_3(PO_4)_2 + 6H^+ + 3SO_4^{2-} \rightarrow 3CaSO_4 + 2H_3PO_{4\,(aq)}$$

The only remaining atom is oxygen. As written, there are 20 oxygen atoms in the reactants and 20 oxygen atoms in the products. The equation is now balanced. ■

GASES

An ideal gas consists of atoms and/or molecules that move completely independently of each other and occupy a volume that is much greater than the total volume of the atoms/molecules; i.e., the actual volume of the atoms/molecules is negligible relative to the total volume. Ideal gases obey two well-known laws that relate pressure and volume (Boyle's law) and temperature and volume (Charles's law). All real gases depart to some extent from ideality.

Experiments carried out on real gases show an inverse relationship between the volume of the gas and the pressure. If the pressure is doubled, the volume decreases by

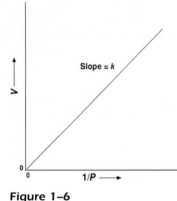

Figure 1–6

A plot of V versus $1/P$ for an ideal gas gives a straight line, and the slope of the line is the Boyle's law constant k.

one-half. A plot of volume versus the reciprocal of pressure yields a straight line (Figure 1–6). **Boyle's law** relates the changes in pressure and volume of a gas and can be stated mathematically as

$$PV = k \qquad (1-13)$$

where P is pressure (force per unit area), V is volume, and k is a constant. In SI units, pressure is measured in Pascals (Pa) and volume is measured in m³. However, atmospheres are often used for pressure and liters are more commonly used for the volume of a gas. The equation of a straight line is

$$y = mx + b \qquad (1-14)$$

With reference to Figure 1–6, y = volume, $x = 1/P$, $m = k$, and $b = 0$ (the line intercepts the y-axis at a volume of 0 liters; i.e., at infinite pressure the gas is compressed to zero volume). Note that the slope k corresponds to volume times pressure, and in commonly used units the constant would be expressed in L atm. For an ideal gas, this value is 22.41 L atm. At low pressures, real gases have constant slopes and k values that are near 22.41 L atm. At higher pressures, real gases do not show a linear relationship between V and $1/P$. However, the deviation with pressure is usually relatively small.

Experiments carried out on real gases show that there is a direct relationship between the volume of the gas and the temperature. As the temperature of the gas decreases so does the volume. This relationship is known as **Charles's law** and can be stated mathematically as

$$V = bT \qquad (1-15)$$

where V is the volume, T is the temperature (in K), and b is a proportionality constant. An interesting result of these experiments is that extrapolation to zero volume for all gases gives a consistent value of $-273.15°C$. At this temperature, the volume of an ideal gas would be zero. At first glance, this result seems ridiculous (a little bit like the Cheshire cat in Alice in Wonderland), because matter cannot contract to zero volume. The key phrase here is "ideal gas." All gases liquify before reaching $-273.15°C$ and therefore clearly do not behave as ideal gases at these low temperatures. The temperature at which the volume of an ideal gas becomes zero is the basis for the absolute (Kelvin) temperature scale: K = 0 corresponds to $-273.15°C$. The Kelvin and Centigrade degrees are the same size; hence, the Kelvin and Centigrade scales can be mathematically related as follows:

$$K = °C + 273.15 \qquad (1-16)$$

Note that in the ideal gas law the temperature is given in Kelvin. Because a gas cannot have a negative volume, 0 K has a special significance and is referred to as **absolute zero**. Although absolute zero has never been achieved, a number of clever experiments have been conducted in which temperatures close to 0 K have been obtained.

Avogadro postulated that equal volumes of gases at the same temperature and pressure contain the same number of particles. **Avogadro's law** relates the volume of a gas to the number of moles and can be stated mathematically as

$$V = an \qquad (1-17)$$

where V is the volume, n is the number of moles, and a is a proportionality constant.

Boyle's, Charles's, and Avogadro's laws can be combined to form the **ideal gas law**, which is stated mathematically as

$$PV = nRT \qquad (1-18)$$

where P is the pressure, V is the volume, n is the number of moles, R is the universal (or ideal) gas constant, and T is the temperature (in K). The value of the universal gas constant depends on the units used to express pressure and volume. For example, if the pressure is in atmospheres and the volume is in liters, $R = 0.08206$ L atm K^{-1} mol^{-1}. The ideal gas law is an equation of state because it defines the condition of the gas at a given time. If any three of the four properties of the gas (pressure, volume, number of moles, and temperature) are known, the fourth can be determined from the ideal gas law. Most gases approximately obey this relationship at pressures less than 1 atm.

EXAMPLE 1–9 A balloon is filled with helium gas. The outside temperature is 20°C, the atmospheric pressure is 1 atm, and the volume of the balloon is 2 L. Calculate the number of moles of helium in the balloon.

From the ideal gas law,

$n = PV/RT = [(1 \text{ atm})(2 \text{ L})]/[(0.08206 \text{ L atm K}^{-1} \text{ mol}^{-1})(293.15 \text{ K})] = 0.083$ mol ∎

The *volume occupied by 1 mole of a gas* is referred to as the **molar volume**. For an ideal gas at 0°C and 1 atm pressure, the molar volume is 22.42 L. Note that this volume would vary as a function of temperature and pressure. The specific conditions of 0°C and 1 atm are referred to as **standard temperature and pressure (STP)**. Molar volumes at STP for a number of common gases are tabulated in Table 1–5.

Table 1–5 Molar Volumes for Various Gases at STP*

Gas	Molar volume (L)
Oxygen (O_2)	22.397
Nitrogen (N_2)	22.402
Hydrogen (H_2)	22.433
Helium (He)	22.434
Argon (Ar)	22.397
Carbon dioxide (CO_2)	22.260
Ammonia (NH_3)	22.079

*From Zumdahl (1989).

EXAMPLE 1–10 Calculate the volume of CO_2 released at STP when 25 g of calcite ($CaCO_3$) are dissolved in hydrochloric acid. The molecular weight of $CaCO_3$ is 100.1 g.

Moles of $CaCO_3$ dissolved = 25 g/100.1 g mol^{-1} = 0.25 mol

During this reaction 0.25 mol of CO_2 is released. At STP, the molar volume of CO_2 is 22.260 L mol^{-1} (Table 1–5).

Volume of CO_2 released = (0.25 mol)(22.260 L mol^{-1}) = 5.565 L ∎

So far, we have considered the behavior of ideal gases. But most gases, particularly at high pressures and low temperatures, deviate from ideal behavior. In fact, experimental studies have shown that real gases most closely approximate ideal gases only at very low pressures and very high temperatures. There are two major reasons for this deviation from ideality: (1) Gas molecules/atoms do have a finite volume, and when the number of gas molecules/atoms increases (as they would at high pressures), the gas molecules/atoms comprise a significant portion of the volume; and (2) gas molecules/atoms do interact with each other, and as the number of molecules/atoms increases, the number of interactions increases. These deviations were taken into account in a model developed in the late 1800s by van der Waals. Two terms were introduced to the ideal gas equation. The first has the form $V - nb$, where n is the number of moles of gas and b is an empirical constant that depends on the gas. This first term corrects for the volume of the gas molecules/atoms. The second term has the form $a(n/V)^2$, where a is an empirical constant, n is the number of moles, and V is the volume. The second term corrects for the interactions between the gas molecules/atoms. Taking into account these two corrections, the observed pressure is described by the following mathematical relationship:

$$P_{\text{obs}} = \frac{nRT}{V - nb} - a\left(\frac{n}{V}\right)^2 \tag{1–19}$$

Table 1–6   Van der Waals Constants for Some Common Gases*

Gas	a (atm L^2 mol^{-2})	b (L mol^{-1})
He	0.0346	0.0238
Ne	0.208	0.0167
Ar	1.355	0.0320
Xe	4.192	0.0516
Kr	5.193	0.0106
H_2	0.245	0.0265
N_2	1.370	0.0387
O_2	1.382	0.0319
Cl_2	6.343	0.0542
CO_2	3.658	0.0429
CH_4	2.303	0.0431
NH_3	4.225	0.0371
H_2O	5.537	0.0305

*From *CRC Handbook of Chemistry and Physics* (2000).

This equation can be rearranged to give the **van der Waals equation**:

$$\left[P_{obs} + a\left(\frac{n}{V}\right)^2 \right](V - nb) = nRT \qquad (1\text{–}20)$$

Van der Waals constants for some of the common gases are tabulated in Table 1–6.

EXAMPLE 1–11 A cylinder of compressed nitrogen has a volume of 100 L and contains 500 mol of N_2. At a temperature of 25°C, calculate the pressure exerted by the gas on the cylinder.

First solve the van der Waals equation for pressure.

$$P_{obs} = \frac{nRT}{V - nb} - a\left(\frac{n}{V}\right)^2$$

25°C = 298.15 K. Substituting the appropriate values into the van der Waals equation,

$$P_{obs} = \frac{(500 \text{ mol})(0.08206 \text{ L atm K}^{-1} \text{ mol}^{-1})(298.15 \text{ K})}{(100 \text{ L}) - (500 \text{ mol})(0.0387 \text{ L mol}^{-1})}$$

$$- (1.37 \text{ atm L}^2 \text{ mol}^{-2})\left(\frac{500 \text{ mol}}{100 \text{ L}}\right)^2 = 117.4 \text{ atm}$$

Gas cylinder pressures are often reported in pounds per square inch (psi); 1 atm = 14.7 lb in^{-2}. Therefore, for this gas cylinder,

$$P = 117.4 \text{ atm} \times 14.7 \text{ lb in}^{-2} \text{ atm}^{-1} = 1726 \text{ psi} \qquad \blacksquare$$

STRUCTURE AND PROPERTIES OF WATER

Water is arguably the most important substance on earth, and without water, life as we know it would not exist. Compared to other liquids, water has a number of unique properties. These are summarized in Table 1–7.

These unique properties are due to the structure of the water molecule (Figure 1–7). The structure of nonmetallic molecules can be described using the **valence shell electron-pair repulsion (VSEPR) model**, which is described in introductory chemistry textbooks. This model assumes that the molecular structure is determined principally by the minimization of electron-pair repulsions. There are eight valence electrons involved in the water molecule. Six electrons are associated with oxygen and one electron is associated

Table 1–7   Properties of Water*

Property	Comparison to other substances
Heat capacity	Highest of all common liquids (except ammonia) and solids
Latent heat of fusion	Highest of all common liquids (except ammonia) and most solids
Latent heat of vaporization	Highest of all common substances
Dissolving ability	Dissolves more substances (particularly ionic compounds), and in greater quantity than any other common liquid
Transparency	Relatively high for visible light
Physical state	The only substance that occurs naturally in all three states at the earth's surface
Surface tension	Highest of all common liquids
Conduction of heat	Highest of all common liquids (Hg is higher)
Viscosity	Relatively low viscosity for a liquid

*Adapted from Sverdrup et al. (1942).

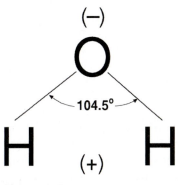

Figure 1–7
Structure of the water molecule.
Water behaves as a polar molecule.

with each of the hydrogens. Electrons exist in pairs. Two pairs are shared between the oxygen and two hydrogens. The other two pairs are nonbonding and are sometimes referred to as *lone pairs*. In order to minimize the electron-pair interactions, the pairs are arranged in a tetrahedral configuration around the oxygen so that the angle between pairs is 109.5°. We see from Figure 1–7 that this angle is somewhat smaller for the water molecule. This is usually explained by assuming that the lone pairs need more space around the oxygen atom than do bonding pairs. From the structure of the water molecule it is apparent that water is a polar molecule (Figure 1–7), and this is the key factor in determining the physical and chemical properties of water. For example, ionic compounds dissolve readily in water because of the polarity of the water molecule. Conversely, covalently bonded compounds are relatively insoluble in water. The hydrogen bonding between water molecules is also responsible for the high latent heats of vaporization (539 cal g^{-1}) and fusion (80 cal g^{-1}). The heat of vaporization varies as a function of temperature and can be calculated using the following equation:

$$L = 596 \text{ cal g}^{-1} - [(0.56 \text{ cal g}^{-1}\,°C^{-1}) \times T\,(°C)] \tag{1–21}$$

Water also has one of the highest heat capacities of any common substance. **Heat capacity** is the *amount of heat energy needed to raise the temperature of 1 g of a substance 1°C*. For water, the heat capacity is 1 cal g^{-1} °C^{-1}.

As noted, water can exist in all three states at the earth's surface. In the vapor state, water exists as separate molecules that move freely past each other. In the liquid state, water is a mixture of clusters of water molecules (many forming six-sided rings) and single water molecules (Figure 1–8b). In the solid state, water molecules form a network of

Figure 1–8
(a) The crystal structure of ice showing the six-sided rings formed by 24 water molecules. (b) The structure of liquid water. In the same volume of liquid water, there are 27 water molecules; hence, liquid water has a greater density than ice. From Gross and Gross (1996).

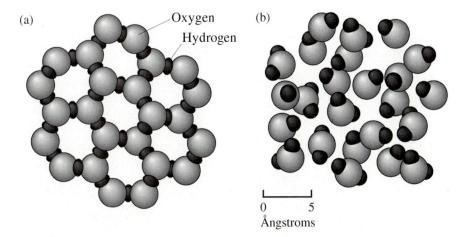

six-sided rings (Figure 1–8a). These variations in the structure of water are responsible for water behaving unlike almost any other substance in terms of density variations. For virtually all substances, the solid state is denser than the liquid state. However, in the case of water, the solid state is less dense. This is because the six-sided ring structures occupy a greater volume than the individual water molecules and clusters of molecules found in the liquid state (Figure 1–8).

Another unique characteristic of water is that its temperature of maximum density is above the freezing point (Figure 1–9). As the temperature of water decreases the molecular motion of the molecules decreases, so they tend to be closer together (i.e., the number of water molecules per unit volume increases). At the same time, the water molecules are forming clusters and six-sided rings that occupy a greater volume than the same number of single water molecules. The increase in volume caused by this ordering eventually becomes more important than the decrease in volume due to the decreasing molecular motion. The crossover point represents the temperature of maximum density (~4°C for freshwater). The freezing-point temperature, and the temperature of maximum density, decrease as the salt content of water increases (Figure 1–10). This becomes an important factor in the oceans. Salinity (defined in Chapter 10) is a measure of the salt content of seawater. Average salinity for seawater is ~35‰. With reference to Figure 1–10, seawater is densest at the freezing-point temperature, unlike the situation for freshwater. These differences in the freezing-point temperature, and temperature of maximum density, play an important role in the circulation of freshwater and seawater bodies, the formation of sea ice, and other processes. The interested student can learn more about this topic in a standard reference on oceanography.

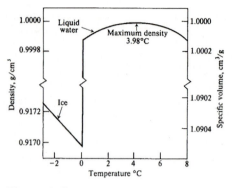

Figure 1–9
Density of pure water near the freezing point. From Duxbury (1971).

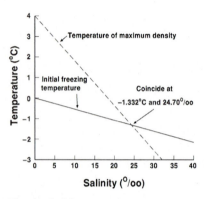

Figure 1–10
Relationship between salinity, decrease in freezing-point temperature, and temperature of maximum density. After Duxbury (1971).

BOX MODELS AND GEOCHEMICAL CYCLES

Box models and the steady-state assumption are widely used in the environmental sciences to trace the passage of various constituents through the bio-geo-hydro-atmo-spheres and to assess the impact of anthropogenic inputs on these natural cycles. A ***box model*** consists *of several boxes showing the reservoirs for a particular substance and the rate at which material is transferred between the reservoirs.* Fluxes between these reservoirs are generally indicated by arrows. In a ***steady-state system***, the *total amount of a substance in each reservoir remains constant*; thus the *rate of addition to, and removal of, a material from a reservoir must be the same.* In a pioneering work, Garrels et al. (1975) used box models to describe geochemical cycles for a number of elements of environmental interest.

A very simple box model for the hydrologic cycle is shown in Figure 1–11. Using the data from this figure, we can calculate the residence time for water in the various

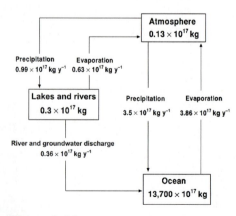

Figure 1–11
Simplified box model of the hydrologic cycle. Modified from Drever (1997).

reservoirs. ***Residence time*** is defined as *the average length of time a particular substance will reside in a reservoir*. The mathematical expression is

$$\text{Residence time} = \frac{\text{Amt. of material in reservoir}}{\text{Rate of addition (removal)}} \tag{1–22}$$

EXAMPLE 1–12 Calculate the residence time for water in the atmospheric reservoir.

Because this is a steady-state model, the rate of addition of water to the reservoir must equal the rate of removal. We can use either set of fluxes. Using the rate of addition, 0.63×10^{17} kg of water are added each year by evaporation from lakes and rivers and 3.86×10^{17} kg are added each year by evaporation from the ocean. The total rate of addition is 4.49×10^{17} kg y^{-1}. The residence time for water in the atmospheric reservoir is

$$\text{Residence time} = \frac{\text{Amt. material in reservoir}}{\text{Rate of addition}} = \frac{0.13 \times 10^{17} \text{ kg}}{4.49 \times 10^{17} \text{ kg y}^{-1}}$$
$$= 0.029 \text{ y} = 10.6 \text{ d}$$

The residence time of water vapor in the atmospheric reservoir is very short. This result suggests that changes in the rate of addition of water to the atmosphere (e.g., increases in the rate of evaporation due to atmospheric warming) would lead to rapid increases in water vapor in the atmosphere and a corresponding increase in the amount of precipitation. This is sometimes referred to as the *speeding up of the hydrologic cycle* due to atmospheric warming. ∎

What happens when we perturb a steady-state system? This problem can be investigated in several ways. The approach illustrated here is a ***first-order kinetics model***. In Chapter 2, we will discuss kinetics at some length and present equations that describe various reaction orders. One of the most common types of reactions is a first-order reaction, an example of which is radioactive decay. In terms of geochemical cycles, we can write the following first-order kinetics equation:

$$dA_i/dt = F_{\text{input}} - F_{\text{output}} = F_{\text{input}} - kA_i \tag{1–23}$$

where dA_i/dt is the rate of change of the amount of substance A in reservoir i, F_{input} is the rate of addition of substance A to the reservoir, F_{output} is the rate of removal of substance A from the reservoir, and k is the rate constant. When the system is in a steady state, $F_{\text{input}} = F_{\text{output}}$. Solving equation 1–23 for the amount of substance A in reservoir i at some particular time gives

$$A_i(t) = \frac{F_{\text{input}}}{k} - \left(\frac{F_{\text{input}}}{k} - A_i^0 \right) \exp(-kt) \tag{1–24}$$

where A_i^0 is the amount of substance A in the reservoir at time zero. When the system is in a steady state, equation 1–23 can be written

$$dA_i/dt = 0 = F_{\text{input}} - kA_i^0 \tag{1–25}$$

and

$$k = F_{\text{input}}/A_i^0 \tag{1–26}$$

Note that in this model the new steady state for a particular reservoir is approached exponentially.

EXAMPLE 1–13 Garrels et al. (1975) present a box model for the prehuman mercury cycle. This model is shown in Figure 1–12. The vast majority of the mercury is contained in the sediments and is very slowly released to the land reservoir by erosion. The small

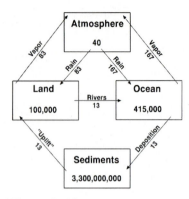

Figure 1–12
Prehuman cycle for mercury. Reservoir masses in units of 10^8 g. Fluxes in units of 10^8 g y^{-1}. From Garrels et al. (1975).

amount of mercury vapor in the atmospheric reservoir suggests that changes in the mercury flux to this reservoir could substantially affect the concentration of mercury in the atmosphere. Garrels et al. (1975) considered the impact of mining, which greatly accelerates the rate at which mercury is transferred from the sediments to the land. Mining results in a 90×10^8 g y^{-1} increase in the flux of mercury to the land reservoir. Emission of mercury vapor to the atmosphere due to anthropogenic activities was estimated to be 102×10^8 g y^{-1}. The other fluxes also increased.

Let us evaluate the impact of the anthropogenic mercury vapor emissions on the mercury content of the atmosphere using the first-order kinetic model. First we calculate the rate constant using equation 1–26 and the steady-state model shown in Figure 1–12. In the prehuman mercury cycle, the total input of mercury vapor to the atmospheric reservoir is 250×10^8 g y^{-1}.

$$k = F_{\text{input}}/A_i^0 = 250 \times 10^8 \text{ g y}^{-1}/40 \times 10^8 \text{ g} = 6.25 \text{ y}^{-1}$$

Adding the anthropogenic mercury vapor input of 102×10^8 g y^{-1} to the prehuman input gives $F_{\text{input}} = 352 \times 10^8$ g y^{-1}. We can now evaluate equation 1–24 for any time t. For example, when $t = 1$ year,

$$A_i(t) = \frac{F_{\text{input}}}{k} - \left(\frac{F_{\text{input}}}{k} - A_i^0\right)\exp(-kt) = \frac{352}{6.25} - \left(\frac{352}{6.25} - 40\right)\exp[(-6.25)(1)]$$

$$= 56.29 \times 10^8 \text{ g}$$

We see that after 1 year the mercury content of the atmosphere has increased by 41%. Are we near the final steady-state value for the atmosphere? Because this is an exponential relationship, the approach to the new equilibrium will be asymptotic. We can easily solve equation 1–24 for multiple time intervals using a spreadsheet. Such a solution is shown in Figure 1–13. From the steady-state model shown in Figure 1–12, we know that the mean residence time for mercury in the atmospheric reservoir is 0.16 year. Given this short mean residence time, we would expect that the atmosphere would quickly achieve a new steady state. From Figure 1–13, we see that the atmosphere has effectively reached its new equilibrium value after 1 year. Note that in doing these calculations we have assumed the rate constant calculated from the steady-state model remains constant when the system is perturbed. This is not necessarily true, and calculations of this type need to be evaluated in terms of the actual behavior of the system. ∎

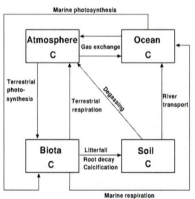

Figure 1–14a
The short-term carbon cycle, excluding anthropogenic inputs. After Berner (1999).

Another important factor in developing and using steady-state models is time. Perhaps the clearest distinction can be made between short-term and long-term cycles. This is illustrated using carbon. In the short-term carbon cycle (Figure 1–14a), the processes are biological and rapid physicochemical processes, such as gas exchange. The time scales are on the order of days to years. These types of cycles are easily disrupted by anthropogenic changes—in the case of carbon by the release of CO_2 by the burning of fossil fuels and deforestation—and are, hence, of significant environmental interest. In the long-term carbon cycle, the processes are largely geological and operate on the time scale of hundreds of thousands to millions of years. The long-term cycle for carbon is illustrated in Figure 1–14b.

In a sense, the traditional steady-state box models are static; i.e., they do not show the various cause-and-effect feedbacks that can occur between the different reservoirs. Berner (1999) proposed a different approach using an interactive model that shows the various cause-and-effect feedbacks. This type of model (a cause-and-effect feedback model), for the long-term carbon cycle, is

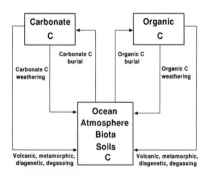

Figure 1–14b
The long-term carbon cycle. After Berner (1999).

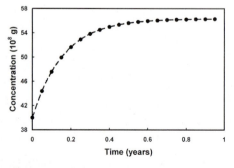

Figure 1–13
Variation in mercury content of the atmosphere as a function of time.

shown in Figure 1–15. In this model, the long-term carbon cycle consists of two sub-cycles—silicate-carbonate and organic. In the silicate-carbonate subcycle, atmospheric CO_2 is used in the weathering of Ca and Mg silicate minerals, the Ca and Mg ions are removed from solution as carbonate minerals, and the CO_2 is ultimately returned to the atmosphere by the thermal breakdown of carbonates. The organic subcycle involves the sequestering of carbon by the burial of organic matter in sediments and the subsequent release of carbon during weathering by oxidation of organic matter in old sediments. Note that both of these subcycles are largely geological in character.

On the cause-and-effect feedback diagram (Figure 1–15), the arrows originate at causes and end at effects. Arrows with small

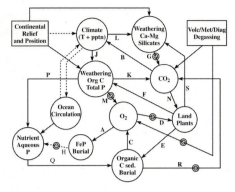

Figure 1–15
Cause-and-effect feedback diagram for the long-term carbon cycle. Arrows originate at causes and end at effects. Arrows with small concentric circles represent inverse responses; arrows without concentric circles represent direct responses. From Berner (1999).

concentric circles indicate inverse responses; those without small concentric circles indicate direct responses. For example, the arrow labeled M shows the cause-and-effect relationship between the weathering of organic carbon and atmospheric oxygen. In this case, the weathering of organic carbon leads to a decrease in atmospheric oxygen, an inverse response as indicated by the small concentric circles on the arrow. The arrow labeled L shows the cause-and-effect relationship between climate and the weathering of Ca-Mg silicates. An increase in temperature and/or precipitation leads to an increase in the rate of weathering, a direct response as indicated by the absence of small concentric circles on this arrow.

We can also look at subcycles in this diagram to see if they have positive or negative feedbacks. If the subcycle contains an even number of concentric circles, the feedback is positive. If the subcycle contains an odd number of concentric circles, the feedback is negative. Positive feedbacks lead to an amplification of an initial increase or decrease, and negative feedbacks lead to a dampening of the initial increase or decrease. For example, consider the subcycle B-L-G, which has an odd number of concentric circles. In this cycle, increasing CO_2 leads to a warmer and wetter climate with a concomitant increase in the weathering of Ca-Mg silicates. This increase in weathering leads to an increase in the uptake of CO_2, a negative feedback. Thus, this cycle tends to dampen increases in CO_2. The subcycle B-J-K contains no concentric circles (by default, an even number). In this cycle, an increase in CO_2 leads to a warmer and wetter climate and increased weathering of organic carbon, releasing CO_2 to the atmosphere, a positive feedback loop. Conversely, a decrease in CO_2 would lead to a cooler and drier climate and a decrease in the weathering of organic carbon, decreasing the amount of CO_2 released to the atmosphere. This is also a positive feedback, because it reinforces the original change, which was a decrease in atmospheric CO_2.

QUESTIONS AND PROBLEMS

1. Distinguish between the *continuous* and *discontinuous* view of matter.
2. What is a *photon*?
3. Distinguish between the *principal*, *azimuthal*, *magnetic*, and *spin* quantum numbers.
4. What is the *Pauli exclusion principle*, and what is its significance?
5. What is the *Heisenberg uncertainty principle*, and what is its significance?
6. Why are the $5s$ orbitals filled before the $4d$ orbitals?

7. Define *valence*.

8. With reference to Appendix I, explain why the rare earth elements (La–Lu) usually have a valence of +3.

9. Why are the elements Sc–Zn sometimes referred to as the transition elements?

10. Distinguish between *cations* and *anions*.

11. Distinguish between, and explain, *emission* and *absorption spectra*.

12. Define *isotope*.

13. What is the *strong force*?

14. Name and describe the four types of chemical bonding.

15. Define *molarity*, *molality*, and *normality*.

16. What are *precipitation*, *acid–base*, and *oxidation–reduction* reactions?

17. Distinguish between an *empirical* and a *molecular formula*.

18. Define and discuss *Boyle's* and *Charles's laws*.

19. What is the significance of *absolute zero*? Describe the experiment that led to the initial concept and estimate of the temperature of absolute zero.

20. Describe the structure of water, and explain why this structure is responsible for many of the unique properties of water, such as its high heat capacity, high latent heats of fusion and vaporization, and the rather remarkable observation that solid water is less dense than liquid water.

21. When we make ice cream, we pack a mixture of ice and salt around the container of ice cream mix. Why do we do this?

22. What do we mean when we say a system is in a *steady state*?

23. Distinguish between short-term and long-term geochemical cycles. What is the key factor used to make this distinction?

24. How does a cause-and-effect feedback model differ from a steady-state box model?

25. For the Bohr atom, calculate the energy that is released when an electron moves from the fourth allowed orbit to the second allowed orbit. Calculate the wavelength of the emitted radiation.

26. Calculate the wavelength of an electron in the third orbit of the Bohr atom.

27. Given the principal quantum number n = 3, how many electrons can be accommodated in the various orbitals? Write the complete set of quantum numbers.

28. Given the principal quantum number, n = 4, how many electrons can be accommodated in the various orbitals? Write the complete set of quantum numbers.

29. Write the electron formula for a neutral atom having 16 electrons. Predict the possible valences for this atom.

30. Write the electron formula for a neutral atom having 51 electrons. Predict the possible valences for this atom.

31. $^{57}_{26}$Fe contains how many protons? how many neutrons?

32. $^{121}_{51}$Sb contains how many protons? how many neutrons?

33. There are three stable isotopes of neon.
    a. Using the following data, calculate the atomic weight of neon.

Isotope	% Abundance	Mass (amu)
^{20}Ne	90.48	19.99244
^{21}Ne	0.27	20.99385
^{22}Ne	9.25	21.99139

    b. If 5.4 mol of neon gas are present in a gas cylinder, calculate the weight of neon gas in the cylinder.

    c. Calculate the number of neon atoms in the gas cylinder.

    d. Calculate the mass (in g) of a single neon atom.

34. There are five stable isotopes of Ti.

    a. Using the following data, calculate the atomic weight of Ti.

Isotope	% Abundance	Mass (amu)
^{46}Ti	8.0	45.95263
^{47}Ti	7.28	46.95176
^{48}Ti	73.8	47.94795
^{49}Ti	5.5	48.94787
^{50}Ti	5.4	49.94479

    b. The chemical formula for the mineral rutile is TiO_2. Calculate the number of grams of Ti in 45.3 g of rutile.

    c. Calculate the number of moles of Ti in the rutile.

    d. Calculate the number of Ti atoms in the rutile.

    e. Calculate the mass (in kg) of a single Ti atom.

35. The $^{51}V/^{12}C$ ratio $= 4.2453$. Calculate the mass (in amu) of ^{51}V.

36. Calculate the molecular weight of kaolinite $[Al_2Si_2O_5(OH)_4]$.

37. 1.25 g of NaCl is dissolved in a 1-L solution. Calculate

    a. The number of moles of NaCl dissolved in the solution.

    b. The molarity of the NaCl solution.

    c. The number of Na^+ ions in the solution.

38. 0.02 g of $SrSO_4$ is dissolved in a 1-L solution. Calculate

    a. The number of moles of $SrSO_4$ dissolved in the solution.

    b. The molarity of the $SrSO_4$ solution.

    c. The number of Sr^{2+} ions in solution.

39. A 1-L solution contains 2.2×10^{-4} mol of $CaCO_3$. Calculate the mass of $CaCO_3$ dissolved in the solution.

40. A $CaCO_3$-$MgCO_3$ solid solution contains 3 wt% Mg. Calculate the mole fraction of $MgCO_3$ in the solid solution.

41. A $CaMgSi_2O_6$-$CaFeSi_2O_6$ solid solution (Ca-rich pyroxene) contains 5 wt% Fe. Calculate the mole fraction of $CaMgSi_2O_6$ in the solid solution.

42. If you have read the book *Civil Action* you know that the organic chemical TCE played an important role in this famous environmental case. An analysis of TCE reveals that it consists of 18.283 wt% C, 0.767 wt% H, and 80.950 wt% Cl. The molecular weight of TCE is 131.389 g.

    a. Determine the empirical formula for TCE.

    b. Determine the molecular formula for TCE.

43. A chemical analysis of a clay mineral reveals that it consists of 20.903 wt% Al, 21.759 wt% Si, 55.776 wt% O, and 1.562 wt% H. The molecular weight of the clay mineral is 258.16 g.

    a. Determine the empirical formula for this clay mineral.

    b. Determine the molecular formula for this clay mineral.

44. A chemical analysis of a carbon compound gives 82.659 wt% C and 17.341 wt% H. The molecular weight of the carbon compound is 58.123 g.

    a. Determine the empirical formula for this organic compound.

    b. Determine the molecular formula for this organic compound.

45. Balance the following chemical reactions:

a. $Al(OH)_{3\ gibbsite} + H^+ \rightarrow Al^{3+} + H_2O$

b. $Al_2Si_2O_5(OH)_{4\ kaolinite} + H^+ \rightarrow Al^{3+} + H_4SiO_{4\ (aq)} + H_2O$

c. $CaAl_2Si_2O_{8\ anorthite} + H^+ \rightarrow Ca^{2+} + Al^{3+} + H_4SiO_{4\ (aq)}$

d. $FeOOH_{\ goethite} + H^+ \rightarrow Fe^{3+} + H_2O$

e. $KFe_3(SO_4)_2(OH)_{6\ jarosite} + H^+ \rightarrow K^+ + Fe^{3+} + SO_4^{2-} + H_2O$

f. $KAlSi_3O_{8\ feldspar} + H_2O + H^+ \rightarrow K^+ + Al^{3+} + H_4SiO_{4\ (aq)}$

g. $Mg_5(CO_3)_4(OH)_2 \cdot 4H_2O_{\ hydromagnesite} + H^+ \rightarrow Mg^{2+} + CO_3^{2-} + H_2O$

h. $SiO_2 + H_2O \rightarrow H_4SiO_{4\ (aq)}$

i. $As_2S_{3\ orpiment} + H_2O \rightarrow H_3AsO_{3\ (aq)} + HS^- + H^+$

j. $Cu_3(OH)_2(CO_3)_{2\ azurite} + H^+ \rightarrow Cu^{2+} + H_2O + HCO_3^-$

46. The German airship *Hindenberg*, filled with hydrogen gas, met an unhappy and dramatic fate at Lakehurst Naval Air Station in New Jersey. Assume that the volume occupied by hydrogen gas was 6×10^7 L, pressure = 1 atm, and temperature = 25°C. Calculate the number of moles of hydrogen gas in the *Hindenberg*. Use the ideal gas equation to solve this problem.

47. The partial pressure of CO_2 gas in a classroom is 0.00036 atm. Partial pressure is the pressure that would be due to CO_2 alone, and partial pressures are often used to express the concentrations of gases. The size of the classroom is 15 m $\times$ 6 m $\times$ 3 m. Remember that 1 L = 1000 cm^3. The classroom temperature is 25°C. Calculate the number of CO_2 molecules in the classroom. Use the van der Waals equation to solve this problem. We will assume that the only interactions are between CO_2 molecules (obviously not true, but we want to give you a fighting chance to solve the problem). At first glance, the algebraic manipulation appears daunting. However, the clever student will recognize that this problem can be set up on a spreadsheet and quickly solved by successive approximations.

48. Calculate the latent heat of vaporization of water at 20°C.

49. With reference to Figure 1–11, calculate

a. The residence time of H_2O in the ocean reservoir.

b. The residence time of H_2O in the lakes and rivers reservoir.

c. Comment on the relative residence times of H_2O in all three reservoirs in terms of which would be most affected by changes in the rate of H_2O addition and removal, and which would be least affected. Why?

50. One predicted consequence of greenhouse warming is a speeding up of the hydrologic cycle. With reference to Figure 1–11, suppose greenhouse warming led to a 10% increase in the rate of evaporation of water from lakes, rivers, and the ocean. Calculate the new steady-state value for water in the atmospheric reservoir using a first-order kinetics model. Select an appropriate time for this calculation based on the residence time of water vapor in the atmosphere.

51. Using the cause-and-effect feedback diagram for carbon (Figure 1–15), for each of the following determine if they are positive or negative feedback cycles. What criteria did you use to decide if they were positive or negative feedback cycles? Describe in words what happens within each cycle.

a. N-S-G

b. A-H-Q-C

c. D-E-C

d. D-F-P-Q-C

e. D-F-M

f. B-J-P-Q-R

Equilibrium Thermodynamics and Kinetics

Equilibrium thermodynamics predicts the concentrations (or, more precisely, activities) of various species and phases if a reaction reaches equilibrium. **Kinetics** *tells us how fast, or if, the reaction will reach equilibrium.* Thermodynamics is an elegant way to deal with problems of chemical equilibria, but it is important to note that kinetics will determine if these equilibrium conditions are actually attained. In the following sections we will consider these topics in the context of the typical conditions found in the surface and shallow subsurface environments.

THE LAWS OF THERMODYNAMICS

Thermodynamic principles are applied to systems. A **system** is *that portion of the universe we wish to study*, e.g., a beaker containing a solution, a room, the ocean, planet earth, the universe. The system can be **open** (*exchanging matter and energy with its surroundings*), **closed** (*not exchanging matter with its surroundings*), or **isolated** (*exchanges neither matter nor energy with its surroundings*). As an example, consider a beaker of water standing on a table. The system is open because it can exchange both heat with the surroundings and gases with the atmosphere. Now seal the top of the beaker. The result is a closed system because it cannot exchange matter (gases) with the atmosphere. If we place the beaker in a thermos bottle, it becomes an isolated system because it can exchange neither heat nor matter with its surroundings.

The properties of a system can be either intensive or extensive. **Intensive properties** are *independent of the magnitude of the system*. Examples are pressure and temperature. **Extensive properties** are *dependent on the magnitude of the system*. Examples are volume and mass.

A system can be described in terms of phases and components. A **phase** is defined as *"a uniform, homogeneous, physically distinct, and mechanically separable portion of a system"* (Nordstrom and Munoz, 1986, p. 67). **Components** are the *chemical constituents (species) needed to completely describe the chemical composition of every phase in a system*. The choice of components is determined by the physical-chemical conditions of the system. For example, consider the three-phase system solid water (ice)–liquid water–water vapor that would exist under normal surface conditions. The composition of each phase can be completely described by a single component—H_2O. Now consider the same system over a much wider range of temperatures so that a fourth phase, plasma, is found. In a plasma, the H_2O would break down into hydrogen and oxygen atoms. To completely describe the composition of this system we would need two components—H and O. The solid, liquid, and vapor phases of H_2O would be formed by combining the two components in the proportions $2H + O \rightarrow H_2O$.

First Law of Thermodynamics

The *first law of thermodynamics* deals with the conservation of energy. One statement of the law is that *energy can be neither created nor destroyed, it can only be changed from one form to another*. The concept of enthalpy (heat flow) arises from the first law.

The internal energy of a system is the sum of the kinetic and potential energies of its constituent atoms. Let us change the internal energy of this system by adding (or subtracting) heat and by doing mechanical work (on or by the system). We can write the following equation:

$$\Delta E = q - w \tag{2-1}$$

where ΔE is the change in internal energy of the system, q is the heat added or removed from the system, and w is the work done on or by the system. By convention, heat added to a system is positive and work done by a system is positive. Thus, the internal energy of a system will increase if heat is added and will decrease if work is done by the system. For infinitesimal changes, equation 2–1 can be written

$$dE = dq - dw \tag{2-2}$$

If the work done by or on a system causes a change in volume at constant pressure (pressure–volume work), then the equation for the change in internal energy can be written

$$\Delta E = q - P\Delta V \tag{2-3}$$

For an infinitesimal change, equation 2–3 can be written

$$dE = dq - P\,dV \tag{2-4}$$

Enthalpy is equal to the *heat flow when processes occur at constant pressure and the only work done is pressure–volume work*. This is the most likely situation in the natural surface environment. For an infinitesimal change, enthalpy can be written

$$dH = dE + P\,dV + V\,dP \tag{2-5}$$

At constant P, $dP = 0$ and

$$dH = dE + P\,dV \tag{2-6}$$

If we substitute for dE (equation 2–4), then dH equals dq at constant P.

$$dH = (dq - P\,dV) + P\,dV = dq \tag{2-7}$$

Note that it is very difficult to determine absolute values for either internal energy (E) or enthalpy (H). Hence, these values are determined on a relative basis compared to standard conditions (see later). **Exothermic reactions** *release heat energy* (i.e., enthalpy is negative for the reaction), and **endothermic reactions** *use heat energy* (i.e., enthalpy is positive for the reaction).

The **heat of formation** (sometimes called the *enthalpy* of formation or the *standard heat* of formation) is the *enthalpy change that occurs when a compound is formed from its elements at particular temperature and pressure* (the **standard state**). It is convenient to use 25°C and 1 bar as the temperature and pressure for the standard state. Hence, the standard state for a gas is the ideal gas at 1 bar and 25°C, for a liquid it is the pure liquid at 1 bar and 25°C, and for a solid it is a specified crystalline state at 1 bar and 25°C. The heat of formation of the most stable form of an element is arbitrarily set equal to zero. For example, the heat of formation for La $_{metal}$ and $N_{2\,(g)}$ equals zero. For dissolved ionic species, the heat of formation of H^+ is set equal to zero.

Second Law of Thermodynamics

The **second law of thermodynamics** deals with the concept of entropy. One statement of the law is that *for any spontaneous process, the process always proceeds in the direction of increasing disorder*. Another way of looking at this law is that during any spontaneous process there is a decrease in the amount of useable energy. As a simple example of the second law consider the burning of coal. In coal the atoms are ordered; i.e., they occur as complex organic molecules. During the combustion process these molecules are broken down, with the concomitant release of energy and the production of CO_2 and H_2O. The atoms are now in a much more disordered (dispersed) state. In order to produce more coal, we need to recombine these atoms, which requires energy—in fact, more energy than was released by the burning of the coal. Hence, there has been a decrease in the amount of useable energy.

The second law has important practical and philosophical implications. During any process there is a decline in the amount of useable energy. This is an important concept in ecology in terms of the efficiency of ecosystems. As a rough rule of thumb, in biological systems about 90% of the energy is lost in going from one trophic level (nourishment level) to another. For example, for every 1000 calories of "grass energy" consumed by a cow, only 100 calories are converted to biomass. If the cow ends up as a steak, only 10% of the energy in the cow's biomass ends up as human biomass. Thus, vegetarians are more efficient users of primary biomass (green plants) than meat eaters. The second law of thermodynamics also predicts that at some point the universe will cease to function. A way of looking at this is to divide the universe into a high-temperature reservoir—the stars—and a low-temperature reservoir—the interstellar medium. As energy is lost from the stars to the interstellar medium, the temperature of the interstellar medium will rise. At some point, the temperatures of these reservoirs will become equal and energy transfers will cease. This has sometimes been referred to as the "heat death of the universe."

A mathematical statement of the second law is

$$\Delta S = \frac{q}{T} \tag{2-8}$$

where ΔS is the change in entropy and T is temperature in Kelvin (the absolute temperature scale). Rearranging equation 2–8 to give $q = T\Delta S$ and substituting into equation 2–1 gives

$$\Delta E = T\Delta S - w \tag{2-9}$$

or, in differential form,

$$dE = T\,dS - dw \tag{2-10}$$

If we consider only pressure–volume work, then equation 2–10 becomes

$$dE = T\,dS - P\,dV \tag{2-11}$$

Substitution of equation 2–11 into equation 2–5 yields

$$dH = (T\,dS - P\,dV) + P\,dV + V\,dP = T\,dS + V\,dP \tag{2-12}$$

EQUILIBRIUM THERMODYNAMICS

In the real world, systems can exist in several states: unstable, metastable, and stable. To illustrate these different states, consider a ball sitting at the top of a hill (Figure 2–1, p. 30). In terms of gravitational energy, the lowest energy state is achieved when the ball is at the bottom of the hill. At the top the ball is unstable with respect to gravitational energy. The least disturbance will cause the ball to start rolling down the hill. Partway down the hill there is a small notch. If the ball does not have enough energy to roll up over the lip of the notch, it will be stuck at this position. Clearly, this is not the lowest possible energy state, which occurs at the bottom of the hill, and the ball is said to be in metastable equilibrium. However, it can remain indefinitely at this position if there isn't enough energy available to push it over the lip of the notch. This is an example of a kinetic impediment to the achievement of equilibrium, and we can regard the energy needed to push the ball out of the notch to be equivalent to the activation energy. Activation energy will be discussed in the latter part of this chapter. If sufficient energy is put into the system to push the ball out

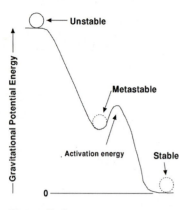

Figure 2–1
Illustration of the states of a system in terms of the gravitational energy of a ball.

Note: In the thermodynamic data table (Appendix II) the tabulated values for ΔG_f^0 and ΔH_f^0 are, respectively, the standard free energies and enthalpies of formation for the compounds from the elements in their standard state. S^0 is also tabulated in the thermodynamic data table. Free energy and enthalpy are in kJ mol^{-1} while entropy is in J mol^{-1} K^{-1}. When doing thermodynamic calculations be sure to convert either enthalpy to joules (1 kJ = 1000 J) or entropy to kilojoules (1 J = 1 × 10^{-3} kJ). Temperature is the absolute temperature in Kelvin. To convert centigrade temperatures to Kelvin, add 273.15 to the centigrade value. Failing to do this is one of the most common mistakes made by students on homework and exam problems. By international convention, the accepted units are SI units. In SI units, heat energy is expressed as joules. For years scientists and engineers used the calorie as the unit of heat measurement. Hence, many older tables of thermodynamic data tabulate heat energy in calories. The following conversion factor is used to convert calories to joules:

$$1 \text{ calorie} = 4.184 \text{ joules}$$

---

of the notch, it will roll to the bottom of the hill. At this position, the gravitational potential energy equals zero and the ball has achieved its equilibrium position in terms of gravitational energy. *In equilibrium thermodynamics it is this lowest energy state that is determined*.

Free Energy

A system at equilibrium is in a state of minimum energy. In chemical thermodynamics this energy is measured either as *Gibbs free energy* (when the reaction occurs at constant T and P) or *Helmholtz free energy* (when the reaction occurs at constant T and V). Here we will use Gibbs free energy, named for J. Willard Gibbs, a Yale University chemist.

For a system at constant T and P, Gibbs free energy can be written

$$G = H - TS \tag{2–13}$$

where H is enthalpy (kJ mol^{-1}), S is entropy (J mol^{-1} K^{-1}), and T is temperature in K (Kelvin).

For changes that occur at constant T and P, the expression for Gibbs free energy becomes

$$\Delta G = \Delta H - T\Delta S \tag{2–14}$$

If ΔG is (−), the process occurs spontaneously. If $\Delta G = 0$, the process is at equilibrium. If ΔG is (+), the reaction does not occur spontaneously. *Note that chemical reactions are written from left to right*. For example, consider the reaction

$$CaSO_{4 \text{ anhydrite}} \rightleftharpoons Ca^{2+} + SO_4^{2-}$$

If the process occurs spontaneously, $CaSO_4$ will dissolve to form Ca^{2+} and SO_4^{2-} ions (i.e., the reaction is running from left to right). If the process is at equilibrium, the concentration of the various species remains constant. If the process does not occur spontaneously, the reaction would actually run from right to left. In the last case, if you rewrote the equation so that the ions were on the left side and the compound on the right side, the free energy would be negative.

We now write equation 2–14 as follows:

$$\Delta G_R^0 = \Delta H_R^0 - T\Delta S_R^0 \tag{2–15}$$

where ΔG_R^0 is the free-energy change, ΔH_R^0 is the enthalpy change, and ΔS_R^0 is the entropy change for the reaction at standard conditions. The enthalpy and entropy changes are calculated, respectively, by subtracting the sum of the enthalpies or entropies of the reactants from the sum of the enthalpies or entropies of the products. *Neither equation 2–14 nor 2–15 can be used to find the ΔG_R^0 value for a single compound or species*.

Chemical Potential

Chemical potential is defined as

$$\mu_i = \left(\frac{\Delta G}{\Delta n_i} \right)_{T,P} \tag{2–16}$$

where μ_i is the chemical potential of a certain component in a system and Δn_i is the change in moles of that component in the system. For a system at equilibrium, μ_i is the same in all phases.

Activity and Fugacity

An important concept in dealing with chemical systems is *activity* (or *fugacity*, for a gas). This is the *apparent (or effective) concentration of a species* as opposed to the actual concentration. Activity and fugacity are a measure of the departure of a system from ideal behavior and need to be taken into account even when dealing with relatively dilute solutions. Activity (or fugacity) is related to concentration through the ***activity coefficient***.

$$\gamma_i = \frac{a_i}{m_i} \tag{2-17}$$

where γ_i is the activity coefficient, a_i is the activity, and m_i is the actual concentration. Rearranging equation 2–17 gives

$$a_i = \gamma_i m_i \tag{2-18}$$

A later section will deal with the calculation of the activity coefficient.

The Equilibrium Constant

We can write the chemical potential for component i as follows:

$$\mu_i = \mu_i^0 + RT \ln a_i \tag{2-19}$$

where μ_i^0 is the chemical potential of component i in its standard state and R is the gas constant ($8.3143 \text{ J mol}^{-1} \text{ K}^{-1}$). For solid solutions, solutions of two miscible liquids, and the solvent in aqueous solutions, the standard state is the pure substance at the same temperature and pressure.

Let us suppose we have a chemical reaction of the form

$$aA + bB \rightleftharpoons cC + dD \tag{2-20}$$

The uppercase letters represent the species and the lowercase letters represent the number of each species (each chemical entity). What follows is an exercise in letter manipulation, a common activity in the sciences and mathematics. To determine the change in free energy for the system, we subtract the free energies of the products (right side of the equation) from the free energy of the reactants (left side of the equation).

$$\Delta G_{\text{reaction}} = \Sigma \, \Delta G_{\text{products}} - \Sigma \, \Delta G_{\text{reactants}} \tag{2-21}$$

From the definition of chemical potential (equation 2–16),

$$\Delta G = n_i \mu_i \tag{2-22}$$

and, on substitution, equation 2–21 becomes

$$\Delta G_{\text{R}} = c\mu_{\text{c}} + d\mu_{\text{d}} - a\mu_{\text{a}} - b\mu_{\text{b}} \tag{2-23}$$

We substitute equation 2–19 for μ_i and write equation 2–23 as follows:

$$\Delta G_{\text{R}} = c\mu_{\text{C}}^0 + d\mu_{\text{D}}^0 - a\mu_{\text{A}}^0 - b\mu_{\text{B}}^0 + RT \ln \left(\frac{a_{\text{C}}^c \cdot a_{\text{D}}^d}{a_{\text{A}}^a \cdot a_{\text{B}}^b} \right) \tag{2-24}$$

The portion of equation 2–24 dealing with the chemical potentials in the standard state is equivalent to ΔG^0, and thus equation 2–24 reduces to

$$\Delta G_{\text{R}} = \Delta G_{\text{R}}^0 + RT \ln \left(\frac{a_{\text{C}}^c \cdot a_{\text{D}}^d}{a_{\text{A}}^a \cdot a_{\text{B}}^b} \right) \tag{2-25}$$

At equilibrium, $\Delta G_{\text{R}} = 0$ and equation 2–25 becomes

$$RT \ln \left(\frac{a_{\text{C}}^c \cdot a_{\text{D}}^d}{a_{\text{A}}^a \cdot a_{\text{B}}^b} \right) = -\Delta G_{\text{R}}^0 \tag{2-26}$$

Dividing both sides of equation 2–26 by R, T, and converting the natural log to an exponent, gives

$$K_{\text{eq}} = \left(\frac{a_{\text{C}}^c \cdot a_{\text{D}}^d}{a_{\text{A}}^a \cdot a_{\text{B}}^b} \right) = \exp\left(\frac{-\Delta G_{\text{R}}^0}{RT} \right) \tag{2-27}$$

For our last manipulation we will rearrange equation 2–27 to yield

$$\ln K_{eq} = \left(\frac{-\Delta G_R^0}{RT} \right) \tag{2-28}$$

Note that in equation 2–28 we are calculating the natural log of K_{eq}. A natural log result can be converted to a base 10 log result by dividing by 2.30259, or by using the proper keystroke sequence on a calculator. If the reactions of interest are occurring at 25°C and 1 bar and the free energy is in kJ mol^{-1}, equation 2–28 can be written in base 10 log form as

$$\log K_{eq} = \frac{-\Delta G_R^0}{5.708} \tag{2-29}$$

Equation 2–29 can only be used for reactions occurring at 25°C and 1 bar. We have now completed our exercise in letter manipulation and everyone should feel very refreshed. The following example illustrates the calculation of an equilibrium constant.

A word about thermodynamic data. There are a number of data compilations in the literature. You may find that for any particular species the different compilations do not give the same value. Thermodynamic data are determined experimentally and are thus subject to error. In compiling a set of thermodynamic data, the compilers attempt to make the data internally consistent; i.e., calculations using the data set give reasonable and consistent answers. Where values have been determined by more than one laboratory, a judgment must be made by the compilers as to which values are more consistent with their data set. In addition, errors can creep into compilations, which can lead to some rather paradoxical answers. The user must be aware of these potential pitfalls. When using a water-chemistry computer model, you should take note of the thermodynamic database used by the model.

EXAMPLE 2–1   Calculate the solubility product for gypsum at 25°C. The solubility product is a special form of an equilibrium constant (i.e., it enables us to calculate the activity of the ions in solution at saturation).

The reaction is

$$CaSO_4 \cdot 2H_2O_{\text{ gypsum}} \rightleftharpoons Ca^{2+} + SO_4^{2-} + 2H_2O$$

The equilibrium equation is written

$$K_{eq} = \frac{[Ca^{2+}][SO_4^{2-}][H_2O]^2}{[CaSO_4 \cdot 2H_2O_{\text{ gypsum}}]}$$

Note that the various species are enclosed in brackets. The convention is that enclosing the *species in brackets indicates activity*, while enclosing the *species in parentheses indicates concentrations*. $CaSO_4 \cdot 2H_2O$ is in its standard state (pure solid) and its activity equals 1. For a dilute solution, the activity of water also equals 1. The activity of water in cases other than dilute solutions is considered in a later section. For the dissolution of gypsum, the equilibrium equation becomes

$$K_{eq} = [Ca^{2+}][SO_4^{2-}] = K_{sp}$$

Selecting the appropriate free energies of formation (Appendix II, source 2) yields (remember, products − reactants)

$$\Delta G_R^0 = (-552.8) + (-744.0) + (2)(-237.14) - (-1797.36) = 26.28 \text{ kJ mol}^{-1}$$

$$\log K_{sp} = \frac{-\Delta G_R^0}{5.708} = \frac{-26.28}{5.708} = -4.60, \qquad K_{sp} = 10^{-4.60} \qquad \blacksquare$$

Henry's Law

This relationship is used in several ways. In solutions, it is used to describe the activity of a dilute component as a function of concentration. In this case, the relationship is written

$$a_i = h_i X_i \tag{2-30}$$

where a_i is the activity of species i, h_i is the Henry's law proportionality constant, and X_i is the concentration of species i.

For gases, Henry's law relates the fugacity of the gas to its activity in solution. At total pressures of 1 bar or less and temperatures near surface temperatures, gases tend to obey the ideal gas law, and hence the fugacity of a gas equals its partial pressure. In this case, we write Henry's law as

$$C_i = K_H P_i \tag{2-31}$$

Table 2–1 Henry's Law Constants for Gases at 1 Bar Total Pressure in Mol L^{-1} Bar^{-1}*

T (°C)	O_2	N_2	CO_2	H_2S	SO_2
0	2.18×10^{-3}	1.05×10^{-3}	7.64×10^{-2}	2.08×10^{-1}	3.56
5	1.91×10^{-3}	9.31×10^{-4}	6.35×10^{-2}	1.77×10^{-1}	3.01
10	1.70×10^{-3}	8.30×10^{-4}	5.33×10^{-2}	1.52×10^{-1}	2.53
15	1.52×10^{-3}	7.52×10^{-4}	4.55×10^{-2}	1.31×10^{-1}	2.11
20	1.38×10^{-3}	6.89×10^{-4}	3.92×10^{-2}	1.15×10^{-1}	1.76
25	1.26×10^{-3}	6.40×10^{-4}	3.39×10^{-2}	1.02×10^{-1}	1.46
30	1.16×10^{-3}	5.99×10^{-4}	2.97×10^{-2}	9.09×10^{-2}	1.21
35	1.09×10^{-3}	5.60×10^{-4}	2.64×10^{-2}	8.17×10^{-2}	1.00
40	1.03×10^{-3}	5.28×10^{-4}	2.36×10^{-2}	7.41×10^{-2}	0.837
50	9.32×10^{-4}	4.85×10^{-4}	1.95×10^{-2}	6.21×10^{-2}	—

*Data are from Pagenkopf (1978).

where C_i is the concentration of the gaseous species in solution, K_H is the Henry's law constant in mol L^{-1} bar^{-1}, and P_i is the partial pressure of gaseous species i. Henry's law constants vary as a function of temperature (Table 2–1). We will use Henry's law in later chapters to calculate the activity of various gases dissolved in water and the partial pressure of various volatile organic solvents.

EXAMPLE 2–2 Calculate the solubility of oxygen in water at 20°C.

At sea level—i.e., a total atmospheric pressure of 1 bar (in terms of the standard atmosphere, precisely 1.0135 bar)—the partial pressure of oxygen is 0.21 bar. At 20°C, the Henry's law constant for oxygen is 1.38×10^{-3} mol L^{-1} bar^{-1}.

$$O_{2 \, (aq)} = K_H P_{O_2} = (1.38 \times 10^{-3} \text{ mol } L^{-1} \text{ bar}^{-1})(0.21 \text{ bar}) = 2.90 \times 10^{-4} \text{ mol } L^{-1}$$

Converting to concentration in mg L^{-1} (equivalent to ppm in freshwater at temperatures near 25°C)

$$\text{Concentration} = (2.9 \times 10^{-4} \text{ mol } L^{-1})(32.0 \text{ g } O_2 \text{ mol}^{-1})$$
$$= 9.28 \times 10^{-3} \text{ g } L^{-1} = 9.28 \text{ mg } L^{-1} \qquad \blacksquare$$

Free Energies at Temperatures Other Than 25°C

So far we have considered reactions that take place at 25°C. Free energy does vary as a function of temperature. If the reaction of interest occurs at a temperature other than 25°C, the free-energy values must be corrected. Unfortunately, this is not a trivial issue. We will briefly consider the problem here. More detailed discussions can be found in Drever (1997), Langmuir (1997), and elsewhere.

If the deviations in temperature from 25°C are small (15° or less, i.e., from 10° to 40°C), we can make the assumption that ΔH_R^0 and ΔS_R^0 are constant. With reference to equation 2–15 and equation 2–28 we can write

$$\Delta G_R^0 = \Delta H_R^0 - T \Delta S_R^0 = -RT \ln K_{eq} \qquad (2–32)$$

Rearranging yields

$$\ln K_{eq} = \frac{-\Delta H_R^0}{RT} + \frac{\Delta S_R^0}{R} \qquad (2–33)$$

We can also write this equation in terms of the equilibrium constant and the standard enthalpy of the reaction (a form of the van't Hoff equation) as follows:

$$\ln K_t = \ln K_r + \frac{\Delta H_R^0}{R}\left(\frac{1}{T_r} - \frac{1}{T_t}\right) \qquad (2–34)$$

where K_t is the equilibrium constant at temperature t, K_r is the equilibrium constant at 25°C, T_t is the temperature t, and T_r is 298.15 K (25°C). $R = 8.314 \times 10^{-3}$ kJ mol^{-1} K^{-1}.

EXAMPLE 2–3 Calculate the solubility product for gypsum at 40°C using equation 2–34. From a previous example (2–1), the solubility product at 25°C is $K_{sp} = 10^{-4.60}$. Calculate ΔH_R^0 for the reaction.

Using the thermodynamic values from Appendix II, source 2,

$$\Delta H_R^0 = (-543.0) + (-909.34) + (2)(-285.83) - (-2022.92) = -1.08 \text{ kJ mol}^{-1}$$

Substituting the appropriate values into equation 2–34 yields

$$\ln K_t = \ln(10^{-4.60}) + \frac{-1.08}{8.314 \times 10^{-3}}\left(\frac{1}{298.15} - \frac{1}{313.15}\right) = -10.61$$

Converting to base 10,

$$\log K_t = -4.61 \quad \text{or} \quad K_t = 10^{-4.61}$$

The solubility of gypsum decreases slightly as temperature changes from 25°C to 40°C. For this particular reaction, the change is small (about 2.0%). However, for other reactions the change can be large (see the problem set). ■

For temperature departures of more than 15°C from standard conditions, the computation becomes more complex. The following equations are easily solved using a spreadsheet (or a computer program). The biggest problem is obtaining the appropriate thermodynamic data, particularly for the ionic species. For enthalpy, we can write the following equation:

$$\int_{H_{298}^0}^{H_T^0} dH = \int_{298}^{T} c_P \, dT \tag{2–35}$$

where T is the temperature of interest and c_P is the heat capacity. **Heat capacity** is defined as *the amount of heat energy required to raise the temperature of 1 gram of a substance 1°C*. There are two different heat capacities, one determined at constant volume (c_v) and the other determined at constant pressure (c_P). At constant volume, changes in heat energy only change the temperature of the system. At constant pressure, changes in heat energy lead to changes in both temperature and volume (pressure–volume work). Thus, c_P is always larger than c_v. The heat capacity varies as a function of temperature. The relationship can be written as follows:

$$c_P = a + bT - \frac{c}{T^2} \tag{2–36}$$

and a, b, and c are experimentally determined constants. Particularly for ionic species, the requisite experiments have not been done. Thus, in many cases it is not possible to calculate c_P as a function of temperature. Substituting equation 2–36 into equation 2–35 and integrating yields

$$H_T^0 = H_{298}^0 + a(T - 298) + \frac{b}{2}(T^2 - 298^2) + c\left(\frac{1}{T} - \frac{1}{298}\right) \tag{2–37}$$

Similarly, for entropy,

$$S_T^0 = \int_{298}^{T} \frac{c_P}{T} \, dT + S_{298}^0 = \left[a \ln T + bT + \frac{c}{2T^2}\right]_{298}^{T} + S_{298}^0 \tag{2–38}$$

which becomes, after inserting limits,

$$S_T^0 = \left[a \ln\left(\frac{T}{298}\right) + b(T - 298) + \frac{c}{2}\left(\frac{1}{T^2} - \frac{1}{298^2}\right)\right] + S_{298}^0 \tag{2–39}$$

Substitution of equations 2–37 and 2–39 into equation 2–15 allows us to calculate Gibbs free energy as a function of temperature. In the context of environmental geochemistry, many processes of interest occur at or near standard (1 atm) pressure. For substantial departures from standard pressure, we would need to include a term to account for changes in free energy due to changes in pressure and volume. This is straightforward for solids, reasonably simple for liquids, but complicated in the case of gases (changes in volume are significant). A detailed account of the effect of pressure on free-energy calculations can be found in Langmuir (1997).

Le Châtelier's Principle

In the preceding sections we have developed several quantitative measures that can be used to determine what happens during equilibrium reactions. We can also make reasonable predictions about the effect a perturbation will have on an equilibrium reaction using *Le Châtelier's principle*, which can be stated: *If a change is imposed on a system at equilibrium, the position of the equilibrium will shift in a direction that tends to reduce the change.* We can consider three possibilities: changes in concentration, changes in pressure, and changes in temperature.

Changes in Concentration   Consider the following reaction that we used in Example 2–1:

$$CaSO_4 \cdot 2H_2O \text{ }_{gypsum} \rightleftharpoons Ca^{2+} + SO_4^{2-} + 2H_2O$$

We have already calculated the equilibrium constant for this reaction—i.e., $K_{eq} = 10^{-4.60}$—and have noted that this is a solubility product. At equilibrium, the concentration of $Ca^{2+} = SO_4^{2-} = 10^{-2.30}$. Suppose we added 0.01 mol of Ca^{2+} ion to the solution. The solution would now be oversaturated and the reaction would go to the left until enough of the added Ca^{2+} ion had been removed for the reaction to return to equilibrium. If you do the calculation, you will find that when the reaction is once again at equilibrium there will be more Ca^{2+} than we started with (and less SO_4^{2-}), but the amount of Ca^{2+} will be less than that immediately after we added Ca^{2+} to the solution. In terms of changes in concentration, we can state Le Châtelier's principle as follows: If a product or reactant is added to a system at equilibrium, the reaction will go in the direction that decreases the amount of the added constituent. If a product or reactant is removed from a system at equilibrium, the reaction will go in the direction that increases the amount of the removed constituent.

Changes in Pressure   There are three possibilities: (1) Add or remove a gaseous reactant or product, (2) add an inert gas (one not involved in the reaction), and (3) change the volume of the container. Case (1) is analogous to what happens when you change the concentration of a constituent. Consider the following reaction, a very familiar one in metamorphic petrology:

$$CaCO_3 \text{ }_{calcite} + SiO_2 \text{ }_{quartz} \rightleftharpoons CaSiO_3 \text{ }_{wollastonite} + CO_2 \text{ }_{(g)}$$

If we added CO_2 to the system, the reaction would move to the left in order to reduce the amount of CO_2. Case (2) has no effect on the system. At first glance this might not seem reasonable because an increase in pressure should favor the solid phases, which occupy a smaller volume. But note that the gas we are concerned with is CO_2, and only changes in the pressure of CO_2 would affect the reaction. If there was a decrease in volume [Case (3)], the reaction would respond by reducing the number of gaseous molecules in the system. This is accomplished by converting some of the CO_2 into solid $CaCO_3$, which occupies a significantly smaller volume. An increase in volume would have the opposite effect.

Changes in Temperature   Changes in temperature are different from the previous cases in that changes in temperature cause changes in the equilibrium constant. However, we can make predictions regarding the effect that changes in temperature will have on the equilibrium constant. With increasing temperature, reactions move in the direction that consumes heat energy. Returning to the reaction

$$CaSO_4 \cdot 2H_2O \text{ gypsum} \rightleftharpoons Ca^{2+} + SO_4^{2-} + 2H_2O$$

in Example 2–3, we found that this reaction was exothermic because ΔH_R^0 for the reaction was negative. At first glance, this may appear counterintuitive. But if you write the equation used to determine ΔH_R^0, you will find that heat is a product. If ΔH_R^0 is positive, the reaction is endothermic (i.e., heat is a reactant). Because heat is consumed when the reaction moves to the left, with increasing temperature we would expect the reaction to move to the left and the concentrations of the products would decrease with respect to the concentration of the reactant. This would lead to a decrease in the equilibrium constant, as confirmed by the calculations in Example 2–3. If the reaction was endothermic, increasing temperature would cause the reaction to shift to the right, leading to an increase in the equilibrium constant.

CALCULATION OF ACTIVITY COEFFICIENTS

In an ideal solution, activity would equal concentration. It is often assumed that in very dilute solutions concentration does equal activity; i.e., the solution is behaving ideally. While this assumption may be justified in special cases, for most real solutions ideality is not achieved. This is particularly true for solutions that contain ionic species. The departure from ideal behavior is caused mainly by two factors:

1. Electrostatic interactions between charged ions.
2. The formation of hydration shells around ions.

The latter factor is easily understood in terms of the structure of the water molecule. Because the bond angle between the H atoms in H_2O is 104.5°, one side of the molecule has a slight positive charge and the other side a slight negative charge. The water molecule is said to be polar. Positive ions (cations) in solution will be surrounded by water molecules with their negative sides facing the cation, while negative ions (anions) will be surrounded by water molecules with their positive sides facing the anion. This tends to shield the cations and anions from each other. For uncharged species, which do not have electrostatic interactions, concentration equals activity in dilute solutions. In concentrated solutions, the uncharged species do show deviations from ideality, and an activity coefficient must be calculated for the uncharged species.

A variety of models (Debye–Hückel, Davies, Truesdell–Jones, Bronsted–Guggenheim–Scatchard specific ion interaction theory, and Pitzer) are used to calculate activity coefficients. Each is effective for a particular range of ionic strengths. Langmuir (1997) gives a detailed description of the different models and their effective concentration ranges. The first step in an activity coefficient calculation is to determine the ionic strength of the solution. The **ionic strength** of a solution is calculated as follows:

$$I = \frac{1}{2} \sum m_i z_i^2 \qquad (2-40)$$

where m_i = the moles per liter of ion i and z_i = the charge of ion i.

Debye–Hückel Model

The simplest form of the model assumes that (1) positive ions are surrounded by a cloud of negative ions and vice versa, (2) interactions between species are entirely electrostatic, (3) the ions can be considered to be point charges, and (4) ions around any particular ion follow a Boltzmann distribution. This simple form of the model fails at relatively low ionic strengths because it does not take into account the finite size of the ions. The more complex form of the model takes into account the size of the ions and is the preferred version of the Debye–Hückel equation,

$$\log \gamma_i = \frac{-A z_i^2 \sqrt{I}}{1 + B a_i \sqrt{I}} \qquad (2-41)$$

Table 2–2 Density of Water, Dielectric Constant, and Debye–Hückel and Truesdell–Jones Constants at 1 Bar Pressure*

Temperature (°C)	Density of water (kg m^{-3})	Dielectric constant	Debye–Hückel Constants	
			A	B
0	0.99984	87.8191	0.4912	0.3248
5	0.99977	85.8838	0.4942	0.3254
10	0.99970	83.9785	0.4976	0.3262
15	0.99896	82.1042	0.5012	0.3270
20	0.99821	80.2618	0.5052	0.3279
25	0.99693	78.4520	0.5094	0.3289
30	0.99565	76.6755	0.5138	0.3299
35	0.99394	74.9326	0.5185	0.3310
40	0.99222	73.2238	0.5235	0.3322
45	0.99013	71.5493	0.5287	0.3334
50	0.98803	69.9094	0.5342	0.3347
55	0.98562	68.3043	0.5399	0.3360
60	0.98320	66.7342	0.5459	0.3373

*Density of water from *CRC Handbook of Chemistry and Physics* (2000).

A word of caution. Compilations of the hydrated radii of ions (a_i) tend to use different units. If the radii are given directly in angstroms, then the value for B can be used as calculated (or as given in Table 2–2). If the radii are tabulated as 10^{-10} m or 10^{-8} cm, then B must be multiplied by 10^{10} or 10^{8}, respectively.

where A and B are constants depending only on T and P, and a_i is the hydrated radius of a particular ion.

At atmospheric pressure, $A = 1.824928 \times 10^6 \rho_0^{0.5}(\varepsilon T)^{-1.5}$ and $B = 50.3(\varepsilon T)^{-0.5}$, where ρ_0 is the density of water, ε is the dielectric constant of water, and T is in Kelvin. At any temperature T (in the range 0° to 100°C), the dielectric constant can be determined from the following relationship:

$$\varepsilon = 2727.586 + 0.6224107T - 466.9151 \ln T - 52000.87/T \qquad (2-42)$$

where T is in Kelvin. At 25°C, $\rho_0 = 0.99693$, $\varepsilon = 78.4520$, $A = 0.5094$, and $B = 0.3289$. Values for the density of water, the dielectric constant of water, and the A and B Debye–Hückel constants at 1 bar, from 0° to 60°C, are tabulated in Table 2–2.

Truesdell–Jones Model

The Truesdell–Jones (Truesdell and Jones, 1974) and Davies (Davies, 1962) equations are extended versions of the Debye–Hückel equation. An additional term is added to the Debye–Hückel equation that takes into account the observation that in high-ionic-strength experimental systems the activity coefficients begin to increase with increasing ionic strength. The Truesdell–Jones equation is written

$$\log \gamma_i = \frac{-Az_i^2\sqrt{I}}{1 + Ba_i\sqrt{I}} + bI \qquad (2-43)$$

where a_i and b are determined from experimental data. Because the a_i values are determined experimentally—i.e., they are selected so that the calculated curves fit the observed data—these a_i values can only be used in the Truesdell–Jones equation (2–43). The A and B constants are the same in both the Truesdell–Jones and Debye–Hückel equations. Selected values for Debye–Hückel a_i and Truesdell–Jones a_i and b are given in Table 2–3.

Table 2–3   Parameters for the Debye–Hückel and Truesdell–Jones Equations at 1 Atm*

Ion	Debye–Hückel a_i (Å)	Truesdell–Jones a_i (Å)	Truesdell–Jones b (L mol^{-1})
H^+	9.0	4.78	0.24
Na^+	4.0	4.32	0.06
K^+	3.0	3.71	0.01
Mg^{2+}	8.0	5.46	0.22
Ca^{2+}	6.0	4.86	0.15
Sr^{2+}	5.0	5.48	0.11
Ba^{2+}	5.0	4.55	0.09
Mn^{2+}	6.0	7.04	0.22
Fe^{2+}	6.0	5.08	0.16
Co^{2+}	6.0	6.17	0.22
Ni^{2+}	6.0	5.51	0.22
Zn^{2+}	6.0	4.87	0.24
Cd^{2+}	5.0	5.80	0.10
Pb^{2+}	4.5	4.80	0.01
Al^{3+}	9.0	6.65	0.19
OH^-	3.5	10.65	0.21
F^-	3.5	3.46	0.08
Cl^-	3.0	3.71	0.01
HCO_3^-	4.0	5.4	0
CO_3^{2-}	4.5	5.4	0
SO_4^{2-}	4.0	5.31	−0.07

*Debye–Hückel radii from Kielland (1937). Truesdell–Jones parameters from Truesdell and Jones (1974) and Parkhurst (1990).

EXAMPLE 2–4   Given the following river water chemistry, calculate the activity coefficient for Ca^{2+} at 25°C using both the Debye–Hückel and Truesdell–Jones equations.

	River Water Concentration (mg L^{-1})						
Ca^{2+}	Mg^{2+}	Na^+	K^+	Cl^-	SO_4^{2-}	HCO_3^-	SiO_2
14.7	3.7	7.2	1.4	8.3	11.5	53	10.4

Water-chemistry data are sometimes reported in ppm, which is a weight/weight measure. For example, 1 ppm = 1 mg kg^{-1}. This is approximately equivalent to 1 mg L^{-1}, but for very precise calculations the density of the water should be used to adjust ppm to mg kg^{-1}; i.e., multiply ppm by the density of the water. To calculate the ionic strength, we need to convert the weight of each ion to moles of each ion. 1 mg = 10^{-3} g. Divide the weight of each ion (in grams) by its gram-molecular weight. For example, $Ca^{2+} =$ 14.7 × 10^{-3} g/40.08 = 3.67 × 10^{-4} mol L^{-1} and $HCO_3^- = $ 53 × 10^{-3} g/61.02 = 8.69 × 10^{-4} mol L^{-1}. Converting all of the ions to mol L^{-1} and solving for I,

$$I = 0.5 \sum m_i z_i^2 = 0.5[(3.67 \times 10^{-4})(2)^2 + (1.52 \times 10^{-4})(2)^2 + (3.13 \times 10^{-4})(1)^2$$
$$+ (3.58 \times 10^{-5})(1)^2 + (2.34 \times 10^{-4})(1)^2 + (1.20 \times 10^{-4})(2)^2$$
$$+ (8.69 \times 10^{-4})(1)^2] = 2 \times 10^{-3} \text{ mol charge L}^{-1}$$

Solving for the activity coefficient using the Debye–Hückel equation,

$$\log \gamma_i = \frac{-A z_i^2 \sqrt{I}}{1 + B a_i \sqrt{I}} = \frac{-(0.5094)(2)^2 \sqrt{2.0 \times 10^{-3}}}{1 + (0.3289)(6.0)\sqrt{2.0 \times 10^{-3}}} = -0.084$$

$$\gamma_i = 0.82$$

Table 2–4 Appropriate Ranges of Ionic
Strengths for Activity-
Coefficient Models

Model	Ionic strength (mol L^{-1})
Debye–Hückel	0 to 0.1
Davies	0 to 0.6
Truesdell–Jones	0 to 2
Specific ion interaction	0 to 4
Pitzer	0 to 6

*From Langmuir (1997).

Solving for the activity coefficient using the Truesdell–Jones equation,

$$\log \gamma_i = \frac{-A z_i^2 \sqrt{I}}{1 + B a_i \sqrt{I}} + bI$$

$$= \frac{-(0.5094)(2)^2 \sqrt{2.0 \times 10^{-3}}}{1 + (0.3289)(4.86)\sqrt{2.0 \times 10^{-3}}} + (0.15)(2.0 \times 10^{-3}) = -0.085$$

At concentrations typical of river water, both models yield essentially the same activity coefficient, and hence either model could be used to calculate the activity coefficients for river and lake waters with low ionic strength. The useable range of ionic strengths for each model is tabulated in Table 2–4. ■

Pitzer Model

For solutions of higher ionic strength, the Pitzer model would be most appropriate. The Pitzer model (Pitzer, 1973, 1979, 1980) takes into account binary interactions between two ions of the same or opposite sign and ternary interactions between three or more ions. This model is most effective for concentrated brines. The solutions are generally complex and best carried out by computer. Further details on the Pitzer model can be found in Langmuir (1997). Several of the commonly used water-chemistry computer codes (SOLMINEQ.88, PHRQPITZ, and PRHEEQC) use the Pitzer model.

Why Do We Care About Activity-Coefficient Models?

The preceding has not been an exhaustive discussion of activity coefficient calculations for ionic species, but it has drawn attention to the types of models and the limitations of the models. A number of computer codes have been developed to calculate speciation in natural waters. These different programs use different activity-coefficient models. The user should be aware of these differences. For example, it would be inappropriate to do speciation calculations for a brine using a computer code based on the Debye–Hückel model. The user should select a computer model appropriate for the system being considered. Table 2–4 summarizes the range of ionic strengths appropriate for each activity-coefficient model.

Calculation of Activity Coefficients for Uncharged Species

There are several activity-coefficient models for uncharged species. Plummer and MacKenzie (1974) calculate the activity coefficient (γ) for an uncharged species as follows:

$$\gamma = 10^{0.1I} \tag{2–44}$$

where I is the ionic strength. The empirical Setchenow equation (Millero and Schreiber, 1982) calculates γ as follows:

$$\log \gamma_i = K_i I \tag{2–45}$$

where K_i is a constant ranging in value from 0.02 to 0.23 at 25°C. For relatively dilute aqueous systems, such as rivers and lakes, which have ionic strengths on the order of 2×10^{-3}, and brackish waters with ionic strengths on the order of 2×10^{-2}, both equations give activity coefficients close to 1. For concentrated solutions, such as seawater, the activity coefficient is greater than 1. Seawater has an ionic strength of about 0.7, which, using the Plummer and MacKenzie (1974) model, yields an activity coefficient of 1.17. Values calculated from the Setchenow equation will vary as a function of the uncharged species. For example, the K_i value for $H_4SiO_{4\ (aq)}$ at 25°C is 0.080 (Marshall and Chen, 1982), giving a calculated activity of 1.14 for this uncharged species. An important point is that with increasing ionic strength, compounds that yield uncharged species on dissolution (such as quartz, which dissolves to form $H_4SiO_{4\ (aq)}$) become less soluble while ionic compounds (such as NaCl, which dissolves to form Na^+ and Cl^-) become more soluble (Figure 2–2). In very high ionic strength solutions (>1.0 mol L^{-1}), the activity coefficients for the ions will also start to increase and the preceding observation will no longer be true.

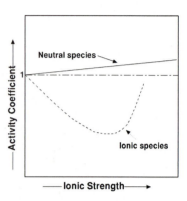

Figure 2–2
Schematic representation of the variation in activity coefficients versus the ionic strength of solutions.

Activity of Water

When water is a solvent, pure liquid water at infinite dilution is used as the standard state; i.e., the activity of $H_2O = 1$ at infinite dilution. The activity of water is related to the mole fraction of pure water, X_{H_2O}, as follows:

$$\mu_{H_2O} = \mu_{H_2O}^0 + RT \ln X_{H_2O} \tag{2–46}$$

In most cases, we are dealing with dilute solutions and we can set the activity of $H_2O = 1$. In more concentrated solutions, such as seawater, the activity will be slightly less than 1.

AQUEOUS COMPLEXES

An ***aqueous complex*** is a *dissolved species formed from two or more simpler species, each of which can exist in aqueous solution* (Drever, 1997, p. 34). In the context of equilibrium calculations, complexes are important because their formation can increase the solubility of various compounds. Consider the following reaction:

$$A^+ + B^- \rightleftharpoons AB_{\ (aq)} \tag{2–47}$$

We can write an equilibrium equation for this reaction in the usual way. In this case, the equilibrium constant is called a stability constant because it is a measure of the stability of the aqueous complex.

$$K_{stab} = \frac{[AB_{\ (aq)}]}{[A^+][B^-]} \tag{2–48}$$

The solubility of compounds whose ions form aqueous species is increased over that predicted from the solubility product for the compound. This is because some of the ions released during the dissolution process are taken up by the aqueous complex. For a solution at saturation, the concentration of the aqueous complex can be determined from

$$[AB_{\ (aq)}] = K_{stab} \cdot [A^+][B^-] = K_{stab} \cdot K_{sp} \tag{2–49}$$

Setting the activity of the solid equal to 1, $K_{sp} = [A^+][B^-]$. Since the aqueous complex is an uncharged species, at low to moderate ionic strengths the activity coefficient is 1, and concentration equals activity.

EXAMPLE 2–5 In pure water the solubility of gypsum is 10.2×10^{-3} mol L^{-1}. Calcium and sulfate ions in solution form an aqueous complex according to the following reaction:

$$Ca^{2+} + SO_4^{2-} \rightleftharpoons CaSO_{4\,(aq)}$$

$$K_{stab} = \frac{[CaSO_{4\,(aq)}]}{[Ca^{2+}][SO_4^{2-}]} = 10^{2.23}$$

$$[CaSO_{4\,(aq)}] = K_{stab}[Ca^{2+}][SO_4^{2-}] = K_{stab} \cdot K_{sp\,(gypsum)}$$
$$= (10^{2.23})(10^{-4.60}) = 10^{-2.37} = 4.3 \times 10^{-3}$$

The activity coefficient for an uncharged species is approximately 1. Therefore, the activity and concentration of the uncharged species is the same. Given that the concentration of $CaSO_{4\,(aq)}$ is 4.3×10^{-3} mol L^{-1}, the solubility of gypsum in pure water has increased from 10.2×10^{-3} mol L^{-1} to 14.5×10^{-3} mol L^{-1}, an increase in solubility of approximately 40% due to the formation of the aqueous complex. ∎

Some elements occur in solution predominantly as complexes rather than free ions. For these elements it is the properties of the complexes rather than the free ions that determine their behavior in natural systems. Certain metals become soluble when the opportunity arises to form a particular type of aqueous complex. For example, Fe^{3+} and Al^{3+} are generally immobile in the weathering environment. However, in the presence of oxalic acid ($H_2C_2O_4$) these ions can become mobile. Oxalic acid dissociates to form H^+ and COO^- (an oxalate ion). This ion can bond to a metal, forming a complex species. The anion is a ***ligand*** *(an anion or neutral molecule that can combine with a cation)*. Given the abundance of dissolved organic matter in the natural environment, this type of aqueous complex can be important in the transport of iron and aluminum.

Ligands can be either monodentate (one pair of shared electrons in a complex) or multidentate (more than one pair of shared electrons in a complex). Inorganic ligands tend to form monodentate complexes, and organic ligands tend to form multidentate complexes. The maximum number of ligands that can bond to a single cation is a function of the relative size of the cation and ligand (see Chapter 7 for a discussion of coordination). In natural waters the number of potential ligands is usually too low for this maximum number to be achieved. Ligands that form multiple bonds with a cation and have a "cagelike" structure are referred to as *chelates*. The metal cation is strongly bonded in these types of structures and chelation is an important process for the removal of metals from the aqueous environment.

MEASUREMENT OF DISEQUILIBRIUM

We have now considered the thermodynamic basis of equilibrium calculations. A remaining question is, How do we measure how close a particular reaction is to equilibrium? Consider the dissolution of gypsum:

$$CaSO_4 \cdot 2H_2O_{\,gypsum} \rightleftharpoons Ca^{2+} + SO_4^{2-} + 2H_2O$$

The solubility product for this reaction is written

$$K_{sp} = [Ca^{2+}][SO_4^{2-}] = 10^{-4.60}$$

If the system is at equilibrium, the concentration of species in solution (*Activity Product* [AP] if both ions and uncharged species are involved; *Ion Activity Product* [IAP] if only charged species are involved) will equal the solubility product. In this example we are dealing with ions, so at equilibrium IAP = $10^{-4.60}$ (the K_{sp} for gypsum). If the IAP is less than $10^{-4.60}$, the solution is undersaturated with respect to gypsum; and if the IAP is greater than $10^{-4.60}$, the solution is supersaturated with respect to gypsum. Quantitatively, the approach to equilibrium can be expressed as AP (or IAP)/K_{sp}, which is 1 at equilibrium, or log [AP (or IAP)/K_{sp}], which is 0 at equilibrium.

EXAMPLE 2–6 In a particular solution $[Ca^{2+}] = 10^{-3}$ mol L^{-1} and $[SO_4^{2-}] = 10^{-2}$ mol L^{-1}. At 25°C, is the solution over- or undersaturated with respect to gypsum? Give a quantitative measure of the degree of over- or undersaturation.

$$\frac{IAP}{K_{sp}} = \frac{(10^{-3})(10^{-2})}{10^{-4.60}} = 10^{-0.40}$$

IAP/K_{sp} is less than 1 so the solution is undersaturated with respect to gypsum. ∎

KINETICS

Equilibrium thermodynamics predicts the final state of the system. Kinetics tells us if the system will actually achieve this state within a reasonable time. In practice, the determination of rates of reaction is not a straightforward exercise. A number of factors affect these rates, and careful experimentation is required to understand the reaction mechanisms. Consider, for example, the dissolution of gypsum in water.

$$CaSO_4 \cdot 2H_2O \text{ gypsum} \rightarrow Ca^{2+} + SO_4^{2-} + 2H_2O$$

This reaction involves several steps. First, the ions need to be freed from the crystal structure, and then they need to be transported away from the crystal surface. The first step requires an energy input to break the bonds in the crystal structure, and the second step is diffusion controlled. The importance of the second step may not be obvious, but consider what will happen if the Ca^{2+} and SO_4^{2-} ions are not removed from the immediate vicinity of the gypsum crystal. The concentration of the ions in solution will increase until the microenvironment surrounding the gypsum crystal becomes saturated with respect to gypsum. At this point the dissolution will stop. Whichever of these steps is the slowest will determine the rate at which the dissolution reaction proceeds. As another example, from metamorphic petrology, consider the apparently simple phase transition sillimanite $\rightleftharpoons$ kyanite. Sillimanite and kyanite are polymorphs of Al_2SiO_5, and the reaction suggests that sillimanite is directly converted to kyanite through the rearrangement of atoms in the crystal structure. Petrographic investigations, however, suggest that this is a much more complex reaction, in which the sillimanite first breaks down to form other minerals and is then reformed as kyanite. Hence, what appears to be a very simple reaction actually involves several steps. The slowest of these steps would determine the rate for the overall reaction.

Reactions are of two types: homogeneous and heterogeneous. **Homogeneous** reactions *only involve one phase (gas, liquid, or solid).* **Heterogeneous** reactions *involve two or more phases.* Consider the condensation of water vapor in the atmosphere. If the condensation process involves water vapor condensing directly from the vapor phase, the reaction is homogeneous (it occurs in the gas phase). If the water vapor condenses onto a particle, the reaction is heterogeneous (a gas and a solid are involved). Direct nucleation turns out to be difficult, and in the homogeneous system significant supersaturation is required before condensation actually occurs. On the other hand, the presence of particles facilitates nucleation and in the heterogeneous system condensation occurs close to saturation. Another example from everyday experience is the crystallization of "rock candy." In practice, this is done by heating water on a stove and dissolving copious amounts of sugar in the water. The water is then slowly cooled to room temperature, and the system is now significantly supersaturated with respect to sugar. A small sugar crystal is introduced into the water and serves as a site of nucleation. Rapid crystal growth occurs as the sugar in the supersaturated system precipitates onto the sugar crystal. Virtually all reactions of interest in nature are heterogeneous, but the best understood reactions in terms of kinetics are homogeneous, a not unusual situation in science.

Order of Reactions

For an elementary reaction, the *order* is defined by the number of individual atoms or molecules involved in the reaction. We can talk about order in terms of an individual species or in terms of the overall reaction. Consider the reaction $A + B \rightarrow AB$. The reaction is first order in terms of A and B and the overall reaction is second order. The reaction $A + 2B \rightarrow C$ is first order with respect to A and C, second order with respect to B, and third order overall. Differential and integrated equations follow for the different types of reactions. Also listed is the equation for the half-life ($t_{1/2}$) of each type of reaction—i.e., the time it will take for half of the reactant to be consumed in the reaction.

Zeroth order. The reaction rate is independent of the concentration of the reactant (A).

$$\frac{dA}{dt} = -k, \qquad A = A_0 - kt, \qquad t_{1/2} = \frac{0.5A_0}{k} \qquad (2\text{--}50)$$

First order. The reaction rate is dependent on the concentration of the reactant, $A \rightarrow B$.

$$\frac{dA}{dt} = -kA, \qquad \ln A = \ln A_0 - kt, \qquad t_{1/2} = \frac{0.693}{k} \qquad (2\text{--}51)$$

The first-order rate equations may look familiar since radioactive decay is a first-order reaction—$t_{1/2}$ in this case being the half-life of the radioactive parent.

Second order. The reaction rate is dependent on the concentration of the reactant, $2A \rightarrow B$.

$$\frac{dA}{dt} = -kA^2, \qquad \frac{1}{A} = \frac{1}{A_0} - kt, \qquad t_{1/2} = -\frac{1}{kA_0} \qquad (2\text{--}52)$$

Note that the units for k depend on the order of the reaction. For example, zeroth-order reaction, $\text{mol cm}^{-1} \text{s}^{-1}$; first-order, s^{-1}, and second-order, $\text{cm}^3 \text{mol}^{-1} \text{s}^{-1}$. Higher-order reactions are possible, but in most geochemical systems of interest the reactions are second order or less. The order of a reaction can be determined by experiment. For example, if the reaction is zeroth order, an arithmetic plot of concentration versus time will yield a straight line. If the reaction is first order, an arithmetic plot of concentration versus time will yield a curved line, because for a first-order reaction the relationship between concentration and time is exponential. A log-log plot of the rate of the reaction versus concentration will yield a straight line whose slope defines the order of the reaction (0 for zeroth order, 1 for first order, 2 for second order, etc.). The various reaction orders are shown graphically in Figure 2–3.

Figure 2–3
Graphical illustration of various reaction orders in terms of the variation in concentration of species A versus time and reaction rate versus concentration of species A. After Appelo and Postma (1996).

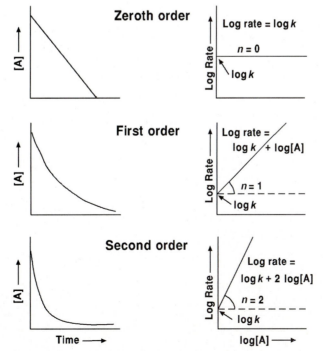

EXAMPLE 2–7 At pH > 4, the oxidation of Fe^{2+} in solution can be represented by the overall reaction

$$Fe^{2+} + \tfrac{1}{4}O_2 + \tfrac{5}{2}H_2O \rightarrow Fe(OH)_3 + 2H^+$$

For this reaction, the empirical rate law (Langmuir, 1997) is

$$\frac{d(Fe(II))}{dt} = -k_+ \frac{(Fe^{2+})}{(H^+)^2} P_{O_2}$$

k_+ indicates that this is the rate constant in the forward direction. For this reaction, at 20°C, $k_+ = 1.2 \times 10^{-11}$ mol^2 bar^{-1} d^{-1}. Under atmospheric conditions, $P_{O_2} = 0.2$ bar. Given that Fe^{2+} in solution $= 1 \times 10^{-3}$ mol L^{-1}, calculate the reaction rate at pH $= 5$ and pH $= 7$.

At pH $= 5$,

$$\text{Rate} = -k_+ \frac{(Fe^{2+})}{(H^+)^2} P_{O_2} = -(1.2 \times 10^{-11}) \frac{(1 \times 10^{-3})}{(10^{-5})^2}(0.2) = -2.4 \times 10^{-5} \text{ mol L}^{-1} \text{ d}^{-1}$$

At pH $= 7$,

$$\text{Rate} = -k_+ \frac{(Fe^{2+})}{(H^+)^2} P_{O_2} = -(1.2 \times 10^{-11}) \frac{(1 \times 10^{-3})}{(10^{-7})^2}(0.2)$$

$$= -2.4 \times 10^{-1} \text{ mol L}^{-1} \text{ d}^{-1}$$

The reaction rate increases by 4 orders of magnitude in going from pH $= 5$ to pH $= 7$. ∎

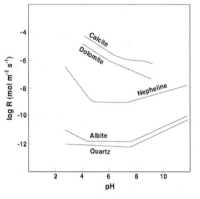

Figure 2–4
Dissolution rates (mol m^{-2} s^{-1}) for common minerals, carbonates, and silicates, as a function of pH. After Lerman (1990). From "Transport and kinetics in surficial processes" by A. Lerman in AQUATIC CHEMICAL KINETICS edited by W. Stumm, pp. 505–534. Copyright © 1990. This material is used by permission of John Wiley & Sons, Inc.

Reaction rates have been determined experimentally for a number of important reactions. The results of some of these experiments are portrayed graphically in Figure 2–4. A more complete discussion of reaction rates and their calculation can be found in standard textbooks on water chemistry. Of particular interest for geoscientists is the text by Langmuir (1997), and the interested student should consult this text for further details regarding the calculation of reaction rates.

The half-life of a reaction is a useful concept in that it enables us to assess the significance of kinetics on the attainment of thermodynamic equilibrium. In Figure 2–5 the residence time of water in various reservoirs is compared to the reaction rate (measured as $t_{1/2}$) for a number of different types of chemical processes. If $t_{1/2}$ for a particular chemical

Figure 2–5
Comparison of half-lives of various reactions and residence times of water in different reservoirs. From Langmuir (1997).

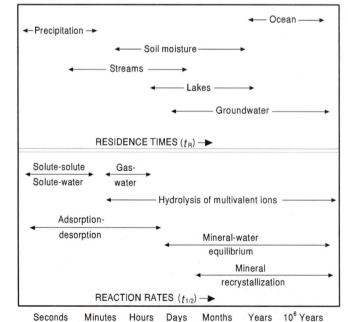

process is less then the residence time, then it would be expected that this process would achieve equilibrium. For example, gas–water reactions are rapid ($t_{1/2}$ = hours to days) compared to the residence time of water in the groundwater and ocean reservoirs ($t_{1/2}$ = days to millions of years). Hence, we would expect that gas–water reactions would be in equilibrium in these reservoirs. Conversely, if $t_{1/2}$ was greater than the residence time, then it would be expected that the process would not achieve equilibrium. The reaction rates for mineral–water equilibria are long ($t_{1/2}$ = days to thousands of years) compared to the residence time for precipitation in the atmosphere ($t_{1/2}$ = seconds to hours). Hence, in atmospheric aerosols we would not expect solid particles and the liquid or vapor phase to be in equilibrium.

The Arrhenius Equation

The **Arrhenius equation** *relates the rate at which a reaction occurs to the temperature*:

$$k = A \exp\left(\frac{-E_a}{RT}\right) \tag{2–53}$$

where A is a pre-exponential factor generally determined by experiment and relatively independent of temperature, E_a is the activation energy for the reaction, R is the ideal gas constant, and T is the temperature in Kelvin. Converting to base 10 logs gives

$$\log k = \log A - \frac{E_a}{2.303RT} \tag{2–54}$$

Activation energies are determined by experiment. If a particular reaction follows the Arrhenius relationship, then a plot of $\log k$ versus $1/T$ yields a straight line with a slope of $-E_a/2.303R$. A rough rule of thumb is that the reaction rate doubles with every 10°C increase in temperature. Measured activation energies (in kJ mol^{-1}) vary from 8 to 500 depending on the process. Simple physical adsorption has low activation energies, and solid phase reactions (such as solid-state diffusion) have high activation energies.

EXAMPLE 2–8 One of the reactions in the carbonate system is

$$CaCO_{3 \text{ calcite}} + H_2CO_{3 \text{ (aq)}} \rightarrow Ca^{2+} + 2HCO_3^-$$

For this reaction, the rate constant at 25°C is 3.47×10^{-5} s^{-1} and $\log k = -4.46$. Determine the value of the pre-exponential term A for this reaction and determine the rate constant for the reaction at 10°C.

 The activation energy for the reaction is 41.85 kJ mol^{-1}. First determine the pre-exponential factor by rearranging equation 2–54 to solve for $\log A$.

$$\log A = \log k + \frac{E_a}{2.303RT} = -4.46 + \frac{41.85 \text{ kJ mol}^{-1}}{(2.303)(8.314 \times 10^{-3} \text{ kJ mol}^{-1} \text{ K}^{-1})(298.15 \text{ K})}$$

$$= -4.46 + 7.33 = 2.87$$

Calculate the reaction rate constant at 10°C using equation 2–54.

$$\log k = \log A - \frac{E_a}{2.303RT} = 2.87 - \frac{41.85 \text{ kJ mol}^{-1}}{(2.303)(8.314 \times 10^{-3} \text{ kJ mol}^{-1} \text{ K}^{-1})(283.15 \text{ K})}$$

$$= 2.87 - 7.72 = -4.85$$

and $k = 1.41 \times 10^{-5}$ s^{-1}. Changing the temperature from 25°C to 10°C leads to an approximately 2.5X decrease in the reaction rate. ∎

Nucleation

There are two types of nucleation—homogeneous and heterogeneous. **Homogeneous** nucleation occurs when a *nucleus forms spontaneously in an oversaturated solution.*

Heterogeneous nucleation occurs when a *nucleus forms in contact with a, usually solid, surface.* Homogeneous nucleation requires a much greater degree of supersaturation than heterogeneous nucleation.

The free energy of formation of a nucleus consists of the energy gained from the formation of bonds and the energy required to create the surface. This can be written mathematically as

$$\Delta G_{nuc} = \Delta G_{bulk} + \Delta G_{surf} \tag{2–55}$$

For an oversaturated solution, ΔG_{bulk} is always negative. For ΔG_{bulk}, we can write the following equation:

$$\Delta G_{bulk} = \frac{4\pi r^3}{3V} k_B T \ln \frac{a}{a_0} \tag{2–56}$$

where $4\pi r^3/3V$ is the volume of a spherical nucleus, V is the *molecular* volume, k_B is the Boltzmann constant (1.3805×10^{-23} J K^{-1}), T is the temperature in Kelvin, a is actual activity, and a_0 is the activity for a saturated solution.

For ΔG_{surf}, we can write

$$\Delta G_{surf} = 4\pi r^2 \overline{\gamma} \tag{2–57}$$

where $\overline{\gamma}$ is the interfacial energy. Interfacial energies vary over several orders of magnitude. For example, amorphous silica has a surface free energy of 46×10^{-3} J m^{-2} and goethite has a surface free energy of 1600×10^{-3} J m^{-2}. Combining equations 2–56 and 2–57 yields

$$\Delta G_{nuc} = -\frac{4\pi r^3}{3V} k_B T \ln \frac{a}{a_0} + 4\pi r^2 \overline{\gamma} \tag{2–58}$$

From equation 2–58 it is obvious that increasing the degree of oversaturation favors nucleation, as does increasing the radius of the particle. For any particular particle size and degree of oversaturation the ease of nucleation is a function of the interfacial energy; e.g., from the data just given it is much more difficult to nucleate goethite than amorphous silica. The rate at which nuclei form can be determined from the standard Arrhenius rate equation,

$$\text{Rate of nucleation} = \overline{A} \exp\left(\frac{-\Delta G^*}{k_B T}\right) \tag{2–59}$$

where $\overline{A}$ is a factor related to the efficiency of collisions of ions or molecules, ΔG^* is the maximum energy barrier (see Figure 2–6), k_B is the Boltzmann constant, and T is the temperature in Kelvin.

Interfacial energies are different for nuclei formed via homogeneous reactions versus nuclei formed via heterogeneous reactions. A discussion of this difference can be found in standard water-chemistry books, such as Stumm and Morgan (1996).

Dissolution and Growth

Once a nucleus has formed, the next question is, how fast will the particle grow? This is a function of several factors: (1) the rate at which the ions (or complex molecules) diffuse through the liquid to the surface of the growing particle and (2) the rate at which the ions or molecules are attached to the surface of the growing particle. One of these will be the

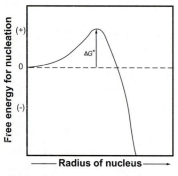

Figure 2–6

Variation of free energy of nucleation as a function of particle radius. The maximum free energy of formation corresponds to the maximum energy barrier. At greater particle radii the free energy of nucleation decreases and eventually becomes negative, and nucleation will proceed spontaneously. Modified from Drever (1997).

rate-limiting step. In addition, during dissolution of a particle, there may be a reaction zone between the surface of the original grain and the solution. For example, feldspars break down by *incongruent dissolution* (the feldspar decomposes to a mineral of different composition and species that enter the solution). Thus, a feldspar grain may become coated with a rim of reaction product through which the ions must diffuse in order for the reaction to continue. As the thickness of this reaction rim increases, the diffusion of ions or molecules through the reaction zone may become the rate-limiting step (Figure 2–7).

The topology of the mineral surface is also important in determining the rate at which both dissolution and particle growth occur. The mineral surface is usually not planar, but consists of *steps* and *kinks*. Atoms that form steps and kinks have higher energy, and dissolution or growth takes place at these locations. **Dislocations** occur *when one part of a crystal structure is offset relative to another part*. The number and types of dislocations are important in determining the rate of dissolution or growth. **Inhibitors**, or **surface poisons**, are *foreign species adsorbed at points of high energy on the crystal surface that may inhibit crystal growth or dissolution*. For example, in seawater the concentration of phosphate ions affects the dissolution rate of calcite. It is believed that the phosphate ion acts as an inhibitor and is preferentially attached to sites of high energy on the crystal surface.

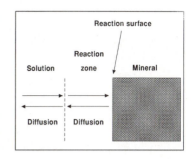

Figure 2–7

Schematic representation of a mineral reacting with a solution during dissolution. The rate-controlling step can be the diffusion of species through the solution, diffusion of species through the reaction zone, or the rate of the surface reaction. Given the slow rate of diffusion of species through the reaction zone, at some point the thickness of this zone will become sufficiently great that the diffusion of species through the reaction zone will become the rate-controlling step. Modified from Drever (1997).

WATER-CHEMISTRY COMPUTER MODELS

The previous discussion of thermodynamics and kinetics has been far from exhaustive but was intended to introduce the basic concepts used in the modeling of water chemistry. Because these calculations can be very laborious, a number of computer models have been developed to do these types of computations. These computer models can be divided into three basic types: speciation, mass transfer, and chemical mass transport. **Speciation models** *calculate the partitioning of elements between aqueous species and determine the degree of saturation with respect to mineral and gas phases.* The calculations that we considered in the section on equilibrium thermodynamics were of this type. The U.S. Geological Survey's WATEQ4F is an example of this type of model. **Mass transfer models** do the *same types of calculations as speciation models but in addition consider the effect of mass transfer processes (dissolution, precipitation, gas exchange, ion exchange, adsorption, etc.).* Examples of these types of models are PHREEQC and PHRQPITZ (U.S. Geological Survey). **Chemical mass transport models** include *speciation, mass transfer processes, and hydrodynamic advection and dispersion.* An example of this type of model is the U.S. Geological Survey's PHREEQM-2D.

The major sources of free computer models are the U.S. Environmental Protection Agency and the U.S. Geological Survey. These models can be accessed from the appropriate agency's web pages. The addresses listed here are current at the time of publication, but the various agencies do occasionally change their web addresses.

The site for the U.S. Environmental Protection Agency software is **http://www.epa.gov/epahome/models.htm**. The programs are developed and maintained by the Center for Exposure Assessment Modeling (CEAM). These are DOS-based programs, and a number of different types are available. The water-chemistry model is MINTEQA2, which can be used to calculate the equilibrium compositions of dilute solutions.

The site for the U.S. Geological Survey software is **http://water.usgs.gov/software**. Most of the programs can be run on UNIX or DOS platforms. Windows versions are available for some of the programs. One of the problems with USGS software is that it has not been very user-friendly. This problem has been addressed through the development of CHEMFORM, which serves as an interface for the various programs. Included in the program inventory are mass balance models (NETPATH), speciation models (WATEQ4F) and mass transfer models (PHREEQC, PHREEQCI, PHRQPITZ). Besides water-chemistry models, a number of other hydrological models are available at this site.

Another site of interest, not only in terms of water-chemistry models but also in terms of geochemical data and analysis in general, is the Geochemical Earth Reference Model (GERM). The address for GERM is **http://earthref.org/GERM/main.htm**. This site not only provides links to a number of water-chemistry models but also has tabulations of thermodynamic data, information on elemental abundances in various reservoirs, and a number of other useful tabulations. The site is under continual development and over time should become a major source of geochemical and environmental data. The Geochemical Society, **http://gs.wustl.edu**, also has links to sites that provide geochemical data. This site, presumably, will also expand with time and the number of linkages will increase.

Among the commercially available software, the most powerful and widely used is the Geochemist's Workbench (Bethke, 1996). This set of computer models was originally developed at the University of Illinois and consists of a number of modules that can be used to carry out calculations involving speciation, plotting of stability diagrams, reaction paths, and a variety of mass transfer processes.

Only a few of the models include kinetics, and a variety of activity-coefficient models are used. The user should consult the documentation that comes with each computer code to determine how that particular code does the computations. Particular attention should be paid to the activity-coefficient models (i.e., are they appropriate for the problem?) and the thermodynamic database. Further information on computer models can be found in Mangold and Tsang (1991), van der Heijde and Elnawawy (1993), and Langmuir (1997).

CASE STUDIES

The following three case studies show how water-chemistry computer models can be applied to environmental problems. The Case Study 2–1 considers the impact of acid mine drainage on the downstream quality of a watercourse. The author of the study used geochemical modeling to determine the chemical processes affecting the concentration of each species of interest. The author was also able to determine the first-order rate constant for the removal of iron from the stream. Such calculations are of interest because they tell us how far a particular contaminant will be transported by a river system.

In Case Study 2–2, the authors used several ionic species as tracers to determine the relative percentage of leachate from a municipal landfill sited on an aquifer. One of the species was found to behave conservatively, and the other was found to show nonconservative behavior when there was a significant leachate component in the aquifer.

In Case Study 2–3, the author investigated the impact of acid deposition on groundwater quality and concluded that the system was not well buffered against acid additions. Increased acidity might result in the release of trace metals, tied up in clay minerals, to the groundwater system.

CASE STUDY 2–1
Geochemical Modeling of Coal Mine Drainage, Summit County, Ohio

A serious problem associated with coal mining is the generation of acid mine drainage (AMD). During mining, sulfur-bearing minerals, such as pyrite (FeS_2), are exposed to oxygen and water, leading to a series of oxidation and hydrolysis reactions that produce sulfuric acid. The resulting waters are strongly acidic (pH of 2 or less is possible) and have high concentrations of SO_4^{2-}, Fe^{2+}, Al^{3+}, and Mn^{2+}. Such waters are toxic to aquatic life and vegetation.

Foos (1997) investigated the downstream changes in the chemistry of coal mine drainage at Silver Creek Metropark, Summit County, Ohio. The first step was to construct a simple mixing model in which AMD and water discharged from Silver Creek lake were the end members. This model successfully predicted the concentrations of Cl^-, PO_4^{3-}, Ca^{2+}, Mg^{2+} and Na^+,

indicating that these species behaved conservatively; i.e., they were not reacting with their surroundings. However, the model did not accurately predict the concentrations of HCO_3^-, SO_4^{2-}, Fe^{3+}, Mn^{2+}, and Si. The model underestimated the concentration of HCO_3^-, indicating that this species was being added to the system, and overestimated the concentrations of the other four species, indicating that they were being removed from the system. Sampling along the length of the discharge stream showed a downstream increase in HCO_3^- and a downstream decrease in the other four species. There was an excellent correlation between Fe^{2+} concentration and distance.

Saturation indices (SI) were calculated, using the water-chemistry program WATEQ4F, for solid phases that could play a role in controlling the concentrations of these species. The results are tabulated here. SI = log(IAP/K_{sp}) and the equilibrium constants were calculated for a temperature of 11.5°C, the temperature of the AMD discharge.

Phase	Formula	SI	Phase	Formula	SI
Hematite	Fe_2O_3	14.93	Quartz	SiO_2	0.53
Goethite	FeOOH	6.96	Chalcedony	SiO_2	0.06
Ferrihydrite	$Fe(OH)_3$	1.07	$SiO_{2 (aq)}$	SiO_2	−0.83
Pyrolusite	MnO_2	−14.27	Aragonite	$CaCO_3$	−2.09
Manganite	MnOOH	−6.44	Calcite	$CaCO_3$	−1.94
Rhodochrosite	$MnCO_3$	−1.17	Dolomite	$CaMg(CO_3)_2$	−4.61

The water was supersaturated with respect to all of the iron-containing phases, and further investigation revealed that about 80% of the iron was removed as ferrihydrite. Thus, precipitation of iron-containing phases was the cause of the decrease in Fe^{2+} in the downstream direction. The water was undersaturated with respect to all Mn-bearing phases, and it was concluded that Mn^{2+} was being removed by adsorption onto the surface of the iron hydroxides. The water was slightly oversaturated with respect to the Si-containing phases, so it is possible that precipitation of these phases was causing the downstream decrease in Si. The waters are undersaturated in terms of the carbonate-containing phases, and it was concluded that the downstream increase in HCO_3^- was due to the addition of or-

ganic carbon where the stream flowed from an area of mowed lawn into a wooded area of dense vegetation.

The author of the study was able to calculate a rate constant for the removal of Fe^{2+} assuming a steady-state model. The model is a first-order rate equation that can be written as follows:

$$V \cdot \frac{dC}{dx} - kC = 0$$

where V is the velocity of the stream, C is the concentration, and x is the distance. Solving this equation for k gives a rate constant of 2.9×10^{-4} s^{-1}.

Source: Foos (1997).

CASE STUDY 2–2
Leachate Geochemistry at a Municipal Landfill, Memphis, Tennessee

Mirecki and Parks (1994) investigated the impact of leachate from a municipal landfill on the Memphis aquifer. The landfill was sited in alluvium of Quaternary age and was believed to be isolated from the Tertiary-age Memphis aquifer. However, it was found that a discontinuity existed in the confining unit that allowed landfill leachate to enter the Memphis aquifer. Chloride (which is usually considered to be a conservative tracer), barium, boron, and strontium were investigated as possible tracers because they occurred in much higher concentra-

tions in the alluvial aquifer downgradient from the landfill. The water-chemistry program PHREEQE was used to construct mixing curves for these four species using the mean concentrations of leachate-contaminated alluvial aquifer samples and the mean concentrations of uncontaminated Memphis aquifer samples as end members. Chloride and boron were ultimately not used as tracers because they did not seem to behave conservatively. Thus, the analysis focused on barium and strontium concentrations. The following table gives the calculated saturation indices ($\log[\text{Ion Activity Product}/K_{sp}]$) for barite ($BaSO_4$), celestite ($SrSO_4$), and strontianite ($SrCO_3$) for different mixtures of the end members.

Mineral	100% contaminated alluvial aquifer water	40%	30%	20%	10%	5%	100% Memphis aquifer water
Barite	0.399	0.136	0.063	−0.031	−0.159	−0.246	−0.360
Celestite	−3.015	−3.287	−3.361	−3.454	−3.581	−3.665	−3.775
Strontianite	−4.355	−4.649	−4.726	−4.821	−4.947	−5.067	−5.140

These calculations indicate that strontium would behave conservatively and that only in mixtures that contained more than 28% water from the contaminated alluvial aquifer would barium behave nonconservatively. In this case, the concentration of barium in the mixture could be predicted from the model calculations. The barium tracer indicated that the leachate component in the Memphis aquifer ranged between 5% and 7%, while the strontium tracer suggested higher amounts of con-

tamination that ranged between 10% and 37%. The authors suggested that the leachate probably flows away from the landfill as discrete pulses in response to precipitation events and that this may account for the variability observed in the chemical date.

Source: Mirecki and Parks (1994).

CASE STUDY 2–3
Ionic Composition and Mineral Equilibria of Acidic Groundwater on the West Coast of Sweden

Halland County, situated on the west coast of Sweden, has some of the most acidified soil and groundwater in the country. This is due to the transport of sulfur and nitrous oxides from Central Europe, leading to high deposition of acidic components. Sjöström (1993) investigated the groundwater chemistry of 14 shallow wells and 10 springs in Halland County. Most of the wells and springs are situated on glacial till or glaciofluvial sediments. The Halland County soils were found to consist of quartz, plagioclase, K-feldspar, amphibole, and chlorite-vermiculite. In the A_2 and B soil subhorizons, goethite, hematite, Ti-oxides, and, possibly, pyrite are common. The groundwaters are dominated by the following major ions: Cl^-, SO_4^{2-}, NO_3^-, HCO_3^-, Ca^{2+}, Mg^{2+}, Na^+, and K^+ but have an overall low ionic strength, presumably due to the short residence times for the ionic species. The water-chemistry model WATEQX was used to calculate the saturation indices (SI) for minerals that might regulate the water chemistry. The results of these calculations are given here for solid phases that were at least 50% over- or undersaturated in the spring and ground waters.

Mineral	Average SI	Mineral	Average SI
Silica gel	−0.75	Vermiculite	−17.46
SiO_2, amorphous	−1.03	Na,K,Mg-beidellite	1.74
Quartz	−0.67	Mg-montmorillonite	4.42
Albite	−4.04	Ca-montmorillonite	5.43
Anorthite	−6.44	Kaolinite	3.14
Sanidine	−1.05	Diaspore	2.98
K-mica	−0.44	Boehmite	1.27
Chlorite	−21.90	Jarosite	−15.89
Hydroxy-Al vermiculite	8.24		

The waters are significantly undersaturated in the mafic silicate minerals, feldspar, K-mica, and chlorite. The soils are dominated by quartz and feldspar, and because these minerals decompose slowly, the buffering capacity of the soils is limited. The waters are oversaturated with respect to a number of the clay minerals, and these could form in the soils as weathering products of the primary aluminosilicates. During increased acidification it is these clay minerals plus, possibly, oxides of Al and Fe that will buffer the H^+ addition (see Chapter 3). These buffering reactions will tend to maintain the pH, but Al and Fe will be released to the groundwater and may, at high enough concentrations, pose a health risk.

Source: Sjöström (1993).

QUESTIONS AND PROBLEMS

1. Why are herbivores more energy efficient than carnivores?

2. What are *intensive variables*? What are *extensive variables*?

3. Distinguish between *open*, *closed*, and *isolated* systems.

4. How can you tell if a reaction is *endothermic* or *exothermic*?

5. Define *enthalpy* and *entropy*.

6. What is the *standard state*?

7. Define *heat capacity*.

8. What is *Le Châtelier's principle* and how can it be used to predict changes in a reaction at equilibrium?

9. When you add HCl to $CaCO_3$, the following reaction occurs:

$$CaCO_3 + 2HCl \rightarrow Ca^{2+} + 2Cl^- + H_2O + CO_2$$

This reaction will proceed differently if it is in a closed container (sealed from the atmosphere) versus open to the atmosphere (for example, on a desktop). How will the reaction differ under these two conditions?

10. What factors are responsible for the nonideal behavior of ions in solution?

11. What is an *aqueous complex*?

12. NaCl dissolves in water to produce Na^+ and Cl^-. Salt is added to a beaker of water until no more salt will dissolve and salt crystals are observed in the beaker.
    a. How many phases are present in this system?
    b. How many components are needed to completely describe the system, and what are the components?
    c. What is the free energy of the system when it becomes saturated in NaCl—i.e., no more salt crystals dissolve in the solution?

13. What is the difference between activity and the measured concentrations of a species in solution?

14. Why and how does the formation of the aqueous complex $CaCO_3$ (aq) affect the solubility of calcite ($CaCO_3$)?

15. Distinguish between *homogeneous* and *heterogeneous* reactions.

16. Why is an understanding of the kinetics of reactions important in dealing with environmental problems?

17. How could you determine if a reaction was zeroth, first, or second order?

18. What is an *inhibitor*?

19. For the reaction Diamond $\rightarrow$ Graphite, $\Delta G_R^0 = -2.9 \, \text{kJ mol}^{-1}$ at 25°C and 1 atm pressure. Thermodynamically, diamond should be the stable form of carbon at the earth's surface. Why don't diamonds invert to graphite at the earth's surface?

20. The rate-determining step in the growth of a particle can be either the rate at which species are added to the surface of the growing particle or the rate at which species diffuse to the surface of the growing particle. Figure 2–8 schematically illustrates the variation in concentration of an ionic species in solution as a function of distance from the surface of a particle. For case A, what is the rate-controlling step, and why? For case B, what is the rate-controlling step, and why?

21. Refer to Case Study 2–2. What is meant by conservative behavior? Why did the authors conclude that barium would only behave conservatively in mixtures that contained less than 28% water from the contaminated aquifer while strontium would behave conservatively in all possible mixtures?

22. With reference to Case Study 2–3, what should happen to anorthite grains when acidified water passes through the soil horizon? Explain.

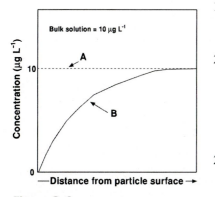

Figure 2–8

Variation in concentration as a function of distance from the surface of a growing particle.

23. During chemical weathering, forsterite is dissolved by the carbonic acid in rainwater. The weathering reaction is as follows:

$$Mg_2SiO_{4\ forsterite} + 4H_2CO_{3\ (aq)} \rightarrow 2Mg^{2+} + 4HCO_3^- + H_4SiO_{4\ (aq)}$$

Use the thermodynamic data from Appendix II, source 2, for the following calculations.
    a. Calculate the K_{eq} for this weathering reaction at 25°C.
    b. If the reaction is at equilibrium, using Le Châtelier's principle, predict what would happen if Mg^{2+} ions were added to the solution.
    c. Using Le Châtelier's principle, predict what would happen to the equilibrium constant if the reaction occurred at a higher temperature.
    d. Calculate the K_{eq} for this reaction at 40°C. Does the solubility of forsterite increase or decrease with increasing temperature? How does this result compare with your prediction in part (c)?

24. a. Calculate the K_{sp} for magnesite ($MgCO_3$) at 25°C. Use the thermodynamic data from Appendix II, source 2, for this calculation.
    b. What is the solubility of magnesite in pure water? How much difference does it make if it is assumed that activity equals concentration versus activity calculated from the Debye–Hückel equation? (*Note*: In order to answer the latter part of the question, you will have to calculate the final solubility by successive approximations. Use the original ion concentrations to determine the ionic strength, then determine the activity coefficients. Use the coefficients to recalculate the ion concentrations. Repeat this process until there is no change in the activity coefficients.)
    c. Mg^{2+} and CO_3^- form the complex $MgCO_{3\ (aq)}$. For this complex, $K_{stab} = 10^{2.98}$. How does the existence of this complex affect the solubility of magnesite? A numerical answer is required.

25. The Hubbard Brook watershed in the White Mountain National Forest in New Hampshire, USA, has been the site of a long-term study on the effect of natural and anthropogenic processes on aquatic ecosystems. Water collected from the brook has the following chemical composition (Likens et al., 1977):

Concentration (mg L^{-1})	
Ca^{2+}	1.7
Mg^{2+}	0.4
Na^+	0.9
K^+	0.3
Cl^-	0.55
SO_4^{2-}	6.3
HCO_3^-	0.9
$SiO_{2\ (aq)}$	4.5
pH	4.9
TDS	19

Note that the pH for Hubbard Brook is much less than 5.7. This is a region in which acid rain is a problem, and the acid rain input is reflected in the water chemistry.
    a. Calculate the concentrations of the ionic species in mol L^{-1}.
    b. Calculate the total negative and positive charge. Do the charges balance; i.e., does the total positive charge equal the total negative charge?
    c. Calculate the activity coefficients for Ca^{2+} and SO_4^{2-} at 25°C using the Debye–Hückel equation.
    d. At 25°C, gypsum has the following solubility product, $K_{sp} = 10^{-4.60}$. Is Hubbard Brook saturated or undersaturated with respect to gypsum? Calculate the IAP and the saturation index.

26. The Rio Grande river at Laredo, Texas, has the following water chemistry (Livingstone, 1963):

Concentration (mg L^{-1})	
Ca^{2+}	109
Mg^{2+}	24
Na^+	117
K^+	6.7
Cl^-	171
SO_4^{2-}	238
HCO_3^-	183
$SiO_{2\ (aq)}$	30
TDS	881

The Rio Grande water chemistry is very different from that of Hubbard Brook (problem 25). Hubbard Brook is located in a forested watershed in the northeastern United States, a region of temperate climate and moderate rainfall; the Rio Grande is on the U.S.–Mexico border, a warm region with low annual precipitation. The differences in climate are at least partly responsible for the differences in water chemistry. Use the thermodynamic data from Appendix II, source 2, for the following calculations.
a. Calculate the concentrations of the ionic species in mol L^{-1}.
b. Calculate the solubility product (equilibrium constant) for gypsum given a water temperature of 30°C.
c. Calculate the activity coefficient for Ca^{2+} and SO_4^{2-} at 30°C using the Debye–Hückel equation.
d. By how much is the water supersaturated or undersaturated with respect to $CaSO_4 \cdot 2H_2O$? Your answer should include a numerical estimate.

27. Fluoride is sometimes found in high concentrations in ground and surface waters. Fluoride, in low concentrations, is often added to drinking water to minimize dental carries, but in higher concentrations fluoride can pose a health risk. The following table gives the composition of relatively fluoride-rich waters from three locations: groundwaters drawn from the Bongo granite, Ghana, groundwaters from Rajasthan, India, and surface waters from Lake Abiata, Ethiopia. Use the thermodynamic data from Appendix II, source 3, for the following calculations.

	Ground and Surface Water Chemistry (mg L^{-1})		
Species	Bongo granites, Ghana[1]	Rajasthan, India[2]	Lake Abiata, Ethiopia[2]
Ca^{2+}	25.3	27.1	1.0
Mg^{2+}	11.5	19.1	0.9
Na^+	23.9	1101	4460
K^+	2.0	5.9	192
Cl^-	6.7	617	1911
F^-	2.3	6.8	119
SO_4^{2-}	4.8	500	14.4
HCO_3^-	156	903	8420

[1]Apambire et al. (1997).
[2]Apello and Postma (1996).

a. Two minerals that may control the concentration of fluoride ion in solution are fluorite (CaF_2) and villiaumite (NaF). For each of the waters listed in the table, determine the saturation index at 25°C for these minerals. (*Note*: Activity is important and must be included in the calculations.) Would either of these minerals control the fluoride concentration of these waters?

b. What is a possible explanation for the very high concentrations of Na^+, Cl^-, F^-, and HCO_3^- in the waters of Lake Abiata? The lake has a pH of 9.62.

28. Knauss et al. (2000) determined various thermodynamic quantities for the aqueous solubilities of two organic liquids of environmental interest, trichloroethene (TCE) and tetracholorethene (PCE). The thermodynamic data at 298 K are tabulated here. Note that the reaction is organic liquid → organic liquid dissolved in water.

	ΔG_{soln} (kJ mol^{-1})	ΔH_{soln} (kJ mol^{-1})	ΔS_{soln} (J mol^{-1} K^{-1})	$\Delta C_{p\,soln}$ (J mol^{-1} K^{-1})
TCE	11.282	−3.35	−49.07	385.2
PCE	15.80	−1.79	−59.00	354.6

a. Calculate the equilibrium constants for both reactions.

b. If 1 kg of each organic liquid was spilled into 1000 kg of water, calculate the concentration of each organic liquid dissolved in the water.

c. Using Le Châtelier's principle, predict what would happen to the solubility of each organic liquid in water with increasing temperature.

d. For each organic liquid calculate the equilibrium constant at 40°C. Were the predictions you made using Le Châtelier's principle, part (c), correct?

29. Asbestos minerals are considered to be a health hazard. The most common type of asbestos is chrysotile, and this mineral comprises about 95% of the asbestos in the United States. Small asbestos fibers can be taken into the lung, where they interact with the lining of the lung. This problem deals with the solubility of chrysotile asbestos in the lung. A later problem deals with the rate at which chrysotile fibers will break down in the lung. The dissolution reaction for chrysotile can be written

$$Mg_3Si_2O_5(OH)_{4\,chrysotile} + 6H^+ \rightarrow 3Mg^{2+} + 2H_4SiO_{4\,(aq)} + H_2O$$

Use the thermodynamic data from Appendix II, source 3, for the following calculations.

a. Calculate the equilibrium constant for this reaction at $T = 37°C$, the average temperature of the human body.

b. For fluid in lung tissues, pH = 4, $Mg^{2+} = 8.7 \times 10^{-4}$ mol L^{-1}, and $H_4SiO_{4\,(aq)} = 1.5 \times 10^{-6}$ mol L^{-1}. Assume activity equals concentration. Are the lung fluids under- or oversaturated with respect to chrysotile? Give a numerical answer.

c. Redo the calculation in part (b) given an ionic strength for the lung fluids of 0.12.

d. Based on these calculations, would you expect chrysotile to persist in the lungs? Explain your answer.

30. Knauss et al. (2000) determined Henry's law constants for TCE and PCE at various temperatures. Henry's law is often expressed as

$$H_c = \frac{\text{Concentration vapor}}{\text{Concentration liquid}}$$

which is the form of Henry's law used by Knauss et al. (2000).

TCE		PCE	
T (K)	H_c (atm m^3 mol^{-1})	T (K)	H_c (atm m^3 mol^{-1})
294	0.006942	295	0.017574
323	0.025183	324	0.057607
348	0.048869	348	0.121925
372	0.067645	374	0.182980
390	0.060473	397	0.189062

a. 1×10^{-4} mol m^{-3} of TCE are dissolved in water at 294 K. Calculate the equilibrium vapor pressure.

b. 1×10^{-5} mol m^{-3} of PCE are dissolved in water at 324 K. Calculate the equilibrium vapor pressure.

31. At 25°C and pH = 5, the following dissolution rates have been obtained for quartz, microcline, albite, diopside, forsterite, and anorthite (data from Lasaga et al., 1994). These dissolution reactions are zeroth-order reactions.

Mineral	Formula	Density g cm^{-3}	Log rate mol m^{-2} s^{-1}
Quartz	SiO_2	2.65	−13.39
Microcline	$KAlSi_3O_8$	2.59	−12.50
Albite	$NaAlSi_3O_8$	2.62	−12.26
Diopside	$CaMgSi_2O_6$	3.22	−10.15
Forsterite	Mg_2SiO_4	3.22	−9.50
Anorthite	$CaAl_2Si_2O_8$	2.76	−8.55

Assuming a planar surface, calculate how long it will take to dissolve a 1-mm-thick layer from each of the minerals. Note that the dissolution rates are determined for a square meter of surface area, so you should do these calculations for 1 m^2 of surface area retreating 1 mm. You will first need to calculate the total moles of each mineral in the volume to be removed. Then calculate the dissolution times, in years, for each mineral. Compare these dissolutions times to the preservation/loss of these minerals during weathering and transport. Is there a relationship between the dissolution times and the persistence of these minerals in the weathering environment?

32. Hume and Rimstidt (1992) investigated the dissolution of chrysotile in lung solutions. They found that the rate-controlling step is the release of Si to solution, and that the reaction is zeroth order. The experimentally determined rate constant is $k = 5.9 \times 10^{-10}$ mol m^{-2} s^{-1}. They modeled the breakdown of a fiber as if it were an infinitely long cylinder dissolving over its lateral surface. The resulting equation is

$$t = \tfrac{3}{4}(d/V_m k)$$

where t is the time in seconds, d is the diameter of the fiber in meters, V_m is the volume occupied by 1 mol of silica in chrysotile (5.4×10^{-5} m^3 mol^{-1}), and k is the rate constant. Calculate the dissolution time for a chrysotile fiber 1 μm in diameter.

33. Foos (1997) determined a first-order rate constant for the removal of Fe^{2+} from a stream (Case Study 2–1). Discharge from a point source adds Fe^{2+} to a river. Using the rate constant determined by Foos (1997) and a stream velocity of 0.5 m s^{-1}, calculate the transport distance required to achieve a 90% reduction in the amount of

Fe^{2+} in solution. To do this problem you will first need to find the amount of time required to reduce the iron concentration by 90%.

34. The breakdown of pyrite (and marcasite) is important in determining the pH of waters draining from coal mines and spoils (acid mine drainage). Aqueous oxidation of pyrite by molecular oxygen can be represented by the following reaction:

$$FeS_{2\ pyrite} + \tfrac{7}{2}O_2 + H_2O \rightarrow Fe^{2+} + 2SO_4^{2-} + 2H^+$$

For this reaction, Williamson and Rimstidt (1994) determined the following rate constant:

$$r = 10^{-8.19} \frac{m_{DO}^{0.5}}{m_{H^+}^{0.11}}$$

where m_{DO} is the amount of dissolved oxygen, m_{H^+} is the hydrogen ion concentration, and r is the rate of pyrite destruction in mol m^{-2} s^{-1}. Assume the activation energy for this reaction is 60 kJ mol^{-1}. Spoils from a coal mine contain small pyrite cubes having an average specific surface area of 0.01 m^2 g^{-1}. Water percolating down through this spoils pile is saturated with respect to atmospheric oxygen ($10^{-3.6}$ mol L^{-1}). The water has a pH of 5.0. Assume that the breakdown of pyrite only occurs according to the preceding reaction.

    a. Calculate the rate constant for the breakdown of pyrite in contact with the water percolating through the spoils pile.
    b. Given that the reaction is pseudo zero order, calculate the time required ($t_{1/2}$) for oxidation of half the pyrite in the spoils pile. You will first need to calculate the number of moles in 1 g of pyrite, and then determine the moles per square meter of specific surface area. The result of this calculation is the value for A_0 in the rate equation.
    c. If the reaction rate was determined at 20°C, calculate the reaction rate at 30°C.

35. Several studies have been done on the rate at which organic matter breaks down in the marine environment. Westerich and Berner (1984) identified three types of organic matter in marine sediments—highly reactive, less reactive, and nonreactive (didn't break down during the course of their experiments). The following equation summarizes the results of their experiments:

$$G_T(t) = G_{01}[\exp(-k_1 t)] + G_{02}[\exp(-k_2 t)] + G_{nr}$$

where G_T is the total organic carbon, G_{01} is the highly reactive fraction, G_{02} is the less reactive fraction, and G_{nr} is the nonreactive fraction. Different rate constants were determined for oxic and anoxic decay. Middelburg (1989) developed a different model that expressed the decay of organic matter in marine sediments in terms of a single first-order rate equation in which k changes with time. The following equation relates the rate constant and its change with time:

$$\log k = -0.95 \log t - 0.81$$

and

$$G_{t1} = G_{t0} \exp 3.2(-t^{0.05})$$

where G_{t1} is the amount of organic carbon remaining at time t, G_{t0} is the amount of organic carbon initially present, and t is time.

   For a number of years New York City has disposed of its sewage sludge in the New York Bight.

    a. Using the model of Westerich and Berner (1984), calculate the amount of sludge remaining 1 year after deposition in the New York Bight. Assume that the sewage consists of 45% highly reactive organic carbon, 45% less reactive organic carbon, and 10% nonreactive organic carbon. For oxic decay, $k_1 = 18 \text{ y}^{-1}$ and $k_2 = 2.3 \text{ y}^{-1}$.

  b. Do the same calculation as in (a) using the model of Middelburg (1989).
  c. Compare the two answers. Why are they different?
  d. Using the equation of Middelburg (1989), calculate the length of time it will take for 50% and 99% of the organic matter to decompose. Are both of these answers realistic? You may want to read the original paper to get a better understanding of the model.

36. The dissolution of quartz in H_2O can be represented by the following reaction:

$$SiO_{2 \, (s)} + 2H_2O \rightarrow H_4SiO_{4 \, (aq)}$$

Rimstidt and Barnes (1980) determined the reaction rate for the dissolution of quartz as a function of temperature. Their data are listed in the following table:

Temperatures and Reaction Rates
for the Dissolution of Quartz

$T \, (°C)$	$k \, (\text{sec}^{-1})$
65	3.81×10^{-9}
105	8.15×10^{-9}
145	1.36×10^{-7}
170	7.68×10^{-8}
213	1.05×10^{-6}
265	1.85×10^{-6}
305	3.69×10^{-6}

  a. Plot log k versus $1/T$. Note that temperature should be in Kelvin.
  b. Using the graph, determine the activation energy for this reaction and the pre-exponential factor.

37. The bacterial reduction of organic matter in marine sediments has been investigated by Berner (1981a). The following data table, showing the variation in sulfate as a function of time, was derived from the experiments of Berner (1981a).

Dissolved Sulfate Versus Time for Marine Sediments

SO_4^{2-} (mmol L^{-1})	Time (days)
20	0
15	5
10	10
5	15
0	20

  a. Graph the data.
  b. What is the order of the reaction?
  c. Calculate the rate constant.

38. If you have read the book *Civil Action* or seen the movie, you may recall that trichloroethene (TCE) was one of the contaminants found in wells G and H. Much of the court case dealt with the source of this contaminant. Knauss et al. (1999) determined the kinetic rate law for the aqueous oxidation of TCE in aerobic, pH neutral waters. The Arrhenius activation energy (E_a) for the reaction was determined to be $108.0 \pm 4.5 \text{ kJ mol}^{-1}$. A series of experiments yielded the following concentration and rate data. Note that the data are given in log form.

Run	Log C_o (mol)	Log initial rate (mol kg^{-1} s^{-1})
TCE-35	−4.330	−9.748
TCE-39	−3.781	−9.724
TCE-41	−4.949	−10.300
TCE-42	−5.662	−11.261
TCE-43	−4.899	−10.195
TCE-51	−4.329	−9.741
TCE-53	−4.401	−9.770

a. Plot the data. Put Log C_o on the *x-axis* and Log rate on the *y-axis*. Fit a straight line through the data points. Note that these are real data and there is scatter. The straight line can be fit either by eye or by linear regression. Linear regression is the preferred method and is most easily done using a spreadsheet. The intercept of this line with the *y*-axis gives the rate constant for this reaction at 25°C. What is the value of the rate constant? What is the slope of the line?

b. Based on the slope of the line, what is the order of the reaction?

c. Using the rate constant from part (a), calculate the pre-exponential factor for the rate equation.

d. Calculate the rate of the reaction at 5°C. By how much does the reaction rate change in going from 25°C to 5°C?

e. Calculate the reaction half-life at 25°C and 5°C.

f. Assuming that the groundwater in the vicinity of wells G and H had a temperature of 5°C, what might you conclude about the transport time for TCE delivered to these wells, i.e., would it be months or years? Explain your answer. In answering this question you need to consider how long TCE would persist in the groundwater under these conditions. If it has already degraded by the time the water reaches the well, it wouldn't be found in the well water.

3

Acid–Base Equilibria

Acids and bases play a key role in a number of environmentally important chemical reactions, including weathering, transport of metals in solution, and CO_2 atmosphere–water equilibria. In this chapter we will develop the concept of an acid and a base, characterize strong and weak acids, develop the pH scale and the concept of buffers, and look at some of the important reactions that take place in the surface environment.

DEFINITION OF ACIDS AND BASES

Acids

According to the **Arrhenius concept**, *acids are substances that produce hydrogen ions* (H^+) *in aqueous solutions*. A more generalized view of acids is provided by the **Brönsted–Lowry model**. In this model *an acid is a proton donor*. Consider the following reaction

$$HCl_{(aq)} + H_2O_{(l)} \rightleftharpoons H_3O^+_{(aq)} + Cl^-_{(aq)}$$

In this reaction a proton is transferred from the HCl molecule to the water molecule to form the hydronium ion (H_3O^+). A more general form of this type of reaction can be written

$$\underset{\text{Acid}}{HA_{(aq)}} \quad + \quad \underset{\text{Base}}{H_2O_{(l)}} \quad \rightleftharpoons \quad \underset{\text{Conjugate acid}}{H_3O^+_{(aq)}} \quad + \quad \underset{\text{Conjugate base}}{A^-_{(aq)}}$$

One way to view this reaction is that it represents a competition between two bases ($H_2O_{(l)}$ and $A^-_{(aq)}$) for the proton. If $H_2O_{(l)}$ is a much stronger base than $A^-_{(aq)}$ the equilibrium position will be far to the right and the $HA_{(aq)}$ will be completely broken down (dissociated). If $A^-_{(aq)}$ is a stronger base, then the reaction will not go far to the right and the $HA_{(aq)}$ will be only partially dissociated. As with any chemical reaction, we can calculate an equilibrium constant for the reaction. This is a particular type of equilibrium constant called an **acid dissociation constant**, which represents the *degree to which the acid has dissociated*.

$$K_a = \frac{[H_3O^+][A^-]}{[HA]} = \frac{[H^+][A^-]}{[HA]} \tag{3–1}$$

These types of reactions are often written using only the $H^+_{(aq)}$ species, rather than the $H_3O^+_{(aq)}$ species, and this is the convention we will use in this book. A **strong acid** is one that *undergoes significant dissociation and has a very large* K_a. A **weak acid** only *partially dissociates and has a relatively small* K_a.

Table 3–1 Dissociation Constants for Acids at 25°C*

Acid	Formula	pK_{a_1}	pK_{a_2}	pK_{a_3}
Hydrochloric	HCl	~ -3		
Sulfuric	H_2SO_4	~ -3	1.99	
Nitric	HNO_3	0		
Oxalic	$H_2C_2O_4$	1.2	4.2	
Phosphoric	H_3PO_4	2.15	7.20	12.35
Hydrofluoric	HF	3.18		
Formic	$HCOOH$	3.75		
Acetic	CH_3COOH	4.76		
Carbonic	H_2CO_3	6.35	10.33	
Hydrosulfuric	H_2S	7.03	>14	
Boric	H_3BO_3	9.27	>14	
Silicic	H_4SiO_4	9.83	13.17	

*Data from Drever (1997), Faure (1998), and Langmuir (1997).

Acids can contain more than one acidic proton. An example is the diprotic (contains two acidic protons) acid H_2SO_4 (sulfuric acid). This acid undergoes a two-step dissociation as follows:

$$H_2SO_4 \rightarrow H^+ + HSO_4^-$$

and

$$HSO_4^- \rightarrow H^+ + SO_4^{2-}$$

A triprotic acid has three acidic protons. An example is H_3PO_4 (phosphoric acid), which is a weak acid. If the acid contains more than one acidic proton, there will be more than one dissociation constant. These are listed in the order of the dissociation reactions. Table 3–1 lists dissociation constants for a number of common and/or environmentally important acids. The dissociation constants are often reported as pK_a values, where $pK_a = -\log K_a$. This is the format used in Table 3–1. Acids that undergo significant dissociation have a negative pK_a, and acids that only partly dissociate have a positive pK_a.

Bases

Bases are defined similarly to acids. Thus, according to the **Arrhenius concept**, *a base is a substance that produces OH^- ions in aqueous solution*, and according to the more general **Brönsted–Lowry model**, *a base is a proton acceptor*. Let us consider two reactions. The first is the dissolution of $NaOH_{(s)}$ in water,

$$NaOH_{(s)} \rightleftharpoons Na^+_{(aq)} + OH^-_{(aq)}$$

in which the OH^- ions are derived from the solid NaOH. The second reaction involves the aqueous ammonia molecule (NH_3),

$$NH_{3\,(aq)} + H_2O_{(l)} \rightleftharpoons NH_4^+_{(aq)} + OH^-_{(aq)}$$

In this reaction, the NH_3 molecule acts as a proton acceptor and water acts as an acid and provides the proton for the aqueous ammonia. **Strong bases** are those that *undergo essentially complete dissociation*, and **weak bases** only *undergo partial dissociation*. As was the case for acids, we can write a general equation for the dissociation of a base and define an equilibrium constant for the dissociation. The generalized equation is

$$\underset{\text{Base}}{B_{(aq)}} + \underset{\text{Acid}}{H_2O} \rightleftharpoons \underset{\substack{\text{Conjugate} \\ \text{acid}}}{BH^+_{(aq)}} + \underset{\substack{\text{Conjugate} \\ \text{base}}}{OH^-_{(aq)}}$$

Table 3–2 *Dissociation Constants for Bases at 25°C**

Base (hydroxide)	Formula	pK_{b_1}	pK_{b_2}	pK_{b_3}
Methylamine	CH_3NH_2	3.36		
Ammonium	$NH_4(OH)$	4.7		
Magnesium	$Mg(OH)_2$	8.6	2.6	
Pyridine	C_5H_5N	8.8		
Manganese	$Mn(OH)_2$	9.4	3.4	
Ferrous	$Fe(OH)_2$	10.6	4.5	
Al, amorphous	$Al(OH)_3$	12.3	10.3	9.0
Al, gibbsite	$Al(OH)_3$	14.8	10.3	9.0
Ferric, amorphous	$Fe(OH)_3$	16.5	10.5	11.8

*Data from Faure (1998) and Zumdahl (1989).

and the equilibrium constant for this reaction is

$$K_b = \frac{[BH^+][OH^-]}{[B]} \tag{3–2}$$

A base such as $Al(OH)_3$ contains three OH molecules and, as was the case for acids that contain more than one H, there are multiple dissociation steps. As was done for acids, we define $pK_b = -\log K_b$. Dissociation constants for some common and/or environmentally important bases are given in Table 3–2.

THE DISSOCIATION OF WATER AND pH

pH is an important and fundamental concept. It arises from the dissociation of water as represented by the following equation:

$$H_2O \rightleftharpoons H^+ + OH^-$$

and

$$K_w = \frac{[H^+][OH^-]}{[H_2O]} = [H^+][OH^-] \tag{3–3}$$

Note that the activity of $H_2O_{(l)} = 1$ in dilute solutions. For this reaction, at 25°C,

$$\Delta G_R^0 = (0) + (-157.2) - (-237.14) = 79.94$$

and

$$\log K_w = \frac{-79.94}{5.708} = -14.00 \tag{3–4}$$

or

$$K_w = 10^{-14} = [H^+][OH^-] \tag{3–5}$$

This is the equilibrium constant for water at 25°C. The equilibrium constant varies as a function of temperature (Table 3–3, p. 62), and at higher temperatures the equilibrium constant is greater than 10^{-14}. Thus, the pH of a neutral solution is a function of temperature. Also note that $K_w = 10^{-14}$ exactly at 24°C (Table 3–3), not at 25°C, but the difference is so minor that for simplicity $K_w = 10^{-14}$ is used at 25°C.

The pH scale is defined, at 25°C, using equation 3–3. By definition, $pH = -\log[H^+]$. When the activity of H^+ and OH^- are equal, the solution is neutral (i.e., the activity of both ions is 10^{-7}, pH = 7). When $[H^+] > 10^{-7}$, the pH is less than 7 and the solution is acidic. When $[H^+] < 10^{-7}$, the pH is greater than 7 and the solution is basic; i.e., there are more OH^- ions than H^+ ions. Remember that we are dealing with a negative exponent, so

Table 3–3 Dissociation Constants (K_w) for Water as a Function of Temperature*

T (°C)	$-\log K_{eq}$	T (°C)	$-\log K_{eq}$
0	14.938	30	13.836
5	14.727	35	13.685
10	14.528	40	13.542
15	14.340	45	13.405
20	14.163	50	13.275
24	14.000	55	13.152
25	13.995	60	13.034

*Data from *CRC Handbook of Chemistry and Physics* (2000).

the larger the number the smaller the activity of the particular ion. pH is referred to as an **environmental parameter**; that is, *pH is theoretically determined by all the equilibria reactions in the system.*

pH OF NATURAL WATERS

The majority of natural waters have pH values between 4 and 10 (Figure 3–1). For most natural waters the carbonic acid–carbonate system exerts the major control on pH. Waters in equilibrium with atmospheric CO_2 would have pH values of around 5.7. The controlling reaction involves the dissociation of the weak acid H_2CO_3. The presence of calcium carbonate, a salt of the weak acid H_2CO_3, results in a buffered system with a pH around 8.2. This buffer is the major control for ocean water pH. Because of its importance, the carbonic acid–carbonate system is discussed at some length in a subsequent section. Additionally, most minerals can be considered to be salts of weak acids and strong bases. For example, $CaAl_2Si_2O_8$ (plagioclase) contains the salt of a weak acid (H_4SiO_4) and two strong bases [$Ca(OH)_2$, $Al(OH)_3$]. The addition of powdered silicate and aluminosilicate minerals, with the exception of quartz, to pure water usually gives rise to a basic pH. pH values less than 4 are usually due to the weathering of sulfide minerals or acid rain. pH values greater than 10 are usually due to the presence of strong bases. The more important pH controlling reactions are discussed in the following sections.

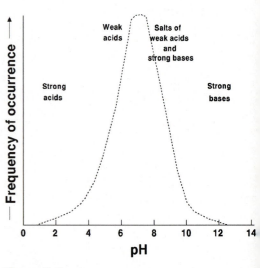

Figure 3–1
Schematic diagram showing the frequency of pH values in natural waters. The major controls for each pH range are indicated on the diagram. After Langmuir (1997).

Strong Acids

The principal strong acids in natural waters are hydrochloric (HCl), nitric (HNO_3), and sulfuric (H_2SO_4). The primary source of nitric acid is acid rain. The nitric acid is produced by the oxidation of various NO_x compounds produced during combustion. The principal sources of sulfuric acid are the weathering of sulfide minerals and acid rain. The source of the sulfur in acid rain is the combustion of fossil fuel, which releases SO_2. Subsequent oxidation in the atmosphere results in sulfuric acid aerosols. The topic of acid rain is considered in Chapter 8, and acid mine drainage is considered in Chapter 9. The effect of strong acids on the pH of natural waters is controlled to some extent by mineral–water reactions. These buffering reactions will be considered in a subsequent section.

Weak Acids

The common weak acids are carbonic acid (H_2CO_3), silicic acid (H_4SiO_4), and various organic acids. The organic acids are discussed in Chapter 5. In addition to these *naturally occurring* acids, other organic acids can be produced by anthropogenic actions. These include acetic acid (CH_3COOH), often found in landfill leachates and sewage, and formic acid ($HCOOH$), often found in groundwater in association with hydrocarbons.

Acetic Acid   Acetic acid is a weak monoprotic acid. The dissociation of acetic acid in water can be written

$$CH_3COOH_{(aq)} \rightleftharpoons CH_3COO^- + H^+$$

and the equilibrium equation, at 25°C, is

$$K_a = \frac{[H^+][CH_3COO^-]}{[CH_3COOH_{(aq)}]} = 10^{-4.76} \qquad (3\text{–}6)$$

Because acetic acid is a weak acid, as is evident from the small equilibrium constant, only a small amount of the dissolved acid will disassociate. Example 3–1 illustrates how to calculate the amount of H^+ in solution.

EXAMPLE 3–1 Calculate the concentration of H^+ ions in solution if 0.1 mol of acetic acid is dissolved in 1 L of water.

Because we start with only acetic acid and pure water, $[H^+] = [CH_3COO^-] = x$, and the amount of acetic acid left after dissociation is $0.1 - x$. We can now write the equilibrium equation (3–6) as

$$K_a = \frac{[x][x]}{[0.1 - x]} = \frac{x^2}{0.1 - x} = 10^{-4.76}$$

The student will recall from introductory algebra that this is a quadratic equation. If we put this equation into the standard form, we get

$$ax^2 + bx + c = 0 = x^2 + 10^{-4.76}x - 10^{-5.76}$$

Substituting the appropriate values, $a = 1$, $b = 10^{-4.76}$, and $c = -10^{-5.76}$, we get

$$x = \frac{-b \pm (b^2 - 4ac)^{1/2}}{2a} = \frac{-10^{-4.76} \pm [(10^{-4.76})^2 - 4(1)(-10^{-5.76})]^{1/2}}{(2)(1)}$$

$$= 1.32 \times 10^{-3}$$

We ignore the negative answer, which is clearly inappropriate. Thus, at equilibrium $[H^+] = [CH_3COO^-] = 1.32 \times 10^{-3}$ mol L^{-1} and $[CH_3COOH_{(aq)}] = 0.1 - 1.32 \times 10^{-3} = 0.0987$ mol L^{-1}. The degree of dissociation is

$$D = \frac{1.32 \times 10^{-3}}{0.0987} \times 100 = 1.34\%$$

a very small amount indicative of the weak acid character of acetic acid. Note that we could simplify the calculation, without introducing a significant error, by assuming that x is small relative to the amount of $CH_3COOH_{(aq)}$. This would immediately give

$$K_a = \frac{x^2}{0.1} = 10^{-4.76}, \qquad x^2 = 10^{-5.76}, \qquad x = 1.32 \times 10^{-3} \qquad ■$$

Carbonic Acid   Carbonic acid (a diprotic acid) is the most abundant acid in natural waters. With its salt, $CaCO_3$, it forms a buffer that plays an important role in regulating the pH of natural waters. Hence, an understanding of the carbonic acid system is essential to the understanding of the pH of natural waters. Three equilibrium relationships are required to describe this system. The first involves the equilibrium between atmospheric CO_2 and H_2CO_3 in solution, which is written

$$CO_{2\ (g)} + H_2O \rightleftharpoons H_2CO_{3\ (aq)}$$

where $H_2CO_{3\ (aq)} = CO_{2\ (aq)} + H_2CO_3^0$ (the true concentration of H_2CO_3, which at 25°C is slightly less than 0.3%). For convenience, the convention is adopted that all the CO_2 in solution exists as $H_2CO_{3\ (aq)}$, and the equilibrium constants used are consistent with this convention. The amount of H_2CO_3 in solution in equilibrium with a particular CO_2 pressure in a coexisting gas phase can be calculated from Henry's law,

$$[H_2CO_{3\ (aq)}] = K_{CO_2}P_{CO_2} \tag{3–7}$$

where K_{CO_2} is the Henry's law constant and P_{CO_2} is the partial pressure of CO_2 in atm. Henry's law constants for various temperatures can be found in Table 2–1.

H_2CO_3 is a weak acid and undergoes a two-step dissociation. The first dissociation is

$$[H_2CO_{3\ (aq)}] \rightleftharpoons [H^+] + [HCO_3^-]$$

and the equilibrium equation is

$$K_{a_1} = \frac{[H^+][HCO_3^-]}{[H_2CO_{3\ (aq)}]} \tag{3–8}$$

The second dissociation is

$$[HCO_3^-] \rightleftharpoons [H^+] + [CO_3^{2-}]$$

and the equilibrium equation is

$$K_{a_2} = \frac{[H^+][CO_3^{2-}]}{[HCO_3^-]} \tag{3–9}$$

Equilibrium constants, as a function of temperature, for the carbonate system are listed in Table 3–4.

Let us rewrite equations 3–8 and 3–9 as follows, and insert the appropriate equilibrium constants for a solution at 25°C.

$$\frac{[H_2CO_{3\ (aq)}]}{[HCO_3^-]} = \frac{[H^+]}{K_{a_1}} = \frac{[H^+]}{10^{-6.35}} \tag{3–10}$$

and

$$\frac{[HCO_3^-]}{[CO_3^{2-}]} = \frac{[H^+]}{K_{a_2}} = \frac{[H^+]}{10^{-10.33}} \tag{3–11}$$

Table 3–4 Equilibrium Constants for the Carbonate System*

| T (°C) | Dissociation constants | | Solubility products | |
	pK_{a_1}	pK_{a_2}	pK_{cal}	pK_{arg}
0	6.58	10.63	8.38	8.22
5	6.52	10.55	8.39	8.24
10	6.46	10.49	8.41	8.26
15	6.42	10.43	8.43	8.28
20	6.38	10.38	8.45	8.31
25	6.35	10.33	8.48	8.34
30	6.33	10.29	8.51	8.37
45	6.29	10.20	8.62	8.49
60	6.29	10.14	8.76	8.64

*Data from Plummer and Busenberg (1982).

Figure 3–2

Relative activity of various carbonate species as a function of pH.

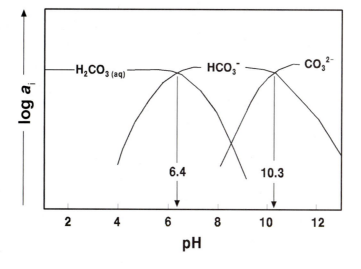

According to equation 3–10, when $[H^+] = 10^{-6.35}$ (pH = 6.35), the ratio $H_2CO_{3\ (aq)}$/HCO_3^- = 1; that is, these two species are present in equal abundance (or more precisely, they are of equal activity). Similarly, from equation 3–11, when pH = 10.33, HCO_3^- and CO_3^{2-} are present in equal abundance. These positions of equal abundance (activity) of species are determined solely by the dissociation constants, which vary only slightly with temperature. Returning to equation 3–10, if the pH is less than 6.35, $[H_2CO_{3\ (aq)}] > [HCO_3^-]$, and if the pH is greater than 6.35, $[HCO_3^-] > [H_2CO_{3\ (aq)}]$. Similarly, from equation 3–11, if pH < 10.33, $[HCO_3^-] > [CO_3^{2-}]$, and if pH > 10.33, $[CO_3^{2-}] > [HCO_3^-]$. These variations in the relative abundances (activities) of the different species are shown qualitatively on Figure 3–2. Note that in waters of intermediate pH, HCO_3^- is the dominant species. In relatively acidic waters $H_2CO_{3\ (aq)}$ is the dominant species, and only in very basic waters is CO_3^{2-} the dominant species.

EXAMPLE 3–2 Calculate the relative activity, at 25°C, of H_2CO_3 to HCO_3^- in a water whose pH = 4.

$$\frac{[H_2CO_{3\ (aq)}]}{[HCO_3^-]} = \frac{[H^+]}{K_{a_1}} = \frac{10^{-4.0}}{10^{-6.35}} = 224$$

The activity of $[H_2CO_{3\ (aq)}]$ is 224 times greater than that of $[HCO_3^-]$ ∎

So far we have considered the abundances of the carbonate species in a qualitative way and have found that the relative abundance of the various species is controlled by pH. We will now consider quantitative solutions to the distribution of the species. There are two end-member cases, an ***open system*** when *the system is in equilibrium with atmospheric CO_2* and a ***closed system*** when the *system is isolated from atmospheric CO_2*.

We will first consider an open system, which is in equilibrium with atmospheric P_{CO_2}. We need four equations to solve this problem: (1) equation 3–7, which relates the partial pressure of CO_2 to the concentration of $H_2CO_{3\ (aq)}$; (2) equation 3–8, which describes the dissociation of $H_2CO_{3\ (aq)}$; (3) equation 3–9, which describes the dissociation of HCO_3^-; and (4) a charge balance equation. ***A basic constraint of solution chemistry is that the positive and negative charges of the species in solution must be equal. This is an important concept and is used repeatedly in solving water-chemistry problems***. For the system CO_2–H_2O, the charge balance equation is written

$$m_{H^+} = m_{HCO_3^-} + 2m_{CO_3^{2-}} + m_{OH^-} \tag{3–12}$$

where m is the molar concentration of each species. We will make the assumption that measured molar concentrations are equal to activity. This is approximately true in very dilute solutions. For more precise calculations, activity coefficients can be calculated as described in Chapter 2 and the molar concentrations adjusted to take into account activity. CO_3^{2-} has two negative charges so the molar concentration of CO_3^{2-} is multiplied by 2 to get the electronic charge. From equation 3–12, note that if there are any HCO_3^- or CO_3^{2-} ions in solution, $[H^+] > [OH^-]$ and the solution is acidic. From Figure 3–2, we see that at acidic pHs the dominant charged species in solution is HCO_3^-. Thus, we can make the simplifying assumption that the relative concentrations of CO_3^{2-} and OH^- are negligible and $m_{H^+} = m_{HCO_3^-}$. Example 3–3 illustrates a calculation of this type.

EXAMPLE 3–3 Calculate the pH of rainwater in equilibrium with atmospheric CO_2.

For the present-day atmosphere, $P_{CO_2} = 10^{-3.5}$ atm. At 25°C, the Henry's law constant is $10^{-1.47}$ (Table 2–1).

$$[H_2CO_{3\,(aq)}] = K_{CO_2}P_{CO_2} = (10^{-1.47})(10^{-3.5}) = 10^{-4.97}$$

Rearranging equation 3–8 gives

$$[H^+][HCO_3^-] = K_{a_1}[H_2CO_{3\,(aq)}] = (10^{-6.35})[10^{-4.97}] = 10^{-11.32}$$

Making the simplifying assumption that $[H^+] = [HCO_3^-]$,

$$[H^+]^2 = 10^{-11.32} \quad \text{and} \quad [H^+] = 10^{-5.66}$$

Therefore,

$$pH = -\log[H^+] = -\log[10^{-5.66}] = 5.66$$

Note that the pH of *pure* rainwater is acidic. The term **acid rain** is used to describe *rainwater that has a pH of less than 5.66.* ∎

Returning to equation 3–12, note that the presence of any anions other than OH^- results in an acidic pH. There are a number of waters in equilibrium with atmospheric CO_2—ocean water, groundwater in contact with limestone, alkaline lake waters, etc.—that are basic. For this to be true, other cations must be present in the system.

In Example 3–3 we calculated the pH of waters in equilibrium with atmospheric CO_2. Changes in the concentration of CO_2 will cause changes in pH. Table 3–5 lists some of the processes that can lead to changes in the pH and CO_2 content of surface and ground waters.

From previous calculations, as schematically summarized in Figure 3–2, we determined that at pH values of less than 10.3, HCO_3^- is the dominant anion in solution. Because this is the case, we can combine equations 3–7 and 3–8 to form an equation that relates the CO_2 partial pressure, pH, and bicarbonate ion activity. Substituting equation 3–7 into equation 3–8, we get

$$K_{a_1} = \frac{[H^+][HCO_3^-]}{K_{CO_2}P_{CO_2}} \tag{3–13}$$

Table 3–5 Examples of Processes That Control the CO_2 Content and pH of Surface and Ground Waters

Process	Reaction	pH
Temperature change	Increase T, decrease solubility of $CO_{2\,(g)}$	Increases
	Decrease T, increase solubility of $CO_{2\,(g)}$	Decreases
Photosynthesis	$6CO_{2\,(g)} + 6H_2O \rightarrow C_6H_{12}O_6 + 6O_{2\,(g)}$	Increases
Respiration	$C_6H_{12}O_6 + 6O_{2\,(g)} \rightarrow 6CO_{2\,(g)} + 6H_2O$	Decreases
Anaerobic decay	$2CH_2O \rightarrow CH_{4\,(g)} + CO_{2\,(g)}$	Decreases
Denitrification	$5CH_2O + 4NO_3^- + 4H^+ \rightarrow 5CO_{2\,(g)} + 2N_{2\,(g)} + 7H_2O$	Increases
Dissolution of carbonate	$CaCO_{3\,calcite} + 2H^+ \rightarrow Ca^{2+} + H_2O + CO_{2\,(g)}$	Increases
Precipitation of carbonate	$Ca^{2+} + H_2O + CO_{2\,(g)} \rightarrow CaCO_{3\,calcite} + 2H^+$	Decreases
Weathering of Al-silicate minerals	$2KAlSi_3O_{8\,feldspar} + 2CO_{2\,(g)} + 11H_2O \rightarrow$	Increases
	$\quad Al_2Si_2O_5(OH)_{4\,kaolinite} + 2K^+ + 2HCO_3^- + 4H_4SiO_{4\,(aq)}$	

Rearranging and solving for P_{CO_2} gives

$$P_{CO_2} = \frac{[H^+][HCO_3^-]}{K_{a_1}K_{CO_2}} \qquad (3\text{--}14)$$

Taking the log, and remembering that $pH = -\log[H^+]$, results in the final form of the equation relating the partial pressure of CO_2, pH, and bicarbonate ion activity.

$$\log P_{CO_2} = -pH + \log\left(\frac{[HCO_3^-]}{K_{a_1}K_{CO_2}}\right) \qquad (3\text{--}15)$$

See also Case Study 3–1 (p. 68).

EXAMPLE 3–4 A groundwater sample has a measured pH of 6.84 and HCO_3^- of 460 mg L^{-1}. We will assume that activity equals concentration. At 25°C, calculate the P_{CO_2} for this groundwater sample.

First, we need to convert the measured concentration of HCO_3^- to moles per liter. The atomic weight of $HCO_3^- = 61.0$ g. Therefore,

$$[HCO_3^-] = \frac{460 \times 10^{-3} \text{ g L}^{-1}}{61.0 \text{ g mol}} = 7.54 \times 10^{-3} \text{ mol L}^{-1}$$

Substituting into equation 3–15, and using the proper constants for $T = 25°C$,

$$\log P_{CO_2} = -6.84 + \log\left(\frac{[7.54 \times 10^{-3} \text{ mol L}^{-1}]}{10^{-6.35}10^{-1.47}}\right) = -6.84 + \log(4.982 \times 10^5)$$

$$= -6.84 + 5.70 = -1.14$$

The calculated partial pressure for CO_2 is significantly greater than that for the atmosphere, hence the water is supersaturated in CO_2 with respect to the atmosphere and the groundwater is behaving as a closed system. The ground water sample was collected near a septic tank, and the elevated P_{CO_2} is presumably due to aerobic and/or anaerobic decay of dissolved organic carbon in the water from the drain field. ∎

Lastly, we will consider a system in which the total carbonate concentration remains constant. What we are interested in are the absolute abundances of the different species as a function of pH. Four equations are required to describe this system, one which represents the total amount of carbonate and the other three the absolute abundances of the three carbonate species. The equation for the total carbonate concentration is

$$C_T = H_2CO_{3\,(aq)} + HCO_3^- + CO_3^{2-} \qquad (3\text{--}16)$$

Rewriting equation 3–8 in terms of HCO_3^- gives

$$[HCO_3^-] = \frac{K_{a_1}[H_2CO_{3\,(aq)}]}{[H^+]} \qquad (3\text{--}17)$$

Substituting in equation 3–9 for HCO_3^- gives

$$[CO_3^{2-}] = \frac{K_{a_1}K_{a_2}[H_2CO_{3\,(aq)}]}{[H^+]^2} \qquad (3\text{--}18)$$

Equations 3–17 and 3–18 are combined with equation 3–16 to give

$$C_T = [H_2CO_{3\,(aq)}]\left(1 + \frac{K_{a_1}}{[H^+]} + \frac{K_{a_1}K_{a_2}}{[H^+]^2}\right) \qquad (3\text{--}19)$$

We now have an expression that relates total carbonate to the amount of $H_2CO_{3\,(aq)}$. As we will see, this is a very useful expression. In order to simplify subsequent equations, we will define a variable, α_H, that encompasses all of the terms inside the parentheses in equation 3–19. Note that this portion of equation 3–19 deals with the activity of the hydrogen ion, i.e., pH.

$$\alpha_H = \left(1 + \frac{K_{a_1}}{[H^+]} + \frac{K_{a_1}K_{a_2}}{[H^+]^2}\right) \qquad (3\text{--}20)$$

CASE STUDY 3–1
Relationship Between Concentration of Bicarbonate Ion, pH, and Partial Pressure of CO₂

Langmuir (1997) summarizes the results of several studies that investigated the relationships between pH, HCO_3^-, and P_{CO_2} (Figure 3–C1–1). The three investigations gave very different trends in terms of these variables, and the reasons for these trends are illustrative of the various factors that can effect the carbonate system. The NJ Coastal Plain is a sand aquifer. As the waters move downdip they react with alumino-silicate minerals. These reactions consume CO_2 (see Table 3–5), leading to an increase in pH, a decrease in CO_2, and an increase in HCO_3^-. In the case of the Floridian aquifer, traces of gypsum are present in the carbonate rock. The Ca released by the dissolution of gypsum causes the precipitation of additional calcite and an increase in CO_2, according to the reaction

$$CaSO_4 \cdot 2H_2O_{\text{ gypsum}} + 2HCO_3^- \rightarrow$$
$$CaCO_{3 \text{ calcite}} + SO_4^{2-} + 3H_2O + CO_2$$

This is referred to as *common-ion driven precipitation*, because the dissolution of gypsum releases Ca^{2+}, which is also involved in the calcite precipitation reaction. Anaerobic decay of organic matter (Table 3–5) and sulfate reduction also occur, and both reactions release CO_2. The result is that CO_2 increases in the downdip direction and this increase is accompanied by both a decrease in pH and an increase in HCO_3^-. In the case of Slab Cabin Run, the HCO_3^- concentration in the stream remains essentially constant. The plotted data represent diurnal variations in the CO_2 content with concomitant changes in pH. CO_2 decreases during the day when photosynthesis is occurring (increasing pH) and increases at night when only respiration is taking place (decreasing pH). Plots such as that shown in Figure 3–C1–1 represent a useful way to summarize relations in the carbonate system and to investigate the factors that lead to variations in pH, P_{CO_2}, and HCO_3^-.

Source: Langmuir (1997).

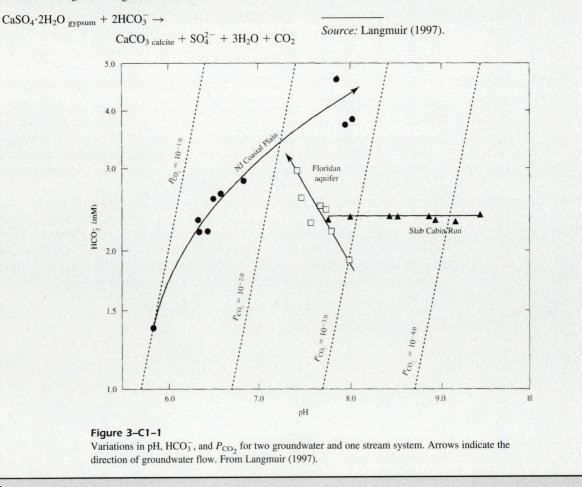

Figure 3–C1–1
Variations in pH, HCO_3^-, and P_{CO_2} for two groundwater and one stream system. Arrows indicate the direction of groundwater flow. From Langmuir (1997).

We now write equation 3–19 in terms of α_H:

$$C_T = [H_2CO_{3 \text{ (aq)}}]\alpha_H \tag{3–21}$$

Solving equation 3–21 for $[H_2CO_{3 \text{ (aq)}}]$ gives

$$[H_2CO_{3 \text{ (aq)}}] = \frac{C_T}{\alpha_H} \tag{3–22}$$

Substituting equation 3–22 into equation 3–17 gives

$$[HCO_3^-] = \frac{C_T K_{a_1}}{[H^+]\alpha_H}$$ (3–23)

Substituting equation 3–22 into equation 3–18 gives

$$[CO_3^{2-}] = \frac{C_T K_{a_1} K_{a_2}}{[H^+]^2 \alpha_H}$$ (3–24)

EXAMPLE 3–5 Calculate the concentration of each carbonate species in solution at 25°C when $C_T = 1 \times 10^{-3}$ mol L^{-1} and pH = 5.7.

Selecting the appropriate dissociation constants from Table 3–4 and setting $[H^+] = 10^{-5.7}$ mol L^{-1},

$$\alpha_H = 1 + \frac{K_{a_1}}{[H^+]} + \frac{K_{a_1} K_{a_2}}{[H^+]^2} = 1 + \frac{(10^{-6.35})}{[10^{-5.7}]} + \frac{(10^{-6.35})(10^{-10.33})}{[10^{-5.7}]^2} = 1.224$$

$$H_2CO_{3\,(aq)} = \frac{C_T}{\alpha_H} = \frac{1 \times 10^{-3}\ \text{mol L}^{-1}}{1.224} = 8.17 \times 10^{-4}\ \text{mol L}^{-1}$$

$$(HCO_3^-) = \frac{C_T K_{a_1}}{[H^+]\alpha_H} = \frac{(1 \times 10^{-3}\ \text{mol L}^{-1})(10^{-6.35})}{(10^{-5.7})(1.224)} = 1.83 \times 10^{-4}\ \text{mol L}^{-1}$$

$$(CO_3^{2-}) = \frac{C_T K_{a_1} K_{a_2}}{[H^+]^2 \alpha_H} = \frac{(1 \times 10^{-3})(10^{-6.35})(10^{-10.33})}{[10^{-5.7}]^2(1.224)} = 4.29 \times 10^{-9}\ \text{mol L}^{-1}$$

As expected, at pH = 5.7, $H_2CO_{3\,(aq)}$ is the dominant species and the abundance of CO_3^{2-} is negligible. ∎

Silicic Acid Silicic acid is produced during the weathering of silicate minerals (Chapter 9). Silicic acid is a tetraprotic acid and has four dissociation steps. Because of the very small dissociation constants (Table 3–1), in natural waters only the first two dissociation steps are important. The first dissociation is

$$H_4SiO_{4\,(aq)} \rightleftharpoons H_3SiO_4^- + H^+$$

and the equilibrium equation, at 25°C, is

$$K_{a_1} = \frac{[H^+][H_3SiO_4^-]}{[H_4SiO_{4\,(aq)}]} = 10^{-9.83}$$ (3–25)

The second dissociation is

$$H_3SiO_4^- \rightleftharpoons H_2SiO_4^{2-} + H^+$$

and the equilibrium equation, at 25°C, is

$$K_{a_2} = \frac{[H^+][H_2SiO_4^{2-}]}{[H_3SiO_4^-]} = 10^{-13.17}$$ (3–26)

Inspection of equation 3–25 reveals that at pH = 9.83, $[H_4SiO_{4\,(aq)}] = [H_3SiO_4^-]$. For pH values less than 9.83, the dominant species is $H_4SiO_{4\,(aq)}$. Similarly, at pH = 13.17 (equation 3–26), $[H_3SiO_4^-] = [H_2SiO_4^{2-}]$. Only at pH values greater than 13.17, which are essentially never encountered in natural waters, is $H_2SiO_4^{2-}$ the dominant species. Thus, $H_4SiO_{4\,(aq)}$ is the dominant species in all but relatively basic waters.

EXAMPLE 3–6 Calculate the relative abundance of $H_4SiO_{4\,(aq)}$ to $H_3SiO_4^-$ at pH = 7.0 and 25°C.

$$\frac{[H_4SiO_4]}{[H_3SiO_4^-]} = \frac{[H^+]}{K_{a_1}} = \frac{[10^{-7.0}]}{10^{-9.83}} = 676$$

H_4SiO_4 (aq) is 676 times more abundant in solution than $H_3SiO_4^-$. Note that we have ignored activities in this calculation. In dilute solutions the activity of the aqueous species, H_4SiO_4 (aq), would be 1, but the activity of $H_3SiO_4^-$ would be less than 1. More precise calculations require the calculation of the activity coefficient, and these more precise calculations would give a slightly lower ratio in terms of the absolute abundance of the two species in solution. ∎

As we did for the carbonate system, we can calculate the total dissolved concentration of silica species as follows:

$$Si_T = H_4SiO_4 \text{ (aq)} + H_3SiO_4^- + H_2SiO_4^{2-} \tag{3–27}$$

The second and third terms would only be significant in relatively basic solutions. We can rewrite equation 3–27 as follows:

$$Si_T = [H_4SiO_4 \text{ (aq)}]\left(1 + \frac{K_{a_1}}{[H^+]} + \frac{K_{a_1}K_{a_2}}{[H^+]^2}\right) \tag{3–28}$$

As we did for the carbonates, we can solve for α_H and the concentrations of the various silicic acid species at a fixed pH and fixed total dissolved silica content.

Carbonic Acid–Carbonate System   When we developed the equations for the carbonic acid system, it was pointed out, with reference to equation 3–12, that in order for natural waters to have a pH greater than 7, other cations in addition to H^+ must be present in solution. In mildly basic waters, the most important equilibria are those involving carbonic acid and the carbonate minerals calcite and aragonite. In this case, Ca^{2+} is the additional cation and waters in equilibrium with carbonate minerals are basic. These carbonate equilibrium reactions are important in determining the pH of the ocean and waters in contact with limestones. There are a number of possible cases involving these equilibria, depending on whether or not we are dealing with an open system (P_{CO_2} constant) or a closed system (ΣCO_2 fixed) and when the system became closed. As an example, we will consider the pH of a system open to the atmosphere and saturated in calcite, for instance, water in a lake located in limestone. Other cases can be found in Garrels and Christ (1965), Drever (1997), and Langmuir (1997).

The following equations describe the system CO_2–$CaCO_3$. The solubility products for the $CaCO_3$ phases (the polymorphs calcite and aragonite) as a function of temperature are listed in Table 3–4. The following three equations were developed previously:

$$[H_2CO_3 \text{ (aq)}] = K_{CO_2}P_{CO_2} \tag{3–29}$$

$$K_{a_1} = \frac{[H^+][HCO_3^-]}{[H_2CO_3 \text{ (aq)}]} \tag{3–30}$$

$$K_{a_2} = \frac{[H^+][CO_3^{2-}]}{[HCO_3^-]} \tag{3–31}$$

Because this system also involves calcite, an additional reaction representing the solubility of calcite is required.

$$CaCO_3 \text{ calcite} \rightleftharpoons Ca^{2+} + CO_3^{2-}$$

and

$$K_{sp} = [Ca^{2+}][CO_3^{2-}] \tag{3–32}$$

A final equation represents the charge balance and is written

$$m_H^+ + 2m_{Ca^{2+}} = m_{HCO_3^-} + 2m_{CO_3^{2-}} + m_{OH^-} \tag{3–33}$$

We now have five equations and six unknowns (P_{CO_2}, [H_2CO_3 (aq)], [HCO_3^-], [CO_3^{2-}], [Ca^{2+}], and [H^+]). If we fix one of these variables, we can solve for the other variables.

The exercise now becomes one of algebraic manipulation, although the manipulations may not always be obvious. In this example, P_{CO_2} is fixed and we are going to solve for pH. We need to emphasize a significant point. In all of the equilibrium equations, we are dealing with the activities of the species. In the charge balance equation, we are dealing with the absolute concentrations. In order to solve this problem, we are going to assume that activity = concentration. A more exact solution requires the calculation of the activity coefficients. This can be done by an iterative technique, as was illustrated in Chapter 2.

EXAMPLE 3–7 Calculate the pH of a solution saturated with respect to calcite and in equilibrium with atmospheric CO_2. For the solution, $T = 25°C$.

As an initial assumption, we will anticipate that the final pH is going to be near neutrality, so equation 3–33 can be simplified to

$$2m_{Ca^{2+}} = m_{HCO_3^-}$$

Rewriting equation 3–31 in terms of CO_3^{2-} and substituting into equation 3–32,

$$K_{sp} = [Ca^{2+}]K_{a_2}\frac{[HCO_3^-]}{[H^+]}$$

Substituting HCO_3^- for Ca^{2+} using the simplified charge balance equation, $Ca^{2+} = 0.5HCO_3^-$,

$$K_{sp} = 0.5[HCO_3^-]K_{a_2}\frac{[HCO_3^-]}{[H^+]}$$

Combining equations 3–29 and 3–30 and solving for HCO_3^- gives

$$[HCO_3^-] = \frac{K_{a_1}K_{CO_2}P_{CO_2}}{[H^+]}$$

Combining this result with the previous equation for calcite solubility gives

$$K_{sp} = 0.5\left(\frac{K_{a_1}K_{CO_2}P_{CO_2}}{[H^+]}\right)^2\frac{K_{a_2}}{[H^+]}$$

Solve for $[H^+]$ by inserting the appropriate constants and P_{CO_2}.

$$[H^+] = \left(\frac{K_{a_1}^2K_{CO_2}^2P_{CO_2}^2K_{a_2}}{2K_{sp}}\right)^{1/3} = \left(\frac{(10^{-6.35})^2(10^{-1.47})^2(10^{-3.5})^2(10^{-10.33})}{2(10^{-8.48})}\right)^{1/3} = 10^{-8.26}$$

pH = $-\log[H^+]$, so the final pH is 8.26. Is our assumption that the dominant ionic carbonate species would be HCO_3^- correct? Rearranging equation 3–31 and solving for the CO_3^{2-}/HCO_3^- ratio gives

$$\frac{[CO_3^{2-}]}{[HCO_3^-]} = \frac{K_{a_2}}{[H^+]} = \frac{10^{-10.33}}{10^{-8.26}} = 10^{-2.07} = 0.009$$

The activity of CO_3^{2-} is less than 1% that of HCO_3^-, so our simplification of the charge balance equation is acceptable. ∎

Case Study 3–2 (p. 72) demonstrates how carbonate equilibria can be used to determine the source(s) of groundwater and the changes that occur along the flow path.

Salts of Weak Acids and Strong Bases

Most, but not all, minerals can be considered to be salts of weak acids and strong bases. Waters in equilibrium with these minerals are normally basic. We will illustrate this by considering two examples, one involving our old friend calcite ($CaCO_3$), a common mineral in limestones, and the other natron ($Na_2CO_3 \cdot 10H_2O$), a mineral that forms in evaporative alkaline lakes, such as those found in the Rift Valley of East Africa.

CASE STUDY 3–2
Groundwater Chemistry of a Plateau near Matsumoto City, Japan

Li and Misawa (1994) investigated the chemistry of groundwater seeping into the Matsumoto tunnel and the chemistry of well, spring, and river waters in the immediate vicinity of the tunnel. The purpose of the study was to determine the source of the groundwater seeping into the tunnel and the changes that occurred along the flow path of the groundwater.

The plateau consists of Miocene andesitic tuff breccia and sandstone overlain by Quaternary river terrace deposits. The plateau is forested and the source of the HCO_3^- ions in the groundwater was thought to be CO_2 gas derived from the soils. The reaction is

$$CO_2 + H_2O \rightarrow HCO_3^- + H^+$$

Solving equation 3–15 at $T = 25°C$,

$$\log P_{CO_2} = -pH + \log[HCO_3^-] + 7.81$$

The source of Ca^{2+} ions in the groundwater was inferred to be dissolution of calcite veins found in the rocks surrounding the tunnel, according to the reaction

$$CaCO_{3 \; calcite} + 2H^+ \rightarrow Ca^{2+} + CO_2 + H_2O$$

For this reaction at 25°C,

$$K_{eq} = 10^{9.8} = \frac{[Ca^{2+}]P_{CO_2}}{[H^+]^2}$$

and

$$\log P_{CO_2} = -2pH - \log[Ca^{2+}] + 9.8$$

Plots of HCO_3^- and Ca^{2+} versus pH are shown in Figure 3–C2–1. Calculated $\log P_{CO_2}$ varies from -1 to -4, with tunnel seepage showing the smaller values. P_{CO_2} values for waters calculated using both equations were generally in agreement. A number of groundwater samples had P_{CO_2} values that exceeded atmospheric P_{CO_2}, supporting the hypothesis that CO_2 had been added by the decay of organic material. The lower CO_2 pressures found in the tunnel seepage were ascribed to reactions between the groundwater and the silicate minerals of

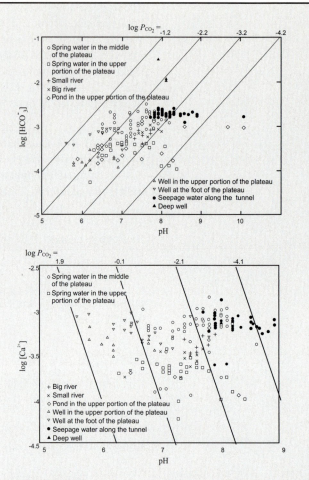

Figure 3–C2–1
The relationship between pH and $\log[HCO_3^-]$ and $\log[Ca^{2+}]$. From Li and Misawa (1994).

the andesitic tuffs as the groundwater flowed to the tunnel. These reactions are of the general type

$$CaAl_2Si_2O_{8 \; anorthite} + 2CO_2 + 3H_2O \rightarrow$$
$$Ca^{2+} + 2HCO_3^- + Al_2Si_2O_5(OH)_{4 \; kaolinite}$$

which consume CO_2.

Source: Li and \isawa (1994).

EXAMPLE 3–8 An excess of pure calcite is placed in distilled water that has been purged of all gas phases—for example, by boiling. Therefore, for the purpose of this example we will assume that $P_{CO_2} = 0$ atm. The following chemical reactions describe what happens in this system:

$$CaCO_{3 \; calcite} \rightleftharpoons Ca^{2+} + CO_3^{2-}$$

$$CO_3^{2-} + H_2O \rightleftharpoons HCO_3^- + OH^-$$

$$HCO_3^- + H_2O \rightleftharpoons H_2CO_{3 \; (aq)} + OH^-$$

All the carbon atoms in solution come from the dissolution of calcite, giving the mass balance equation

$$m_{Ca^{2+}} = m_{CO_3^{2-}} + m_{HCO_3^-} + m_{H_2CO_3}$$

We can also write a charge balance equation:

$$2m_{Ca^{2+}} + m_{H^+} = 2m_{CO_3^{2-}} + m_{HCO_3^-} + m_{OH^-}$$

We will once again assume that concentration equals activity, bearing in mind our standard warning about this simplification. We anticipate that the solution will be sufficiently basic so that $[H_2CO_{3\ (aq)}]$ is significantly smaller than $[HCO_3^-]$ and can be neglected in the following calculations. At 25°C (equilibrium constants from Table 3–4), the following equations describe this system:

$$[Ca^{2+}][CO_3^{2-}] = 10^{-8.48}$$

$$\frac{[H^+][HCO_3^-]}{[H_2CO_{3\ (aq)}]} = 10^{-6.35}$$

$$\frac{[H^+][CO_3^{2-}]}{[HCO_3^-]} = 10^{-10.33}$$

$$[H^+][OH^-] = 10^{-14}$$

We will write the mass balance equation substituting the activities of the various species for concentrations and ignoring $H_2CO_{3\ (aq)}$.

$$[Ca^{2+}] = [CO_3^{2-}] + [HCO_3^-]$$

Similarly, for the charge balance equation, making the assumption that H^+ will be small relative to Ca^{2+},

$$2[Ca^{2+}] = 2[CO_3^{2-}] + [HCO_3^-] + [OH^-]$$

Substituting from the mass balance equation for Ca^{2+} into the charge balance equation,

$$[HCO_3^-] = [OH^-]$$

What we are interested in is the pH of the solution. Therefore, it is more useful to solve this problem in terms of the H^+ rather than the OH^- ion. Hence, we want to write the equations in what is called the *proton condition*. As a first step we write OH^- in terms of H^+.

$$[OH^-] = \frac{K_w}{[H^+]} = \frac{10^{-14}}{[H^+]}$$

We can now write

$$[HCO_3^-] = [OH^-] = \frac{10^{-14}}{[H^+]}$$

We solve for CO_3^{2-} in terms of H^+ by rearranging the $HCO_3^- \rightleftharpoons H^+ + CO_3^{2-}$ equilibrium equation and solving for CO_3^{2-}:

$$[CO_3^{2-}] = \frac{[HCO_3^-]10^{-10.33}}{[H^+]} = \frac{10^{-14}10^{-10.33}}{[H^+][H^+]} = \frac{10^{-24.33}}{[H^+]^2}$$

Substituting for HCO_3^- in the calcite solubility equation,

$$[Ca^{2+}] = \frac{10^{-8.48}}{[CO_3^{2-}]} = \frac{10^{-8.48}[H^+]^2}{10^{-24.33}} = 10^{15.85}[H^+]^2$$

Substituting into the modified mass balance equation, $[Ca^{2+}] = [CO_3^{2-}] + [HCO_3^-]$,

$$10^{15.85}[H^+]^2 = \frac{10^{-24.33}}{[H^+]^2} + \frac{10^{-14}}{[H^+]}$$

Multiplying through by $[H^+]^2$ and rearranging,

$$10^{15.85}[H^+]^4 - 10^{-14}[H^+] = 10^{-24.33}$$

This equation can be solved by successive approximations. The final value is $[H^+] = 10^{-9.9}$ and pH = 9.9. We can compare this result with the calculation we did in Example 3–7, in which the system was in equilibrium with atmospheric CO_2. In the case of the open system, pH = 8.26, a less basic pH than that calculated here, indicating the effect of P_{CO_2} on carbonate equilibria.   ∎

EXAMPLE 3–9 Evaporative alkaline lakes contain a number of strange and exotic minerals. One such mineral is natron, which dissolves according to the following reaction:

$$Na_2CO_3 \cdot 10H_2O \,_{natron} \rightarrow 2Na^+ + CO_3^{2-} + 10H_2O$$

releasing Na^+ ions to solution and increasing the alkalinity of the lake water.

Calculate the pH for an alkaline lake in Tanzania in which the concentration of Na = 0.1 mol L^{-1}. The lake waters are in equilibrium with atmospheric CO_2, and the water temperature is 30°C.

We expect that this lake will have a very basic pH, hence we will ignore the H^+ ion. In addition, we will assume that Na^+ is the only other cation in the water. We can now write the combined mass balance and charge balance equation as follows:

$$m_{Na^+} = m_{HCO_3^-} + 2m_{CO_3^{2-}} + m_{OH^-} = 10^{-1}$$

As we have done in the other examples, we now solve the various carbonate equilibria equations for the activity of HCO_3^- and CO_3^{2-}. The appropriate equilibrium constants are from Table 2–1 (Henry's law constants) and Table 3–4.

$$[HCO_3^-] = \frac{[H_2CO_{3\,(aq)}]K_{a_1}}{[H^+]} = \frac{K_{CO_2}P_{CO_2}K_{a_1}}{[H^+]} = \frac{10^{-1.53}10^{-3.5}10^{-6.33}}{[H^+]} = \frac{10^{-11.36}}{[H^+]}$$

$$[CO_3^{2-}] = \frac{[HCO_3^-]K_{a_2}}{[H^+]} = \frac{10^{-11.36}10^{-10.29}}{[H^+]^2} = \frac{10^{-21.65}}{[H^+]^2}$$

As in the case of Example 3–8, OH^- is replaced by $10^{-14}/[H^+]$. Substituting into the charge balance equation,

$$\frac{10^{-11.36}}{[H^+]} + (2)\left(\frac{10^{-21.65}}{[H^+]^2}\right) + \frac{10^{-14}}{[H^+]} = 10^{-1.0}$$

Multiplying through by $[H^+]^2$,

$$10^{-11.36}[H^+] + 10^{-21.35} + 10^{-14}[H^+] = 10^{-1.0}[H^+]^2$$

Rearranging and grouping terms,

$$[H^+]^2 - 10^{-10.36}[H^+] - 10^{-20.35} = 0$$

This is a quadratic equation and can be solved in the standard way (Example 3–1), giving pH = 10.04. In doing this problem we have once again made the simplifying assumption that activity equals concentration. The alkaline lake water has a significantly higher ionic charge than the solutions in the preceding examples, and a more precise calculation taking into account activity would give a slightly different pH value.   ∎

Strong Bases

pH values greater than 10 are rarely found in nature, and when they are found they are inevitably due to the presence of strong bases such as $Ca(OH)_2$ and $Mg(OH)_2$. Possible natural sources for such bases are the weathering products of ultramafic rocks.

AMPHOTERIC HYDROXIDES

Amphoteric hydroxides are *hydroxides that can behave as either an acid or a base*. This behavior varies as a function of pH. As an example, consider the following reaction, which represents the dissolution of $Cd(OH)_{2\ (s)}$ when a hydroxyl ion is added:

$$Cd(OH)_{2\ (s)} + OH^- \rightleftharpoons Cd(OH)_3^-$$

The equilibrium equation for this reaction is

$$K_A = \frac{[Cd(OH)_3^-]}{[OH^-]} = 10^{-5.3} \tag{3-34}$$

where K_A is the equilibrium constant for the amphoteric reaction. Equilibrium constants for other amphoteric compounds are found in Table 3–6. When $Cd(OH)$ is partially dissolved in water, OH^- ions are taken up to form $Cd(OH)_3^-$ and a hydrogen ion is released through the dissociation of water. In this case, $Cd(OH)_2$ acts as an acid because H^+ (protons) are released. Le Châtelier's principle predicts that in a basic solution the reaction will move to the right, using up OH^- ions, while in an acidic solution the reaction will move to the left, releasing OH^- ions. Thus, in basic solutions amphoteric compounds act as acids, and in acidic solutions they act as bases. If this does not seem obvious, consider the dissociation reaction for water. If OH^- ions are removed from solution, in order for the equilibrium constant to stay constant, the activity of H^+ ions must increase. The reverse is true for an acidic solution.

Table 3–6 *Equilibrium Constants for Amphoteric Compounds**

Hydroxide	Formula	$-\log K_A$
Aluminum, amorphous	$Al(OH)_3$	−1.1
Aluminum, gibbsite	$Al(OH)_3$	1.4
Cadmium	$Cd(OH)_2$	5.3
Cobalt	$Co(OH)_2$	5.2
Copper	$Cu(OH)_2$	2.9
Ferrous	$Fe(OH)_2$	5.1
Ferric	$Fe(OH)_3$	4.4
Manganese	$Mn(OH)_2$	5.1
Nickel	$Ni(OH)_2$	4
Thorium, amorphous	$Th(OH)_4$	5.8
Uranium	$UO_2(OH)_2$	3.6
Zinc, amorphous	$Zn(OH)_2$	1.9

*Data from Krauskopf and Bird (1995).

ACIDITY AND ALKALINITY

Acidity is the *capacity of water to donate protons*; *alkalinity* is the *capacity of water to accept protons*. What do we mean by the capacity of water to donate or accept protons? Let us suppose we have an acidic solution whose acidity is due to the presence of hydrochloric acid. In this solution, HCl will have dissociated to H^+ and Cl^- ions. If we add hydroxyl ions to the solution, the following reaction will occur:

$$H^+ + OH^- \rightarrow H_2O$$

A hydronium ion in solution has donated a proton (i.e., an H^+), which combines with the OH^- to produce H_2O. For a weak acid, such as H_2CO_3, the reaction would be

$$H_2CO_{3\ (aq)} + OH^- \rightarrow HCO_3^- + H_2O$$

and once again an H^+ ion has combined with an OH^- ion to produce water. The preceding are examples of acidity reactions. The acidity of a solution can be determined by titrating the solution with a strong base, such as NaOH. Examples of alkalinity reactions are those involving OH^- or HCO_3^- ions, such as

$$OH^- + H^+ \rightarrow H_2O \quad \text{and} \quad HCO_3^- + H^+ \rightarrow H_2CO_{3\ (aq)}$$

Alkalinity is determined by titrating a solution with a strong acid, such as HCl.

Recall that a fundamental principle of aqueous chemistry is that solutions must be electrically neutral. Suppose that we have a solution of distilled water in equilibrium with atmospheric CO_2. For this solution we can write the following charge balance equation:

$$m_H^+ = m_{HCO_3^-} + 2m_{CO_3^{2-}} + m_{OH^-}$$

Note, as we already know, this solution is acidic, because m_{H^+} must be greater than m_{OH^-}. If we add acid to the solution, the H^+ ions will combine with OH^- to form H_2O, with CO_3^{2-} to form HCO_3^-, and with HCO_3^- to form $H_2CO_{3\ (aq)}$. During this set of reactions, the overall electrical neutrality of the solution must be maintained, but the concentration of the individual species will change. Hence, these species are considered to be **nonconservative species** because their *abundances vary as a function of pH or some other intensive variable (such as pressure or temperature)*.

Now let us consider a more complicated solution in which other species occur. As an example, consider a typical groundwater. For this solution, an appropriate charge balance equation is

$$2m_{Ca^{2+}} + 2m_{Mg^{2+}} + m_{Na^+} + m_{K^+} + m_{H^+} = m_{Cl^-} + 2m_{SO_4^{2-}} + m_{HCO_3^-} + 2m_{CO_3^{2-}} + m_{OH^-}$$

If we add acid to this solution, the concentrations of HCO_3^-, CO_3^{2-}, and OH^- will change, but the concentrations of the other species (as long as precipitation or complex species formation doesn't occur) will remain constant. These species are referred to as **conservative species** because their *abundances do not vary as a function of pH or some other intensive variable (such as pressure or temperature)*. Rearranging the equation so the conservative species are on the left and the nonconservative species are on the right (a reversal of the political spectrum) gives

$$2m_{Ca^{2+}} + 2m_{Mg^{2+}} + m_{Na^+} + m_{K^+} - m_{Cl^-} - m_{SO_4^{2-}} = m_{HCO_3^-} + 2m_{CO_3^{2-}} + m_{OH^-} - m_{H^+}$$

The term on the right now represents the *excess of nonconservative bases with respect to hydrogen* and is a quantity known as **total alkalinity**. A final point with regard to conservative and nonconservative species is that during water mixing the total abundance of each conservative species will remain constant but the total abundance of each nonconservative species may change because of changes in pH, temperature, etc.

For waters in which acidity or alkalinity are largely determined by the carbonate system and completely dissociated acids or bases, we can define acidity (C_A) and alkalinity (C_B) as follows:

$$C_A = 2H_2CO_{3\ (aq)} + HCO_3^- + H^+ - OH^- \tag{3-35}$$

$$C_B = HCO_3^- + 2CO_3^{2-} + OH^- - H^+ \tag{3-36}$$

Systems in which other species are important contributors to the acidity or alkalinity would require additional terms. For example, weathering of pyrite and other sulfide minerals in coal leads to acid mine drainage that has pH values of 2 to 3. Under these conditions, sulfuric acid, bisulfate ion (HSO_4^-), and acid cations are produced. In this case, total acidity might be defined as

$$C_A = H^+ + HSO_4^- + 3Fe^{3+} + 2Fe(OH)^{2+} + Fe(OH)_2^+$$

Fe^{3+} is considered an acid cation because it can combine with OH^- to form Fe hydroxides.

Acidity and alkalinity are often reported in meq L^{-1}. Recall from Chapter 1 that a gram-equivalent weight is the molecular or atomic weight divided by the valence, or, in the case of acids and bases, the number of H^+ or OH^- ions that can be produced when the acid or base is dissolved in water. Hence, in the equation for total acidity, the number of moles of H_2CO_3 is multiplied by 2 because this acid can produce two H^+ ions. In the total alkalinity equation, the number of moles of CO_3^{2-} is multiplied by 2 because there are two gram-equivalents in one mole of CO_3^{2-}.

EXAMPLE 3–10 In Example 3–5 we calculated the abundances of the various carbonate species in a solution in equilibrium with the atmosphere at pH = 5.7. Calculate the total acidity and alkalinity of this solution.

From equation 3–35, total acidity is

$$C_A = 2H_2CO_{3\,(aq)} + HCO_3^- + H^+ - OH^-$$
$$= (2)(10^{-3.09}) + 10^{-3.74} + 10^{-5.7} - 10^{-8.3}$$
$$= 1.81 \times 10^{-3} \text{ eq } L^{-1} = 1.81 \text{ meq } L^{-1}$$

From equation 3–36, the total alkalinity is

$$C_B = HCO_3^- + 2CO_3^{2-} + OH^- - H^+ = 10^{-3.74} + (2)(10^{-8.37}) + 10^{-8.3} - 10^{-5.7}$$
$$= 1.80 \times 10^{-4} \text{ eq } L^{-1} = 0.18 \text{ meq } L^{-1} \qquad \blacksquare$$

Another common way to report total alkalinity is as equivalent weight of calcium carbonate (in mg L^{-1}). In this calculation one must use the equivalent weight of calcium carbonate. The equation is

$$\text{Total alkalinity} = \frac{C_A V_A \times \text{eq. wt } CaCO_3(0.050 \text{ g/meq}) \times 1000 \text{ mg/g}}{V_s} \qquad (3\text{–}37)$$

where C_A is the concentration of the acid (in meq L^{-1}) used in the alkalinity titration, V_A is the volume (in L) of acid used in the alkalinity titration, and V_s is the volume (in L) of the solution.

Acidity and alkalinity are important environmental parameters for natural waters. Acidity affects the weathering of silicate and carbonate minerals, the transport of metal ions in solution, and the functioning of aquatic ecosystems. High-acidity waters are generally considered to be environmentally detrimental. Conversely, alkalinity is important in determining the capacity of a water to ameliorate acid additions and is generally considered to be a positive characteristic.

ACIDITY AND ALKALINITY TITRATIONS

If we know the concentrations of all the acidic (or basic) species in a solution, we can calculate the acidity (or alkalinity). This is generally a straightforward problem for very dilute (low ionic strength) solutions. However, in more concentrated (high ionic strength) solutions, uncertainties arise regarding the formation of complexes and the calculation of activity coefficients. Thus, total acidity or alkalinity is often determined by titrating the solution with a strong base (acidity determination) or a strong acid (alkalinity determination). At the endpoint pH (discussed next), the acidity (or alkalinity) can be calculated using the following equation:

$$C_t \times V_t = C_s \times V_s \qquad (3\text{–}38)$$

where C_t is the concentration (meq L^{-1}) of acid or base in the titrant, V_t is the volume of the titrant (L^{-1}), C_s is the acidity or alkalinity (meq L^{-1}) of the unknown solution, and V_s is the volume (L^{-1}) of the unknown solution.

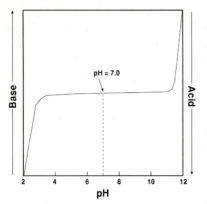

Figure 3–3

Schematic representation of an acid–base titration curve for a strong acid (base). The inflection point is at pH = 7.0, where all the excess H^+ (acid) or OH^- (base) has been consumed. Titrant is a base for an acidity determination; an acid for an alkalinity determination.

We will first consider a solution that contains only a strong acid. We take a fixed volume of this solution and titrate with a strong base. As we add the base we monitor the change in the pH of the solution (Figure 3–3). In this case, we would find that there is an inflection in the curve at pH = 7.0, i.e., at neutrality. The reaction taking place in the solution is

$$H^+ + OH^- \rightarrow H_2O$$

and at neutrality all the excess H^+ has been consumed. If we repeat this experiment using a solution that contains only a strong base and titrate with an acid, we once again find that there is an inflection point at pH = 7.0. We are only dealing with strong acids and bases, so the definitions of acidity and alkalinity reduce to the following simple equations:

$$C_A = H^+ - OH^- \quad \text{and} \quad C_B = OH^- - H^+$$

Because the activity of the OH^- ions in acidic solutions is negligible, and the activity of H^+ ions in basic solution is negligible, $C_A = H^+$ and $C_B = OH^-$.

EXAMPLE 3–11 100 mL of an acidic solution is titrated with a 100 meq L^{-1} NaOH solution. Neutrality (pH = 7) is achieved after 50 mL of titrant have been added to the acid solution. Calculate the acidity of the solution.

From equation 3–38,

$$C_s = \frac{C_t \times V_t}{V_s} = \frac{(100 \text{ meq L}^{-1})(50 \times 10^{-3} \text{ L})}{100 \times 10^{-3} \text{ L}} = 50 \text{ meq L}^{-1}$$

Total acidity is 50 meq L^{-1}. ∎

If both strong and weak acids are present in a solution, the titration curve is more complex. A schematic titration curve is shown in Figure 3–4 for a solution that contains a strong acid and carbonic acid. You will note that there are a number of inflection points on this curve, so the question arises as to what we are measuring at each inflection point. The total acidity of this solution is

$$C_A = 2H_2CO_{3 \text{ (aq)}} + HCO_3^- + H^+ - OH^-$$

Figure 3–4

Schematic acidity titration curve for a solution that contains a strong acid and carbonic acid. See text for details.

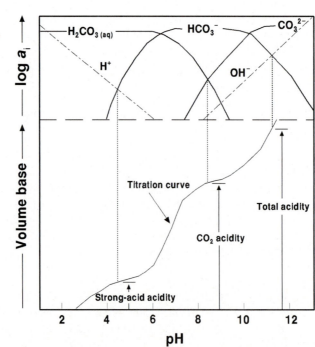

We titrate this solution with a strong base and measure the pH changes. The first inflection point occurs at pH ≈ 4.5. This is the point at which the reaction $H^+ + OH^- \rightarrow H_2O$ is essentially complete and $H^+ = HCO_3^-$. A point to note here is that while the position of the H^+ and OH^- curves on the top part of Figure 3–4 are fixed relative to pH, the positions of the curves representing the various carbonate equilibria change as a function of total carbonate content. Thus, the crossing points can vary by several tenths of a pH unit as a function of carbonate concentration. This first inflection point corresponds to the *strong-acid acidity*. The next segment of the titration curve is relatively steep; i.e., large additions of base result in only small changes in pH. In this region there is a buffer reaction involving $H_2CO_{3\ (aq)}$ and HCO_3^-, which tends to inhibit pH changes. Buffers will be discussed in the next section. At pH ≈ 8.35, there is another inflection point. At this point, 99% of the $H_2CO_{3\ (aq)}$ has been converted to HCO_3^-, as demonstrated by the following calculation, with which you should be very familiar:

$$\frac{[HCO_3^-]}{[H_2CO_{3\ (aq)}]} = \frac{K_{a_1}}{[H^+]} = \frac{10^{-6.35}}{10^{-8.35}} = 10^{2.00}$$

This inflection point corresponds to the *CO_2 acidity*. The final inflection point occurs when $HCO_3^- = OH^-$. This inflection point corresponds to the *total acidity*.

If our solution contained a strong base and a weak acid, the titration curve would have the same shape as in Figure 3–4, except we would titrate with a strong acid and the pH would decrease as the titration proceeded. The point at which $OH^- = HCO_3^-$ represents the *caustic alkalinity* (the free OH^- ions from strong bases), the inflection point at pH ≈ 8.3 represents the *carbonate alkalinity*, and the inflection point at pH ≈ 4.5 represents the *total alkalinity*.

Alkalinity titrations are done using a strong acid and two endpoint indicators—phenolphthalein (pH = 8.3) and methyl orange (pH = 4.5). The first endpoint corresponds to the carbonate alkalinity and the second endpoint to the total alkalinity. Note from Figure 3–4 that near the two endpoints pH changes rapidly with small additions of acid. Hence, a precise determination of pH is not required. If a more precise measurement is desired, the pH change of the solution, as a function of the volume of acid added, can be monitored with a pH meter and a titration curve plotted from the data.

EXAMPLE 3–12 In Example 3–9 we calculated the pH of water from an alkaline lake in Tanzania. A 100 mL water sample from the lake is titrated with 100 meq L^{-1} nitric acid. Inflection points are noted in the titration curve at about pH = 8.34 (after 54 mL of titrant have been added) and at about pH = 3.0 (after 102 mL of titrant have been added). Calculate the carbonate alkalinity and total alkalinity for this water sample in equivalent weight of calcium carbonate. Carbonate alkalinity corresponds to the inflection point at pH = 8.34.

$$\frac{\text{Carbonate}}{\text{alkalinity}} = \frac{C_A V_A \times \text{eq. wt } CaCO_3(0.050\ \text{g/meq}) \times 1000\ \text{mg/g}}{V_s}$$

$$= \frac{(100\ \text{meq L}^{-1})(54 \times 10^{-3}\ \text{L}^{-1})(0.050)(1000)}{(100 \times 10^{-3}\ \text{L}^{-1})} = 2700\ \text{mg L}^{-1} \quad \text{as } CaCO_3$$

Total alkalinity corresponds to the inflection point at pH = 3.0.

$$\frac{\text{Total}}{\text{alkalinity}} = \frac{C_A V_A \times \text{eq. wt } CaCO_3(0.050\ \text{g/meq}) \times 1000\ \text{mg/g}}{V_s}$$

$$= \frac{(100\ \text{meq L}^{-1})(102 \times 10^{-3}\ \text{L}^{-1})(0.050)(1000)}{(100 \times 10^{-3}\ \text{L}^{-1})} = 5100\ \text{mg L}^{-1} \quad \text{as } CaCO_3$$

BUFFERS

A weak acid and its salt (or a weak base and its salt) form what is known as a *buffer*. A *buffered solution* is *a solution that resists changes in pH when either hydrogen or hydroxyl ions are added to the solution.* Let us first illustrate what we mean by a buffer by considering once again the carbonic acid system. For all acidic waters, H_2CO_3 (aq) and HCO_3^- (Figure 3–2) will be the dominant carbonate species. Equation 3–8 relates the activity of these carbonate species to the H^+ activity. We can rewrite equation 3–8 as follows:

$$\frac{1}{[H^+]} = \frac{1}{K_{a_1}} \frac{[HCO_3^-]}{[H_2CO_{3\,(aq)}]} \tag{3–39}$$

Taking the logs of both sides, and remembering that $pH = -\log [H^+]$, gives

$$pH = -\log K_{a_1} + \log \frac{[HCO_3^-]}{[H_2CO_{3\,(aq)}]} \tag{3–40}$$

We can use equation 3–40 to calculate the effect that the addition of hydrogen ions will have on the pH.

EXAMPLE 3–13 Let us suppose that we have 1 L of pure water with pH = 7.0. If we add 10^{-4} mol of H^+ ions to the pure water, the pH will drop to 4 (10^{-7} mol + 10^{-4} mol = 10^{-4} mol L^{-1} of H^+). If carbonic acid is present in the solution, what will be the change in pH when the H^+ ions are added? At pH = 7 and $T = 25°C$, assume that $[HCO_3^-] = 10^{-3}$ mol L^{-1}.

From equation 3–8 we get $[H_2CO_{3\,(aq)}] = 10^{-3.65}$ mol L^{-1}. When we add H^+ ions to the solution, the following reaction occurs:

$$H^+ + HCO_3^- \rightarrow H_2CO_{3\,(aq)}$$

an example of LeChâtelier's principle. According to this reaction, the added H^+ ions combine with the HCO_3^- to form $H_2CO_{3\,(aq)}$. Thus, the activity of HCO_3^- decreases by 10^{-4} mol L^{-1} and the activity of $H_2CO_{3\,(aq)}$ increases by 10^{-4} mol L^{-1}. Solving equation 3–40 for pH gives

$$pH = -\log K_{a_1} + \log\left(\frac{[HCO_3^-]}{[H_2CO_{3\,(aq)}]}\right)$$

$$= 6.35 + \log\left(\frac{[10^{-3.0} - 10^{-4.0}]}{[10^{-3.65} + 10^{-4.0}]}\right) = 6.35 + 0.44 = 6.79$$

Due to the presence of carbonic acid, the pH only decreases by 0.21 units rather than 3.0 units.   ∎

The calculation we have just done used the ***Henderson–Hasselbalch equation***. Consider the following reaction:

$$H^+ + A^- \rightarrow HA$$

In this reaction, a H^+ ion has combined with the salt of a weak acid (a conjugate base) to form a weak acid. We can write equation 3–40 in a general form (the Henderson–Hasselbalch equation) as

$$pH = -\log K_a + \log\left(\frac{[A^-]}{[HA]}\right) \tag{3–41}$$

Buffers are important in the natural environment because they control the impact of acid or base additions on natural waters. A widely discussed example of this type of impact is acid rain (see Chapter 8), which can lead to the acidification of streams and lakes. Even in the same geographic area, lakes can have widely different pH values, depending upon the available buffers. There are two types of reactions that can control pH—those due to dissolved species in water and those that require interactions between water and mineral species. The carbonic acid–bicarbonate ion reaction considered in Example 3–13 is illustrative of the first type of pH-controlling reaction. An example of the second type of pH-controlling reaction is the calcite–carbonate system, which controls the long-term pH of the oceans. If we add H^+ ions to the ocean, in addition to the carbonate equilibria illustrated in Example 3–13, the following reaction can occur, which will remove H^+ ions from solution:

$$CaCO_{3 \, calcite} + H^+ \rightarrow Ca^{2+} + HCO_3^-$$

If we add OH^- ions, the following reaction may occur:

$$Ca^{2+} + HCO_3^- + OH^- \rightarrow CaCO_{3 \, calcite} + H_2O$$

As a second example, consider the following reaction between the clay minerals illite and kaolinite:

$$2KAl_3Si_3O_{10}(OH)_{2 \, illite} + 2H^+ + 3H_2O \rightarrow 3Al_2Si_2O_5(OH)_{4 \, kaolinite} + 2K^+$$

In this reaction, hydrogen ions are consumed during the conversion of illite to kaolinite.

The **buffering capacity** is *a measure of the amount of H^+ or OH^- ions a solution can absorb without a significant change in pH.* The capacity of a buffered solution is determined by the amount of [HA] and [A$^-$] in the solution. For an acidity titration, the buffering index, B, is defined as

$$B = \frac{dC_B}{d\text{pH}} \tag{3–42}$$

where dC_B is the increment of strong base added in eq L^{-1} (or meq L^{-1}) and dpH is the corresponding change in solution pH. For an alkalinity titration, the buffering index is written

$$B = \frac{dC_A}{d\text{pH}} \tag{3–43}$$

where dC_A is the incremental amount of strong acid added.

The ability of a particular system to buffer changes in pH varies as a function of pH. For weak acids and bases, the maximum buffering capacity occurs at pH values that equal the dissociation constants, pK_a or pK_b, of the weak acid or base. At pH = pK_a (or pK_b), the ratio [A$^-$]/[HA] = 1 and is most resistant to change. The following example will illustrate this point.

EXAMPLE 3–14 We will return to Example 3–13, the carbonate buffer. In this case, the initial pH is 6.35, equivalent to the first dissociation constant for carbonic acid. The total carbonate content is 2×10^{-3} mol L^{-1}. At pH = 6.35, $H_2CO_{3 \, (aq)}$ and HCO_3^- will be present in equal abundance. The activity of the CO_3^{2-} ion is negligible.

Hence, we can write

$$\text{pH} = -\log K_{a_1} + \log\left(\frac{[HCO_3^-]}{[H_2CO_{3 \, (aq)}]}\right) = 6.35 + \log\left(\frac{[10^{-3.0}]}{[10^{-3.0}]}\right) = 6.35 + 0.0 = 6.35$$

We now add 10^{-5} mol of H^+ ions to 1 L of solution. After the addition we recalculate the pH:

$$\text{pH} = -\log K_{a_1} + \log\left(\frac{[HCO_3^-]}{[H_2CO_{3 \, (aq)}]}\right)$$

$$= 6.35 + \log\left(\frac{[10^{-3.0} - 10^{-5.0}]}{[10^{-3.0} + 10^{-5.0}]}\right) = 6.35 + (-0.009) = 6.34$$

Because of the buffering capacity of the system, the pH has only changed by -0.01 units. Let us continue to add acid until the pH of the solution is reduced to 5.35. At this pH, $H_2CO_{3 (aq)}$ is 10 times more abundant in the solution than HCO_3^-. Under these conditions, what is the effect of a 10^{-5} mol addition of H^+?

$$pH = -\log K_{a_1} + \log\left(\frac{[HCO_3^-]}{[H_2CO_{3 (aq)}]}\right)$$

$$= 6.35 + \log\left(\frac{[1.818 \times 10^{-4.0} - 10^{-5.0}]}{[1.818 \times 10^{-3.00} + 10^{-5.0}]}\right) = 6.35 + (-1.03) = 5.32$$

In this case, the pH has changed by -0.03 pH units. Although not a large difference from the previous calculation, we can see that there has been a decrease in the buffering capacity of the solution. Repeating the calculation when pH = 4.35, with the same addition of acid, gives pH = 4.04, a change of -0.31 pH units, clearly showing the declining effectiveness of the buffer. ∎

The derivation of the buffering index for various systems is not always straightforward. Several examples will be considered next. A more complete discussion can be found in Langmuir (1997). We will start with the simple system H_2O and then extend our analysis to include weak acids.

Buffering Index for H_2O

We start with pure water, which we titrate with a NaOH solution; i.e., we are determining the total acidity of water. In this example, and all the following examples, we will assume that activity equals concentration. The charge balance equation is

$$Na^+ + H^+ = OH^-$$

and $C_B = Na^+$, which represents the total amount of base added to the solution. Substituting into the charge balance equation we can write

$$C_B = OH^- - H^+ = \frac{K_w}{H^+} - H^+ \tag{3–44}$$

Taking the derivative of both sides gives

$$dC_B = \left[-\frac{K_w}{(H^+)^2} - 1\right]dH^+ \tag{3–45}$$

$pH = -\ln(H^+)/2.3$, so $dH^+ = dpH[-2.3(H^+)]$. Substituting for dH^+ gives the final form of the equation:

$$B_{H_2O} = \frac{dC_B}{dpH} = 2.3\left[\frac{K_w}{H^+} + H^+\right] \tag{3–46}$$

Equation 3–46 is easily solved on a spreadsheet for a range of pH values, and the result of such a computation is shown graphically in Figure 3–5. The buffering capacity of water is greatest at very low and very high pHs and quickly declines to essentially zero at intermediate pH values. Hence, in most natural waters, H_2O is not an important pH buffer.

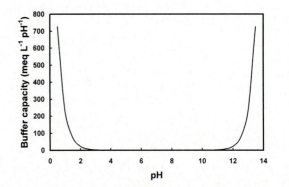

Figure 3–5

Variations in buffer capacity as a function of pH for water. Only at very low and very high pH values is water an important buffer.

Buffering Index for a Weak Monoprotic Acid

We start with a solution that contains a weak monoprotic acid (HA) and we determine its acidity by titrating with NaOH. The total acidity due to the weak acid is

$$C_A = HA + A^-$$

The charge balance equation for the system monoprotic weak acid plus water plus NaOH is

$$Na^+ + H^+ = A^- + OH^-$$

Because $C_B = Na^+$, we can write the charge balance equation as

$$C_B = A^- + OH^- - H^+$$

What we now want to do is express A^- in terms of hydrogen ions and the acidity (C_A). We do this by rearranging the equation for total acidity to give

$$HA = C_A - A^-$$

and then substituting for HA in the equilibrium equation:

$$K_a = \frac{(H^+)(A^-)}{(HA)} = \frac{(H^+)(A^-)}{(C_A - A^-)} \qquad (3\text{–}47)$$

Solving equation 3–47 for A^- gives

$$A^- = \frac{K_a C_A}{(K_a + H^+)} \qquad (3\text{–}48)$$

We now substitute into the charge balance equation:

$$C_B = A^- + OH^- - H^+ = \frac{K_a C_A}{(K_a + H^+)} + \frac{K_w}{H^+} - H^+ \qquad (3\text{–}49)$$

Taking the derivative of both sides,

$$dC_B = \left[-\frac{K_a C_A}{(K_a + H^+)^2} - \frac{K_w}{(H^+)^2} - 1 \right] dH^+ \qquad (3\text{–}50)$$

and rearranging gives

$$-\frac{dC_B}{dH^+} = \left[\frac{K_a C_A}{(K_a + H^+)^2} + \frac{K_w}{(H^+)^2} + 1 \right] \qquad (3\text{–}51)$$

Recalling that $dH^+ = dpH[-2.3(H^+)]$ and substituting for dH^+ gives the final form of the buffering index equation for a weak monoprotic acid:

$$B = \frac{dC_B}{dpH} = 2.3 \left[\frac{K_a C_A (H^+)}{(K_a + H^+)^2} + \frac{K_w}{H^+} + H^+ \right] \qquad (3\text{–}52)$$

The first term in this equation describes the buffering index for the weak monoprotic acid; the second and third terms describe the buffering index for water. This equation is plotted for acetic acid in Figure 3–6 with $C_A = 1$ mol. As noted previously, acetic acid has its maximum buffering capacity at a pH equal to the pK_a for acetic acid.

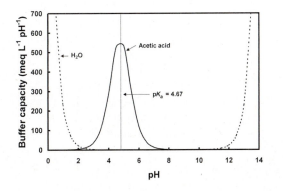

Figure 3–6

Variations in buffering capacity as a function of pH for water and acetic acid when $C_A = 1$ mol.

Buffering Index for Weak Polyprotic Acids

As long as the dissociation constants differ by a factor of 20 or more, the solution for the buffering index of polyprotic acids is analogous to equation 3–52. Additional terms are added to represent the additional dissociation steps. Equation (3–53) is for a diprotic acid. Note that there are now two terms at the beginning of the equation that describe the first and second dissociation for the diprotic acid.

$$B = \frac{dC_B}{d\text{pH}} = 2.3 \left[\frac{K_{a_1} C_A(\text{H}^+)}{(K_{a_1} + \text{H}^+)^2} + \frac{K_{a_2} C_A(\text{H}^+)}{(K_{a_2} + \text{H}^+)^2} + \frac{K_w}{\text{H}^+} + \text{H}^+ \right] \qquad (3\text{–}53)$$

As previously mentioned, carbonic acid is one of the most important acids in the natural waters. Because carbonic acid is a diprotic acid, it will have two pH regions in which it acts as a buffer. The equation for carbonic acid is plotted in Figure 3–7 for $C_A = 1 \times 10^{-3}$ mol L^{-1}. Note the two regions of highest buffering capacity corresponding to $pK_{a_1} = 6.35$ and $pK_{a_2} = 10.33$, the first and second dissociation constants, respectively, of carbonic acid. At pH values less than 4.7 and greater than 10.4, the buffering capacity of water is greater than that of carbonic acid.

Figure 3–7

Variations in buffering capacity as a function of pH for the system carbonic acid–water when $C_A = 1 \times 10^{-3}$ mol L^{-1}.

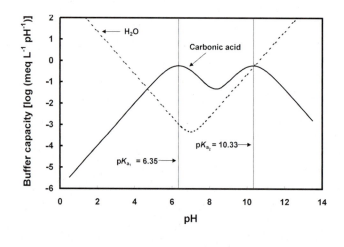

MINERAL–WATER REACTIONS

So far we have considered buffering reactions that utilize dissolved species. On a longer time frame, we are also interested in acidity and alkalinity due to interactions between minerals and water. Most minerals can be considered to be salts of weak acids and strong bases, so these interactions contribute to the alkalinity of the water. There are several exceptions to this generality, particularly in the case of acid mine drainage, where minerals may be present that are salts of a weak base and a strong acid. Such minerals would contribute to the acidity of the water.

Consider what happens when acidic waters come in contact with K-feldspar. This reaction can be written as follows:

$$2\text{KAlSi}_3\text{O}_8 \text{ K-feldspar} + 2\text{H}_2\text{CO}_3 \text{ (aq)} + 9\text{H}_2\text{O} \rightarrow \text{Al}_2\text{Si}_2\text{O}_5(\text{OH})_4 \text{ kaolinite}$$
$$+ 2\text{K}^+ + 4\text{H}_4\text{SiO}_4 \text{ (aq)} + 2\text{HCO}_3^-$$

In the course of this reaction, carbonic acid is consumed and bicarbonate ion (HCO$_3^-$) is formed. Also formed is silicic acid, but this acid is essentially undissociated in acidic waters. The net result of this reaction is the consumption of carbonic acid and an increase in bicarbonate ion. Thus, the weathering of feldspar tends to offset acid additions (reduces acidity). Carbonate minerals play a significant role in buffering acid additions, particularly because they react with water much more quickly than do the silicate minerals. A representative reaction is

$$\text{CaCO}_3 \text{ calcite} + \text{H}_2\text{CO}_3 \text{ (aq)} \rightarrow \text{Ca}^{2+} + 2\text{HCO}_3^-$$

in which the carbonic acid reacts with calcite to form Ca^{2+} and bicarbonate ions. As in the case of the silicate minerals, carbonic acid is consumed and there is an increase in the bicarbonate ion.

Buffering Index for the Calcite–Carbonic Acid System

As has been noted, this important system controls the long-term pH of the ocean. It also exerts a significant influence on the pH of fresh waters in contact with limestone or in soils containing carbonate minerals. The full derivation will not be done here (it can be found in Langmuir, 1997). This calculation is for a calcite-saturated solution with fixed total carbonate. The solution is titrated with HCl; therefore, $C_A = Cl^-$. The charge balance equation is

$$2Ca^{2+} + H^+ = HCO_3^- + 2CO_3^{2-} + Cl^- + OH^-$$

Substituting C_A for Cl^- and rearranging the charge balance equation in terms of C_A gives

$$C_A = H^+ - OH^- + 2Ca^{2+} - HCO_3^- - 2CO_3^{2-}$$

As we have done in the previous cases, we now carry out a set of substitutions so that the equation is written in terms of the H^+ ion and then differentiate the resulting equation. After completion of these operations, the equation becomes

$$B = 2.3\left[(H^+) + \frac{K_w}{(H^+)}\right] + 2.3\left[\frac{C_T K_{a_2}^{-1}(H^+)[K_{a_1}^{-1}K_{a_2}^{-1}(H^+)^2 + 4K_{a_2}^{-1}(H^+) + 1]}{[K_{a_1}^{-1}K_{a_2}^{-1}(H^+)^2 + K_{a_2}^{-1}(H^+) + 1]^2}\right]$$

$$+ 2.3\left[\frac{(H^+)}{K_{cal}C_T}[2(H^+) + K_{a_1}]\right] \tag{3-54}$$

where C_T is the total carbonate concentration, K_{a_1} and K_{a_2} are the first and second dissociation constants for carbonic acid, and K_{cal} is the solubility constant for calcite. If we wanted to solve this equation for aragonite, we would use the aragonite solubility constant, K_{arag}. The equation has been written in three parts to illustrate the various components of the calculation. The first part represents the buffering capacity of water, the second the buffering capacity of carbonic acid, and the third the buffering capacity due to the dissolution of calcite. This system reacts relatively rapidly to changes in pH because calcite readily dissolves in acidic waters. Note that the dissolution of calcite tends to offset pH decreases. The precipitation of calcite is more difficult, so this system would not be as effective in mitigating pH increases. The buffering curve for the system calcite–carbonic acid, at 25°C and $C_T = 1 \times 10^{-3}$ mol L^{-1}, is shown in Figure 3–8.

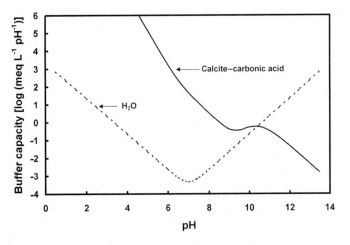

Figure 3–8
Variations in buffering capacity as a function of pH for the calcite–carbonic acid system. $C_T = 1 \times 10^{-3}$ mol L^{-1}.

EXAMPLE 3–15 Calculate the buffering index for the system calcite–carbonic acid at pH = 6, total carbonate = 1×10^{-2} mol L^{-1}, and $T = 25°C$. Select the appropriate equilibrium constants from Table 3–3 (for water) and Table 3–4 (for the carbonates). Solve for the buffering index using equation 3–54.

$$B = 2.3 \left[(H^+) + \frac{K_w}{(H^+)} \right] + 2.3 \left[\frac{C_T K_{a_2}^{-1}(H^+)[K_{a_1}^{-1}K_{a_2}^{-1}(H^+)^2 + 4K_{a_1}^{-1}(H^+) + 1]}{[K_{a_1}^{-1}K_{a_2}^{-1}(H^+)^2 + K_{a_2}^{-1}(H^+) + 1)]^2} \right]$$

$$+ 2.3 \left[\frac{(H^+)}{K_{cal}C_T}[2(H^+) + K_{a_1}] \right]$$

$$= 2.3 \left[10^{-6} + \frac{10^{-13.995}}{10^{-6}} \right]$$

$$+ 2.3 \left[\frac{10^{-2}10^{10.33}10^{-6}[10^{6.35}10^{10.33}(10^{-6})^2 + 4 \times 10^{6.35}10^{-6} + 1]}{[10^{6.35}10^{10.33}(10^{-6})^2 + 10^{10.33}10^{-6} + 1]^2} \right]$$

$$+ 2.3 \left[\frac{10^{-6}}{10^{-8.48}10^{-2}} [2 \times 10^{-6} + 10^{-6.35}] \right]$$

$$= 2.32 \times 10^{-6} + 4.91 \times 10^{-3} + 0.16994 = 0.17485 \text{ eq } L^{-1} \text{ pH}^{-1}$$

$$= 174.9 \text{ meq } L^{-1} \text{ pH}^{-1} \qquad \blacksquare$$

Buffering Index for Water–Silicate Mineral Systems

Various water–silicate mineral reactions can occur that tend to mitigate the effect of the acid addition. These reactions proceed much more slowly than those just discussed because of kinetic constraints. As examples of these types of buffers, consider the following reactions:

$$KAl_3Si_3O_{10}(OH)_2 \text{ muscovite} + H^+ + 1.5H_2O \rightarrow 1.5Al_2Si_2O_5(OH)_4 \text{ kaolinite} + K^+$$

and

$$KAlSi_3O_8 \text{ K-feldspar} + H^+ + 4.5H_2O \rightarrow 0.5Al_2Si_2O_5(OH)_4 \text{ kaolinite} + K^+ + 2H_4SiO_4 \text{ (aq)}$$

In the first reaction, hydrogen ions are consumed in the conversion of muscovite to the clay mineral kaolinite. In the second reaction, a common weathering reaction in humid temperate climates, hydrogen ions are consumed when K-feldspar is weathered to kaolinite. Both reactions are essentially irreversible, so they will only buffer hydrogen ion additions, i.e., pH decreases. We start by calculating the equilibrium constants (at 25°C) for both reactions. In each case,

$$K_{eq} = \frac{[K^+]}{[H^+]}$$

because activity = 1 for the other species in both equations. We calculate the equilibrium constants as was done in Chapter 2. For this set of calculations, we will use the free-energy values of Robie et al. (1978) (Appendix II, source 3), an internally consistent set of thermodynamic data for calculations involving common minerals. For the muscovite–kaolinite reaction, $\Delta G_R^0 = -25.25$ kJ mol^{-1} and log $K_{eq} = 4.42$. For the K-feldspar–kaolinite reaction, $\Delta G_R^0 = 11.05$ kJ mol^{-1} and log $K_{eq} = -1.94$.

The buffering index will have the same form for both mineral pairs. Here we will derive the buffering index equation for the muscovite–kaolinite pair. We start by assuming that we are titrating a water–mineral mixture with HCl ($C_A = Cl^-$). The charge balance equation is

$$H^+ + K^+ = OH^- + Cl^-$$

The total acidity equation is

$$C_A = H^+ + K^+ - OH^-$$

Because $[K^+] = K_{eq}[H^+]$, substitution into the total acidity equation for K^+ and OH^- gives

$$C_A = K_{eq}[H^+] + [H^+] - \frac{K_w}{[H^+]} \tag{3-55}$$

Taking the derivative and substituting for dH^+, as in previous examples, gives

$$B = 2.3\left[(H^+) + \frac{K_w}{(H^+)} + K_{eq}(H^+)\right] = 2.3\left[(H^+) + \frac{10^{-14}}{(H^+)} + 10^{4.42}(H^+)\right] \tag{3-56}$$

For the K-feldspar–kaolinite reaction, we have a similar equation—the only difference is the value for the equilibrium constant. Silicic acid is not included in the buffer calculation because only at very basic pH values would there be any significant dissociation to $H_3SiO_4^-$ (first dissociation constant $= 10^{-9.9}$). The buffering index equation for the K-feldspar–kaolinite pair is

$$B = 2.3\left[(H^+) + \frac{10^{-14}}{(H^+)} + 10^{-1.94}(H^+)\right] \tag{3-57}$$

Equations 3–56 and 3–57 are plotted in Figure 3–9.

Buffering Capacities of Natural Waters

Figure 3–9 summarizes the results of our calculations of buffering reactions for natural waters. Several important generalizations can be made from this diagram. Except at very low or very high pH, water has little buffering capacity. In most natural waters carbonic acid is the most important system for buffering pH at intermediate values. If calcite is present in the system, there is a significant increase in buffering capacity. Hence, waters in contact with limestone, or soils containing carbonate minerals, will be very resistant to pH changes, particularly decreasing pH changes. Although they operate more slowly because of kinetic factors, clay-type minerals have significant buffering capacity for acid additions. Conversely, weathering reactions involving feldspar have very low buffering capacity (note from Figure 3–9 that the buffering capacity of the feldspar–kaolinite reaction is less than that of water). These are important observations in terms of the effect of acid additions on natural waters. In terrains where carbonates (and/or clay minerals) are present, there will be significant buffering of acid additions. In terrains underlain by rocks such as granite (quartz and feldspar minerals), buffering capacity will be nil. Hence, we would expect that lakes and streams draining limestones areas would be much less affected by acid rain than those draining granitic terrains.

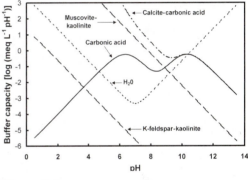

Figure 3–9

Summary of buffering reactions for natural waters containing carbonic acid and/or in contact with various minerals. Note the importance of the carbonate and clay minerals for buffering acid additions to natural waters.

Mineral Reactivity

A final factor to consider is the rate at which buffering reactions will occur. Carbonate minerals are generally considered to be highly reactive, particularly when present as minute grains. Using the reactivity of carbonate minerals as a reference point, Table 3–7 (p. 88) lists the relative reactivity, at pH = 5, for a number of common minerals. What is meant by relative reactivity is, under the given set of conditions, compared to calcite, how readily the mineral will react with the fluid. For minerals with low relative reactivity, there

Table 3–7 Relative Reactivity of Common Minerals at pH = 5*

Mineral group	Typical minerals	Relative reactivity at pH 5
Dissolving	Calcite, aragonite, dolomite, magnesite, brucite	1.00
Fast weathering	Anorthite, nepheline, forsterite, olivine, garnet, jadeite, leucite, spodumene, diopside, wollastonite	0.40
Slow weathering	Plagioclase feldspars (albite, oligoclase, labradorite), clays (vermiculite, montmorillonite)	0.01
Very slow weathering	K-feldspars (orthoclase, microcline), muscovite	0.01
Inert	Quartz, rutile, zircon	0.004

*From Lawrence and Scheske (1997).

is little interaction between the fluid and the mineral. Even if these minerals were effective buffers, we would not expect them to have a significant effect on pH.

Lawrence and Scheske (1997) used relative reactivities to calculate the neutralization potential of mining wastes. This was done by using the normative mineralogy (weight abundance of ideal end-member minerals calculated from the chemical composition of the rock) and the relative reactivity of the various minerals to calculate a neutralization potential for the rock that could be compared to that of limestone. From a strictly kinetic point of view, rocks composed of reactive minerals would be more effective at neutralizing acid waters than those containing minerals of low relative reactivity.

EXAMPLE 3–16 A rock contains 24% (24 g/100 g) normative forsterite (Mg_2SiO_4). Calculate the contribution of the forsterite to the neutralization potential (NP) of the rock. NP values are expressed as kg $CaCO_3$-equivalent per 1000 kg of rock (1 ton).

In doing the calculation, we must take account of the differences in molecular weights.

$$\text{NP contribution} = \frac{\text{Normative wt mineral (g)}}{100 \text{ g}} \times \frac{\text{mol wt calcite}}{\text{mol wt mineral}}$$

$$\times \frac{1000 \text{ kg}}{\text{ton}} \times \text{relative reactivity}$$

$$= \frac{24}{100} \times \frac{100.1}{140.7} \times 1000 \text{ kg ton}^{-1} \times 0.4$$

$$= 68.3 \text{ kg CaCO}_3\text{-equivalent ton}^{-1} \qquad \blacksquare$$

ACID–BASE EQUILIBRIA CASE STUDIES

We conclude our discussion of acid–base equilibria with two case studies that illustrate the application of several concepts developed in this chapter to "real-world" problems. In Case Study 3–3, a model is developed for the attenuation of acid mine drainage by natural processes. In Case Study 3–4, the potential of deep saline aquifers as disposal sites for carbon dioxide and other acid gases is investigated.

CASE STUDY 3–3
A Process Model for the Natural Attenuation of Acid Mine Drainage

Berger et al. (2000) investigated the chemistry of acid mine drainage from an abandoned Cu-Pb-Zn mine located in the southern Sangre de Cristo mountains in New Mexico. Waste rock from the mining operation was piled on top of an existing drainage channel, and acid waters discharged directly from this waste pile. The waste rock is largely composed of quartz-chlorite and biotite schists. The underlying bedrock consists of a Pennsylvanian-age conglomeratic unit composed of siliclastic and carbonate components. Flow through the waste pile is essentially vertical, and atmospheric O_2 + rainwater react with the waste pile to produce acid waters and metal ions in solution. In the immediate vicinity of the waste pile, waters have a pH of 4.1 to 4.2. Willow Creek runs along the western edge of the site, and the contaminated waters enter this creek. Upstream from the site, Willow Creek waters have pH = 7.2. Given the relatively high pH of Willow Creek, simple mixing with Willow Creek water would lead to a pH increase and a dilution in the concentrations of the dissolved metals in the acid mine drainage. This pH increase may also cause the precipitation of metal ions as insoluble hydroxides, oxides, and sulfates. An additional factor affecting pH and the precipitation of metals is the interaction between the acid mine drainage and the carbonate component of the underlying bedrock. Field observations indicated that there had been acid water–carbonate interactions, as shown by the cracked and discolored nature of the limestone. The calcite dissolution reaction can be written

$$CaCO_{3\ calcite} + 2H^+ \rightarrow Ca^{2+} + H_2O + CO_{2\ (aq)}$$

This reaction will lead to an increase in pH and will release Ca^{2+} ions to the water.

The authors investigated these questions by collecting water samples from several drainage channels and measuring the change in chemistry of the water samples. The assumption was made that SO_4^{2-} was a conservative component; i.e., its concentration was not affected by processes occurring in the water. Thus, sulfate concentration varies simply as a function of the ratio of contaminated water to uncontaminated water. For example, contaminated waters from the site had maximum sulfate concentrations of approximately 8400 mg L^{-1}, while uncontaminated Willow Creek water had sulfate concentrations on

the order of 120 mg L^{-1}. A 50/50 mixture of contaminated and uncontaminated water would have a sulfate concentration of approximately 4260 mg L^{-1}. Plots of various components versus sulfate showed that Zn, Mg, and Mn behaved conservatively; Fe, Al, Pb, and to some degree Cu showed concentrations less than that expected by simple mixing. Ca concentrations exceeded those due to simple mixing. The elevated Ca concentrations led the authors to conclude that the dissolution of limestone was a factor in determining the pH of the acid mine drainage. The authors developed a reaction path model, illustrated in Figure 3–C3–1, to explain the variations in Ca^{2+}. In this model, calcite dissolution and mixing lead to an increase in pH to a value of 6; mixing alone leads to a final pH value of 7.2. Under these conditions Al, Cu, and Fe precipitate directly from solution. The authors concluded that the chemistry of the acid mine drainage was controlled by two processes: mineral buffering and mixing with ambient waters.

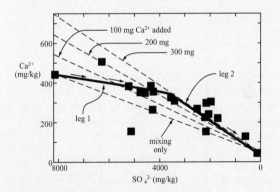

Figure 3–C3–1

Trend in Ca concentration predicted by the reaction path model. Dashed lines show simple mixing for fluids with different amounts of added Ca. Filled squares represent individual water samples. The reaction pathway is shown by the heavy line. In leg 1, Ca is added by dissolution of limestone. In leg 2, Ca concentrations change only by simple dilution. In leg 1, 2.5 kg of uncontaminated water and 250 mg of Ca are added to the solution. In leg 2, 12.5 kg of uncontaminated water is added. From Berger et al. (2000).

Source: Berger et al. (2000).

CASE STUDY 3–4
Aquifer Disposal of Acid Gases

Gunter et al. (2000) investigated the possibility of disposing of acid gases (CO_2, H_2SO_4 and H_2S) in deep saline aquifers in the Alberta sedimentary basin. Models were developed for the chemical reactions that would occur between these acid gases and two sedimentary reservoirs: the Glauconitic Sandstone aquifer, consisting of 87% quartz, 2% K-feldspar, 1% plagioclase, 5% glauconite, 2% kaolinite, 1% calcite, 1% dolomite, and 1% siderite; and the Nisku carbonate aquifer, consisting predominantly of dolomite and/or calcite with or without accessory anhydrite, quartz, pyrite, and illite. The modeling was done using the reaction path computer model PATHARC.94 and a single injection of acid gases. The reaction between the acid gases and the carbonate aquifer was essentially instantaneous, while reactions with the Glauconitic Sandstone took hundreds of years to run to completion (Figure 3–C4–1). In the carbonate aquifer, the reaction involving CO_2 is

$$CaCO_{3 \text{ calcite}} + H_2O + CO_2 \rightarrow Ca^{2+} + 2HCO_3^-$$

This reaction quickly reaches equilibrium (Figure 3–C4–1), but little CO_2 is removed from solution. In the model calculation, the reduction in CO_2 pressure is from 87 to 86 bars. Thus, carbonate rocks have little trapping capacity for CO_2. For the Glauconitic Sandstone, hundreds of years were required to reach equilibrium (Figure 3–C4–1), but in this case essentially all of the CO_2 is trapped by mineral reactions of the following types. In the calculations, muscovite was used as a proxy for illite.

(1) $3KFe_3AlSi_3O_{10}(OH)_{2 \text{ annite}} + 11CO_2 \rightarrow$ $KAl_3Si_3O_{10}(OH)_{2 \text{ muscovite}} + 9FeCO_{3 \text{ siderite}}$ $+ 6SiO_{2 \text{ quartz}} + H_2O + 2K^+ + 2HCO_3^-$

(2) $3NaAlSi_3O_{8 \text{ albite}} + 2H_2O + 2CO_2 + K^+ \rightarrow$ $KAl_3Si_3O_{10}(OH)_{2 \text{ muscovite}} + 6SiO_2 + 3Na^+ + 2HCO_3^-$

(3) $3KAlSi_3O_{8 \text{ K-feldspar}} + 2H_2O + 2CO_2 \rightarrow$ $KAl_3Si_3O_{10}(OH)_{2 \text{ muscovite}} + 6SiO_2 + 2K^+ + 2HCO_3^-$

The calculations showed that the carbonate aquifer would not be a good trap for CO_2, but the Glauconitic Sandstone would be an effective trap.

Similar calculations were done for the disposal of SO_2 (as H_2SO_4). In the carbonate aquifer, the reaction is

$$CaCO_{3 \text{ calcite}} + H_2SO_4 \rightarrow CaSO_{4 \text{ anhydrite}} + H_2O + CO_2$$

In this process CO_2 pressure builds up to substantial levels, in the model run to 80 bars. This is a significant overpressure that can fracture the rocks, thus permitting the escape of CO_2. Because the Glauconitic Sandstone also contained carbonate minerals, a similar reaction would occur. However, in this case reactions 1–3 would also occur, leading to the reduction of the CO_2 pressure to ambient values. Thus, the Glauconitic Sandstone is a potential trap for sulfuric acid.

Source: Gunter et al. (2000).

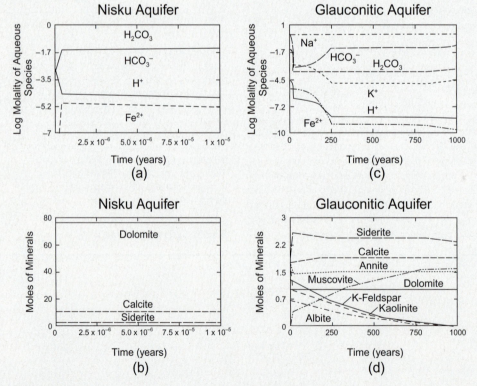

Figure 3–C4–1

Water–rock reactions as a function of time. Initial input is 1 mol of CO_2 into 1 kg of formation water. (a) Variation in amounts of aqueous species and (b) minerals for the Nisku carbonate aquifer. (c) Variation in amounts of aqueous species and (d) minerals for the Glauconitic Sandstone aquifer. From Gunter et al. (2000).

QUESTIONS AND PROBLEMS

1. Distinguish between the *Arrhenius concept* and the *Brönsted–Lowry model* of acids and bases.

2. How does a strong acid differ from a weak acid?

3. Does a pH of 7 always indicate neutrality? Explain.

4. At 250°C, the equilibrium constant for the dissociation of water is $10^{-11.16}$ (Ohmoto et al., 1994). At this temperature, what is the pH of a neutral solution?

5. pH is referred to as an *environmental parameter*. What does this mean?

6. Most minerals can be considered to be salts of weak acids and strong bases. Identify the weak-acid and strong-base components in each of the following minerals:
   a. Olivine [Mg_2SiO_4]
   b. K-feldspar [$KAlSi_3O_8$]
   c. Diopside [$CaMgSi_2O_6$]
   d. Kaolinite [$Al_2Si_2O_5(OH)_4$]

7. In terms of the carbonic acid system, distinguish between an *open* system and a *closed* system.

8. Explain why the pH of surface waters increases when green plants in the water are actively photosynthesizing.

9. With reference to Case Study 3–1, explain the trends shown by waters from the NJ Coastal Plain and Floridian aquifers.

10. With reference to Case Study 3–2, explain why interactions between groundwaters and silicate minerals would lead to a reduction in P_{CO_2}.

11. What is an *amphoteric hydroxide*?

12. Define *acidity* and *alkalinity*.

13. Why do titration curves for weak acids show regions in which large additions of base result in very small pH changes while in other regions small additions of base lead to large pH changes?

14. What is a *buffer*?

15. Define *buffering capacity*.

16. Refer to Figure 3–9 to answer this question. For waters of pH 4, 5.7, 7, 8.5, and 10, identify the most important buffer(s). Why did you choose these particular buffers?

17. What is meant by *mineral reactivity*?

18. In Case Study 3–3, the authors assumed that SO_4^{2-} was a conservative component. What does this mean?

19. With reference to Case Study 3–4, why would a limestone formation be a poor choice as a disposal site for CO_2? Why would a siliciclastic unit be a better choice?

20. With reference to Case Study 3–4, in terms of the various chemical reactions explain the observed variations in the mineral abundances with time (Figure 3–C4–1).

21. At 60°C, the pH of a water sample is 5.1. Calculate the concentration of the hydroxyl ion.

22. Calculate the concentration of H^+ ions in solution if 0.2 mol of formic acid (Table 3–1) is dissolved in 1 L of pure water. The water temperature is 25°C.

23. Calculate the concentration of OH^- ions in solution if 0.1 mol of ammonium (Table 3–2) is dissolved in 1 L of pure water. The water temperature is 25°C.

24. Water from a septic tank drainage field enters an aquifer. The water contains a significant amount of dissolved organic carbon. Oxidation of this dissolved organic carbon adds CO_2 to the groundwater. The system is effectively isolated from the atmosphere and after the breakdown of the dissolved organic carbon, $P_{CO_2} = 10^{-2.5}$ atm. The temperature of the groundwater is 15°C. Calculate the pH of the groundwater.

25. A sandstone aquifer underlies an agricultural area. Extensive use of nitrogen fertilizer has led to a significant nitrogen burden in the groundwater of the sandstone aquifer. Denitrification has caused an increase in dissolved CO_2 in the groundwater. The measured $P_{CO_2} = 10^{-2}$ atm. The temperature of the groundwater is 20°C. Calculate the pH of the groundwater.

26. A water sample from a lake has a measured pH = 8.0 and $HCO_3^- = 22$ mg L^{-1}. The water temperature is 15°C. Is the lake water in equilibrium with atmospheric CO_2 (a calculation is required)? What type of rock serves as the basin for the lake, and why did you chose this type of rock?

27. Calculate the concentration of each carbonate species in a solution at 20°C when $C_T = 3 \times 10^{-3}$ mol L^{-1} and pH = 6.2.

28. Water samples were collected from a series of wells drilled into a sandstone aquifer. Bicarbonate ion concentration and pH were measured for each water sample. The following table lists the data for the wells measured in the downdip $(1 \rightarrow 7)$ direction. Plot pH versus HCO_3^- in millimoles L^{-1}. You will first have to convert the concentration of HCO_3^- to millimoles L^{-1}. Draw a smooth curve through the data. For each water sample, calculate the P_{CO_2}. Assume the water temperature is 25°C. Interpret the data. You may want to refer to Case Study 3–1 when you do this problem.

Well	pH	HCO_3^- (mg L^{-1})	HCO_3^- $(1 \times 10^{-3}$ mol $L^{-1})$	P_{CO_2} (atm)
1	5.7	92		
2	6.0	122		
3	6.4	171		
4	6.9	220		
5	7.3	244		
6	7.7	262		
7	8.0	275		

29. Calculate the relative abundance of $H_4SiO_{4\,(aq)}$ to $H_3SiO_4^-$ at pH = 8.2. $T = 25$°C.

30. A river contains 10.5 mg L^{-1} of dissolved silica (as SiO_2). The measured pH = 6.5 and the water temperature is 25°C. Calculate the concentration of each silica species in solution.

31. Calculate the pH of a solution saturated in calcite and in equilibrium with atmospheric CO_2. The temperature of the solution is 5°C.

32. A water sample collected from a well has $T = 25$°C, pH = 7.9, $Ca^{2+} = 96$ mg L^{-1}, and $HCO_3^- = 30$ mg L^{-1}. With reference to Case Study 3–2, calculate the P_{CO_2} for the water sample using both the HCO_3^- and Ca^{2+} concentrations. Do the calculations agree? Is this sample in equilibrium with atmospheric CO_2?

33. A water sample collected from a well has $T = 12$°C, pH = 7.9, $Ca^{2+} = 61$ mg L^{-1}, and $HCO_3^- = 48$ mg L^{-1}. With reference to Case Study 3–2, calculate the P_{CO_2} for the water sample using both the HCO_3^- and Ca^{2+} concentrations. Note that the equations in Case Study 3–2 are for 25°C. These will have to be modified because of the temperature difference. The equation for the bicarbonate ion can be modified using Henry's law (Table 2–1) and the carbonate equilibria constants in Table 3–4. The equation for the Ca ion can be modified by using Henry's law (Table 2–1) and the solubility products for calcite from Table 3–4. Do the calculations agree? Is this sample in equilibrium with atmospheric CO_2?

34. $Cu(OH)_{2\,(s)}$ is an amphoteric compound and dissolves according to the following reaction:

$$Cu(OH)_{2\,(s)} + OH^- \rightarrow Cu(OH)_3^-$$

For the following calculations assume that the water temperature is 25°C.

a. Excess copper hydroxide is added to beaker of water; i.e., solid copper hydroxide is present in the water. Using a base, the solution pH is adjusted to 10. Calculate the concentration of $Cu(OH)_3^-$ in the solution at equilibrium. Equilibrium constants for amphoteric compounds are found in Table 3–6.

b. A strong acid is added to the solution until the pH is reduced to 4. At this pH, calculate the concentration of $Cu(OH)_3^-$ in the solution. Will the amount of $Cu(OH)_{2\ (s)}$ in the beaker increase or decrease? Explain.

35. Calculate the total acidity and total alkalinity of a carbonic acid solution for which $C_T = 3 \times 10^{-3}$ mol L^{-1}, $T = 20°C$, and pH = 6.2. If you did problem 27 you have already calculated the abundance of the various carbonate species.

36. Calculate the buffering index for a solution containing 1×10^{-2} mol L^{-1} of acetic acid when the solution pH = 5 and $T = 25°C$.

37. Calculate the buffering index, as a function of pH, for an aqueous solution containing 1×10^{-2} mol L^{-1} of acetic acid at 25°C. This problem is easily done on a spreadsheet.

38. Calculate the buffering index for a solution containing 1×10^{-2} mol L^{-1} of silicic acid at pH = 7, $T = 25°C$. Assume that only the first two dissociation steps are significant.

39. Calculate the buffering index, as a function of pH, for an aqueous solution containing 1×10^{-2} mol L^{-1} silicic acid at 25°C. Assume that only the first two dissociation steps are significant. This problem is easily done on a spreadsheet.

40. Calculate the buffering index for the system calcite–carbonic acid at pH = 8.2, total carbonate = 1×10^{-2} mol L^{-1}, and $T = 15°C$. Select the appropriate equilibrium constants from Table 3–3 (for water) and Table 3–4 (for the carbonates).

41. Calculate the buffering index, as a function of pH, for the system calcite–carbonic acid with total carbonate = 1×10^{-2} mol L^{-1} and $T = 15°C$. Select the appropriate equilibrium constants from Table 3–3 (for water) and Table 3–4 (for the carbonates). This problem is easily done on a spreadsheet.

42. Calculate the buffering index for the reaction

$$KAl_3Si_3O_{10}(OH)_{2\ muscovite} + H^+ + 1.5H_2O \rightarrow 1.5Al_2Si_2O_5(OH)_{4\ kaolinite} + K^+$$

at pH = 5.7. $T = 25°C$.

43. Derive the buffering index equation for the reaction

$$Al_2Si_2O_5(OH)_{4\ kaolinite} + 6H^+ \rightarrow 2Al^{3+} + 2H_4SiO_{4\ (aq)} + H_2O$$

and at pH = 3.5, calculate the buffering index for the reaction. Use the thermodynamic data from Appendix II, source 3. $T = 25°C$.

44. Derive the buffering index equation for the reaction

$$Mg_2SiO_{4\ forsterite} + 4H^+ \rightarrow 2Mg^{2+} + H_4SiO_{4\ (aq)}$$

and at pH = 5.0, calculate the buffering index for the reaction. Use the thermodynamic data from Appendix II, source 3. $T = 25°C$.

45. Use the model of Lawrence and Scheske (1997) to calculate the acid neutralization potential of waste rock from a sulfide ore deposit that has the following mineralogical composition:

Quartz (SiO_2)—26%   Orthoclase ($KAlSi_3O_8$)—10%
Albite ($NaAlSi_3O_8$)—13%   Anorthite ($CaAl_2Si_2O_8$)—11%
Diopside ($CaMgSi_2O_6$)—4%   Hypersthene ($MgSiO_3$)—26%
Calcite ($CaCO_3$)—10%

Oxidation–Reduction Reactions

Oxidation–reduction reactions (often referred to as *redox* reactions) involve the gain or loss of electrons. The species involved in the reactions may be atoms, molecules, or ions. A **cation** is a *positively charged ion* (fewer electrons than the number of protons in the nucleus), and an **anion** is a *negatively charged ion* (more electrons than the number of protons in the nucleus). For many elements, the oxidation state is an important factor in determining their behavior in the natural environment. For example, Fe^{2+} is more soluble in water than Fe^{3+}. If water in which iron occurs as the Fe^{2+} cation is exposed to higher oxygen concentrations, as might occur when groundwater is pumped to the surface, some of the iron would be oxidized to Fe^{3+} and an iron hydroxide would precipitate from solution. Hence, the oxidation–reduction characteristics of the natural environment play a key role in the transport and ultimate fate of various contaminants.

BASIC PRINCIPLES

Electrochemical Cells

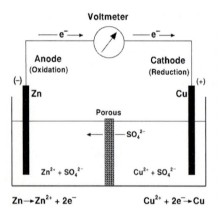

Figure 4–1
Diagram of a Zn-Cu electrochemical cell. Zn and Cu metal electrodes are immersed in a $CuSO_4$ solution. Electrons flow from left to right and a potential is recorded by the voltmeter. With time, this potential decreases to zero, the concentration of Zn^{2+} increases in the left-hand half of the cell, and the concentration of Cu^{2+} decreases in the right-hand half of the cell. After Faure (1998).

Consider the following simple experiment (Figure 4–1). Zinc and copper metal bars are connected by wires to a voltmeter. The bars are immersed in a container that is divided into two compartments by a porous partition. The zinc bar is immersed in a $ZnSO_4$ solution and the copper bar is immersed in a $CuSO_4$ solution. The porous partition very cleverly allows SO_4^{2-} anions to pass from one side to the other, but does not allow metal cations to pass from one side to the other. The experiment we have set up is an electrochemical (or galvanic) cell. After the metal bars are immersed in the solution, several changes occur. The Zn^{2+} cations increase in the left-hand side of the container and the concentration of Cu^{2+} cations decreases in the right-hand side of the container. SO_4^{2-} anions move from right to left through the porous partition, thus maintaining the charge balance in both halves of the container. Immediately after immersing the bars, a voltage is recorded by the voltmeter. With time, this voltage decreases to zero. What has occurred is an oxidation–reduction reaction that with time achieved equilibrium.

Let us consider the various components of this reaction. In the left-hand side of the container, Zn metal has gone into solution as Zn^{2+} cations. Because there has been a loss of electrons, we say that the Zn has been oxidized. In the right-hand side of the container, Cu^{2+} cations have precipitated onto the Cu electrode as copper metal. Because there has been a gain in electrons, we say that the Cu^{2+} cations have been reduced. The oxidation of Zn is the source of electrons for the system and Zn is referred to as a *reducing agent*. During reduction of the copper cations, electrons are removed from the system and copper is referred to as an *oxidizing agent*. This nomenclature is sometimes a major stumbling block for students. Remember **oxidation** is the *loss of electrons* and **reduction** is the *gain of electrons*. A **reducing agent** is an *electron donor* (it loses electrons and is oxidized), and an **oxidizing agent** is an *electron acceptor* (it gains electrons and is reduced). In terms of our example, the complete electrochemical reaction is

$$Zn + Cu^{2+} \rightleftharpoons Zn^{2+} + Cu$$

Table 4–1   Oxidation–Reduction
                  Reactions

Reaction	ΔG_R^0 (kJ mol^{-1})
$Zn + Fe^{2+} \rightleftharpoons Zn^{2+} + Fe$	-68.4
$Fe + Cu^{2+} \rightleftharpoons Fe^{2+} + Cu$	-144.4
$Cu + 2Ag^+ \rightleftharpoons Cu^{2+} + 2Ag$	-88.7

Zn is oxidized and acts as a reducing agent, and Cu is reduced and acts as an oxidizing agent. This reaction can be divided into two half-reactions

$$Zn \rightarrow Zn^{2+} + 2e^- \quad \text{and} \quad Cu^{2+} + 2e^- \rightarrow Cu$$

Half-reactions (often called *electrodes*) will be further considered in subsequent sections.

Electromotive Series

Three oxidation–reduction reactions, and their free energies, are listed in Table 4–1. Each reaction has a negative free energy. Thus, the reactions proceed spontaneously to the right. In the first reaction, Zn acts as a reducing agent and Fe acts as an oxidizing agent. In the second reaction, Fe acts as a reducing agent and Cu acts as an oxidizing agent. In the third reaction, Cu acts as a reducing agent and Ag acts as an oxidizing agent. For these four elements, in terms of their strength as reducing agents from strongest to weakest, the order is Zn, Fe, Cu, and Ag. The electromotive series is a listing of the elements (as half-reactions) in terms of their decreasing strengths as reducing agents. The electromotive series for a number of elements, and some of their oxidation states, is given in Table 4–2 (p. 96).

Electromotive Force

Returning to the experiment illustrated in Figure 4–1, the voltage generated by the electrons flowing from the zinc bar to the copper bar is referred to as the *electromotive force (emf)*. The electromotive force is related to the free energy of an oxidation–reduction reaction as follows:

$$\Delta G_R = -n\mathscr{F}E \tag{4–1}$$

where ΔG_R is the free energy of the reaction in any state, E is the corresponding electromotive force, n is the number of electrons transferred in the reaction, and $\mathscr{F}$ is Faraday's constant (96,489 coulombs mol^{-1}, 23.06 kcal volt^{-1} gram equivalent^{-1}, or, in SI units, 96.42 kJ volt^{-1} gram equivalent^{-1}).

EXAMPLE 4–1 Calculate the *standard emf (E°)* for the oxidation–reduction reaction illustrated in Figure 4–1. In the standard state $[Zn^{2+}] = [Cu^{2+}] = 1.0$, $P = 1$ atm, and $T = 25°C$. The reaction is written

$$Zn + Cu^{2+} \rightleftharpoons Zn^{2+} + Cu$$

Calculating the free energy using the values in Appendix II, source 3, gives

$$\Delta G_R^0 = [(-147.3) + (0.00)] - [(0.00) + (65.5)] = -212.8 \text{ kJ mol}^{-1}$$

Calculating the emf gives

$$E° = \frac{-\Delta G_R^0}{n\mathscr{F}} = \frac{-(-212.8)}{(2)(96.42)} = 1.10 \text{ V}$$

The number of electrons transferred in the reaction is 2; therefore, $n = 2$. ∎

Table 4–2 Electromotive Series for Selected Half-Reactions Arranged in Order of Decreasing Strengths as Reducing Agents*

Oxidizing agent			Reducing agent	Standard electrode potential, V
Li^+	$+1e^-$	$\rightarrow$	Li	-3.045
Na^+	$+1e^-$	$\rightarrow$	Na	-2.714
Pu^{3+}	$+3e^-$	$\rightarrow$	Pu	-2.07
Th^{4+}	$+4e^-$	$\rightarrow$	Th	-1.90
Np^{3+}	$+3e^-$	$\rightarrow$	Np	-1.86
Al^{3+}	$+3e^-$	$\rightarrow$	Al	-1.66
U^{4+}	$+4e^-$	$\rightarrow$	U	-1.38
Mn^{2+}	$+2e^-$	$\rightarrow$	Mn	-1.18
V^{3+}	$+3e^-$	$\rightarrow$	V	-0.87
Zn^{2+}	$+2e^-$	$\rightarrow$	Zn	-0.763
S	$+2e^-$	$\rightarrow$	S^{2-}	-0.44
Fe^{2+}	$+2e^-$	$\rightarrow$	Fe	-0.41
Cd^{2+}	$+2e^-$	$\rightarrow$	Cd	-0.403
Co^{2+}	$+2e^-$	$\rightarrow$	Co	-0.277
Ni^{2+}	$+2e^-$	$\rightarrow$	Ni	-0.250
Sn^{2+}	$+2e^-$	$\rightarrow$	Sn	-0.140
Pb^{2+}	$+2e^-$	$\rightarrow$	Pb	-0.126
$2H^+$	$+2e^-$	$\rightarrow$	H_2	0.00
Ti^{4+}	$+1e^-$	$\rightarrow$	Ti^{3+}	0.04
Sn^{4+}	$+2e^-$	$\rightarrow$	Sn^{2+}	0.15
Cu^{2+}	$+2e^-$	$\rightarrow$	Cu	0.337
Cu^+	$+1e^-$	$\rightarrow$	Cu	0.521
Se	$+2e^-$	$\rightarrow$	Se^{2-}	0.67
Fe^{3+}	$+1e^-$	$\rightarrow$	Fe^{2+}	0.771
Ag^+	$+1e^-$	$\rightarrow$	Ag	0.799
Pu^{4+}	$+1e^-$	$\rightarrow$	Pu^{3+}	0.97
Au^{3+}	$+3e^-$	$\rightarrow$	Au	1.50
Co^{3+}	$+1e^-$	$\rightarrow$	Co^{2+}	1.82

*Data from Daniels and Alberty (1967) and Faure (1998).

Note: By international convention half-cell reactions are written so that the electrons appear on the left-hand side of the equation; i.e., the reduced form of the element appears on the right and the oxidized form appears on the left. Not all textbooks (or professional papers) follow this convention, an added difficulty for the student attempting to understand a conceptually difficult subject.

It is impossible to measure only the electromotive force of a half-reaction. In order to measure the potential of a half-reaction (electrode), another metallic electrode is required. This is done by arbitrarily assigning a potential to one electrode (half-reaction) and comparing all other electrodes to this electrode. The universally accepted reference electrode is the ***standard hydrogen electrode (SHE)*** for which the value of $E°$ is arbitrarily set equal to 0.00 V (at $P = 1$ atm and $T = 25°C$). This electrode is written

$$H^+ + e^- \rightarrow \tfrac{1}{2}H_{2\,(g)}$$

If $E° = 0.00$ V, then $\Delta G_R^0 = 0.00$, from which it follows that $G°(H^+) = G°(e^-) = 0.00$. Also, for a hydrogen half-cell in the standard state, $[H^+] = [H_2] = 1$. It is now possible to determine the emf of any half-reaction by comparison to the standard hydrogen electrode. The results of these determinations are tabulated in Table 4–2 (the electromotive series).

EXAMPLE 4–2 Calculate the standard electrode potential for the oxidation of scandium. During oxidation, scandium gives up three electrons. Remember the convention is that the electrons appear on the left-hand side of the equation. The reaction with the hydrogen electrode is written

$$Sc^{3+} + 1.5H_{2\,(g)} \rightleftharpoons Sc + 3H^+$$

Calculating the free energy using the values in Appendix II, source 4, gives

$$\Delta G_R^0 = [(0.00) + (3)(0.00)] - [(-586.6) + (1.5)(0.00)] = 586.6 \text{ kJ mol}^{-1}$$

Calculating the emf gives

$$E^\circ = \frac{-\Delta G_R^0}{n\mathscr{F}} = \frac{-(586.6)}{(3)(96.42)} = -2.03 \text{ V}$$ ■

A hydrogen electrode is difficult to prepare and maintain. In its stead, a calomel electrode is often used. The reaction for this electrode is

$$Hg_2Cl_2 + 2e^- \rightleftharpoons 2Hg + 2Cl^-$$

For a normal calomel electrode (one molar solution of potassium chloride) the potential is 0.2802 V, and for a saturated calomel electrode (saturated in potassium chloride) the potential is 0.2444 V, both relative to the standard hydrogen electrode. The saturated calomel electrode is the one most frequently used when making environmental measurements of oxidation–reduction potentials.

The electromotive force for an oxidation–reduction reaction can be determined by combining the appropriate half-reactions. Consider the reaction between Zn and Cu electrodes that we used as an example at the beginning of this chapter. Referring to Table 4–2, we see that Zn is the stronger reducing agent. Thus, Zn will be oxidized and Cu will be reduced (as we observed in the experiment). We can write the half-reactions and combine them as follows:

$$
\begin{array}{ll}
Zn \rightarrow Zn^{2+} + 2e^- & E^\circ = 0.763 \\
\underline{Cu^{2+} + 2e^- \rightarrow Cu} & \underline{E^\circ = 0.337} \\
Zn + Cu^{2+} \rightleftharpoons Zn^{2+} + Cu & E^\circ = 1.100 \text{ V}
\end{array}
$$

Because Zn is oxidized, we have reversed the Zn half-reaction and reversed the sign for the electromotive force. Thus, the Zn half-reaction has a positive value.

Balancing Oxidation–Reduction Equations

The oxidation–reduction reactions we have considered so far are relatively simple. Most reactions are more complicated and we will need to balance complex equations. The following rules are used when balancing oxidation–reduction equations:

1. Determine the valence numbers (charge) for all elements involved in the reaction.
2. Determine the number of electrons given off and taken up in the reaction. Balance the transfer of electrons.
3. Balance the elements, except oxygen and hydrogen, on both sides of the equation.
4. Balance the number of oxygen atoms by adding H_2O.
5. Balance the number of hydrogens by adding H^+.

EXAMPLE 4–3 Acid mine drainage is caused by the oxidation of pyrite when it comes in contact with water containing dissolved oxygen. The elements of interest are Fe and S. The reaction is written as follows:

$$FeS_{2 \text{ pyrite}} + O_2 \rightarrow Fe(OH)_{3 \text{ (ppt)}} + SO_4^{2-}$$

Both Fe and S are oxidized. Fe is oxidized from Fe^{2+} to Fe^{3+}, releasing 1 electron in the process. S is oxidized from S^- in pyrite to S^{6+} in the sulfate ion, releasing 7 electrons in the process. We have now determined the valences for the elements involved in the reaction (rule 1). The breakdown of one pyrite molecule releases a total of 15 electrons, 1 from Fe oxidation and 14 (2×7) from S oxidation. Each oxygen receives 2 electrons to become O^{2-}. Thus, 4 electrons are consumed by the O_2 molecule. In summary, 15 electrons are released by FeS_2 and 4 are consumed by O_2. We could balance the transfer of electrons by setting $FeS_2 = \frac{4}{15}$. However, this would leave us with some very nasty fractions in the equation. It is preferable to work with simple whole numbers. We accomplish this by cross-multiplying these two numbers, which gives the following (rule 2):

$$4FeS_2 + 15O_2 \rightarrow Fe(OH)_3 + SO_4^{2-}$$

The electrons released and consumed are now in balance, $4 \times 15 = 15 \times 4$. We now balance all the elements except hydrogen and oxygen (rule 3).

$$4FeS_2 + 15O_2 \rightarrow 4Fe(OH)_3 + 8SO_4^{2-}$$

We now balance the number of oxygens by adding H_2O (rule 4).

$$4FeS_2 + 15O_2 + 14H_2O \rightarrow 4Fe(OH)_3 + 8SO_4^{2-}$$

Lastly, we balance the number of hydrogens by adding H^+ (rule 5).

$$4FeS_{2\ pyrite} + 15O_2 + 14H_2O \rightarrow 4Fe(OH)_{3\ (ppt)} + 8SO_4^{2-} + 16H^+$$

Note that during the oxidation of pyrite a substantial number of H^+ ions are produced; i.e., for each mole of pyrite that is oxidized 4 mol of H^+ ions are released. It is these H^+ ions that are responsible for the very low pH of waters draining from coal and base-metal mines (acid mine drainage).   ∎

The Nernst Equation and Eh

Consider the following generalized oxidation–reduction reaction:

$$aA_{ox} + bB_{red} \rightleftharpoons cC_{red} + dD_{ox} \qquad (4\text{--}2)$$

As we did in Chapter 2 (equation 2–25), we can write this reaction in terms of free energy.

$$\Delta G_R = \Delta G_R^0 + RT \ln \frac{[C_{red}]^c[D_{ox}]^d}{[A_{ox}]^a[B_{red}]^b} \qquad (4\text{--}3)$$

Substituting for $\Delta G_R = -n\mathscr{F}E$ and $\Delta G_R^0 = -n\mathscr{F}E°$ gives

$$E = E° - \frac{RT}{n\mathscr{F}} \ln \frac{[C_{red}]^c[D_{ox}]^d}{[A_{ox}]^a[B_{red}]^b} = E° - \frac{RT}{n\mathscr{F}} \ln K \qquad (4\text{--}4)$$

Equation 4–4 is referred to as the **Nernst equation** and can be used to calculate the emf of an oxidation–reduction reaction under any conditions. $E°$ is the emf of the reaction in the standard state, R is the gas constant (8.314×10^{-3} kJ deg^{-1} mol^{-1}), T is the temperature (K), n is the number of electrons transferred in the reaction, $\mathscr{F}$ is the Faraday constant (96.42 kJ volt^{-1} gram equivalent^{-1}), and K is the equilibrium constant.

We usually work in base 10, rather than base e, so it is useful to write this equation in terms of base 10 logarithms. Assuming that the reaction occurs at 25°C, the Nernst equation can be written

$$E = E° - \frac{RT}{n\mathscr{F}} \ln K = E° - \frac{(8.314 \times 10^{-3})(298.15)(2.303)}{(n)(96.42)} \log K$$

$$= E° - \frac{0.0592}{n} \log K \qquad (4\text{--}5)$$

If we are measuring the emf of a half-cell relative to the hydrogen electrode (or a proxy, such as the calomel electrode), the Nernst equation is usually written so that the activity product of the oxidized species is in the numerator. What this means is that the sign is reversed in the Nernst equation, giving

$$E = E° + \frac{0.0592}{n} \log\left(\frac{\text{Activity product of oxidized species}}{\text{Activity product of reduced species}}\right) \qquad (4\text{--}6)$$

This has the potential to be a very confusing complication. How the calculation is carried out is illustrated in Example 4–4.

EXAMPLE 4–4 Calculate the emf for the half-reaction

$$Mn^{3+} + e^- \rightarrow Mn^{2+}$$

Combining this half-reaction with the hydrogen electrode yields

$$Mn^{3+} + \tfrac{1}{2}H_{2\,(g)} \rightleftharpoons Mn^{2+} + H^+$$

For this reaction,

$$\Delta G_R^0 = [(-228.1) + (0.00)] - [(-84.8) + (\tfrac{1}{2})(0.00)] = -143.3 \text{ kJ mol}^{-1}$$

$$E° = \frac{-\Delta G_R^0}{n\mathscr{F}} = \frac{-(-143.3)}{(1)(96.42)} = 1.49 \text{ V}$$

The Nernst equation for the reaction is

$$E = E° + \frac{0.0592}{n} \log K = 1.49 + 0.0592 \log\frac{[Mn^{3+}][H_{2\,(g)}]^{1/2}}{[Mn^{2+}][H^+]}$$

For the hydrogen half-cell in the standard state, $[H^+] = [H_2] = 1$. Thus, the $[Mn^{3+}]/[Mn^{2+}]$ ratio is determined by the electromotive force of the system. For example, groundwater has $E = 0.00$ V, and surface water has $E = 0.80$ V. What would happen when groundwater is pumped to the surface? At $E = 0.00$ V,

$$\log\frac{[Mn^{3+}]}{[Mn^{2+}]} = \frac{E - 1.49}{0.0592} = \frac{0.00 - 1.49}{0.0592} = -25.2$$

and at $E = 0.80$ V,

$$\log\frac{[Mn^{3+}]}{[Mn^{2+}]} = \frac{E - 1.49}{0.0592} = \frac{0.80 - 1.49}{0.0592} = -11.7$$

When the groundwater is exposed to the oxidizing surface environment, Mn^{3+} increases relative to Mn^{2+}. Because the solubility of Mn^{3+} is less than that of Mn^{2+}, this may lead to the precipitation of manganese. ■

We can define a special type of electromotive force, Eh. **Eh** is *the electromotive force of any reaction measured relative to the standard hydrogen electrode*. What was determined in Example 4–4 was Eh. Eh is an environmental parameter that reflects the overall oxidation–reduction potential of a natural system relative to the hydrogen electrode. Because the hydrogen electrode is not easily transported, measurements in the field are usually made between a Pt electrode and a reference electrode. The measurements are then corrected, using equation 4–7, to the value that would be observed if the reference electrode was a hydrogen electrode.

$$Eh = E_{meas} - E_{ref} \qquad (4-7)$$

The most commonly used reference electrode is the calomel electrode, for which $E_{ref} = 244.4$ mV (25°C). Ionic equilibria calculated from the Nernst equation using the measured Eh often are not in good agreement with the measured ionic equilibria. Thus, except in certain circumstances, Eh measurements are only used in a qualitative sense. The reasons for this disagreement are numerous, and the student should consult standard reference works, such as Langmuir (1997) and Stumm and Morgan (1996), for further details. In practice, it is best to determine Eh based on specific oxidation–reduction reactions (see Example 4–5).

EXAMPLE 4–5 Calculate the Eh for a water sample at 25°C for which $[Mn^{2+}] = 10^{-3}$ mol L^{-1} and $[Mn^{3+}] = 10^{-14}$ mol L^{-1}. Using the data from Example 4–4,

$$Eh = 1.49 + 0.0592 \log[Mn^{3+}] - 0.0592 \log[Mn^{2+}]$$

$$= 1.49 + 0.0592 \log[10^{-14}] - 0.0592 \log[10^{-3}] = 0.84 \text{ V}$$

A direct Eh measurement is made using a calomel electrode as the reference electrode. The measured Eh is 901 mV. The corrected Eh is

$$Eh = E_{meas} - E_{ref} = 901 - 244 = 657 \text{ mV} = 0.657 \text{ V}$$

The measured environmental Eh suggests that there should be significantly less Mn^{3+} present than the amount measured. Mn is an example of an electroactive species, and electrons are easily exchanged at the surface of an Eh electrode. If the water sample was largely composed of electroactive species, there should be good agreement between the measured Eh and the Eh calculated from the Mn reaction. The lower measured Eh suggests that there are also nonelectroactive species present that cannot easily exchange electrons at the surface of the Eh electrode. C, N, O, H, and oxidized sulfur are examples of such species. The presence of nonelectroactive species gives rise to mixed potentials because these species are not in equilibrium with the Eh electrode. In these cases, the Eh measurement is not environmentally significant. ∎

Oxidation–Reduction Reactions and pe

An alternative way to look at oxidation–reduction is through the concept of electron activity, pe. This approach is often used by chemists and engineers and is algebraically simpler than the Nernst equation. Geo- and environmental scientists still tend to use Eh. Consider the half-reaction

$$Fe^{3+} + e^- \rightleftharpoons Fe^{2+}$$

We can write this reaction as follows:

$$K = \frac{[Fe^{2+}]}{[Fe^{3+}][e^-]} \tag{4–8}$$

The electrons appear explicitly in this equation. However, this should not be interpreted as a concentration of electrons, but rather as the tendency to accept or release electrons. Similar to pH,

$$pe = -\log[e^-] \tag{4–9}$$

Rewriting equation 4–8 in logarithmic form and substituting for e^- using equation 4–9 gives

$$\log K = \log[Fe^{2+}] - \log[Fe^{3+}] - \log[e^-] = \log[Fe^{2+}] - \log[Fe^{3+}] + pe \tag{4–10}$$

Solving equation 4–10 for pe gives

$$pe = \log K + \log[Fe^{3+}] - \log[Fe^{2+}] \tag{4–11}$$

Note that in the standard state, $[Fe^{2+}] = [Fe^{3+}] = 1$ and $pe° = \log K$. For the reaction illustrated by equation 4–11, only one electron is transferred. A more general statement of the relationship between $pe°$ and K is

$$pe° = \frac{1}{n} \log K \tag{4–12}$$

where $pe°$ is the standard state and n is the number of electrons transferred in the reaction. For the general case when n numbers of electrons are transferred at conditions other than the standard state,

$$pe = pe° + \frac{1}{n} \log\left(\frac{\text{Activity product of oxidized species}}{\text{Activity product of reduced species}}\right) \tag{4–13}$$

Note that in terms of pe this is analogous to the general form of the Nernst equation (4–6).

EXAMPLE 4–6 At 25°C, calculate $pe°$ for the reaction

$$Zn + Fe^{2+} \rightleftharpoons Zn^{2+} + Fe$$

From Table 4–1, $\Delta G_R^0 = -68.4$ kJ mol^{-1}.

$$\log K = \frac{-\Delta G_R^0}{5.708} = \frac{-(-68.4)}{5.708} = 11.98$$

and

$$pe^\circ = \frac{1}{n} \log K = \frac{1}{2} \times 11.98 = 5.99$$

If $[Fe^{2+}] = 10^{-3}$ and $[Zn^{2+}] = 10^{-6}$, calculate the pe.

$$pe = pe^\circ + \frac{1}{n} \log[Zn^{2+}] - \frac{1}{n} \log[Fe^{2+}] = pe^\circ + \frac{1}{2} \log[10^{-6}] - \frac{1}{2} \log[10^{-3}]$$

$$= 6 - 3 + 1.5 = 4.5 \qquad \blacksquare$$

Eh and pe are related as follows:

$$Eh = \frac{2.303RT}{\mathscr{F}} pe \qquad (4–14)$$

and at 25°C,

$$Eh = 0.059\, pe \qquad (4–15)$$

OXIDATION–REDUCTION DIAGRAMS

The stability of many species in solution and the solubility of solid phases is controlled by the oxidation–reduction potential of the environment. In many cases, H^+ ions are involved in these reactions; hence, they are also pH sensitive. Diagrams that represent the range of stability for various species in terms of Eh and pH can be constructed from thermodynamic data. These graphical representations are used to make first-order inferences about the distribution of species in various natural environments. The pioneering work in this area was that of Garrels and Christ (1965). More recently, Brookins (1988) has prepared a number of Eh–pH diagrams for geologically and environmentally important systems. The basic principles used to construct these diagrams and their interpretation are considered in this section.

Stability of Water

Although one does not often think of water in terms of oxidation–reduction, limits for the breakdown of water can be derived in terms of Eh and pH. Consider the following reaction (referred to as the *water electrode*):

$$2H^+ + \tfrac{1}{2}O_{2\,(g)} + 2e^- \rightleftharpoons H_2O_{\,(l)} \qquad (4–16)$$

In this reaction, the oxygen in the water molecule is converted to a gas, the valence number of oxygen changes from -2 to 0, and two electrons are released to solution. This is an oxidation reaction and its emf can be determined by reference to the standard hydrogen electrode. The emf of the standard hydrogen electrode is, by definition, zero, so we only need to calculate the emf for the water half-reaction. The Nernst equation for this reaction is

$$Eh = E^\circ + \frac{0.0592}{2} \log([O_2]^{1/2}[H^+]^2) \qquad (4–17)$$

We calculate E°, at 25°C, from the standard free energy for reaction 4–16.

$$\Delta G_R^0 = [-237.14] - [(2)(0.00) + (\tfrac{1}{2})(0.00) + (2)(0.00)]$$

$$= -237.14 \text{ kJ mol}^{-1} \qquad (4–18)$$

and

$$E° = \frac{-(-237.14)}{(2)(96.42)} = 1.23 \text{ V} \qquad (4\text{–}19)$$

The Eh equation is

$$\text{Eh} = 1.23 + \frac{0.0592}{2} \log([O_2]^{1/2}[H^+]^2) \qquad (4\text{–}20)$$

However, we would like to write this equation in terms of Eh and pH because pH is what we determine when we measure $[H^+]$ in solution. We can rewrite equation 4–20 as follows:

$$\text{Eh} = 1.23 + \frac{0.0592}{2}\left(\frac{1}{2}\right)\log[O_2] + \frac{0.0592}{2}(2)\log[H^+] \qquad (4\text{–}21)$$

Recall that the definition of pH is $\text{pH} = -\log[H^+]$. Substituting for pH in equation 4–21, and carrying out the appropriate multiplications, gives

$$\text{Eh} = 1.23 + 0.0148 \log[O_2] - 0.0592 \text{ pH} \qquad (4\text{–}22)$$

which is the Eh–pH equation for the water electrode.

The limits of the natural Eh–pH environment are determined by the conditions under which water will break down into its gaseous components. We can write the following equilibrium reaction:

$$2H_2O_{(l)} \rightleftharpoons O_{2\,(g)} + 2H_{2\,(g)}$$

The equilibrium constant for this reaction is

$$K = [O_2][H_2]^2 = 10^{-83.1} \qquad (4\text{–}23)$$

Clearly, the pressure of neither gas can exceed 1 atm. If we set $H_2 = 1$ atm, then $O_2 = 10^{-83.1}$ atm; and if we set $O_2 = 1$ atm, then $H_2 = 10^{-41.65}$ atm. Substituting the two extreme values for O_2 yields the limits of the natural environment. At $O_{2\,(g)} = 1$ atm,

$$\text{Eh} = 1.23 + 0.0148 \log[1] - 0.0592 \text{ pH} = 1.23 - 0.0592 \text{ pH} \qquad (4\text{–}24)$$

and at $O_{2\,(g)} = 10^{-83.1}$,

$$\text{Eh} = 1.23 + 0.0148 \log[10^{-83.1}] - 0.0592 \text{ pH} = -0.0592 \text{ pH} \qquad (4\text{–}25)$$

If we use the present-day atmospheric partial pressure of $O_2 = 0.2$ atm, $E°$ in equation 4–24 is 1.22. The limits for the natural environment, in terms of Eh and pH, are shown in Figure 4–2. These natural limits are the starting point for the construction of Eh–pH diagrams. Also shown on Figure 4–2 are the Eh–pH ranges for some natural waters.

We can also plot this diagram in terms of pe, and the pe units are shown on the right-hand side of Figure 4–2. To further illustrate the relationship between Eh and pe, we will derive the pe–pH equations for the limits of the natural environment. For the water electrode (equation 4–16), $\Delta G_R^0 = -237.14$ kJ mol^{-1}, from which we can calculate the equilibrium constant for the reaction at 25°C.

$$\log K = \frac{-\Delta G_R^0}{5.708} = \frac{-(-237.14)}{5.708} = 41.55 = \frac{[H_2O_{(l)}]}{[H^+]^2[O_{2\,(g)}]^{1/2}[e^-]^2} \qquad (4\text{–}26)$$

Writing the equilibrium equation 4–26 using pe gives

$$\log K = -\tfrac{1}{2} \log P_{O_2} + 2 \, \text{p}e + 2 \, \text{pH} = 41.55 \qquad (4\text{–}27)$$

Note that $[e^-]$ has been replaced by pe and $[H^+]$ has been replaced by pH. Rewriting equation 4–27 in terms of pe gives

$$\text{p}e = \tfrac{1}{4} \log P_{O_2} - \text{pH} + 20.775 \qquad (4\text{–}28)$$

Substituting $P_{O_2} = 1$ atm gives

$$\text{p}e = 20.775 - \text{pH} \qquad (4\text{–}29)$$

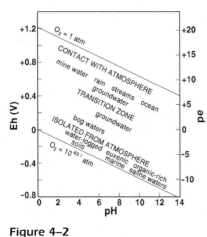

Figure 4–2

Stability limits for natural waters at the earth's surface in terms of Eh and pH at 25°C. The limits are based on partial pressures of oxygen of 1 and $10^{-83.1}$ atm. Also shown is the emf in pe units. The range of Eh and pH conditions for various natural environments is modified from Garrels and Christ (1965).

and substituting $P_{O_2} = 10^{-83.1}$ atm gives

$$pe = -pH \tag{4–30}$$

In pe units, equations 4–29 and 4–30 are analogous to equations 4–24 and 4–25, respectively.

In the previous equations we have dealt with oxygen in terms of its partial pressure in the atmosphere. At 25°C the relationship between the partial pressure of oxygen, in atm, and the oxygen dissolved in water, in mol L^{-1}, is (Stumm and Morgan, 1996)

$$\frac{[O_{2\,(aq)}]}{[O_{2\,(g)}]} = 10^{-2.9} \tag{4–31}$$

Solving equation 4–31 for oxygen dissolved in water gives

$$[O_{2\,(aq)}] = [O_{2\,(g)}](10^{-2.9}) \tag{4–32}$$

EXAMPLE 4–7 A groundwater sample, at 25°C, has Eh = 0.6 and pH = 6.0. Assuming the system is at equilibrium, calculate the amount of dissolved oxygen in the groundwater sample.

Rearrange equation 4–22 to solve for $[O_{2\,(g)}]$.

$$0.0148 \log[O_{2\,(g)}] = Eh - 1.23 + 0.0592\, pH = 0.6 - 1.23 + (0.0592)(6)$$

$$[O_{2\,(g)}] = 10^{-18.57} \text{ atm}$$

The amount of dissolved oxygen in the groundwater is

$$[O_{2\,(aq)}] = [O_{2\,(g)}](10^{-2.9}) = [10^{-18.57}](10^{-2.9}) = 10^{-21.47} \text{ mol } L^{-1}$$

In ppm, the amount of dissolved oxygen is

$$(10^{-21.47} \text{ mol } L^{-1})(32 \text{ g mol}^{-1}) = 1.08 \times 10^{-20} \text{ g } L^{-1}$$
$$= 1.08 \times 10^{-17} \text{ mg } L^{-1} = 1.08 \times 10^{-17} \text{ ppm}$$

This is an incredibly small amount of dissolved oxygen given that we have an oxidizing environment. Thus, even in oxidizing environments the amount of dissolved oxygen can be very low. To put this calculation into perspective, fish require on the order of 5 to 7 ppm dissolved oxygen, so this environment would not support fish or other higher forms of aquatic life. ∎

Constructing Eh–pH Diagrams

This topic is covered in detail in a number of textbooks, and one of the most complete descriptions is found in the classic text by Garrels and Christ (1965). In this section we will consider two types of Eh–pH interactions—those between minerals and dissolved species, as illustrated by the Fe system, and those between dissolved species and gases, as illustrated by the nitrogen system. Further examples are found in the problem set at the end of the chapter.

The Iron Eh–pH Diagram The iron Eh–pH diagram will be constructed for surface conditions and $T = 25°C$. Iron can exist in the natural environment in three oxidation states, 0, +2, and +3. Native iron, Fe, has 0 valence. The two stable oxides are Fe_3O_4, in which iron has both +2 and +3 valence, and Fe_2O_3, in which iron has +3 valence. We will start by plotting the boundaries for these three solid phases on the Eh–pH diagram.

Native iron is oxidized to magnetite according to the following half-reaction:

$$Fe_3O_4 + 8e^- \rightarrow 3Fe$$

A total of 8 electrons are released because in the magnetite molecule two of the Fe atoms are in the +3 valence state and one is in the +2 valence state. In order to balance the number of oxygens on the left-hand side of the equation, we add 4 water molecules to the right-hand side, which results in $8H^+$ on the left-hand side of the equation.

$$Fe_3O_4 + 8e^- + 8H^+ \rightleftharpoons 3Fe + 4H_2O$$

Relative to the hydrogen electrode, we can write the following Nernst equation. In this, and all subsequent calculations, we will use the hydrogen electrode as the reference electrode. This will not be explicitly stated.

$$Eh = E° + \frac{0.0592}{8} \log\left(\frac{[Fe_3O_4][H^+]^8}{[Fe]^3[H_2O]^4}\right) \qquad (4-33)$$

The activities of the solid phases and water are unity, assuming that the water is nearly pure. We will make this assumption in subsequent calculations, and the Nernst equations will include only species that have an activity other than unity.

$$Eh = E° + \frac{0.0592}{8} \log[H^+]^8 \qquad (4-34)$$

For this reaction, $\Delta G_R^0 = 64.2$ kJ mol^{-1} and $E° = -0.083$ V. Substituting into equation 4–34 and converting $[H^+]$ to pH gives

$$Eh = -0.083 - 0.0592 \text{ pH} \qquad (4-35)$$

The half-reaction for the oxidation of Fe_3O_4 to Fe_2O_3 is

$$3Fe_2O_3 + 2e^- \rightarrow 2 Fe_3O_4$$

Only 2 electrons are released in this reaction because in the Fe_3O_4 molecule two of the Fe atoms already have a +3 valence. Adding water to balance the number of oxygens gives

$$3Fe_2O_3 + 2e^- + 2H^+ \rightleftharpoons 2Fe_3O_4 + H_2O$$

For this reaction, $\Delta G_R^0 = -34.2$ kJ mol^{-1} and $E° = 0.18$ V. The Nernst equation is

$$Eh = E° + \frac{0.0592}{2} \log[H^+]^2 = 0.18 - 0.0592 \text{ pH} \qquad (4-36)$$

Equations 4–35 and 4–36 are plotted in Figure 4–3. Note that the boundary between metallic iron and magnetite falls outside of the stability field of water. Hence, metallic iron is not stable in the surficial environment. What this means in practice is that native iron is inherently unstable in the presence of water and readily oxidizes. You are familiar with one of these oxidation products, rust. With reference to Figure 4–2, one would expect that magnetite would be the stable form of iron oxide only in reducing environments isolated from the atmosphere.

Magnetite and hematite are slightly soluble in the surficial environment and coexist in equilibrium with Fe^{2+} and Fe^{3+} ions in solution. We will first consider equilibrium reactions involving magnetite, hematite, and Fe^{3+}. For hematite, we can write

$$Fe_2O_3 + 6H^+ \rightleftharpoons 2Fe^{3+} + 3H_2O$$

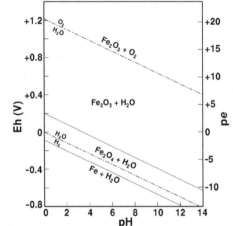

Figure 4–3

Eh–pH diagram showing the stability limits of hematite, magnetite, and metallic iron at 25°C in the presence of water.

Note that in hematite, iron has a +3 valence. There is no change in valence in going from hematite to Fe^{3+} and this reaction is only pH dependent; i.e., it is not an oxidation–reduction reaction. This often comes as a surprise to students because we are plotting Eh–pH diagrams and the presumption is made that both variables must be involved in all calculations. Later we will write a reaction that is dependent only on Eh; i.e., pH is not a variable. We can write the reaction in terms of an equilibrium constant:

$$K_{eq} = \frac{[Fe^{3+}]^2}{[H^+]^6} \tag{4-37}$$

For this reaction, $\Delta G_R^0 = 22.2$ kJ mol^{-1} and $K = 10^{-3.89}$. Writing equation 4–37 in logarithmic form, we get

$$2 \log[Fe^{3+}] - 6 \log[H^+] = -3.89 \tag{4-38}$$

or

$$\log[Fe^{3+}] + 3 \text{ pH} = -1.95 \tag{4-39}$$

Note that the boundary between hematite and Fe^{3+} is fixed either by the concentration of Fe^{3+} ions in solution or the pH. This boundary has been plotted in Figure 4–4 for various values of $[Fe^{3+}]$. For magnetite, the reaction is

$$3Fe^{3+} + 4H_2O + e^- \rightleftharpoons Fe_3O_4 + 8H^+$$

$\Delta G_R^0 = -50.4$ kJ mol^{-1} and $E^\circ = 0.52$ V. The Nernst equation for this reaction is

$$Eh = 0.52 + \frac{0.0592}{1} \log\left(\frac{[Fe^{3+}]^3}{[H^+]^8}\right) = 0.52 + 0.177 \log[Fe^{3+}] + 0.473 \text{ pH} \tag{4-40}$$

This boundary is also plotted in Figure 4–4 for various values of $[Fe^{3+}]$.

The boundaries we have drawn in Figure 4–4 are actually activity contour lines. Each represents a particular activity of Fe^{3+} in solution coexisting with either hematite or magnetite. Note that only under very acidic conditions is a significant amount of Fe^{3+} found in solution and that the equilibrium concentration of Fe^{3+} in solution increases with decreasing pH (increased acidity). Under any Eh–pH conditions there will be some Fe^{3+} in a solution coexisting with hematite or magnetite, but at other than very low pH the Fe^{3+} in solution will be infinitesimal. For any Eh–pH calculation that involves the activity of ions other than H^+, an activity value must be selected in order to draw the Eh–pH boundary.

We now repeat the preceding calculations, but in this case for Fe^{2+} in equilibrium with hematite or magnetite. For hematite, we can write the following reaction:

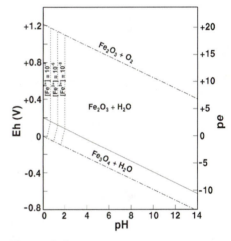

Figure 4–4
Eh–pH diagram showing the variation in $[Fe^{3+}]$, in mol L^{-1}, for a solution coexisting with hematite or magnetite. Only at very low pH values does Fe^{3+} have a significant activity.

$$Fe_2O_3 + 2e^- \rightleftharpoons 2Fe^{2+}$$

Adding appropriate amounts of H_2O and H^+ gives

$$Fe_2O_3 + 6H^+ + 2e^- \rightleftharpoons 2Fe^{2+} + 3H_2O$$

For this reaction, $\Delta G_R^0 = -126.4$ kJ mol^{-1} and $E^\circ = 0.66$ V. The resulting Nernst equation is

$$Eh = 0.66 + \frac{0.0592}{2} \log\left(\frac{[H^+]^6}{[Fe^{2+}]^2}\right) = 0.66 - 0.177 \text{ pH} - 0.0592 \log[Fe^{2+}] \tag{4-41}$$

Similarly for magnetite,

$$Fe_3O_4 + 8H^+ + 2e^- \rightleftharpoons 3Fe^{2+} + 4H_2O$$

For this reaction, $\Delta G_R^0 = -172.5$ kJ mol^{-1} and $E^\circ = 0.89$ V. The Nernst equation is

$$Eh = 0.89 + \frac{0.0592}{2} \log\left(\frac{[H^+]^8}{[Fe^{2+}]^3}\right) = 0.89 - 0.237\, pH - 0.0888 \log[Fe^{2+}] \quad (4\text{–}42)$$

As was the case for $[Fe^{3+}]$, in Figure 4–5 the Eh–pH boundaries for these equilibria are plotted using $[Fe^{2+}] = 10^{-4}$, 10^{-6}, and 10^{-8} mol L^{-1}.

In both Figure 4–4 and Figure 4–5 we have in essence drawn contour lines showing the activity of Fe^{3+} and Fe^{2+} for solutions in equilibrium with hematite or magnetite. Note that these are for solutions in equilibrium with the solid phases. For example, within the field of hematite as shown in Figure 4–3, hematite is always present. Consider Figure 4–5. To the left of the activity contour labeled $[Fe^{2+}] = 10^{-8}$, Fe^{2+} is present in solution and has an activity of greater than 10^{-8} mol L^{-1}. These solutions are in equilibrium with either hematite or magnetite.

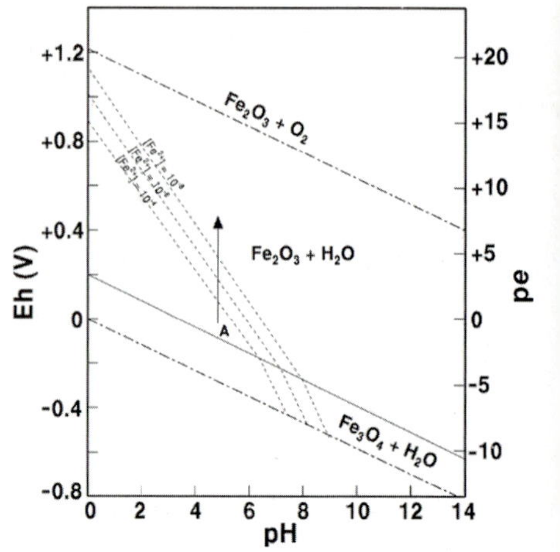

What would happen to the activity of Fe^{2+} if the oxidation–reduction potential of an environment was changed? Referring to Figure 4–5, let us suppose that local groundwater had the Eh and pH indicated by point A and was in equilibrium with hematite. For this sample, $[Fe^{2+}] = 10^{-2}$ mol L^{-1}. The groundwater is pumped to the surface and exposed to atmospheric oxygen. The Eh of the sample would change in the direction shown by the arrow. Note that the activity of Fe^{2+} decreases in this direction and the solution would be supersaturated in Fe^{2+} with respect to hematite. Thus, ignoring kinetic factors, we would expect that Fe would precipitate as hematite. Note that this is an oxidation–reduction reaction, because the valence of Fe must change from +2 to +3.

We can now combine Figures 4–3, 4–4, and 4–5 into a single diagram, referred to as a *composite diagram* (Figure 4–6). With reference to Figures 4–4 and 4–5, you will note that in the upper left-hand corner of the diagram, $[Fe^{3+}] > [Fe^{2+}]$. Thus, there is a field in which Fe^{3+} is the dominant iron cation. The boundary for this field is represented by the reaction

$$Fe^{3+} + e^- \rightleftharpoons Fe^{2+}$$

For this reaction, $\Delta G_R^0 = -74.3$ kJ mol^{-1} and $E^\circ = 0.77$ V. The Nernst equation is

$$Eh = 0.77 + \frac{0.0592}{1} \log\left(\frac{[Fe^{3+}]}{[Fe^{2+}]}\right) \quad (4\text{–}43)$$

When $[Fe^{3+}] = [Fe^{2+}]$, the Nernst equation reduces to

$$Eh = 0.77 \text{ V} \quad (4\text{–}44)$$

This is only an oxidation–reduction reaction, so the boundary is independent of pH. The $[Fe^{3+}]$ field occupies a small area in the upper left corner of the water stability field (Figure 4–6). To reemphasize an earlier point, for the Eh–pH conditions represented by the field labeled Fe^{3+}, waters that

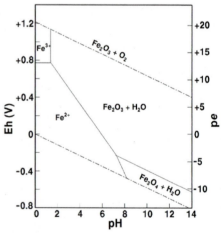

Figure 4–5
Eh–pH diagram showing the variation in $[Fe^{2+}]$, in mol L^{-1}, for a solution coexisting with hematite or magnetite. The arrow indicates increasing oxidation. See text for discussion.

Figure 4–6
Composite diagram showing the stability fields for hematite and magnetite as a function of Eh and pH. The boundary for the ionic species is drawn with activity = 10^{-6} mol L^{-1}. In the field labeled Fe^{2+}, $[Fe^{2+}] > [Fe^{3+}]$. In the field labeled Fe^{3+}, $[Fe^{3+}] > [Fe^{2+}]$. For a system containing hematite or magnetite, the solution is in equilibrium with the solid phase.

plot in this field and that are in equilibrium with hematite would contain both Fe^{2+} and Fe^{3+}, but $[Fe^{3+}] > [Fe^{2+}]$. Waters that plot in the field labeled Fe^{2+} and that are in equilibrium with either hematite or magnetite would have $[Fe^{2+}] > [Fe^{3+}]$. By convention, the diagrams are constructed with ionic activities $= 10^{-6}$ mol L^{-1}.

Iron also occurs in the natural environment as the iron carbonate mineral siderite ($FeCO_3$). This introduces another complication into our analysis in that the fugacity of CO_2 gas or the activity of carbonate species must also be considered.

For a solution in equilibrium with $CO_{2\,(g)}$ and magnetite, we can write the following reaction:

$$Fe_3O_4 + 3CO_{2\,(g)} + 2H^+ + 2e^- \rightleftharpoons 3FeCO_3 + H_2O$$

For this reaction, $\Delta G_R^0 = -41.4$ kJ mol^{-1} and $E^\circ = 0.215$ V. The Nernst equation is

$$Eh = 0.215 + \frac{0.0592}{2} \log([H^+]^2[CO_2]^3) = 0.215 - 0.0592\,pH + 0.0888 \log[CO_2]$$
(4–45)

The partial pressure of CO_2 in the earth's atmosphere is $10^{-3.5}$ atm. Substituting this value into equation 4–45 gives

$$Eh = -0.096 - 0.0592\,pH \tag{4–46}$$

This Eh–pH equation yields values that are slightly less than that of the water breakdown boundary. Thus, siderite is not stable at the earth's surface in water in equilibrium with atmospheric CO_2; i.e., if siderite is present, it is a metastable phase. Two points to consider are that the difference is very small and that kinetic factors always need to be taken into account. In practice, siderite is found in natural waters and thus persists despite the thermodynamic limitations. Most authors set $CO_2 = 10^{-2}$ atm, a pressure greater than that of atmospheric CO_2, in order to plot siderite on the iron Eh–pH diagram. Using this value for CO_2, the Eh–pH equation for the siderite–magnetite boundary becomes

$$Eh = 0.0374 - 0.0592\,pH \tag{4–47}$$

The dissolution of siderite to yield Fe^{2+} ions in solution can be written

$$FeCO_3 + 2H^+ \rightleftharpoons Fe^{2+} + CO_2 + H_2O$$

This is not an oxidation–reduction reaction. The equilibrium constant is written

$$K = \frac{[CO_2][Fe^{2+}]}{[H^+]^2} \tag{4–48}$$

For this reaction, $\Delta G_R^0 = -43.7$ kJ mol^{-1}, which gives $K = 10^{7.656}$. Taking the logs and substituting pH for $-\log[H^+]$ gives

$$7.656 = \log[CO_2] + \log[Fe^{2+}] + 2\,pH \tag{4–49}$$

Setting $[CO_2] = 10^{-2}$ atm and $[Fe^{2+}] = 10^{-6}$ mol L^{-1} gives pH $= 7.83$. The siderite field under these conditions is shown in Figure 4–7.

If the solution is not in equilibrium with $CO_{2\,(g)}$—i.e., it is a closed system—we must take into consideration the more complex equilibria of the carbonate system. This topic was covered in Chapter 3. In brief, we have three carbonate species, $H_2CO_{3\,(aq)}$, HCO_3^-, and CO_3^{2-}, and the relative abundances of these species is a function of the pH. Thus, the reactions involving magnetite, hematite, and the dissolved carbonate species will be dependent upon pH. For example, at low pH the dominant species is $H_2CO_{3\,(aq)}$ and the reactions used to construct the boundaries between magnetite, hematite, and siderite would involve this particular carbonate species. This topic will not be considered further here. The interested student can find a complete discussion in Garrels and Christ (1965) and Faure (1998).

Another important group of iron-containing minerals is the sulfides. As was the case for the iron carbonates, sulfur species such as $H_2S_{(aq)}$, HS^-, HSO_4^-, and SO_4^{2-} must be considered when writing Eh–pH equations involving iron sulfides. Under normal surface

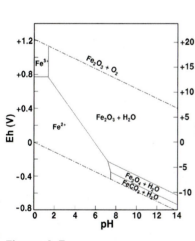

Figure 4–7

Composite Eh–pH diagram showing the stability fields of hematite, magnetite, and siderite as a function of Eh and pH. For ionic species, activity $= 10^{-6}$ mol L^{-1}. $P_{CO_2} = 10^{-2}$ atm, a partial pressure greater than that of the earth's atmosphere. At $P_{CO_2} = 10^{-3.5}$ atm, the actual partial pressure, siderite would plot below the stability limit of liquid H_2O.

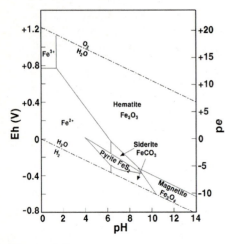

Figure 4–8
Composite Eh–pH diagram showing the stability relations for the iron oxides, carbonates, and sulfides in water at 25°C. Activity of Fe species = 10^{-6} mol L^{-1}, total dissolved carbonate = 1 mol L^{-1}, and total dissolved sulfur = 10^{-6} mol L^{-1}. After Garrels and Christ (1965).

conditions, pyrite (FeS_2) is the stable iron sulfide and its stability field is controlled by the activity of the sulfur-containing species. We will not consider these calculations here. The interested student can consult Garrels and Christ (1965) or Faure (1998) for the solutions to the stability limits in the iron sulfide system. A final composite diagram for the iron system, including the minerals hematite, magnetite, siderite, and pyrite, is shown in Figure 4–8. Note that pyrite is stable over a fairly broad range of pH values at low Eh.

The Nitrogen Eh–pH Diagram As a second example of an Eh–pH diagram, this one containing only gases and aqueous species, we will draw the diagram for the nitrogen system under surface conditions at 25°C. The extensive use of nitrogen in fertilizers has led to an interest in the fate of nitrogen species in ground and surface waters. Nitrogen can show a variety of valences from +5 (NO_3^-) to −3 (NH_4^+). The major nitrogen species in surface and ground waters are $N_{2\,(g)}$, $NH_{3\,(g)}$, NO_2^-, NO_3^-, and NH_4^+. Because nitrogen minerals are readily soluble, they are rarely found in nature and will, therefore, not be included in our Eh–pH calculations.

Oxidation of N_2 to NO_3^- is represented by the following reaction:

$$2NO_3^- + 10e^- \rightarrow N_2$$

Balancing the equation gives

$$2NO_3^- + 12H^+ + 10e^- \rightleftharpoons N_2 + 6H_2O$$

For this reaction, $\Delta G_R^0 = -1199.6$ kJ mol^{-1} and $E° = 1.24$ V. The Nernst equation is

$$Eh = 1.24 + \frac{0.0592}{10} \log\left(\frac{[H^+]^{12}[NO_3^-]^2}{[N_2]}\right) \tag{4–50}$$

The partial pressure of nitrogen in the earth's atmosphere is 0.77 atm, thus $[N_2] = 0.77$ atm. In the following calculations we will assume that the waters are in equilibrium with atmospheric nitrogen. Note that in many groundwaters N_2 partial pressure is less than 0.77 atm. This will change the position, but not the slopes, of the various stability boundaries calculated here. Nitrate concentrations in groundwater are on the order of 10^{-3} mol L^{-1}, thus $[NO_3^-] = 10^{-3}$ mol L^{-1}. Substituting these values gives the following Eh–pH equation:

$$Eh = 1.21 - 0.071 \text{ pH} \tag{4–51}$$

Oxidation of NH_4^+ to N_2 is represented by the following reaction:

$$N_2 + 8H^+ + 6e^- \rightleftharpoons 2NH_4^+$$

For this reaction, $\Delta G_R^0 = -158.8$ kJ mol^{-1} and $E° = 0.27$ V. The resulting Nernst equation is

$$Eh = 0.27 + \frac{0.0592}{6} \log\left(\frac{[N_2][[H^+]^8}{[NH_4^+]^2}\right) \tag{4–52}$$

Setting $[NH_4^+] = 10^{-3}$ mol L^{-1} and $N_2 = 0.77$ atm gives

$$Eh = 0.33 - 0.0791 \text{ pH} \tag{4–53}$$

Oxidation of NH_3 to N_2 is represented by the following reaction:

$$N_2 + 6H^+ + 6e^- \rightleftharpoons 2NH_3$$

For this reaction, $\Delta G_R^0 = -53.2$ kJ mol^{-1} and $E° = 0.09$ V. The resulting Nernst equation is

$$Eh = 0.09 + \frac{0.0592}{6} \log\left(\frac{[N_2][H^+]^6}{[NH_3]^2}\right) \tag{4–54}$$

Setting $[NH_3] = 10^{-3}$ mol L^{-1} and $N_2 = 0.77$ atm gives

$$Eh = 0.15 - 0.0592 \text{ pH} \tag{4-55}$$

Lastly, consider the relationship between NH_4^+ and NH_3. This is not an oxidation–reduction reaction because there is no change in the valence of nitrogen. The reaction can be written

$$NH_4^+ \rightleftharpoons NH_3 + H^+$$

For this reaction, $\Delta G_R^0 = 52.8$ kJ mol^{-1} and $K = 10^{-9.25}$. Writing this reaction in terms of the equilibrium constant,

$$K = \frac{[NH_3][H^+]}{[NH_4^+]} = 10^{-9.25} \tag{4-56}$$

Setting $[NH_3] = [NH_4^+] = 1$ gives pH = 9.25 for the boundary between these two species when they are at equal activity.

We can now plot an Eh–pH diagram (Figure 4–9) for the nitrogen species. The relationship between N_2 and NO_3^- is given by equation 4–51, the relationship between N_2 and NH_4^+ is given by equation 4–53, and the relationship between N_2 and NH_3 is given by equation 4–55. The boundary between NH_4^+ and NH_3 is independent of Eh and is at a constant pH = 9.25 (equation 4–56). We first plot equation 4–51. We then plot equations 4-53 and 4-55. Note that the slopes of the lines calculated from the latter two equations are different. The change in slope corresponds to the change from the NH_4^+-dominant field to the NH_3-dominant field.

NO_2^- is also found in groundwater. We will now calculate the Eh–pH relationship for this species. NO_2^- is intermediate in the oxidation–reduction sequence between NO_3^- and N_2. We can write the following equation to represent the oxidation of NO_2^- to NO_3^-:

$$NO_3^- + 2H^+ + 2e^- \rightleftharpoons NO_2^- + H_2O$$

For this reaction, $\Delta G_R^0 = -160.6$ kJ mol^{-1} and $E° = 0.83$ V. The Nernst equation is

$$Eh = 0.83 + \frac{0.0592}{2} \log\left(\frac{[NO_3^-][H^+]^2}{[NO_2^-]}\right) \tag{4-57}$$

Setting $[NO_3^-] = 10^{-3}$ mol L^{-1} and solving equation 4–57 for Eh, given $[NO_3^-]/[NO_2^-]$ ratios of 1 and 10^4, gives, respectively, the following Eh–pH equations:

$$Eh = 0.83 - 0.0592 \text{ pH} \tag{4-58}$$

and

$$Eh = 0.95 - 0.0592 \text{ pH} \tag{4-59}$$

These two curves are plotted in Figure 4–9. The significance of these results is that the stability field for NO_2^- falls within the field of N_2 stability, a more reduced form of nitrogen. The conclusion is that NO_2^- is a metastable species. It occurs in water as a kinetic intermediary species formed during the reduction of NO_3^- to N_2.

To reemphasize an earlier point, consider the boundary in Figure 4–9 between N_2 and NO_3^-. On the more oxidizing side of this boundary, NO_3^- activities are 10^{-3} mol L^{-1} or greater (as the system becomes more oxidizing). However, N_2 is also present in the system under these conditions. This boundary does not represent the disappearance of N_2, but rather the occurrence of NO_3^- with activites of 10^{-3} mol L^{-1} or greater.

Case Study 4–1 illustrates an environmental application of Eh–pH diagrams. In the case study we are concerned with acidic runoff from a coal mine and the factors that affect the concentrations of dissolved iron and manganese in the runoff.

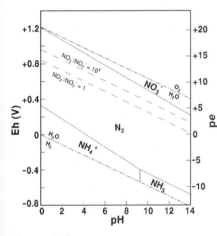

Figure 4–9

Eh–pH diagram for the nitrogen system at 25°C $N_2 = 0.77$ atm and activity of the dissolved species = 10^{-3} mol L^{-1}. The dashed lines indicate the boundary between NO_3^- and NO_2^- for activity ratios of 1 and 10^4. Because N_2 is a more reduced form of nitrogen, NO_2^- must exist metastably in this region. Only at very low concentrations of NO_2^- would it be the stable species relative to N_2.

CASE STUDY 4–1
Sources of High Dissolved Manganese Concentrations in Mildly Acidic Runoff from a Coal Mine in Eastern Tennessee

Larsen and Scarbrough (2000) investigated the sources of dissolved manganese in mildly acidic runoff from a coal mine. The mined coal is low in sulfur, and the presence of overburden with a large neutralization capacity (as a consequence of the common occurrence of siderite cement) indicated that acid mine drainage should not be a problem at this site. However, within a year of opening the mine the mildly acidic (pH = 4) mine waters contained significant quantities (up to 60 mg L^{-1}) of dissolved manganese. Because these manganese concentrations exceeded those permitted for the mine effluent, a study was undertaken to determine the source of the manganese and the steps that might be taken to reduce its concentration in the effluent.

Mineralogical and extraction studies were conducted on samples from the mine site to identify the source of the extractable manganese. Two sources were identified: exchangeable manganese on clay minerals (mainly illite + muscovite and chlorite) in the shales and mudstones and manganese in siderite concretions and cement. The study suggested that the siderite was the major source of the dissolved manganese. The proposed mechanism was that sulfuric acid formed by the oxidation of pyrite in the mine spoils reacted with the siderite. The acid was neutralized by the hydrolysis of the siderite, with the concomitant release of Ca, Mg, Mn, and Fe to the solutions percolating through the mine spoils. Because these are electroactive species (refer to Example 4–5), an environmental Eh measurement should be meaningful in this case. The Eh–pH range of the spoil solution is plotted in the iron Eh–pH diagram (Figure 4–C1–1) and the manganese Eh–pH diagram (Figure 4–C1–2). Note that the spoil solution plots within the $Fe(OH)_{3\ ppt}$ field in Figure 4–C1–1, indicating that iron should be removed as an iron-hydroxide flocculate. Manganese, on the other hand, is soluble in the Eh–pH range of the spoil solution (Figure 4–C1–2) and will remain in solution.

Potential methods to control the manganese concentration in the spoil solution are (1) limiting oxidation of the pyrite in the spoils in order to minimize the acid contribution to the spoil solution, (2) keeping the siderite-rich spoils separate and up the hydraulic gradient from the pyrite-rich spoils thus limiting siderite dissolution, and (3) increasing the pH and Eh of the effluent to induce Mn-hydroxide flocculation.

Source: Larsen and Scarbrough (2000).

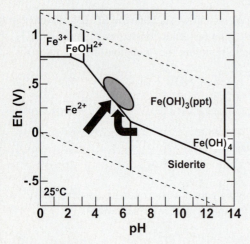

Figure 4–C1–1
Iron Eh–pH diagram at 25°C and 1.013 bars. The ellipsoidal field is the range in Eh–pH values for the spoil solution. The straight and curved arrows show hypothetical reaction paths for waters reacting with the pyritic and sideritic spoils, respectively. The spoil waters fall within the Fe-hydroxide field, and iron is apparently removed from solution as an Fe-hydroxide flocculate. Personal communication, D. Larsen, 2000.

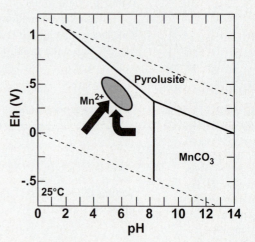

Figure 4–C1–2
Manganese Eh–pH diagram at 25°C and 1.013 bars. Field and arrows same as Figure 4–C1–1. The spoil waters fall within the Mn^{2+} field, and manganese remains in solution. If the pH and Eh were increased, the manganese would be removed from solution as a Mn-hydroxide flocculate. Personal communication, D. Larsen, 2000.

THE ROLE OF MICROORGANISMS IN OXIDATION–REDUCTION REACTIONS

So far, we have considered inorganic processes that can lead to changes in the oxidation state of various species. In the natural environment, microorganisms play an important role in facilitating a number of chemical reactions. In this section we will consider the role of microorganisms in mediating oxidation–reduction reactions. A more complete discussion of the role of microorganisms in the natural environment can be found in Ehrlich (1996) and Banfield and Nealson (1997). Consider the following reaction in which ammonium is oxidized to nitrate:

$$NH_4^+ + 2O_2 \rightarrow NO_3^- + H_2O + 2H^+$$

For this reaction, $\Delta G_R^0 = -266.5$ kJ mol^{-1}. Given the large negative free energy, one would expect that this reaction would readily occur. However, an aqueous solution containing ammonium ions and exposed to the air is very stable. In contrast, this reaction proceeds at a measurable rate in natural waters that contain microorganisms. In this case, the bacteria *Nitrosomonas* sp. and *Nitrobacter* sp. use the ammonium ion as a substrate for their metabolism. In effect, the microorganisms act as a catalyst. In the natural environment many oxidation–reduction reactions are mediated by microorganisms.

It is important to understand that the microorganisms simply act as catalysts; i.e., they facilitate the transfer of electrons. The types of reactions that occur are still controlled by what is thermodynamically possible. For example, the oxidation of ammonium to nitrate can only occur in the Eh–pH range in which nitrate is the stable form of nitrogen. Hence, all the thermodynamic calculations we have done in the preceding section, and the Eh–pH diagrams constructed from these calculations, are still an accurate description of what reactions are possible under any given conditions of Eh, pH, and concentration of individual species.

Microorganisms and Energy Pathways

Microorganisms belong to one of three groups: procaryotes, eucaryotes, and viruses. Viruses are important in terms of the transmission of diseases, but they do not play a role in facilitating chemical reactions and will not be considered further. *Procaryotes* have a *simple cellular structure with rigid cell walls and lack a nucleus.* These are the most primitive life-forms and are believed to have been the first life-forms to arise on earth. The procaryotes include all types of bacteria. *Archaebacteria* (a subgroup of the procaryotes) *inhabit extreme environments, similar to those that may have existed early in earth history, and are considered to represent ancient life-forms.* The archaebacteria include *methanogens (bacteria that produce methane), halophiles (bacteria that exist in high-salinity environments), thermophiles (bacteria that exist in high-temperature environments),* and *thermoacidophiles (bacteria that exist in high-temperature and low-pH environments). Eucaryotes* have a *true nucleus and a more complex structure than procaryotes.* This more complex structure is represented by the presence of mitochondria, chloroplasts, and vacuoles and the compartmentalization of some key metabolic processes. Algae, fungi, and protozoa are members of the eucaryote group.

Microorganisms are classified in various ways depending on how they obtain their carbon and energy. *Autotrophs obtain their carbon from CO_2, HCO_3^-, or CO_3^{2-} and use external energy for the synthesis of organic compounds. Photosynthesizers use sunlight as the external energy source*, and *chemosynthesizers use chemical reactions involving various inorganic molecules as the external energy source. Photolithotrophs are photosynthetic autotrophs*, and *chemolithotrophs are chemosynthetic autotrophs. Heterotrophs derive their energy from the oxidation of organic compounds and use previously synthesized carbon as their source of carbon.* Microorganisms can be either *aerobes, microorganisms that directly use molecular oxygen as the electron acceptor for their oxidation reactions*, or *anaerobes, microorganisms that use other electron-poor species (SO_4^{2-}, NO_3^-) as the electron acceptor for their oxidation reactions. Obligate anaerobes* can *only function in the absence of oxygen; facultative anaerobes can use either oxygen or other electron acceptors for oxidation.*

Organisms themselves use catalysts, called *enzymes*, to facilitate the transfer of electrons. *Enzymes* are *proteins that facilitate a reaction by forming a complex with the reactants that brings the reactants into close proximity.* Enzymes are usually specific to a particular substrate. *Oxygenases* are *enzymes that facilitate oxidation reactions*, and *reductases* are *enzymes that facilitate reduction reactions.* An important characteristic of microorganisms is their ability to synthesize a new enzyme that can utilize a previously unavailable substrate. This characteristic is particularly important in microbial remediation of pollutants. The pollutant is often a new organic chemical not previously found in the environment. The synthesis of an enzyme that can utilize this new substrate is the key step in the remediation process.

Five groups of microorganisms, important in mediating oxidation–reduction reactions, are briefly described here.

Bacteria are single-celled organisms ranging in size from 0.2 to 50 μm in diameter, but commonly less than 5 μm. They are the most abundant microorganisms in water and soil. The shape of bacterial cells is usually spherical, straight-rod, or curved-rod. Because of their small size, they have a large surface-to-volume ratio and usually have a negative surface charge. Bacteria can exist in aerobic and/or anaerobic environments. Bacteria obtain their carbon either directly from carbonate species (autotrophs) or from organic compounds (heterotrophs). Most bacteria are chemoautotrophs and derive their energy from chemical oxidation reactions. Bacteria prefer neutral to slightly alkaline environments.

Actinomycetes are a class of uniceluar organisms that show similarities to both fungi and bacteria. They typically form branched filamentous colonies and are found in both aquatic and terrestrial environments. Their principle role is the degradation of existing organic compounds. Like bacteria, actinomycetes prefer alkaline environments.

Fungi are multinucleate and do not have internal cell boundaries. The living mass of a fungus is bounded externally by a rigid wall composed of cellulose or chitin. The basic structure is a tubular, often branched filament. Fungi use organic compounds as their carbon source and play a key role in the degradation of litter in the soil. Fungi prefer acidic environments.

Algae range from unicellular to complex multicellular varieties and vary in size from microscopic to large, plant-like structures. The main differences between the types of algae are biochemical: (1) types of chlorophyll and other pigments, (2) chemical composition of the cell wall, and (3) chemical nature of stored foods. Algae convert inorganic carbon into organic compounds.

Protozoa are mostly unicellular organisms with animal-like characteristics and are commonly 5–50 μm in size. They are found either in water or in thin films of water on the surface of particles and are most abundant in warm, well-oxygenated environments of intermediate pH.

Examples of Oxidation–Reduction Reactions Mediated by Microorganisms

Aerobic degradation is the decomposition of dead plant and microbial material in the presence of oxygen. The reaction can be written

$$CH_2O_{(aq)} + O_{2(g)} \rightarrow CO_{2(g)} + H_2O$$

and at 25°C, $\Delta G_R^0 = -501.8$ kJ mol^{-1}. Thermodynamically, the reaction is highly favored. However, in practice, without the presence of microorganisms this reaction occurs very slowly. Microorganisms that facilitate aerobic degradation are heterotrophic species of bacteria, fungi, protozoa, and actinomycetes. As an example, consider the secondary treatment of sewage effluent. At a sewage treatment facility you may have seen large circular or rectangular tanks filled with gravel. Following primary treatment (which removes particulate matter), sewage effluent is trickled onto this gravel bed. The community of microorganisms that exist on the gravel mediate the aerobic oxidation of the dissolved organic matter in the effluent. A problem that occasionally rises at sewage treatment facilities is the destruction of this microorganism community by harmful chemicals in the effluent. When this happens, the microbial community has to be reestablished in order for aerobic oxidation of dissolved organic matter to occur.

Anaerobic degradation occurs in an environment without measurable oxygen or nitrate and sulfate (other possible oxidizing agents). The reaction can be written

$$2CH_2O_{(aq)} \rightarrow CH_{4(g)} + CO_{2(g)}$$

and at 25°C, $\Delta G_R^0 = -185.7 \, \text{kJ mol}^{-1}$. Thermodynamically, the reaction is not as strongly favored as aerobic oxidation, but the reaction should proceed spontaneously. Species of actinomycetes are the major anaerobes responsible for mediating this reaction. This type of biomass degradation typically occurs in swamps and other wetlands and releases methane to the atmosphere. Methane is a greenhouse gas (Chapter 8), and anaerobic degradation is one of the major sources of atmospheric methane.

Nitrification is the oxidation of ammonium ion to nitrate ion. It is a two-step process, and both steps are mediated by autotrophic bacteria.

$$NH_4^+ + 1.5O_2 \rightarrow NO_2^- + H_2O + 2H^+ \quad \text{and} \quad NO_2^- + 0.5O_2 \rightarrow NO_3^-$$

The net reaction is

$$NH_4^+ + 2O_2 \rightarrow NO_3^- + H_2O + 2H^+$$

For the first step, $\Delta G_R^0 = -190.0 \, \text{kJ mol}^{-1}$, and for the second step, $\Delta G_R^0 = -76.5 \, \text{kJ mol}^{-1}$. The nitrification of ammonium ion is thermodynamically favored. However, as in the previous examples, it is the presence of microorganisms (the bacteria *Nitrosomonas* and *Nitrobacter*, respectively) that facilitates these reactions.

Denitrification is the reduction of nitrogen in nitrate to nitrogen gas, nitrous oxide, or ammonium ion. Which pathway occurs depends on the Eh and pH of the environment. In an aerobic environment, denitrification is represented by the following reaction:

$$4NO_3^- + 5CH_2O_{(aq)} + 4H^+ \rightarrow 2N_2 + 7H_2O + 5CO_2$$

For this reaction, $\Delta G_R^0 = -2548.4 \, \text{kJ mol}^{-1}$, indicating that the reaction is thermodynamically strongly favored. The reaction is mediated by facultative heterotrophic bacteria, such as species of *Pseudomonas* and *Achromobacter*. When small amounts of oxygen are present, the reaction follows a different pathway, leading to the production of nitrous oxide. This reaction is written

$$2NO_3^- + 2CH_2O_{(aq)} + 2H^+ \rightarrow N_2O + 3H_2O + 2CO_2$$

For the reaction, $\Delta G_R^0 = -919.1 \, \text{kJ mol}^{-1}$. From an environmental point of view, this is an interesting reaction because there has been a recent increase in the N_2O content of the atmosphere. This increase may be due, in part, to an increase in the use of fertilizers, which leads to more denitrification.

Sulfide oxidation occurs under aerobic conditions. This oxidation is usually microbially mediated. These sulfide oxidation reactions are important in a number of environmental settings. Consider the following reaction in which sulfide formed during the decomposition of organic matter is oxidized:

$$HS^- + 2O_2 \rightarrow SO_4^{2-} + H^+$$

For this reaction, $\Delta G_R^0 = -756.6 \, \text{kJ mol}^{-1}$. The reaction is mediated by chemoautotrophic bacteria, such as *Thiobacillus thiooxidans*.

In organic-rich environments, such as mangrove swamps and other types of swamps, the large quantity of organic matter leads to a strongly reducing environment and the precipitation of metal sulfides. Soils formed under these conditions contain abundant sulfide minerals. If these soils are drained, thus exposing them to atmospheric oxygen, the sulfide minerals are oxidized, with the release of metal ions, sulfate, and hydrogen ions. The release of hydrogen ions leads to decreasing pH, i.e., acidification of the soils. Similar oxidation reactions occur when sulfide minerals in spoils, either from coal mining or the mining of sulfide ore deposits, are exposed to the atmosphere. The oxidation of FeS_2 (pyrite) can be written

$$2FeS_{2 \, \text{pyrite}} + 7O_2 + 2H_2O \rightarrow 2Fe^{2+} + 4SO_4^{2-} + 4H^+$$

For this reaction, $\Delta G_R^0 = -2327.8$ kJ mol^{-1}. The Fe^{2+} is oxidized to Fe^{3+} as follows:

$$4Fe^{2+} + O_2 + 4H^+ \rightarrow 4Fe^{3+} + 2H_2O$$

For this reaction, $\Delta G_R^0 = -177.4$ kJ mol^{-1}. The reaction is biologically mediated by the acid-tolerant, iron-oxidizing bacteria *Thiobacillus ferrooxidans*. The Fe^{3+} acts as an oxidizing agent and oxidizes additional pyrite according to the reaction

$$FeS_{2\ pyrite} + 14Fe^{3+} + 8H_2O \rightarrow 15Fe^{2+} + 2SO_4^{2-} + 16H^+$$

For this reaction, $\Delta G_R^0 = -543.0$ kJ mol^{-1}. For the oxidation sequence consisting of the latter two reactions, the rate-limiting step is the oxidation of Fe^{2+} to Fe^{3+}. Because of the positive feedback loop, the oxidation of pyrite is a self-accelerating process.

Sulfate reduction occurs in organic-rich reducing environments. An example of a sulfate reduction reaction is

$$SO_4^{2-} + 2CH_2O_{(aq)} + H^+ \rightarrow HS^- + 2CO_2 + 2H_2O$$

For this reaction, $\Delta G_R^0 = -247.0$ kJ mol^{-1}. The reaction is mediated by sulfate-reducing bacteria, such as *Desulfovibrio desulfuricans*, an obligate anaerobe that grows at pH values greater than 5.5.

As an example of a biologically mediated oxidation–reduction reaction, Case Study 4–2 investigates the role of bacteria in the reduction of arsenate to arsenite in a hypersaline lake.

OXIDATION–REDUCTION PROCESSES IN NATURAL SYSTEMS

In the previous sections we developed the thermodynamic basis for oxidation–reduction reactions and looked at the role of microorganisms in oxidation–reduction reactions. In the natural environment the major oxidizing agent is atmospheric oxygen and the major reducing agent is organic carbon. Most carbon compounds are unstable in water; i.e., their stability field falls below the H_2O-H_2 boundary. The oxidation of organic matter first utilizes the dissolved oxygen in the system. If the amount of dissolved oxygen is limited, either because the system is isolated from the atmosphere or the rate of replenishment of dissolved oxygen is less than its rate of consumption by organic matter, the decomposition of organic matter proceeds with the reduction of NO_3^-, NO_2^-, and SO_4^{2-}. A further decrease in oxidation–reduction potential can lead to the bacterial reduction of carbonate species and fermentation, resulting in the production of methane.

Table 4–3 lists reduction and oxidation reactions that may be combined to produce biologically mediated redox reactions. These reactions are combined in different ways as the

Table 4–3 Reduction and Oxidation Reactions That May Be Combined to Produce Biologically Mediated Redox Reactions*

	Reduction		Oxidation
A	$O_{2\ (g)} + 4H^+ + e^- \rightarrow 2H_2O$	L	$CH_2O_{(aq)} + H_2O \rightarrow CO_{2\ (g)} + 4H^+ + 4e^-$
B	$2NO_3^- + 12H^+ + 10e^- \rightarrow N_{2\ (g)} + 6H_2O$	L-1	$HCOO^- \rightarrow CO_{2\ (g)} + H^+ + 2e^-$
C	$MnO_{2\ (s)} + HCO_3^- + 3H^+ + 2e^- \rightarrow MnCO_{3\ (s)} + 2H_2O$	L-2	$CH_2O_{(aq)} + H_2O \rightarrow HCOO^- + 3H^+ + 2e^-$
D	$NO_3^- + 10H^+ + 8e^- \rightarrow NH_4^+ + 3H_2O$	L-3	$CH_3OH_{(aq)} \rightarrow CH_2O_{(aq)} + 2H^+ + 2e^-$
E	$FeOOH_{(s)} + HCO_3^- + 2H^+ + e^- \rightarrow FeCO_{3\ (s)} + 2H_2O$	L-4	$CH_{4\ (g)} + H_2O \rightarrow CH_3OH_{(aq)} + 2H^+ + 2e^-$
F	$CH_2O_{(aq)} + 2H^+ + 2e^- \rightarrow CH_3OH_{(aq)}$	M	$HS^- + 4H_2O \rightarrow SO_4^{2-} + 9H^+ + 8e^-$
G	$SO_4^{2-} + 9H^+ + 8e^- \rightarrow HS^- + 4H_2O$	N	$FeCO_{3\ (s)} + 2H_2O \rightarrow FeOOH_{(s)} + HCO_3^- + 2H^+ + e^-$
H	$CO_{2\ (g)} + 8H^+ + 8e^- \rightarrow CH_{4\ (g)} + 2H_2O$	O	$NH_4^+ + 3H_2O \rightarrow NO_3^- + 10H^+ + 8e^-$
J	$N_{2\ (g)} + 8H^+ + 6e^- \rightarrow 2NH_4^+$	P	$MnCO_{3\ (s)} + 2H_2O \rightarrow MnO_{2\ (s)} + HCO_3^- + 3H^+ + 3e^-$

*Modified from Stumm and Morgan (1996).

CASE STUDY 4–2
Bacterial Dissimilatory Reduction of Arsenate and Sulfate in Mono Lake, California

Mono Lake, California, is an alkaline and hypersaline lake with high concentrations ($\sim$200 μmol L^{-1}) of dissolved inorganic arsenic. For Mono Lake, pH = 9.8, salinity = 75 to 90 g L^{-1} dissolved solids, and dissolved carbonates = 0.4 mol L^{-1}. Oremland et al. (2000) studied the change in speciation from As^{5+} (arsenate) to As^{3+} (arsenite) that occurred in the transition from oxic (oxygen-rich) surface waters to anoxic (oxygen-depleted) bottom waters. The distinction between arsenic species is important because arsenate is strongly adsorbed to mineral surfaces, whereas arsenite is much more mobile and toxic. Arsenic is derived from both anthropogenic sources, such as drainage from mines and mine tailings, pesticides, and biocides and from natural sources, such as hydrothermal leaching or solution of arsenic-containing minerals in rocks. Numerous studies have documented the change from arsenate to arsenic species, with depth, in stratified lakes. Two possible biochemical reduction pathways have been identified: (1) the reduction of As by a reductase present in the cytoplasm of certain bacteria that leads to the rapid expulsion of As from the cell and (2) a respiratory (*dissimilatory*) As^{5+} reductase present in certain anaerobes that enables them to conserve energy produced by the oxidation of organic substrates.

For Mono Lake the authors measured the change in a number of parameters as a function of depth (Figure 4–C2–1). The rapid decrease in oxygen between 11 and 17 m represents the transition from the oxic surface layer to the anoxic deep layer. In the surface waters, arsenic occurs as As^{5+} and sulfide and methane are absent. In the deep waters, arsenic occurs predominantly as As^{3+} and sulfide and methane are present in significant amounts. Total cell counts show a significant increase in the deep waters. Relative to shallow waters (5 m depth), deep waters (24 m depth) contained significant amounts of sulfate-respiring and arsenate-respiring bacteria ($\sim$50X increase for sulfate-respiring bacteria and $\sim$100X increase for arsenate-respiring bacteria).

Previous studies found that the arsenic-respiring bacteria *B. selenitireducens* and *B. arsenicoselenatis* occurred in the Mono Lake sediments. The first species was used to develop a radioassay technique to measure the reduction of As^{5+} to As^{3+}. A radioassay technique was also used to measure the reduction of sulfate. In the surface waters, the rate constants for the reduction of arsenate are essentially zero, but in the deepest waters, the rate constants vary between 0.277 and 0.290 d^{-1}. Similar observations were made for the reduction of sulfate, with the surface waters having rate constants of zero and the deepest waters having rate constants that varied between 5×10^{-6} and 2×10^{-5} d^{-1}. For the 1999 measurements, the integrated reduction rates were sulfate, 12.6 mmol m^{-2} d^{-1}

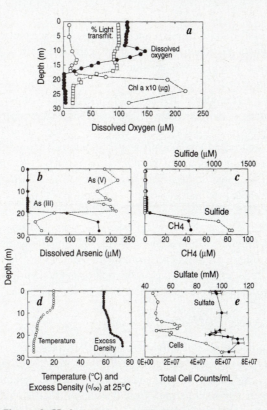

Figure 4–C2–1
Variations with depth of various physical, chemical, and biological parameters for Mono Lake in July, 1999. From Oremland et al. (2000).

(continued)

CASE STUDY 4–2 (continued)

between 21 and 28 m depth, and arsenate, 17.3 mmol m^{-2} d^{-1} between 18 and 28 m depth.

The authors considered two reactions for the oxidation of organic carbon, one involving arsenate reduction and the other sulfate reduction. The two reactions are

$$CH_2O + 2H_2AsO_4^- \rightarrow HCO_3^- + 2H_2AsO_3^- + H^+$$

and

$$2CH_2O + SO_4^{2-} \rightarrow 2HCO_3^- + HS^- + H^+$$

Using the integrated daily reduction values for arsenate and sulfate, the authors concluded that these two processes could account for the oxidation of 32 to 55% of the organic matter produced annually by photosynthesis in the surface waters.

The authors also made the provocative suggestion that arsenate-reducing bacteria might be found on Mars. They noted that in hypersaline environments there is a high energy requirement for the maintenance of osmotic pressure. In a situation such as this, arsenate reduction would be favored over sulfate reduction. For example, consider the following reactions calcu-

lated for an ion concentration of 1 μmol L^{-1}, pH = 7, and $P_{H_2} = 10^{-6.6}$ atm:

$$SO_4^{2-} + H^+ + 4H_2 \rightarrow HS^- + 4H_2O$$
$$(\Delta G' = -0.42 \text{ kJ mol electrons}^{-1})$$

and

$$H_2AsO_4^- + H^+ + H_2 \rightarrow H_3AsO_3 + H_2O$$
$$(\Delta G' = -23.0 \text{ kJ mol electrons}^{-1})$$

The free energy for arsenate reduction is much greater than that for sulfate reduction. Hence, this reaction would be energetically favored. During the late stages of the "drying up" of Mars, hypersaline lakes would represent the last biome. Hydrothermal activity associated with Martian shield volcanoes would introduce arsenic into these lakes. If some oxygen was present, a biogeochemical cycle might develop in which the cycling of arsenic between the +3 and +5 oxidation states would provide the energy for anaerobic decomposition of organic matter.

Source: Oremland et al. (2000).

overall Eh of the environment changes. The sequence of biologically mediated oxidation–reduction reactions is shown graphically in Figure 4–10. The calculations required to construct this diagram were done at pH = 7 and HCO$_3^-$ = 1 × 10^{-3} mol L^{-1}. The box in Figure 4–10 lists examples of various redox reactions that are combinations of the reduction and oxidation reactions listed in Table 4–3. With decreasing oxygen content, and a corresponding decrease in Eh, reactions will proceed in the order aerobic respiration → denitrification → nitrate reduction → fermentation → sulfate reduction → methane fermentation. If MnO$_2$ is present, it should be reduced to Mn^{3+} in the same Eh range as nitrate reduction, and if FeOOH $_{(s)}$ or Fe(OH)$_3$ $_{(s)}$ is present, it will be reduced to Fe^{2+} at Eh's similar to those for fermentation. The oxidation–reduction reaction sequence is paralleled by an ecological succession of microorganisms: aerobic heterotrophs → denitrifiers → fermentors → sulfate reducers → methane bacteria. The sequence illustrated here might occur in a lake undergoing eutrophication. As the organic content of the lake increases, and oxidizing agents are utilized in the decomposition of the organic matter, the series of reactions outlined here would occur in the deep waters of the lake. Under extreme conditions, the production of methane might be possible. Such a result might be expected when a lake evolves to a swamp or marsh.

Redox Buffering

Similar to pH buffering discussed in Chapter 3 is the concept of ***redox buffering***. When oxygen is present in water, the Eh is controlled by the oxygen–water half-reaction. During the oxidation of organic matter, as long as measurable oxygen is present, there is very little change in Eh. Recall from our earlier discussion that relatively large variations in dissolved oxygen lead to only small variations in Eh. When dissolved oxygen is no longer significant, there is a rapid drop in Eh until the sulfate and nitrate oxidation–reduction reactions become important. The sulfate and nitrate serve as electron acceptors for the oxi-

Figure 4–10
Sequence of microbially mediated redox processes. Letters refer to reactions in Table 4–3. From AQUATIC CHEMISTRY, 3rd Edition by W. Stumm and J. J. Morgan. Copyright © 1996. This material is used by permission of John Wiley & Sons, Inc.

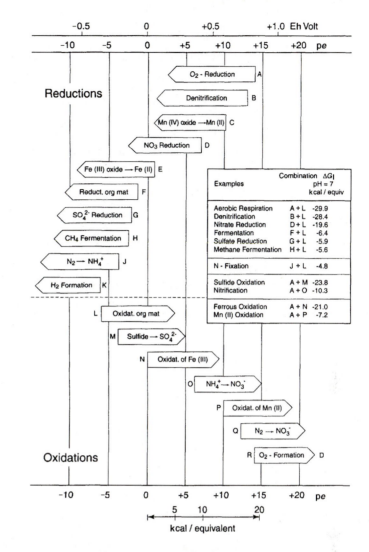

dation of organic matter. In these reactions sulfate is reduced to HS^- and nitrate to nitrite, ammonium ion, or N_2, depending on the environmental conditions. After the sulfate and nitrate are exhausted (i.e., there are no additional oxidants), there is a further small decline in Eh until fermentation (anaerobic biodegradation) becomes important. Because of these processes, natural waters tend to fall into two main Eh ranges, referred to as high and low Eh (or pe) environments (Figure 4–11).

Figure 4–11
Schematic representation of redox buffering regions in acidic waters. Modified from vanLoon and Duffy (2000).

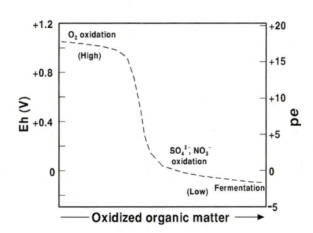

Classification of Oxidation–Reduction Environments

When the Eh concept was first introduced earlier in this chapter, the comment was made that there was often a discrepancy between the measured Eh and the observed oxidation–reduction equilibria. Because of the difficulty of measuring a thermodynamically meaningful Eh, Berner (1981b) suggested a simplified scheme. The initial distinction is between environments that contain *significant dissolved oxygen* (**oxic**) and those that have *no significant dissolved oxygen* (**anoxic**). Subsequent modification led to the recognition of a third environment, **suboxic**, which *contains some dissolved oxygen*. The anoxic environment is subdivided into two environments, one that *contains significant dissolved sulfide* (**sulfidic**) and the other with *minimal dissolved sulfide* (**nonsulfidic**). The nonsulfidic environment is subdivided on the basis of whether or not sulfate reduction and methane production can occur. These different oxidation–reduction environments and their characteristics are summarized in Table 4–4. Also included in the table are phases that characterize each of these environments.

Table 4–4 Oxidation–Reduction Classification of Natural Environments*

Environment	Dissolved gases $(10^{-6} \text{ mol L}^{-1})$	Characteristic phases
Oxic	$O_2 > 30$	Hematite, goethite, ferrihydrite, MnO_2-type phases, no organic matter
Suboxic	$O_2 < 30$ and $\geqslant 1$	Hematite, goethite, ferrihydrite, MnO_2-type phases, minor organic matter
Anoxic	$O_2 < 1$	
Sulfidic	$H_sS \geqslant 1$	Pyrite, marcasite, rhodocrosite, organic matter
Nonsulfidic	$H_2S < 1$	
Postoxic		Low-temperature Fe^{2+} and Fe^{3+} silicates, siderite, vivianite, rhodocrosite, no sulfide minerals, minor organic matter
Methanic		Siderite, vivanite, rhodocrosite, earlier formed sulfide minerals, organic matter

*After Berner (1981b) and Anderson et al. (1994) as modified by Langmuir (1997).

Oxidative and Reductive Capacity

Oxidative and reductive capacity was defined by Scott and Morgan (1990). It is essentially a measure of the net oxidation or reduction potential of an environment. The mathematical expression is

$$\text{OXC} = \Sigma \, n_i[\text{Ox}]_i - \Sigma \, n_i[\text{Red}]_i = -\text{RDC} \qquad (4\text{–}60)$$

where OXC is the oxidative capacity, RDC is the reductive capacity, $[\text{Ox}]_i$ and $[\text{Red}]_i$ are the molar concentrations of the oxidants and reductants, respectively, including both dissolved and solid species, and $\Sigma \, n_i$ is the number of electrons transferred in the oxidation–reduction reactions. Given these units, capacity is expressed in moles of electrons per liter; i.e., capacity is a measure of the excess or deficiency of electrons in the environment.

EXAMPLE 4–8 A groundwater sample is collected from an aquifer overlain by cultivated fields. For this sample $[O_{2 \, (aq)}] = 2 \times 10^{-4} \text{ mol L}^{-1}$, $[NO_3^-] = 1 \times 10^{-3} \text{ mol L}^{-1}$, colloidal $Fe(OH)_{3 \, (s)} = 1 \times 10^{-4} \text{ mol L}^{-1}$, and dissolved organic carbon (DOC) as $CH_2O = 1 \times 10^{-3} \text{ mol L}^{-1}$. We did not consider the iron hydroxides when we constructed the iron Eh–pH diagram, but they are common in natural systems as metastable phases intermediate to the iron oxides.

We can write four possible reactions that control the concentrations of these species:

$$4H^+ + O_{2\,(g)} + 4e^- \rightleftharpoons 2H_2O$$

$$NO_3^- + 6H^+ + 5e^- \rightleftharpoons \tfrac{1}{2}N_2 + 3H_2O$$

$$Fe(OH)_3 + 3H^+ + e^- \rightleftharpoons Fe^{2+} + 3H_2O$$

$$HCO_3^- + 5H^+ + 4e^- \rightleftharpoons CH_2O \text{ (organic matter)} + 2H_2O$$

Substituting into the OXC equation (4–60) gives

$$
\begin{aligned}
OXC &= (n_i)[O_{2\,(aq)}] + (n_i)[NO_3^-] + (n_i)[Fe(OH)_3] - (n_i)[CH_2O] \\
&= (4)(2 \times 10^{-4}) + (5)(1 \times 10^{-3}) + (1)(1 \times 10^{-4}) - (4)(1 \times 10^{-3}) \\
&= 1.9 \times 10^{-3} \text{ mol electrons L}^{-1}
\end{aligned}
$$

The environment has excess oxidizing capacity; i.e., all the dissolved organic matter will be oxidized and there will still be some oxidative capacity left in the groundwater system. ∎

Oxidation–Reduction Ladders

An *oxidation–reduction ladder* is *a graphical representation of oxidation–reduction reactions arranged in order of decreasing oxidation potential*; i.e., species of couples higher on the ladder can oxidize species of couples lower on the ladder. Eh depends on both pH and concentrations of individual species, so this information must be specified when constructing an Eh ladder. Figure 4–12 is an example of an Eh ladder on which has been plotted the Eh for the various reactions described in the section on the construction of Eh–pH diagrams. These calculations are done at pH = 5.7 (the pH of surface waters in equilibrium with atmospheric CO_2) and activity of ionic species = 1×10^{-3} mol L^{-1}. Plotted on this ladder are two reactions that were not previously described, i.e., the oxidation reaction between dissolved methane and bicarbonate ion and the oxidation reaction between dissolved CH_2O (a proxy for dissolved organic carbon, DOC) and bicarbonate ion. For dissolved methane the reaction is

$$HCO_3^- + 9H^+ + 8e^- \rightleftharpoons CH_{4\,(aq)} + 3H_2O$$

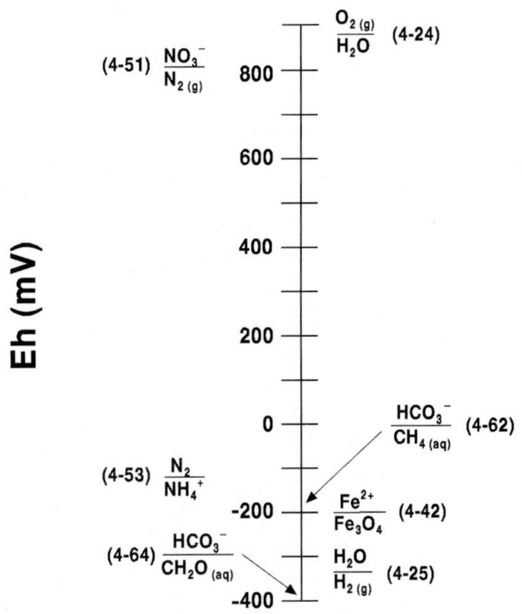

Figure 4–12
Eh ladder plot at pH = 5.7 for some of the reactions previously described in the chapter. $O_{2\,(g)}$ = 0.21 atm, $N_{2\,(g)}$ = 0.77 atm, activities of aqueous species = 1×10^{-3} mol L^{-1}. Numbers in parentheses refer to equations in the text.

For this reaction, $\Delta G_R^0 = -158.8$ kJ mol^{-1} and $E° = 0.206$ V. The Nernst equation is

$$E = 0.206 + \frac{0.0592}{8} \log\left(\frac{[HCO_3^-][H^+]^9}{[CH_{4\,(aq)}]}\right) \qquad (4\text{–}61)$$

If we set $[HCO_3^-] = [CH_{4\,(aq)}]$, the Eh–pH equation becomes

$$Eh = 0.206 - 0.0666 \text{ pH} \qquad (4\text{–}62)$$

The reaction between $CH_2O_{(aq)}$ and bicarbonate ion is

$$HCO_3^- + 5H^+ + 4e^- \rightleftharpoons CH_2O_{(aq)} + 2H_2O$$

For this reaction, $\Delta G_R^0 = -17.1$ kJ mol^{-1} and $E° = 0.044$ V. The Nernst equation is

$$E = 0.044 + \frac{0.0592}{4} \log\left(\frac{[HCO_3^-][H^+]^5}{[CH_2O_{(aq)}]}\right) \qquad (4\text{–}63)$$

If we set $[HCO_3^-] = [CH_2O_{(aq)}]$, the Eh–pH equation becomes

$$Eh = 0.044 - 0.0740 \text{ pH} \qquad (4\text{–}64)$$

At pH = 5.7, Eh = -378 mV for reaction 4-61. At this pH, the Eh for the breakdown of water to H_2 gas is -337 mV. Thus, $CH_2O_{(aq)}$ is only stable under conditions more reducing than those that lead to the breakdown of water. What this means is that $CH_2O_{(aq)}$ exists metastably in water and should readily oxidize. Whether or not complete oxidation of DOC will take place depends on the total oxidizing capacity of the environment.

We can use the Eh ladder to predict what would happen during changes in the natural system. For example, if oxygen was added to the system, the first reaction to occur would be the oxidation of $CH_2O_{(aq)}$ to HCO_3^- because the greatest difference in oxidation–reduction potential exists between these two couples. If dissolved oxygen was still present after the completion of this reaction, the next step would be the oxidation of Fe^{2+}, resulting in the precipitation of Fe_3O_4 (magnetite). If the dissolved oxygen was exhausted during this process, the next step would be the reduction of NO_3^- to N_2. Thus, reactions would proceed in a stepwise fashion, moving down the ladder as each oxidant was exhausted and moving up the ladder as each reductant was exhausted. The oxidation–reduction processes would be completed when the only remaining couples had the same Eh. If DOC was added to the system, the first step would be the removal of dissolved O_2, then reduction of N_2, and so on. Thus, additions of oxygen (or other oxidants) to the system or the addition of organic matter (or other reductants) leads to changes in the species present in a system.

Because there is also a kinetic factor—i.e., reactions don't occur instantaneously—one can envision that the Eh of a system can change with time. For example, groundwater entering an aquifer will initially be in equilibrium with atmospheric oxygen. Once the groundwater is isolated from the atmosphere (we will assume a closed system), various oxidation reactions will begin to utilize dissolved oxygen. As these reactions proceed, moving down the Eh ladder, the Eh of the system will change toward some final equilibrium value. Thus, with distance from the recharge area, which is proportional to the time since the water was in equilibrium with the atmosphere, the Eh of the groundwater will change.

The changes in Eh-sensitive species in a system can be used to determine the impact of an anthropogenic input. For example, let us suppose that landfill leachate is entering a stream. This organic-rich leachate will cause a series of reduction reactions to occur in the stream. In the immediate vicinity of the landfill, there will be significant reduction and $CH_{4\,(aq)}$ and NH_4^+ may occur in the stream given sufficient organic input. In the downstream direction, the addition of atmospheric oxygen will lead to a series of oxidation reactions that will, at some distance, return the stream to its more or less natural conditions. Monitoring the Eh-sensitive species would allow us to construct a model of this stream rejuvenation process. Similarly, Case Study 4–3 describes the reactions that took place when septic system effluent was discharged into a groundwater system.

CASE STUDY 4–3
Effect of Septic System Effluent on Groundwater

Robertson and Blowes (1995) investigated the impact of septic system effluent discharged to a silt–fine-sand aquifer. Given the nature of the aquifer material, there was not a great deal of buffering capacity. Hence, it was possible to achieve acidic conditions in the effluent plume. Partial oxidation of sulfide minerals in the aquifer sediments and the oxidation of DOC and NH_4^+ released H^+ ions to solution. The oxidation of organic carbon can be represented by the following equation:

$$CH_2O + O_2 \rightarrow CO_2 + H_2O \rightarrow H_2CO_3 \rightarrow H^+ + HCO_3^-$$

and the oxidation of NH_4^+ by

$$NH_4^+ + 2O_2 \rightarrow NO_3^- + 2H^+ + H_2O$$

The effluent had pH = 6.4, and pH values in the effluent plume ranged down to pH = 4.4. The latter value can be compared with the natural groundwater pH of 5.6 to 5.9. NH_4^+ and NO_3^- occurred in approximately equal amounts in the effluent plume. The persistence of NH_4^+, relatively high DOC levels compared to aerobic septic system plumes, and high Fe concentrations, presumably as the more soluble Fe^{2+} species, led the authors to conclude that the plume must be anaerobic.

Assuming that Cl^- is a conservative species, i.e., a species whose concentration is not changed by processes occurring in the system, the authors found that there was a significant decrease with time (distance) in the total nitrogen content of the effluent plume; i.e., the N/Cl ratio decreased. This decrease could be due to reduction reactions involving either dissolved organic carbon and/or sulfide minerals present in the aquifer.

For DOC,

$$5CH_2O + 2NO_3^- \rightarrow 5CO_2 + 2N_2 + 3H_2O + 4OH^-$$

Note that this reaction would tend to raise the pH. For sulfide minerals,

$$FeS_2 + 2NO_3^- + 2H_2O \rightarrow Fe^{2+} + 2SO_4^{2-} + N_2 + 4H^+$$

Sulfate concentrations were elevated in the NO_3^--depleted zone and also increased along the flow path, suggesting that the latter reaction might be important in the reduction of nitrate.

Fe, Mn, Ni, Cr, Zn, and Co concentrations were found to be greater in the effluent plume than in the noncontaminated groundwater. The authors suggested that these metals were added to the system either during oxidation of sulfide minerals (previous equation) or during reduction reactions, as illustrated by the following equations:

$$MnO_{2\ birnesite} + CH_2O \rightarrow Mn^{2+} + CO_2 + H_2O$$

and

$$Fe(OH)_3 + CH_2O \rightarrow Fe^{2+} + CO_2 + 2H_2O + H^+$$

The introduction of DOC and dissolved ammonia by septic system discharge to a groundwater system resulted in a number of oxidation–reduction reactions that significantly modified groundwater chemistry. These modifications included a decrease in pH, an increase in dissolved sulfate ion, and an increase in dissolved metals. These changes were monitored by measuring variations in Eh-sensitive species.

Source: Robertson and Blowes (1995).

The Redox Interface

The *redox interface* or *redox front* is a *zone of rapidly changing Eh*. This zone may be sharp and extend over a distance of only several millimeters (see Case Study 4–4, p. 122) or it may extend over a distance of many meters or tens of meters. Redox interfaces occur where environments of very different oxidation–reduction potentials come into contact. As one example, consider the interface between seawater and the seafloor. Ocean waters are generally aerobic at all depths, whereas pore waters in bottom sediments, depending on the organic carbon content of the sediments, are often suboxic to anoxic. This change can occur over a distance of several to 10 millimeters and is reflected by an increase in reduced species in the pore waters relative to the overlying seawater. As a second example, consider the discharge of organic-rich waste to the environment, as described in Case Study 4–3. In the immediate vicinity of the discharge, the high DOC produces anoxic waters. As one moves away from the discharge site, mixing with groundwater adds oxygen to the system and the waters become oxic. This transition occurs over a distance of several meters, and the transition zone is marked by several subzones of varying Eh. As a third example, consider groundwater flow through an aquifer. Where the water enters the aquifer, the aquifer is near the surface and unconfined, and the waters are well aerated and oxic. As the waters move through the aquifer, the aquifer deepens and becomes confined; i.e., impermeable layers prevent exchange of oxygen with the atmosphere, and the waters become anoxic due to O_2-consuming redox reactions. In terms of Eh-sensitive species and Eh, Figure 4–13 schematically shows the changes that might be expected in crossing a redox

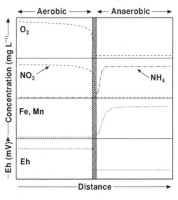

Figure 4–13

Schematic representation of changes that would occur across a redox interface between aerobic and anaerobic groundwaters.

CASE STUDY 4–4
Corrosion of Brass in a Marine Environment

An interesting example of the significance of the redox front is a study on the corrosion of brass shells done by Stoffyn-Egli et al. (1998). In 1942 a British munitions ship caught fire in Halifax Harbour and was sunk by naval gunfire using sand-filled shells. A most unusual, but effective, way to put out a fire. In 1994 brass (30% Zn, 70% Cu) ammunition shells were retrieved from the seafloor around the vessel. These shells were 10 cm in diameter and 30 cm long and had a wall thickness of approximately 1 mm. The shells were partially embedded in the soft, organic-rich clayey-silts that comprise the bottom sediments of this portion of Halifax Harbour. A sharp redox front occurs between the oxygenated marine waters of Halifax Harbour and the pore waters of the anaerobic bottom sediments. The redox front extends vertically over a distance of 10 cm and the partially buried brass shells straddled the redox front. The surface of the brass shell embedded in the bottom sediments had a coating of native copper (Cu) and djurleite ($Cu_{1.96}S$). The portion of the shell protruding above the bottom had a coating of cuprite, atacamite, and a new copper sulfate mineral that appeared to have a stability field similar to that of connellite. Thus, the sequence of minerals on the surface of the shell matched the changing Eh–pH conditions of the surrounding environment (Figure 4–C4–1). The arrow in Figure 4–C4–1 indicates the changing Eh–pH conditions in going from the sediment pore waters to marine waters. A similar pattern was observed for the corrosion coating on the inside of the hollow shells: Native copper and djurlite were found immediately above the brass while the surface layers consisted of atacamite and the new copper sulfate mineral. The more oxidized coatings tended to act as a protective cover on the brass shells, and the more reduced coatings tended to peel off the brass shells, thus

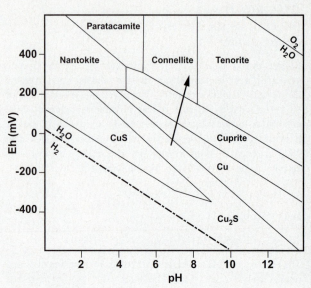

Figure 4–C4–1
Eh–pH diagram for copper. Arrow shows Eh–pH trend for redox front. From Stoffyn-Egli et al. (1998).

exposing more material to corrosion. The authors suggest that this may be an important observation because Cu-based canisters have been considered as containers for high-level nuclear wastes. The results of this study suggest that reducing, sulfur-rich waters could accelerate corrosion of these types of canisters.

Source: Stoffyn-Egli et al. (1998).

---

interface. Note the increase in Fe and Mn in solution. The lower Eh would favor Fe^{2+} and Mn^{2+}, species that are more soluble. Mn- and Fe-containing minerals in the aquifer would begin to decompose, releasing these ions to solution.

In this chapter the basic principles needed to understand and describe oxidation–reduction reactions have been developed. Oxidation and reduction processes play a role in a number of different environments, and further descriptions of redox processes will be found in the chapters on fresh and marine waters. This chapter concludes with case studies that illustrate the importance of redox processes in the distribution and degradation of contaminants.

Case Study 4–5 considers changes that occur across the oxic/postoxic boundary in marine sediments. The significant observation is that under certain conditions Hg can be immobilized in selenide minerals, thus inhibiting its release to the environment.

Case Study 4–6 concerns the effect of a hydrocarbon spill that entered an aquifer on the geochemistry of the aquifer sediments and the implications of the resulting changes in terms of hydrocarbon degradation and migration.

CASE STUDY 4–5
A Coupled Natural Immobilization Mechanism for Mercury and Selenium in Deep-Sea Sediments

Mercone et al. (1999) investigated the distribution of Hg and Se, among other elements, across the oxic/postoxic boundary (Berner's classification of redox environments) in marine sediments. Three different deep-sea sediment sequences (turbidite, sapropel, and glacial/interglacial transition) were investigated. Sapropels are dark, very organic-rich (>2 wt% organics) sediments. In each of these sequences the oxic/postoxic boundary has persisted in a narrow depth zone for thousands of years. Sediment samples collected across this boundary showed an enrichment in Hg, Se, Cd, V, Sb, and Tl just below (on the postoxic side) the boundary. Hg maxima were either coincident with, or very close to, the Se maxima, and in several cases a similar relationship was noted for Cd and Pt. The underlying

postoxic sediments contained sulfide minerals that are absent above the redox boundary. The authors suggested that oxidation of sulfide minerals was the source of the trace metals. The sulfide released from the sulfide minerals was oxidized to sulfate, while the metals migrated downward across the redox boundary. In the reducing, sulfide-poor postoxic environment, these metals were reprecipitated as selenide minerals such as tiemannite (HgSe). These minerals will not form in the presence of sulfide, but the sulfide-deficient pore waters are a suitable environment. The minerals will persist under oxidizing conditions, thus immobilizing Hg and other associated metals. The authors suggest that this mechanism may be important during the weathering of sulfide wastes that contain significant Hg. The presence of Se in these wastes can retard the loss of Hg to the environment as long as sulfide is oxidized to sulfate.

Source: Mercone et al. (1999).

CASE STUDY 4–6
The Effect of a Hydrocarbon Spill on the Sediments of the Contaminated Aquifer

A glacial-outwash, sandy aquifer near Bemidji, Minnesota (Figure 4–C6–1) was contaminated in 1979 by a high-pressure oil pipeline burst. Oil was sprayed onto the land surface and a 1-m-thick oil slick was found floating on the water table in the vicinity of the spill. This is now a USGS Toxic Substance Hydrology study site. Previous studies have outlined the extent of the contaminant plume and its redox conditions (Figure 4–C6–1). These studies have largely focused on the groundwater chemistry. The degradation of the organic compounds largely occurs by microbial activity, during which Fe^{3+} is reduced to Fe^{2+}. This inference is supported by the observations that within the anoxic part of the plume Fe^{2+} and Mn^{2+} are abundant and the DOC is a complex mixture containing notable amounts of low-molecular-weight aliphatic and aromatic organic acids, metabolic intermediates in the degradation of hydrocarbons. The source of the Fe^{3+} is iron-containing minerals in the aquifer.

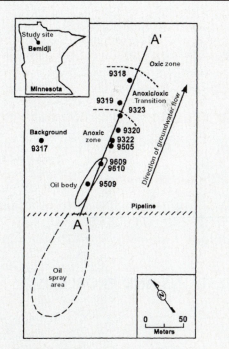

Figure 4–C6–1
Location of study site, redox zones, and locations of core samples. From Tuccillo et al. (1999).

(continued)

CASE STUDY 4–6 (continued)

Tuccillo et al. (1999) investigated the impact of these processes on the sediment geochemistry. A series of sediment core samples were collected along the plume axis (Figure 4–C6–2). A series of selective iron extractions were done and sediment mineralogy was determined. The purpose of the selective iron extractions was to determine the relative importance of the various Fe-containing phases that were contributing iron to the oxidation–reduction process. It was found that there was a rough inverse correlation between the Fe^{2+} concentration in the pore waters and the Fe^{3+} loss from the sediment. The samples collected closest to the oil spill had the highest Fe^{2+} in the pore waters and the greatest Fe^{3+} loss from the sediments. These data supported the microbial degradation of hydrocarbons according to reactions such as the following for the degradation of toluene:

$$C_7H_8 + 36Fe(OH)_3 + 65H^+ \rightarrow 7HCO_3^- + 36Fe^{2+} + 87H_2O$$

The continued availability of Fe^{3+} is required for this reaction. Once Fe^{3+} is exhausted, other degradation reactions become important.

The authors also noted that there was an increase in Fe^{2+} in the affected sediments relative to background values. This was ascribed to the precipitation of an iron-containing carbonate. One possible reaction is

$$0.2Ca^{2+} + 0.8Fe^{2+} + HCO_3^- \rightarrow Fe_{0.8}Ca_{0.2}CO_3 + H^+$$

There may also be some precipitation of Fe^{2+} as an iron hydroxide.

Finally, it was observed that there was a significant increase in the iron content of the sediments in the anoxic/oxic transition zone. This was believed to be due to the reoxidation (Fe^{2+} to Fe^{3+}) and precipitation of iron as Fe^{3+} oxyhydroxides. These oxyhydroxides may be efficient adsorbers of organic compounds and thus act as a barrier to the migration of the hydrocarbons.

Source: Tuccillo et al. (1999).

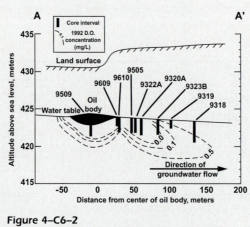

Figure 4–C6–2
Cross-section showing the extent of the anoxic plume, dissolved oxygen content, and location of the core samples. From Tuccillo et al. (1999).

QUESTIONS AND PROBLEMS

1. Define *anion* and *cation*.

2. Define *oxidation* and *reduction*.

3. Define *reducing agent* and *oxidizing agent*.

4. For the following oxidation–reduction reaction, identify the reducing agent, the oxidizing agent, the species that is oxidized, and the species that is reduced.

$$Ce + Pu^{3+} \rightarrow Ce^{3+} + Pu$$

5. What is the *electromotive series*?

6. Define *standard hydrogen electrode* and *saturated calomel electrode*. How are the two electrodes related?

7. Define *Eh*.

8. What is the relationship between Eh and p*e*?

9. Why must we select a value for the activity of ionic species when we draw Eh–pH diagrams?

10. What is an *electroactive* species?

11. Case Study 4–1 dealt with the release of Mn in a spoil solution draining a coal mine waste dump. One suggested solution to the problem was to increase the pH so that Mn-hydroxide would precipitate from solution. With reference to this case study, explain why increasing pH would produce this result. Note that a field for Mn-hydroxide is not shown in Figure 4–C1–2. Reference to Figure 4–C1–1, and the construction of the Fe Eh–pH diagram described in the text, might give you a hint about how to answer this question.

12. Microorganisms play an important role in mediating chemical reactions. Consider the following reaction:

$$NO_3^- + H_2O + H^+ \rightarrow NH_4^+ + 2O_2$$

For this reaction, $\Delta G_R^0 = 266.5 \text{ kJ mol}^{-1}$. Will the presence of microorganisms facilitate this reaction? Explain.

13. What are the characteristics that distinguish *procaryotes* from *eucaryotes*?

14. Distinguish among *methanogens*, *halophiles*, *thermophiles*, and *thermoacidophiles*.

15. Distinguish between *autotrophs* and *heterotrophs*.

16. Distinguish between *photolithotrophs* and *chemolithotrophs*.

17. How do *obligate anaerobes* differ from *facultative anaerobes*?

18. What is the role of enzymes?

19. How does *aerobic degradation* of biomass differ from *anaerobic degradation* of biomass?

20. Soils in the Mekong delta have been continuously submerged and used for rice growing. These soils developed on pyrite-rich former marine deposits. What might happen if these soils were drained so that they could be used for other types of agriculture?

21. With reference to Case Study 4–2, what is *dissimilatory reduction*?

22. What is a *radioassay*? To answer this question, you should refer to the article used for Case Study 4–2.

23. Refer to Figure 4–C2–1. From the plots of dissolved oxygen, As(III), As(V), sulfide, and methane concentrations as a function of depth, what can you conclude about the processes that determine the oxidation state of arsenic and sulfur in Mono Lake?

24. Relate the reaction order shown in Figure 4–10 to the presence of different types of microorganisms.

25. Define *Eh buffering* and give an example of an Eh buffer.

26. Describe and discuss Berner's oxidation–reduction classification of natural environments.

27. Define *oxidative capacity*. What is the significance of this concept?

28. What are *oxidation–reduction* ladders, and why are they useful when investigating oxidation–reduction reactions in natural environments?

29. What is a *redox interface*, and why is it important?

30. Describe what happens, in terms of Eh and species in solution, when an organic-rich effluent enters aerobic groundwater.

31. What might happen to a Cu-based canister used to store radioactive materials if it came in contact with reducing saline groundwater? Refer to Case Study 4–4 to answer this question.

32. With reference to Case Study 4–5, explain how the presence of Se in sulfide wastes might retard the loss of Hg to the environment.

33. With reference to Case Study 4–6, explain why the highest Fe^{2+} concentrations in the pore waters corresponded with the lowest Fe^{3+} concentrations in the sediment.

34. Balance the following oxidation–reduction reactions:

$$PbO_{2\ (s)} \rightarrow Pb^{2+} + O_2$$

$$Fe^{2+} + MnO_{2\ (s)} \rightarrow Fe^{3+} + Mn^{2+}$$

$$As_2S_{3\ (s)} + NO_3^- \rightarrow HAsO_4^{2-} + S^0 + NO_2^-$$

$$CH_2O_{\ (aq)} + NO_3^- \rightarrow N_2 + HCO_3^- + CO_{2\ (g)}$$

$$FeS_{2\ pyrite} + NO_3^- \rightarrow N_{2\ (g)} + FeOOH_{\ goethite} + SO_4^{2-}$$

$$C_2HCl_3 + O_2 \rightarrow CO_2 + Cl^-$$

$$Mn^{2+} + O_2 + Co^{2+} \rightarrow MnO_{2\ (s)} + Co_3O_{4\ (s)}$$

35. For the following reaction, at 25°C,

$$Ce_{\ metal} + Pu^{3+} \rightarrow Ce^{3+} + Pu_{\ metal}$$

calculate:
a. The standard-state free energy. Use the thermodynamic data from Appendix II, source 4 or source 5, as appropriate.
b. The standard-state electromotive force.
c. The K_{eq} value for the reaction and the $[Ce^{3+}]/[Pu^{3+}]$ ratio at equilibrium.

36. a. Calculate the standard electrode potentials, at 25°C, for each of the following half-reactions. Use the thermodynamic data from Appendix II, source 4 or source 5, as appropriate.

$$Np^{3+} + 3e^- \rightarrow Np \qquad Ce^{3+} + 3e^- \rightarrow Ce$$
$$Pu^{3+} + 3e^- \rightarrow Pu \qquad Th^{4+} + 4e^- \rightarrow Th$$

b. Arrange the half-reactions in the order strongest reducing agent to weakest reducing agent.

37. Using the results from problem 36,
a. Calculate the standard electromotive force for the reaction $Pu + Np^{3+} \rightarrow Pu^{3+} + Np$.
b. Calculate K_{eq} for this reaction.
c. If $[Np^{3+}] = 10^{-6}$ mol L^{-1}, calculate the activity of Pu^{3+} at equilibrium.

38. In order for fish to survive, the dissolved oxygen content of a lake or stream must be at least 3 μg mL^{-1}. A particular lake has pH = 6.5 and $T = 25°C$.
a. Calculate the minimum Eh required for this lake to support a fish population.
b. There are a number of summer homes around this lake and each has its own septic system. The effluent from the septic systems drains into the lake and this supports a summer algal bloom. In early spring, the lake has an Eh = 0.90 V, but by late summer Eh = 0.4 V. What would you conclude about the health of the fish in this lake?

39. a. Derive the Eh–pH equation at standard conditions for the reaction $Co^{3+} + e^- \rightarrow Co^{2+}$. Use the thermodynamic data from Appendix II, source 4.
b. Given $[Co^{2+}] = 10^{-4}$ mol L^{-1}, calculate $[Co^{3+}]$ when Eh = 0.95.

40. Construct an Eh–pH diagram, at standard conditions, based on the progressive oxidation of Co to form oxides, starting with metallic Co. The appropriate reactions are

$$CoO + 2H^+ + 2e^- \rightarrow Co + H_2O \quad and \quad Co_3O_4 + 2H^+ + 2e^- \rightarrow 3CoO + H_2O$$

For each of these reactions derive the Eh–pH equation and plot the resulting equation. Use the thermodynamic data from Appendix II, source 4.

41. Derive the Eh–pH equations, at standard conditions, for the solution of the oxide minerals of Co to form Co^{2+}. Plot these equations on the same Eh–pH diagram you used for problem 40. Set $[Co^{2+}] = 10^{-6}$ mol L^{-1}. Use the thermodynamic data from Appendix II, source 4.

$$Co^{2+} + 2e^- \rightarrow Co$$
$$CoO + 2H^+ \rightarrow Co^{2+} + H_2O$$
$$Co_3O_4 + 8H^+ + 2e^- \rightarrow 3Co^{2+} + 4H_2O$$

42. Calculate the Eh for acid mine drainage (pH = 4) and ocean water (pH = 8.3) in equilibrium with atmospheric oxygen ($P_{O_2} = 0.2$ atm) at $T = 25°C$.

43. In Case Study 4–3, the impact of septic system effluent on groundwater was described. Assume the groundwater temperature is 25°C. Use the thermodynamic data from Appendix II, source 4.
   a. Considering the portion of the plume that had pH = 4.4, calculate the Eh of the water. $[NH_4^+] = [NO_3^-]$, and the Eh–pH equation relating these two species is $NO_3^- + 8e^- + 10H^+ \rightarrow NH_4^+ + 3H_2O$.
   b. At this Eh, calculate the concentration of dissolved oxygen in the water.
   c. A direct measurement of the oxidation–reduction potential of the plume water, using a calomel electrode as the reference electrode, gives Eh = 974 mV. Calculate the corrected Eh value. How does this compare to the Eh determined in part (a)?

44. The iron hydroxides were ignored when we constructed the iron Eh–pH diagram. This was done because the hydroxides are metastable phases that will convert with time to hematite and magnetite. However, they can and do occur in natural systems in a metastable state. Construct an Eh–pH diagram, at standard conditions, for the Fe hydroxides, $Fe(OH)_2$ and $Fe(OH)_3$. Use the thermodynamic data from Appendix II, source 4. Set the activity of dissolved species $= 10^{-6}$ mol L^{-1}. The aqueous species $Fe(OH)^{2+}$ can occur in this system, but for this problem ignore this species. Using the equations developed for the diagram plus the diagram itself, what will happen at constant pH = 7 if the Eh of the environment increases from 0 V to 0.6 V?

45. Construct an Eh–pH diagram, at standard conditions, for the sulfur species. The species are $S_{(s)}$, SO_4^{2-}, HSO_4^-, HS^-, and $H_2S_{(aq)}$. Set the activity of all the dissolved species $= 10^{-3}$ mol L^{-1}. Use the thermodynamic data from Appendix II, source 2.

46. The oxidation of pyrite can be represented by the reaction $4FeS_{2\ pyrite} + 15O_2 + 14H_2O \rightarrow 4Fe(OH)_{3\ ferrihydrite} + 8SO_4^{2-} + 16H^+$.
   a. Write the pe equation for this reaction at 25°C. Remember that when calculating the free energy the reaction is written so that the oxidized forms are on the left and the reduced forms are on the right. Use the thermodynamic data from Appendix II, source 4.
   b. Calculate the pe when pH = 4, $SO_4^{2-} = 10^{-3}$ mol L^{-1}, and the reaction is in equilibrium with atmospheric oxygen ($P_{O_2} = 0.2$ atm).

47. Fredrickson et al. (2000) investigated the microorganism (*Shewanella putrefaciens*)-mediated reduction of U^{6+} to U^{4+}. U^{6+} is soluble and uranium is transported in solution in the oxidized form, whereas U^{4+} is relatively insoluble. Hence, the oxidation state of uranium is an important factor in determining the mobility of uranium in surface and ground waters. One of the investigated reactions was $0.5UO_2(CO_3)_3^{4-} + H^+ + e^- \rightarrow 0.5UO_{2\ uraninite} + 1.5HCO_3^-$.
   a. Calculate $E°$ for this reaction at 25°C. Use the thermodynamic data from Appendix II, source 4 or source 5, as appropriate.
   b. Calculate the Eh for this reaction at pH = 5.7, $UO_2(CO_3)_3^{4-} = 1 \times 10^{-5}$ mol L^{-1}, and total carbonate, $C_T = 3 \times 10^{-2}$ mol L^{-1}. You will first have to calculate the amount of C_T that exists as at the given pH.
   c. Based on these calculations, comment on the solubility of uranium in acidic waters such as those found in swamps and marshes.

48. One of the microorganism-mediated reduction reactions considered in Case Study 4–2 is $CH_2O_{(aq)} + 2H_2AsO_4^- \rightarrow HCO_3^- + 2H_2AsO_3^- + H^+$.
   a. Calculate $E°$ for this reaction at 25°C. Use the thermodynamic data from Appendix II, source 4.
   b. Write the Nernst equation for this reaction and calculate the ratio of $H_2AsO_4^-$ to $H_2AsO_3^-$ when Eh = 0.57 V and Eh = 0.28 V. In order to do this calculation, set $CH_2O = 1 \times 10^{-4}$ mol L^{-1}. Use the pH and total carbonate, C_T, given in the case study. Remember to determine the proportion of total carbonate that occurs as HCO_3^-.
   c. Based on these two calculations, would you expect As^{5+} or As^{3+} to be the dominant ionic form of arsenic in a bog? Test your hypothesis by calculating the Eh when the two arsenic species are in equal abundance given pH = 4, $C_T = 1 \times 10^{-3}$ mol L^{-1}, and $CH_2O = 1 \times 10^{-2}$ mol L^{-1}.

49. Oxidation of organic matter can be represented by the reaction $CH_2O_{(aq)} + O_{2(g)} \rightarrow CO_{2(g)} + H_2O$. It was suggested earlier that as long as measurable amounts of oxygen are present in a system, the Eh will remain relatively constant during oxidation of organic matter (redox buffering). Suppose the system becomes isolated from the atmosphere and the oxygen content decreases by roughly 10 orders of magnitude. Calculate the Eh of a system when $P_{O_2} = 0.2$ atm and $P_{O_2} = 10^{-10}$ atm. Initially, $P_{CO_2} = 10^{-3.5}$ atm. At the lower partial pressure of oxygen, $P_{CO_2} = 10^{-2.5}$ atm. Set $CH_2O_{(aq)} = 1 \times 10^{-4}$ mol L^{-1}. Use the thermodynamic data from Appendix II, source 4. Based on these calculations, does the system tend to act as a buffer?

50. A lake is in equilibrium with atmospheric oxygen ($P_{O_2} = 0.2$ atm). A chemical analysis of a water sample from the lake gives $NO_3^- = 1 \times 10^{-3}$ mol L^{-1}, $SO_4^{2-} = 1 \times 10^{-4}$ mol L^{-1}, and $CH_2O_{(aq)} = 1 \times 10^{-3.5}$ mol L^{-1}. Four possible reactions that control the concentrations of these species are

$$O_{2(g)} + 4H^+ + 4e^- \rightarrow 2H_2O$$
$$NO_3^- + 6H^+ + 5e^- \rightarrow \tfrac{1}{2}N_2 + 3H_2O$$
$$SO_4^{2-} + 9H^+ + 8e^- \rightarrow HS^- + 4H_2O$$
$$CO_{2(g)} + 4H^+ + 4e^- \rightarrow CH_2O_{(aq)} + H_2O$$

Calculate the oxidative capacity of the lake water. You will need to use Henry's law to determine the amount of dissolved O_2. Assume $T = 25$°C. Will all of the organic matter in the lake be oxidized? Explain.

5

Carbon Chemistry

Scientists practicing in the environmental field are constantly faced with the problem of organic contaminants. Hence, a familiarity with the nomenclature and basic principles of organic chemistry is essential. This is a vast and complex field, as illustrated by the fact that more than 9 million carbon-based compounds have been identified. Only the rudiments of organic chemistry will be covered here, with particular emphasis on those organic compounds that are environmentally important. Additional topics involving organic chemistry are considered in Chapters 8 (atmospheric environment), 9 (freshwater environment), and 10 (marine environment). Although the naming of organic compounds may seem a dry topic, it is useful to be able to relate the names of compounds to their structure. For this reason, the chapter starts with a discussion of the basic nomenclature for organic compounds.

BASIC STRUCTURES AND NAMING OF ORGANIC COMPOUNDS

Organic compounds are primarily composed of carbon and hydrogen atoms. Other elements—e.g., oxygen, nitrogen, sulfur, phosphorous, and the halogens—may also occur in organic compounds. The fundamental building block of organic compounds is the carbon-carbon bond. The bonding between carbon atoms is covalent. **Single bonds** are formed when *one pair of electrons is shared between two carbon atoms*. **Double bonds** involve the *sharing of two pairs of electrons*, and **triple bonds** the *sharing of three pairs of electrons*. *If multiple bonds occur between carbon atoms, the compound is* referred to as **unsaturated**. *If there are only single bonds between the carbon atoms, the compound is* **saturated**. The carbon atoms are joined together to form chains, branched structures, or rings. *A functional group is an atom or a group of atoms that can be attached to the carbon atoms*. In most cases, a functional group contains at least one noncarbon atom. Also important is the order in which the functional groups are attached and the geometric position of the attached groups. All of this variability means that the naming of organic molecules is not an easy task.

Organic compounds have been named in a variety of ways, which, for the nonspecialist, introduces additional complexity. Organic compounds are often identified by trade names, common names, or acronyms that bear no relationship to the actual chemistry of the compound. **Systematic names** *contain structural information*, and the International Union of Pure and Applied Chemistry (IUPAC) naming system will be described in this section. All organic compounds listed in the Chemical Abstracts System (CAS) are given a unique number. In order to search for a particular compound in Chemical Abstracts, one must know the CAS number of the compound.

Formulas

The three-dimensional geometry of an organic molecule is important because it partly determines the properties of the molecule. **Structural formulas** *show the orientation of the*

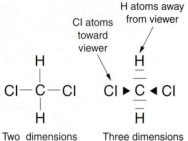

Structural formulas

Figure 5–1

Two- and three-dimensional structural formulas for dichloromethane, CH_2Cl_2. In a two-dimensional representation, single, normal-weight lines are used to connect atoms. In a three-dimensional representation, broken lines indicate bonds extending away from the viewer and arrows indicate bonds extending toward the viewer.

Butane
Condensed structural formula

$$CH_3 - CH_2 - CH_2 - CH_3$$

Structural formula

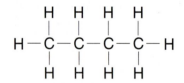

Figure 5–2

Relationship between the condensed structural formula and the structural formula of butane.

atoms in a molecule in either two or three dimensions. **Condensed structural formulas** *are used to save space and are written in terms of carbon units.* **Molecular formulas** *give the numbers of each type of atom in the molecule.* As one example of the types of formulas, consider the compound dichloromethane. The molecular formula is CH_2Cl_2. Because this molecule contains only one carbon atom, the condensed structural formula would be the same. The two- and three-dimensional structural formulas are illustrated in Figure 5–1. Figure 5–2 illustrates the relationship between the condensed structural formula and the structural formula for butane. Note that in the condensed structural formula the H subscript indicates the number of hydrogen atoms attached to each carbon atom.

Isomers

Isomers *are compounds with the same molecular formula but different structural formulas.* Isomers are analogous to the polymorphs of inorganic chemistry. The two isomers of C_4H_{10} are illustrated in Figure 5–3. Note that in this case we are using condensed structural formulas.

If the carbon atoms have double bonds, another type of isomer, called a *geometric* isomer, is possible. The presence of the double bond prevents free rotation between the two carbon atoms. If hydrogen atoms are replaced by substituent groups, there are two possible arrangements for these substituent groups. The two groups may occur on the same side of the double bond, the "cis" orientation, or they may occur on opposite sides of the double bond, the "trans" orientation. The physical and chemical properties of these two isomers are different. Thus, the distinction is important. As an example of the two types of geometric isomers, the condensed structural formulas for *cis*-2-butene and *trans*-2-butene are shown in Figure 5–4.

Hydrocarbons

Hydrocarbons contain only carbon and hydrogen and are the most common group of organic compounds. The major types of hydrocarbons are alkanes, alkenes, alkynes, and aromatic (or aryl) compounds. In **alkanes** *the C atoms have single bonds and occur as straight or branched chains and as rings.* **Alkenes** *have a double bond between two of the carbons,* and **alkynes** *have a triple bond between two of the carbons.* The type of bonding, single versus multiple, affects the *reactivity* of the molecule. Alkanes, because of their strong single carbon bond, tend to be chemically nonreactive except at high temperatures, such as those found during combustion. Alkanes can undergo **substitution reactions** in which *one or more hydrogens are replaced by atoms of another element.* For example,

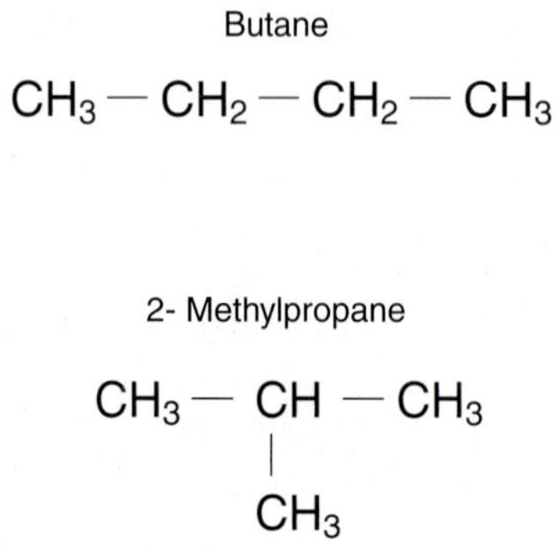

Butane

$$CH_3 - CH_2 - CH_2 - CH_3$$

2- Methylpropane

$$CH_3 - CH - CH_3$$
$$|$$
$$CH_3$$

Figure 5–3

Isomers of C_4H_{10}. Both butane and 2-methylpropane have the same number of C and H atoms, but 2-methylpropane has a branched chain.

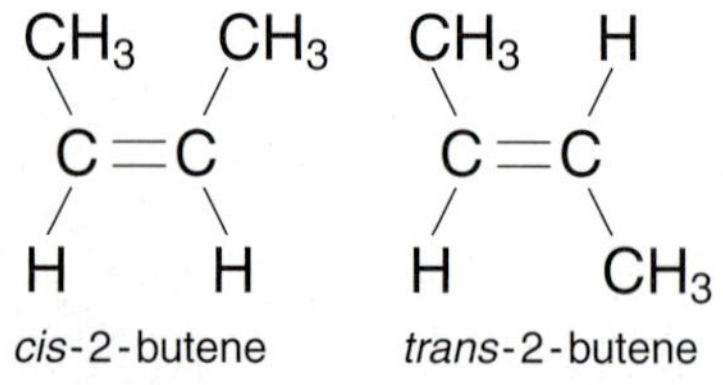

cis-2-butene   *trans*-2-butene

Figure 5–4

Examples of geometric isomers, *cis*-2-butene and *trans*-2-butene. In the first isomer, the two substituent groups are on the same side of the double bond. In the second isomer, the two substituent groups are on opposite sides of the double bond.

consider the photochemical reaction shown in Figure 5–5. In this reaction one of the hydrogen atoms in methane is replaced by chlorine. Hence, this is a substitution reaction. Alkenes and alkynes, because they have double and triple carbon bonds (they are unsaturated compounds), are chemically more reactive. For these types of compounds, ***addition reactions***, *in which some of the electrons in the unsaturated bonds are used to bind with additional atoms*, are common. Thus, atoms are added to the original organic molecule. An example of this type of reaction is the addition of hydrogen to ethylene to produce ethane (Figure 5–6).

Figure 5–5
Example of a substitution reaction in which a hydrogen atom is replaced by a chlorine atom to produce chloromethane.

$$Cl_2 + UV \rightarrow Cl^{\cdot} + Cl^{\cdot} \qquad (Cl^{\cdot} \text{ is a free radical})$$

$$Cl^{\cdot} + CH_4 \rightarrow HCl + CH_3^{\cdot} \qquad (\text{methyl radical})$$

$$CH_3^{\cdot} + Cl_2 \rightarrow CH_3Cl + Cl^{\cdot} \qquad (\text{chloromethane})$$

Figure 5–6
Example of an addition reaction in which hydrogen is added to ethylene to produce ethane (a hydrogenation reaction).

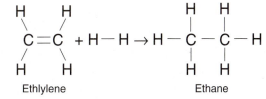

Ethlylene   Ethane

Naming and Types of Hydrocarbon Compounds

Alkanes   Many organic compounds can be derived from alkanes, so it is useful to know the names of the more common alkanes and how these names are derived. The naming process is relatively straightforward and is done according to the following rules:

1. The name of the compound is based on the longest continuous chain of carbon atoms. The root name indicates the number of carbon atoms in the chain, and the suffix *ane* is added to the root name (Table 5–1). For example, a chain with three carbon atoms is propane.

Table 5–1   Prefixes for the Number of Carbon Atoms

# C	Prefix	# C	Prefix	# C	Prefix
1	Meth	11	Undec	21	Henicos
2	Eth	12	Dodec	22	Docos
3	Prop	13	Tridec	23	Tricos
4	But	14	Tetradec	24	Tetracos
5	Pent	15	Pentadec	25	Pentacos
6	Hex	16	Hexadec	26	Hexacos
7	Hept	17	Heptadec	27	Heptacos
8	Oct	18	Octadec	28	Octacos
9	Non	19	Nonadec	29	Nonacos
10	Dec	20	Eicos	30	Triacont

2. The carbon atoms in the longest continuous chain are numbered sequentially from one end. If substituent groups are added to the chain, the numbering starts at the end of the chain that will give the lowest numbers for the substituent groups.
3. All groups attached to the longest continuous chain are identified by the number of the carbon atom to which they are attached and the name of the substituent group. If the substituent group is derived from an alkane, it is named by removing the terminal *ane* and replacing it with *yl*. For example, methane → methyl group. If two or more groups are present, their names are listed in alphabetical order.
4. A prefix is used to denote multiple substitutions by the same kind of group. This prefix is ignored when deciding the alphabetical order (rule 3).

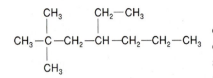

Figure 5–7
Condensed structural formula for the alkane 4-ethyl-2,2-dimethylheptane.

As an example, consider the naming of the alkane shown in Figure 5–7. The longest continuous chain consists of seven carbon atoms; hence, the root name is heptane. The chain is numbered from the left because this will give the lowest numbers for the substituent groups. There are two types of substituent groups: CH_3- is methyl (one carbon atom is methane, substitute *yl* for *ane* to get methyl) and CH_2-CH_3- is ethyl. There are two methyl groups; hence we will use the prefix *di* to give dimethyl. Ethyl is earlier in the alphabet than methyl and will be listed first. The methyl groups are attached to the second carbon, and the ethyl group is attached to the fourth carbon. The complete name for this compound is 4-ethyl-2,2-dimethylheptane, where "4-ethyl" indicates that the ethyl group is attached to the fourth carbon, "2,2-dimethyl" indicates that both methyl groups are attached to the second carbon, "di" indicates two methyl groups, and "heptane" indicates that the longest continuous chain is heptane (i.e., it has seven carbon atoms). At first glance this may seem to be a complicated name, but when considered in light of the rules, the name accurately represents the structural formula of the hydrocarbon.

Alkenes and Alkynes Because of the presence of double or triple bonds, the naming of alkenes and alkynes is somewhat more complicated than the naming of alkanes, but the rules are still relatively straightforward. Suffixes are used to designate double (*ene*) or triple (*yne*) bonds, and prefixes are used to locate the bonds. The initial phase of the naming process is similar to that for the alkanes. The rules are as follows:

1. Find the longest chain containing the double and/or triple bonds. The stem name is based on the number of carbon atoms, as is the case for the alkanes. A prefix number is used to indicate the position of the carbon atom immediately before the double or triple bond, and a suffix (*ene* or *yne*) is used to indicate whether the compound is an alkene or alkyne.
2. The chain is numbered so that the double/triple bond will have the lowest location number. If both types of bonds are present, the double bond takes precedence in establishing the start of the numbering sequence.
3. If more than one double bond is present, the location numbers are indicated first and the ending is given as *diene*, *triene*, etc. If a triple bond is also present, the double bonds are designated in the main stem, dropping the *e*, and the triple bond is designated by an extra suffix.
4. Unsaturated substituents are named by using the prefix of the chain after replacing the terminal *e* with *yl*. For example, propene becomes propenyl.

The following examples illustrate the application of these rules to the naming of alkenes and alkynes.

$$CH_3-CH_2-CH_2-CH=CH-CH_3$$
$$\quad 6 \quad\;\; 5 \quad\;\; 4 \quad\;\;\; 3 \quad\; 2 \quad\; 1$$

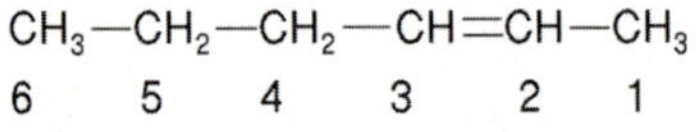

Figure 5–8
Condensed structural formula for 2-hexene.

EXAMPLE 5–1 Given the condensed structural formula shown in Figure 5–8, name the compound.

The double bond is located toward the right end of the chain, so we number from right to left. There is only one double bond and no triple bonds, so the suffix for the stem name is *ene*. The chain consists of six carbons, giving a stem name of hexene. The double bond starts with the second carbon atom, so the locator number is 2. The complete name of the compound is 2-hexene. If the compound contained a single triple bond, rather than a single double bond, the name would be 2-hexyne. ∎

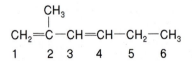

Figure 5–9
Condensed structural formula for 2-methyl-1,3-hexdiene.

EXAMPLE 5–2 Given the condensed structural formula shown in Figure 5–9, name the compound.

The first double bond is located toward the left end of the chain, so we number from left to right. There are two double bonds, so the suffix for the stem name is *diene*, and there are six carbons, giving a stem name of hexdiene. The double bonds begin with carbon atoms 1 and 3, the location numbers. There is one methyl group attached to carbon 2. The complete name of the compound is 2-methyl-1,3-hexdiene. ∎

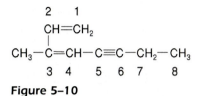

Figure 5–10
Condensed structural formula for 3-methyl-1,3-octdien-5-yne.

EXAMPLE 5–3 Given the condensed structural formula shown in Figure 5–10, name the compound.

We start numbering with the double bond closest to the end of the chain. There are two double bonds and one triple bond in the chain, and there are eight carbon atoms, so the stem name becomes 1,3-octdien-5-yne. Note that the *e* is dropped from *diene*. "Oct" indicates there are eight carbons, and "dien" indicates that there are two double bonds. The 1,3 indicates the positions of the double bonds, and the "5-yne" indicates the position of the triple bond (the fact that it's a triple bond is indicated by the "yne" suffix). The substituent group is a methyl group, and this group is attached to the third carbon atom. The complete name of the compound is 3-methyl-1,3-octdien-5-yne.   ■

Cyclic Hydrocarbons   In cyclic hydrocarbons the carbon atoms are arranged in a ring. The cyclic hydrocarbons do not form flat structures like aromatic hydrocarbons and are thus distinguished from them. For simple rings, without attached substituent groups, the prefix *cyclo* is used to distinguish a ring structure. For example, cyclohexane would consist of six singly bonded carbon atoms in a ring structure, cyclohexene would consist of six carbon atoms with one double bond in a ring structure, and cyclohexyne would consist of six carbon atoms with one triple bond in a ring structure. If substituent groups are attached, the carbon atoms are numbered so as to give the substituents the lowest possible location numbers. If a double bond is present, the numbering of the carbon atoms starts with the double bond.

Aromatic (Aryl) Hydrocarbons   The basic building block of the aromatic hydrocarbons is the benzene ring, which consists of six carbon atoms and six hydrogen atoms. It was originally thought that the ring consisted of alternating single and double bonds, but it was eventually realized that the ring has six identical bonds, neither single nor double. Modern representations of the benzene ring show it as a hexagon enclosing a circle, indicating that it consists of a ring of delocalized (π) electrons. Thus, one can find two different symbols (Figure 5–11) for the benzene ring in the literature. In the IUPAC system, the benzene-ring carbon atoms are numbered from 1 to 6, starting with one substituent and continuing either clockwise or counterclockwise so that the lowest number for each substituent is obtained. Benzene derivatives are also identified in the literature by specific names based on long usage or on the basis of the ortho-, meta-, and para-system. An example of a name based on long usage is *xylene*, which is used for a benzene ring with two attached methyl groups. In the ortho-, meta-, para-system, the positions are defined relative to a particular constituent with ortho- (*o*) one carbon atom removed from the constituent, meta- (*m*) two carbon atoms removed, and para- (*p*) three carbon atoms removed. An example of this naming convention, for a benzene ring with attached methyl groups, is shown in Figure 5–12. Aromatic hydrocarbons are characterized by their low H:C atomic ratio, relatively strong C-C bonds, a tendency to undergo substitution reactions, and the delocalization of π electrons over several carbon atoms, which adds substantial stability to the ring structure.

Benzene ring shown as alternating single and double bonds

Benzene ring shown as a ring of delocalized electrons

Figure 5–11
Representations of the benzene ring found in the literature.

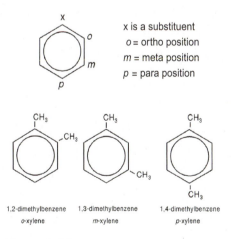

x is a substituent
o = ortho position
m = meta position
p = para position

1,2-dimethylbenzene
o-xylene

1,3-dimethylbenzene
m-xylene

1,4-dimethylbenzene
p-xylene

Figure 5–12
Illustration of the naming conventions for the benzene ring using attached methyl groups. Note that while not shown on the diagram, the other sites are occupied by hydrogen atoms.

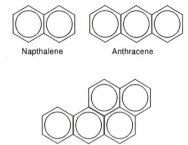

Napthalene   Anthracene

Benzo(a)pyrene

Figure 5–13
Examples of polycyclic aromatic hydrocarbons (PAHs) including the known carcinogen benzo(a)pyrene. Where the benzene rings do not share edges, hydrogen atoms are attached to the carbon atoms.

Polycyclic Aromatic Hydrocarbons (PAH)   This important group of compounds consists of two or more fused benzene rings, which form a planar structure (Figure 5–13). These compounds are formed by the incomplete combustion of other hydrocarbons. Because there are many partial combustion and pyrolysis processes that favor the production of PAHs, they are widely distributed in the environment. Some PAHs are precursors to cancer-causing metabolites and are of significant environmental interest. For example, benzo[a]pyrene is a particularly potent carcinogen.

Functional Groups

So far, we have considered the basic structures of organic molecules that consist only of carbon and hydrogen. However, organic molecules can contain other elements, such as oxygen, nitrogen, and the halogens (F, Cl, Br). These atoms occur in functional groups that are attached to the main carbon skeleton and lead to the great diversity in properties noted for the organic compounds. ***Functional groups*** *consist of specific bonding configurations of atoms in organic molecules and usually, but not always, contain at least one element other than carbon or hydrogen.* Note that two carbon atoms connected by either a double bond or a triple bond can also be considered a functional group. The functional groups of most interest in environmental geochemistry are carboxylic acid (COOH), hydroxyl (OH), aldehyde (COH), amine (NH_2), nitro compounds (NO, NO_2), sulfonic acids (SOOOH), and organohalides (Cl, F, Br). Examples of compounds containing these different types of functional groups are shown in Figure 5–14.

Organohalide Compounds

This particular group contains some of the most useful, most toxic, and most environmentally sensitive organic compounds. The chlorofluorocarbons (CFCs) that have achieved notoriety for their effect on stratospheric ozone are organohalides, as are DDT, PCB, and dioxins. The fundamental structures are the same as those described in the preceding sections on hydrocarbons except that each molecule contains at least one atom of F, Cl, Br, or I. Thus, we can have saturated, unsaturated, cyclic, and aromatic halides, each of which starts with the basic hydrocarbon structure. The naming is relatively straightforward in that the halogen atoms are listed as substituents using prefixes such as fluoro (for F), chloro (for Cl), bromo (for Br), and iodo (for I). As an example, consider the nam-

Figure 5–14
Examples of important functional groups. For each molecule the functional group is enclosed in a box and the name of the group is indicated by the arrow. The name of the compound is listed under each molecule.

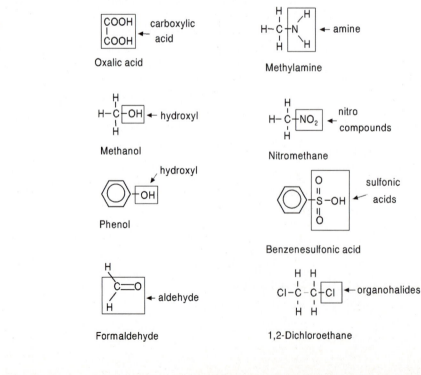

ing of the compound 1,2-dichloroethane (shown in Figure 5–14). This is a halogenated hydrocarbon because it contains two chlorine atoms that have replaced two hydrogen atoms. The basic hydrocarbon structure is a straight chain consisting of two carbon atoms; hence the root name is ethane. There are two substituent chlorine atoms, giving the prefix dichloro (di = two, chloro = Cl). One chlorine atom is attached to each carbon atom, Thus, they are in positions 1 and 2. We now have the complete name for the compound, 1,2-dichloroethane. If more than one type of halogen atom appears in the molecule, then multiple halogen prefixes are used. The numbers of each type of halogen atom are indicated by a prefix added to the halogen prefix (sometimes used, mono = 1, di = 2, tri = 3, tetra = 4, penta = 5, hexa = 6, etc.). For example, a compound having the formula CCl_2F_2 (this is a methane molecule in which all the hydrogen atoms have been replaced by halogen atoms) would be named dichlorodifluoromethane. This is a rather long name, but it does accurately describe the chemical formula. For unsaturated chains, naming is done in a similar way. For example, a compound having the condensed structural formula CH_2–CCl_2 would be named 1,1-dichloroethene. The root chain is ethene (two carbon atoms doubly bonded), and there are two chlorine atoms (dichloro) attached to the first carbon atom. A similar practice is followed for the aromatic hydrocarbons. For example, consider a benzene ring with two attached chlorine atoms, one in position 1 and the other in position 3. The name of this compound would be 1,3-dichlorobenzene. The root structure is the benzene ring, and the two chorine atoms (dichloro) are attached to carbon atoms 1 and 3. A listing of representative halogenated hydrocarbons is given in Table 5–2.

Polychlorinated biphenyls (PCBs) *consist of two benzene rings joined by a single bond with attached chlorine atoms.* Both benzene rings have five available carbon atoms that can bond to the substitutent chlorine atoms. For these molecules the root name is biphenyl and the positions of the substituents are indicated, for example, by 2 and 2′ (2 indicating the first ring and 2′ the second ring), etc. The standard prefixes are used to indicate the type and number of each of the substituent atoms. PCB compounds have very high chemical, thermal, and biological stability; low vapor pressures; and high dielectric constants. For this reason, they have wide industrial applications. However, because of their high stability, PCBs accumulate in the environment and subsequently in the fatty tissues of birds and fish. They are viewed as a significant environmental and health hazard and are no longer manufactured in the United States.

The *chlorinated phenols*, particularly pentachlorophenol and the trichlorophenol isomers, are significant hazardous wastes. These compounds have been widely used as wood preservatives. With reference to Figure 5–14, a phenol is a benzene ring that has an attached OH functional group. In pentachlorophenol all the hydrogens attached to the benzene ring have been replaced. At one site the OH functional group is attached to the benzene ring, and at the other five sites chlorine atoms are attached to the ring. In the case of trichlorophenol, three of the five hydrogen atoms have been replaced by chlorine atoms.

Table 5–2   Examples of Halogenated Hydrocarbons

Name	Formula	Name	Formula
Chloromethane	CH_3Cl	1,1,2,2-Tetrachloroethane	Cl_2CH–$CHCl_2$
Bromomethane	CH_3Br	1,1,1,2-Tetrachloroethane	Cl_3C–CH_2Cl
Dichloromethane	CH_2Cl_2	Pentachloroethane	Cl_3C–$CHCl_2$
Trichloromethane	$CHCl_3$	Hexachloroethane	Cl_3C–CCl_3
Tetrachloromethane	CCl_4	2-Chloropropane	CH_3–$CHCl$–CH_3
Bromodichloromethane	$CHCl_2Br$	1,2-Dibromo-3-chloropropane	$BrCH_2$–$BrCH$–CH_2Cl
Trichlorofluoromethane	CCl_3F	Chloroethene	CH_2=$CHCl$
Chlorodifluoromethane	$CHClF_2$	1,1-Dichloroethene	CH_2=CCl_2
Dichlorodifluoromethane	CCl_2F_2	Trichloroethene	$ClCH$=CCl_2
Chloroethane	CH_3–CH_2Cl	Tetrachloroethene	Cl_2C=CCl_2
1,2-Dichloroethane	$ClCH_2$–CH_2Cl	1-Chloropropene	$ClCH$=CH–CH_3
1,1,1-Trichloroethane	CH_3–CCl_3	1,3-Dichloropropene	$ClCH$=CH–CH_2Cl

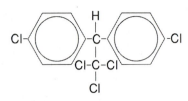

Figure 5–15
Structure of DDT. The molecule consists of two benzene rings joined by a chlorinated ethane group. Each ring has a Cl atom substituting for a H atom. The IUPAC name for this molecule would be dichlorodiphenyltrichloroethane.

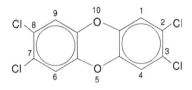

Figure 5–16
Structure of 2,3,7,8-tetrachlorodibenzo-*p*-dioxin (TCDD). The rings are numbered as shown. Note that the chlorine atoms occur in positions 2, 3, 7, and 8. The tetra prefix indicates that there are four chlorine atoms.

DDT (dichlorodiphenyltrichloroethane) is a halogenated insecticide that was once widely used. It is known as a "hard pesticide" because of its persistence in the environment. This persistence was seen as a plus because after application the pesticide remained active for a relatively long period of time and did not have to be reapplied. It also has a low toxicitiy in mammalian systems. However, as Carson (1962) so eloquently described in her book *Silent Spring*, the unforeseen hazards, due to biomagnification, to other biosystems were significant. DDT breaks down to DDD (dichlorodiphenyldichloroethane) and DDE (dichlorodiphenyldichloroethene). DDE has been shown to be hazardous to male animals and affects reproduction. Because of these problems, the use of DDT has been banned in the United States and many other countries. The structure of DDT is shown in Figure 5–15.

Dioxins are a byproduct of pesticide manufacture and are often found as residues in pesticides. They can also be produced by the incomplete combustion of PCBs during waste incineration. The dioxins have a high thermal stability and do not decompose until the temperature exceeds 700°C. Of this group of compounds the most hazardous is TCDD (Figure 5–16), often referred to simply as *dioxin*. For some animals, this compound is one of the most toxic of all synthetic substances. The basic structure of dioxin is two benzene rings connected by an oxygen-containing cyclical hydrocarbon. The benzene rings have from one to eight chlorine atoms substituting for hydrogen atoms, giving a total of 75 possible chlorinated derivatives.

Organooxygen Compounds

These compounds have an oxygen-containing functional group. Some of these chemicals are hazardous and/or toxic. *Alcohols* contain a hydroxyl functional group attached to an alkane. They are named by changing the *e* ending of the alkane to *ol*—hence, methanol (Figure 5–14). *Phenols* (aromatic alcohols) are benzene rings in which one hydrogen is replaced with an OH functional group (Figure 5–14). *Ketones* contain a carbonyl group (C=O) that occurs in the alkane molecule bonded to two other carbon atoms. In the IUPAC system, the ketones are named by changing the *e* ending of the alkane to *one*. However, because of common usage, older names are often retained. For example, propanone is usually referred to as acetone. *Carboxylic acids* are organic acids that contain a hydroxyl group attached to a carbonyl group (the carboxylic acid functional group −COOH). The acids of alkanes are named by replacing the *e* of the root alkane with *oic acid*. For example, propane with a substituent carboxylic acid group would be called *propanoic* acid. *Ethers* consist of two carbon groups connected by an oxygen. In traditional naming schemes, if the two groups are the same, then the group name is only used once with the word *ether* added. If the two groups are different, then both groups are named. For example, CH_3-O-CH_3 is dimethyl ether and $CH_3-O-CH_2-CH_3$ is ethyl methyl ether (the alkane on the left of the oxygen is a methyl group and the alkane on the right of the oxygen is an ethyl group). In the IUPAC system, the naming of ethers is different. One group (the smaller if the groups are of different size) is combined with the oxygen to form a substituent, and the other group becomes the root name of the compound. An *oxy* ending is added to the group serving as the substituent. For example, methane becomes methoxy and phenol becomes phenoxy. For the examples given, in the IUPAC system methyl ether ≡ methoxymethane and ethyl methyl ether ≡ methoxyethane.

Organonitrogen Compounds

Nitrogen occurs in a number of functional groups in organic compounds. Nitrogen can form single, double, or triple covalent bonds, and nitrogen functional groups are found as substituents on alkanes, alkenes, alkynes, and aromatic hydrocarbons. Two important classes of organonitrogen compounds are the amino acids and amines. *Amino acids* are characterized by the presence of an amino group (NH_2, Figure 5–14). *Amines* can be considered as derivatives of ammonia (NH_3), in which one or more of the hydrogen atoms are

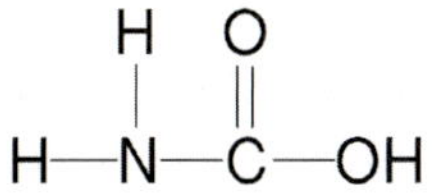

H O
│ ‖
H—N—C—OH

Figure 5–17
Carbamic acid.

replaced by alkyl groups (functional groups derived from alkanes are referred to as *alkyl* groups) or an aromatic ring. The naming of the organonitrogen compounds will not be discussed here, but a few examples will be given. *TNT* (2,4,6-trinitrotoluene) consists of a benzene ring with one attached methyl group (this is usually called *toluene* from long usage) and three NO_2 functional groups. *Carbamates* are extensively used as pesticides and herbicides. These are complex organic molecules derived from carbamic acid (Figure 5–17). The resulting carbamate is combined with methyl and other functional groups, and aromatic or heterocyclic compounds, to produce a wide variety of pesticides and herbicides. The carbamates are referred to as *soft pesticides* because they quickly break down in the natural environment. They are, however, neurotoxins and at high concentrations are toxic to mammals. *Aminocarboxylic acids* ($-CH_2CO_2H$ groups bonded to nitrogen atoms) form a variety of organonitrogen compounds that are chelating agents. An example is the monohydrate of trisodium nitrilotriacetate (NTA) that is used as a substitute for phosphate detergent and in metal plating formulations. When released to the environment, this chelating agent can mobilize heavy metals.

Organosulfur Compounds

Sulfur can occur in a number of oxidation states (-2, $+4$, $+6$), leading to a variety of organic sulfur compounds. There are three major groups of organosulfur compounds: (1) compounds that contain a functional group in which oxygen is replaced by sulfur, (2) compounds that contain a functional group consisting of sulfur and oxygen (Figure 5–14), and (3) sulfur heterocyclics. Examples of the various types of sulfur compounds are shown in Figure 5–18. *Thiols* (called *mercaptans* in the older literature) are analogous to alcohols in which the oxygen atom has been replaced by sulfur (R–SH, R is the root molecule). *Sulfides* contain a single sulfur atom that connects hydrocarbon groups (e.g., dimethyl sulfide), and *disulfides* contain two sulfur atoms joined by a single bond (e.g., dimethyl disulfide). *Thiourea* compounds contain both sulfur and nitrogen. One example is 1-napthylthiourea (ANTU), which is an excellent rodenticide. *Sulfoxides* and *sulfones* contain both sulfur and oxygen. Sulfoxides are analogous to ketones, in which the carbonyl carbon has been replaced by sulfur. In sulfones, sulfur has six bonds, four with oxygen and two with carbon. *Sulfonic acids and salts* are organic derivatives of sulfurous acid (H_2SO_3), and *sulfates* are organic derivatives of inorganic sulfuric acid (H_2SO_4).

Figure 5–18
Examples of organosulfates. Thiols—methanethiol and benzenethiol, sulfides—dimethyl sulfide and dimethyl disulfide, thiourea—thiourea, sulfoxides—dimethylsulfoxide, sulfonic acids—benzenesulfonic acid, sulfates—methylsulfuric acid.

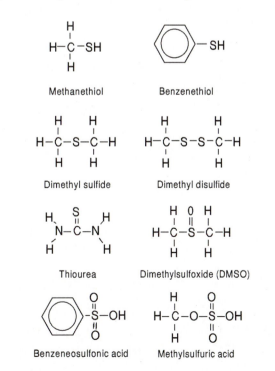

Organophosphorus Compounds

Phosphorus can occur in organic compounds in either the trivalent (3+) or pentavalent (5+) oxidation state. Compounds containing trivalent phosphorus are indicated by the suffix *ous* (acid) or *ite*; compounds containing pentavalent phosphorus are indicated by the suffix *ic* (acid) or *ate*. Organophosphorus compounds can also be distinguished on the basis of their linkages, i.e., number of P–C bonds relative to the number of P–O–C linkages. Compounds that contain only P–C linkages can be viewed as having been derived from phosphine (PH_3) or phosphane (PH_5). Note the *ine* and *ane* suffixes, indicating the trivalent and pentavalent oxidation states, respectively. In the naming of organophosphorus compounds, the prefix (*phosph*) indicates that the compound contains phosphorus atoms, the center part of the name indicates the number of hydroxyl groups ("in" = 1, "on" =, 2 and "or" = 3), and the suffix indicates the oxidation state of the phosphorus. Many of the important nerve gases and insecticides belong to this class of organic compounds. Examples of various types of organophosphorus compounds are shown in Figure 5–19. Phosphinous acid contains phosphorus in the trivalent state and has one hydroxyl group. Parsing the name, "phosph" indicates a phosphorus compound, "in" indicates one hydroxyl group, and "ous" indicates a trivalent phosphorus atom. Phosphinic acid has an "ic" suffix, indicating that the phosphorus is pentavalent. Dimethylphosphinous acid has two methyl groups ("dimethyl"), one hydroxyl group ("in"), and phosphorus in the trivalent state ("ous"). Trimethyl phosphate has three methyl groups ("trimethyl"), and no hydroxyl groups, and the phosphorus is in the pentavalent oxidation state ("ate").

Esters

Esters are a special group of compounds formed by a condensation reaction involving carboxylic acids and alcohols. Esters are named by listing the alcohol portion with a *yl* ending and, separated by a space, the acid portion with an *oate* ending. Consider the reaction shown in Figure 5–20 in which methanol (an alcohol) and butanoic acid (a carboxylic acid) are combined. The resulting compound is methyl butanoate (methyl is the alcohol

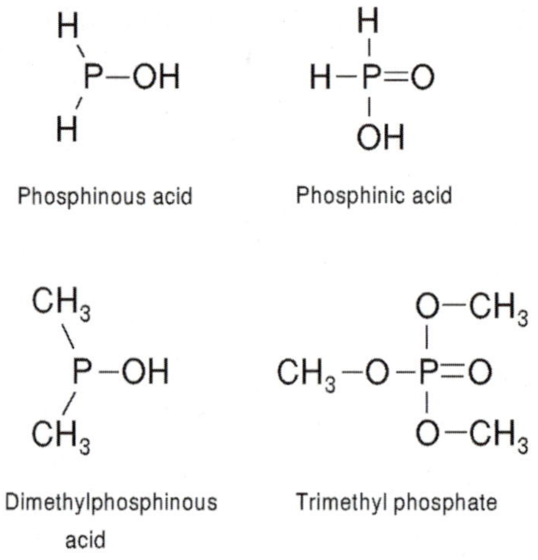

Figure 5–19
Examples of organophosphorus compounds. See text for description of naming conventions.

Figure 5–20
Condensation reaction between an alcohol (methanol) and a carboxylic acid (butanoic acid) to form an ester (methyl butanoate). Note that water is expelled during this reaction. The ester functional group is enclosed by the box.

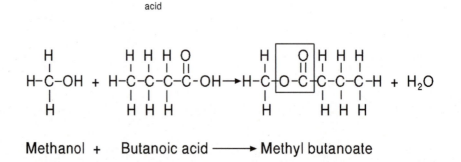

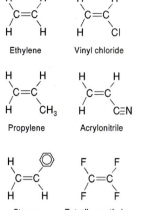

Figure 5–21
Examples of monomers used to build the common synthetic polymers. See text for examples of the uses of each synthetic polymer.

portion of the ester and butanoate is the carboxylic acid portion of the ester). In the condensation reaction, H_2O is expelled.

Polymers

The last group of organic compounds that are of environmental interest are the polymers. Synthetic polymers have a variety of uses and are widely dispersed in the environment. Many are resistant to degradation, and these materials have become an important component of landfills. During combustion, some of these polymers can release toxic gases. *Polymers* are *high-molecular-weight compounds that are composed of a large number of simple repeating units (called monomers)*. The conversion process from monomer to polymer is known as *polymerization*. The monomers are linked together to form straight or branching chains. The polymer may contain only a single monomer (homopolymer) or two or more different monomers (copolymers). Some of the common monomers are shown in Figure 5–21. Typical polymers formed from these monomers are polyvinylchloride (water and sewer pipes), polyethylene (plastic bags and milk bottles), polypropylene (impact-resistant plastics), polyacrylonitrile (carpets), polystyrene (foam insulation), and polytetrafluoroethylene (Teflon coatings).

CARBON COMPOUNDS IN THE ENVIRONMENT

Carbon-based compounds, both of natural and anthropogenic origin, are widely dispersed in the environment. All known life is carbon based and, hence, carbon is the most important element in the biosphere. In Chapter 8 we will consider the impact of carbon-based compounds—notably, CO_2, CH_4, and chlorofluorocarbons—on the atmosphere. In this chapter we will emphasize the role of carbon in the hydrosphere and geosphere, and we will look at the interactions between synthetic carbon-based compounds and the biosphere, atmosphere, hydrosphere, and geosphere.

Natural Sources

The ultimate source of all carbon at the earth's surface is the mantle of the earth. This carbon (usually referred to as *primordial carbon*) was released to the atmosphere during volcanic processes that led to the partial degassing of the mantle. This release of carbon is the initial step in the carbon cycle (Figure 5–22). Atmospheric carbon (in the form of CO_2) is

Figure 5–22
Major carbon reservoirs and interactions between these reservoirs. Carbon concentrations are in grams of carbon. Data from Schidlowski (1988).

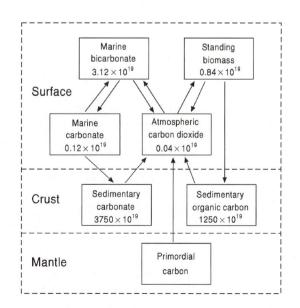

removed either by photosynthetic processes or by dissolution in fresh and marine (by far the most important) waters. Photosynthesis is often represented by the following simplified reaction:

$$6CO_2 + 6H_2O \xrightarrow[\text{sunlight}]{\text{chlorophyll}} C_6H_{12}O_6 + 6O_2$$

where $C_6H_{12}O_6$ is a sugar molecule. During cellular respiration, this process is reversed and sugars and oxygen are consumed while CO_2 and water are released. The production of biomass via photosynthesis is referred to as *primary* production, and photosynthetic organisms are primary producers. Note that a minor source of primary production is the sulfur-reducing bacteria. The biomass produced by the primary producers is subsequently consumed by higher-trophic-level organisms (consumers). The cycling of organic matter in the surface environment is complex and will not be considered in detail here. However, the excretions of organisms and the breakdown of biomass through decay and decomposition releases carbon to the surficial environment (i.e., soil, streams, groundwater). Much of this carbon is recycled to produce new organic matter, but some of it is transferred to the sedimentary environment. The other major reservoir for surficial carbon is the ocean. Given the pH (~8.2) of ocean water, the bulk of the dissolved carbon occurs as bicarbonate ion. Carbon is ultimately removed from the ocean as carbonate shell material and is preserved in the carbonate sediments.

Killops and Killops (1993) suggest that a useful way to look at the cycling of organic carbon is to define two subcycles: geochemical and biochemical. The geochemical subcycle contains the bulk of the carbon, 99.95%, and this carbon is found in the sedimentary rocks both as organic carbon (peat, coal, petroleum) and as sedimentary carbonate. This cycle operates on a time scale of millions of years. The biochemical subcycle consists of DOC (dissolved organic carbon), POC (particulate organic carbon), biota biomass, and soil organics and operates on a time scale of tens of years. These two cycles are linked by a two-way flux, which represents the transfer of organic carbon between sedimentary rocks and the surface environment. It would normally be expected that this transfer would approximate a steady state, but given the current significant use of fossil fuel, the cycle is not in balance. More carbon is currently transferred to the surficial environment than is returned to the sedimentary rocks.

In these various cycles, organic carbon is released to the environment. For example, the excretion of urea by organisms contributes organonitrogen compounds to streams and lakes. This is a natural addition (nonanthropogenic). However, if this release is due to large concentrations of animals being raised as livestock, then the release is often considered anthropogenic. Similarly, petroleum released to the environment by natural oil seeps is not considered anthropogenic, but petroleum released during an oil spill *is* considered anthropogenic. PAHs are considered to be a significant air pollutant produced during the incomplete combustion of fossil fuels. They are generated under conditions of local oxygen deficiency and can be viewed as pyrolysis products. Interestingly, PAHs are found in the sedimentary record and are apparently the result of wildfires initiated by lightning strikes. They can also be caused by igneous intrusions that locally heat coal and wood to sufficiently high temperatures to cause incomplete combustion. Chemically, there is no difference between these types of releases and anthropogenic releases.

Anthropogenic Sources

A vast number of organic chemicals have been synthesized for various uses. Among these are a wide variety of compounds used as pesticides, insecticides, and herbicides. Many of these have a significant environmental impact. Other organic compounds have various uses in industrial processes. Because most of these organic compounds have not previously existed in the natural environment, their impact can be significant. In addition, there are releases of organic compounds that are naturally present, but in amounts far above those normally encountered (e.g., oil spills, leakage from underground storage tanks, etc.). In terms of environmental impact, anthropogenic releases of organic compounds are typically the most serious and require the greatest expenditure of effort for cleanup and remediation.

HUMIC SUBSTANCES

Humic substances can be defined as *"a general category of naturally occurring, biogenic, heterogeneous organic substances that can generally be characterized as being yellow to black in color, of high molecular weight, and refractory"* (Aiken et al., 1985). Humic substances are found in soils, brown coal, fresh and marine waters, and marine and lacustrine sediments. There are three types of humic substances: humin, humic acid, and fulvic acid (Figure 5–23). *Humin* is *that fraction of the humic material that is insoluble in water at all pH values.* *Humic acid* is *soluble in water at pH values greater than 2,* and *fulvic acid* is *soluble in water at all pH values.* Humic and fulvic acids are the major components of both freshwater and marine DOC, although there are significant differences in the types of humic substances that occur in these two environments.

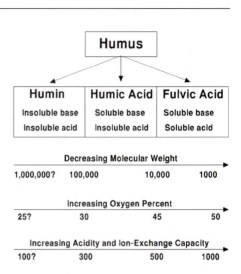

Figure 5–23

Types of humic substances and general characteristics. Note that the distinction between the types of substances is based on their solubility in acids and bases. After Oades (1989).

Attempts to structurally characterize humic substances have generally proven unsuccessful. However, compositional information can be obtained from elemental and functional group analysis. Chemical compositions for soil humic and fulvic acids and characteristic elemental ratios are given in Table 5–3.

The character of the humic substances varies as a function of the chemical environment, i.e., soil vs. freshwater vs. marine. Some of these differences are summarized in Table 5–4 (p. 142). Of note is the higher aromatic content of soil humic substances compared to marine humic substances, which suggests that humic substances from these two environments may be differentiated on the basis of their aromaticity. The H/C atomic ratio is of significance in that it is an indirect indicator of the relative proportion of aliphatic to aromatic compounds. An increase in aromatic compounds is marked by a decrease in the H/C ratio. Also of note is the lower molecular weight of freshwater humic substances compared to soil humic substances.

Table 5–3   Elemental Compositions, Atomic Ratios, and Atomic Weights for Soil Humic and Fulvic Acids[1]

	Humic acids	Fulvic acids
Carbon (wt%)	53.6–58.7	40.7–50.6
Hydrogen (wt%)	3.2–6.2	3.8–7.0
Nitrogen (wt%)	0.8–5.5	0.9–3.3
Oxygen (wt%)	32.8–38.3	39.7–49.8
Sulfur (wt%)	0.1–1.5	0.1–3.6
Mol wt (daltons[2])	2000–5000	500–2000
H/C (atomic)	~0.8	~1.3
O/C (atomic)	~0.5	~0.8

[1]Sources of data: Schnitzer (1978), Thurman (1985a), Killops and Killops (1993).
[2]The dalton is a unit of molecular weight. The unit is defined so that the weight of ^{12}C = 12.000 daltons.

Table 5–4 *Composition of Humic Substances in Various Environments**

	Soil	Groundwater	Surface water	Lake sediments	Seawater	Marine sediments
Aromatic hydrocarbons	20–35%	—	17–30%	< 15%	< 15%	< 15%
H/C	0.5–1.0	0.7–1.2	0.7–1.1	1.0–1.6	1.6	1.0–1.5
Molecular weights	10^3–10^6	500–10^4	< 10^4	10^3–10^6	< 10^3	—

*Sources of data: Hatcher et al. (1980), Harvey and Boran (1985), Ishiwatari (1985), Killops and Killops (1993), Malcolm (1985), Thurman (1985b).

Humic and fulvic acids are macromolecules that consist of flexible extended chains with only limited branching and cross-linking. These chains form the backbone of the macromolecule to which smaller molecules are attached. Humic substances form stable complexes with polyvalent (metal) cations. These complexes are important in the mobilization and transport of metal ions. This ability to form complexes with metal ions is due to the high abundance of oxygen-containing functional groups. Infrared spectroscopy studies have shown that COOH groups play a predominant role in the complexing of metal ions by fulvic and humic acids. There is some evidence to suggest that OH, C=O, and NH_2 groups may also play a role. In general, the maximum amount of any metal ion that can be complexed by humic or fulvic acid is approximately equal to the content of acidic functional groups (in particular, COOH). For most humic substances, the amount of metal cation that can be complexed falls in the range 1.5 to 5.0 meq g^{-1} of humic substance. Factors that affect the actual amount of metal ion bound to humic substances are the pH, ionic strength of the solution, and the functional group content. At any particular pH and ionic strength, the trivalent cations are bound in greater amounts than the divalent cations. For any particular metal cation, stability constants tend to increase with pH and then at some point begin to decrease with increasing pH. Increasing ionic strength of the solution leads to a decrease in the stability constants for all metal-humic complexes. Case Studies 5–1 and 5–2 explore the relationships between humic substances and the mobility of various trace metals.

CASE STUDY 5–1
Chemical and Physical Characteristics of Heavy Metals in Humus in the Vicinity of the Base Metal Smelter at Flin Flon, Manitoba, Canada

Henderson et al. (1998) investigated the distribution of heavy metals in humus from the area surrounding the Cu-Zn smelter at Flin Flon. Concentrations of a variety of trace metals were found to diminish with distance from the smelter, and background values were reached at 70–104 km from the smelter. For example, Zn samples collected within 5 km of the smelter have concentrations 94X background, 40X background at 10 km, 16X background at 20 km, 5X background at 40 km, and 1.6X background at 80 km, a logarithmic variation with distance. Sequential extraction analysis was used to determine the occurrence of the trace metals in the humus: soluble organics (humic and fulvic acid complexes), labile phases in the insoluble residue (humin), and nonlabile phases in the insoluble residue (mineral matter). For the smelter-derived elements, Zn, Pb, Cd, and Cu were found to correlate with the soluble organic phase,

and As and Hg were correlated with the nonlabile insoluble residue. In the insoluble residue, the concentration of Zn increased in the labile fraction with distance from the smelter. This was ascribed to a decrease in the particulate component with increasing distance from the smelter and secondary processes, such as weathering and alteration, becoming relatively more important. In the case of Hg, the nonlabile component of the insoluble residue contained greater than 95% of the total Hg, indicating that this element largely occurs with mineral components and is relatively insoluble in the natural environment. Investigation of a soil horizon revealed that more than 70% of the Zn was found in the soluble organic and labile fractions, indicating that this element would be readily mobile in the soil horizon, but only 16% of the Hg was found in these fractions. Thus, the association of Zn with humic substances leads to a high mobility for this element in the natural environment.

Source: Henderson et al. (1998).

SOIL AND MARINE ORGANIC MATTER

In this section we will briefly consider some of the fundamental differences between organic matter found in soils and organic matter found in marine sediments. A more complete discussion of this topic can be found in Hedges and Oades (1997).

Soil Organic Matter

Soil can be defined as *"a complex system of air, water, decomposing organic matter, living plants, and animals, in addition to the residues of rock weathering, organized into definite structural patterns as dictated by the environmental conditions"* (Loughnan, 1969). Most soils exhibit well-developed soil horizons, as illustrated in Figure 5–24 for a spodosol. This type of soil forms in response to strong leaching and acidity and is typical of soils found in temperate regions. The O horizon is rich in accumulated plant material. The underlying A1 horizon contains organic matter, and the A2 horizon is a leached zone. The B1 horizon is a zone of net accumulation of organic material; the B2 horizon is a zone of net accumulation of metal oxyhydroxides. The B3 horizon is a transition zone to the C horizon, which consists of weathered rock unaffected by the processes of soil formation.

The major determining factor in soil formation is climate, and the two most important variables are temperature and precipitation. Although the bedrock that is weathered to form the soil is of secondary importance, it should be noted that lithology does exert a fundamental influence; i.e., soils produced from the weathering of quartz sandstone cannot be rich in calcium, and those produced from the weathering of limestone cannot be rich in silica. A simple classification of soils as a function of climate is shown in Figure 5–25. With reference to Figure 5–25, *pedalfers* are *soils that are high in Al and Fe and are found in humid regions*, *pedocals* are *soils that*

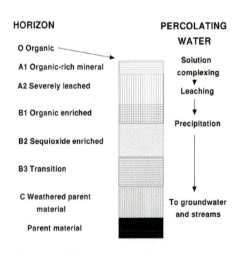

Figure 5–24
Generalized soil profile for a spodosol. After Stevenson (1985). From HUMUS CHEMISTRY, GENESIS, COMPOSITION, REACTIONS by Stevenson. Copyright © 1994. This material is used by permission of John Wiley & Sons, Inc.

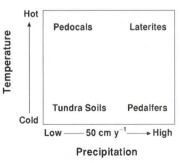
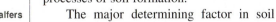

Figure 5–25
Soil types as a function of the climate variables temperature and precipitation.

are high in Ca and are found in deserts or semiarid regions, **laterites** *are* *highly leached* *soils that largely consist of iron and aluminum oxides and are found in the tropics,* and **tundra soils** are *found in cold, dry climates.* Table 5–5 gives a modern classification of soil types.

Table 5–5   Soil Classification*

Soil type	Characteristics and climatic conditions
Oxisols	Extensive alteration of silicate minerals to form clays, such as kaolinite, and iron oxides; wet equatorial regions
Calcisols	Precipitation of calcium carbonate as nodules and layers; hot, dry regions
Gypsisols	Calcium sulfate forms in soil as gypsum or anhydrite; very arid regions
Gleysols	Waterlogged soils with poor drainage and reducing conditions; temperate and polar regions
Argillisols	Layers and grain coating of clay are common; wet midlatitude regions
Spodosols	Grain coatings of organic material and iron oxides; wet midlatitude regions
Vertisols	Poorly developed layering but strong vertical structures due to repeated desiccation of expandable clays; relatively dry regions
Histisols	High concentrations of organic material forming peat layers; moist temperate regions and occasionally humid equatorial regions
Protosols	Poorly developed soils; most common in polar regions but can occur anywhere

*Modified from Nichols (1999).

Soils are a dynamic mixture of mineral particles, organic material, organisms, water, and air. Porosity varies from 30 to 60%, with 50% porosity being an average value. The pore spaces are filled with either water or air and the water/air ratio is continuously changing. Temperature changes can be significant on a daily or monthly basis. pH values range from 4.0 to 8.5. The major input of organic matter to soils is litterfall or root invasion by vascular plants. These sources have high C/N ratios (wood, 175–400; tree leaves, 20–50; grasses and herbaceous plants, 25–80) (Hedges and Oades, 1997). Herbivores and detritivores feed on the plant parts. The total microbial biomass accounts for 2–3% of the soil organic matter. This microbial biomass is composed of bacteria, actinomycetes, and fungi. These organisms are nitrogen-rich, with C/N ratios ranging from about 5 (bacteria dominant) to 15 (fungi dominant). Thus, two distinct types of organic matter are found at the earth's surface, organic matter associated with vascular plants, which has high C/N ratios, and organic matter associated with microbial biomass, which has low C/N ratios (see Figure 5–26).

Marine Organic Matter

The marine sedimentary environment differs from the soil environment in a number of important ways. Marine sediments tend to have greater porosity than soils, with the porosity of fine-grained sediments ranging from 70 to 90% at the surface and from 50 to 60% at depths on the order of a meter or more. The pH of the pore waters varies between 7 and 8, a much narrower range than that shown by soils. Surface layers are oxic, but with increasing depth the pore waters become anoxic.

Vascular plants are essentially absent from the ocean, and the major source of organic material is phytoplankton. C/N ratios in phytoplankton are around 6. Bacteria are the primary decomposers. In terms of carbon isotopes (see Chapter 6), marine plankton tend to be isotopically heavier than either vascular plants or soil organic matter. Thus, it is possible to distinguish the sources of organic carbon on the basis of their C/N ratios and $\delta^{13}C$ values (Figure 5–26). Case Study 5–3 describes how various chemical and isotopic signatures can be used to determine the sources of organic matter in a freshwater system.

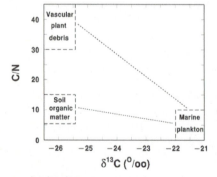

Figure 5–26
Atomic C/N versus $\delta^{13}C$ for various reservoirs of organic carbon. Mixtures of the three end members will lie within the field defined by the dashed lines. Modified from Hedges and Oades (1997).

CASE STUDY 5–3
Using Sedimentary Organic Matter to Record Environmental Changes in a River–Lake Ecosystem

Lake George is located in the St. Marys River system, which flows from Lake Superior to Lake Huron. The lake acts as a trap for sediments transported by the river. Previous work showed that sediments accumulated in Lake George at a relatively high rate of 0.5 cm y^{-1} and that the sediments were not disturbed by mixing. The region around the St. Marys River was clear-cut for lumber in the late 1800s and industrialization began in the 1900s with the construction of paper mills, steel mills, and tanneries. In 1986, a 64-cm sediment core was collected by a SCUBA diver from the northern end of the lake. This core was used by Tenzer et al. (1999) to document the impact of anthropogenic activities on the watershed of the St. Marys River.

The sediment core was divided into 1-cm sections for the top 20 cm and into 2-cm sections for the lower part of the core. The chronology of the layers was determined by ^{210}Pb geochronology and confirmed by comparing the calculated age for the bomb fallout maximum with the ^{137}Cs peak in the core (see Chapter 6). The geochronology indicated that the deposition rate decreased from 0.265 g cm^{-2} y^{-1} for the top part of the core to 0.165 g cm^{-2} y^{-1} for pre-1950 deposition.

The authors determined total organic carbon (TOC), total nitrogen (TN), δ^{13}C, the concentrations of total C^{16}-C^{36} n-alkanes (THC, total extractable hydrocarbons), and several other parameters. The results of some of the analyses are shown graphically in Figure 5–C3–1. Starting from the present, TOC increases until about 1970 and then begins to decrease. The biggest change occurs around 1925, and from about 1920 on TOC is uniform and relatively low. The C/N ratio shows a similar pattern, with C/N maxima roughly corresponding to TOC maxima. Similar to TOC, C/N shows low and relatively constant ratios prior to 1920. δ^{13}C values become less negative with time, and the maximum negative value corresponds with the TOC and C/N maxima. Typical Great Lakes plankton have a C/N ratio of about 7 compared to C/N ratios from 17 to 42 for leaves and needles of vascular plants from the St. Marys water-

shed. Vascular plants had δ^{13}C values between −25.1‰ and −29.8‰; organic material produced by algae were inferred to have δ^{13}C = −22‰. The increased C/N ratios and more negative δ^{13}C values were interpreted to reflect the increased contribution of land-derived organic matter that accompanied industrial development along the St. Marys River. The decrease in C/N ratios and increase in δ^{13}C after 1970 reflect a decrease in anthropogenic inputs following the United States–Canada Great Lakes Water Quality Agreement of 1972. Assuming end-member δ^{13}C values of −22‰ for pristine (algae) organic matter and −28‰ for average vascular plant material, and average δ^{13}C = −24.6‰ for sediments deposited around 1970, gives an approximately 40% greater contribution of land plant material to the sediments relative to 1900.

Hydrocarbons are particularly useful in determining the source of organic carbon. Aquatic algae and photosynthetic bacteria are dominated by C$_{17}$ n-alkanes; the waxy coatings of vascular plants are dominated by C$_{27}$ to C$_{31}$ n-alkanes. The terrigenous/aquatic ratio, (TAR$_{HC}$) = (C$_{27}$ + C$_{29}$ + C$_{31}$)/(C$_{15}$ + C$_{17}$ + C$_{21}$), and variations in this ratio are interpreted to indicate changes in the land-derived component of the organic matter. Petroleum hydrocarbons can be distinguished from biological hydrocarbons on the basis of two characteristics: (1) odd-carbon chain lengths are not dominant in petroleum hydrocarbons (biogenic n-alkanes have odd/even ratios = 5 to 10, while petroleum n-alkanes have odd/even ratios ~ 1), and (2) a more diverse range of molecular structures results in an *unresolved complex mixture* (UCM) that cannot be separated by high-resolution gas chromatography. During the 1900s, the change with time of TAR$_{HC}$ suggests that vascular plants were a less important component of the organic carbon, an observation at variance with other lines of evidence. The authors suggest that this is because the short-chain hydrocarbons were better preserved in the younger sediments. The odd/even ratio = 1 and UCM is at a maximum around 1970, indicating a significant input of petroleum-derived hydrocarbons. Changes in these values in more recent sediments indicate a reduction in anthropogenic inputs.

Source: Tenzer et al. (1999).

Figure 5–C3–1
Organic carbon concentrations (C$_{org}$), organic matter C/N ratios, δ^{13}C$_{org}$, total extractable hydrocarbons (THC), concentration ratios of terrigenous/aquatic n-alkanes (TAR$_{HC}$), odd/even n-alkanes and unresolved complex mixture (UCM) relative to sediment age. Shaded area represents time of greatest environmental change as determined from the sediment record. Modified from Tenzer et al. (1999).

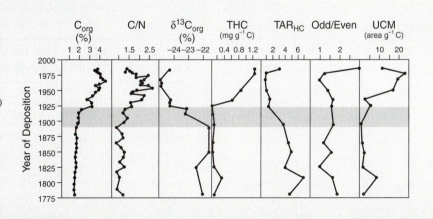

FOSSIL FUELS

Coals

Coals are classified as **humic coals** if they are *derived from humic substances via a peat stage* and as **sapropelic coals** if they are *formed from fairly fine-grained organic muds in quiet, oxygen-deficient shallow waters.* The humic coals are lustrous, dark-brown to black, exhibit stratification, and often contain the remains of woody tissue. The sapropelic coals are dull, unstratified, and contain allochthonous organic and mineral matter. The sapropelic coals are subdivided into cannel and boghead coals. The conversion of organic matter to coal involves two stages: biochemical and geochemical. The biochemical phase involves biological processes; the geochemical stage involves physicochemical processes due to increasing pressure and temperature. These processes lead to a release of volatiles, primarily methane, water, and carbon dioxide; the elimination of oxygen-containing functional groups; and an increase in the percent of aromatic units. These changes are illustrated in Figure 5–27, which shows the relative increase in carbon with increasing degree of coalification.

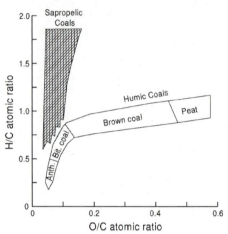

Figure 5–27
van Krevelen (1963) diagram indicating the changes in hydrogen, oxygen, and carbon content during the coalification process. After Killops and Killops (1993).

Of major environmental interest is the metal, sulfur, and fly ash content of coals. Metals that were incorporated in the original organic material tend to remain during the coalification process. Thus, coal is often high in a number of trace elements of environmental concern. Lyons et al. (1989) determined the concentrations of trace elements in a variety of coals, ranging from high-volatile bituminous coals to meta-anthracites. Concentration ranges determined for elements of environmental interest are given in Table 5–6.

Lyons et al. (1989) also investigated the distribution of these trace elements between organic and inorganic (i.e., mineral) components of the coal. For the elements listed in Table 5–6, they concluded that As, Co, Cu, Ni, Sc, and Th generally showed an inorganic association, whereas Cr, Sb, U, V, W, and Zn generally showed an organic association. Se and Pb did not correlate with either association. Metals associated with inorganic components would presumably remain with the fly ash and could be successfully removed from the effluent stream by electrostatic precipitation of the fly ash. Metals associated with

Table 5–6 Concentration Ranges in Coal for Some Elements of Environmental Interest*

Element	Concentration range (ppm)	Element	Concentration range (ppm)
Sc	0.6–10.3	As	0.7–31
V	6–109	Se	0.4–3.3
Cr	2.6–25.4	Sb	0.1–7.3
Co	1.1–24.1	W	0.2–1.3
Ni	2–50	Pb	3–20
Cu	6–54	Th	0.5–3.7
Zn	3–65	U	0.2–3.8

*From Lyons et al. (1989).

CASE STUDY 5–4
Distribution of Metals and Other Elements in Coals

A number of studies have been carried out on the distribution of metals and other elements in coal. Several of these are cited in this case study to indicate the variability of these elemental associations in coal. Christanis et al. (1998) found that in the Philippi peat of eastern Macedonia, Greece, Ge, Mo, Pb, Se, Ta, Tl, U, and W were correlated with organic content and the other trace elements were correlated with mineral phases. For lignites from central Anatolia, Turkey, Querol et al. (1997) found that Mn, B, Be, Ge, Y, Zr, Nb, Hf, W, and U showed positive correlation with organic matter and Mn, U, Th, Sc, Ge, and HREE showed positive correlation with phosphate, suggesting that apatite or another phosphate mineral is the major host. Co, Mo, Ta, Pb, and lREE showed positive correlation with both Fe and S, indicating that Fe-sulfide minerals are the hosts for these elements. As and V had mixed affinities. For United Kingdom coal fields, Spears and Zheng (1999) found that Rb, Cr, Th, Ce, Zr, Y, Ga, La, Ta, Nb, and V were associated with the clay minerals; As, Mo, Sb, Tl, Se, Bi, and Pb with pyrite; Sr and Ba with phosphate minerals. Germanium was the only element that had a major organic association. For medium- to high-volatile bituminous coals from Nova Scotia, Canada, Mukhopadhyay et al. (1998) found that most lithophile elements, including B, Cr, U, V, and the REE, were associated with the clay minerals; Cu, Pb, and Zn with the sulfide minerals; and Cl, Ca, As, and Mn with organic matter. Of note is the marked variability of the element associations. Major factors affecting trace element concentrations are the abundance and sources of detrital minerals, the presence of volcanic material, and seawater or brine incursions into the basins of deposition.

organic components would be released to the effluent stream during combustion and dispersed into the atmosphere. These associations would be expected to vary on a coal-by-coal basis, so it is necessary to determine the distribution of elements in a particular coal in order to determine the potential trace metal contribution of this coal to the atmosphere during combustion (see Case Study 5–4).

Sulfur occurs in coal either combined with organic molecules or as physically separate sulfide minerals (of which pyrite is the most common). Sulfur occurs in reduced form (S^-, S^{2-}) because the organic-rich depositional environment is reducing (see Chapter 4). If Fe^{2+} (or other metal ions) is present, it will combine with the reduced sulfur to produce minerals such as marcasite (FeS) and pyrite (FeS_2). If there is a low abundance of metal ions, free hydrogen sulfide (H_2S) and polysulfides (HS_4^-, HS_5^-, S_4^{2-}, S_5^{2-}) are formed. These species can combine with various organic molecules to form a variety of sulfur-containing compounds. In terms of "cleaning up" coal, the way in which sulfur occurs in the coal is of key importance. If the sulfur occurs in separate sulfide minerals, then in principle it can be physically separated by flotation techniques. If the sulfur is bound to organic molecules, there is no effective way to remove the sulfur before burning the coal.

Petroleum

Petroleum is *any hydrocarbon-rich fluid (liquid or gas) derived from kerogen by increases in pressure and temperature.* *Kerogen* is *a polymeric organic material that occurs in sedimentary rocks in the form of finely disseminated organic macerals (the preserved remains of plant material).* Four types of kerogen (I, II, III, and IV) have been distinguished on the basis of H/C and O/C atomic ratios, types of organic molecules (aliphatics versus aromatics), functional groups, and other criteria. Type II kerogen is the most common precursor for petroleum. Type I has the highest petroleum-forming potential but is present in low abundance. Type IV kerogen is not a petroleum precursor. Kerogen is converted to petroleum through a series of reactions starting in the *diagenetic environment (the environment in which processes occur at pressures and temperatures greater than those of the weathering environment but below those required to produce metamorphism)* and ending at temperatures of about 225°C. These reactions result in the expulsion of CO_2 and CH_4 and a reduction in the size of the hydrocarbon molecules. The hydrocarbon formation process for a typical source rock is schematically illustrated in Figure 5–28 (p. 148). Note that natural gas is generated throughout the P–T range, but the formation of oil is confined to a narrow P–T range. Thus, the thermal history of a basin is an important factor in terms of its oil-producing capability. Also note the reduction in the size of the mean hydrocarbon molecule with increasing temperature.

Figure 5–28
Schematic representation of production of petroleum from a typical reservoir rock assuming an average geothermal gradient. Note the narrow P–T range over which oil is produced (called the *oil window*). With increasing temperature, the size of the mean hydrocarbon molecule decreases. During catagenesis, hydrocarbon chains break off and are expelled from the kerogen. During metagenesis, only methane is released. Modified from Killops and Killops (1993).

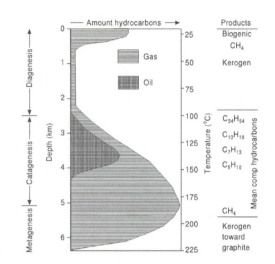

Table 5–7 Elemental Composition of Crude Oil*

Element	Abundance
	In wt %
C	82.2–87.1
H	11.8–14.7
S	0.1–5.5
O	0.1–4.5
N	0.1–4.5
Others	< 0.1
	In ppm
Ni	0.3–200
V	0.3–1000

*From Baker (1964), Hodgson (1954), Levorsen (1967).

The elemental composition of crude oil is given in Table 5–7. Nitrogen, sulfur, and oxygen are often associated with the resins and asphaltenes (fragments of polymeric material with molecular weight > 500) and are referred to as *polar NSO compounds*. Trace metals, such as nickel and vanadium, are also associated with polar NSO compounds. Note the range in sulfur and nitrogen concentrations in crude oil. These are important elements because the SO_2 and NO_x gases produced during combustion are precursors to the sulfuric and nitric acids of acid rain. *Crude oil that is low in sulfur* is referred to as **sweet**, and *crude oil that is high in sulfur* is referred to as **sour**. As might be anticipated, sweet crude is of lesser abundance, and hence more costly, so that reducing sulfur emissions through fuel substitution incurs additional cost. The *average oil* contains 57% aliphatic hydrocarbons, 29% aromatic hydrocarbons, and 14% resins and asphaltenes. Sulfur, as thiophenic compounds, accounts for 2% of the aromatic hydrocarbon fraction (Killops and Killops, 1993). Crude oils can be classified according to the relative amounts of acyclic (normal and branched) alkanes, cycloalkanes, and aromatic hydrocarbons plus NSO compounds (Tissot and Welte, 1984). The main classes of normal crude oils, according to the classification of Tissot and Welte (1984), are (1) paraffinic oils containing mainly acyclic alkanes and < 1% suflur, (2) paraffinic-naphthenic oils containing mainly acyclic alkanes and cycloalkanes and < 1% sulfur, and (3) aromatic-intermediate oils containing < 50% aromatic hydrocarbons and > 1% sulfur. Class 1 crude oil is derived from terrestrial organic matter, class 2 from terrestrial or marine organic matter, and class 3 from marine organic matter.

NATURAL CARBON INPUTS TO SURFACE AND GROUND WATERS

Carbon in water is generally reported as dissolved organic carbon (DOC), particulate organic carbon (POC), or total organic carbon (TOC). **DOC** is *the carbon remaining in a water sample after it has passed through a 0.45-μm filter.* By convention, any material that passes through a 0.45-μm filter is considered to be in solution. **POC** is *the particulate matter retained by a 0.45-μm filter.* **TOC** is the *total carbon content of the sample (DOC + POC).* All of these measurements are made by converting the organic matter to CO_2 and then measuring the amount of CO_2 produced. The distinction between DOC and POC is considered significant because carbon that occurs in the dissolved state is chemically more reactive than carbon that occurs as plant and animal organic matter and as organic coatings on silt and clay.

The sources of the carbon can either be **allochthonous** *(from outside the aquatic system)* or **autochthonous** *(from within the aquatic system)*. The two allochthonous sources are soil and plant material, and a major difference between these two sources is the much longer decomposition time for the soil organic material (> 100 years). Leaching experiments on fresh plant litter using distilled water have shown that a significant proportion (25–40%) of the plant organic matter can be solubilized in 24 hours. This material is similar to fulvic acid but differs from soil and aquatic fulvic acids in that it has a lower carboxylic acid content and a higher percentage of carbohydrates. A number of studies have shown that plant material is a major contributor to the DOC of streams, a conclusion largely based on the young ^{14}C age of the DOC. Because stream fulvic acid differs from plant leachates, it is generally assumed that rapid oxidation of the plant material in the upper soil layers results in an increase in carboxyl content and a decrease in carbohydrate content. Autochthonous organic carbon is largely derived from algae, either as direct excretions or by the bacterial and chemical breakdown of POC. The relative significance of these two types of inputs, allochthonous vs. autochthonous, depends on the size of the body of water. In streams and small lakes, allochthonous sources are dominant, whereas in large lakes, and notably the ocean, autochthonous sources become dominant.

The concentration of dissolved and particulate organic carbon varies with the aquatic environment. For example, both seawater and groundwater have low organic carbon concentrations (Figure 5–29) and insignificant particulate organic carbon. At the other extreme are marsh and bog waters, which have both high total organic carbon and significant particulate organic carbon (Figure 5–29). The primary source of seawater carbon is photosynthetic activity, which is confined to the top 200 meters of the ocean. Hence, surface waters have higher carbon content than deep waters. Coastal waters also have a higher carbon content due to land inputs. Groundwater has generally low carbon content because organic carbon is used as a food supply by heterotrophic microbes and converted to CO_2 (in aerobic groundwater) or CH_4 (in anaerobic groundwater) and is also adsorbed onto grain surfaces, where it is chemically and biochemically de-

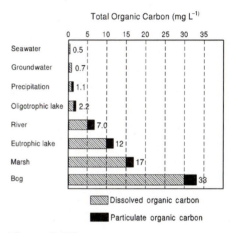

Figure 5–29
Average organic carbon concentrations for various aquatic systems. Note that the actual range in TOC can be substantial and depends on a variety of factors. See text. Taken from Thurman (1985a).

graded to CO_2. However, the carbon content of any particular aquifer can be very different from the average depending on the type of aquifer (Table 5–8). Marsh and bog waters have the highest organic carbon concentrations for the obvious reason that they are in contact with an organic-rich environment. Stream organic carbon content varies as a function of both the climate (which determines biological productivity) and the stream discharge. In general, the DOC content of a stream first increases during increased flow and then declines. The increase is due to the initial flushing of organic carbon from the soil into the

Table 5–8   Median Concentrations of Organic Carbon in Various Types of Aquifers*

Aquifer	DOC (mg C L^{-1})
Sand and gravel	0.7
Limestone	0.7
Sandstone	0.7
Igneous	0.5
Oil shale	3
Humic colored	10
Petroleum associated	100

*Data from Leenheer et al. (1974) and Feder and Lee (1981).

stream, and the decline reflects the subsequent dilution effect of the increased stream flow. Stream POC shows a logarithmic correlation with the sediment concentration. The difference in carbon content between oligotrophic and eutrophic lakes (Figure 5–29) is directly related to the much higher biological productivity of the eutrophic lake. Precipitation carbon content is, on average, 1.1 mg L^{-1}. However, precipitation samples collected within the canopy zone of a forest, or as drip from the canopy, have higher carbon contents. This is due to the washing of carbon from the leaves and limbs of the canopy. From this discussion it is clear that a number of factors control the concentration of organic carbon in aquatic systems. A more detailed discussion of this topic can be found in Thurman (1985a).

ANTHROPOGENIC CARBON INPUTS TO SURFACE AND GROUND WATERS

A tremendous variety of organic chemicals occur in surface and ground waters. In this section we will look at some of these inputs and their impact and remediation. This will not be an exhaustive discussion and the student should access the literature for a broader perspective.

Organic Carbon

The TOC of surface and ground waters is an important parameter, and at high concentrations organic carbon can be considered a pollutant. The oxidation of organic matter leads to the extraction of dissolved oxygen from the water. This process can be represented by the following simple reaction:

$$CH_2O_{(aq)} \text{ [carbohydrate]} + O_{2\,(aq)} \rightarrow CO_{2\,(g)} + H_2O_{(aq)}$$

If there is a significant amount of organic matter, the oxygen may be totally depleted in the water column, leading to anoxic conditions. This is what happens during eutrophication, when excessive production of organic matter leads to essentially total depletion of dissolved oxygen. As an example of eutrophication, consider a lake surrounded by homes, each having its own septic system. If phosphorus is present in the effluent (phosphate detergents are used by the homeowners), this phosphorus can enter the lake. For any biological system there is a limiting factor that controls productivity, and this limiting factor is often a nutrient element (N, K, P). If in the case of the lake phosphorus is the limiting element, the addition of the sewage effluent to the lake will result in increased production of plant material. The subsequent decay of this plant material leads to oxygen depletion and a reduction, or elimination, of fish and other oxygen users in the lake ecosystem. The lake "dies" from *overnourishment (eutrophication). The capacity of the organic matter in a sample of natural water to consume oxygen* is referred to as the **biological (or biochemical) oxygen demand (BOD)**. BOD is measured experimentally by holding a sealed water sample at 25°C for 5 days and measuring the change in oxygen content (mg L^{-1}). For unpolluted surface waters in the United States, the median BOD is 0.7 mg L^{-1} O$_2$. This is much less than the saturation O$_2$ value at 25°C of 8.7 mg L^{-1}. Under anaerobic (oxygen-free) conditions, organic matter will decompose if the appropriate bacteria are present. This decomposition can be represented by the following simple equation:

$$2CH_2O \text{ [organic matter]} \rightarrow CH_4 \text{ [methane]} + CO_2$$

This is a reducing environment and elements and compounds occur in their reduced state (see Chapter 4). Also note that methane gas is released, a common product of anaerobic decay of organic material both in natural waters and sanitary landfills.

Pesticides

Pesticides are used to control the growth of unwanted organisms. They can be divided into three types depending on their purpose: *insecticides*, which *control insects*; *herbicides*, which *control plant growth*; and **fungicides**, which *control the growth of various types of fungus*. Pesticides are often divided into two groups, *hard* and *soft*. **Hard pesticides** are *typically organochlorine compounds that are persistent in the environment but have a low toxicity in mammalian systems*. **Soft pesticides**, *typically organophosphorus compounds and carbamates, have a short residence time in the environment but are often highly toxic* (Table 5–9). Toxicity is usually expressed as **LD$_{50}$**, *the amount of the chemical per unit of body mass required to produce death in 50% of an exposed animal population*. This measurement is specific to a particular group of animals, often rats and guinea pigs, and there is some debate about the universal application of this measure to other animal populations. However, on a relative basis it can be used to rank the toxicity of various chemicals. The organochlorines have a low solubility in water (Table 5–9) but are much more soluble in organic solvents. This means that they are concentrated in the fatty tissue of animals. DDT is an example of an organochlorine insecticide. Because of its low mammalian toxicity and persistence in the environment, thus minimizing the number of repeat applications, it was considered an ideal pesticide. Many animals metabolize DDT by the elimination of HCl, forming a derivative called DDE (referred to as a *metabolite* of DDT). This metabolite is also persistent in the environment and can interfere with reproductive functions. In the case of some birds, this interference caused the birds to produce shells that were deficient in calcium carbonate and thus easily broken.

The tendency of a particular organic chemical in water to concentrate in the fatty tissue can be expressed as the bioconcentration factor (**BCF**). The BCF is based on the partitioning of a particular solute between water and fish. The BCF can be estimated using the following relationship (Kenaga and Goring, 1980):

$$\log \text{BCF} = 0.935 * \log K_{ow} - 1.495 \qquad (5–1)$$

where K_{ow} is the octanol/water partition coefficient, which is defined as follows:

$$K_{ow} = [S]_{octanol}/[S]_{water} \qquad (5–2)$$

where $[S]_{octanol}$ is the concentration of the chemical in the alcohol 1-octanol and $[S]_{water}$ is the concentration of the chemical in water. Because this is usually a large ratio, it is commonly reported as the log of the partition coefficient. K_{ow} values for some common pesticides are given in Table 5–9. The higher the K_{ow} value, the more strongly the chemical is partitioned into the fatty tissue. Chemicals with log K_{ow} values of 7 to 8 or greater tend to be strongly adsorbed to sediments and are thus unlikely to enter living tissue.

Because of their persistence and partitioning into fatty tissue, the concentration of organochlorine compounds increases as one moves higher in a food chain. This is called

Table 5–9   Data for Selected Pesticides[1]

Pesticide	Solubility in H$_2$O (mg kg^{-1})	LD$_{50}$ (mg kg^{-1})	log K_{ow}	log BCF[2]
Hard				
HCB	0.0062	3,500–10,000	5.3	3.5
DDT	0.0034	115	3.9–6.2	2.2–4.3
Toxaphene	n/a	85	2.9–3.3	1.2–1.6
Dieldrin	0.20	46	5.1–6.2	3.3–4.3
Mirex	0.20	700	5.8	3.9
Soft				
Malathion	145	1,375–2,800	2.7	1.0
Parathion	24	3.6–13	n/a	—
Atrazine	30	1,870–3,080	2.3	0.7

[1]From Baird (1995).
[2]BCF values calculated from the relationship of Kenaga and Goring (1980).

biomagnification. The efficiency of an ecological process is about 10%. In moving from one trophic level to another, only about 10% of the total biomass is transferred to the next trophic level, but effectively *all* of the pesticide is transferred to the next trophic level. As an example, consider the following simple coastal ocean food chain:

$$\text{Plankton} \rightarrow \text{Minnows} \rightarrow \text{Needlefish} \rightarrow \text{Cormorants}$$
$$(0.05) \qquad (0.5) \qquad (5) \qquad (50)$$

The values in parentheses, assuming 10% efficiency, are the DDE concentrations in ppm. Relative to the plankton, there is a 1000X increase in DDE in the cormorants. The water may have significantly lower DDE concentrations than the plankton, so the total biological magnification from water to cormorant may be on the order of millions of times.

Dioxins, PCBs, and Dibenzofurans (DFs)

These compounds are all chlorinated hydrocarbons that have been shown to be toxic in animal experiments. The basic structure of dioxin and PCB has been described in the section on nomenclature. The basic furan ring contains five atoms, one is oxygen and the other four are carbon. In dibenzofuran, benzene rings are attached to each side of the furan ring (Figure 5–30). There are 75 different varieties of dioxins, called *congeners—members of the same chemical family that differ only in the number and position of the same substituent*.

The toxicity of dioxin varies as a function of the number and location of the chlorine atoms. TCDD (Figure 5–16) is considered to be the most toxic dioxin. Note that in this molecule there are an intermediate number of Cl atoms and they are located at the positions farthest from the central ring. Dioxins with fewer or greater numbers of Cl, or with the Cl atoms located closer to the central ring, are found to be less toxic, and a fully chlorinated dioxin is virtually nontoxic. Dioxins are found as contaminants in herbicides and fungicides. Perhaps the most famous example of this contamination was "Agent Orange," used as a defoliant in Vietnam. Dioxin concentrations were around 10 ppm in Agent Orange, far above the values now found in herbicides (< 0.1 ppm). PCBs are chemically inert liquids, have low vapor pressures, high combustion temperatures, and are excellent electrical insulators. For this reason they were widely used as coolants for power transformers and capacitors. Other industrial uses were subsequently found. PCBs are virtually insoluble in water but very soluble in fatty or oily substances and thus tend to concentrate in fatty tissue. There are 209 congeners in the PCB family. Dibenzofurans are produced when PCBs are heated in the presence of oxygen. There are 135 dibenzofuran congeners. Furans with intermediate amounts of Cl have a toxicity similar to that of TCDD.

The relative toxicities of a number of dioxins and furans (compared to TCDD) are given in Table 5–10. Note the variation in toxicity with the numbers and positions of the Cl atoms. The *toxicity equivalence factor (TEQ)* for a mixture of dioxins and furans is *calculated on the basis of the relative abundance of each of the dioxins*.

Dibenzofuran

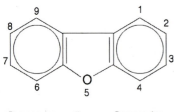

Benzene ring   Furan   Benzene ring

Figure 5–30
Basic structure of dibenzofuran. Positions 1 through 4 and 6 through 9 have attached hydrogens that can be replaced by chlorines in chlorodibenzofurans.

EXAMPLE 5–4 Suppose that an individual ingested 20 pg (picograms) of 2,3,7,8-tetrachlorodibenzo-*p*-dioxin, 200 pg of octachlorodibenzo-*p*-dioxin, and 40 pg of 1,2,3,4,7,8, 9-heptachlorodibenzofuran. Using the toxicity factors listed in Table 5–10,

$$\text{TEQ} = (20 \text{ pg})(1) + (200 \text{ pg})(0.001) + (40 \text{ pg})(0.01) = 20.6$$

The individual has ingested a total of 260 pg of dioxin but this is equivalent to only 20.6 pg of TCDD. Thus, not only the total exposure to dioxins but the actual composition of the dioxins is important in determining the health or environmental hazard. ∎

PAHs

Polycyclic aromatic hydrocarbons (PAHs) consist of fused benzene rings. Representative structures are shown in Figure 5–13. The only commercially produced PAH is napthalene, whose vapor is toxic to some insects. PAHs are common air pollutants and are the result

Table 5–10 Toxicity Equivalence Factors (TEQ) for Some Important Dioxins and Furans*

Dioxin or furan	TEQ
2,3,7,8-Tetrachlorodibenzo-*p*-dioxin	1
1,2,3,7,8-Pentachlorodibenzo-*p*-dioxin	0.5
1,2,3,4,7,8-Hexachlorodibenzo-*p*-dioxin	0.1
1,2,3,7,8,9-Hexachlorodibenzo-*p*-dioxin	0.1
1,2,3,6,7,8-Hexachlorodibenzo-*p*-dioxin	0.1
1,2,3,4,6,7,8-Heptachlorodibenzo-*p*-dioxin	0.01
Octachlorodibenzo-*p*-dioxin	0.001
2,3,7,8-Tetrachlorodibenzofuran	0.1
2,3,4,7,8-Pentachlorodibenzofuran	0.5
1,2,3,7,8-Pentachlorodibenzofuran	0.05
1,2,3,4,7,8-Hexachlorodibenzofuran	0.1
1,2,3,7,8,9-Hexachlorodibenzofuran	0.1
1,2,3,6,7,8-Hexachlorodibenzofuran	0.1
2,3,4,6,7,8-Hexachlorodibenzofuran	0.1
1,2,3,4,6,7,8-Heptachlorodibenzofuran	0.01
1,2,3,4,7,8,9-Heptachlorodibenzofuran	0.01
Octachlorodibenzofuran	0.001

*From Baird (1995).

of incomplete combustion, particularly of wood and coal. PAHs are found in creosote, which is used as a wood preservative. They can also be introduced into the aquatic environment through oil spills. Certain PAHs are known carcinogens and their distribution in the environment is of concern. In the aquatic environment they tend to be adsorbed on sediment particles because their solubility in H_2O is low. Like other hydrocarbons, they tend to concentrate in fatty tissue, and various aquatic organisms have relatively high burdens of PAHs. Case Study 5–5 describes the distribution and sources of PAHs found in Lake Erie sediments.

NAPLs

Nonaqueous-phase liquids (NAPLs) are *hydrocarbons that have very low solubility in water.* They can be very important contaminants in groundwater systems. The NAPLs are divided into two groups on the basis of density. *LNAPLs (light nonaqueous-phase liquids)* are *less dense than water and will float on the water table.* *DNAPLs (dense nonaqueous-phase liquids)* are *more dense than water and can sink deep into the saturated zone.*

CASE STUDY 5–5
Distribution of PAHs in Lake Erie Sediments

Smirnov et al. (1998) investigated the distribution of PAHs in the sediments of Lake Erie. Among the PAHs found in the sediments were anthracene, benz(a)anthracene and its isomers, and benzo(a)pyrene. Principal component analysis (PCA) revealed three distinct groups of sediments in terms of PAH content and $\delta^{13}C$. These were identified as the City Zone (average TPAH = 4997 ng g^{-1}, $\delta^{13}C$ = −25.4‰), the South Zone (average TPAH = 1914 ng g^{-1}, $\delta^{13}C$ = −26.2‰), and the North Zone (average TPAH = 764 ng g^{-1}, $\delta^{13}C$ = −27.0‰). The City

Zone was interpreted to be the result of direct input of fluvial sediments from Detroit, Cleveland, and Buffalo. The South Zone represented a region of small isolated inputs, and the North Zone was essentially uncontaminated by fluvial inputs. The authors concluded that approximately 80% of the total PAHs in the sediments were deposited by direct fluvial input of contaminated material. The other 20% represented atmospheric deposition.

Source: Smirnov et al. (1998).

Table 5–11 Solubility of BTEX Aromatic Hydrocarbons in Water

BTEX hydrocarbon	Solubility in distilled water (ppm)	Solubility in seawater (ppm)	Reference
Benzene	1696	201	1
Toluene	580	50	1
Ethylbenzene	161	111	2
o-Xylene	171	130	2
m-Xylene	148	106	2
p-Xylene	156	111	2

[1]Keeley et al. (1988).
[2]Sutton and Calder (1975).

LNAPLs are predominantly petroleum and gasoline-related compounds that are less dense than water and hence float on the groundwater table. Thus, they are rapidly dispersed over a wide area and pose a significant environmental problem. Among these compounds are a group of aromatic compounds that are often referred to as BTEX (benzene, toluene, ethylbenzene, and xylenes). The BTEX compounds have a high solubility in freshwater (Table 5–11) and are potentially serious health hazards. The solubility of BTEX compounds is much lower in seawater, a high-ionic-strength solution. Toluene [methylbenzene] consists of a benzene ring with one attached methyl group. Ethylbenzene consists of a benzene ring with one attached ethyl group. The xylenes—o-xylene [1,2-dimethylbenzene], m-xylene [1,3-dimethylbenzene], and p-xylene [1,4-dimethylbenzene]—have two methyl groups attached at different positions on the benzene ring. The names in brackets are the corresponding IUPAC names. Of concern are those LNAPLs that have a relatively high solubility (benzene and toluene) in freshwater. These can enter the drinking water supply and pose a health hazard. For example, benzene is a known carcinogen.

DNAPLs are mostly halogenated hydrocarbons and commonly occur as contaminants in groundwater. Sources are facilities that use chlorinated solvents, manufacture pesticides, and treat wood. Spills of transformer oils are also an important source. Chlorinated aliphatic hydrocarbons, PAHs, and PCBs are the compounds of most concern because they are relatively soluble in water. DNAPLs commonly found in the environment are listed in Table 5–12. Because these compounds do penetrate the zone of saturation, they are widely dispersed in groundwater. They are also not readily biodegradable. Hence, DNAPL releases are generally of much greater environmental concern than LNAPL releases.

NAPL concentrations are affected by four processes: volatilization, hydrolysis, in situ biodegradation, and dissolution. ***Volatilization*** occurs when NAPLs come in contact with air. For any particular chemical, standard reference works give equilibrium vapor pressures. For example, for benzene the equilibrium vapor pressure is 0.10 atm. For a mixture of hydrocarbons, the vapor pressure for a particular compound is proportional to its relative molar abundance in the mixture. The volatility of organic compounds is utilized in the sparging remediation process. In this process large volumes of air are passed through the contaminated zone and the organic compounds volatilize into the air stream. ***Hydrolysis*** occurs when an organic compound reacts with water to form a derivative compound. This type of reaction is particularly important for the chlorinated hydrocarbons that are often found in DNAPLs. The reaction can be written

$$R\text{-}X + H_2O \rightarrow R\text{-}OH + X^- + H^+$$

where R-X is an organic compound and X is an attached halogen, carbon, phosphorous, or nitrogen group. The attached group is replaced by OH^-. ***Biodegradation*** is a biological process involving aerobic bacteria. The bacteria utilize dissolved oxygen to directly consume the organic compound. This process is particularly important for LNAPLs. ***Dissolution*** occurs when the organic compounds dissolve in water. The effectiveness of this process is directly related to the solubility of the particular organic compound and the volume of water through which it can be dispersed. Case Study 5–6 describes the natural attenuation of hydrocarbons in a groundwater system.

Table 5–12   Common DNAPLs*

Halogenated volatiles	Nonhalogenated semivolatiles
Chlorbenzene	2-Methyl naphthalene
1,2-Dichloropropane	o-Cresol
1,1-Dichloroethane	p-Cresol
1,1-Dichloroethylene	2,4-Dimethylphenol
1,2-Dichloroethane	m-Cresol
trans-1,2-Dichloroethylene	Phenol
cis-1,2-Dichloroethylene	Naphthalene
1,1,1-Trichloroethane	Benzo[a]anthracene
Methylene chloride	Fluorene
1,1,2-Trichloroethane	Acenaphthene
Trichloroethylene	Anthracene
Chloroform	Dibenzo[a,h]anthracene
Carbon tetrachloride	Fluoranthene
1,1,2,2-Tetrachloroethane	Pyrene
Tetrachloroethylene	Chrysene
Ethylene dibromide	2,4-Dinitrophenol
Halogenated semivolatiles	**Miscellaneous**
1,4-Dichlorobenzene	Coal tar
1,2-Dichlorobenzene	Creosote
Aroclor 1242, 1254, 1260	
Chlordane	
Dieldrin	
2,3,4,5-Tetrachlorophenol	
Pentachlorophenol	

*From Bedient et al. (1994).

CASE STUDY 5–6
Natural Attenuation of Aromatic Hydrocarbons in Groundwater Near a Former Manufactured-Gas Plant

Manufactured-gas plants produced gas from either coal or oil. These plants disappeared with the advent of interstate gas distribution pipelines, but pollution associated with these plants still exists. Contaminants from these sites include trace elements, volatile organic compounds (VOCs), monoaromatic hydrocarbons (MAHs), and PAHs. Landmeyer et al. (1998) investigated the breakdown of organic pollutants in groundwater from one of these sites located in South Carolina. The transport of organic contaminants through the groundwater system is a function of the rate of groundwater flow and the rate of degradation. If degradation rates are relatively high compared to groundwater flow, transport of the organic contaminants will be limited.

Biodegradation rates are often determined by laboratory experiments in which a sample of aquifer material is spiked with ^{14}C-labeled compounds. In this study, the authors used ^{14}C-napthalene (a PAH), ^{14}C-benzene, and ^{14}C-toluene

(MAHs). Experiments were carried out under both aerobic and anaerobic conditions. It should be noted that in a groundwater system contaminated with hydrocarbons, conditions are usually anaerobic. First-order biodegradation rate constants under aerobic conditions were toluene = -0.84 d^{-1}, benzene = -0.03 d^{-1}, and napthalene = -0.88 d^{-1}. Under anaerobic conditions, the rate constants were toluene = -0.002 d^{-1}, benzene = -0.00014 d^{-1}, and napthalene = -0.000046 d^{-1}. Under aerobic conditions, which are unlikely to be attained in the aquifer, the breakdown of all three aromatic hydrocarbons is rapid. Under anaerobic conditions, the breakdown is much slower.

The authors also determined the adsorption coefficients for toluene and napthalene [the partitioning of a contaminant between an aqueous phase (groundwater) and a solid phase (aquifer material)]. For napthalene, $K_d = 137$ L kg^{-1} and for toluene, $K_d = 9.9$ L kg^{-1}. The difference in the two partition coefficients indicates that napthalene would be retained in the solid phase relative to toluene as groundwater moved through the system.

(continued)

GEOCHEMICAL AND BIOCHEMICAL PROCESSES THAT CONTROL THE CONCENTRATION OF ORGANIC CHEMICALS IN WATER

In the previous sections we have considered the types of organic compounds, their distribution in the environment, and some of the factors that control their distribution and decomposition. Here we will look at the specific physical and biological processes that control the concentrations of carbon compounds in natural waters. These are important processes because they will determine the rate at which a contaminant is removed from the system, the rate at which it is dispersed through the system, and the techniques that can be used to remove the pollutant. Physical/chemical processes are sorption/partition, precipitation, volatilization, and oxidation–reduction. Biological processes are direct consumption by primary producers (i.e., photo- and chemosynthesis), decomposition by bacteria under aerobic or anaerobic conditions, and oxidation–reduction.

Sorption/Partition

Particles are present in natural systems, and dissolved organic carbon is partitioned between the aqueous phase and the particles. The particle mixture is complex, consisting of both the products of erosion and biological activity. The possibilities are a variety of silicate (usually dominated by the clays), carbonate, oxide, and sulfide minerals; organic particles; and organic and oxide coatings on pre-existing particles. Given the complexity of natural systems, it is usually best to experimentally determine the sorption/partition characteristics (Case Study 5–6). What is usually determined is the distribution coefficient, i.e., the amount adsorbed per unit mass of solid divided by the concentration of the substance in the aqueous phase. The amount adsorbed per unit mass may change with variations in absolute concentration. When this occurs, the adsorption process is represented by an adsorption isotherm. We will consider adsorption isotherms in Chapter 7. If the distribution of the substance is independent of concentration, we have the simple relationship

$$C_{ads} = K_d * C_{soln}$$

where C_{ads} is the concentration of the substance adsorbed by the particles, C_{soln} is the concentration of the substance in water, and K_d is the distribution coefficient.

___

EXAMPLE 5–5 In Case Study 5–6 it was found that the adsorption of napthalene onto sedimentary particles followed a linear relationship, with $K_d = 137 \text{ L kg}^{-1}$. If the napthalene concentration in the groundwater was 3.2 mg L^{-1}, calculate the concentration of napthalene adsorbed onto the sedimentary particles in the aquifer.

$$C_{ads} = K_d \times C_{soln} = (137 \text{ L kg}^{-1})(3.2 \text{ mg L}^{-1}) = 438 \text{ mg kg}^{-1} = 438 \text{ ppm}$$

Assuming that 1 L of groundwater has a mass of 1 kg, the concentration of napthalene in the groundwater is 3.2 ppm compared to a concentration of 438 ppm for napthalene adsorbed onto the sedimentary particles. ■

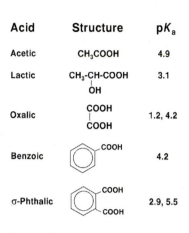

Acid	Structure	pK_a
Acetic	CH₃COOH	4.9
Lactic	CH₃-CH-COOH OH	3.1
Oxalic	COOH COOH	1.2, 4.2
Benzoic	⬡—COOH	4.2
σ-Phthalic	⬡—COOH COOH	2.9, 5.5

Figure 5–31
Structures of some common organic acids and their dissociation constants, pK_a. Oxalic and σ-phthalic acid have two dissociation constants because both acids have two-step dissociations. After Drever (1997).

In the case of organic acids and bases, pH plays an important role in the adsorption process. For acids, the pH of the solution has to be 2 units less than the pK_a for complete adsorption, and when the pH is 2 units greater than the pK_a, maximum desorption occurs. The desorption is related to the ionization of the organic acid, which is less at lower pH. In the case of bases, the opposite effect is observed, and bases do not adsorb in acid solutions because H^+ ions are combined with OH^- (protonation), leading to ionization of the base.

Because organic acids are important in the natural environment, we will briefly consider their properties. The equilibrium reaction for an organic acid, in this case one containing the carboxylic acid group, can be written

$$RCOOH \rightleftharpoons H^+ + RCOO^-$$

where R is the root molecule and COOH is carboxylic acid. The equilibrium constant for this reaction is written

$$K_a = \frac{[H^+][RCOO^-]}{[RCOOH]}$$

Analagous to pH, the equilibrium constant is often expressed as

$$pK_a = -\log_{10} K_a$$

For a given pK_a, as the solution becomes more acidic (lower pH equals higher H^+ concentration) the amount of dissociated acid decreases; i.e., $RCOO^-$ becomes smaller. At higher pH, more of the acid is dissociated. Adsorption is inversely related to the degree of dissociation, so increasing pH leads to a decrease in adsorption. Some common organic acids are illustrated in Figure 5–31 along with their pK_a's. Note that acids with more than one carboxylic acid group have multiple-step dissociations and, thus, have more than one pK_a value.

In the case of the partitioning of an organic chemical between soil and water, the partition coefficient is often normalized to the organic carbon content of the soil. The relationship becomes

$$K_{oc} = K_d \times \frac{100}{\%oc} = \frac{\mu g \text{ chemical/g organic carbon}}{\mu g \text{ chemical/g water}} \qquad (5–3)$$

K_{oc} values for selected compounds are found in Table 5–13.

Table 5–13 K_{oc} and Henry's Law Constants for Selected Organic Chemicals*

Compound	log K_{oc}	H (unitless)
DDT	5.18	0.00213
Hexachlorobutadiene	4.46	186.797
1,2-Dichlorobenzene	3.23	0.07889
Lindane	3.11	0.00333
Napthalene	3.11	0.04701
Ethylbenzene	3.04	0.26282
m-Xylene	2.99	0.43736
o-Xylene	2.92	0.20846
4-Chloro-m-cresol	2.69	0.00010
Tetrachloroethene	2.56	1.05865
Toluene	2.48	0.26037
Benzene	1.92	0.22849
2,4-D	1.78	7.7E-9
Dichlorodifluoromethane	1.76	121.39765
Bromobenzene	2.18	0.07848

*Data from Hounslow (1995).

EXAMPLE 5–6 A field has been treated with DDT (obviously, not in the United States) and rainwater percolating down through the soil contains 3 ppb DDT. Organic carbon makes up 3% of the soil. Calculate the amount of DDT adsorbed per gram of soil.

For DDT, $\log K_{oc} = 5.18$ (Table 5–13).

$$C_{ads} = K_{oc} * C_{soln} = (1.514 \times 10^5)(3 \text{ ng g}^{-1})$$
$$= 4.542 \times 10^5 \text{ ng g}^{-1} \text{ C} = 454 \ \mu\text{g g}^{-1} \text{ C}$$

Only 3% of the soil is organic carbon, so the amount of DDT adsorbed per gram of soil is

$$454 \times 0.03 = 13.6 \ \mu\text{g DDT g}^{-1} \text{ of soil} \equiv 13.6 \text{ ppm} \qquad \blacksquare$$

Precipitation

Precipitation of organic matter generally occurs when there is a significant change in the ionic strength of a solution, with an increase in ionic strength leading to decreased solubility. Recall from Chapter 2 that for neutral species increasing ionic strength leads to activity coefficients greater than 1 and a decrease in solubility. The most likely place for organic precipitation to occur is where freshwater meets seawater, for example, in an estuary.

Volatilization

Many organic compounds have low volatility; i.e., they do not readily go into the vapor state (evaporate). However, evaporation is an important process for a number of organic compounds of anthropogenic origin. These include the hydrocarbons, ketones, aldehydes, and esters. In general, low-molecular-weight nonpolar molecules have the greatest vapor pressures and are most easily evaporated. As common, everyday examples, nail polish remover (essentially acetone) and gasoline spills (when you overfill your gas tank) evaporate relatively quickly. The first example is a ketone and the second is a hydrocarbon.

The partitioning of an organic compound between liquid and vapor is often expressed in terms of a Henry's law constant, which is written

$$H_i = \text{Concentration in vapor state/Concentration in water} \qquad (5–4)$$

A number of different units have been used for this constant, including atm-liter g^{-1}, atm-liter mol^{-1}, atm (mole fraction)$^{-1}$, and a dimensionless unit. Note that the units depend on how the concentrations of the organic compound in the vapor and liquid are measured. For example, in the case of atm-liter g^{-1}, the concentration in the vapor is measured in atmospheres and the concentration in the liquid is measured in g L^{-1}. For the dimensionless unit, the concentrations in both the vapor and liquid phase are generally measured in mol L^{-1}. Dimensionless Henry's law constants for selected organic compounds are found in Table 5–13.

The volatility of organic compounds is important for one type of remediation called *sparging*. This technique is used to remove volatile organics, such as LNAPLs, from soil and groundwater. The process involves passing large volumes of air through the soil. The volatile organic compound partitions into the air and is exhausted from the site. You should recognize the practical difficulties of this type of remediation, in which the interconnectivity of the pore space (permeability) is an important factor.

EXAMPLE 5–7 Benzene (a BTEX hydrocarbon) is released during an industrial accident and enters the groundwater system. The average concentration of benzene in the water is 1000 ppm. If the water is in contact with air, calculate the concentration of benzene in the air in mol L^{-1}. First convert 1000 mg kg^{-1} to mol L^{-1}. Assume that 1 kg of water is equivalent to a volume of 1 L. The molecular weight of benzene (C_6H_6) is 78.

$$1000 \times 10^{-3} \text{ g L}^{-1} \text{ benzene/78 g mol}^{-1} = 1.28 \times 10^{-2} \text{ mol L}^{-1} \text{ benzene}$$

The Henry's law constant for benzene is 0.22849 (Table 5–13).

$$C_v = H_i * C_w = (0.22849)(1.28 \times 10^{-2}\ mol\ L^{-1}) = 2.92 \times 10^{-3}\ mol\ L^{-1}$$

It is decided to remediate this spill by sparging. The goal is to reduce the benzene contamination to 10% of its original concentration in the groundwater. Assuming that we are dealing with an ideal situation, and that saturation of the vapor space is instantaneous, how many liters of air must we pass through the system for each liter of contaminated water? Note that the rate at which the concentration of benzene decreases with time is not constant; i.e., the concentration decrease is a first-order process (see Chapter 2). Using the dimensionless Henry's law constant and setting 1 cycle = 1 liter of air,

$$Cycles = \ln(C_f/C_i)/(-H_i) = \ln(0.1/1)/(-0.22849) = 10\ cycles = 10\ liters\ of\ air$$

If we decide to reduce the concentration to 1% of the original, we would need to pass 20 liters of air through the system per liter of water. Note that this calculation assumes an ideal situation. In the real world much larger volumes of air would be required.   ■

Oxidation–Reduction

In Chapter 4 we considered the basic principles of electrochemistry. The abundance of organic matter in the natural environment is important in terms of decomposition reactions that deplete the waters in oxygen and, thus, lower the overall Eh. In addition, it has been shown that humic ($E° = 0.7$ V) and fulvic ($E° = 0.5$ V) acids can act as reducing agents.

Biological Processes

Particulate organic matter (POC) is removed from the water column by consumers and decomposers (bacteria and fungi). Dissolved organic carbon (DOC) is removed by microbial action. In both cases, microbial decomposition appears to be most important. The presence of sediment increases the rate of microbial decomposition. The rate of microbial decomposition is also related to the size of the organic molecule, with increasing size leading to decreasing rates of decomposition. Microbial decomposition under aerobic conditions can be represented by the following simple reaction:

$$CH_2O + O_2 \rightarrow CO_2 + H_2O + Energy$$

This is the dominant reaction in the aerated portions of streams, lakes, and the ocean. Anaerobic decomposition can be represented by the following simple reaction:

$$CH_2O + A(oxidized) \rightarrow CO_2 + A(reduced)$$

where A is an electron acceptor (an oxidizing agent, see Chapter 4). If A is an organic compound, then the product is an organic compound. Hence, during anaerobic decay a number of intermediate organic compounds are produced that contribute to the total DOC. The final products of anaerobic decay are usually CH_4 and CO_2. Other types of microbial actions involve transformations of organic compounds. For example, through microbial action an organohalide compound can be degraded by *dehalogenation*, in which a halogen atom is replaced by another atom or group, such as OH. Case Study 5–7 (p. 160) deals with the biodegradation of chlorinated and nonchlorinated organic contaminants.

Degradation Half-Life

A useful concept is that of **degradation half-life**, i.e., *the length of time it takes for 50% of an organic compound to be degraded*. Note that this is an exponential relationship just as in the case of radioactive decay (a first-order process). Thus, for each half-life the existing concentration is reduced by 50%. The half-life of organic compounds in soil and surface and ground water is a function of the initial concentration of the compound, the temperature, and whether the decomposition occurs under aerobic or anaerobic conditions.

CASE STUDY 5–7

Biotransformations of Organics in Soil Columns and an Infiltration Area

Bosma et al. (1996) investigated the biodegradation of chlorinated and nonchlorinated organic contaminants in sand sediments from the Rhine River and from a dune infiltration site at the Amsterdam municipal water works. Under aerobic conditions all the nonchlorinated compounds were transformed, but most of the chlorinated compounds were not transformed. This situation was reversed under anaerobic conditions in which the chlorinated compounds were transformed by reductive dechlorination (H replaces Cl), but of the nonchlorinated organic con-

taminants only toluene showed significant transformation. In all of the experiments the authors observed that there was a lag time before the transformations occurred. They interpreted these time delays to reflect the time required to increase the microbial population to a level where the transformations became measurable, or the time required to induce enzymes needed for the contaminant breakdown. In the anaerobic experiments the reductive dechlorination was slower at 4°C than at 20°C.

Source: Bosma et al. (1996).

CASE STUDY 5–8

Aerobic Biodegradation of Alkylated Aromatic Hydrocarbons

Budzinski et al. (1998) investigated the aerobic degradation of aromatic compounds using a bacterial community from marine sediments. The aromatic fraction was prepared from an Arabian light-crude residue. The experiments were conducted over a period of seven days and measurements were made at 1, 2, 3, and 7 days. At the end of seven days, the napthalenes had completely degraded and there were significant reductions in the amounts of dibenzothiophenes and phenanthrenes. The degradation was found to follow first-order kinetics, and there were no differences between sulfur-containing aromatic compounds (dibenzothiophenes) and non-sulfur-containing aromatic com-

pounds (phenanthrenes). For the naphthalenes, $t_{1/2} \approx 17$ h; dibenzothiophenes, $t_{1/2} = 32$–50 h; and phenanthrenes, $t_{1/2} = 32$–51 h. Given these data, the nonalkylated aromatic compounds are degraded first. Among the alkylated aromatic compounds, there were important differences in biodegradation rates depending on the position of the alkyl substituents on the aromatic rings. Remember, alkyl groups are derived from alkanes. The half-lives determined by these experiments indicate that all the aromatic compounds would degrade rapidly in marine sediments. For the aromatic hydrocarbon with the longest half-life (51 h), 99% of the hydrocarbon would be degraded after 14 days.

Source: Budzinski et al. (1998).

In general, the half-life increases with increasing concentration and decreases with increasing temperature. Half-lives are determined by experiment or by measuring changes with time in natural systems (Case Study 5–8).

EXAMPLE 5–8 Assume an organic contaminant is introduced into an aquifer. Right after the spill the concentration of the organic contaminant is 12 mg L^{-1}. A measurement made 20 days later reveals that the concentration of the contaminant has decreased to 1 mg L^{-1}. Assuming that this is a first-order reaction, what is the half-life for this particular organic contaminant?

$$C_i = C_o e^{-kt}$$

Rearranging this equation and solving for k gives

$$k = \ln C_o/C_i \times 1/t = \ln 12 \times 1/20 \text{ d} = 0.124 \text{ d}^{-1}$$

The half-life is the time it takes 50% of the organic contaminant to degrade; i.e., $C_o/C_i = 2$.

$$t_{1/2} = \ln 2/k = 0.693/0.124 \text{ d}^{-1} = 5.6 \text{ d} \qquad \blacksquare$$

Half-lives for various organic compounds can be found in the literature (Table 5–14). These values are often determined for a particular system and may not be appropriate for other systems. The data, however, do give us an idea of the relative rates at which various organic chemicals break down under aerobic and anaerobic conditions in the natural environment.

Table 5–14   *Half-Lives for Selected Organic Chemicals**

| | Half-Life in Days | | | |
| | Aerobic decomposition | | Anaerobic decomposition | |
Compound	Minimum	Maximum	Minimum	Maximum
Cresol(s)	< 1	29	10	49
Phenol	< 1	4	8	28
Napthalene	1	20	25	258
Toluene	4	22	56	210
Benzene	5	16	112	720
Xylene(s)	7	28	180	360
Methyl parathion	15	70	1	7
Aldecarb	20	361	62	635
Lindane	31	413	6	31
Dieldrin	175	1080	1	7
Tetrachloroethene	180	360	98	1653
Chlordane	238	1386	1	7
DDT	730	5708	16	100

*Data from Hounslow (1995).

QUESTIONS AND PROBLEMS

1. What is the difference between a *saturated* and an *unsaturated* organic compound?

2. What is a *functional group*?

3. How do *molecular* and *structural* formulas differ?

4. What is an *isomer*?

5. Distinguish between *alkanes*, *alkenes*, and *alkynes*.

6. Give an example of a *substitution* reaction.

7. Give an example of an *addition* reaction.

8. What are the distinguishing characteristics of a *PAH*?

9. Distinguish between *hard* and *soft* pesticides.

10. Define *polymer*.

11. What is *primordial carbon*?

12. Compare and contrast the *geochemical* and *biochemical* carbon cycles.

13. Define *humic substances*.

14. Distinguish between *humin*, *humic acid*, and *fulvic acid*.

15. Define the terms *labile* and *nonlabile* as used in Case Study 5–1.

16. What role do humic and fulvic acids play in the mobility of radionuclides in the surface environment (Case Study 5–2)?

17. Define *soil*.

18. What are the major factors in soil formation?

19. How could you distinguish organic carbon from a marine source from organic carbon from a land source?

20. How can sulfur occur in coal, and why is this important in the context of the environmental impact of coal combustion?

21. Summarize the associations of trace metals in coal with mineral and organic fractions (Case Study 5–4).

22. What factors are important in the formation of petroleum, and how do these factors affect the final petroleum product?

23. Distinguish among *POC*, *DOC*, and *TOC*.

24. Distinguish between *allochthonous* and *autochthonous* sources of carbon.

25. What is *BOD*, and how is it determined?

26. What is LD_{50}, and what is its significance?

27. What is *BCF*, and how is it determined?

28. What are the major factors that determine the toxicity of a *dioxin*?

29. What are *NAPLs*, *LNAPLs*, and *DNAPLs*?

30. Name and describe the four processes that affect NAPL concentrations.

31. Discuss the relationship between pH and the adsorption of organic acids.

32. Why do dissolved organic compounds tend to precipitate when the ionic strength of a solution increases?

33. What are the factors that favor high vapor pressures for organic compounds?

34. What is *sparging*, and why is the volatility of an organic compound important when using this process?

35. What is *BTEX*?

36. The famous Woburn toxics case (*A Civil Action*) involved the transport of various chlorinated hydrocarbons via the groundwater system to two water supply wells. Referring to Case Study 5–7, what variables in the groundwater system are important in the breakdown of chlorinated hydrocarbons?

37. Define *degradation half-life*.

38. Draw the structural formula for the benzene ring (remember the ring consists of six carbon atoms and six hydrogen atoms). Given this structure, and charge balance considerations, why was the carbon bonding in the benzene ring once thought to consist of alternating single and double bonds?

39. Draw the structural formula for 2-methylbutane.

40. Draw the structural formula for 2,5-dimethylhexane, and write the molecular formula.

41. Draw the structural formula for 3-ethyl-2-methylpentane, and write the molecular formula.

42. Draw the structural formula for 1,4-dimethylcyclohexane, and write the molecular formula.

43. Write the condensed structural formula for 2,4-octadiene.

44. Write the condensed structural formula for 2-ethyl-1,3-hexadiene.

45. Write the condensed structural formula for 2-methyl-1-penten-3-yne.

46. Draw the structural formula for 1,2-dichlorobenzene.

47. Draw the structural formula for 1,1-dichloroethylene.

48. Draw the structural formula for 1,2,3,6,9-pentachlorodibenzo-*p*-dioxin.

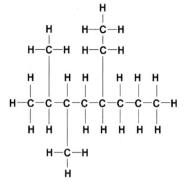

Figure 5–P50
Structural formula for problem 50.

49. Draw the structural formula for methylamine.

50. Name the compound illustrated in Figure 5–P50.

51. A chemical analysis of dissolved organic carbon (DOC) in a water sample yields the following:

Partial Water Analysis	
Carbon (wt%)	45.0
Hydrogen (wt%)	6.0
Oxygen (wt%)	46.0
Molecular weight (daltons)	800

a. Calculate the atomic H/C and O/C ratios.
b. Is the DOC fulvic or humic acid? List the criteria used to make the identification.
c. Is this a saltwater or freshwater sample? List the criteria used to make the identification.

52. Keil et al. (1994) analyzed organic material extracted from the Columbia River, the Columbia River estuary, and the continental shelf sediments off Washington. The sediments were separated into sand, silt, and clay size fractions. Bulk sediment analyses were also done. A partial listing of their data is given here.

Data for Columbia River and Continental Shelf Sediments				
Location	Size	C (wt%)	N (wt%)	$\delta^{13}C‰$
Columbia River sedimentary plume	Bulk	1.34	0.12	−22.5
	Clay	3.48	0.43	−22.9
	Sand	32.1	0.77	−25.6
Columbia River estuary suspended sediment	Silt	3.93	0.37	−26.0
Midshelf	Bulk	1.02	0.09	−22.5
	Clay	3.99	0.47	−23.1
	Sand	25.8	0.80	−24.7
Continental slope	Bulk	2.70	0.26	−22.0
	Clay	3.21	0.38	−22.1
	Sand	0.27	0.01	−23.7

a. Calculate the atomic C/N ratio for each sample.
b. Plot the data on a copy of Figure 5–26. Indicate the type of material you are plotting, i.e., bulk, clay, silt, or sand, and use different symbols to indicate each environment.
c. Calculate the relative percentage of vascular plant debris, soil organic matter, and marine plankton in each sample. The problem is solved graphically using the lever rule. (See the section in Chapter 6 dealing with mixing diagrams.)
d. What conclusions can you draw about the source of the organic material in the different sediment size fractions?

53. A harbor is currently the site of an oil refinery, and a number of ships enter the harbor to unload crude oil and to load gasoline. A sediment core was obtained from the floor of the harbor. Using the dating techniques discussed in Chapter 6, the ages of the various layers in the core were determined. The organic material in the sediment was analyzed and the ratio of odd chain lengths to even chain lengths was determined for the alkanes. Also determined was the unresolved complex mixture (UCM), which represents molecular structures that cannot be resolved with high-resolution gas chromatography. Some of these data are listed here. You will need to refer to Case Study 5–3 in order to answer these questions.

Data for Harbor Sediments		
Age of Sediment layer	Odd/even	UCM (area g^{-1} C)
1990	5	2
1980	3	4
1970	1	20
1960	5	1
1950	5	1
1940	5	1

a. When did the oil refinery first begin to operate? Why did you select this date?

b. What might have happened around 1970? Why?

c. What has happened since 1970? List some reasons for the changes observed in the chemical data after 1970.

54. The following chemical data were obtained for four coal samples. Plot the data on the van Krevelen diagram (Figure 5–27), and classify each of the samples.

Data for Some Coal Samples				
Sample	1	2	3	4
C (wt%)	55.44	72.95	84.24	93.50
H (wt%)	5.60	5.24	5.55	2.81
O (wt%)	36.56	20.50	8.67	2.72

55. For Carbaryl (a carbamate), log K_{ow} = 2.81 (Karickhoff, 1981; note that this paper tabulates K_{ow} values for a number of organic compounds). Calculate the BCF for this insecticide.

56. An individual has the following dioxin exposure: 10 pg 1,2,3,7,8-pentachlorodibenzo-p-dioxin, 40 pg 1,2,3,7,8,9-hexachlorodibenzo-p-dioxin, 120 pg 2,3,4,7,8-pentachlorodibenzofuran, and 200 pg 1,2,3,4,6,7,8-heptachlorodibenzofuran. Calculate the TEQ for this exposure.

57. In Case Study 5–6, the first-order rate constant for the breakdown of toluene under aerobic conditions was -0.84 d^{-1}, and under anaerobic conditions the rate constant was -0.002 d^{-1}. Calculate the time required to degrade 90% of the toluene under aerobic and anaerobic conditions.

58. In Case Study 5–6 it was found that the adsorption of toluene onto sedimentary particles followed a linear relationship, with K_d = 9.9 L kg^{-1}. If the toluene concentration in the groundwater is 4.5 mg L^{-1}, calculate the concentration of toluene adsorbed onto the sedimentary particles in the aquifer.

59. Log K_{oc} = 2.36 for the carbamate pesticide Carbaryl (Karickhoff, 1981). Rainwater percolating down through the soil contains 5 ppb Carbaryl. Organic matter makes up 2% of the soil. Calculate the amount of Carbaryl adsorbed per gram of soil.

60. m-Xylene enters the groundwater system. The average concentration of m-Xylene in the groundwater is 100 ppm. If the groundwater at the top of the saturated zone is in equilibrium with air, calculate the concentration of m-Xylene in the air in mol L^{-1}.

61. In a particular system, the half-life of Lindane undergoing aerobic decomposition is 210 days. Calculate the length of time required for 90% of the Lindane to decompose.

62. Because of an oil spill a marine sediment becomes contaminated with napthalene. With reference to Case Study 5–8, calculate the length of time required for 99% of the napthalene to decompose.

6

Isotopes

Atoms consist of a nucleus and a surrounding cloud of electrons. The nucleus contains positively charged *protons* and anywhere from 0 to over 100 *neutrons* (particles that have nearly the same mass as a proton, but no charge). The number of protons in the nucleus uniquely determines the element—i.e., hydrogen has 1 proton, helium 2 protons, uranium 92 protons, etc. *Isotopes* are *atoms that have the same number of protons but a different number of neutrons*. The **atomic number** of an element is the *number of protons in the nucleus*, and the **atomic mass number** is the *number of protons plus the number of neutrons in the nucleus*. For example, there are three isotopes of hydrogen, $_1^1H$, $_1^2H$, and $_1^3H$, all of which have 1 proton in the nucleus. $_1^1H$ contains 1 proton and no neutrons, $_1^2H$ contains 1 proton and 1 neutron, and $_1^3H$ contains 1 proton and 2 neutrons. By international convention, the number at the bottom left of the element symbol is the atomic number and the number at the upper left of the element symbol is the atomic mass. Often, the atomic number is not attached to the symbol because it is implied by the element name.

It is convenient to divide the isotopes into two groups: radioactive and stable. **Radioactive isotopes** are *atoms that undergo spontaneous breakdown of their nuclei to form other isotopes*. **Stable isotopes** do not spontaneously break down to form other isotopes. Both types of isotopes have a wide range of applications in the physical and biological sciences. In this chapter we will focus on those isotopes that have been found to be particularly useful in describing and characterizing the surface and near-surface environment.

RADIOACTIVE ISOTOPES

Basic Principles

Radioactivity is defined as *the spontaneous breakdown of a nucleus*. This breakdown can follow several pathways: alpha (α) or beta (β^-) particle emission, positron (β^+) particle emission, and K-electron capture (ε). As an example of the different types of decay processes, consider the radioactive decay of $_{92}^{238}U$ (Table 6–1, p. 166). This uranium isotope contains 92 protons and 146 neutrons. The isotope decays through a series of alpha and beta emissions, ultimately arriving at a stable form of lead ($_{82}^{206}Pb$). The first step in the decay chain is the emission of an alpha particle. An **alpha particle** *consists of 2 protons and 2 neutrons* (equivalent to a helium nucleus). The resulting nucleus (a thorium atom) has 90 protons and an atomic mass of 234 ($_{90}^{234}Th$). This nucleus is radioactive (unstable). The next step in the decay chain is the emission of a **beta particle** (*an electron*). The beta particle is produced by the breakdown of a neutron to yield a proton and an electron. The result is that the atomic number of the nucleus increases by 1, but there is no change in the atomic mass number. The resulting nucleus is a protactinium atom ($_{91}^{234}Pa$). This decay process continues until a stable nuclear configuration (in this case, $_{82}^{206}Pb$) is achieved. As an example of K-electron capture, consider the decay of $_{36}^{81}Kr$. This isotope decays by capturing an electron from the K-shell. The resulting nucleus ($_{35}^{81}Br$) has the same atomic mass

165

Table 6–1   Uranium Decay Series*

	Uranium 238			Uranium 235	
Isotope	Emitted particle	Half-life	Isotope	Emitted particle	Half-life
$^{238}_{92}$U	α	4.47×10^9 y	$^{235}_{92}$U	α	7.038×10^8 y
$^{234}_{90}$Th	β^-	24.1 d	$^{231}_{90}$Th	β^-	1.063 d
$^{234}_{91}$Pa	β^-	1.17 min	$^{231}_{91}$Pa	α	3.248×10^4 y
$^{234}_{92}$U	α	2.48×10^5 y	$^{227}_{89}$Ac	β^-	21.77 y
$^{230}_{90}$Th	α	7.52×10^4 y	$^{227}_{90}$Th	α	18.72 d
$^{226}_{88}$Ra	α	1.60×10^3 y	$^{223}_{88}$Ra	α	11.435 d
$^{222}_{86}$Rn	α	3.8235 d	$^{219}_{86}$Rn	α	3.96 s
$^{218}_{84}$Po	α	3.10 min	$^{215}_{84}$Po	α	1.78×10^{-3} s
$^{214}_{82}$Pb	β^-	27 min	$^{211}_{82}$Pb	β^-	36.1 min
$^{214}_{83}$Bi	β^-	19.9 min	$^{211}_{83}$Bi	α	2.14 min
$^{214}_{84}$Po	α	1.64×10^{-4} s	$^{207}_{81}$Tl	β^-	4.77 min
$^{210}_{82}$Pb	β^-	22.3 y	$^{207}_{82}$Pb		Stable
$^{210}_{83}$Bi	β^-	5.01 d			
$^{210}_{84}$Po	α	138.38 d			
$^{206}_{82}$Pb		Stable			

*Data source: Chart of the Nuclides (1989).

number, but the atomic number has decreased by 1 due to the conversion of a proton to a neutron. **Gamma rays** (*high-energy electromagnetic radiation*) are also emitted during the decay process. Following the emission of an alpha or beta particle, the nucleus may be left in an excited state. The return of the nucleus to the ground state is accomplished by the emission of the excess energy as gamma rays. Similar to the case with electronic transitions in the atom yielding characteristic X rays, these gamma rays are characteristic of the nucleus and are the basis of a branch of analytical nuclear chemistry known as gamma ray spectroscopy. Both uranium and thorium are important elements in environmental problems, and various parts of their decay chains have been used in a variety of analytical methods. The complete decay chains for uranium are listed in Table 6–1, and the decay chain for thorium is listed in Table 6–2.

Note that in the thorium decay chain (Table 6–2), $^{212}_{83}$Bi has two decay paths (branching decay), one by α decay to $^{208}_{81}$Tl and the other by β^- decay to $^{212}_{84}$Po. The bismuth isotopes in the uranium decay chains (Table 6–1) show similar branching decays, but in these chains the alternative path comprises only 0.04% (U-238 chain) and 0.3% (U-235 chain). Thus, the resulting radiogenic progeny are not shown in Table 6–1.

Table 6–2   Thorium Decay Series*

Isotope	Emitted particle	Half-life	Isotope	Emitted particle	Half-life
$^{232}_{90}$Th	α	1.40×10^{10} y	$^{216}_{84}$Po	α	0.145 s
$^{228}_{88}$Ra	β^-	5.76 y	$^{212}_{82}$Pb	β^-	10.64 h
$^{228}_{89}$Ac	β^-	6.15 h	$^{212}_{83}$Bi	α (33.7%) β^- (66.3%)	1.009 h
$^{228}_{90}$Th	α	1.913 y	$^{208}_{81}$Tl	β^-	3.053 min
$^{224}_{88}$Ra	α	3.66 d	$^{212}_{84}$Po	α	2.98×10^{-7} s
$^{220}_{86}$Rn	α	55.6 s	$^{208}_{82}$Pb		Stable

*Data source: Chart of the Nuclides (1989).

Radioactive Decay and Growth   The decay of a radioactive isotope is a first-order reaction and can be written

$$\frac{dN}{dt} = -\lambda N \qquad (6\text{--}1)$$

where N is the number of unchanged atoms at time t and λ is the radioactive decay constant (instantaneous fraction of atoms decaying per unit time). The negative sign indicates that the number of atoms is decreasing with time. Rearranging this equation,

$$\frac{dN}{N} = -\lambda\, dt \qquad (6\text{--}2)$$

and integrating from $t = 0$ to t and from N_0 to N (N_0 is the number of atoms present at $t = 0$), we get

$$\ln \frac{N}{N_0} = -\lambda t \qquad (6\text{--}3)$$

Taking the antilogs and rearranging gives

$$N = N_0 e^{-\lambda t} \qquad (6\text{--}4)$$

the basic form of the radioactive decay equation.

The **half-life** of a radioactive isotope is the *length of time it takes for half of the atoms to spontaneously decay*. Note that this is an exponential relationship. For example, if we start with 1000 atoms, after one half-life 500 will be left, after the second half-life 250 will remain, after the third half-life 125 will remain, etc. The relationship between the decay constant and the half-life can be determined from equation (6–4). After one half-life, $N = N_0/2 = 0.5$ when $N = 1$. Substituting into equation 6–4 gives

$$\tfrac{1}{2} = 1 e^{-\lambda t_{1/2}} \qquad (6\text{--}5)$$

Taking the antilog and rearranging gives

$$t_{1/2} = \frac{\ln 2}{\lambda} = \frac{0.693}{\lambda} \qquad (6\text{--}6)$$

The radioactive decay equation can also be written in terms of activity, i.e., the number of alpha particles or beta particles or gamma rays emitted per unit time.

$$A = A_0 e^{-\lambda t} \qquad (6\text{--}7)$$

where A is the activity at some time t, A_0 is the activity at time zero, and λ is the decay constant. Rearranging this equation, taking the log and solving for t, we get

$$t = \frac{1}{\lambda} \ln \frac{A_0}{A} \qquad (6\text{--}8)$$

Thus, the ratio of the initial activity to the observed activity enables us to determine the radiometric age of the system.

Consider the radioactive decay of $^{238}_{92}\text{U}$ to $^{234}_{90}\text{Th}$. $^{238}_{92}\text{U}$ is the *radioactive parent* and $^{234}_{90}\text{Th}$ is the *radiogenic progeny*. For any closed system at time t, the number of progeny atoms produced plus the number of parent atoms remaining must equal the total number of parent atoms at the start. Returning to equation 6–4 and substituting P for the number of progeny atoms produced, we get

$$N = (N + P)e^{-\lambda t} \qquad (6\text{--}9)$$

Rearranging and solving for t gives

$$t = \frac{1}{\lambda} \ln \left[1 + \frac{P}{N} \right] \qquad (6\text{--}10)$$

another form of the radiometric age equation that is used when both parent and progeny are measured. Note that N and P are numbers of atoms.

Returning to equation 6–4, we can also describe radioactive decay in terms of the growth of the radiogenic progeny. At any time t the number of radiogenic progeny atoms produced by the decay of the parent is

$$P = N_0 - N \qquad (6\text{–}11)$$

Substituting for N, using equation 6–4, gives

$$P = N_0 - N_0 e^{-\lambda t} = N_0(1 - e^{-\lambda t}) \qquad (6\text{–}12)$$

The relationships between the growth of the radiogenic progeny and the decay of the radiogenic parent is shown graphically in Figure 6–1. In this example it is assumed that the radiogenic progeny is not itself radioactive.

If the radiogenic progeny is radioactive (as is the case for all but the last of the radiogenic progeny in the U and Th decay series,

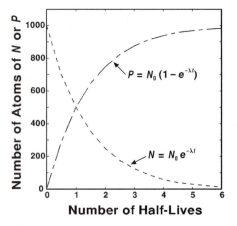

Figure 6–1
Graphical representation of decay of radioactive parent (N) and growth of radiogenic progeny (P).

Tables 6–1 and 6–2), we need to use a form of the radioactive decay equation that not only considers the growth of the radiogenic progeny but also its decay to another radiogenic progeny. We can rewrite the radioactive decay equation 6–1 in terms of the radiogenic progeny.

$$\frac{dN_2}{dt} = \lambda_1 N_1 - \lambda_2 N_2 \qquad (6\text{–}13)$$

where N_1 is the number of parent atoms, N_2 is the number of radiogenic progeny atoms, and λ_1 and λ_2 are the respective decay constants. Substituting for N_1 and N_2 using equation 6–4, integrating, and taking the antilogs gives

$$N_2 = \frac{\lambda_1}{\lambda_2 - \lambda_1} N_1^0 (e^{-\lambda_1 t} - e^{-\lambda_2 t}) + N_2^0 e^{-\lambda_2 t} \qquad (6\text{–}14)$$

where N_1^0 and N_2^0 are, respectively, the number of radioactive parent and radiogenic progeny atoms present at $t = 0$. If there are no radiogenic progeny atoms initially present, the last term of equation 6–14 is zero. If $\lambda_1 > \lambda_2$, the amount of the progeny will first increase, and then after the parent has disappeared, the amount of the progeny will decrease at a rate determined by its half-life. A more interesting case is when $\lambda_1 < \lambda_2$. In this case, a transient equilibrium will be achieved. This occurs because as the decay time increases, the terms involving $e^{-\lambda_2 t}$ become negligible. Equation 6–14 is now written

$$N_2 = \frac{\lambda_1}{\lambda_2 - \lambda_1} N_1^0 e^{-\lambda_1 t} \qquad (6\text{–}15)$$

Substituting $N_1^0 = \dfrac{N_1}{e^{-\lambda_1 t}}$ (by rearranging equation 6–4) gives

$$\frac{N_1}{N_2} = \frac{\lambda_2 - \lambda_1}{\lambda_1} \qquad (6\text{–}16)$$

If the half-life of the parent is much longer than that of the progeny, i.e., $\lambda_1 \ll \lambda_2$, equation 6–16 becomes

$$\frac{N_1}{N_2} = \frac{\lambda_2}{\lambda_1} \qquad (6\text{–}17)$$

This last case is an example of secular equilibrium in which the ratio of the number of parent to the number of progeny atoms remains constant. In effect, what has happened is that the rate of decay of the progeny is balanced by the rate of formation of the progeny by the decay of the parent.

Table 6–3 RBE Values for Various Types of Radiation*

Radiation	RBE
X and γ rays	1
Beta rays and electrons	1
Thermal neutrons	2
Fast neutrons	10
Protons	10
Alpha particles	20
Heavy ions	20

*From Cember (1983).

Measurement of Radioactivity   The *becquerel (Bq)* is the basic unit of measurement for radioactivity, *1 Bq = 1.000 disintegrations s^{-1}*. Radioactivity is often expressed in *curies (Ci)*, *1 Ci = 3.700 $\times$ 10^{10} disintegrations per second*. A picocurie is 1×10^{-12} curies. In the quantitative study of the chemical and biological effects of radiation, a measure of the amount of energy absorbed (the dose) is required. The unit used for the dose is the *gray (Gy)*, and *a dose of 1 Gy deposits 1 joule of energy per kilogram of material*. Another often-used unit is the *rad*: 1 Gy = 100 rad. The biological effects of radiation are due to changes in the chemistry of the cells. These changes are caused by ionization, excitation, dissociation, and atomic displacement due to the passage of the radiation through the cells. The *sievert (Sy)* is the *unit of radiation dosage used to measure the biological effects of radiation*. In the past, the *rem (roentgen equivalent man)* was the commonly used unit: 1 Sy = 100 rem. The dosage in sieverts is equal to the dosage in grays multiplied by the *relative biological effectiveness (RBE)*, *a measure of the ability of various types of radiation to cause ionization in biological materials*. RBE values for different types of radiation are given in Table 6–3.

Radioactive Isotopes Used in Environmental Studies   A number of radioactive isotopes (Table 6–4) have been used in environmental studies to determine the age of groundwater, sediments, and ice cores; to measure groundwater flow rates; and as tracers for groundwater movement. Which isotopic system to use is determined by the nature of the problem. Important parameters are the half-life of the radioisotope and the reactivity of the isotope in the system of interest.

Tritium Dating

There are three isotopes of hydrogen: ^1_1H, ^2_1H (deuterium), and ^3_1H (tritium), with average terrestrial abundances (in atomic %) of 99.985, 0.015, and $< 10^{-14}$, respectively. The first two isotopes are stable but tritium is radioactive ($t_{1/2} = 12.43$ y) and is produced in the upper atmosphere by the bombardment of nitrogen with cosmic-ray-produced neutrons.

$$^{14}_7N + {}^1_0n \rightarrow {}^3_1H + {}^{12}_6C \tag{6–18}$$

With the onset of atmospheric testing of fusion bombs in 1952, bomb-produced tritium became the major source of this isotope. Since the signing of the Atmospheric Test Ban Treaty in 1963, there has been a continual decline in tritium due to the cessation of atmospheric testing and radioactive decay. Tritium abundances are measured in several ways: *tritium unit (TU) = 1 tritium atom per 10^{18} hydrogen atoms*, dpm L^{-1} = disintegrations

Table 6–4   Radiometric Isotopes Used in Environmental Studies

Radioactive parent	Radiogenic progeny	Type of decay	Half-life (y)	Decay constant (y^{-1})
^3_1H	^3_2He	β^-	12.43	5.575×10^{-2}
$^{14}_6C$	$^{14}_7N$	β^-	5.73×10^3	1.209×10^{-4}
$^{36}_{17}Cl$	$^{36}_{18}Ar$	β^-	3.01×10^5	2.302×10^{-6}
$^{81}_{36}Kr$	$^{81}_{35}Br$	ε	2.10×10^5	3.300×10^{-6}
$^{234}_{92}U$	$^{230}_{90}Th$	α	2.48×10^5	2.794×10^{-6}
$^{230}_{90}Th$	$^{226}_{88}Ra$	α	7.52×10^4	9.217×10^{-6}
$^{226}_{88}Ra$	$^{222}_{84}Rn$	α	1.622×10^3	4.272×10^{-4}
$^{210}_{82}Pb$	$^{210}_{83}Bi$	β^-	22.26	3.11×10^{-2}
$^{231}_{91}Pa$	$^{227}_{89}Ac$	α	3.248×10^4	2.134×10^{-5}
$^{87}_{37}Rb$	$^{87}_{38}Sr$	β^-	4.88×10^{10}	1.42×10^{-11}
$^{232}_{90}Th$	$^{208}_{82}Pb$	α, β^-	1.401×10^{10}	4.948×10^{-11}
$^{235}_{92}U$	$^{207}_{82}Pb$	α, β^-	7.038×10^8	9.849×10^{-10}
$^{238}_{92}U$	$^{206}_{82}Pb$	α, β^-	4.468×10^9	1.551×10^{-10}

ε = K-electron capture.

per minute per liter, and pCi L^{-1} = picocuries per liter (of water). The units are related as follows:

$$1 \text{ TU} = 7.1 \text{ dpm L}^{-1} = 3.25 \text{ pCi L}^{-1} \qquad (6\text{--}19)$$

Prior to the atmospheric testing of fusion devices, the tritium content of precipitation was probably between 2 and 8 TU (Thatcher, 1962). A peak concentration of several thousand TU was recorded in northern hemisphere precipitation in 1963 (Weiss et al., 1979).

EXAMPLE 6–1 Assume that in 1951, before the first atmospheric tests of fusion bombs, the concentration of tritium in rainwater was 8 TU. Calculate the maximum activity (in pCi L^{-1}) for prebomb tritium in a present-day (2003) groundwater reservoir.

$$3.25 \text{ pCi L}^{-1} \times 8 \text{ TU} = 26 \text{ pCi L}^{-1}$$

Fifty-two years have elapsed since 1951 (2003 − 1951), and the decay constant for tritium is $5.575 \times 10^{-2} \text{ y}^{-1}$. Using equation 6–7, we can calculate the present-day activity.

$$A = A_0 e^{-\lambda t} = (26 \text{ pCi L}^{-1})e^{-(0.05575)(52)} = 1.4 \text{ pCi L}^{-1}$$

For any groundwater sample that had a tritium activity of greater than 1.4 pCi L^{-1}, we would conclude that at least some of the water was added to the reservoir since 1951. ■

The major use of tritium has been to determine the age, flow rates, and mixing of groundwaters. If the water moves rapidly through the vadose zone (the zone in which the pores are filled with air), it will have a tritium content reflecting that of the rainwater. Once it becomes part of the groundwater system, the water is isolated from the atmospheric reservoir and the tritium activity decreases according to the radiometric decay law. Waters that were added to the groundwater reservoir between 1952 and 1963 will have distinctly elevated tritium abundances. These waters form an age marker that can be followed through the groundwater system, an example of what is called ***event dating***. Other nuclides produced during nuclear bomb testing ($^{36}_{17}$Cl, $^{89}_{38}$Sr, $^{90}_{38}$Sr, $^{133}_{55}$Cs, $^{137}_{55}$Cs) can be used in a similar manner. The circulation of water masses formed at the surface of lakes or oceans between 1952 and 1963 can also be traced using their inventory of bomb-created isotopes.

Tritium concentration in rainwater shows a strong geographical and seasonal variation. Tritium content in water vapor over the oceans is low because of exchange with oceanic surface water, which has a low tritium concentration. As the air moves across a continent, it acquires tritium both by evapotranspiration and from the stratosphere. There is also a latitudinal and a seasonal effect each spring that is caused by the breakup of the tropopause (the boundary, at about 10 km, between the lowermost layer of the atmosphere, the troposphere, and the overlying stratosphere) between 30° and 60°N latitude. Stratospheric air, which contains a higher abundance of tritium, mixes with tropospheric air, resulting in a seasonal spike in tritium concentrations in midlatitude precipitation. This combination of factors means that there is significant seasonal and latitudinal variation in tritium abundances in rainwater, which makes it difficult to use tritium alone for precise dating and mixing studies. The most common approach is to use any increased tritium (over the preatmospheric testing background) as an indication that some of the water in the system must have been added since 1952. Other issues involve the modeling of water flow through an aquifer (i.e., piston-flow approaches, reservoir models, compartment models, and advection–dispersion models). A further discussion of these issues can be found in Plummer et al. (1993).

Many of the problems discussed in the previous paragraph can be avoided by using the ***tritium-helium-3 dating method***. In this method, both ^{3}H and its radiogenic progeny, ^{3}He, are measured. In a confined system, one that is unaffected by dispersion, measurement of both these isotopes allows us to calculate the amount of tritium originally present in the water sample of interest. Potential complications are the presence of sources and sinks for ^{3}He in the system. Schlosser et al. (1989) suggested the following mass balance equation for ^{3}He in a groundwater system:

$$^3\text{He}_{\text{tot}} = {}^3\text{He}_{\text{trit}} + {}^3\text{He}_{\text{eq}} + {}^3\text{He}_{\text{exc}} + {}^3\text{He}_{\text{nuc}} \tag{6-20}$$

where $^3\text{He}_{\text{tot}}$ is the total ^{3}He, $^3\text{He}_{\text{trit}}$ is the tritogenic component, $^3\text{He}_{\text{eq}}$ is the ^{3}He added when the water moves through the vadose zone and equilibrates with the air, $^3\text{He}_{\text{exc}}$ is derived from air entering the groundwater during recharge, and $^3\text{He}_{\text{nuc}}$ is derived from decay processes, other than the breakdown of tritium, occurring within the system. We will be using TU units for both the radioactive parent and the radiogenic progeny, so we can solve for the age using equation 6–10.

$$t = \frac{1}{\lambda} \ln\left[1 + \frac{P}{N}\right] = \frac{1}{5.575 \times 10^{-2}} \ln\left[1 + \frac{{}^3\text{He}_{\text{trit}}}{{}^3\text{H}}\right] = 17.937 \ln\left[1 + \frac{{}^3\text{He}_{\text{trit}}}{{}^3\text{H}}\right] \tag{6-21}$$

The amount of tritogenic ^{3}He in a water sample is calculated from the measured ^{3}He/^{4}He ratio and the concentration of ^{4}He in the water sample. A correction for the nucleogenic component is done using the amount of neon in the sample, which is assumed to have only an atmospheric source. For further details, the student should read the papers by Schlosser et al. (1988, 1989). Both ^{3}H and $^3\text{He}_{\text{trit}}$ concentrations are given in TU (or TR, the tritium ratio: 1 TR = ^{3}H/H$_{\text{tot}}$ = 1×10^{-18}). Remember that $^3\text{He}_{\text{trit}}$ is produced by the decay of ^{3}H and the TU for this isotope represents the original amount of ^{3}H that gave rise to the measured amount of $^3\text{He}_{\text{trit}}$.

EXAMPLE 6–2 Water samples are taken from two wells, 100 m apart. Both samples are collected 1 m below the water table. For well #1, ^{3}H = 25 TU and $^3\text{He}_{\text{trit}}$ = 0.8 TU. For well #2, ^{3}H = 20.5 TU and $^3\text{He}_{\text{trit}}$ = 6.2 TU. Calculate the ages of the water samples. Given that groundwater moves from well #1 to well #2, calculate the rate of groundwater flow.

Using equation 6–21,

$$\text{Well \#1: } t = 17.937 \ln\left[1 + \frac{0.8}{25}\right] = 0.6 \text{ y}$$

$$\text{Well \#2: } t = 17.937 \ln\left[1 + \frac{6.2}{20.5}\right] = 4.7 \text{ y}$$

Rate of flow = 100 m/(4.7 − 0.6) = 100 m/4.1 y = 24 m y^{-1} ∎

In Case Study 6–1 (p. 172) the ^{3}H/^{3}H dating method is used to determine the time and location of a contaminant release to a groundwater system.

Carbon-14 Dating

The tritium dating system can only be used for young (i.e., less than 50-year-old) samples. Thus, there is a need for other isotopic systems that can be used for older samples. One of these is carbon-14, which has been widely used in the geosciences and archaeology. It is the most commonly used system for dating samples less than 50,000 years old. Using accelerator mass spectrometry (AMS), the method can be extended to samples that are greater than 100,000 years old, although care must be taken to ensure that the samples are not contaminated by modern carbon. There are three isotopes of carbon: $^{12}_6$C, $^{13}_6$C, and $^{14}_6$C, with average terrestrial abundances (in atomic %) of 98.90, 1.10, and $< 10^{-10}$, respectively. The first two isotopes are stable but $^{14}_6$C is radioactive ($t_{1/2}$ = 5730 years) and is produced in the upper atmosphere by the bombardment of nitrogen with cosmic-ray-produced neutrons.

$$^{14}_7\text{N} + {}^1_0n \rightarrow {}^1_1\text{H} + {}^{14}_6\text{C} \tag{6-22}$$

The decay of ^{14}C occurs as follows:

$$^{14}_6\text{C} \rightarrow {}^{14}_7\text{N} + \beta^- + \bar{\nu} + Q \tag{6-23}$$

CASE STUDY 6–1
Determination of the Source and Time of Contaminant Release to an Unconfined Gravel Aquifer Using ^{3}H/^{3}He Dating

High concentrations of benzene and ethylene dibromide (EDB) were encountered in exploratory water wells at the FS-12 field site on Cape Cod, Massachusetts (Figure 6–C1–1). This site is adjacent to the Massachusetts Military Reservation. A road runs through the site and an underground pipeline follows this road. The pipeline was used to transport both aviation gasoline and JP-4 jet fuel. Approximately 2000 gallons of fuel leaked from this pipeline during 1972, and this was thought to be the source of the organic contaminants. However, monitoring wells installed along the roadway and a soil gas survey did not detect contamination.

Solomon et al. (1995) used the ^{3}H/^{3}He dating method to determine the ages of groundwater at the site and from these ages developed a model that allowed them to determine the location and date of the contaminant release. The site lies within a glacial outwash plain, and groundwater flow occurs within an unconfined gravel aquifer. Five multilevel sampling systems (MLS) were installed on the site (Figure 6–C1–1). In each well, water was collected at various depths and an age determined for each sample. For the leading edge of the contaminant plume, the groundwater age was 16 years. The measurements were made in 1993, so the time at which this groundwater was at the water table would be 1977. Because of dispersion, the leading edge of the contaminant plume moves at a different rate than the groundwater. This difference was estimated to correspond to an age difference of two years, giving the time at which the contaminant was at the water table as 1975. The vertical age profiles were used to determine recharge rates, and these results were then used to estimate horizontal groundwater flow rates. Given the flow rates and the age of the leading edge of the contaminant plume, an upgradient source for the organic contaminant was identified. Site investigations found fuel floating on the water table in the area identified as the source of the contaminant.

Source: Solomon et al. (1995).

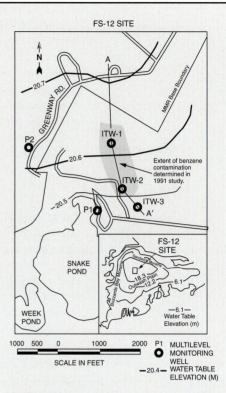

Figure 6–C1–1
Location and extent of spill. From "Site Characterization using 3H/3He Ground Water Ages, Cape Cod, MA" by D. K. Solomon, R. J. Poreda, P. G. Cook and A. Hunt, GROUND WATER; Vol. 33, No. 6, p. 989 (Figure 1, FS 12). November/December 1995. Reprinted from GROUND WATER with permission of the National Ground Water Association. Copyright 1995.

where $\bar{\nu}$ is the antineutrino and Q is the maximum decay energy, 0.156 MeV (million electron-volts) for this reaction. No gamma (γ) rays are emitted, so the abundance of the radioactive isotope is determined by beta counting. This is done either by gas counting or liquid scintillation counting. Even for a relatively young sample, the number of beta emissions is small, so great care has to be taken to shield the counter from background radiation. A number of laboratories now do ^{14}C determinations by AMS. In this method, the atoms in the sample are converted to a beam of fast-moving ions. The appropriate mass is then selected through the application of electric and magnetic fields. The advantages of this method are increased sensitivity and much smaller sample size than that required for conventional beta counting methods.

In order to use the ^{14}C system for geochronology, it is assumed that the atmosphere is in ***secular equilibrium*** with respect to ^{14}C. *By this we mean that the rate at which ^{14}C is produced by the cosmic ray flux is balanced by the rate of decay of ^{14}C so that the abundance of ^{14}C in the atmosphere remains constant.* As long as an organism is alive, or water is in contact with the atmosphere, there is a constant exchange of CO_2 between the systems. When the organism dies, or the water becomes isolated from the atmosphere, the ^{14}C content begins to decrease through radioactive decay. The measured activity of a sample at any time t compared to the activity when the system became closed to atmospheric exchange yields the age of the sample. Using equation 6–8, with $\lambda = 1.209 \times 10^{-4} \, y^{-1}$

$$t = \frac{1}{1.209 \times 10^{-4}} \ln\left(\frac{A_0}{A}\right) = 8.271 \times 10^3 \ln\left(\frac{A_0}{A}\right) \tag{6–24}$$

where A_0 is the initial activity of the sample at $t = 0$ and A is the activity of the sample at some time t.

EXAMPLE 6–3 The ^{14}C activity of modern-day carbon is 13.56 dpm g^{-1} of carbon. Charcoal from an ancient campfire has an activity of 6.22 dpm g^{-1} of carbon. Assuming that the wood was only a year or so old when it was burned in this fire, what was the age of the fire?

$$t = 8.271 \times 10^3 \ln\left(\frac{A_0}{A}\right) = 8.271 \times 10^3 \ln\left(\frac{13.56}{6.22}\right) = 6446 \text{ years before present} \quad \blacksquare$$

In practice, the atmospheric concentration of ^{14}C does not remain constant. There are a number of reasons for these variations: (1) addition of significant amounts of ^{14}C to the atmosphere since 1952 due to the testing of nuclear weapons and the operation of reactors and particle accelerators; (2) addition of ^{14}C-depleted carbon through the burning of fossil fuel; (3) changes in the cosmic-ray proton flux; (4) changes in the earth's magnetic field, which modulates proton flux; and (5) climatic changes causing changes in the carbon reservoirs of the earth. A scheme to correct for these ^{14}C variations has been devised using dendrochronology (the study of tree rings). The correction is done by taking long-lived species, such as the bristlecone pine, and determining the ^{14}C content of the various rings. It is assumed that after each ring is formed it is isolated from the atmosphere. Using living trees and preserved logs, it has been possible to extend this correction back about 10,000 years. After making the ^{14}C measurements, ages in this time span are corrected for the atmospheric variations in ^{14}C.

Another problem with ^{14}C dating is isotopic fractionation, a topic that will be more fully developed in a later section of this chapter. Consider the formation of carbonate shell material by a mollusk. CO_2 in the atmosphere is dissolved in seawater and subsequently taken up, along with Ca, by the mollusk to form the shell material. There are a variety of physical, chemical, and biological processes that will cause a change in the relative proportions of the three C isotopes. This is known as *isotope fractionation*. Thus, the shell material will not have the same amount of ^{14}C relative to ^{12}C and ^{13}C as is observed in the atmosphere. In order to minimize this problem, the same species of organism is used for radiocarbon dating. The present-day ^{14}C activity is determined from a living organism, and this activity is compared to that of the ancient shell. This approach is still not without problems because variations in atmospheric ^{14}C must also be taken into account, plus the possibility that there may have been introduction of nonradiogenic carbon into the aquatic system.

There are also several pitfalls in groundwater dating. The starting assumption is that the groundwater was in equilibrium with atmospheric CO_2, yielding an initial ^{14}C activity. This activity, however, can be changed by various processes within the groundwater system. These include the addition of dissolved organic carbon and inorganic carbon released by the dissolution of limestone. Consider the latter possibility. Suppose the groundwater comes in contact with 100-million-year-old limestone. The limestone contains no ^{14}C. If some of the limestone is dissolved, releasing carbon to the groundwater, the result is a dilution in the concentration of ^{14}C relative to the other carbon isotopes. Consequently, the ^{14}C age of this groundwater would be older than the true age (less ^{14}C per gram of carbon). Conversely, the addition of modern carbon to a sample would increase the amount of ^{14}C relative to the total carbon and hence yield an age younger than the true age.

Despite the possible complications in ^{14}C dating, this is a widely used technique. In many cases, the measured ages must be corrected in various ways, but careful assessment of the variables that can affect the ages often permits an accurate age determination.

U-Series Disequilibrium Methods of Dating

A number of geochronological methods have been developed using various parts of the U (Table 6–1) and Th (Table 6–2) decay chains. We will consider a few of these methods here. More detailed and complete discussions can be found in isotope geochemistry textbooks (Faure, 1986; Dickin, 1997). For a system closed for a sufficiently long time—i.e., secular equilibrium has been achieved—the relative abundance of each isotope in the decay chain will be constant. The basis of the disequilibrium methods of dating is the separation of various isotopes in the decay chain due to differences in their geochemical behavior. This separation can either be a progeny from its parent, with subsequent decay of the progeny, or separation of a parent from its progeny, with subsequent growth of the progeny. The result is that the decay chain is now in disequilibrium and the reestablishment of equilibrium can be used as a dating method.

^{230}Th Dating of Marine Sediments ^{230}Th occurs in the ^{238}U decay chain. Its immediate radioactive parent is ^{234}U. ^{230}Th decays to ^{226}Ra. Due to its oxidation state, U is nonreactive in the marine environment and has a residence time (average length of time an element remains in the ocean) on the order of 500,000 years. Thorium is very reactive in the marine environment and has a residence time of about 300 years. It is removed by adsorption onto the surfaces of solids and by incorporation into authigenic minerals. Given that the addition and removal of U to the ocean is in balance (a steady state has been achieved with respect to this element), ^{230}Th is produced at a constant rate. ^{230}Th activity is measured as a function of depth in a marine sediment core (the actual measurements are done on leachates so that contributions from silicate minerals are eliminated). As long as there has been no disruption of the sediment layers, the uppermost layer represents the present-day ^{230}Th deposition to the sediments. Comparison of the ^{230}Th activity at some depth compared to the surface-layer ^{230}Th activity gives the age of the sediment layer. For ^{230}Th, $t_{1/2} = 75,200$ years and $\lambda = 9.217 \times 10^{-6}$ y^{-1}. The ^{230}Th age can be determined using equation 6–8.

$$t = 108,495 \ln\left(\frac{^{230}\text{Th}_{\text{initial}}}{^{230}\text{Th}_{\text{meas}}}\right) \tag{6–25}$$

EXAMPLE 6–4 The ^{230}Th activity is measured for a marine sediment core. The top layer of the core has a ^{230}Th activity of 62 dpm. At a depth of 1 m the activity is 28 dpm. Calculate the age of the sediment at a depth of 1 m.

$$t = 108,495 \ln\left(\frac{^{230}\text{Th}_{\text{initial}}}{^{230}\text{Th}_{\text{meas}}}\right) = 108,495 \ln\left(\frac{62}{28}\right) = 86,246 \text{ y}$$

Assuming a constant sedimentation rate, calculate the rate of sediment accumulation.

$$\text{Rate} = \left(\frac{\text{Sediment thickness}}{\text{Time}}\right) = \left(\frac{1 \text{ m}}{86,246 \text{ y}}\right) = \left(\frac{100 \text{ cm}}{86.246 \times 10^3}\right) = 1.16 \text{ cm}/1000 \text{ y}$$

Note that marine sedimentation rates are usually given in cm/1000 years. ∎

^{230}Th/^{232}Th Dating of Marine Sediments ^{232}Th is the radioactive parent for the ^{232}Th–^{208}Pb decay chain. The first step in the chain is the decay of ^{232}Th to ^{228}Ra via an alpha emission with $t_{1/2} = 1.40 \times 10^{10}$ years. It is assumed that both isotopes of Th are removed simultaneously from seawater and that the ^{230}Th/^{232}Th ratio has been constant for the last several hundred thousand years—both reasonable assumptions. The two isotopes should chemically and physically show the same behavior, so variations in Th removal to the sediments will equally affect both isotopes. Thus, while there may be differences in the absolute concentrations of Th scavenged to the sediments, the initial

^{230}Th/^{232}Th activity ratio will be maintained. Because the half-life of ^{232}Th is orders of magnitude greater than the half-life of ^{230}Th, changes in the activity ratio will be solely due to changes in the abundance of ^{230}Th. Using equation 6–7, we can write

$$\left(\frac{^{230}\text{Th}}{^{232}\text{Th}}\right) = \left(\frac{^{230}\text{Th}}{^{232}\text{Th}}\right)^0 e^{-\lambda t} \qquad (6\text{--}26)$$

Setting $R_0 = (^{230}\text{Th}/^{232}\text{Th})^0$ and $R = (^{230}\text{Th}/^{232}\text{Th})$, the initial and measured activity ratios, and using equation 6–8, we can write

$$t = \frac{1}{\lambda} \ln\left(\frac{R_0}{R}\right) = \frac{1}{9.217 \times 10^{-6}} \ln\left(\frac{R_0}{R}\right) = 108{,}495 \ln\left(\frac{R_0}{R}\right) \qquad (6\text{--}27)$$

This system is used exactly the same way as the ^{230}Th system for the dating of sediments except that we are now using activity ratios instead of the activity of ^{230}Th alone.

So far, we have considered what is called *unsupported* ^{230}Th. By this we mean that the isotope has been separated from its radioactive parent (^{234}U) and the change in ^{230}Th activity can be described simply by the decay of this isotope. This is often the case, but there is another possibility that needs to be considered. If some U was also taken up by the sediment, there is a source of new ^{230}Th, and we refer to this as *supported* ^{230}Th (produced by radioactive decay). In this case, we would have a complex relationship in which ^{230}Th would initially decrease and then reach a constant activity when secular equilibrium was achieved. If we can determine the ^{238}U/^{232}Th ratio for the sediment, and if this ratio remained constant throughout the depositional interval, it is possible to correct the measured activities for the addition of supported ^{230}Th. At secular equilibrium, the activity of ^{234}U and ^{238}U would be the same; for each atom of ^{238}U that decays one atom of ^{234}U must decay. If this wasn't the case, then the concentration of ^{234}U or ^{230}Th would change and the system wouldn't be in secular equilibrium. The half-life of ^{238}U is so much longer than that of ^{230}Th that for the purpose of this correction the concentration of ^{238}U can be assumed to remain constant. We rewrite equation 6–26 with a second term on the right that accounts for the addition of ^{230}Th through the decay of ^{238}U.

$$\left(\frac{^{230}\text{Th}}{^{232}\text{Th}}\right) = \left(\frac{^{230}\text{Th}}{^{232}\text{Th}}\right)^0 e^{-\lambda t} + \left(\frac{^{238}\text{U}}{^{232}\text{Th}}\right)(1 - e^{-\lambda t}) \qquad (6\text{--}28)$$

Because the ^{238}U/^{232}Th ratio is essentially constant over a million-year time frame, and the half-life of ^{238}U is orders of magnitude greater than that of ^{230}Th, the growth of ^{230}Th is solely controlled by the decay constant of ^{230}Th. Hence, in equation 6–28, $\lambda = 9.217 \times 10^{-6}$ y^{-1}. As an illustration of the effect of supported ^{230}Th on age calculations, the ^{230}Th/^{232}Th ratio is calculated (Table 6–5) in 100,000-year increments, assuming that initial ^{230}Th/^{232}Th $= 42$ and ^{238}U/^{232}Th $= 2$.

Table 6–5   Calculation of ^{230}Th/^{232}Th for Unsupported and Supported ^{230}Th and Total ^{230}Th

Age	^{230}Th/^{232}Th Ratio		
10^5 years	Unsupported	Supported	Observed
0	42	0	42
1	16.7	1.20	17.9
2	6.65	1.68	8.33
3	2.64	1.87	4.51
4	1.05	1.95	3.00
5	0.42	1.98	2.40
6	0.17	1.99	2.16
7	0.07	2.00	2.07
8	0.03	2.00	2.03
9	0.01	2.00	2.01
10	0	2.00	2.00

Except for $t = 0$, the observed ^{230}Th/^{232}Th ratio is greater than the unsupported ratio. Hence, an age calculated on the assumption that only unsupported ^{230}Th was present would be younger than the actual age. At about 1 million years, the unsupported ^{230}Th has decayed away while the supported ^{230}Th has achieved secular equilibrium. For sediments older than 1 million years, the ^{230}Th/^{232}Th ratio would be constant.

^{230}Th/^{231}Pa Dating of Marine Sediments This method is very similar to the ^{230}Th/^{232}Th dating technique described in the previous section. ^{231}Pa is produced in the ^{235}U decay chain and has $t_{1/2} = 3.248 \times 10^4$ y and $\lambda = 2.134 \times 10^{-5}$ y^{-1}. Like thorium, protactinium is rapidly removed from seawater. If the removal proceeds with equal efficiency, then the activity ratio of ^{230}Th/^{231}Pa varies with time as a function of the difference in the decay constants. For unsupported ^{230}Th and ^{231}Pa,

$$\left(\frac{^{230}\text{Th}}{^{231}\text{Pa}}\right) = \left(\frac{^{230}\text{Th}}{^{231}\text{Pa}}\right)^0 \frac{e^{-\lambda_{230}t}}{e^{-\lambda_{231}t}} \tag{6–29}$$

If the uranium concentration of the leachable fraction of the sediment is negligible, the effective decay constant becomes $\lambda_e = \lambda_{230} - \lambda_{231} = 9.217 \times 10^{-6}$ y^{-1} $- 2.134 \times 10^{-5}$ y$^{-1} = -1.2123 \times 10^{-5}$ y^{-1} and equation (6–29) can be written

$$\left(\frac{^{230}\text{Th}}{^{231}\text{Pa}}\right) = \left(\frac{^{230}\text{Th}}{^{231}\text{Pa}}\right)^0 e^{\lambda_e t} \tag{6–30}$$

where λ_e is now positive and equals 1.2123×10^{-5} y^{-1}. Because the half-life of ^{230}Th is greater than that of ^{231}Pa, the activity ratio will actually increase as a function of age. If uranium is present in the leachable fraction, then corrections must be applied for supported ^{230}Th and ^{231}Pa, a more complicated problem that will not be discussed here.

Activity and Sedimentation-Rate Relationships Because radioactive decay is a first-order reaction—i.e., the relationship is exponential—even when the sedimentation rate is constant a plot of activity versus sediment depth will yield a curved line (Figure 6–2). However, for constant sedimentation rates a plot of logarithmic activity versus arithmetic sediment depth will yield a straight line (Figure 6–3). In the case of a constant sedimentation rate, the rate can be calculated from the following relationship:

$$a = \frac{\lambda}{-2.303m} \tag{6–31}$$

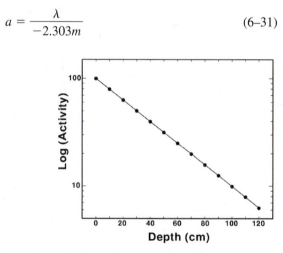

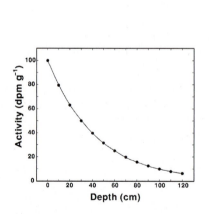

Figure 6–2

Variation of activity with depth at a constant sedimentation rate. Because radioactivity is a first-order reaction, the data define a curved line.

Figure 6–3

Plot of log activity versus depth (arithmetic). For a constant sedimentation rate, the data define a straight line and the sedimentation rate can be determined from the slope of the line (see text). In this case, $m = -0.01003$. If $\lambda = 9.24 \times 10^{-6}$, the sedimentation rate (a) is 4×10^{-4} cm/y = 0.4 cm/1000 y.

where a is the sedimentation rate, m is the slope of the straight line, and λ is the decay constant. If the sedimentation rate is not constant, then a logarithmic activity versus arithmetic depth plot will not yield a straight line (or straight-line segments).

^{230}Th/^{234}U Dating of Calcium Carbonate

This dating system is based on the observation that calcium carbonate formed in the ocean and saline lakes is essentially free of Th but contains significant concentrations (0.1 to 5 ppm) of U. In the ^{238}U decay chain, ^{234}U decays to ^{230}Th via an alpha emission. The half-life for this decay is 2.48×10^5 y with a corresponding decay constant of 2.794×10^{-6} y^{-1}. If the ^{234}U is in secular equilibrium with its parent ^{238}U, the activity of ^{234}U will remain constant with time while the activity of ^{230}Th will increase. Hence, we can use the change in the ^{230}Th/^{234}U as a measure of the age of the calcium carbonate. In this case, we are looking at the growth of a radiogenic progeny, rather than its decay, so the decay equation is written

$$\left(\frac{^{230}\text{Th}}{^{234}\text{U}}\right) = 1 - e^{-\lambda t} \tag{6–32}$$

which on rearranging and substituting for λ becomes

$$t = \frac{1}{\lambda} \ln\left[1 - \left(\frac{^{230}\text{Th}}{^{234}\text{U}}\right)\right] = 108{,}495 \ln\left[1 - \left(\frac{^{230}\text{Th}}{^{234}\text{U}}\right)\right] \tag{6–33}$$

In the ocean there is actually a slight excess of ^{234}U; i.e., the ^{234}U/^{238}U ratio is 1.15 rather than 1, as would be observed if the two isotopes were in secular equilibrium. If a correction is not made for this excess, the calculated ages are slightly older than the true ages. Once the system has been isolated from seawater, the excess uranium begins to decrease, and after a sufficiently long time the 1:1 ratio expected for secular equilibrium is achieved. Given the 2.48×10^5 y half-life of ^{234}U, it will take more than a million years for secular equilibrium to be reestablished, and the correction to be applied will be a function of the age of the sample.

Why Do We Want to Date Marine Sediments and Carbonate Minerals?

Both ^{14}C and the U-series disequilibrium systems can be used to date marine sediments. Besides our intrinsic interest in earth processes, the dating of marine sediments is essential for climate change studies. At least in the deep ocean, vertical sedimentation is the main depositional process. Typical sediment accumulation rates are on the order of 0.1 to 1 cm/1000 years. Thus, we have a simple stratigraphic record and the potential for a large amount of information to be contained in a relatively short core section of seafloor sediment. Processes at the earth's surface affect the sedimentary record in the deep ocean. For example, variations in the amount of water stored on the continents as ice and surface temperature variations affect a variety of isotopic systems in the ocean (to be discussed later in this chapter). These variations can be used to infer climatic conditions. Additional examples include changes in coarse-fractions deposited in deep-sea sediments due to variations in continental exposure and glacial action and changes in the distribution of planktonic species in the surface waters (subsequently preserved in the sediment record) in response to changes in the surface temperatures of the ocean. It is generally believed that an understanding of past climates will be important in assessing the potential impact of anthropogenic changes to the earth's present atmosphere. Thus, this is an important area of current research, and it is essential to have good chronological control when interpreting the seafloor sediment record.

^{210}Pb Dating

As a final example of U-series disequilibrium dating, we will consider the ^{210}Pb system, which can be used to date a variety of processes occurring during the last 100 years. ^{210}Pb occurs in the ^{238}U–^{206}Pb decay chain. ^{226}Ra decays to ^{222}Rn, which subsequently decays through a series of short-lived daughters to ^{210}Pb. Rn (radon) is a noble gas and rapidly escapes into the atmosphere. Radon decays to ^{210}Pb in a matter of days. The ^{210}Pb is rapidly removed from the atmosphere (mean residence time of about 10 days) by rain or snow. ^{210}Pb is removed from river and ocean water by a combination of inorganic and biochemical reactions. Its mean residence time in the aquatic systems is short, usually less than one year. ^{210}Pb is a short-lived isotope with $t_{1/2} = 22.26$ y and $\lambda = 3.11 \times 10^{-2}$ y^{-1}. Assuming that all the ^{210}Pb is unsupported, equation 6–8 can be used to determine the age:

$$t = \frac{1}{3.11 \times 10^{-2}} \ln\left(\frac{^{210}\text{Pb}^0}{^{210}\text{Pb}}\right) = 32.15 \ln\left(\frac{^{210}\text{Pb}^0}{^{210}\text{Pb}}\right) \qquad (6\text{–}34)$$

where ^{210}Pb^0 = the initial activity of ^{210}Pb and ^{210}Pb = the measured (present-day) activity of ^{210}Pb. In principle, this method can be used to determine the age of water samples, the accumulation rate for snow and ice, and sedimentation rates. The presence of U in the sediment can complicate the determination of the sedimentation rate because the decay of ^{238}U will add ^{210}Pb to the sediment (i.e., supported ^{210}Pb). The contribution of supported ^{210}Pb to the total ^{210}Pb activity can be determined by analyzing a sediment sample (a deeper layer in the core) in which all the unsupported ^{210}Pb has decayed or by measuring the activity of ^{226}Ra in each sample and using this activity to calculate the amount of supported ^{210}Pb. Case Study 6–2 describes an application of the ^{210}Pb dating method.

To illustrate the variety of applications of radioisotopes to environmental problems, Case Study 6–3 (p. 180) describes how ^{234}U and ^{238}U isotopes were used to determine the impact of uranium-mill effluent on near-surface groundwater.

Radiogenic Isotopic Tracers

Radiogenic isotopic tracers are widely used in the geosciences. Several of these isotopic systems have environmental applications. For example, the isotopic composition of Sr in groundwaters can be used to infer the source(s) of the groundwaters. Similarly, the isotopic composition of Pb, a widespread and significant contaminant, can be used to infer the source of the Pb. In this section we will briefly consider the Rb–Sr and U–Th–Pb isotopic systems.

Rb–Sr System   There are two naturally occurring isotopes of Rb: $^{85}_{37}$Rb (72.654%) and $^{87}_{37}$Rb (27.8346%). $^{87}_{37}$Rb is radioactive and decays to $^{87}_{39}$Sr via a beta emission. There are four naturally occurring isotopes of Sr: $^{88}_{38}$Sr (82.53%), $^{87}_{38}$Sr (7.04%), $^{86}_{38}$Sr (9.87%), and $^{84}_{38}$Sr (0.56%). Note that these are only approximate percentages because the amount of $^{87}_{38}$Sr produced by radioactive decay will vary from sample to sample. This isotopic system is used in the geosciences to determine the age of igneous and metamorphic rocks, but given the long half-life, it is not a useful system for dating environmental processes. Our interest here is in the variations in the amount of $^{87}_{38}$Sr in various reservoirs. Mass spectrometers can very precisely measure mass ratios, so we generally measure the amount of the radioactive parent and radiogenic progeny relative to a stable isotope of the element of interest. Thus, the change in radiogenic Sr with time can be represented by the following equation:

$$\left(\frac{^{87}\text{Sr}}{^{86}\text{Sr}}\right)_{\text{meas}} = \left(\frac{^{87}\text{Sr}}{^{86}\text{Sr}}\right)_{\text{initial}} + \left(\frac{^{87}\text{Rb}}{^{86}\text{Sr}}\right)(e^{\lambda t} - 1) \qquad (6\text{–}35)$$

where $(^{87}\text{Sr}/^{86}\text{Sr})_{\text{meas}}$ is the measured Sr isotopic ratio for the sample, $(^{87}\text{Sr}/^{86}\text{Sr})_{\text{initial}}$ is the Sr isotopic ratio for Sr present in the sample when it was formed, $(^{87}\text{Rb}/^{86}\text{Sr})$ is the present-day isotopic ratio, λ is the decay constant, and t is the time that has elapsed since the sample was formed. Note that the two major factors determining the present-day Sr isotopic ratio for the sample are the ^{87}Rb/^{86}Sr ratio and the time that has elapsed since the sample was formed.

CASE STUDY 6–2
^{210}Pb Dating of Sediments in Lake Constance, Germany, and Inputs of Pb and Zn to Lake Constance

Bollhöfer et al. (1994) determined the ^{210}Pb ages for two cores from Lake Constance in Germany and used the age distribution to determine the timing of inputs of Pb and Zn to the lake sediments. They also investigated the distribution of radioactive ^{134}Cs, ^{137}Cs, and ^{241}Am in the sediment. The radioactive Cs had two sources: (1) the Chernobyl reactor accident in 1986 and (2) aboveground nuclear weapons testing that produced a Cs maximum in 1963–1964. ^{241}Am is also the result of weapons testing and is produced by the decay of ^{241}Pu to ^{241}Am by emission of a β^- particle. Two sediment cores, which had been previously dated by lamination counting, were used in the study.

The activity of the various radionuclides was determined by gamma ray spectroscopy. The activity due to supported ^{210}Pb was determined by measuring the activity of ^{214}Pb and ^{214}Bi. The source of the supported ^{210}Pb activity is ^{238}U associated with the detrital material in the sediment. ^{214}Pb and ^{214}Bi are both found in the ^{238}U decay chain (Table 6–1), and hence their activity can be used to estimate the supported ^{210}Pb activity. The calculation is done using the equation

$$Act(^{210}Pb_{excess}) = Act(^{210}Pb_{total}) - Act(^{210}Pb_{supported})$$

The ages determined by ^{210}Pb dating, compared to the lamination ages, are shown in Figure 6–C2–1. In general, there is good agreement between the two methods. A logarithmic plot of ^{210}Pb activity versus the compaction-corrected depth gave a mean accumulation rate of $0.16 \, g \, cm^{-2} \, y^{-1}$. Two ^{137}Cs maxima were found in the cores, one at a depth of approximately 2 cm, corresponding to the time of the Chernobyl reactor accident, and the other at a depth of approximately 16 cm, corresponding to a ^{210}Pb age of 29 ± 4 y before present ($1990_{present} - 29$ years = 1961), in agreement, within error, with the time of the Cs maximum from nuclear bomb testing. ^{241}Am also showed a maximum at 16 cm, corresponding to the time of bomb testing. Hence, in this system both bomb-produced radioisotopes proved to be reliable time markers.

The distribution of Pb and Zn in the lake sediments, versus time, is show in Figure 6–C2–2. Preindustrial input of these two metals was determined by measuring the Pb (20 ppm) and Zn (50 ppm) content of the deepest sediment layers, which were deposited before the input of anthropogenic heavy metals. The Pb maximum of approximately 150 ppm and Zn maximum of approximately 330 ppm greatly exceed the preindustrial values. Both maxima are found below the layer with the maximum inventory of bomb-produced radioisotopes. The ^{210}Pb ages of the maxima are 1957 for Pb and 1960 for Zn. Pb released by the burning of leaded gasoline reached a peak (in Germany) in 1971. The Pb maximum in the Lake Constance sediment is 14 years earlier, and at the time of the Pb atmospheric maxi-

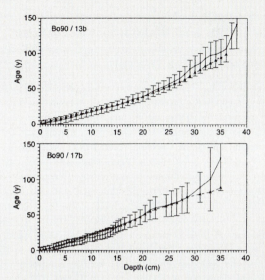

Figure 6–C2–1
^{210}Pb ages for the two cores using a constant-flux model (solid line) compared to lamination counting (dashed line). From Bollhöfer et al. (1994).

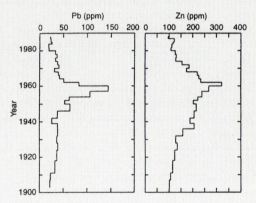

Figure 6–C2–2
Lead and zinc concentrations versus ^{210}Pb age in sediment core 13b. From Bollhöfer et al. (1994).

mum the Pb in the Lake Constance cores is significantly less than the peak value in 1957. Hence, the Pb (and Zn) input to Lake Constance is from other sources, most likely from the Alpenrhein River, which is the main inlet to Lake Constance, and perhaps from localized atmospheric inputs that reached a maximum in 1957. These sources may be coal burning and the inflow of sewage or sewage sludge through lead pipes.

Source: Bollhöfer et al. (1994).

CASE STUDY 6–3
Use of ^{234}U and ^{238}U Isotopes to Evaluate Contamination of Near-Surface Groundwater with Uranium-Mill Effluent

Using the ^{234}U/^{238}U alpha activity ratio (AR), Zielinski et al. (1997) mapped the extent of uranium contamination in near-surface groundwaters in the vicinity of an abandoned uranium mill. The major aquifers are Quaternary alluvial-terrace deposits and Paleocene sandstones and fluvial conglomerate of the Poison Canyon Formation. Liquid waste (raffinate) and tailings solids stored on the site are the source of the uranium contamination in the aquifers. Besides high concentrations of uranium, liquid wastes from the site also have high concentrations of molybdenum and selenium. Because of the high concentrations of these elements in some off-site water wells, the site was added to the Superfund National Priority List in 1984.

For the original uranium ore, secular equilibrium was achieved. Thus, the alpha activity of ^{234}U and ^{238}U was the same, giving a theoretical AR = 1. Measured values for liquid waste ranged from 0.98 to 1.05. Most natural groundwater has AR > 1, ranging up to 3 and in some cases up to > 10. The excess ^{234}U is believed to be due to the fact that alpha-recoil displacement occurs when ^{238}U decays to ^{234}U. This displacement damages the crystal structure in the vicinity of the ^{234}U atom and this site is more easily leached during mineral/water interactions. Three background samples from the alluvial aquifer had AR values ranging from 1.32 to 1.41. The groundwaters were found to have a high oxygen content. Thus, the waters are oxidizing and U (and Mo) are expected to remain in solution.

The AR and U concentration for various groundwater samples from the area are shown in Figure 6–C3–1. The field labeled "Range of raffinate activity ratio" shows the range in uranium concentrations and AR expected for raffinate (liquid waste) from the abandoned uranium plant. The field labeled "Range of background activity ratio" shows the range in uranium concentrations and AR expected for uncontaminated groundwater. The field between represents groundwater samples that consist of a mixture of raffinate and uncontaminated groundwater. The dashed lines are mixing lines for various combinations of raffinate and uncontaminated groundwater. The relative volumes of the two end members in the mixed water can be determined from the following relationship:

$$V_1/V_m = (S_1/S_m) \times [(AR_2 - AR_m)/(AR_2 - AR_1)]$$

where V_1 is the volume of end member 1, V_m is the volume of the mixed water (this ratio gives the proportion of end member 1 in the mixed water), $S = 1$/uranium concentration, AR is the alpha activity ratio, and subscripts 1, 2, and m represent end member 1, 2, and the mixed water, respectively. An areal plot of the AR values from the various groundwater samples showed a plume of dissolved uranium that originated in the vicinity of the old tailings ponds and extended northward into the alluvial aquifer. The uranium plume as defined by AR values did not extend as far downgradient as the molybdenum plume (defined by elevated Mo concentrations in the groundwater), suggesting that uranium may be less mobile than molybdenum in the alluvial aquifer.

_____

Source: Zielinski et al. (1997).

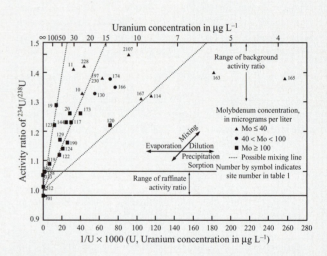

Figure 6–C3–1
Relationship between AR and U content of groundwater samples and raffinate. The effect of various processes on these values is illustrated in the figure. Evaporation and dilution (or sorption) will change the total uranium concentration. Only mixing will change the AR values. From Zielinski et al. (1997).

During the process of crust formation, melts are extracted from the mantle of the earth and added to the crust. Because of differences in the partitioning of Rb and Sr between solid mantle phases (i.e., minerals) and melt, Rb is preferentially enriched with respect to Sr in the melt. The higher Rb/Sr ratio for the melt ultimately leads to higher ^{87}Sr/^{86}Sr ratios for the crust, which is formed either directly by crystallization of the melt or indirectly by rock weathering. For the same Rb/Sr ratio, the greater the age of the crust, the greater the ^{87}Sr/^{86}Sr ratio. Conversely, for the mantle from which the melt was extracted (depleted mantle), the growth in ^{87}Sr with time is less than that for the mantle from which a melt has not been extracted (undepleted mantle). The result is a number of Sr isotopic reservoirs with different ^{87}Sr/^{86}Sr ratios.

As another example of Sr isotopic variability, let us briefly consider what might happen to the Sr isotopic composition of ocean waters throughout geologic time. The Sr found in the oceans comes from rock weathering and submarine volcanism (see Chapter 10 for a discussion of marine chemistry). As the relative proportions of these contributions vary, and as time progresses, the Sr isotopic composition of the ocean will change. This change is preserved in marine carbonates that formed at different times throughout earth history. Relative to Rb, calcite is highly enriched in Sr (this is because of the crystal structure of calcite, see Chapter 7), and the Sr isotopic ratio of the marine carbonate is approximately the same as the seawater from which it formed. Thus, it has been possible to determine the ^{87}Sr/^{86}Sr ratios of ocean waters as a function of time. This has been, and continues to be, an active area of research because the marine Sr isotopic curve can be used for correlation of marine carbonates. Our interest here, however, is the observation that marine carbonates formed at different times in earth history have different Sr isotopic ratios. Thus, there are a number of ways to produce isotopically distinct reservoirs. Sr found in waters in equilibrium with these different reservoirs would have different ^{87}Sr/^{86}Sr ratios, thus fingerprinting the source(s) of the waters.

Th–U–Pb Isotopic Systems   We have already discussed the U (Table 6–1) and Th (Table 6–2) decay schemes. All three decay schemes end with a lead isotope. Of the naturally occurring Pb isotopes, only ^{204}Pb is nonradiogenic. As we did for Sr, we can write equations that represent the change in radiogenic Pb isotopic composition with time, and in this case we will use ^{204}Pb as the reference lead isotope.

$$\left(\frac{^{206}Pb}{^{204}Pb}\right)_{meas} = \left(\frac{^{206}Pb}{^{204}Pb}\right)_{initial} + \left(\frac{^{238}U}{^{204}Pb}\right)(e^{\lambda_{238}t} - 1) \tag{6–36}$$

$$\left(\frac{^{207}Pb}{^{204}Pb}\right)_{meas} = \left(\frac{^{207}Pb}{^{204}Pb}\right)_{initial} + \left(\frac{^{235}U}{^{204}Pb}\right)(e^{\lambda_{235}t} - 1) \tag{6–37}$$

and

$$\left(\frac{^{208}Pb}{^{204}Pb}\right)_{meas} = \left(\frac{^{208}Pb}{^{204}Pb}\right)_{initial} + \left(\frac{^{232}Th}{^{204}Pb}\right)(e^{\lambda_{232}t} - 1) \tag{6–38}$$

where λ_{238} is the decay constant for ^{238}U, λ_{235} is the decay constant for ^{235}U, and λ_{232} is the decay constant for ^{232}Th. As was the case for Sr, the Pb isotopic ratios for a particular sample depend upon the U/Pb and Th/Pb ratios and the time that has elapsed since the sample was formed. Hence, isotopic reservoirs with different U/Pb and Th/Pb ratios formed at different times will have different Pb isotopic ratios.

Our interest here is anthropogenic lead. Anthropogenic lead is derived from lead ore deposits, and the lead ore mineral is galena (PbS). U and Th do not occur in galena because they are excluded by the crystal structure (see Chapter 7). Thus, the galena freezes-in the Pb isotopic ratios that existed at its time of formation. As was the case for the Sr isotopic system, the Pb isotopic ratios increase with time. Thus, ore deposits formed at different times in earth history have different Pb isotopic ratios. These differences can be used to fingerprint the source of the Pb. This topic is revisited in some detail in Chapters 8 and 10.

STABLE ISOTOPES

Stable isotopes do not spontaneously break down to form other isotopes. They are useful in environmental and geological studies because physical, chemical, and biological processes can lead to changes in the isotopic ratios. These changes arise because of the mass differences between the isotopes. The effects are most noticeable for isotopes that have large relative mass differences. Hence, most of the elements used for stable isotope studies have a low atomic number and atomic mass (Table 6–6). For isotopes with an atomic mass of greater than 40, the relative mass differences are too small for any measurable isotopic fractionation. Most elements used in radiometric dating have high atomic mass, and mass fractionation is not significant. Note, however, that in the previous section we did describe several dating systems that involve low-atomic-number elements. Particularly in the case of ^{14}C dating using carbonate shells, there can be significant fractionation of ^{14}C

Table 6–6   *Average Terrestrial Abundances of Stable Isotopes Used in Environmental Studies**

Element	Isotope	Average terrestrial abundance (atom %)
Hydrogen	$^{1}_{1}H$	99.985
	$^{2}_{1}H$	0.015
Carbon	$^{12}_{6}C$	98.9
	$^{13}_{6}C$	1.1
Nitrogen	$^{14}_{7}N$	99.63
	$^{15}_{7}N$	0.37
Oxygen	$^{16}_{8}O$	99.762
	$^{17}_{8}O$	0.038
	$^{18}_{8}O$	0.2
Sulfur	$^{32}_{16}S$	95.02
	$^{33}_{16}S$	0.75
	$^{34}_{16}S$	4.21
	$^{36}_{16}S$	0.014

*Data source: IUPAC (1992).

between the shell material and the water from which the carbon is extracted. To partly compensate for this effect, ^{14}C analyses should only be done on shells from the same species.

Basic Principles

Stable Isotope Fractionation   Isotopic fractionation is the partitioning of isotopes during physical (evaporation, condensation, melting, crystallization, absorption and desorption, diffusion), chemical, or biological processes. This partitioning is proportional to the difference in the masses of the isotopes. The processes can either be ***equilibrium reactions***, in which *forward and backward reaction rates are equal for each isotope*, or ***kinetic reactions***, which are *unidirectional reactions in which reaction rates are dependent on the masses of the isotopes and their vibrational energies.*

As an example of an equilibrium reaction, consider the evaporation of water into a closed space. At equilibrium, the forward and backward rates of the reaction will be the same, but the isotopic ratios in each phase (liquid and vapor) will be different. There are six possible isotopic combinations for H_2O molecules: $^{1}H_2{}^{16}O$, $^{1}H^{2}H^{16}O$, $^{2}H_2{}^{16}O$, $^{1}H_2{}^{18}O$, $^{1}H^{2}H^{18}O$, and $^{2}H_2{}^{18}O$, with atomic masses, respectively, of 18, 19, 20, 20, 21, and 22. We have ignored ^{17}O because of its low abundance. Water molecules move from the liquid to vapor phase when their energy is sufficiently high to overcome the electrostatic attractions that tend to hold the molecules together in the liquid phase. From basic physics: kinetic energy $= \frac{1}{2}$mass $\times$ velocity^2. For any given amount of energy, the isotopically lighter molecule will have a greater velocity than the isotopically heavier molecule. Thus, we would expect the isotopically lighter molecule to preferentially escape to the vapor phase. The result is that the vapor phase becomes enriched in the lighter isotopes (^{1}H and ^{16}O) relative to the liquid.

As an example of a kinetic reaction, consider the breakdown of limestone ($CaCO_3$), on the addition of acid, to Ca^{2+} and CO_2 gas. This reaction is unidirectional—the CO_2 gas escapes and thus does not equilibrate with the solid phase. During a chemical reaction the lighter isotope is more reactive and is concentrated in the reaction products. The relative loss of the lighter isotope from the reactants leads to an enrichment of the reactants in the heavier isotope. The lighter isotope is more reactive because it has a higher vibrational energy and thus a weaker bond. In this example, the gas phase (product) is enriched in the lighter isotopes of C and O, and the enrichment is significantly greater than that which would occur during an equilibrium reaction.

Fractionation Factor The partitioning of stable isotopes between two substances, A and B, is described by the *isotopic fractionation factor, α*. The fractionation factor is written

$$\alpha = \frac{R_A}{R_B} \tag{6-39}$$

where R_A is the ratio of the heavy to the light isotope in molecule or phase A and R_B is the ratio of the heavy to the light isotope in molecule or phase B. The fractionation factor varies as a function of temperature. As an example, consider the evaporation of water (Figure 6–4). ^{16}O is enriched in the vapor phase relative to the liquid. Conversely, ^{18}O is enriched in the liquid relative to the vapor, and thus the fractionation factor is greater than 1. With increasing temperature, the fractionation factor decreases, and it becomes 1 at infinite temperature. This happens because at high temperatures the isotopic species are well mixed. Isotopic fractionation factors can be determined either experimentally or calculated from spectroscopic data. In practice, most isotopic fractionation factors are determined empirically from a large set of observational data or by comparisons of natural materials to experimental results obtained in the laboratory.

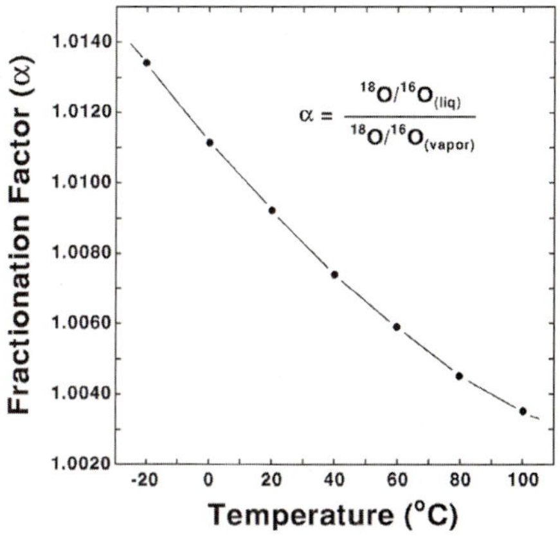

Figure 6–4

Variation of the isotope fractionation factor for oxygen, as a function of temperature, during the evaporation of water. Note that with increasing temperature the fractionation factor approaches 1.0000. Values from Dansgaard (1964).

The isotopic fractionation factor is related to the equilibrium constant of a reaction as follows:

$$\alpha = K^{1/n} \tag{6-40}$$

where K is the equilibrium constant and n is the number of atoms exchanged. Consider the carbonate exchange reaction between solid calcium carbonate and water. This particular reaction is of interest because it is used as a paleothermometer in the marine system. Writing the reaction in terms of a single oxygen, so $n = 1$ in equation 6–40,

$$\tfrac{1}{3}CaC^{16}O_3 + H_2{}^{18}O \rightleftharpoons \tfrac{1}{3}CaC^{18}O_3 + H_2{}^{16}O \tag{6-41}$$

Writing this equation in terms of the equilibrium constant, K,

$$K = \frac{[CaC^{18}O_3]^{1/3}[H_2{}^{16}O]}{[CaC^{16}O_3]^{1/3}[H_2{}^{18}O]} = \frac{([CaC^{18}O_3]/[CaC^{16}O_3])^{1/3}}{[H_2{}^{18}O]/[H_2{}^{16}O]} \tag{6-42}$$

and

$$K = \alpha = \frac{R_c}{R_w} = \frac{^{18}O/^{16}O \text{ ratio of calcium carbonate}}{^{18}O/^{16}O \text{ ratio of water}} \tag{6-43}$$

Because the isotopic fractionation factor varies as a function of temperature, the measured fractionation factor can be compared to the experimental values to determine the equilibrium temperature for the calcium carbonate–water reaction. In practice, this calculation is a bit more complicated, as will be discussed in a later section.

The δ (Delta) Notation Because isotopic variations are very small, isotopic ratios are written using the delta notation.

$$\delta = \left(\frac{\text{Isotopic ratio of sample} - \text{Isotopic ratio of standard}}{\text{Isotopic ratio of standard}} \right) \times 1000 \quad (6\text{--}44)$$

A variety of standards are used for comparison (see Table 6–7): **SMOW** (standard mean ocean water), **V-SMOW** (Vienna standard mean ocean water), and **SLAP** (standard light Antarctic precipitation) for oxygen and hydrogen isotope ratios in water; **PDB** (PeeDee beleminite) for oxygen and carbon isotope ratios in carbonates; **AIR** (air) for nitrogen; and **CDT** (Canyon Diablo troilite) for sulfur. PDB is now exhausted so a new standard, **NBS-19** (National Bureau of Standards, now NIST), has been established for carbon and oxygen in carbonates. The new standards are related to PDB as follows: $\delta^{18}O_{\text{NBS-19/V-PDB}} = -2.20$ and $\delta^{13}C_{\text{NBS-19/V-PDB}} = 1.95$ (O'Neil, 1986). If the sample is enriched in the heavier isotope relative to the standard, the delta values are positive. If the sample is depleted in the heavier isotope relative to the standard, the delta values are negative.

Table 6–7 *Stable Isotope Ratios for Standards**

Element	Standard	Ratio
Hydrogen	V-SMOW	$^2H/^1H = 155.76 \times 10^{-6}$
Carbon	PDB	$^{13}C/^{12}C = 1123.75 \times 10^{-5}$
Oxygen	V-SMOW	$^{18}O/^{16}O = 2005.2 \times 10^{-6}$
	PDB	$^{18}O/^{16}O = 2067.2 \times 10^{-6}$
Nitrogen	NBS-14	$^{15}N/^{14}N = 367.6 \times 10^{-5}$
Sulfur	CDT	$^{34}S/^{32}S = 449.94 \times 10^{-4}$

*Data from Kyser (1987).

EXAMPLE 6–5 The isotopic ratio of $^{18}O/^{16}O$ in V-SMOW is 0.0020052. A rainwater sample collected in Boston, Massachusetts, has an $^{18}O/^{16}O$ ratio of 0.0019750. Calculate the delta value for this rainwater sample.

$$\delta = \left(\frac{\text{Ratio}_{\text{sample}} - \text{Ratio}_{\text{standard}}}{\text{Ratio}_{\text{standard}}} \right) \times 1000 = \left(\frac{0.0019750 - 0.0020052}{0.0020052} \right) \times 1000$$

$$= -15.1\text{‰}$$

The delta value is reported in parts-per-thousand (‰), and the negative value means that the sample is isotopically lighter than the standard. ∎

In practice, isotopic ratios are usually measured with respect to the appropriate standard and the results are reported directly in delta values. Because isotopic measurements are reported in delta values, it is useful to derive a relationship that relates the measured delta values to the isotopic fractionation factor. From equation 6–44,

$$R_a = \frac{\delta_a R_{\text{std}}}{10^3} + R_{\text{std}} = \frac{R_{\text{std}}(\delta_a + 10^3)}{10^3} \quad (6\text{--}45)$$

Substituting for R_A and R_B in equation 6–39 gives

$$\alpha_b^a = \frac{R_a}{R_b} = \frac{10^3 R_{\text{std}}(\delta_a + 10^3)}{10^3 R_{\text{std}}(\delta_b + 10^3)} = \frac{\delta_a + 1000}{\delta_b + 1000} \quad (6\text{--}46)$$

EXAMPLE 6–6 The oxygen isotopic fractionation factor for the evaporation of water under equilibrium conditions at 25°C is 1.0092 (Craig and Gordon, 1965). Calculate the

δ-value for vapor in equilibrium with lake water having a δ-value $= -5.0$‰.
From equation 6–46,

$$\alpha_v^l = 1.0092 = \frac{R_l}{R_v} = \frac{\delta_l + 1000}{\delta_v + 1000}$$

Rearranging and solving for δ_v,

$$\delta_v = \frac{\delta_l + 1000}{\alpha_v^l} - 1000 = \frac{-5 + 1000}{1.0092} - 1000 = -14.1‰ \qquad \blacksquare$$

Oxygen and Hydrogen Isotopes in Water

Water molecules consist solely of hydrogen and oxygen so both isotopes usually show similar fractionation patterns, and hence we will consider both elements in the following discussion. Because these elements *are* water, they make ideal tracers for water. Different waters have different isotopic signatures, and these can be used to trace the source(s) of a water mass and/or to provide information on processes within the hydrologic cycle. The source of water and its movement through surface and ground water systems is essential information for studies involving water contamination. A number of factors are responsible for the variations in water isotopic chemistry and these are considered next.

Factors Affecting the Isotopic Composition of Water As described in a previous section and illustrated in Figure 6–4, isotopic fractionation factors vary as a function of temperature. For seawater at 25°C, the isotopic fractionation factors for oxygen and hydrogen are $\alpha_{18} = 1.0092$ and $\alpha_D = 1.074$ (Craig and Gordon, 1965). The ^2H isotope of hydrogen is called deuterium, and the isotopic ratio is often written D/H, where H stands for common hydrogen (^1H) and D stands for deuterium. The isotopic fractionation factor for hydrogen is often written with a D subscript. Given that seawater has a delta value of zero for both isotopes, solving equation 6–46 for the vapor phase yields $\delta_{18} = -9.1$‰ and $\delta_D = -69$‰ for vapor in equilibrium with seawater. At 10°C, $\delta_{18} = -10.1$‰ and $\delta_D = -84$‰, illustrating the effect temperature has on isotopic fractionation. Also note that both the absolute delta values and the change in delta values are much greater for hydrogen. This is because the relative mass differences are much greater for hydrogen than oxygen. Thus, one factor that will affect isotopic ratios for water vapor and precipitation is temperature variation. This temperature variation can be either latitudinal (higher latitudes equal lower temperatures) or altitudinal (higher altitudes equal lower temperatures). Note that these calculations assume an equilibrium reaction. Actual values for water vapor over the ocean are on the order of -13‰ to -14‰, indicating that the evaporative process is a kinetic reaction rather than an equilibrium reaction (see later).

When raindrops form in a cloud they are enriched in ^{18}O and D relative to the coexisting vapor phase. The first droplets will have an isotopic composition similar to that of ocean water, the exact value depending on the temperature at which the condensation process occurs. Conversely, during condensation the water vapor will become enriched in ^{16}O and ^1H. As condensation continues, the water vapor will become progressively isotopically more negative and there will be a corresponding change in the isotopic composition of the water droplets. This process can be described by the Rayleigh distillation equation (Broecker and Oversby, 1971)

$$\frac{R}{R_0} = f^{(\alpha - 1)} = \frac{\delta^{18}O + 1000}{\delta^{18}O_0 + 1000} \qquad (6\text{–}47)$$

where R is the $^{18}O/^{16}O$ ratio of the water vapor at any point during the condensation process, R_0 is the $^{18}O/^{16}O$ ratio of the vapor before condensation begins, f is the fraction of water vapor remaining, and α is the fractionation factor R_l/R_v. We can rewrite equation 6–47 in terms of the $\delta^{18}O$ value of the water vapor:

$$\delta^{18}O_v = [\delta^{18}O_0 + 1000] f^{(\alpha - 1)} - 1000 \qquad (6\text{–}48)$$

To calculate the $\delta^{18}O$ value for the condensate in equilibrium with the vapor at any instant,

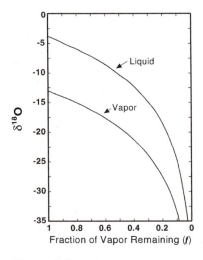

Figure 6–5

Fractionation of oxygen isotopes during Rayleigh distillation of water vapor at 25°C. The initial $\delta^{18}O$ value of the vapor is $-13‰$.

we solve equation 6–46 for the $\delta^{18}O$ value of the liquid by setting R_a = the liquid ratio and R_b = the vapor ratio:

$$\delta^{18}O_l = \alpha(\delta^{18}O_v + 1000) - 1000 \quad (6\text{–}49)$$

The changes in isotopic compositions for both vapor and liquid, assuming a Rayleigh distillation model, are illustrated in Figure 6–5. The initial water vapor has $\delta^{18}O = -13‰$ and the reaction takes place at 25°C ($\alpha = 1.0092$).

As condensation proceeds, both the vapor and the liquid become isotopically lighter. Thus, the isotopic composition of rainwater changes during a precipitation event; i.e., earlier precipitation is isotopically heavier than later precipitation. One outcome of this process is that the delta values of rain tend to decrease as one moves inland from a continental margin. With subsequent reevaporation and reprecipitation, there is a continuing decrease in the delta values for rainwater. Thus, as one moves poleward the rain becomes isotopically lighter, a latitudinal effect that is also enhanced by the decline in ambient atmospheric and seawater temperatures. On the basis of a large number of analyses of meteoric waters collected at different latitudes, Craig (1961) was able to show there was a relationship between the $\delta^{18}O$ and δD values that could be represented by the equation

$$\delta D = 8\delta^{18}O + 10 \quad (6\text{–}50)$$

This relationship defines the mean global meteoric water line (Figure 6–6). Also note that in equation 6–50 there is an excess in δD. Dansgaard (1964) found that there was a good correlation between the $\delta^{18}O$ values of average annual precipitation and the average annual air temperature. More generally, we can write equation 6–50 as

$$\delta D = 8\delta^{18}O + d \quad (6\text{–}51)$$

where d = the deuterium excess.

Equation 6–50 defines isotopic variations on a global basis; on a local basis each area has its own meteoric water line (Figure 6–6). These plots are usually made by setting the slope of the straight line to 8, although regression of the actual data may yield slopes other than 8. In these plots excess deuterium varies from 0‰ to +22‰ and cool and humid

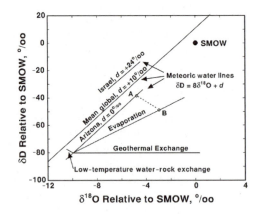

Figure 6–6

Plot of δD versus $\delta^{18}O$ illustrating the mean global meteoric water line and local meteoric water lines. Other processes that affect the isotopic ratios—e.g., low-temperature water–rock exchange, geothermal exchange, and evaporation—are also illustrated. A and B are two water masses and the dashed line represents the possible isotopic compositions of water produced by the simple mixing of these two end members. The diagram is modified from "Uses of Environmental Isotopes" by T. B. Coplen in REGIONAL GROUND WATER QUALITY edited by W. M. Alley, pp. 227–254. Copyright © 1993. This material is used by permission of John Wiley & Sons, Inc.

areas have the lower values. Thus, it may be possible to use differences in the *d*-value to tag water masses. For example, during Pleistocene glaciation meteoric waters had a lower deuterium excess (Merlivat and Jouzel, 1979), which might be used to identify waters recharged to an aquifer during the Pleistocene.

Evaporation of water from surface-water bodies is a nonequilibrium process and the $\delta D/\delta^{18}O$ slope is usually between 3 and 6. This process can be described using the Craig–Gordon evaporation model (Craig and Gordon, 1965). The model takes into account the effect of relative (temperature normalized) humidity on isotopic fractionation during evaporation. If the vapor space is saturated, the process is an equilibrium reaction. For relative humidities of less than 100%, the process is a kinetic reaction in which diffusive isotope fractionation is an important factor. The greater the degree of undersaturation, the greater is the kinetic effect.

Water–rock interactions also affect isotopic compositions. Interactions between geothermal waters and rocks lead to changes in isotopic compositions. Because rocks have very little hydrogen, the exchange only affects the oxygen isotopes. Rocks have $\delta^{18}O$ values in the range of +2‰ to +14‰. Geothermal waters interacting with country rocks will thus become enriched in $\delta^{18}O$, the degree of enrichment depending in part on the temperature at which the interaction is taking place. Recall that fractionation factors approach 1 as temperature increases. At sufficiently high temperatures, and assuming that isotopic equilibrium has been achieved, the $\delta^{18}O$ value of the water should approach that of the country rock. At low temperatures a common diagenetic reaction, the hydrolysis of silicate minerals, also leads to changes in isotopic compositions. These diagenetic reactions usually lead to a decrease in the $\delta^{18}O$ value of the water and an increase in the δD.

The sum total of these various isotopic interactions is to produce waters with different isotopic signatures. Thus, it may be possible to determine the contributions of various sources to ground or surface waters using stable isotopes (Case Study 6–4).

CASE STUDY 6–4
Sources of Groundwater in the Albuquerque, New Mexico, Area

A major area of concern in arid and semiarid regions is the source and recharge rate of groundwaters because groundwater is often a major water source. Lambert and Balsley (1997) determined the hydrogen and oxygen isotopic composition of water from municipal wells in Albuquerque, New Mexico. The results of this study were compared to δD values determined in an earlier study. The δD and $\delta^{18}O$ values plot in two clusters along the meteoric water line (MWL) (Figure 6–C4–1) and define two end members for the groundwater system. The less negative cluster was named the *eastern domain* and was inferred to be derived from runoff from the Sandia and Manzano mountains along the eastern edge of the basin. The other cluster was named the *central basin domain*, which may partly consist of seepage from the Rio Grande river. Samples from two of the wells (Figure 6–C4–1) do not fall on the MWL and have excess deuterium of 5‰. The authors concluded that these waters were emplaced during climatic conditions different from the present day and may represent a separate groundwater reservoir. The authors also found that δD values along the eastern and western margins of the basin had become more negative over a ten-year period, suggesting an expansion of the *central basin domain* waters. Pumping at multiple depths in two wells showed that there were significant variations in isotopic compositions between different levels, indicating that the vertically stacked groundwaters have very little interconnection. It was

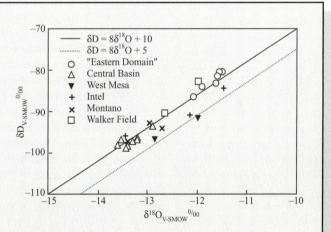

Figure 6–C4–1
δD and $\delta^{18}O$ values for groundwaters from the Albuquerque Basin showing the various groundwater domains. From Lambert and Balsley (1997).

suggested that long-term pumping tests could be used to determine the degree of interconnectivity between the various groundwater intervals; i.e., because the various levels have different isotopic compositions, changes in isotopic composition during pumping would indicate the movement of water between the different intervals.

Source: Lambert and Balsley (1997).

EXAMPLE 6–7 A dam and reservoir were constructed on a river in Arizona. River water sampled from a point 40 km downstream from the dam has $\delta D = -44.6‰$ and $\delta^{18}O = -3.6‰$. The river water has two sources, groundwater recharge and overflow from the dam. The isotopic composition of the groundwater is $\delta D = -38‰$ and $\delta^{18}O = -4.5‰$ (A in Figure 6–6). Because of evaporative water loss from the reservoir behind the dam, the isotopic composition of the water discharged from the reservoir is $\delta D = -49‰$ and $\delta^{18}O = -3.0‰$ (B in Figure 6–6). Because these are the only two sources of water for the river, the resulting river water will be a simple mixture of the two end members. Calculate the groundwater component of the river flow.

Because we are using two isotopes, we can write two mass balance equations:

$$\text{For deuterium: } X\,\delta D_{\text{groundwater}} + Y\,\delta D_{\text{reservoir}} = \delta D_{\text{river}}$$

and

$$\text{For oxygen: } X\,\delta^{18}O_{\text{groundwater}} + Y\,\delta^{18}O_{\text{reservoir}} = \delta^{18}O_{\text{river}}$$

where X is the contribution from the groundwater system and Y is the contribution from the reservoir. Substituting the appropriate values gives two simultaneous equations that can be solved for X and Y.

$$\text{For deuterium: } X(-38) + Y(-49) = -44.6$$

$$\text{For oxygen: } X(-4.5) + Y(-3.0) = -3.6$$

Solving gives $X = 0.40$ and $Y = 0.60$; 40% of the current stream flow represents groundwater input. ∎

Climate Change Because the fractionation of hydrogen and oxygen isotopes is temperature sensitive, it has been possible to reconstruct past climatic conditions from stable isotope measurements on ice cores. Ice cores have been obtained from Greenland and Antarctica, and the results of these studies are reported in a number of papers (e.g., Dansgaard et al., 1969, 1971; Johnson et al., 1972). In brief, the isotopic composition of snow at high latitudes reflects the prevailing air temperature (Figure 6–7). Declining air temperatures lead to more negative δD and $\delta^{18}O$ values, and increasing air temperature leads to less negative δD and $\delta^{18}O$ values. During Southern Hemisphere summers (dates in Figure 6–7), ambient temperature increases are marked by less negative $\delta^{18}O$ and δD values. During Southern Hemisphere winters, $\delta^{18}O$ and δD values become more negative. The snow ultimately forms glacial ice, and if the ice stratigraphy is not disrupted, it is possible to map past temperature changes (Figure 6–8). A confounding variable is the isotopic composition of the ocean water from which the original water vapor is derived. During an ice age a significant amount of water is removed from the ocean and stored on the continents. This leads to an increase in the δD and $\delta^{18}O$ values for ocean waters (they become less negative), and thus the water vapor becomes less negative. In this case, at constant air temperature, we would expect to see an increase in the δD and $\delta^{18}O$ values for snow, which would be interpreted as atmospheric warming. In practice, temperature changes generally outweigh this effect so that the direction, if not the absolute magnitude, of the temperature change is correct. Another point of contention is the meaning of the relative temperature changes; e.g., do they sim-

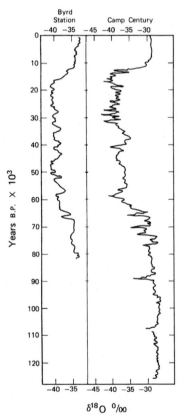

Figure 6–8
Variation in $\delta^{18}O$ in ice cores from Byrd Station and Camp Century. From PRINCIPLES OF ISOTOPE GEOLOGY, 2nd Edition by G. Faure. Copyright © 1986. This material is used by permission of John Wiley & Sons, Inc.

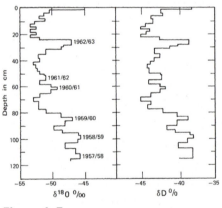

Figure 6–7
Seasonal variations in $\delta^{18}O$ and δD in snow and firn at the South Pole. From PRINCIPLES OF ISOTOPE GEOLOGY, 2nd Edition by G. Faure. Copyright © 1986. This material is used by permission of John Wiley & Sons, Inc.

ply reflect local conditions, such as a change in the prevailing wind direction, or do they reflect global changes? If a reliable geochronology is available for ice cores collected in the Northern and Southern hemispheres, the presence of the same temperature variation in cores from both hemispheres is interpreted to represent a global variation, whereas the presence of a temperature variation in cores from only one hemisphere is interpreted to represent more local changes. In Figure 6–8, the Byrd Station (South Pole) and Camp Century (Greenland) ice cores are seen to show similar variations in $\delta^{18}O$ for the last 80,000 years, suggesting that the temperature changes are global. The increase in $\delta^{18}O$ during the last 10,000 years represents the global warming following the end of the last ice age. The lower $\delta^{18}O$ values found between 10,000 and 70,000 B.P. reflect the colder global temperatures during the last ice age. Climate change will be considered further in Chapter 8.

Carbon

Carbon occurs in a number of reservoirs (Figure 6–9), each with a characteristic range of $\delta^{13}C$ values. With respect to the PDB standard, most of the carbon at or near the earth's surface is enriched in ^{12}C relative to the standard; i.e., $\delta^{13}C$ values are negative. For the modern atmosphere, $\delta^{13}C = -7‰$. Both coal (petroleum is similar to coal) and methane are isotopically more negative than the present-day atmosphere. Thus, the addition of CO_2 to the atmosphere by the burning of fossil fuels leads to more negative values for $\delta^{13}C$. The $\delta^{13}C$ values for plants depend on the biochemical pathway used for carbon fixation (Deines, 1980). Most plants use the Calvin cycle (C_3) and have $\delta^{13}C$ in the range -24 to $-37‰$. Tropical grasses, marine aquatics, and desert and salt marsh plants use the Hatch–Slack cycle (C_4) and have $\delta^{13}C$ in the range -9 to $-24‰$. Some algae, cacti, and succulents use the CAM (Crassulacean acid metabolism) cycle and have $\delta^{13}C$ values in the range -12 to $-30‰$. In principle, it is possible to use differences in $\delta^{13}C$ to distinguish between organic matter produced by the C_3 or C_4 cycles. Carbon occurs in the aquatic environment as DIC (dissolved inorganic carbon) and DOC (dissolved organic carbon).

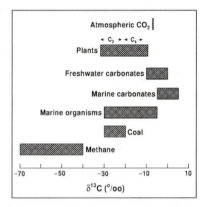

Figure 6–9
Range of $\delta^{13}C$ values for different carbon reservoirs.

DIC in Aqueous Systems The isotopic behavior of carbon in aqueous systems is complex. The major sources of carbon in water are carbonate minerals, organic matter, and atmospheric CO_2. The relative importance of these sources is a function of the system; e.g., atmospheric CO_2 is generally a minor source for groundwater systems but an important source for rivers, lakes, and the ocean. The isotopic composition of the total dissolved carbon in solution, and the isotopic composition of $CaCO_3$ in equilibrium with the solution, is a function of the temperature and pH (which affects the relative abundance of carbonate and bicarbonate ions) of the solution.

A number of authors (Deines et al., 1974; Grootes et al., 1969; Mook et al., 1974; Turner, 1982) have investigated the isotopic fractionation of carbon between CO_2 gas and CO_2 dissolved in solution, bicarbonate and carbonate ions, and solid calcium carbonate. Each species has its own fractionation factor, and hence the isotopic composition of the solution is a function of the relative proportions of the different species. The empirical equations, based on experimental results, of Deines et al. (1974) are given in Table 6–8 and graphically illustrated in Figure 6–10 (p. 190).

For $CO_{2\,(aq)}$ in equilibrium with $CO_{2\,(g)}$, the offset in the delta value can be calculated from the following equation (Mook et al., 1974):

$$\varepsilon = -0.373 \times 10^3/T + 0.19‰ \tag{6–52}$$

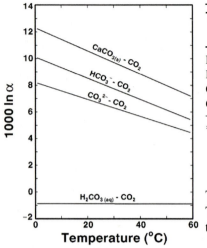

Figure 6–10
Isotopic fractionation factors, relative to CO_2 gas, for carbonate species as a function of temperature. Deines et al. (1974).

Table 6–8   Fractionation Factors for Carbonate Species Relative to Gaseous CO_2*

H_2CO_3	$1000 \ln \alpha = -0.91 + 0.0063 \times 10^6/T^2$
HCO_3^-	$1000 \ln \alpha = -4.54 + 1.099 \times 10^6/T^2$
CO_3^{2-}	$1000 \ln \alpha = -3.4 + 0.87 \times 10^6/T^2$
$CaCO_{3\,(s)}$	$1000 \ln \alpha = -3.63 + 1.194 \times 10^6/T^2$

*From Deines et al. (1974).

The calculated value (ε) is the difference between the $\delta^{13}C$ value for $CO_{2\,(g)}$ and $CO_{2\,(aq)}$. This offset is always negative, so the CO_2 in solution is always isotopically heavier than the coexisting CO_2 gas. In all of these equations, temperatures are in K.

EXAMPLE 6–8 Calcium carbonate is precipitating at 25°C in equilibrium with atmospheric CO_2 ($\delta^{13}C = -7.0‰$). Calculate the $\delta^{13}C$ for the carbonate.
   From Table 6–8,

$$1000 \ln \alpha = -3.63 + 1.194 \times 10^6/T^2 = -3.63 + 1.194 \times 10^6/(298.15)^2 = 9.8$$

$$\ln \alpha = 9.8 \times 10^{-3}$$

$$\alpha = 1.00985$$

Using equation 6–46 and rearranging,

$$\delta^{13}C_{\,(carbonate)} = 1.00985(-7.0 + 1000) - 1000 = 2.8‰$$

The carbonate is enriched by 2.8‰ in ^{13}C.   ∎

As was discussed in Chapter 3, the distribution of carbonate species in solution is a function of the pH of the solution (Figure 3–2). Because isotopic fractionation factors are different for each carbonate species, the relative abundance of the species will determine the $\delta^{13}C$ for the solution. For example, at pH = 4, $H_2CO_{3\,(aq)}$ is the dominant species in solution and the $\delta^{13}C$ of the DIC will be determined by the fractionation of carbon between $H_2CO_{3\,(aq)}$ and $CO_{2\,(g)}$. At pH = 8, the dominant species in solution is HCO_3^-, and the $\delta^{13}C$ of the DIC will be determined by the fractionation of carbon between HCO_3^- and $CO_{2\,(g)}$.

Another important factor controlling the isotopic composition of DIC is whether the system is open or closed to CO_2. In a closed system, there is no mass transfer across boundaries, whereas in an open system, there is mass transfer across boundaries. A confined aquifer would be an example of a closed system and the ocean is an example of an open system (there is free exchange of CO_2 between the atmosphere and the ocean).

EXAMPLE 6–9 The ocean is an open system. For the present-day atmosphere, $\delta^{13}C = -7.0‰$. Average ocean water has a pH ≈ 8.2. With reference to Figure 3–2, at this pH bicarbonate ion is the dominant species; thus, the $\delta^{13}C$ for oceanic DIC will be the same as the bicarbonate $\delta^{13}C$. Given a surface temperature of 10°C, calculate the $\delta^{13}C$ value for the bicarbonate ion.
   Using the equation for the bicarbonate ion in Table 6–8, at 10°C $\alpha = 1.00921$ and using equation 6–46 and rearranging,

$$\delta^{13}C_{\,(bicarbonate)} = 1.00921(-7.0 + 1000) - 1000$$

$$= 2.1‰ \text{ (for oceanic DIC)}$$

∎

EXAMPLE 6–10 A confined limestone aquifer (closed system) has a pH of 8. For the limestone, $\delta^{13}C = 0‰$; for the CO_2, $\delta^{13}C = -25‰$ (the soil CO_2 is from C_3 vegetation). At this pH the dissolution reaction for limestone can be written

$$CaCO_3 + CO_2 + H_2O \rightarrow Ca^{2+} + 2HCO_3^-$$

In a closed system, $\delta^{13}C$ is a function of the relative amounts of CO_2 and $CaCO_3$ and their $\delta^{13}C$ values. In this case, the relative proportions of CO_2 and $CaCO_3$ are 1:1, which yields a final DIC value of $-12.5‰$. Because bicarbonate ion is the dominant species in solution, it will also have $\delta^{13}C = -12.5‰$.   ■

Sources of Methane Methane sources include petroleum, methane produced from kerogen at elevated temperatures (thermogenic), methane produced by bacterial action (either fermentation of acetate or reduction of CO_2), methane produced by the pyrolytic decomposition of organic material using heat energy derived from a magmatic source (geothermal), and abiogenic methane derived from the mantle. In terms of carbon and hydrogen isotopes, each of these potential sources has a different isotopic signature (Figure 6–11). Hence, it would, for example, be possible to distinguish between methane derived from petroleum and methane derived by bacterial reduction. With reference to Figure 6–11, it is seen that methane from petroleum would be isotopically heavier, in terms of carbon, than methane from bacterial reduction. If the methane in groundwater was derived from both sources, then the observed isotopic composition would fall on a mixing line between the two end members. As another example, consider the source(s) of methane in deep ocean waters. The isotopically heaviest methane is derived from the mantle. This type of methane has been

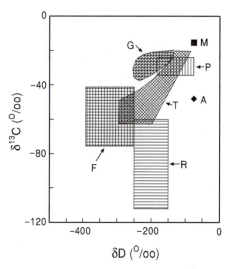

Figure 6–11
^{13}C and deuterium isotopic values for methane from various sources and reservoirs. M—abiogenic (from the mantle), P—petroleum, A—atmosphere, G—geothermal (pyrolitic from interaction with magmatic heat), T—thermogenic (from kerogen at elevated temperatures), F—acetate fermentation (bacterial), and R—CO_2 reduction (bacterial). After Schoell (1984, 1988).

observed in hydrothermal systems associated with the midocean ridges. In deep ocean waters, which are isolated from the atmosphere and are therefore not in equilibrium with atmospheric methane, it would be possible to determine the amount of mantle-derived methane contributing to the total methane content of the deep waters. Case Study 6–5 (p. 192) illustrates how carbon isotopes can be used to distinguish leachate water from a landfill from natural (uncontaminated) water.

DOC in Aqueous Systems Dissolved organic carbon has multiple sources (see Chapter 5), including natural sources such as plant and animal remains and humic substances and manufactured organic compounds. To the extent that these compounds have unique isotopic signatures, they can be traced in the environment.

Tracers in Food Chains Because C_3 and C_4 plants have different isotopic characteristics, it is possible to trace the food sources for different organisms in a food chain. When plant material is consumed by a herbivore or an omnivore, the carbon is incorporated into the consumer with minimal isotopic fractionation (e.g., about $+1‰$). For example, maize

Use of Isotopes to Characterize Landfill Leachates and Gases

Hackley et al. (1996) describe how isotopic systems can be used to characterize gases and leachates from landfills. The student should refer to this paper for a more complete discussion. During methanogenesis (the production of methane gas, CH_4), the hydrogen and carbon isotopes are fractionated. The two main processes are thermal decomposition and microbial decomposition of organic matter. Microbial decomposition can either proceed by acetate–fermentation or CO_2–reduction. These processes are described in Chapter 5. Each pathway produces isotopically distinct products (Figure 6–C5–1). Landfill methane gas has isotopic compositions typical of methane produced by acetate–fermentation. During methanogenesis there is a significant enrichment of deuterium in the leachate waters. This enrichment can be used to distinguish leachate water from natural waters (Figure 6–C5–2). The degree to which these isotopic distinctions are developed is a function of the time during which the processes have been active and the isolation of the landfill. During the early stages of biodegradation, the oxidation phases would be isotopically light ($\delta^{13}C = -10$ to -35‰), reflecting the isotopic composition of the original organic matter. As anaerobic decomposition proceeds, the values become more negative. Hence, gas and leachate from a relatively young landfill, or a young portion of a landfill, will not be distinctively different from natural sources. If the landfill is open to water, the shift in deuterium values due to methanogenesis will be overwhelmed by the input of natural waters. In Figure 6–C5–2 these two cases are illustrated by the points labeled C2 (where isotopically light groundwater was infiltrating and mixing with the leachate) and point C4 (a leachate sample from a new portion of the landfill).

The radiogenic isotopes of hydrogen and carbon may also be useful in identifying landfill gases and leachates. Both of these isotopes were produced during atmospheric testing of nuclear weapons and reached a maximum in 1962–1963. ^{14}C is enriched in landfill methane and leachates because most of the organic materials in landfills are young and have been influenced by the increased ^{14}C content of the atmosphere due to weapons testing. Tritium values are often much higher than those to be expected from the input of local contemporaneous precipitation. These higher tritium values (ranging up to 8000 or more TU) may be due to the disposal of luminescent paints, which contain tritiated hydrocarbons, in landfills. Application of these isotopes to three different landfills in Illinois is illustrated in Figure 6–C5–3. Background-I was from surface water and groundwater near the surface and reflects the present-day input of these radiogenic isotopes. Backgrounds II and III were from confined aquifers approximately 30 meters below the surface. These aquifers presumably contain old waters that were not affected by the weapons-produced ^3H and ^{14}C. The low background radioactivity for these two aquifers would increase the sensitivity of the isotope tracers.

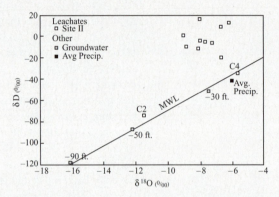

Figure 6–C5–2

δD versus $\delta^{18}O$ for leachates from a municipal landfill in Illinois. The leachates are relatively enriched in deuterium compared to meteoric water. The two leachate samples (C2 and C4) that fall on the meteoric water line (MWL) represent leachate significantly diluted by precipitation (C2) or from a recent portion of the landfill (C4).*

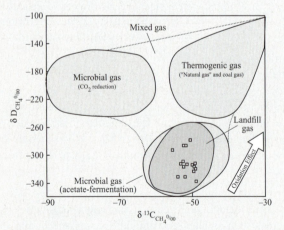

Figure 6–C5–1

δD versus $\delta^{13}C$ for methane from different sources. Landfill methane plots in the field of acetate–fermentation.*

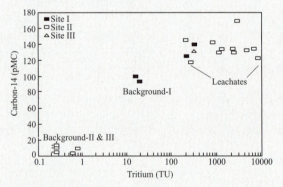

Figure 6–C5–3

^{14}C and ^3H activities for leachates from three Illinois landfills. Background-I was determined from surface and shallow groundwater. Backgrounds-II and III were determined for confined aquifers at a depth of 30 m. Background-I shows the influence of present-day bomb-derived radiogenic isotopes.*

*Source: From "Environmental Isotope Characteristics of Landfill Leachates and Gases" by K. C. Hackley, C. L. Lui and D. D. Coleman, GROUND WATER; Vol. 34, No. 5, pp. 831–834 (Figures 5, 8, 9). September/October 1996. Reprinted from GROUND WATER with permission of the National Ground Water Association. Copyright 1996.

is a C^4 plant ($\delta^{13}C \approx -15‰$) and wheat is a C_3 plant ($\delta^{13}C \approx -30‰$). This difference has been used by archaeologists to determine if the staple grain of an ancient population was maize or wheat. In combination with age-dating, $\delta^{13}C$ isotopic data have enabled archaeologists to determine when corn was introduced into the Americas.

Nitrogen

The main uses of nitrogen isotopes in the environmental field have been to trace the sources of nitrate pollution in ground and surface waters and to trace the sources of NH_3 and HNO_3 in rainfall (cars and factories versus agricultural). The processes controlling the isotopic composition of nitrogen are generally biological, rather than inorganic. Nitrogen isotopic values ($\delta^{15}N$) are reported relative to atmospheric air (AIR).

Nitrates in Surface and Ground Waters High abundances of nitrates in natural waters can cause adverse health effects and, if nitrogen is a limiting nutrient, can lead to eutrophication of rivers and lakes. Hence, there is an interest in the source of nitrates. The main sources of nitrates in surface and ground waters are fertilizer, animal or human waste, natural soil organic matter, nitrogen fixation, and rain. Nitrogen isotopes, in combination with other isotopes, have been used to elucidate which of these source(s) is contributing nitrates to the aqueous environment. These studies have been particularly useful in the identification of fertilizer or human and animal waste sources. Because the nitrogen in fertilizer is extracted from the atmosphere, it has $\delta^{15}N = 0‰ \pm 2‰$. Animal and human waste has $\delta^{15}N = 15‰ \pm 10‰$. Thus, in principle, it is possible to distinguish between fertilizer- and waste-derived nitrates (Case Study 6–6, p. 194). During denitrification, the $\delta^{15}N$ value of the residual nitrate increases, and it eventually becomes indistinguishable from the $\delta^{15}N$ value of manure. For this reason, better separation of sources is achieved if the oxygen in the nitrate is also analyzed. The oxygen in fertilizer will have $\delta^{18}O = +23‰$ (the $\delta^{18}O$ value for atmospheric oxygen), and for manure $\delta^{18}O = -10‰ \pm 10‰$ (oxygen derived from water). Denitrification is represented by the following equation:

$$4NO_3^- + 5CH_2O \rightarrow 2N_{2\,(g)} + 5HCO_3^- + H^+ + 2H_2O$$

Both nitrogen and oxygen isotopes are fractionated during denitrification, although the effect is greater for nitrogen. Because nitrogen is going into the gas phase, we would expect the residual nitrate to be enriched in the heavier isotope and the $\delta^{15}N$ value of the residual nitrate to increase. Oxygen shows a similar pattern, with an increase in the $\delta^{18}O$ value. These processes are graphically illustrated in Figure 6–12. If both sources are undergoing denitrification, it is possible to determine the relative contribution of each source to the water because the process is one of simple mixing. This is illustrated in Figure 6–12, in which both fertilizer and manure are contributing nitrates to a groundwater system. At some point, the fertilizer has an isotopic composition represented by point A and the manure has an isotopic composition represented by point B. A mixture of these two components would lie along the line A–B. In this case, the groundwater has a nitrate isotopic composition represented by point C. The relative percentage of each end member in the groundwater can be calculated from the lever law—e.g., the amount of A equals the distance C–B divided by the distance A–C. In this case, A provides 60% of the total nitrate to the groundwater.

NH_4^+ and NO_3^- in Rain Anthropogenic sources—fossil fuel combustion and biomass burning—account for 65% of the total NO_x input to the atmosphere, and 56% of this 65% is due to fossil fuel combustion in automobile engines and power plants (Berner and Berner, 1996). NO_x has a residence time of about 6 days in the atmosphere and is then removed as HNO_3, a locally important component of acid rain. NH_4^+ (ammonium ion) is produced by the interaction of ammonia gas (NH_3) with water. The major continental sources of atmospheric ammonia are (1) bacterial decomposition of animal and human waste (45%), (2) bacterial decomposition of nitrogenous organic matter in soils (20%), (3)

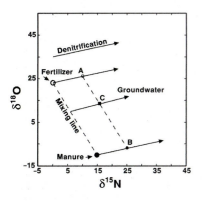

Figure 6–12

Determination of the relative importance of nitrate sources to a groundwater system. Two sources for nitrates are fertilizer and manure. Both are undergoing denitrification. A and B represent each source at a particular stage in the denitrification process. C is the isotopic composition of the nitrate in the groundwater due to simple mixing. In this example, approximately 60% of the nitrate is contributed by the fertilizer.

CASE STUDY 6–6
Nitrogen Isotopes as Indicators of Nitrate Sources in Minnesota Sand-Plain Aquifers

Sand-plain aquifers, consisting of Quaternary-age glacial outwash sands with interlayered clay, silt, silty sand, and gravel, underlie approximately 32,000 km^2 of central Minnesota and are important sources of water for irrigation and drinking. The aquifers are generally less than 30 m thick and are mainly recharged by the infiltration of precipitation. Thus, they are vulnerable to contaminants derived from the surface. In about 20% of the wells, nitrate concentrations exceed the recommended limit of 10 mg L^{-1} (as N). Komor and Anderson (1993) used nitrogen isotopes to identify the source of nitrates in groundwaters underlying five types of land-use areas: feedlot, cultivated-irrigated, cultivated-nonirrigated, residential with septic system, and natural. In general, various nitrate sources can be characterized as follows: nitrates from animal wastes, $\delta^{15}N$ = +10 to +22‰; nitrate from natural organic material in soil, $\delta^{15}N$ = +4 to +9‰; and nitrate from commercial fertilizers, $\delta^{15}N$ = −4 to +4‰. These values can be changed during transport through the zone of aeration and the aquifer. Many of these processes lead to an increase in the $\delta^{15}N$ value of the residual nitrate. A total of 51 wells from the five land-use areas were sampled. The results of the nitrogen isotopic analyses are shown in Figure 6–C6–1. $\delta^{15}N$ values of greater than +22‰ found in some feedlot wells were attributed to denitrification. The samples from the residential area that showed high $\delta^{15}N$ values were believed to reflect leakage from septic systems. In general, cultivated areas in which both manure and fertilizer were applied had higher $\delta^{15}N$ values. For the cultivated-nonirrigated areas, the $\delta^{15}N$ values did not reflect the type of fertilizer (inorganic versus manure) applied to the fields. The authors suggested that this difference was due to the lower amount of water received by the nonirrigated fields. The nitrogen fertilizer added to the field would spend a longer period of time in contact with the natural soil material and would tend to homogenize with the soil nitrates. For the irrigated fields, the greater amount of water delivered during the growing season would wash much of the nitrate out of the soil horizon, thus di-

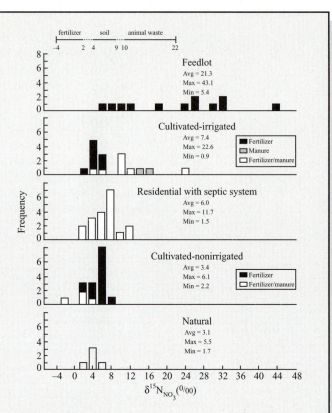

Figure 6–C6–1

$\delta^{15}N$ values for groundwaters from various land-use areas. Forms of nitrogen applied to cultivated areas are shown on the diagram. From "Nitrogen Isotopes as Indicators of Nitrate Sources in Minnesota Sand-Plain Aquifers" by S. C. Komor and H. W. Anderson, GROUND WATER; Vol. 31, No. 2, p. 266 (Figure 3). March/April, 1993. Reprinted from GROUND WATER with permission of the National Ground Water Association. Copyright 1993.

minishing the residence time of nitrates in the soil. On a qualitative basis, it should be possible to identify the source(s) of nitrates occurring in groundwater from the sand-plain aquifers.

Source: Komor and Anderson (1993).

fertilizer (8%), (4) burning of coal (<<1%), and (5) biomass burning (3%). The remaining 24% is contributed by sea surface release (Berner and Berner, 1996). NO$_x$ derived from fossil fuel combustion has $\delta^{15}N$ values in the range −10 to +5‰, and ammonia derived from the bacterial decomposition of animal and human wastes has $\delta^{15}N$ values in the range −15 to +30‰. In favorable circumstances, it may be possible to distinguish between combustion and waste-decomposition sources of rainwater N.

Sulfur

Sulfur isotope compositions are reported in terms of the $^{34}S/^{32}S$ ratio in parts-per-thousand relative to the CDT (Canyon Diabolo troilite) standard. Two major processes are responsible for variations in $\delta^{34}S$: reduction of sulfate to hydrogen sulfide by anaerobic bacteria, which results in the enrichment of hydrogen sulfide in ^{32}S, and various exchange reactions, which result in ^{34}S being concentrated in the compound with the highest sulfur oxidation state. The major sources of sulfate are ocean water, the dissolution of evaporites, oxidation of sulfide minerals, and atmospheric precipitation of sulfate. The major sources of sulfide are ore deposits, volcanic emanations, reducing environments, and the reduction of sulfate. Each of these sources has a characteristic range of δ^{34}S. Sulfur is the major contributor to acid rain, so the source of the sulfur in the atmosphere has been an area of ongoing interest. Figure 6–13 shows the range of δ^{34}S values for sources of atmospheric sulfur. Under certain circumstances, the difference in the δ^{34}S values for the different sources can be used to determine the relative importance of each source to the total atmospheric burden of sulfur. This is often done by determining the δ^{34}S values of local sulfur sources and then comparing these to the measured atmospheric δ^{34}S values. An example of such an approach is given in Case Study 6–7.

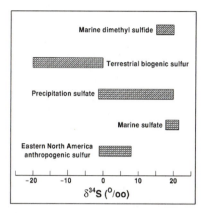

Figure 6–13
Range of δ^{34}S values for sulfur sources that contribute to atmospheric sulfur.

CASE STUDY 6–7
Sources of Sulfur Gases in Salt Lake City Atmosphere

Jensen (1972) identified three major sources for the sulfur gases found in Salt Lake City air:

  Sulfur from oil refineries + automobiles: δ^{34}S = +16‰ (S1)

  Sulfur from copper smelters: δ^{34}S = +1‰ (S2)

  Bacteriogenic sulfur (H$_2$S) from Great Salt Lake: δ^{34}S = +5.3‰ (S3)

When the smelters were operating, δ^{34}S = +1.5‰ near the smelters and δ^{34}S = +3.1‰ in Salt Lake City proper. This suggests that the major source of atmospheric sulfur was the smelters (δ^{34}S is near 1‰), but the significance of the other two sources for Salt Lake City was unknown. During a strike by smelter workers, δ^{34}S = +5.3‰ near the smelters and +6.4‰ in Salt Lake City. Because the measured δ^{34}S value near the smelters is identical to the bacteriogenic sulfur value, the data indicate that the smelters and bacteriogenic sulfur are the only sources of sulfur for the atmosphere in the vicinity of the smelters. When the smelters are operating,

$$x \, S2 + (1 - x) \cdot S3 = +1.5‰$$

$$(x)(1) + (1 - x)(5.3) = 1.5$$

where x is the percent of sulfur contributed by the smelters. Solving for x, we get 0.88 (i.e., 88% of the sulfur is from the smelters). The remaining 12% represents the bacteriogenic input. For Salt Lake City, while the smelter workers were on strike, the only sulfur sources were the oil refineries + automobiles and bacteriogenic sulfur. Setting x = the oil refinery + automobile component,

$$x \, S1 + (1 - x) \cdot S3 = +6.4‰$$

$$(x)(16) + (1 - x)(5.3) = 6.4$$

Solving for x, we get 0.10 (e.g., 10% of the sulfur is from the refineries + automobiles). The remaining 90% represents the bacteriogenic input. When the smelters are operating, all three sources contribute to the sulfur gases in the Salt Lake City atmosphere. From the observations made when the smelters weren't operating, we know that the relative contributions of S1 and S3 gave δ^{34}S = 6.4‰. Setting ST = %S1 + %S3 = 6.4, we can write the following equation for the sulfur content of the Salt Lake City air when the smelters are operating:

$$x \, ST + (1 - x) \cdot S2 = +3.1‰$$

$$(x)(6.4) + (1 - x)(1) = 3.1$$

Solving for x gives 0.39 (e.g., 39% of the total sulfur in the Salt Lake City atmosphere comes from the combination of refineries + automobiles and bacteriogenic sulfur). The remaining 61% comes from the smelters. Refineries + automobiles contribute 10% of ST, so the percent contribution of refineries and automobiles to the Salt Lake City atmospheric sulfur is 4%. The other 35% (of ST) comes from bacteriogenic sulfur. Thus, for Salt Lake City 4% of the atmospheric sulfur comes from the oil refineries and automobiles, 35% from bacteriogenic sulfur, and 61% from the copper smelters. From this analysis it is clear that any significant reduction in the total sulfur burden of the Salt Lake City atmosphere can only be achieved by controlling the smelter emissions. This has, in fact, happened through the closure of some of the smelters and the installation of emission control devices on the remaining smelters.

Source: Jensen (1972).

For the United States, the major sources of anthropogenic sulfur are electricity generation (67%) and industry (smelting and refining) (14%) (Berner and Berner, 1996). There are regional variations in the relative contributions of these sources. For example, in the western United States, metal smelters, particularly those in Arizona and New Mexico, contribute about 70% of the total anthropogenic sulfur.

The sulfur isotopic characteristics of the principal fossil fuels, petroleum and coal, vary on a deposit-by-deposit basis. In the case of petroleum, Thode and Monster (1965) concluded that petroleum is enriched in ^{32}S by about 15‰ relative to contemporaneous marine evaporites. They attributed this enrichment to the action of sulfur-reducing bacteria at the time of deposition of the sediment and during the subsequent formation of the petroleum. The source of the sulfur is marine sulfate. The isotopic composition of marine sulfate has varied during the Phanerozoic from +10 to +30‰ (Holser and Kaplan, 1966). Hence, petroleum formed at different times will have different $\delta^{34}S$ values, and given the 15‰ enrichment in ^{32}S, these values would be in the range of −5 to 15‰. In the case of coal, sulfur is present as sulfide minerals, sulfate, elemental sulfur, and organically bound. Each of these forms has different isotopic characteristics, and the $\delta^{34}S$ values of coal are highly variable and depend on the relative amounts of each component. For these reasons, it is difficult to draw any general conclusions as to the sulfur isotopic compositions of the fossil fuels. Therefore, $\delta^{34}S$ values need to be determined for each possible anthropogenic source.

Mixing

Binary Isotopic Mixing   If we mix materials from two different isotopic reservoirs to form a material of intermediate composition, there are several possible relationships. For example, if we mix leachate from a dump with uncontaminated groundwater to produce a contaminated groundwater, in terms of hydrogen and oxygen isotopes, this mixing process can be described by the following equation:

$$\delta_M = \delta_A f_A + \delta_B(1 - f_A) \tag{6–53}$$

where δ_A and δ_B are the isotopic values for the end members, δ_M is the isotopic value for the mixture, and f_A is the fraction of end member A in the mixture. This simple equation applies when the elemental concentrations are the same in the end members and the mixture. This would be the case for both hydrogen and oxygen in this example. However, note that if one of the end members was a brine, the concentrations of hydrogen and oxygen would not be the same in each end member; i.e., on a weight/weight basis there would be less water in a kg of brine and hence less hydrogen and oxygen.

Continuing with the leachate example, if the element of interest is sulfur and the concentration of sulfur is different in the two end members, then concentration must be taken into account when doing the mixing calculations. In this case, we would use the following equation:

$$\delta_M = \delta_A f_A(A/M) + \delta_B(1 - f_A)(B/M) \tag{6–54}$$

where A, B, and M are the concentrations of the element (or ion) in end member A or B or the mixture M; δ_A, δ_B, and δ_M are the corresponding isotopic values; and f_A is the fraction of end member A in the mixture. If we solve equation 6–54 for various mixtures of A and B, a plot of the sulfur isotopic ratio versus sulfur abundance will yield a hyperbola, characteristic of simple binary mixing. In the case of simple binary mixing, a plot of the isotopic ratio versus the reciprocal of the species concentration will yield a straight line (see problem 6–53).

Multi-End-Member Mixing of Chemical Species   If the mixture involves more than two end members, the problem becomes a bit more complicated. However, if the end members can be identified, the composition of the resulting mixture can be calculated for any combination of end members. For example, if there were three end members—A, B,

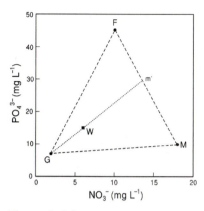

Figure 6–14
Plot of PO_4^{3-} versus NO_3^- in water samples from feedlot runoff (M), cultivated fields (F), uncontaminated groundwater (G), and contaminated well water (W). The sample of contaminated well water falls within the triangle defined by compositions M, G, and F, indicating that this sample is a mixture of these three compositions. The relative proportions of each end member can be determined by applying the lever rule (see Example 6–11).

and C—that were mixed in the proportion 20A:30B:50C, one would first solve for a mixture of A and B (M′) and then for a mixture of M′ and C. The mixture of A and B (M′) would consist of 40% A and 60% B. The final mixture (M) would consist of 50% M′ and 50% C. On a graphical plot any mixture formed from these three end members would fall within a compositional triangle defined by the three end members (Example 6–11, Figure 6–14). The relative proportions of the end members can be determined by the application of the *lever rule*, a useful tool for determining relative proportions when data are presented graphically. Example 6–11 illustrates how the lever rule is applied to a multi-end-member mixing problem.

EXAMPLE 6–11 Elevated concentrations of phosphate and nitrate are found in a water well. There are two potential sources of phosphate and nitrate in the region, runoff from a feedlot and runoff from fertilized cultivated fields. Water samples collected in the immediate area of each of these sources give the following results: feedlot runoff (M), $PO_4^{3-} = 10$ mg L^{-1} and $NO_3^- = 18$ mg L^{-1}; and runoff from cultivated fields (F), $PO_4^{3-} = 45$ mg L^{-1} and $NO_3^- = 10$ mg L^{-1}. For uncontaminated groundwater (G), $PO_4^{3-} = 7$ mg L^{-1} and $NO_3^- = 2$ mg L^{-1}. For the well water (W), $PO_4^{3-} = 15$ mg L^{-1} and $NO_3^- = 6$ mg L^{-1}. These four compositions are plotted in Figure 6–14. Note that the contaminated well water falls within the composition triangle defined by the three end members. We can determine the proportion of each end member in the sample by applying the lever rule. The proportion of uncontaminated groundwater in the sample is determined by measuring the distance Wm′ and dividing this distance by Gm′. The ratio is 0.64 and the proportion of groundwater in the sample is 64%. Note that the relative proportion of a particular component in a mixture is the distance from the mixture to the opposite component divided by the total distance between the two components. This follows from the observation that W is closer to component G than component m′. Thus, component G must be the major contributor to the mixture. The remaining 36% consists of runoff from the feedlot and the cultivated fields. This proportion can be determined as follows: measure the distance Fm′ and divide by the total distance FM. The resulting ratio is 0.44. Thus, of the remaining 36%, 44% is feedlot runoff, and feedlot runoff constitutes 16% (36% $\times$ 0.44) of the well water sample. The remaining 20% is runoff from the cultivated fields. To check your result, calculate the well water composition using the graphically determined proportions: $PO_4^{3-} = (0.64)(7) + (0.16)(10) + (0.20)(45) = 15$ mg L^{-1}, and $NO_3^- = (0.64)(2) + (0.16)(18) + (0.20)(10) = 6$ mg L^{-1}. ■

Paleothermometry

In a previous section we considered the reconstruction of past climates (e.g., temperatures) using ice cores. These measurements have been successfully extended back several hundred thousand years (see Chapter 8). The oldest obtainable information is ultimately limited by the depth at which the ice begins to flow, and the internal stratigraphy of the ice record is therefore lost. In order to look further back into the climatic record, we need other types of paleothermometers.

Urey (1947) suggested that the distribution of oxygen isotopes between water and calcium carbonate could be used to determine the temperature of ancient oceans, and such measurements were subsequently done by Urey et al. (1951). Although not without problems, as will be described, this has become a widely used paleothermometer. The basic equilibria have already been described in equations 6–41 to 6–43. The isotopic relationship between the temperature of the water and the coexisting calcite was experimentally determined by Epstein et al. (1953), and the subsequently modified version of the equation (Craig, 1965) is given here:

$$t\,(°C) = 16.9 - 4.14(\delta_c - \delta_w) + 0.13(\delta_c - \delta_w)^2 \qquad (6–55)$$

where δ_c is the $\delta^{18}O$ value for calcite (relative to PDB) and δ_w is the $\delta^{18}O$ value for water (relative to SMOW). Problems with the paleothermometer include isotopic disequilibrium between the shell material and the water, postdeposition changes, and variations in the isotopic composition of the water. These problems have been addressed by using only species that seem to achieve isotopic equilibrium for isotopic measurements and by independently determining the isotopic composition of the water. Freshwater shows such a wide range in isotopic ratios that freshwater organisms are not suitable for paleotemperature determinations. Ocean isotopic ratios are less variable, but recall from our earlier discussion that removal (glaciation) or addition (melting of glaciers) of water to the oceans will change the $\delta^{18}O$ values of seawater. Because the temperature of the deep ocean remains relatively constant (i.e., there is no variation in isotopic fractionation due to temperature changes), isotopic measurements done on benthic organisms can be used (assuming a particular temperature) to determine the $\delta^{18}O$ value of the coexisting water. Given the sum total of the corrections and assumptions that must be made in order to use this paleothermometer, most investigators use the results of such measurements to indicate relative, rather than absolute, changes in ocean temperatures. Cast Study 6–8 illustrates the application of paleothermometry to an environmental question.

CASE STUDY 6–8
Stable Isotope and Heavy Metal Variations in the Shells of Mussels as an Indicator of Pollution in Lake Erie

Al-Asasm et al. (1998) analyzed the shells of *Dreissena polymorpha*, an exotic freshwater bivalve species found in the Great Lakes, for stable isotopes (carbon and oxygen) and heavy metals (Pb, Cd, Mn, Fe, Mg, V, and Cu). The organisms were collected from the western end of Lake Erie, near the mouth of the Detroit River. The organisms have a 3- to 9-year lifespan and show rapid growth. The shells have distinct growth layers due to seasonal variations—rapid growth in the summer, slow growth in the winter. Mollusks tend to concentrate heavy metals in their shells. Thus, they are potentially sensitive indicators of changes in heavy metal concentrations in the aquatic environment.

Calculations were done to determine if the aragonite shells of *Dreissena polymorpha* were formed in isotopic equilibrium with Lake Erie waters. For oxygen, the equation of Anderson and Arthur (1983) was used:

$$t\,(°C) = 19.0 - 3.52(\delta^{18}O_a - \delta^{18}O_w) + 0.13(\delta^{18}O_a - \delta^{18}O_w)^2$$

where t is the temperature, $\delta^{18}O_a$ (PDB) is the isotopic composition of the shell, and $\delta^{18}O_w$ (SMOW) is the isotopic composition of the water. For Lake Erie, average water temperature is 25.8°C and $\delta^{18}O_w$ is −7.03‰ (SMOW). For the aragonite shells, the authors calculated an equilibrium of

$\delta^{18}O_w$ = −7.36‰ (PDB), in good agreement with the measured values. Substitution of the authors' values into the equation yields a temperature of 20.2°C, suggesting that there may be an error in the paper with respect to the water temperature used in the calculation. A similar calculation was done for carbon using the equation of Rubinson and Clayton (1969):

$$\delta^{13}C_{(aragonite)} = \delta^{13}C_{(HCO_3)} + 1.85 + 0.035(t - 25°C)$$

where t is temperature and the isotopic ratios are reported relative to PDB. The authors' estimated $\delta^{13}C$ equilibrium value is −0.92‰. This value is 1 to 1.5‰ heavier than the values measured for the shells. Thus, the shells are depleted in ^{12}C with respect to the equilibrium value. The authors considered a number of possible reasons for this difference. One of the more significant factors might be the influx of isotopically lighter organic carbon from domestic and industrial sewage.

It was found that the heavy metal concentrations in the shells tended to follow changes in the isotopic ratios. When the shells were enriched in the heavier isotope, there was an increase in the total metal concentration. The isotopic changes were correlated with changes in temperature, the enriched values reflecting higher temperatures, higher metabolic activity for the organisms, and increased mobility of the metals. Shells collected near the mouth of the Detroit River had higher total heavy metal concentrations, a response to their proximity to an anthropogenic source.

Source: Al-Asasm et al. (1998).

QUESTIONS AND PROBLEMS

1. The $^{84}_{38}Sr$ nucleus contains how many protons? how many neutrons?
2. Distinguish between *radioactive*, *radiogenic*, and *stable* isotopes.
3. Define *half-life*.
4. What is *RBE*?
5. What is meant by the term *event dating*?

6. Define *tritium unit* (TU).

7. Discuss the corrections that must be made in the tritium-helium-3 dating method. You may want to refer to the paper of Schlosser et al. (1989) in order to give a complete answer for this question.

8. With reference to Case Study 6–1, why did the authors use multilevel sampling wells?

9. Groundwaters isolated from the atmosphere prior to 1952 can't be dated by the tritium method. Why not?

10. Why is the tritium content of air over the oceans lower than that of air over the continents?

11. How and why will the burning of fossil fuel affect the ^{14}C content of the atmosphere?

12. What factors affect the atmospheric concentration of ^{14}C? Explain how we can correct ^{14}C ages in order to account for these variations.

13. Why would contamination of a sample by modern carbon be a major problem when dating *old* samples using the ^{14}C method? How would the modern carbon affect the measured age (i.e., would it be younger or older)?

14. In terms of radioactive isotopes, what is *secular equilibrium*?

15. Explain the basic concept of a *disequilibrium dating method*. Give an example.

16. What do we mean when we say the activity of a particular radioactive isotope is *supported*?

17. In Case Study 6–2, the authors concluded that the Pb maximum found in the sediments of Lake Constance was not due to the burning of leaded gasoline. Why did the authors reach this conclusion, and what sources did they suggest for the Pb and why?

18. Explain the reasons for the variation in strontium isotopic ratios in different rock samples. How can these variations be used to identify the source(s) of groundwaters?

19. Why would Pb in galena formed at different times in earth history have different isotopic ratios?

20. In terms of the fractionation of stable isotopes, distinguish between *equilibrium* and *kinetic* reactions.

21. Explain why, during the evaporation of water, the vapor is enriched in ^{16}O relative to the liquid.

22. What is the *isotopic fractionation factor*?

23. Why and how does the isotopic fractionation factor vary as a function of temperature?

24. Why do we use isotopic standards?

25. Explain why snow falling at the North Pole is isotopically lighter, in terms of hydrogen and oxygen, than rain falling at the equator.

26. Distinguish between the *mean global meteoric water line* and *local meteoric water lines*.

27. With reference to Case Study 6–4, how could changes in isotopic compositions during the pumping of a well be used to determine the degree of interconnectivity of various groundwater intervals?

28. How can carbon and oxygen isotopes be used to determine temperature variations during the last 100,000 years of earth history? Answer this question in the context of data obtained from Greenland ice cores.

29. How does the addition of CO_2 to the atmosphere by the burning of fossil fuels affect the $\delta^{13}C$ values and why?

30. Why is the pH of a solution an important factor in determining the $\delta^{13}C$ of DIC?

31. An archaeologist uncovers prehistoric human remains in the U.S. southwest. Isotopic analysis of these remains yields $\delta^{13}C = -14$‰. Was the staple food crop for people living at this time wheat or maize? How did you determine your answer?

32. With reference to Case Study 6–5, how could leachate waters from a landfill be distinguished from natural waters?

33. What is the effect of denitrification on the $\delta^{15}N$ value of manure?

34. With reference to Case Study 6–6, how can nitrogen isotopes be used to distinguish between groundwaters contaminated by runoff from feedlots versus groundwaters contaminated by leakage from residential septic systems?

35. List and describe the major processes that affect the isotopic composition of sulfur. How do each of these processes affect the $\delta^{34}S$ value?

36. List and discuss the factors that affect the carbonate paleothermometer.

37. Exposure to radon gas leads to a dose of 1×10^{-4} rad being deposited in a basement dweller's lungs. Calculate the biological exposure. Be sure to take into account the nature of the ionizing radiation. For small doses, a millirem (mrem $= 1 \times 10^{-3}$ rem) is often used. This may be a more appropriate unit for this calculation than the rem.

38. A groundwater sample is collected from a deep aquifer. No tritium activity is observed in the sample. The activity of ^{14}C in the atmosphere is 13.56 dpm g^{-1} of carbon. The observed ^{14}C activity in the water sample is 4.54 dpm g^{-1} of carbon.
    a. What is the significance of the absence of tritium in the sample? Why is it important to check for the presence of tritium before using the ^{14}C method to date a water sample?
    b. Calculate the age of the water sample.
    c. Suppose the water sample had been in contact with Devonian limestone and that 60% of the total carbon was derived by dissolution of this limestone. What effect would this have on the calculated age of the water? Recalculate the age of the water sample accounting for the *limestone* carbon.

39. Schlosser et al. (1989) measured tritium and ^3He abundances in a well at Liedern/Bocholt, West Germany. Their data are reproduced in the accompanying table. The data were all decay-corrected to a sampling date of May 6, 1987. Column 1 gives the depth below the groundwater table. When answering the questions, don't forget that all these measurements are corrected to 1987—i.e., all ages are relative to 1987.

Depth (m)	^3H (TU)	^3He (TU)	$^3H + {}^3He$ (TU)
0.5	23.1	2.7	25.8
1.5	26.7	5.9	32.6
2.5	37.8	20.2	58.0
4.5	55.3	42.2	97.5
5.5	53.0	76.6	129.6
10.5	29.7	66.1	95.8
12.5	30.5	51.7	82.2
16.5	20.5	41	61.5
27.5	3.0	13.6	16.6

   a. The highest tritium values for rainfall occurred in 1963. At what depth is this tritium found in the groundwater? What is the age of the groundwater at this depth?
   b. For each depth, calculate the tritium age. Does the calculated tritium age at the depth that corresponds to the tritium maximum agree with the age you inferred for this depth in part (a)? If it doesn't, what are some possible reasons for the discrepancy?
   c. Plot depth versus the tritium age. What can you conclude from this plot?

40. Stute et al. (1997) determined the ages of groundwater samples at various distances from the Danube River. Recharge to the surficial aquifer is dominated by bank infiltration. Part of their data compilation, for deep wells presumably unaffected by mixing with locally recharged groundwater, is given in the accompanying table.

Distance Danube (km)	^{3}H (TU)	^{3}He (TU)	^{3}H + ^{3}He (TU)
2.0	23.1	7.9	31.0
3.5	24.1	5.5	29.6
8.6	46.5	84.2	130.7
11.2	68.5	164.5	233.0
12.0	79.5	277.1	356.6
14.2	63.6	204.4	268.0
14.8	74.8	298.5	373.3
18.3	3.9	19.4	23.3
29.5	< 1	4.3	< 5.3

This problem is most easily done on a spreadsheet.
a. Calculate the age of each of the groundwater samples.
b. Plot the ages of the groundwater samples versus their distance from the Danube River.
c. Calculate the horizontal flow velocity for the groundwater away from the Danube River.

41. ^{14}C ages for carbonate and organic samples need to be corrected for isotopic fractionation. The following equation is used to calculate the correction:

$$A_{corr} = A_{meas}\left[1 - \frac{2(25 + \delta^{13}C_{PDB})}{1000}\right] \text{dpm g}^{-1}$$

where $\delta^{13}C_{PDB}$ is the delta value for the sample. A mollusk shell was collected from Pleistocene lake deposits in southern New England. For this specimen, $\delta^{13}C_{PDB} = -9.0‰$ and ^{14}C activity = 0.9 dpm g^{-1}. Use the present-day ^{14}C activity of the atmosphere as the activity at $t = 0$. Calculate the age of the shell.

42. Rember et al. (1993) measured the concentrations of ^{137}Cs in the bottom sediments of Medicine Lake, northern Idaho. Heavy-metal-laden tailings from the Coeur d'Alene mining district were deposited in this lake. Tailings were deposited during floods as well-stratified, heavy-metal, contaminated silts on the lake bottom. Atmospheric testing of nuclear bombs, the source of ^{137}Cs, began in 1951 and reached a maximum in 1964. ^{137}Cs is first detected in the varved sediments at a depth of 30–32 cm and the ^{137}Cs maximum occurs at a depth of 16–18 cm.
a. Calculate the average sedimentation rate for the varved sequence.
b. The last layer of varved sediment is encountered at a depth of 10–12 cm. What is the age of this sediment layer?
c. Tailings ponds were installed in the mining district in 1968. Have these ponds reduced the input of tailings to Medicine Lake? Explain.

43. The following ^{230}Th activities were determined for a deep-sea sediment core:

Depth (cm)	^{230}Th (dpm g^{-1})
0	65.5
20	33.6
40	17.4
60	9.6
80	5.7
100	3.5
120	2.4

a. Calculate the sedimentation rate.
b. If the ^{231}Pa activity is 6.9 dpm g^{-1} at 10 cm, what would the activity be at 50 cm?

44. Das et al. (1994) used the activity of ^{210}Pb in lake sediments to determine the sedimentation rate for a number of lakes in northern India. The data for two of the lakes are reproduced here along with the average ^{226}Ra activity due to the presence of ^{238}U in the sediments.

Core no.	Depth (cm)	Total ^{210}Pb (dpm g^{-1})	^{210}Pb excess (dpm g^{-1})
Nainital	0–6	2.628	2.468
Lake NT-1	6–12	2.117	
	12–18	2.000	
	Average ^{226}Ra = 0.16 dpm g^{-1}		
Sattal Lake	0–6	17.916	
ST-1	6–12	11.090	
	12–18	4.510	
	18–24	3.622	
	24–30	0.822	
	Average ^{226}Ra = 0.822 dpm g^{-1}		

a. Complete the data table.

b. For each lake, plot the log of ^{210}Pb excess versus arithmetic depth. For the depth use the midpoint of each interval.

c. From the activity versus depth plots, calculate the average sedimentation rate for each lake. If you do this problem on a spreadsheet, you can use the curve-fitting function of the spreadsheet program to determine the slope of the activity versus depth curve.

d. The catchment area for one of these lakes has a relatively dense population; the other is sparsely populated. Which lake catchment has the low population density and why did you pick this lake?

45. With reference to Case Study 6–3, the average values for uncontaminated groundwater are AR = 1.5 and U = 15 μg L^{-1}, and for dilute raffinate AR = 1.05 and U = 100 μg L^{-1}. A groundwater sample collected peripheral to the site has AR = 1.2 and U = 35 μg L^{-1}. Calculate the percent raffinate in this groundwater sample.

46. For Greenland, average summer temperatures are 0°C and average winter temperatures are -30°C. The fractionation factors for $\delta^{18}O$ are: at -30°C, $\alpha = 1.0155$; and at 0°C, $\alpha = 1.0112$.

a. For Rayleigh fractionation, after 90% condensation ($f = 0.1$), calculate the $\delta^{18}O$ for the water vapor when condensation is occurring at 0°C and -30°C. At the start of the condensation process $(\delta^{18}O)_0 = -10$‰ for the water vapor.

b. Calculate the isotopic composition of snow formed during the summer and winter seasons.

47. Hubbard Brook, in the White Mountain National Forest, New Hampshire, has a pH of 4.9. Assuming that the waters of Hubbard Brook are in equilibrium with atmospheric CO_2 ($\delta^{13}C = -7.0$‰), for a water temperature of 20°C calculate $\delta^{13}C$ for the dissolved inorganic carbon (DIC).

48. The delta values for sulfur from different sources are (1) biogenic hydrogen sulfide = -2.0‰, (2) seawater sulfate = $+20.0$‰, and (3) sulfur in coal = $+5.0$‰. Case Study 6–7 may be of use in solving this problem.

a. An atmospheric aerosol collected over city X has $^{32}S/^{34}S$ = 21.78. Calculate the delta value for this aerosol. What can you conclude about the source of the aerosol and the location of city X?

b. A coastal city has only two major sources of sulfur, rain derived from the ocean and sulfur derived from a coal-burning power plant. A sample of city air yields $\delta^{34}S = +14.0$‰. Calculate the percent sulfur in the city air that is derived from the power plant.

49. Grasby et al. (1997) investigated the hydrogen, oxygen, and sulfur isotope geochemistry of Nose Creek, a tributary to the Bow River in Alberta, Canada. Nose Creek flows southward through the town of Airdrie (25,000) and the city of Calgary (700,000) and enters the Bow River south of Calgary. The Bearspaw Reservoir, located on the Bow River, is the source of the municipal water supply for Calgary and Airdrie. Water samples were collected during the fall of 1993 and the spring of 1996. The results of the isotopic analyses are given here. For oxygen and hydrogen, SMOW was used as the standard; for sulfur, CDT was used as the standard.

Sample	Location	$\delta^{18}O_{(H_2O)}$	$\delta D_{(H_2O)}$	$\delta^{34}S_{(SO_4)}$	$\delta^{18}O_{(SO_4)}$
		Nose Creek (Fall 1993)			
NC1-0893	North Airdrie	−12.51	−105.9	16.66	7.02
NC2-0893	Airdrie	−13.70	−120.3	7.88	3.97
NC3-0893	North crossing	−14.00	−125.9	5.53	3.33
NC4-0893	South crossing	−15.18	−123.0	6.15	3.40
NC5-0893	Country Hills	−15.05	−122.2	4.61	0.63
NC6-0893	32nd Ave. NW	−17.50	−139.5	−5.89	−7.03
NC7-0893	Calgary Zoo	−18.19	−140.6	−6.25	−2.29
NC8-0893	Centre Street	−19.02	−141.4	−7.44	−3.66
		Nose Creek (Spring 1996)			
NC1-0696	North Airdrie	−15.20	−121.9	—	3.90
NC2-0696	Airdrie	−15.40	−121.5	12.56	1.41
NC3-0696	North crossing	−15.40	−124.2	10.82	4.39
NC4-0696	South crossing	−15.50	−124.6	9.68	4.30
NC5-0696	Country Hills	−15.10	−120.3	7.21	1.08
NC6-0696	32nd Ave. NW	−15.20	−115.4	0.33	2.40
NC7-0696	Calgary Zoo	−15.30	−120.9	−1.02	0.37
NC8-0696	Centre Street	−17.80	−131.9	−0.83	0.68
		Below Bearspaw Reservoir			
BR7-0893	Bowness	−18.80	−146.0	10.48	—

a. Plot the δD and $\delta^{18}O$ values listed in the table. Use different symbols to distinguish the spring and fall samples. Draw the local meteoric water line (LMWL) on the graph. The coordinates of the LMWL are $\delta D = -100‰$, $\delta^{18}O = -13.2‰$ and $\delta D = -160‰$, $\delta^{18}O = -20.5‰$. Also plot the Bowness sample on the graph using a distinguishing symbol.

b. From the graphical relationships, what can you conclude about the major source of the Nose Creek waters during the spring and the fall? In making these interpretations keep in mind that evaporation will probably not be an important process for Nose Creek and that the Bow River (Bearspaw Reservoir) is the source of the municipal water supplies.

c. Plot $\delta^{34}S_{(SO_4)}$ versus $\delta^{18}O_{(H_2O)}$. Plot the spring and fall samples on the same graph using different symbols.

d. The two most likely sources of sulfur for Nose Creek are weathering of till and/or bedrock and an anthropogenic input. The Crossfield gas plant is located north of Airdrie. For the sour gas emitted from this plant, $\delta^{34}S = +25‰$. Between Airdrie and urban Calgary, A-horizon soils typically have $\delta^{34}S = +3$ to $+10‰$, and this is an agricultural area. Total S (pyrite + organic S) for the Balzac till, which underlies Calgary, has $\delta^{34}S = -12.5‰$. Based on the isotopic characteristics of the different sulfur sources, describe and explain the variations and patterns seen on the plot constructed in part (c).

50. For a stream draining a region underlain by granites, $Sr = 0.06$ mg L^{-1} and $^{87}Sr/^{86}Sr = 0.7200$. A tributary that drains a region underlain by carbonate rocks enters the stream. For the tributary, $Sr = 0.4$ mg L^{-1} and $^{87}Sr/^{86}Sr = 0.7092$. Immediately downstream from the tributary the following values are measured for the river water: $Sr = 0.128$ mg L^{-1} and $^{87}Sr/^{86}Sr = 0.7178$. Calculate the amount of tributary water added to the main river.

51. Fritz et al. (1994) measured the sulfur isotopic composition of leachate seeping from a closed landfill. For the leachate, $\delta^{34}S = +17‰$. Gypsum mined in the United States that is used for plaster board would have $\delta^{34}S = +20$ to $+30‰$, leading the authors to conclude that the breakdown of plaster board in the landfill was the source of the sulfate in the leachate. Deep and shallow groundwaters in the area have an average of $\delta^{34}S = +0.5‰$. Water from a shallow well downgradient from the landfill had $\delta^{34}S = +10.6‰$. If the concentrations of the SO_4^{2-} ions in the various solutions are as follows: leachate $= 14$ mg L^{-1}, uncontaminated groundwater $= 2.4$ mg L^{-1}, and contaminated groundwater $= 9.5$ mg L^{-1}, calculate the percent leachate in the groundwater.

52. With reference to Case Study 5–5, Smirnov et al. (1998) identified three distinct groups of Lake Erie bottom sediments in terms of PAH content and $\delta^{13}C$. Two end members were identified, fluvial input from Detroit, Cleveland, and Buffalo and uncontaminated bottom sediments (North Zone). A sample of Lake Erie bottom sediment has TPAH = 2351 ng g^{-1} and $\delta^{13}C = -26.4‰$. Calculate the percent fluvial component in this sample of bottom sediment.

53. In the upper reaches of a river, the drainage basin is underlain by granites. In the lower reaches of the river, the drainage basin is underlain by carbonate rock. The carbonate rock is first encountered at a distance of 30 km from the headwaters of the river. Measurements made at various downstream locations are given in the accompanying table.

   a. Plot the $^{87}Sr/^{86}Sr$ ratio versus Sr concentration. Describe the shape of the resulting curve.

   b. Calculate 1/Sr. Plot the $^{87}Sr/^{86}Sr$ ratio versus 1/Sr. Describe the shape of the resulting curve. Do the data fit a simple binary mixing model? Explain.

   c. For waters in equilibrium with the granites, $^{87}Sr/^{86}Sr = 0.7200$ and Sr $= 0.06$ mg L^{-1}. For waters in equilibrium with the carbonate rocks, $^{87}Sr/^{86}Sr = 0.7092$ and Sr $= 0.4$ mg L^{-1}. Assuming simple binary mixing, calculate the proportion of carbonate-derived waters in the water sample collected at 70 km.

Distance (km)	Sr (mg L^{-1})	$^{87}Sr/^{86}Sr$
0	0.060	0.7200
10	0.060	0.7200
20	0.060	0.7200
30	0.060	0.7200
40	0.094	0.7154
50	0.128	0.7133
60	0.162	0.7120
70	0.196	0.7112
80	0.230	0.7106
90	0.264	0.7102
100	0.298	0.7099

54. Runoff from a feedlot and fertilized cultivated fields is entering a groundwater aquifer. Unpolluted groundwater samples have $PO_4^{3-} = 3$ mg L^{-1} and $NO_3^- = 1$ mg L^{-1}. For the runoff from the feedlot, $PO_4^{3-} = 6$ mg L^{-1} and $NO_3^- = 18$ mg L^{-1}, and for runoff from the cultivated fields, $PO_4^{3-} = 42$ mg L^{-1} and $NO_3^- = 13$ mg L^{-1}. For the contaminated groundwater, $PO_4^{3-} = 9$ mg L^{-1} and $NO_3^- = 11$ mg L^{-1}. Plot the data and determine the relative proportions of the three end members using the lever rule. Check your answer by using these proportions to calculate the composition of the contaminated groundwater.

55. The relationship between $\delta^{18}O$ and temperature for calcite in equilibrium with water has been determined experimentally. This relationship is represented by equation 6–55. This geothermometer has been used to track temperature changes in the world's oceans. One of the problems with the geothermometer is that changing climatic conditions can change the isotopic composition of seawater. One way to correct for this is to determine the isotopic composition of a benthic organism and use this value to estimate the seawater isotopic composition, assuming a constant temperature for the deep ocean. This seawater isotopic composition can then be used to determine surface seawater temperature using a pelagic organism.

a. A benthic organism has $\delta_c = +5.25‰$. Assuming that the temperature of the deep ocean remains constant at $1°C$, calculate the isotopic composition of the seawater in equilibrium with this shell. You will need to use the quadratic formula in order to solve this part of the problem.

b. A pelagic organism has $\delta_c = +2.5‰$. Calculate the temperature of the surface water in equilibrium with this shell.

56. Price et al. (2000) determined the carbon and oxygen isotopic characteristics of belemnite genera from the early Cretaceous interval of the Speeton Clay Formation, Filey Bay, England. A subset of their data compilation is given here. The first column gives the height of the sampled interval above the base of the section, arranged in normal stratigraphic order (youngest at the top). The approximate age of this section is 142 to 133 million years. The data for two belemnite species (*Hibolites* sp. and *Acroteuthis* sp.) are listed in the table. An analysis of their results led the authors to conclude that the isotopic data were consistent across these two species.

a. Calculate the paleotemperature for each belemnite sample using equation 6–55. Price et al. (2000) assumed that Early Cretaceous seawater had $\delta^{18}O = -1‰$.

b. Plot stratigraphic height (on the y-axis) versus $\delta^{13}C$ and $\delta^{18}O$. Draw smooth curves through the data points.

c. For each stage, plot the calculated temperature versus $\delta^{13}C$. If you've done these calculations on a spreadsheet, draw a linear regression line for each data subset.

d. From the plots constructed to answer parts (b) and (c), what conclusions can you draw about the relationship between paleotemperature and $\delta^{13}C$? Interpret the results. You may want to refer to the original paper.

Height from base (m)	Stage	Species	$\delta^{13}C$ (‰) (PDB)	$\delta^{18}O$ (‰) (PDB)	T (°C)
26.91	Late Hauterivian	*Hibolites* sp.	1.78	0.63	_____
25.53	Late Hauterivian	*Hibolites* sp.	1.76	0.50	_____
25.50	Late Hauterivian	*Hibolites* sp.	−0.13	0.10	_____
23.49	Early Hauterivian	*Hibolites* sp.	1.62	0.72	_____
22.00	Early Hauterivian	*Hibolites* sp.	1.18	0.46	_____
20.42	Early Hauterivian	*Hibolites* sp.	0.60	0.62	_____
19.14	Early Hauterivian	*Hibolites* sp.	1.60	−0.27	_____
19.02	Early Hauterivian	*Hibolites* sp.	1.23	0.59	_____
19.00	Early Hauterivian	*Hibolites* sp.	1.49	0.87	_____
18.71	Early Hauterivian	*Hibolites* sp.	1.30	−0.89	_____
18.57	Early Hauterivian	*Hibolites* sp.	1.89	0.21	_____
17.96	Early Hauterivian	*Hibolites* sp.	1.00	−0.46	_____
16.71	Early Hauterivian	*Hibolites* sp.	0.25	0.24	_____
16.26	Early Hauterivian	*Hibolites* sp.	0.81	0.40	_____
14.52	Early Hauterivian	*Hibolites* sp.	1.23	0.56	_____
12.33	Early Valanginian	*Acroteuthis* sp.	−0.27	−0.25	_____
11.81	Early Valanginian	*Acroteuthis* sp.	0.14	−0.60	_____
11.21	Early Valanginian	*Acroteuthis* sp.	−0.71	−0.29	_____
11.00	Early Valanginian	*Acroteuthis* sp.	1.30	−0.42	_____
10.56	Early Valanginian	*Acroteuthis* sp.	−1.38	−0.86	_____
9.84	Early Valanginian	*Acroteuthis* sp.	0.31	−0.08	_____
9.09	Early Valanginian	*Acroteuthis* sp.	0.79	0.47	_____
1.68	Late Ryazanian	*Acroteuthis* sp.	0.34	0.44	_____
1.63	Late Ryazanian	*Acroteuthis* sp.	−1.05	−0.06	_____
0.80	Late Ryazanian	*Acroteuthis* sp.	−0.22	0.37	_____

57. Jones and Young (1998) used laser ablation to measure the $\delta^{18}O$ isotopic composition of tooth enamel from the molar of a modern African elephant in Amboseli National Park, Kenya. Ten analyses were done through a 1750-μm section of tooth enamel parallel to the growth direction. The enamel growth rate was estimated to be 150 to 200 μm per month. Thus, the traverse represents approximately 1 year in the life of the elephant. For the local drinking water (largely snowmelt from Kilimanjaro), $\delta^{18}O = -5‰$.

Ayliffe et al. (1994) give the following relationship for the fractionation of oxygen isotopes between water and tooth enamel: $\delta^{18}O_p = 0.94 \pm 0.1\delta^{18}O_w + 23.3 \pm 0.7$.

a. Calculate the $\delta^{18}O$ for the elephant's enamel.

b. Several segments of tooth enamel had $\delta^{18}O = 23‰$. These segments were formed during the wet season when meteoric water is the dominant source for drinking water. Calculate $\delta^{18}O$ for the meteoric water.

c. If only a bulk analysis had been done on this tooth, how would the estimated $\delta^{18}O$ for meteoric water in equatorial Africa differ from the actual $\delta^{18}O$?

Environmental Mineralogy

Minerals are the basic building blocks of the earth's crust and as such are of significant interest in environmental geochemistry. Some minerals, such as asbestos, are considered to be hazardous to humans. Other minerals, such as the clay minerals, play an important role in the transport of ionic species through groundwater systems. The types of atoms, and how they are arranged within the mineral, are the important factors that determine the physical and chemical properties of minerals. Minerals that at first glance may appear to be similar, such as chrysotile and crocidolite, both considered to be asbestos minerals, behave very differently when lodged in the lungs due to differences in their crystal structures. Thus, an understanding of mineral chemistry and crystal structure is important in assessing the role of minerals in the environment.

BASIC MINERALOGY

Definition of a Mineral

A **mineral** is commonly defined as *a naturally occurring, inorganic substance with a characteristic internal structure and a chemical composition that is either fixed or varies within certain limits.* By "naturally occurring" we mean that the substance is found in nature. Compounds made in the laboratory that have no natural analogue are not considered to be minerals. By way of comparison, fewer then 4000 minerals have been identified, but hundreds of thousands of compounds have been made in the laboratory. The "inorganic substance" part of the definition is often troublesome for students. The intent is to convey that minerals do not have a biological origin. However, this definition becomes somewhat blurred in that carbon compounds, such as graphite and diamond, which result from physiochemical processes, and certain substances produced by organisms, such as carbonate shell material, are also considered to be minerals. A "characteristic internal structure" means that the compound must be crystalline; i.e., the atoms are arranged in a periodic array. Therefore, glasses, which are supercooled liquids, are not considered to be minerals because their atoms are not arranged in a regular array. Two crystalline compounds that have the same chemical composition, but in which the atoms are arranged in different ways, form two different minerals (called *polymorphs*). For example, both graphite and diamond consist solely of carbon, but the arrangement of the carbon atoms is different, giving rise to two minerals with vastly different physical properties. Lastly, "a chemical composition that is either fixed or varies within certain limits" means that a particular mineral has a relatively unique chemical composition. For example, in the minerals halite and sylvite the atoms are arranged in a cubic array. Thus, both minerals have the same crystal structure. Halite consists of Na and Cl atoms, whereas sylvite consists of K and Cl atoms, giving rise to two minerals that differ in terms of their chemistry. For mineral groups that form solid solution series, the distinction between mineral names becomes somewhat arbitrary. For example, plagioclase forms a complete solid solution series with the end members albite ($NaAlSi_3O_8$) and anorthite ($CaAl_2Si_2O_8$). The series is formed by the coupled substitution NaSi $\rightleftharpoons$ CaAl. Albite is defined as any plagioclase that contains 90–100% of the albite component, oligoclase as any plagioclase that contains 70–90% of the albite component, etc.

Types of Minerals

The two major elements of the earth's crust are oxygen and silicon. Hence, these elements form the basic building block, the silica tetrahedron, of the common silicate minerals. In this chapter we will mostly be concerned with the silicate minerals. However, there are a number of other mineral groups that are common at the earth's surface. None of these are as abundant as the silicate minerals, but many are economically or environmentally important. These different mineral classes are summarized in Table 7–1.

Crystal Chemistry

In Chapter 1 we discussed the structure of the atom. Depending on the number of electrons in the outer shell, atoms either gain or lose electrons to complete their outer shell. This is referred to as *ionization* and the resulting charged atom is called an *ion*. **Ionization potential** is a *measure of the energy required to remove an electron from an atom and place it at an infinite distance from the nucleus*. In general, as we move from left to right across any row in the periodic table the ionization potential increases; i.e., it becomes more difficult to remove an electron from the atom. **Cations** are *atoms that have lost electrons and have a net positive charge*. **Anions** are *atoms that have gained electrons and have a net negative charge*. Because ionization potential increases from left to right across the periodic table, elements on the left side tend to lose electrons and form cations. These elements are sometimes referred to as *metals*. Elements on the right side tend to gain electrons and form anions. These elements are sometimes referred to as *nonmetals*.

Types of Bonding   Atoms can combine (bond) in four different ways: covalent, ionic, metallic, and hydrogen. In **covalent bonding** *electrons are shared between atoms*. During

Table 7–1   Mineral Classes

Class	Chemical characteristics	Examples
Borates	Various elements in combination with boron	Borax [$Na_2B_4O_7 \cdot 10H_2O$]
Carbonates	Metals in combination with carbonate (CO_3^{2-})	Calcite [$CaCO_3$] Cerrusite [$PbCO_3$]
Halides	Alkali metals or alkaline earths in combination with halogens (F, Cl, Br, I)	Halite [$NaCl$] Fluorite [CaF_2]
Hydroxides	Metals in combination with hydroxyls (OH^-)	Brucite [$Mg(OH)_2$]
Native elements	Pure compound of a metallic or nonmetallic element	Gold [Au] Graphite [C]
Oxides	Metals in combination with oxygen	Hematite [Fe_3O_4]
Phosphates, arsenates, vanadates, chromates, tungstates, and molybdates	Various elements in combination with the ZO_4 radical where Z = P, As, V, Cr, W, Mo	Apatite [$Ca_5(PO_4)_3(F,Cl,OH)$] Carnotite [$K_2(UO_2)_2(VO_4)_2 \cdot 3H_2O$] Scheelite [$CaWO_4$]
Silicates	Metals in combination with silica tetrahedra (SiO_4^{4-}) forming three-dimensional networks, sheets, chains, and isolated tetrahedra	Quartz [SiO_2] Forsterite [Mg_2SiO_4] Orthoclase [$KAlSi_3O_8$]
Sulfates	Alkaline earths or metals in combination with sulfate (SO_4^{2-})	Barite [$BaSO_4$] Epsomite [$MgSO_4 \cdot 7H_2O$]
Sulfides	One or more metals in combination with reduced sulfur or chemically similar elements (As, Se, Te)	Pyrite [FeS_2] Galena [PbS] Skutterudite [$CoAs_3$]

ionic bonding electrons are transferred from one atom to another, thus forming a cation–anion pair that is held together by electrostatic attraction. **Metallic bonding** occurs in the case of *pure metals in which electrons are freely shared among all the atoms.* **Hydrogen bonding**, of which there are many types, involves *polar molecules (such as water) that have a slight charge bias and form a weak electrostatic bond with other ions.* Except for compounds consisting of a single element, pure bonds do not exist. Most bonds are of a mixed ionic-covalent character. The component of ionic bonding can be determined from the difference in electronegativity between the two atoms. **Electronegativity** (as defined by Pauling, 1960) is a *measure of the ability of an atom in a molecule to attract electrons.* The smaller the electronegativity, the smaller is the likelihood of the atom attracting an electron; i.e., elements with small electronegativities are electron donors. Electronegativity is a nondimensional number determined from the relative strengths of bonds between various elements and ranges from 0.7 (Cs) to 4.0 (F). Electronegativities are listed in Table 7–2. Electronegativities do depend on the charge of the ion, and an appropriate ion for each element is listed in Table 7–2. The ionic character of a particular bond is determined by the difference in electronegativities between the two elements involved in the bond. The ionic character corresponding to this difference is tabulated in Table 7–3 (p. 210). Note that bonds between the same element, e.g., C–C, do not have an ionic character; i.e., the electronegativity difference is zero.

Table 7–2   Electronegativities*

Z	Ion	Electro-negativity	Z	Ion	Electro-negativity	Z	Ion	Electro-negativity
1	H^+	2.20	33	As^{5+}	2.18	65	Dy^{3+}	1.22
3	Li^+	0.98	34	Se^{2-}	2.55	67	Ho^{3+}	1.23
4	Be^{2+}	1.57	35	Br^-	2.96	68	Er^{3+}	1.24
5	B^{3+}	2.04	37	Rb^+	0.82	69	Tm^{3+}	1.25
6	C^{4+}	2.55	38	Sr^{2+}	0.95	70	Yb^{3+}	—
7	N^{5+}	3.04	39	Y^{3+}	1.22	71	Lu^{3+}	1.0
8	O^{2-}	3.44	40	Zr^{4+}	1.33	72	Hf^{4+}	1.3
9	F^-	3.98	41	Nb^{5+}	1.6	73	Ta^{5+}	1.5
11	Na^+	0.93	42	Mo^{6+}	2.16	74	W^{6+}	1.7
12	Mg^{2+}	1.31	43	Tc^{2+}	2.10	75	Re^{7+}	1.9
13	Al^{3+}	1.61	44	Ru^{2+}	2.2	76	Os^{6+}	2.2
14	Si^{4+}	1.90	45	Rh^{2+}	2.28	77	Ir^{6+}	2.2
15	P^{5+}	2.19	46	Pd^{2+}	2.20	78	Pt^{4+}	2.2
16	S^{2-}	2.58	47	Ag^+	1.93	79	Au^+	2.4
17	Cl^-	3.16	48	Cd^{2+}	1.69	80	Hg^{2+}	1.9
19	K^+	0.82	49	In^{3+}	1.78	81	Tl^{3+}	1.8
20	Ca^{2+}	1.00	50	Sn^{2+}	1.96	82	Pb^{2+}	1.8
21	Sc^{3+}	1.36	51	Sb^{5+}	2.05	83	Bi^{3+}	1.9
22	Ti^{4+}	1.54	52	Te^{2-}	2.1	84	Po^{4+}	2.0
23	V^{3+}	1.63	53	I^-	2.66	85	At^{5+}	2.2
24	Cr^{3+}	1.66	55	Cs^+	0.79	87	Fr^+	0.7
25	Mn^{2+}	1.55	56	Ba^{2+}	0.89	88	Ra^{2+}	0.9
26	Fe^{2+}	1.83	57	La^{3+}	1.10	89	Ac^{3+}	1.1
27	Co^{2+}	1.88	58	Ce^{3+}	1.12	90	Th^{4+}	1.3
28	Ni^{2+}	1.91	59	Pr^{3+}	1.13	91	Pa^{4+}	1.5
29	Cu^+	1.90	60	Nd^{3+}	1.14	92	U^{6+}	1.7
30	Zn^{2+}	1.65	62	Sm^{3+}	1.17	93	Np^{3+}	1.3
31	Ga^{3+}	1.81	64	Gd^{3+}	1.20	94	Pu^{4+}	1.3
32	Ge^{4+}	2.01						

*From *CRC Handbook of Chemistry and Physics* (2000).

Table 7–3 Percent Ionic Character of a Single Chemical Bond*

Difference in electronegativity	Ionic character, %	Difference in electronegativity	Ionic character, %
0.1	0.5	1.7	51
0.2	1	1.8	55
0.3	2	1.9	59
0.4	4	2.0	63
0.5	6	2.1	67
0.6	9	2.2	70
0.7	12	2.3	74
0.8	15	2.4	76
0.9	19	2.5	79
1.0	22	2.6	82
1.1	26	2.7	84
1.2	30	2.8	86
1.3	34	2.9	88
1.4	39	3.0	89
1.5	43	3.1	91
1.6	47	3.2	92

*From Sargent-Welch (1980).

EXAMPLE 7–1 The mineral fluorite has the chemical composition CaF_2. Calculate the ionic character of the bond between Ca-F.

From Table 7–2, the difference in electronegativity = 3.98 (F^-) − 1.00 (Ca^{2+}) = 2.98. From Table 7–3, the bond is 89% ionic. ■

Coordination Numbers Most crystal structures can be viewed as polyhedra of anions surrounding cations. The number of anions that can "fit" around a cation is a function of the relative size of the cations and anions. In a close-packed array, all the ions are in contact. If we consider the ions to be spheres, then the number of anions that can fit around a cation can be determined by simple geometry. To do this, we calculate the **radius ratio**, the *radius of the cation divided by the radius of the anion*. The smaller the radius ratio, the fewer the number of anions that can fit around the cation. The **coordination number** is the *number of anions that surround a cation in an ionic crystal*. For each coordination number there is a minimum radius ratio. If you have access to styrofoam balls of different diameters, you can determine the coordination number empirically by seeing how many large balls you can fit around a smaller ball. We can calculate

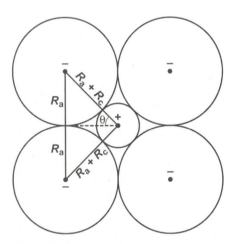

Figure 7–1
Packing of anions around a cation for a coordination number of 4. The minimum radius ratio can be calculated from the geometry of the packing. R_a and R_c are the radii of the anion and cation, respectively. In this case, $\theta = 45°$. See text for calculation.

the minimum radius ratio for any particular coordination number. In Figure 7–1 four large anions (CN = 4) are arranged in close packing around a smaller cation. We calculate the

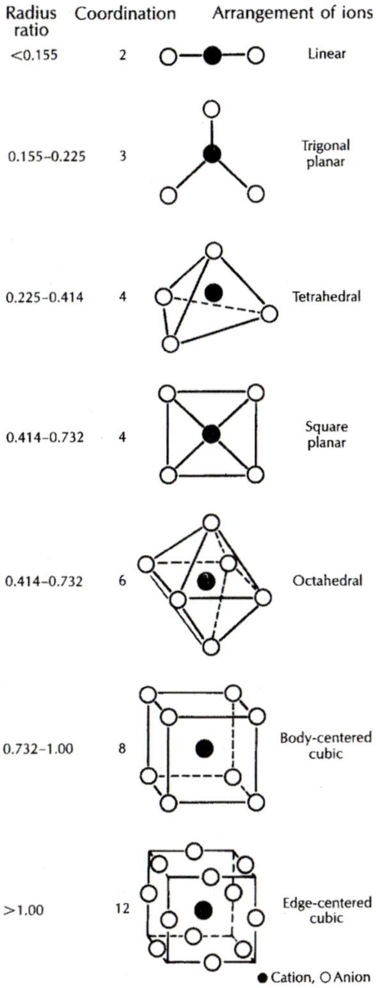

Radius ratio	Coordination	Arrangement of ions	
<0.155	2		Linear
0.155–0.225	3		Trigonal planar
0.225–0.414	4		Tetrahedral
0.414–0.732	4		Square planar
0.414–0.732	6		Octahedral
0.732–1.00	8		Body-centered cubic
>1.00	12		Edge-centered cubic

● Cation, ○ Anion

Figure 7–2

Possible arrangement of ions in crystals for particular coordination numbers (CN). Radius ratio range is listed for each CN. From Faure (1998).

minimum radius ratio for this arrangement as follows. With reference to Figure 7–1, we have a right triangle with hypotenuse $R_a + R_c$ and side R_a. Angle $\theta = 45°$ and

$$\sin \theta = \frac{R_a}{R_a + R_c} \qquad (7\text{–}1)$$

The sin of $45° = 0.7071$, and substituting this value in equation 7–1 gives

$$0.7071 = \frac{R_a}{R_a + R_c} \qquad (7\text{–}2)$$

Rearranging in terms of R_c/R_a gives

$$\frac{R_c}{R_a} = \frac{0.293}{0.707} = 0.414 \qquad (7\text{–}3)$$

This is the minimum radius ratio for fourfold coordination.

In the preceding example the ions lay in a plane. In real crystals ions are arranged in three-dimensional arrays. Examples of possible three-dimensional arrays for various coordination numbers are shown in Figure 7–2. Also shown are the range of radius ratios for each of these coordination numbers. If we know the radii of various ions, it is possible to predict the coordination number for these ions in a crystal structure. This is usually done in terms of the number of anions around a particular cation. The ionic radii for various ions are tabulated in Appendix III. The ionic radii depend on the ionic charge of the ion and the coordination number of the ion in a particular structure. With increasing positive charge ions get smaller, and with increasing negative charge ions get bigger. Some ions can occur in more than one coordination and for these, multiple radii are given. If an ion can occur in more than one coordination, the ionic size in the higher coordination is greater due to the "stretching" of the ion. For Cr, Mn, and Fe, additional ionic radii are given as a function of low (L) or high (H) spin state. In the *low spin state* electrons in the outer orbitals are paired, in the **high spin state** electrons in the outer orbitals exist singly. Hence, an ion in the high spin state has a larger ionic radius than an ion in the low spin state.

EXAMPLE 7–2 Calculate the coordination number for Ca and F in the mineral fluorite (CaF_2). F is twice as abundant as Ca in fluorite so the coordination number for F will be 0.5 that of Ca. As a first guess, we will predict that Ca is in sixfold coordination in fluorite. For sixfold coordination, the ionic radius of $Ca^{2+} = 1.08$ (Appendix III). For threefold coordination, the ionic radius of $F^- = 1.22$. The radius ratio is 0.89. With reference to Figure 7–2, this radius ratio corresponds to an eightfold cubic coordination. To verify that Ca is in eightfold coordination, we recalculate the radius ratio using $Ca^{2+} = 1.20$ (eightfold ionic radius), and $F^- = 1.23$ (fourfold ionic radius). The radius ratio is 0.98, confirming that Ca^{2+} is in eightfold coordination in the fluorite structure. The actual structure, as determined from X-ray crystallography, is a face-centered cube in which each Ca^{2+} is surrounded by eight F^-, and each F^- is surrounded by four Ca^{2+}, a result in agreement with the coordination number calculation. ∎

Ionic Substitutions When minerals crystallize, certain minor or trace elements that are present in the environment can enter the structure of the crystallizing mineral. This topic is of considerable importance in high-temperature geochemistry but also plays a role here in determining the likely repository of trace metals in minerals and the uptake of trace metals by minerals crystallizing from aqueous solutions. There are four *rules* that predict, with many exceptions, the uptake of trace elements by crystallizing minerals.

1. Ions of one element can substitute for those of another in a crystal structure if their radii differ by less than about 15%.
2. Ions that differ by one charge unit substitute readily for each other as long as charge neutrality is maintained. Note that this requires a coupled substitution, such

as occurs in the plagioclase solid solution series. If the charges differ by more than one unit, substitution is limited.

3. When two ions can occupy the same site in a crystal structure, the ion with the higher ionic potential preferentially enters the site.

4. Even if the size and charge of the minor and major ion are similar, substitution may be limited for the minor ion if it has a very different electronegativity and forms a bond of very different character from that of the major ion.

The first three rules were proposed by Goldschmidt and the last rule was later added by Ringwood. Although not perfect, these rules do provide some guidance as to the likely substitution of various ions into crystal structures.

EXAMPLE 7–3 In galena (PbS) lead is in sixfold coordination with sulfur. Will any of the following trace elements—Mn, Ag, or Np—likely be found in galena?

The following table summarizes the pertinent information for these elements. Ionic radii for the ions in sixfold coordination are from Appendix III, and the electronegativities and percent ionic character of the bonds are from Tables 7–2 and 7–3, respectively.

Ion	Ionic radii (sixfold coordination)	Electronegativity	% Ionic character of metal-sulfur bond
Pb^{2+}	1.26	1.8	14
Mn^{2+}	0.75	1.55	23
Ag^+	1.23	1.93	11
Np^{3+}	1.10	1.3	33

Mn^{2+} differs by more than 15% in size from Pb^{2+} and will not substitute in the crystal structure. Ag^+ has a similar ionic radius and bond type, differs by only one charge unit, and will substitute in the crystal structure. Although Np^{3+} meets the size and charge criteria, it will not substitute in the crystal structure because of its very different bond type compared to the Pb-S bond (33% ionic versus 14% ionic). ∎

X-Ray Crystallography

Atoms have radii on the order of 1 Å (1 angstrom = 1×10^{-10} m). Given the small size of atoms, how do we determine their arrangement in crystal structures? Our knowledge of the crystalline state is gained by studies utilizing X rays (the field of X-ray crystallography). Let us suppose that a beam of monochromatic X rays is incident on a crystal (Figure 7–3). If the wavelength of the X rays is similar to that of the spacing between the planes of atoms (for example, let us assume we are using Cu K_α radiation for which $\lambda = 1.54$ Å), the X rays will be scattered by the atoms. For certain incident angles and spacings between the planes of atoms, constructive interference will occur and a diffracted beam will leave the crystal at an angle equal to that of the incident beam. The relationship between X-ray wavelength, angle of incidence, and spacing between planes of atoms is known as Bragg's law. The relationship can be derived as follows. In order for waves to interfere constructively, the differences in travel path must equal multiples of the wavelength. For example, in Figure 7–3 the distance *BCD* must equal one wavelength or a multiple of one wavelength. The angle

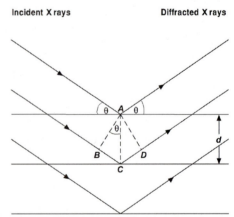

Figure 7–3
Diagram illustrating Bragg's law. θ = angle of incidence and diffraction when Bragg's law conditions are met. d = interplanar spacing.

of the incident X rays is θ. Note that ABC is a right triangle. AC is the hypotenuse and BC is the opposite side. Hence, we can write

$$\sin \theta = \frac{BC}{AC} \qquad (7\text{--}4)$$

$AC = d$ and solving for BC gives

$$BC = d \sin \theta \qquad (7\text{--}5)$$

Remember that for constructive interference to occur the distance BCD must equal one wavelength or a multiple of one wavelength. Thus, for constructive interference

$$\lambda = 2BC = 2d \sin \theta \qquad (7\text{--}6)$$

The more general form of Bragg's law, taking into consideration path lengths that are integers of the wavelength, is

$$n\lambda = 2d \sin \theta \qquad (7\text{--}7)$$

Although a very simple equation, this relationship represents a powerful way to investigate the structure of crystalline materials. A number of different methods are used to acquire X-ray diffraction data. The interested student can refer to any textbook on X-ray crystallography for further information. Perhaps the most common method used in the environmental field is X-ray powder diffractrometry. An X-ray tube provides a monochromatic X-ray beam. Thus, the wavelength is known. The sample is in a fixed geometry, usually powder on a glass slide, and a scintillation counter is scanned through a wide 2θ angle. When constructive interference occurs, a diffracted beam is sent to the scintillation counter, giving the 2θ angle. The remaining unknown is the d spacing. For any particular mineral there are multiple paths with different d spacings, and each of these will be recorded. In addition, for a mineral that consists of more than one element, the areal distribution of the different elements will vary as a function of the plane that is diffracting the X-ray beam. Because different ions interact differently with the X-ray beam, the intensity of the diffracted signal varies as a function of the particular atomic plane that is responsible for the diffracted beam. Thus, we obtain a second important piece of information, the intensity of the diffracted X-ray beam, which can be related to the composition of the crystalline solid. Tens of thousands of crystalline compounds have been characterized by X-ray diffraction. Taking into account the d spacing and relative intensities of the diffracted beam, virtually every crystalline compound has a unique X-ray fingerprint. This information is available in the American Society for Testing and Materials powder data file, which is frequently updated and contains information for all known crystalline materials. Thus, we have a powerful method for identifying crystalline solids even if individual particles are very small.

EXAMPLE 7–4 A beam of monochromatic X rays is focused on a crystalline material. A diffracted beam is found at $\theta = 23.4°$. Calculate the d spacing for the plane of atoms that is causing this diffracted beam. For this experiment we used an iron X-ray tube. Fe $K_\alpha = 1.54$ Å.

Solving for d,

$$d = \frac{\lambda}{2 \sin \theta} = \frac{1.54 \text{ Å}}{2 \sin 23.4} = 1.94 \text{ Å} = 1.94 \times 10^{-10} \text{ m} \qquad \blacksquare$$

EXAMPLE 7–5 Montmorillonite, an expandable clay, is often used as a liner for sanitary landfills. If water percolates down through the landfill, it will encounter the liner and the absorption of the water by the clay will expand the clay, forming an impermeable barrier. In a particular region both kaolinite (a nonexpandable clay) and montmorillonite deposits are found. The contractor constructing the landfill begins to quarry a clay deposit. An X-ray diffraction analysis of this clay yields the following data. Also included in the table are the five most intense peaks for kaolinite and montmorillonite. I/I_1 is the ratio of the peak intensity to that of the strongest peak in the diffraction pattern. For the most intense peak in the diffraction pattern, $I/I_1 = 100$; d spacings are in angstroms (10^{-10} m). Is this a suitable clay to use for the liner of the landfill? The comparative data are from Berry (1974), and the number behind each mineral name is the ASTM powder diffraction file designation.

X-Ray Powder Diffraction Data for Unknown Mineral and Kaolinite and Montmorillonite

Unknown clay					
I/I_1	100	92	80	72	49
d	1.47	2.31	4.39	4.51	7.22
Kaolinite (12–447)					
I/I_1	100	95	85	75	50
d	1.48	2.33	4.34	4.46	7.15
Montmorillonite (13–135)					
I/I_1	100	80	60	60	50
d	15.0	4.50	5.01	3.02	1.50

Based on the X-ray diffraction pattern, the unknown clay is kaolinite and would not be suitable for the landfill liner. Note that there are measurement errors, so the patterns are not expected to be exactly the same. ∎

BASIC SILICATE STRUCTURES

Silicate minerals are the building blocks of the earth's crust and mantle and are the earth's most abundant minerals. The fundamental structural unit of silicate minerals is the silica tetrahedron, an Si^{4+} ion tetrahedrally coordinated with $4 O^{2-}$. Each O^{2-} anion shares one negative charge with the Si^{4+} cation, leaving a negative charge of -1 at each corner of the tetrahedron. The negatively charged silica tetrahedron can bond with various cations to form a three-dimensional crystal structure. The silica tetrahedra can also combine with other silica tetraheda by sharing corners, i.e., O^{2-} anions. The way the silica tetrahedra are combined leads to the various classes of silicate minerals illustrated in Figure 7–4.

The **nesosilicates** consist of *isolated silica tetrahedra that bond with various metal cations to form silicate minerals. In the **sorosilicates**, two tetrahedra are combined by sharing an oxygen. The resulting structural unit has a net negative charge of -6 and the structural units are connected via cations. In the **cyclosilicates**, the tetrahedra are arranged in a ring. In these structural units, the outer corners of the tetrahedra have a -1 charge, as do the apices of the tetrahedra pointing out of the page. In the **inosilicates**, the

Figure 7–4
Arrangement of silica tetrahedra in the various classes of silicate minerals. See text for further discussion. From Brownlow (1996).

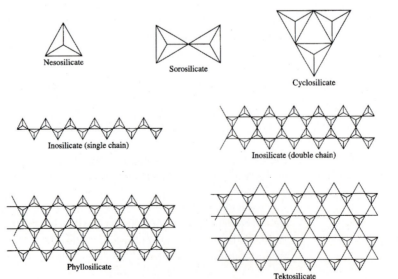

Table 7–4   *Properties of the Silicate Crystal Classes**

Class	Tetrahedral arrangement	# Shared corners	Chemical unit	Si:O	Example
Nesosilicate	Independent tetrahedra	0	SiO_4^{4-}	1:4	Olivine
Sorosilicate	Two tetrahedra sharing a corner	1	$Si_2O_7^{6-}$	1:3.5	Melilite
Cyclosilicate	Three or more tetrahedra sharing two corners, forming a ring	2	SiO_3^{2-}	1:3	Beryl
Inosilicate	Single chain of tetrahedra sharing two corners	2	SiO_3^{2-}	1:3	Augite
	Double chain of tetrahedra alternately sharing two or three corners	2.5	$Si_4O_{11}^{6-}$	1:2.75	Hornblende
Phyllosilicate	Sheet of tetrahedra sharing three corners	3	$Si_2O_5^{2-}$	1:2.5	Kaolinite
Tektosilicate	Framework of tetrahedra sharing all four corners	4	SiO_2	1:2	K-feldspar

*After Brownlow (1996).

silica *tetrahedra are joined in either single or double chains*. The inosilicates are the structural units for two important groups of silicate minerals, the pyroxenes (single chains) and the amphiboles (double chains). In the ***phyllosilicates***, the *tetrahedra are joined to form a layer*. Only the corners of tetrahedra along the outer edge of the layer and the apices of the tetrahedra pointing out of the page have net negative charges. Micas and clay minerals are phyllosilicates. In the ***tektosilicates***, the *tetrahedra are connected in a three-dimensional framework through the sharing of all tetrahedral oxygens*. Quartz (SiO_2) is an example of a tektosilicate. The characteristics of each of these structural groups are summarized in Table 7–4.

In the quartz structure, all the oxygens are bonded to silica and there are no net negative charges. Hence, there seem to be no other chemical possibilities. However, Al can replace Si in the silica tetrahedron. The Al cation has a charge of $+3$, so for any tetrahedron in which this replacement has occurred there is a net negative charge of -1. This substitution, which must be coupled with another cation substitution in order to maintain charge balance, gives rise to one of the important groups of silicate minerals, the feldspars. For K-feldspar ($KAlSi_3O_8$) the substitution is $KAl \rightleftarrows Si$. The plagioclase feldspars form a complete solid solution series with the end members albite ($NaAlSi_3O_8$) and anorthite ($CaAl_2Si_2O_8$). In this series the coupled substitution is $CaAl \rightleftarrows NaSi$.

CLAY MINERALS

In terms of environmental mineralogy, the clays are one of the most important groups of minerals. They are a common product of the weathering of silicate minerals, play an important role in the sequestering of metals via ion exchange, and have a number of engineering uses, such as liners for landfills. This is a very complex group of minerals, made even more so by their fine grain size, about which literally books have been written. A ***clay mineral*** is a *fine-grained hydrous silicate composed of layers of tetrahedrally and octahedrally coordinated cations*. A ***clay*** is any *material less than 2 µm in diameter*. Clays are often composed of clay minerals, but this is not always the case. Hence, a distinction needs to be made between the term *clay*, referring to size, and the term *clay mineral*, referring to a particular type of crystalline substance.

Clay minerals consist of two types of layers, octahedral and tetrahedral, which are arranged in different ways to produce the wide variety of clay minerals (the layered clay mineral groups). The one exception to this basic structural arrangement is the palygorskite group minerals, which have chainlike crystal structures. A simple way of looking at clay structures is to consider the octahedral layer as representing a sandwich spread and the tetrahedral layer as representing a piece of bread. These layers can be put together to form an open-faced sandwich, in which there is one slice of bread covered with sandwich spread, or a closed-face sandwich, with the sandwich spread between two pieces of bread. The sandwiches are then stacked to form the clay mineral.

The octahedral layer consists of two sheets of close-packed hydroxyl ions with Mg^{2+} or Al^{3+} ions between the sheets. The two sheets of hydroxyl ions are offset from each other, as would be two layers of billiard balls. Tetrahedral and octahedral holes (sites) are found between the layers. If Mg^{2+} ions occupy the octahedral holes, all the sites are filled (tri-octahedral). If Al^{3+} ions occupy the octahedral holes, only two out of three sites are filled (di-octahedral). In the case of Al^{3+}, the presence of vacant sites is necessitated by the need to charge balance the octahedral layer. Thus, we have two basic types of sandwich filling: brucite (tri-octahedral), in which the octahedral layers contain Mg^{2+}, and gibbsite (di-octahedral), in which the octahedral layers contain Al^{3+} (Figure 7–5). The tetrahedral layer consists of silicate tetrahedrons arranged in a sheet (Figure 7–5). In the tetrahedral layer, Al^{3+} can substitute for Si^{4+} and this substitution is charge balanced by the addition of Na^+ or K^+. Various combinations of these two types of layers give rise to the main groups of clay minerals (Table 7–5).

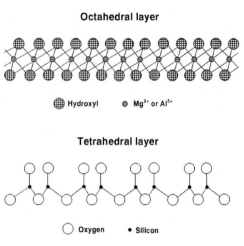

Figure 7–5
Structure of the octahedral and tetrahedral layer. Mg^{2+} in the octahedral layer = brucite. Al^{3+} in the octahedral layer = gibbsite. Al^{3+} can substitute for Si^{4+} in the tetrahedral layer.

Table 7–5 Summary of the Principal Characteristics of the Layered Clay Mineral Groups*

	Kaolinites	Illites	Smectites	Vermiculites
Structure, tetrahedral: octahedral	1:1	2:1	2:1	2:1
Octahedral layer	Di-octahedral	Mostly di-octahedral	Di- or tri-octahedral	Mostly tri-octahedral
Interlayer cations	Nil	K	Ca, Na	Mg
Interlayer water	Only in halloysite	Some in hydromuscovite	Ca, two layers Na, one layer	Two layers
Basal spacing	7.1 Å	10 Å	Variable, most ~ 15 Å	Variable, 14.4 Å when fully hydrated
Ethylene glycol	Only taken up by halloysite	No effect	Two glycol layers, 17 Å	One glycol layer, 14 Å
Cation exchange capacity (CEC) in meq/100 g clay	Nil, 3–15	Low, 10–40	High, 80–150	High, 100–150
Formula	$Al_2Si_2O_5(OH)_4$, little variation	$K_{0.5-0.75}Al_2(Si, Al)_4 O_{10}(OH)_2$	$M^+_{0.7}(Y^{3+}, Y^{2+})_{4-6} (Si, Al)_8O_{20}(OH)_4 \cdot nH_2O$	$M^{2+}_{0.66}(Y^{2+}, Y^{3+})_6 (Si, Al)_8O_{20}(OH)_4 \cdot 8H_2O$
Acids	Scarcely soluble	Readily attacked	Attacked	Readily attacked
Heating 200°C	Except halloysite, unchanged	No marked change	Collapse to approximately 10 Å	Exfoliation, shrinkage of layer spacing
Examples	Kaolinite, dickite, nacrite, halloysite	Illite, hydrous micas, phengite, brammallite, glauconite, celadonite	Montmorillonite, beidellite, nontronite, hectorite, saponite, sauconite	Vermiculite

*Modified from Deer et al. (1992).

Clay Mineral Structures

1:1 Clays . The basic building block of the 1:1 clays consists of a tetrahedral layer and an octahedral layer. The structure of kaolinite, a representative two-layer clay, is shown in Figure 7–6. The octahedral sites in the octahedral layer are occupied by Al^{3+} (this is a gibbsite layer). Substitutions are limited. In the octahedral layer, Fe^{2+} may occasionally substitute for Al^{3+}; and in the tetrahedral layer, there is limited substitution of Al^{3+} for Si^{4+}. The ideal chemical formula for kaolinite is $Al_2Si_2O_5(OH)_4$. Because of the limited substitution in the kaolinite structure, the net charge is minimal. Hence, kaolinite group minerals have a very low cation-exchange capacity (Table 7–5). The structural units (sheets) are attached to other

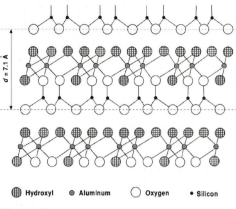

⊕ Hydroxyl   ⊙ Aluminum   ◯ Oxygen   • Silicon

Figure 7–6

Structure of kaolinite. Each structural unit consists of a gibbsite layer and a tetrahedral layer. Note that only two out of three octahedral sites in the octahedral layer are occupied.

sheets by weak van der Waals bonds. Because of this weak bonding, as we know from common experience, clays are soft and easily deformed. The sheets can be stacked in different ways, which gives rise to the polymorphs (dickite and nacrite) of kaolinite. With reference to Table 7–5, note that kaolinite does not take up ethylene glycol (a universal measure of the expandability of a clay) and contains minimal interlayer cations and water.

The ideal di-octahedral (gibbsite) layer would have a slightly smaller horizontal dimension than the tetrahedral layer. In kaolinite, dickite, and nacrite, this small difference in size is accommodated by distorting the tetrahedral layer. This difference can also be accommodated by forming a "tubelike" structure in which the slightly smaller di-octahedral layers are on the inside of the tube. The tubular variety of kaolinite is called *haloysite*.

The tri-octahedral (brucite layer) analogues of the kaolinite group are the serpentine minerals. We will consider the serpentine minerals in a subsequent section.

2:1 Clays   The basic building block of the 2:1 clays is two tetrahedral layers with an intervening octahedral layer. The octahedral layer can be either di- or tri-octahedral, and a variety of substitutions occur in both the octahedral and tetrahedral layers. These substitutions are charge balanced by placing cations in the interlayer positions. The three groups of 2:1 clays (Table 7–5) are illites, smectites, and vermiculites.

In *illites*, the octahedral layer is gibbsite, and Al^{3+} substitutes for Si^{4+} in the tetrahedral layer. This substitution is charge balanced by the addition of K^+ ions between the 2:1 sheets (the interlayer positions). The large number of K^+ cations in the interlayer position results in a strong bond between sheets. These interlayer cations are not readily exchangeable. Thus, illite has a low cation-exchange capacity (CEC) and illites are not expandable clays. Illite is very similar to muscovite, and the original definition of illite (Grim et al., 1937) was that it was all minerals belonging to the mica group with a size of $< 2\mu m$ and a basal spacing of 10 Å. The major difference between muscovite and illite is that illite has fewer interlayer cations, and the subsequently weaker bonding means that there is less regularity in the stacking sequence of illite sheets. Most natural samples of illite contain smectite layers, which may be regularly or randomly intermixed. Such mixed clays show properties between those of illite and smectite, depending on the relative proportion of each mineral. Glauconite is an iron-rich illite.

By definition, **smectites** are *clay minerals whose basal spacing expands to 17 Å when treated with ethylene glycol.* A number of substitutions are possible in this group of minerals—in particular, Fe^{2+} and Mg^{2+} for Al^{3+} in the octahedral layer and Al^{3+} for Si^{4+} in the tetrahedral layer. These substitutions lead to a net negative charge on the 2:1 sheets on

Table 7–6 Substitutions for Smectite-Group Clay Minerals*

Mineral	Tetrahedral cations	Octahedral cations	Exchangeable cations
		Di-octahedral	
Montmorillonite	Si_8	$Al_{3.3}Mg_{0.7}$	$(0.5Ca,Na)_{0.7}$
Beidellite	$Si_{7.3}Al_{0.7}$	Al_4	$(0.5Ca,Na)_{0.7}$
Nontronite	$Si_{7.3}Al_{0.7}$	Fe_4^{3+}	$(0.5Ca,Na)_{0.7}$
		Tri-octahedral	
Saponite	$Si_{7.2}Al_{0.8}$	Mg_6	$(0.5Ca,Na)_{0.8}$
Hectorite	Si_8	$Mg_{5.3}Li_{0.7}$	$(0.5Ca,Na)_{0.7}$
Sauconite	$Si_{6.7}Al_{1.3}$	$Zn_{4-6}(Mg,Al,Fe^{3+})_{2-0}$	$(0.5Ca,Na)_{0.7}$

*Modified from Deer et al. (1992).

the order of 0.2 to 0.5 units per formula unit $[O_{10}(OH)_2]$. This negative charge is balanced by adding cations to the interlayer position. The possible substitutions, and the corresponding mineral names, are given in Table 7–6.

The structure of montmorillonite, a representative example of the smectites, is shown in Figure 7–7. The general formula for montmorillonite is $M_{0.67}(Mg_{0.33}Al_{1.67})Si_4O_{10}(OH)_2 \cdot nH_2O$, where M is the exchangeable cation (Na^+, K^+, $0.5Ca^{2+}$, etc.) in the interlayer position. The ions in the interlayer position are hydrated. If the interlayer cation is predominantly Na, the mineral is Na-montmorillonite and there is generally one water layer in the interlayer position. If Ca is the predominant cation, the mineral is Ca-montmorillonite and there are generally two water layers in the interlayer position. These differences in the number of water layers lead to differences in the basal spacing: 15.4 Å for Ca-montmorillonite and 12.6 Å for Na-montmorillonite. In the case of Na-montmorillonite, the clay mineral will expand as the water content of the immediate environment increases. Thus, these clays will expand and

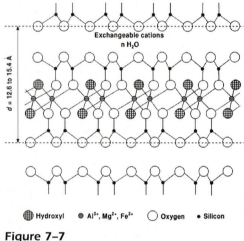

Figure 7–7
Structure of montmorillonite, a 2:1 clay. The octahedral layer is a gibbsite layer. Substitution of Mg^{2+} for Al^{3+} in the octahedral layer is charge balanced by the addition of Na^+ or Ca^{2+} cations (exchangeable cations) in the interlayer position.

contract with changes in water content of the environment. This presents an engineering challenge in designing structures sited on expandable clays. The response of the clay to changes in water content has also been used in the design of sanitary landfills. A layer of Na-montmorillonite is used as one of the barriers in the system. If groundwater seeps through the site and penetrates the impermeable barrier (often polyethylene), it encounters the clay layer. Water entering the clay layer causes the clay mineral to expand, thus producing an impermeable barrier. The interlayer cations are readily exchangeable, leading to a high cation-exchange capacity for montmorillonite.

The ***vermiculites*** have a higher net negative charge of 0.5 to 0.7 unit per $O_{10}(OH)_2$. In this case, the electrostatic force holding the sheets together is sufficiently strong that the interlayer space is no longer easily expandable by ethylene glycol (Table 7–5). When Mg^{2+} is the interlayer cation there are generally two water layers. When K^+ is the interlayer cation there may be one to no water layers. In terms of cation-exchange capacity, vermiculites are similar to smectites.

CASE STUDY 7–1
Mineralogy of Mixed-Layer Clays from Southern and Central Tunisia

Ben Rhaiem et al. (2000) determined the mineralogy of three clay samples from Tunisia using XRD, TG/DTA, CEC, and specific surface measurements. The clay size fraction was separated from each sample and air-dried. In all samples the air-dried XRD patterns showed reflections at 7.2 Å, ~10 Å, and ~15 Å, which are characteristic basal reflections for kaolinite, illite, and Ca-montmorillonite, respectively. The relative intensities of the peaks were different for each sample, indicating that each sample had different relative amounts of these clay minerals. Samples treated with ethylene glycol also had a reflection at ~17 Å, typical of smectites. However, this reflection was relatively weak in one of the three samples, suggesting that smectite was less abundant in this sample. Based on the XRD data, the authors reported the following mineralogical compositions for the three clay samples:

Using TG/DTA analysis, which involves determining the temperatures at which water is lost from the clays, the authors inferred that beidellite was present in the samples. They also made estimates of the amount of kaolinite in each sample, which were in general agreement with those determined by XRD. Specific surface area measurement (refer to the paper of Ben Rhaiem et al., 2000, for an explanation of this type of measurement) was used to estimate the amount of smectite in each sample. These results were in rough agreement with the XRD estimates. CEC determinations were made for each sample. Assuming CEC values for pure kaolinite (1–3 meq/100 g), illite (10–40 meq/100 g), and smectite (80–150 meq/100 g), measured CEC values for the bulk samples were in agreement with those determined using the relative proportions of the three clay minerals in each sample. The authors concluded that for these inhomogeneous mineral compositions several different analytical techniques had to be used in order to produce satisfactory results.

Source: Ben Rhaiem et al. (2000).

	Composition of the Clay Fraction		
Sample	% Kaolinite	% Total illite	% Total smectite
JZ1	30	25	45
A402	10	36	54
22/2	30	50	20

Mixed-layer clays arise when more than one structural type occurs in a clay. For example, it is possible to mix both illite and smectite structures in a single clay mineral (as previously discussed). In addition, gibbsite or brucite layers may occur between layers of smectite and vermiculite. The distribution of structural types is often completely random. This leads to a very complicated clay structure, the characterization of which requires significant effort. Cation exchange and other properties of such clays are best determined by experiment (Case Study 7–1).

The *palygorskite group* minerals [*sepiolite* and *palygorskite (attapulgite)*] consist of 2:1 layers that are arranged in chains rather than sheets. In sepiolite, the chains are six silicon tetrahedra wide; in palygorskite, they are four silicon tetrahedra wide. Channels exist in the octahedral layers that can accommodate water and organic molecules. Simplified formulas are $Mg_4Si_6O_{15}(OH)_2 \cdot 6H_2O$ for sepiolite and $(Mg, Al, Fe)_4Si_8O_{20} \cdot nH_2O$ for palygorskite. In both minerals, the octahedral layer is brucite. There is Al^{3+} substitution in the octahedral layers in both minerals, and there are usually some exchangeable cations.

Ion-Exchange Properties—Origin

Clay minerals can exchange cations (and in special cases, anions) with an aqueous phase. This is an important property of the clay minerals that can have a significant impact on the distribution of metals in the environment. The ranges of cation-exchange capacities for the various groups of clay minerals are listed in Table 7–5. *Cation-exchange capacity (CEC)* is generally determined by measuring *the uptake and release of ammonium ions (NH_4^+) from a 1 M ammonium acetate solution at pH = 7.0*. This very specific definition is used because CEC varies as a function of pH, particle size, and the types of ions involved in the exchange. CEC is normally measured in meq/100 g. Cation-exchange capacity arises when the surface of a mineral has a net negative charge. If the surface has a net positive charge, then the mineral exhibits anion-exchange capacity (AEC). Clay minerals generally have

Table 7–7 Permanent Negative Surface Charge of 2:1 Clay Minerals[a]

Mineral group	Charge (mole sites kg^{-1})[b]
Kaolinite	0.02–0.2
Illites	0.1–0.9
Smectites	0.7–1.7
Vermiculites	1.6–2.5

[a]Data from Sposito (1989) and Langmuir (1997).
[b]Charge in moles of monovalent sites per kg of clay.

negatively charged surfaces, except at very low pH ($<$ 2 to 3 for smectites and vermiculites, $<$ 2 to 4.6 for kaolinite).

Surface Charge   The surface charge of a clay mineral is due to three causes: (1) substitutions in the tetrahedral and octahedral layers (i.e., Al^{3+} for Si^{4+}, Mg^{2+} for Al^{3+}) that cause a decrease in the total positive charge, resulting in a net negative charge; (2) defects or imperfections in the crystal structure (i.e., missing cations) that lead to a net negative charge; and (3) unsatisfied or broken bonds at the corners and edges of a crystal surface, in which case O^{2-} or OH^- ions are normally exposed. The importance of this last mechanism is demonstrated both by the decrease in adsorption with decreasing pH (H^+ ions compete with cations at the crystal surface) and an increase in CEC with decreasing particle size (the smaller the particle, the greater the relative surface area and the greater the number of broken bonds). For the 2:1 clays, the negative surface charge is largely due to causes (1) and (2). Hence, these minerals show relatively constant permanent negative surface charge (Table 7–7). For the 1:1 clays (kaolinites), the permanent surface charge is small and the surface charge is largely controlled by the number of broken bonds.

The relationship between the number of sites per unit surface area (N_{SA}), the total surface area (S_A), and the mole sites per unit mass (N_{SM}) is

$$N_{SA} \text{ (sites nm}^{-2}) = \frac{N_{SM} \text{ (mole sites g}^{-1})}{(1.661 \times 10^{-6})S_A \text{ (m}^2 \text{ g}^{-1})} \tag{7–8}$$

Cation-exchange capacity (in meq/100 g) can be related to charge and surface area according to the following expression:

$$\text{CEC (meq/100 g)} = N_{SA} \text{ (sites nm}^{-2}) \times S_A \text{ (m}^2 \text{ g}^{-1}) \times 0.1661 \tag{7–9}$$

where N_{SA} is the number of monovalent sites per unit surface area and S_A is the surface area per unit mass. Because CEC is reported on a per 100 g basis, the mass unit used in the calculation is grams. If the number of monovalent sites is given on a per unit weight basis, substitution of equation 7–8 into equation 7–9 gives

$$\text{CEC (meq/100 g)} = N_{SM} \text{ (mole sites g}^{-1}) \times 1.0 \times 10^5 \tag{7–10}$$

EXAMPLE 7–6 A montmorillonite has a permanent negative surface charge of 0.8 mole sites kg^{-1}. Assuming that the permanent surface charge is the only factor in determining the cation-exchange capacity, calculate the CEC for this mineral.

$$\text{CEC} = 0.8 \text{ mole sites g}^{-1} \times 10^{-3} \text{ g kg}^{-1} \times 1.0 \times 10^5 = 80 \text{ meq/100 g} \quad \blacksquare$$

Surface Area   The surface area per unit mass can be very large for clay-size minerals. A simple example, using illite, illustrates this point. Illite has a density of 2600 kg m^{-3}. Start with an illite cube 1 m on a side. The total volume is 1 m^3 and the surface area is 6 m^2, which gives a surface area to unit mass of 2.3×10^{-3} m^2 kg^{-1} = 2.3×10^{-6} m^2 g^{-1}. Now subdivide this large cube into 1 cm cubes (1×10^{-2} m). The volume of each

Table 7–8   *Surface Area Per Unit Mass of Illite with a Density of 2600 kg m^{-3}*

Length of side (m)	Number of cubes	Surface area of cube (m^2)	Surface area (m^2 g^{-1})
1	1	6	2.3×10^{-6}
1×10^{-2} (cm)	1×10^6	6×10^{-4}	2.3×10^{-4}
1×10^{-6} (1 μm)	1×10^{18}	6×10^{-12}	2.31
1×10^{-7} (0.1 μm)	1×10^{21}	6×10^{-14}	23.1
1×10^{-8} (0.01 μm)	1×10^{24}	6×10^{-16}	231

cube is 1×10^{-6} m^2 and there are 1×10^6 cubes in the 1-m^3 block. The total surface area is now 1×10^6 cubes $\times$ 6×10^{-4} m^2 cube^{-1} = 600 m^2, which gives a surface area to unit mass of 0.23 m^2 kg^{-1} = 0.00023 m^2 g^{-1}. With decreasing size, the surface area per unit mass increases substantially (Table 7–8). Hence, clay-size particles have a very large surface-to-mass ratio.

Ion-Exchange Properties—Determination

So far we have considered the factors that affect the cation-exchange capacity of clay minerals. Although CEC is a useful measure of the overall ability of a clay to adsorb cations, we are often interested in the adsorption properties of a clay for particular cations. What we want to determine is the partitioning of a particular cation between an aqueous solution and a clay. Because of the number of variables involved in the adsorption of cations from aqueous solutions, the usual approach is to determine the appropriate values by experiment. The two methods that are commonly used are the batch method and the column test experiment.

Batch Method   In the batch method, a sample of the clay of interest is placed in a solution with a known concentration of a particular cation. The pH of the solution is set to an appropriate value. The solution is then agitated for a period of time (minutes, hours, days, etc.) until equilibrium is achieved. The concentration of the cation in the solution is measured after equilibration, and the change in concentration is used to calculate the distribution of the cation between the solution and the clay. The experiments are done with different cation concentrations, different ionic strengths for the solutions, and different pH values in order to assess the effect of these variables on the adsorption of the ion(s) of interest. The results of the experiments are then fitted to adsorption isotherms.

Adsorption Isotherms   Adsorption isotherms represent the partitioning of a particular species between an aqueous phase and solid particles (sorbate). At very low concentrations, this partitioning may follow a linear relationship, but generally more complex relationships are observed in which the distribution of the species between the aqueous and solid phases varies as a function of concentration (Figure 7–8). The dashed line labeled K_d (Figure 7–8) is the tangent to the isotherm at the origin. At very low concentrations, the system behaves ideally, and the partitioning of the species between the aqueous solution and the sorbate is constant. Under these conditions, we can calculate a unique partition coefficient (K_d) that describes the partitioning behavior. As the concentration of the species increases, the partitioning behavior deviates from ideality, as shown by the adsorption isotherm. At sufficiently high concentrations, the species may be removed by precipitation. When this occurs the concentration of the species in solution will not change; i.e., saturation has been achieved. The observed adsorption characteristics may fit one of several mathematical models. These fits are usually done by trial-and-error. The two most common models are the Freundlich isotherm and the Langmuir isotherm.

The ***Freundlich equation*** can be written

$$C_{ads} = KC_{soln}^n \qquad (7-11)$$

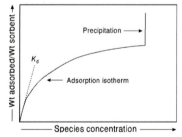

Figure 7–8

Representation of a typical adsorption isotherm showing the distribution of a species between an aqueous phase and a solid (sorbent). At very low concentrations, the distribution behaves ideally and can be represented by a unique value, K_d. At higher concentrations, the partitioning deviates from ideality. If precipitation occurs, the concentration of the species in solution will remain constant; i.e., the solution is saturated with respect to the particular species.

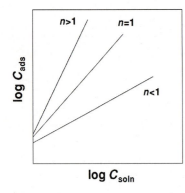

Figure 7–9
Plot of the log of the concentration of the substance adsorbed on a particle (C_{ads}) versus the log of the concentration of the substance in water (C_{soln}). Linear plots indicate that the partitioning can be represented by a Freundlich isotherm. The slope of the line gives the exponent, n, and the intercept gives the value for log K.

where C_{ads} is the weight of the substance (ion of interest) adsorbed by the sorbent (the clay), K is a constant, C_{soln} is the concentration of the ion in solution, and n is a constant. When $n = 1$, $K = K_d$ (the distribution coefficient, which is a measure of the partitioning of the ion between the solution and the clay). In order to determine if a set of experimental data obeys the K_d model or the Freundlich isotherm, the isotherm is linearized by taking the log of equation 7–11.

$$\log C_{ads} = \log K + n \log C_{soln} \tag{7–12}$$

If a plot of log C_{ads} versus log C_{soln} yields a straight line, then the data obey the Freundlich isotherm (Figure 7–9). The y-intercept gives the K value and the slope of the line gives n. If $n = 1$, then the K_d model applies. These models assume an infinite number of unreacted sorbent sites.

The ***Langmuir equation*** can be written

$$C_{ads} = \frac{Q°KC_{soln}}{1 + KC_{soln}} \tag{7–13}$$

where C_{ads} is the weight of the substance adsorbed divided by the weight of sorbent, $Q°$ is the maximum sorptive capacity for the surface, K is a partition coefficient reflecting the extent of sorption, and C_{soln} is the concentration of the ion in solution. The Langmuir equation takes into account the finite nature of the number of sorbent sites. Given an adequate range of experimental values, the maximum number of sorbent sites can be inferred from the C_{ads} value found when increases in C_{soln} don't result in increases in C_{ads}. Examples of Langmuir isotherms for various values of $Q°$ and K are shown in Figure 7–10a. In linear form, the Langmuir equation is written

$$\frac{1}{C_{ads}} = \frac{1}{Q°} + \frac{1}{KQ°} \times \frac{1}{C_{soln}} \tag{7–14}$$

If the data can be described by a Langmuir isotherm, a plot of $1/C_{ads}$ versus $1/C_{soln}$ yields a straight line (Figure 7–10b). The y-intercept is $1/Q°$ and the slope of the line is $1/KQ°$.

The uptake of ions from solution is affected by several factors. Among these are pH, Eh, and the precipitation of phases that may incorporate the ions of interest. The result is

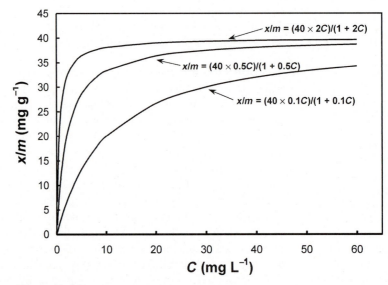

Figure 7–10a
Langmuir isotherms for $Q° = 40$ mg g^{-1} and $K = 0.1, 0.5$, and 2.

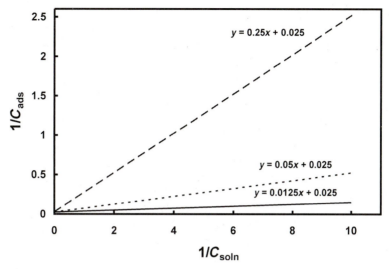

Figure 7–10b
Linearized form of the isotherms plotted in Figure 7–10a. The *y*-intercept is $1/Q°$ and the slope $=$ $1/KQ°$.

that laboratory batch tests are often poor approximations of real-world behavior. At the very least, the experiments need to be done for a relatively wide range of concentrations (in order to determine the shape of the isotherms) and a variety of pH values that encompass those that might be expected in the natural environment.

Column Test Method In the column test method the sorbent is packed into a column and a volume of solution is passed through the column. The concentration of the ion of interest in the original solution is compared to that in the eluent (the solution that has passed through the column) and the K_d is calculated using the following equation:

$$K_d = \left(\frac{C_i - C_f}{C_f}\right)\frac{V}{M}$$ (7–15)

where K_d is the partition coefficient, C_i is the initial concentration of sorbate, C_f is the final concentration of sorbate, V is the volume of solution passed through the column, and M is the mass of sorbent in the column.

EXAMPLE 7–7 Ten grams of montmorillonite are placed in a column and 100 mL of solution are passed through the column. The initial solution had a zinc concentration of 20 mg L^{-1}. After passing through the column, the solution had a zinc concentration of 14.1 mg L^{-1}. Calculate the K_d for zinc between the solution and the montmorillonite.
From equation 7–15,

$$K_d = \left(\frac{C_i - C_f}{C_f}\right)\frac{V}{M} = \left(\frac{20 - 14.1}{14.1}\right)\frac{100 \text{ cm}^3}{10 \text{ g}} = 4.18 \text{ cm}^3 \text{ g}^{-1} \qquad \blacksquare$$

Case Study 7–2 illustrates how the concepts of cation-exchange capacity and partitioning of metals between aqueous solutions and clays can be applied to a specific problem. In this case study the authors investigated the use of clay minerals as adsorbents for heavy metals present in industrial wastewater. The clays used in the study were from various commercial deposits in Spain.

CASE STUDY 7–2
Adsorption Characteristics of Clays for Various Heavy Metals

Sanchez et al. (1999) investigated the adsorbent characteristics of a number of commercially available clay minerals. The purpose of the study was to determine if these clay minerals could be used as adsorbents for heavy metals present in industrial wastewater. The study involved both the determination of CEC for the various clays and investigation of partitioning behavior for specific heavy metals as a function of concentration and pH. The clays were collected from various commercial deposits in Spain. The samples did contain other minerals, notably quartz, calcite, chlorite, and feldspars. The bentonite (weathered volcanic ash) sample was essentially 100% montmorillonite. The CEC for the various clays was determined by the ammonium acetate method (Tan, 1996). The percent of clay minerals in each sample, the results of the CEC measurements, and the specific surface areas determined for each clay are listed in the accompanying table.

Clay	% Clay mineral	Specific surface area $(m^2\ g^{-1})$	CEC (meq/100 g)
Sepiolite (Orera)	87	272	9.1
Sepiolite (Vallecas)	95	271	5.2
Palygorskite (Bercimuelle)	80	157	15.5
Palygorskite (Torrejón)	66	93	9.1
Bentonite	100	57	89.0
Illite	31	41	20.6
Kaolin	78	39	14.8

Adsorption studies for the various metals were done by the batch method. The authors determined adsorption isotherms and investigated the effect of particle size, pH, and ionic strength on adsorption. The adsorption isotherms for sepiolite (Orera) are shown in Figure 7–C2–1. The experimental data were fitted to Freundlich and Langmuir equations. The authors concluded that the Langmuir equation provided the best description of the experiments, i.e., best correlation. The mineral sepiolite was found to have the highest adsorption capacity, and the adsorption capacities varied (from high to low) in the order $Cd^{2+} > Cu^{2+} > Zn^{2+} > Ni^{2+}$. Not surprisingly, decreasing particle size led to higher x/m (mass adsorbed to mass of sorbent) values, i.e., greater adsorption. For sepiolite, increases in pH led to increased adsorption (Figure 7–C2–2), which the authors thought was due to cation hydrolysis because hydroxy complexes are preferentially adsorbed with respect to uncomplexed cations. Increasing ionic strength led to a slight decrease in the adsorption of Cd^{2+} on sepiolite. This decrease was ascribed to competition for nonspecific adsorption sites between Cd^{2+} and K^+ (the cation in the added electrolyte). For Zn^{2+}, the authors found that the adsorption capacity of the various clays followed the order sepiolite (Orera) > sepiolite (Vallecas) > bentonite > palygorskite (Bercimuelle) > illite > kaolin ≈ palygorskite (Torrejón). With reference to the preceding table, we see that the clays that have the greatest specific surface area or greatest CEC are the ones that show the greatest adsorption capacity. The authors concluded that sepiolite might be a good adsorbent for industrial wastes.

Source: Sanchez et al. (1999).

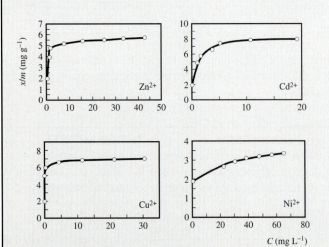

Figure 7–C2–1
Adsorption isotherms for sepiolite. Adsorbent dose = 10 g L⁻¹, agitation time = 3 h, pH = 4, and T = 22°C. The Langmuir equation provided the best fit to the adsorption isotherms. From Sanchez et al. (1999).

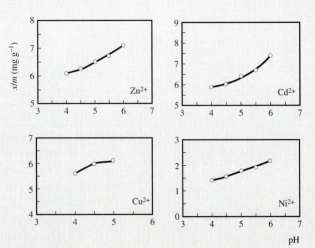

Figure 7–C2–2
Effect of pH on the adsorption of metal cations by sepiolite. Adsorbent dose = 5 g L⁻¹, initial cation concentration = 50 mg L⁻¹, agitation time = 2 h. With increasing pH the adsorption of the metal ions increases. From Sanchez et al. (1999).

ZEOLITES

A *zeolite mineral* is *a crystalline substance with a structure characterized by a framework of linked tetrahedra, each consisting of four O atoms surrounding a cation. This framework contains open cavities in the form of channels and cages. These are usually occupied by H_2O molecules and extra-framework cations that are commonly exchangeable. The channels are large enough to allow the passage of guest species. In the hydrated phases, dehydration occurs at temperatures mostly below about 400°C.* (Coombs et al., 1997). Fifty-two zeolite species are currently recognized, so this is a very complex group of minerals. Because of their structure, they have been used as molecular sieves, ion-exchange resins, and catalysts.

The general chemical formula for the zeolites is

$$M_xD_y[Al_{x+2y}Si_{n-(x+2y)}O_{2n}] \cdot mH_2O$$

where M = Na, K, or other monovalent cations, D = Mg, Ca, Sr, Ba, and other divalent cations, and *m* is the number of water molecules. The portion of the formula within the brackets corresponds to the tetrahedral framework. M and D are the extra-framework exchangeable cations and mH_2O are the water molecules, which often coordinate the extra-framework cations. The atomic ratio (Al + Si)/O is always 1:2. It is the substitution of the trivalent cation Al^{3+} for the tetravalent Si^{4+} that provides the excess negative charge needed to charge balance the M and D cations. Because of the substitutions in the tetrahedral sites, the Si:(Si + Al) ratio is one of the major discriminating factors between various types of zeolites. Some tend toward the Si end and others tend toward the Al end (Table 7–9). The molecular ratio (Ca, Sr, Ba, Na$_2$, K$_2$, etc.)/Al$_2$O$_3$ always equals 1.

Structurally, the zeolites consist of an aluminosilicate framework composed of (Si, Al)O$_4$ tetrahedra. Each oxygen is shared between two tetrahedra. This is the basic feldspar and feldspathoid structure. The zeolites differ in that the structure is more open and contains large cavities and channels (Table 7–9; Figure 7–11, p. 226). The cations and H_2O molecules occur in these channels. Because the removal of cations does not require

Table 7–9 *Chemical Formulas and Various Properties of Representative Natural Zeolites**

Mineral	Representative chemical formula	Si:(Si + Al)	Void space	Free aperture size
Erionite	$NaK_2MgCa_{1.5}(Al_8Si_{28})O_{72} \cdot 28H_2O$	0.75–0.78	0.36	3.6 Å × 5.2 Å 8-rings ⊥ (001)
Mordenite	$Na_3KCa_2(Al_8Si_{40})O_{96} \cdot 8H_2O$	0.80–0.85	0.26	6.7 Å × 7.0 Å 12-rings ∥ (001) 2.9 Å × 5.7 Å 8-rings ∥ (010)
Clinoptilolite	$(Na,K)_6(Al_6Si_{30})O_{72} \cdot 20H_2O$	0.74–0.84	0.34	3.5 Å × 7.9 Å 10-rings ∥ (001) 3.0 Å × 4.4 Å 8-rings
Phillipsite	$K_2(Ca_{0.5},Na)_4(Al_6Si_{10})O_{32} \cdot 12H_2O$	0.54–0.75	0.30	4.2 Å × 4.4 Å 8-rings ∥ (100) 2.8 Å × 4.8 Å 8-rings ∥ (010) 3.3 Å diameter 8-rings ∥ (001)
Chabazite	$Ca_2(Al_4Si_8)O_{24} \cdot 12H_2O$	0.59–0.80	0.48	3.7 Å × 4.1 Å 8-rings

*Modified from Bish and Guthrie (1993).

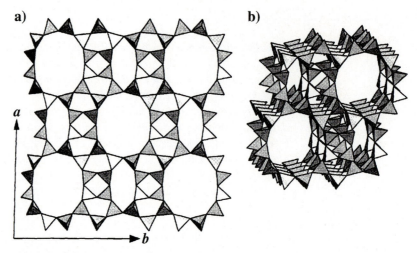

Figure 7–11
The structure of mordenite viewed (a) down the c-axis and (b) obliquely down the c-axis. Note the presence of two sets of channels, one consisting of 12-membered rings and the other of 8-membered rings (count the tetrahedra around the large and small channels). From Bish and Guthrie (1993).

substitutions in the crystal structure, cation exchange occurs readily in zeolites and these minerals have large CECs (100–400 meq g^{-1}). Monovalent cations are, in general, more readily exchanged than divalent cations because they are held by weaker electrostatic bonds. The size of the channel is determined by the way in which the rings of silica tetrahedran are linked. Thus, the size of the channel varies in the different types of zeolites, giving rise to their molecular-sieve properties.

Because of their cation-exchange properties, zeolite minerals have a number of applications in the environmental field. Among these uses are the treatment of nuclear wastewaters, treatment of municipal and industrial wastewaters, remediation of acid mine drainage, and remediation of soils contaminated with heavy metals. The Si/Al (or trivalent cation) ratio determines the theoretical cation-exchange capacity (Table 7–10) for a particular zeolite because the substitution of trivalent cations for Si^{4+} controls the number of negative charges available for cation adsorption.

Table 7–10 Theoretical Cation-Exchange Capacity for Selected Zeolite Minerals*

Zeolite	Typical unit-cell formula	CEC (meq/100 g)
Analcime	$Na_{16}(Al_{16}Si_{32}O_{96})\cdot16H_2O$	450
Chabazite	$Ca_2(Al_4Si_8O_{24})\cdot12H_2O$	390
Clinoptilolite	$(Na,K)_6(Al_6Si_{30}O_{72})\cdot20H_2O$	220
Erionite	$NaK_2MgCa_{1.5}(Al_8Si_{28}O_{72})\cdot28H_2O$	280
Faujasite	$Na_{20}Ca_{12}Mg_8(Al_{60}Si_{132}O_{384})\cdot235H_2O$	360
Ferrierite	$(Na,K)Mg_2Ca_{0.5}(Al_6Si_{30}O_{72})\cdot20H_2O$	230
Heulandite	$(Na,K)Ca_4(Al_9Si_{27}O_{72})\cdot24H_2O$	320
Laumontite	$Ca_4(Al_8Si_{16}O_{48})\cdot16H_2O$	430
Mordenite	$Na_3KCa_2(Al_8Si_{40}O_{96})\cdot28H_2O$	220
Natrolite	$Na_{16}(Al_{16}Si_{24}O_{80})\cdot16H_2O$	530
Phillipsite	$K_2(Ca_{0.5},Na)_4(Al_6Si_{10}O_{32})\cdot12H_2O$	450
Wairakite	$Ca_8(Al_{16}Si_{32}O_{96})\cdot16H_2O$	460

*From Pabalan and Bertetti (2001).

CASE STUDY 7–3
Using Zeolites to Contain Radioactive Wastes

Dyer (2000) describes the use of zeolites for radioactive waste removal and sequestration. ^{90}Sr and ^{137}Cs are two common fission products produced during the operation of fission reactors. They are also common constituents of radioactive fallout. During the Chernobyl nuclear disaster these radionuclides were widely distributed throughout western Europe. Hence, the removal of these isotopes is of considerable interest to both the nuclear industry and environmental scientists.

The selectivity of a zeolite mineral for these radioactive species (Cs^+ and Sr^{2+}) can be determined by doing a batch experiment. In these experiments the distribution coefficient (K_d) is defined as

$$K_d = \frac{(A_o - A_f)V}{A_f W}$$

where A_o is the original radioactivity of the solution, A_f is the final radioactivity of the solution, V is the volume of the solution, and W is the weight of the zeolite used in the experiment. The zeolite mineral is usually converted to a homoionic form before doing the experiment. Liquid nuclear wastes are often acidic, and because of this, they are neutralized using sodium or ammonium hydroxide. Thus, depending on the exact condition of the waste, competing cation species may be H^+, Na^+, or NH_4^+. For this reason, zeolites used in the experiments are converted to the homoionic form of one of these cation species.

Because of its high Si:Al ratio (5:1), clinoptilolite is one of the zeolite minerals that shows a good selectivity for Sr and Cs. The selectivity series for this mineral was listed earlier. Note the high selectivity for both Sr and, particularly, Cs. It has been found that the acid form of this mineral has greater selectivity for both Sr and Cs. Boric acid, another common constituent of nuclear wastes, also has little effect on selectivity. Hence, this mineral has been widely used for removal of these isotopes from radioactive waste. Clinoptilolite has been used to remove Cs and Sr from liquid waste, soil, and animals. Clinoptilolite added to soil sequesters Cs and Sr, significantly reducing the uptake of these radionuclides by plants. It has also been used in animals (including man) to remove these radionuclides. The proposed U.S. high-level nuclear waste repository at Yucca Mountain will be sited within a vitric tuff approximately 100 m above a thick zeolitic tuff consisting predominantly of clinoptilolite. In case of leakage from the repository, it is believed that the zeolitic tuff will act as a significant barrier to the migration of Cs and Sr.

Source: Dyer (2000).

In general, zeolites with high Si/Al ratios tend to be selective for monovalent cations and those with lower Si/Al ratios are selective for divalent cations. The reason is presumably related to the charge distribution on the surface of the crystals. In the case of zeolites with high Si/Al, there are fewer negative charges and they are widely distributed across crystal surfaces. This favors the large monovalent cations. With increasing surface charge density, the smaller divalent ions are favored. For example, for clinoptilolites, which have high Si/Al, the selectivity series for the monovalent cations is

$$Li < Na < NH_4 < K < Rb < Cs$$

and for the divalent cations the selective series is

$$Mg < Ca < Sr < Ba$$

Taking into consideration these selectivity series, zeolites used in various applications are often preconditioned to favor the ion(s) of interest. For example, if one wanted to selectively extract Cs from a solution, one might convert the zeolite to a Na^+ or NH_4^+ homoionic cation form (all the exchangeable cations are the same). Case Study 7–3 describes how zeolite minerals can be used to sequester radioactive wastes.

THE ASBESTOS MINERALS

Probably no group of minerals has attracted greater environmental interest, nor spawned a larger environmental job market, than the asbestos minerals. There are two types of asbestos minerals, the serpentine minerals and the fibrous amphiboles. In this section we will look at the mineralogy of both types and consider their environmental impact.

The definition of asbestos has been a source of controversy. It is generally agreed that the asbestos minerals are the serpentine mineral chrysotile and the amphibole minerals crocidolite, amosite, anthophyllite, actinolite, and tremolite. Crocidolite is the fibrous form of the amphibole riebeckite, and amosite is the fibrous form of the amphibole grunerite.

For regulatory purposes these minerals are considered to be fibrous if they have a length of greater than 5 μm, a diameter of less than 5 μm, and an aspect ratio (length to width or diameter) of 3:1 or greater. The mineralogical community uses a greater aspect ratio of 10:1 or 20:1 to define fibrous minerals. Approximately 95% of the commercial asbestos is chrysotile. Because these minerals belong to two different mineral groups, despite their similarity in terms of fibrous habit, they have very different crystal structures.

Crystal Structures of Asbestos Minerals

The chrysotile structure (Figure 7–12) consists of a layer of silicate tetrahedra bonded to a layer of octahedrally coordinated magnesium ions (1:1 layer silicates). In the octahedral layer each Mg^{2+} ion is surrounded by four hydroxyl molecules and two oxygens. The two oxygens in the octahedron are the apical oxygens of the polymerized sheet of silica tetrahedra. The distance between the oxygens in the octahedral layer is slightly greater than the distance between the apical oxygens of the silica tetrahedra. Thus, when the two sheets are combined the octahedral layer curls around the tetrahedral layer, forming a scrolled tube parallel to the *a* crystallographic axis, which is the long axis of the fiber.

The basic amphibole structure consists of a strip of octahedrally coordinated cations sandwiched between two double silicate chains (the double silicate chain is illustrated in Figure 7–4). The chains extend for an infinite distance in the *c* crystallographic direction. Viewed end on, this structure has the appearance of an I-beam, and simplified representations of amphibole structure make use of this graphical analogy (Figure 7–13). These I-beams are then stacked, as shown in Figure 7–13, to form the amphibole structure.

The amphiboles are a chemically complex group of minerals. The general amphibole formula can be written

$$A_{0-1}B_2C_2T_8O_{22}(OH, F, Cl, O)_2$$

where T refers to the tetrahedrally coordinated sites in the silicate chain, C to the fairly regular octahedral sites, B to the less regular octahedral or 8-coordinated sites, and A to the relatively large irregular cation sites having coordination numbers of 6 to 12. The cations commonly found in the four sites are: A = Na, K; B = Na, Li, Ca, Mn, Fe^{2+}, Mg; C = Mg, Fe^{2+}, Mn, Al, Fe^{3+}, Ti; and T = Si, Al. In the crystal structure the M1, M2, and M3 sites, found in the octahedral layer, are occupied by the C cations, and the M4 site is occupied by the B cations. The large possible number of substitutions gives rise to a number of amphibole minerals. As mentioned, the fibrous form of five of these minerals is considered to be asbestos. Ideal chemical formulas and properties of the asbestos minerals are listed in Table 7–11.

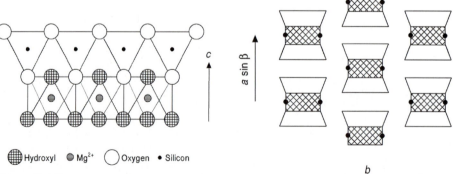

Hydroxyl ⊕ Mg^{2+} ⊜ Oxygen ◯ Silicon •

Figure 7–12
Crystal structure of chrysotile, a 1:1 layer silicate consisting of a tetrahedral layer and an octahedral layer in which Mg^{2+} ions are surrounded by four OH molecules and two oxygens.

Figure 7–13
Simplified clinoamphibole structure looking down the *c*-axis. The cross-hatched area represents the octahedral layer. The filled circles indicate the M4 cation positions.

Table 7–11   *Chemical Formulas and Properties of Asbestos Minerals**

Mineral	Chemical formula	Crystal system	Hardness	Specific gravity
Chrysotile	$Mg_3[Si_2O_5](OH)_4$	Monoclinic	2.5	≤ 2.55
Actinolite	$Ca_2(Mg,Fe^{2+})_5[Si_8O_{22}](OH)_2$	Monoclinic	5–6	3.0–3.2
Amosite	$(Fe^{2+},Mg)_7[Si_8O_{22}](OH)_2$	Monoclinic	5.5–6.0	3.1–3.25
Anthophyllite	$(Mg,Fe^{2+})_7[Si_8O_{22}](OH)_2$	Orthorhombic	5.5–6.0	2.85–3.1
Crocidolite	$Na_2Fe_2^{3+}(Fe^{2+},Mg)_3[Si_8O_{22}](OH)_2$	Monoclinic	5	3.15–3.5
Tremolite	$Ca_2Mg_5[Si_8O_{22}](OH)_2$	Monoclinic	5–6	2.9–3.2

*Data from Deer et al. (1992) and Veblen and Wylie (1993).

Types of Asbestos and Their Uses

Due to its resistance to heat, friction, and acid conditions, asbestos has a large number of commercial uses. The precise use is governed to some extent by the character of the fiber. Chrysotile forms a tubular fibril that has an average diameter of 250 Å. Because there is a limit to the curvature, chrysotile fibrils have a central tube with a diameter of ~70 Å. Because of this structure, chrysotile is a good *spinning fiber* and is used in the production of heat-resistant clothing. Amphibole asbestos minerals tend to be columnar to fibrous in habit and do not make good spinning fibers. Thus, the use of amphibole asbestos is more limited. Asbestos used in the United States is approximately 95% chrysotile (*white* asbestos) and 5% amphibole asbestos—crocidolite (*blue* asbestos) and amosite (*brown* asbestos). Among the common uses of asbestos are fireproof clothing, gloves, face masks, and stage curtains; roofing products; spray-on insulation; pipe insulation; as a binding agent in floor tiles and cement pipes; and as a friction agent in brake linings. Asbestos has been used for these purposes for over a century and, hence, these minerals are common in our modern industrial society.

Health Effects of Asbestos Exposure

It is now known that exposure to asbestos can lead to three pulmonary diseases—asbestosis, mesothelioma, and lung cancer. ***Asbestosis*** is a lung disease caused by asbestos particles deposited in the lung through inhalation. Over time, the lung encapsulates these fibers and hardens, leading to a decrease in the efficiency of the O_2/CO_2 exchange. The result is increased stress on the heart and ultimately death due to heart failure. Inhalation of other types of particulate matter, such as coal dust and silica dust, lead to similar illnesses. ***Mesothelioma*** is a rare, diffuse, malignant cancer of the lining of the lung and stomach. It has a long latency period of 35 to 40 years, and once it appears death usually occurs in 1 to 2 years. About 80% of the mesotheliomas have been linked to asbestos exposure. However, it appears that the type of asbestos is a critical factor in determining the onset of mesothelioma. ***Lung cancer*** is usually linked to cigarette smoking. Other cases are often attributed to radon, second-hand cigarette smoke, or exposure to asbestos. There is also evidence that a synergistic relationship exists between cigarette smoking and exposure to asbestos. The risk associated with various degrees of asbestos exposure is a very controversial area, and we will return to this topic after a consideration of the physiochemical interactions to be expected for the various asbestos minerals.

Much of the following discussion is summarized from Veblen and Wylie (1993) and Hochella (1993). The interested student should consult these papers and others in Guthrie and Mossman (1993) for further details on this topic. Reactions between fluids in the body and solids occurs at the surface of the solid. Therefore, the chemistry of this surface is an important, perhaps key, factor in determining the effect of the solid on cellular processes. A secondary consideration is the length of time the solid remains in the body.

Dissolution of Fibers   In Chapter 2, problem 32 dealt with the dissolution rate of chrysotile fibers. For a fiber 1 μm in diameter, the dissolution time is 9 ($\pm$4.5) months.

Although similar experiments have not been done to determine the dissolution rates for amphibole fibers, experimental studies on rats have shown that the concentration of crocidolite in the lung tissue of exposed animals increases with time, whereas chrysotile initially increases with time but then reaches a constant value. Similarly, studies of asbestos workers with long-term exposure to asbestos fibers have shown an increase in the concentration of tremolite fibers in the lung as a function of duration of exposure, but an essentially constant concentration for chrysotile irrespective of the duration of exposure. This is particularly significant because tremolite comprised only a small fraction of the total asbestos burden. Thus, the evidence to date indicates that chrysotile has a much shorter lifetime in the lung than the amphibole asbestos minerals.

Chemistry of Exposed Fiber Surface   In terms of the surface exposed to the lung fluids, chrysotile presents a much different surface than the amphibole asbestos minerals. Recall that chrysotile occurs as rolled fibers. The outer surface of the roll is the octahedrally coordinated layer and the hydroxyl ions are exposed. At the edges of the roll the silicate tetrahedral layer and Mg ions are exposed, but this is only a small fraction of the entire exposed surface. For the amphibole minerals, a variety of exposed surfaces are possible, including the tetrahedral layers and surfaces that contain the various cations in the octahedral layer. Also, Fe^{2+} may be exposed at the surface of the amphiboles, and the significance of this will be discussed in a subsequent section.

Surface Charge   An important property of particles is their surface charge. This is normally reported as a zeta potential. The *zeta potential* is *a measure of the potential energy difference between a bulk solution and the boundary between the free solvent and the solvent adhering to the surface*. The zeta potential varies as a function of pH. Consider a silica tetrahedron exposed at the mineral surface. The Si–O–Si reacts with water according to the following reaction:

$$Si–O–Si + H_2O \rightarrow 2(Si–OH)$$

which produces two silanol groups (Si–OH). These groups can react in two different ways, depending on the pH of the solution:

$$SiOH \rightarrow SiO^- + H^+ \quad \text{or} \quad SiOH + H^+ \rightarrow SiOH_2^+$$

The first reaction decreases the pH of the solution and the second reaction increases the pH of the solution. At low pHs, the second reaction is favored because H^+ is adsorbed (remember LeChâtelier's principle) and the surface becomes positive. At high pHs, the first reaction is favored because H^+ is released and the surface becomes negative. There is some pH, referred to as the *critical* pH (pH_{crit}), at which the surface is electrically neutral. This is the point of zero net proton charge (PZNPC), and as one moves away from this pH the surface charge (either positive or negative) increases. The attraction of ions to the mineral surface depends on the sign and absolute magnitude of the surface charge. Measurements over a wide range of pH values yield positive zeta potentials for chrysotile and negative zeta potentials for the amphibole asbestos minerals.

Biological Interactions   Foreign objects, such as asbestos fibers, are attacked by the body's defense system. This attack consists of phagocytic cells that attempt to engulf and ultimately remove the foreign object. The phagocytic cells produce highly reactive compounds, such as hydrogen peroxide and hydroxyl radicals, that are designed to kill foreign microorganisms. The production of these short-lived oxygen species is promoted by the presence of ferrous iron. One such reaction can be written

$$H_2O_2 + Fe^{2+} \rightarrow OH + OH^- + Fe^{3+}$$

The increased production of these active oxygen species increases inflammation in the area of the fiber and may cause DNA damage in the surrounding cells. It is important to note that the amphibole asbestos minerals can contain significant amounts of ferrous iron, but chrysotile asbestos contains minor amounts of ferrous iron. Hence, this type of reaction would be much more important for the amphibole asbestos minerals.

Considering the difference in crystal structure, shape of the fiber, surface chemistry, and resistance to dissolution of chrysotile compared to the amphibole asbestos minerals, one might expect these fibers to react differently in the human body. The relationship between the type of fiber and disease is complex. As mentioned, exposure to asbestos has been linked to asbestosis, mesothelioma, and lung cancer. However, a variety of studies have suggested that chrysotile does not play a significant role in mesothelioma. This is presumably due to the difference in surface chemistry between chrysotile and the amphibole asbestos minerals and the relatively rapid dissolution and removal of chrysotile from the lung compared to the amphibole asbestos minerals. A related question involves the issue of whether or not there is a *safe level* of exposure to the asbestos minerals.

Determining the Effect of Asbestos Exposure Determining the effect of a particular substance on humans is approached in a variety of ways: *in vitro* testing, *in vivo* testing, and case studies. ***In vitro*** testing *involves the exposure of individual cells to a particular substance.* An example of this type of test is the Ames test, in which individual cells are exposed to chemicals or other possible mutagens. If cellular mutations are observed, then the substance is a possible carcinogen. This is an inexpensive test that is often used to screen substances for possible health effects. However, the test often gives false positives. ***In vivo*** testing *involves the exposure of laboratory animals to a particular substance.* In order to obtain a significantly meaningful response in a relatively short period of time, the animals are exposed to high concentrations of the substance, concentrations far in excess of what would be encountered in the environment. In addition, there is often significant uncertainty as to how reliably the animal response mimics what might happen in humans exposed to the same substance. In a ***case study*** *a cohort (group of individuals with similar traits) of exposed individuals is compared to a reference cohort, which as nearly as possible matches the exposed individuals except that the reference cohort was not exposed to the substance of interest.* These types of studies are also referred to as *epidemiological* studies. The cohort of exposed individuals is often composed of individuals who have had an occupational exposure and thus a significantly higher exposure than the general public. Using larger populations, it may be possible to identify a subset of the general population that has had a low-level exposure to the substance compared to the population as a whole. At low levels of exposure, the statistical significance of a response becomes an important issue.

The assessment of risk due to exposure to a particular substance is a topic of considerable debate. A simplified model of how risk assessment is carried out is shown in Figure 7–14. Animal studies are done at very high doses and elicit a high response rate. One approach to risk assessment is to linearly extrapolate the animal studies to zero dose, assuming that any dose will lead to a response. There is always an uncertainty in the animal studies in terms of how well the animal system models the response to be expected for humans. If there are results from occupational studies, this extrapolation can be constrained at lower dose–response levels. General population studies may be able to give insight into what happens at very low dose–response levels. As we go to lower dose–response levels, there is a greater statistical uncertainty. What is important to note here, and is of considerable interest from a regulatory point of view, is whether or not the dose–response curve extrapolates linearly to zero. Another possibility is that there is a threshhold dose below which there is no change in response

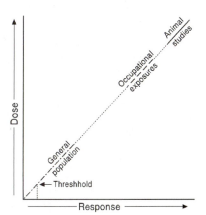

Figure 7–14

Schematic representation of the relationship between dose and response for a particular contaminant. Linear extrapolation to zero is often done based on animal studies and occupational exposures. The threshold dose implies that there is some dose below which no change in response is observed.

(Figure 7–14). This occurs because there are other causes of the particular response. From a regulatory point of view, if there is a dose–response threshhold, there is little point to reducing the concentration of the substance to values less than the threshhold value because you are receiving no benefit from further reductions in the concentration of the substance. Case Study 7–4 addresses the question of whether or not there is a safe level of chrysotile exposure.

CASE STUDY 7–4
Is There a "Safe" Level of Chrysotile Exposure?

Johnson and Mossman (2001) review the risk factors associated with chrysotile exposure. The pulmonary effects of inorganic particles and fibers depends on a number of factors (Figure 7-C4-1). It is the combination of these factors that determines the response to a particular dose. As noted, dose–response effects are determined from animal studies and, to a lesser extent, from occupational exposures. In the case of chrysotile asbestos, there is a marked difference in exposure levels between those used for animal experiments and the actual exposure of human populations (accompanying table, from Johnson and Mossman, 2001). This difference raises questions concerning the applicability of animal studies to human populations and the corollary question of whether or not the currently low exposures are a public health concern. This is an important question both from the standpoint of asbestos remediation and the continued use of asbestos in applications for which there are no really good substitutes.

Exposure setting	Exposure concentration (fibers mL^{-1})
Rodent inhalation	102,000
1936 occupational	5,760
1987 occupational	5.5
Indoor nonoccupational	0.00024
Outdoor	0.000003

Johnson and Mossman (2001) cite several studies of long-term exposure to chrysotile fibers. For example, South Carolina textile workers exposed to chrysotile fibers did not show a significant increase in lung cancer until cumulative exposures of 1000–2500 fibers mL^{-1} d. Other studies have suggested that similarly high burdens are required. These observations can be compared to exposure levels for the general public.

Fiber dimensions are important in determining the effect of chrysotile on the pulmonary system. There appears to be a correlation between the length of the fiber and lung disease, with longer fibers (>20 μm) exerting a more significant effect. Aspect ratios are also important. Short, thick fibers have a low surface-to-volume ratio and are less biologically active than long, thin fibers with large surface-to-volume ratios.

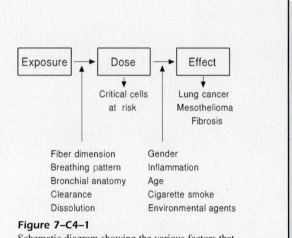

Figure 7–C4–1
Schematic diagram showing the various factors that affect the relationship among exposure–dose–effect. Adapted from Johnson and Mossman (2001).

Chrysotile is cleared much more quickly from the lungs than the amphibole asbestos minerals. This has been demonstrated by a number of studies, both in the laboratory and in human systems. Populations exposed to both kinds of asbestos will see a continued buildup of amphibole asbestos fibers in the lung but a steady-state condition with respect to chrysotile fibers. Magnesium is readily leached from the chrysotile fibers by lung fluids. This leads to a weakening of the structure and subsequent breakage and removal by macrophages. Hence, the biopersistence of chrysotile fibers is low, an important observation.

All these factors need to be considered when assessing the potential health effects of asbestos exposure for the general population. The authors make the point that asbestos clearance occurs on a longer time frame than that of the typical *in vitro* and short-term inhalation study. Long-term exposure studies are required to properly assess the mechanism of chrysotile-induced lung damage and to determine the significance of low-level exposures.

Source: Johnson and Mossman (2001).

CRYSTALLINE AND AMORPHOUS SILICA

Crystalline compounds that have the chemical composition SiO_2 are referred to as *silica phases*. There are six silica **polymorphs**, *minerals with the same chemical composition but different crystal structures*, and their stabilities, in terms of pressure and temperature, are shown in Figure 7–15. The crystalline silica phases are widely distributed and are found in all the major rock types: igneous, metamorphic, and sedimentary. After the feldspars, they are the most common minerals in the rocks of the earth's crust. The two quartz polymorphs are the most abundant silica phases. Tridymite and cristobalite are found in some volcanic rocks that were rapidly quenched from high temperatures. Coesite is found in rocks that have undergone high-pressure metamorphism. Coesite and stishovite are also found in short-lived high-pressure environments, such as mete-

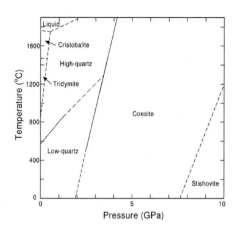

Figure 7–15

Phase diagram for the silica system. Adapted from MANUAL OF MINERALOGY, 21/e by C. Klein and C. S. Hurlbut, p. 681. Copyright © 1993. This material is used by permission of John Wiley & Sons, Inc.

orite craters. Under most conditions, quartz is the mineral most resistant to weathering and it is, therefore, a common constituent of soil dust. In addition, various processes involving the crushing of rock material release the silica phases to the environment. Because quartz is now considered to be a carcinogen (see below), the silica phases are of environmental interest.

Amorphous silica (opal, $SiO_2 \cdot nH_2O$) is also widespread. It is found as siliceous oozes in the seafloor sediments and on land as preserved deposits of marine sediments. Gem-quality opal occurs in veins in volcanic rocks, and its occurrence is generally ascribed to deposition from hydrothermal (hot, water-rich) solutions. Opaline material, called geyserite, can also precipitate from geyser fluids that contain high amounts of dissolved silica.

Dissolution of Silica Minerals

In Chapter 3 we discussed the dissociation of the weak acid $H_4SiO_{4\,(aq)}$. At pH < 9.9, the dominant species is $H_4SiO_{4\,(aq)}$. Hence, for reactions that occur at pH < 9.9, we can write the dissolution of a silica phase as

$$SiO_{2\,(s)} + 2H_2O \rightarrow H_4SiO_{4\,(aq)}$$

Rimstidt and Barnes (1980) give equations that relate the solubility of silica solids to temperatures of up to about 200°C. For quartz,

$$\log K_{sp} = 1.8814 - 2.028 \times 10^{-3}T - \frac{1560.46}{T} \tag{7–16}$$

and for amorphous silica,

$$\log K_{sp} = 0.338037 - 7.8896 \times 10^{-4}T - \frac{840.075}{T} \tag{7–17}$$

where T is in kelvin.

CASE STUDY 7–5
Relationships Between Respiratory Diseases and Quartz-Rich Dust

Norton and Gunter (1999) investigated the occurrence of the respiratory diseases lung cancer and chronic obstructive pulmonary diseases (COPD)—specifically bronchitis, emphysema, and asthma—as a function of various levels of PM10 particle exposure. The study focused on rural Idaho, an area with a high burden of soil dust but a low burden of combustion products. Various air samplers were used to collect particles in the PM10 ($\leq$ 10 μm) and PM2.5 ($\leq$ 2.5 μm) size ranges. X-ray diffraction was used to quantify the concentrations of inorganic constituents. Inorganic constituents were not found in the PM2.5 population, a size fraction that normally consists of the organic products of combustion. On average, the PM10 population consisted of 10% quartz, 30% feldspar, and 60% Mount St. Helens volcanic ash. A subset of the data collected in the Moscow, Idaho, area is reported in the accompanying data table.

PM10 Data Collected for the Moscow Area (μg m^{-3})

Date	PM2.5	PM2.5–10	PM10	Quartz (%)	Feldspar (%)
September 1995	6	28	34	8.6	28
October 1995	6	15	21	11.0	23
November 1995	3	6	10	8.8	33

Note that the total mass in the PM2.5 fraction is very small and that approximately 40% of the PM2.5–10 material is composed of quartz and feldspar. The remainder was amorphous and was inferred to be volcanic glass. Also note the decline in total particulate load associated with the end of the growing season and the onset of winter.

In order to carry out the comparisons, the standard mortality ratio (SMR) must be calculated and corrections applied for smoking and age distribution. The SMR is a statistic that is routinely used to determine if death rates for certain populations are above or below the averages of other populations. The SMR is defined as

$$SMR = \left(\frac{O}{E} \right) * 100$$

where O is the number of observed deaths in the population and E is the number of expected deaths in the reference population. O and E can be reported in deaths per fixed number of individuals (which must be the same for both the investigated population and the reference population, e.g., deaths per thousand) or percent of the population. Values over 100 indicate death rates higher than expected; values below 100 indicate death rates lower than expected. The standard error (SE) for the SMR populations is calculated as follows:

$$SE(SMR) \cong \frac{SMR}{\sqrt{O}}$$

where SE(SMR) is the standard error for the SMR. Note that the greater the number of observations, i.e., deaths, the smaller the standard error. Because there is a documented causal link between smoking and lung cancer, the SMR values must be adjusted for smoking. In addition, for both classes of respiratory diseases corrections must be made for age differences in the populations. The results, after applying the corrections, for 14 of the 44 counties in Idaho for which adequate data were available are shown in Figure 7–C5–1. There is no apparent relationship between the incidence of lung cancer or COPDs and the concentration of PM10-size particles. There is, however, more scatter in the case of the COPD data, an observation that reflects the difficulty of conducting such causal-relationship studies when the response rate is low.

Source: Norton and Gunter (1999).

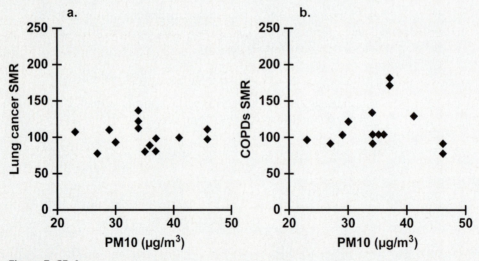

Figure 7–C5–1
Plots of PM10 versus (a) adjusted (age and smoking) lung cancer SMR and (b) adjusted (age) COPDs. From Norton and Gunter (1999).

EXAMPLE 7–8 Calculate the K_{sp} for quartz and amorphous silica at pH $= 5.5$ and $T = 25°C$. Which form of silica is more soluble?

For quartz,

$$\log K_{sp} = 1.8814 - 2.028 \times 10^{-3}T - \frac{1560.46}{T}$$

$$= 1.8814 - (2.028 \times 10^{-3})(298.15) - \frac{1560.46}{298.15} = -3.96$$

For amorphous silica,

$$\log K_{sp} = 0.338037 - 7.8896 \times 10^{-4}T - \frac{840.075}{T}$$

$$= 0.338037 - (7.8896 \times 10^{-4})(298.15) - \frac{840.075}{298.15} = -2.71$$

Amorphous silica has the larger K_{sp} and is the more soluble form of solid silica. ∎

Health Effects of Silica Exposure

Three diseases have been associated with exposure to silica dust: silicosis, silicotuberculosis, and cancer. *Silicosis* is similar to asbestosis in that the inhalation of small quartz particles leads to scarring of the lung tissue and a diminution in respiratory capacity. *Silicotuberculosis* is due to a synergistic effect between silica and *M. Tuberculosis* in which the presence of silica promotes the growth of the pathogen. Over the past 20 years, various studies have implicated quartz as a factor in *lung cancer*, and the International Agency for Research on Cancer now classifies quartz as a Group I human carcinogen.

Saffiotti et al. (1993) have investigated the mechanisms responsible for the carcinogenic properties of quartz. For quartz, the PZNPC occurs at a pH of less than 2, and at most pHs the quartz surface is negatively charged and quartz exhibits negative zeta potentials. Trace-metal impurities are attracted to these sites, and an observed reaction was the oxidation of Fe^{2+} to Fe^{3+}. The result is the production of OH^-. Crystalline silica surfaces also produce hydrogen peroxide (H_2O_2) and the hydroxyl radical. The presence of these oxygen free radicals leads to damage and breakage of DNA strands.

The silica-caused diseases listed are most often the result of occupational exposures. Is the general public at risk from quartz exposure? Recall that most beaches are composed largely of quartz sand, so is going to the beach hazardous to your health beyond the widespread concern regarding melanoma? An important factor is that the silica dust particles must be small enough to enter the lungs. This size is generally considered to be 10 μm (PM10) and smaller. Given quartz particles in this size range, what is the risk associated with nonoccupational exposures? This type of question is difficult to answer because it involves very small exposures, usually well below those that cause readily measurable responses. Case Study 7–5 illustrates one approach to answering this type of question.

MINERAL–MICROORGANISM INTERACTIONS

In Chapter 4 we discussed the role of microorganisms in oxidation–reduction reactions. Besides acting as catalysts for redox reactions, microorganisms also play a role in a number of other reactions involving minerals and aqueous solutions. In this section we revisit the topic of microorganisms from a slightly different perspective and consider their role in mineral reactions.

Microorganisms can be divided into two groups: procaryotes, whose cells lack a nucleus, and eucaryotes, whose cells contain a well-defined nucleus. The procaryotes include all types of bacteria (eubacteria, filamentous and coryneform bacteria, and cyanobacteria) and are the oldest living organisms on earth. The eucaryotes include the fungi, protozoa, and algae. These organisms obtain the energy needed to form new organic matter through the oxidation and reduction of organic and inorganic substances. These reactions can take place both under oxygen-rich (aerobic) and oxygen-deficient (anaerobic) conditions. Besides mediating reactions that serve as energy sources, the microorganisms also release gaseous, liquid, and solid compounds to their environment. Microorganisms can be distinguished on the basis of the energy sources they use and the source of their carbon. *Photolithotrophs use photons as a source of energy and inorganic compounds as electron donors and carbon sources. The green plants are photolithotrophs. Photoorganotrophs use photons as a source of energy and organic compounds as electron donors and carbon sources. Chemolithotrophs utilize the oxidation of inorganic components (Fe^{2+}, Mn^{2+}, S^{2-}, etc.) as their energy source and CO_2 as the source of carbon. Chemoorganotrophs use organic compounds as both a source of energy and carbon.* The chemolithotrophs and chemoorganotrophs represent the majority of soil and sediment microorganisms. Figure 7–16 illustrates the various energy pathways and products released by these microorganisms.

Berthelin et al. (2000) distinguish three ways in which microorganisms play a role in the dissolution and precipitation of minerals:

1. Dissolution and release of metal ions by acid and complexing compounds produced by autotrophic (primary producers) and heterotrophic (primary consumers) microorganisms.
2. Oxidation and reduction reactions promoted by microorganisms (see Chapter 4) that can lead to either increased or reduced solubility of particular metals.
3. Changes in Eh or pH due to the uptake and release of different compounds or ions, which change the solubility of various compounds.

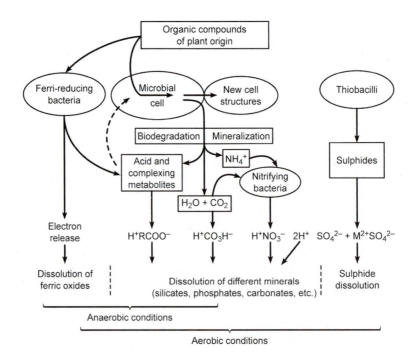

Figure 7–16

Various pathways for the dissolution and weathering of minerals by soil microorganisms. Note both the energy sources and the products of the various pathways. See text for further discussion. From Berthelin et al. (2000).

When heterotrophic organisms break down organic matter, they may produce carbonic or organic acids. Examples of such reactions are

$$\text{Organic matter } (CH_2O)_n \rightarrow CO_2 + H_2O \rightarrow H_2CO_3$$

and

$$\text{Organic matter } (CH_2O)_n \rightarrow \text{Organic acids (RCOOH)}$$

These acids may then react with minerals according to the simplified reaction

$$M^+(\text{Mineral})^- + H^+R^- \rightarrow H^-(\text{Mineral})^- + M^+R^-$$

where $M^+(\text{Mineral})^-$ represents silicate, phosphate, carbonate, etc. minerals, M^+ is the metal cation, and R^- is HCO_3^-, $RCOO^-$, NO_3^-, SO_4^{2-}, etc. produced by the action of the microorganism. Oxalic and citric acids are particularly effective in forming soluble and stable organometallic compounds. Simplified representations of these types of reactions are

$$M^+(\text{Mineral})^- + HL \rightarrow H^+(\text{Mineral})^- + ML$$

and

$$HL + LM \rightarrow L_2M + H^+$$

where L is the organic ligand (oxalic acid, citric acid, etc.; see Chapter 5).

The reaction between a mineral and the acid (and/or complexing compound) produced by the microorganisms occurs at the mineral surface. This reaction can be represented as a two-step process; a fast reaction

$$\text{Surface sites + Reactants } (H^+, OH^-, \text{ligand}) \rightarrow \text{Surface species}$$

and a slow reaction

$$\text{Surface species} \rightarrow \text{Aqueous species}$$

The second reaction, which represents the transfer from the mineral surface to the surrounding solution, controls the rate at which the reaction proceeds. Several factors affect the reaction rate.

1. The zeta potential of the mineral surface is important because the net charge of the surface affects the ability of the surface to attract charged species. Dissolution rates are at a minimum at the PZNPC. Recall that the zeta potential varies as a function of pH; therefore, pH plays a role in determining the reaction rate.
2. The size of the particle. A smaller particle has a larger surface-to-volume ratio and more broken bonds at the surface.
3. The atomic nature of the mineral surface. The presence of dislocations and defect structures, points of high negative free energy, increases the reaction rate.
4. The presence of other ions in solution that may compete for the charged sites on the mineral surface.
5. The attachment of microorganisms to the mineral surface. For example, heterotrophic bacteria can accumulate cations from solution and, therefore, effectively act as organic ligands.

As was discussed in Chapter 4 in some detail, microorganisms can mediate oxidation–reduction reactions. These reactions can lead either to the dissolution of minerals or to precipitation. For example, in an aerobic environment, *Thiobacillus ferrooxidans* (an acidophilic, chemolithotrophic bacteria) can oxidize both sulfur and iron according to the following reactions:

$$FeS_2 + \tfrac{7}{2}O_2 + H_2O \rightarrow FeSO_4 + H_2SO_4$$

and

$$2FeSO_4 + H_2SO_4 + \tfrac{1}{2}O_2 \rightarrow Fe_2(SO_4)_3 + H_2O$$

In anaerobic environments, chemoorganotrophic bacteria are able to reduce Mn^{4+} or Mn^{3+} to Mn^{2+} and Fe^{3+} to Fe^{2+}. A representative reaction is

$$Fe(OH)_{3\,(s)} + 3H^+ + e^- \rightarrow Fe^{2+} + 3H_2O$$

These reactions lead to mineral dissolution.

In aerobic environments, bacterially mediated reactions can lead to the oxidation of Fe and Mn and their subsequent precipitation as oxyhydroxides. A representative oxidation reaction is

$$Fe^{2+} + \tfrac{1}{4}O_2 + H^+ \rightarrow Fe^{3+} + \tfrac{1}{2}H_2O$$

The Fe^{3+} is subsequently precipitated as $Fe(OH)_{3\,(s)}$. In anaerobic environments, sulfate-reducing bacteria use fermentation products as an energy source and sulfate as an electron receptor, a reaction that produces a reduced sulfur species:

$$2CH_3CHOHCOOH + SO_4^{2-} \rightarrow 2CH_3COOH + 2CO_2 + S^{2-} + 2H_2O$$

The S^{2-} can then combine with Fe, or other metals, to form a metal-sulfide precipitate.

QUESTIONS AND PROBLEMS

1. Define *mineral*.
2. What is a *solid solution*? Give an example.
3. What is a *polymorph*? Give an example.
4. Define *ionization potential*.
5. What is the distinction between *metals* and *nonmetals*?
6. What are the distinguishing characteristics of *ionic*, *covalent*, *metallic*, and *hydrogen* bonding?
7. Define *electronegativity*.
8. Define *coordination number*.
9. Name and describe the basic silicate structures.
10. Distinguish between a *clay mineral* and a *clay*.
11. How does a *brucite layer* differ from a *gibbsite layer*?
12. How do kaolinites differ from smectites?
13. What is a *smectite?*
14. Define *cation-exchange capacity*.
15. Name and describe the factors that are responsible for the surface charge of a clay mineral.
16. Define *adsorption isotherm* and discuss the differences between K_d and K determined from adsorption isotherms.
17. The *batch method* and the *column test method* are used to determine the adsorption of cations from aqueous solutions by clays. Describe and distinguish between these two experimental methods.
18. Define *zeolite mineral*.
19. Why can zeolite minerals act as molecular sieves?
20. What determines the theoretical cation-exchange capacity of a zeolite, and why is this the determining factor?
21. Why are zeolite minerals with high Si/Al ratios selective for monovalent cations?
22. Define *asbestos mineral* using both the regulatory and the mineralogical definitions.
23. Describe the structure of chrysotile asbestos and compare this structure to that of the amphibole asbestos minerals.

24. Name and describe the three diseases commonly associated with exposure to asbestos.

25. Compare and contrast the surfaces of chrysotile and amphibole asbestos.

26. Define *zeta potential* and explain why it varies as a function of pH.

27. Name and describe the mechanisms that are responsible for cell damage by asbestos minerals. How would this cell damage vary as a function of the type of asbestos mineral?

28. How are dose–response curves determined?

29. In terms of the regulation of an environmental contaminant, discuss the significance of a dose–response threshold.

30. With reference to Case Study 7–4, discuss the various factors that determine the pulmonary response to an exposure to chrysotile fibers.

31. Name and briefly describe the three respiratory diseases associated with exposure to silica dust.

32. What is meant by *PM10* and *PM2.5*?

33. Define and distinguish between *photolithotrophs*, *photoorganotrophs*, *chemolithotrophs*, and *chemoorganotrophs*.

34. Discuss three ways in which microorganisms play a role in the dissolution and precipitation of minerals. Give some examples.

35. Name and describe five factors that play a role in the rate at which a mineral dissolves.

36. For each of the following calculate the ionic character of the bond:
    a. Pb–S bond in galena (PbS)
    b. Fe–O bond in goethite (FeO·OH)
    c. Si–O, Al–O, and Na–O bonds in albite ($NaAlSi_3O_8$)
    d. P–O bond in apatite [$Ca_5(PO_4)_3(OH,F,Cl)$]
    e. Na–Cl bond in halite (NaCl)
    f. Mg–O bond in forsterite (Mg_2SiO_4)

37. Calculate the coordination number for Si and Al in the mineral K-feldspar ($KAlSi_3O_8$).

38. Calculate the coordination number for Na in the mineral halite (NaCl).

39. Calculate the coordination number for Mg in the mineral forsterite (Mg_2SiO_4).

40. Calculate the coordination number for Mg in the mineral brucite [$Mg(OH)_2$]. Remember that the radius of the hydrogen ion is extremely small and does not need to be considered when doing this type of calculation.

41. Calculate the coordination number for K in the mineral illite [$KAl_4(Si_7AlO_{20})(OH)_4$]. Remember that the radius of the hydrogen ion is extremely small and does not need to be considered when doing this type of calculation.

42. In pyrite (FeS_2), iron is in sixfold coordination. Pyrite is a common mineral found in coal, and the breakdown of pyrite during combustion is responsible for the emission of sulfur dioxide. Pyrite can also contain trace amounts of other metals that could be released to the atmosphere. Assuming that Fe in pyrite is in the high spin state, which of the following elements—Sc, Cr, V, Co, Ni, As, and Sb—might be found in trace amounts in pyrite? If an element can exist in more than one oxidation state, answer the question for each oxidation state. Explain your answers.

43. Goethite (FeOOH) can precipitate from solution when iron is oxidized from Fe^{2+} to Fe^{3+}. In goethite, the iron is in sixfold coordination. Assume that Fe is in the low spin state. Which of the following elements—Sc, Cr, V, Co, Ni, As, and Sb—might enter the crystal structure of goethite when it precipitates from solution? If an element can exist in more than one oxidation state, answer the question for each oxidation state. Explain your answers.

44. An unknown sample is analyzed by X-ray diffraction. An iron X-ray tube is used ($K_\alpha = 1.54$ Å). A diffracted beam occurs at 19.4°. Calculate the d spacing.

45. An unknown sample is analyzed by X-ray diffraction. An iron X-ray tube is used ($K_\alpha = 1.54$ Å). A diffracted beam occurs at 10.2°. Calculate the d spacing.

46. An unknown sample is analyzed by X-ray diffraction. An iron X-ray tube is used ($K_\alpha = 1.54$ Å). The following data are obtained. Complete the table.

θ	Intensity (cps)	d (Å)	I/I_0
2.9	12,481		
8.8	12,541		
9.8	9,406		
13.3	15,677		
18.2	9,249		
20.2	6,270		
26.6	3,135		

47. An unknown sample is analyzed by X-ray diffraction. An iron X-ray tube is used ($K_\alpha = 1.54$ Å). The following data are obtained. Complete the table.

θ	Intensity (cps)	d (Å)	I/I_0
11.0	772		
15.5	15,448		
20.6	4,788		
25.5	4,557		

The sample was collected from a carbonate rock. X-ray powder diffraction data for calcite and dolomite are reproduced here. Identify the mineral.

Calcite (5–586)				
I/I_1	100	18	18	12
d	3.04	2.29	2.10	3.86
Dolomite (11–78)				
I/I_1	100	30	30	5
d	2.89	2.19	1.79	4.03

48. A montmorillonite sample has a mean surface area of 600 m² g⁻¹ and a negative surface charge of 1.1 mole sites kg⁻¹. Calculate the number of monovalent sites per unit surface area (in nm²).

49. A kaolinite sample has a negative surface charge of 0.1 mole sites kg⁻¹. Calculate the CEC for this sample.

50. A vermiculite sample has a negative surface charge of 2.0 mole sites kg⁻¹ and 10 monovalent sites nm⁻². Calculate the surface area of the vermiculite per gram of sample.

51. Sanchez et al. (1999) obtained the following data for three commercially available clays. Calculate the number of monovalent sites per unit surface area (in nm²) for each clay.

Clay	S_A (m² g⁻¹)	CEC (meq/100 g)	N_{SA} (sites nm⁻²)
Kaolin	39	14.8	
Illite	41	20.6	
Bentonite	57	89.0	

52. A batch experiment was done to investigate the adsorption of Ni^{2+} onto kaolin. The experiment was done at 20°C and pH = 5. Agitation time was 3 h. The following data were obtained:

C_{soln} (mg L^{-1})	C_{ads} (mg g^{-1})
0.1	0.15
0.5	0.75
1	1.5
2	3
4	6
8	12
10	15
12	18

a. Plot C_{ads} versus C_{soln}. Based on the arithmetic plot, do you think the data obey a Freundlich or a Langmuir isotherm?
b. Plot log C_{ads} versus log C_{soln}. If the result is a straight line, calculate K and n.
c. For this set of data, does $K = K_d$? Explain.

53. A batch experiment was done to investigate the adsorption of Cd^{2+} onto kaolin. The experiment was done at 25°C and pH = 4.5. Agitation time was 3 h. The following data were obtained:

C_{soln} (mg L^{-1})	C_{ads} (mg g^{-1})
0.1	0.47
0.5	1.06
1	1.50
2	2.12
4	3.00
8	4.24
10	4.74
12	5.20

a. Plot C_{ads} versus C_{soln}. Based on the arithmetic plot, do you think the data obey a Freundlich or a Langmuir isotherm?
b. Plot log C_{ads} versus log C_{soln}. If the result is a straight line, calculate K and n.
c. For this set of data, does $K = K_d$? Explain.

54. A batch experiment was done to investigate the adsorption of Cd^{2+} onto sepiolite. The experiment was done at 22°C and pH = 4. Agitation time was 3 h. The following data were obtained:

C_{soln} (mg L^{-1})	C_{ads} (mg g^{-1})
0.1	0.9
0.5	3.2
1	4.5
2	5.8
3	6.4
4	6.7
6	7.1
8	7.3
10	7.4
15	7.6
20	7.7
25	7.8

    a. Plot C_{ads} versus C_{soln}. Based on the arithmetic plot, do you think the data obey a Freundlich or a Langmuir isotherm?

    b. If the data obey a Langmuir isotherm, plot the linearized version of the Langmuir equation.

    c. What is the maximum sorptive capacity of the surface?

    d. Calculate K.

55. Ten grams of vermiculite are placed in a column and 100 mL of solution is passed through the column. The initial solution had a cadmium concentration of 15 mg L^{-1}. After passing through the column the solution had a cadmium concentration of 11.6 mg L^{-1}. Calculate the K_d.

56. With reference to Case Study 7–3, a batch experiment is done to determine the uptake of ^{137}Cs by a zeolite mineral. The zeolite mineral is first converted to the acid form. Ten grams of the zeolite is added to a 100 mL solution. For this solution, the initial ^{137}Cs activity is 50 dpm (decays per minute). After equilibrium is achieved, the measured activity of ^{137}Cs in the solution is 10 dpm. Calculate the distribution coefficient.

57. Given pH = 4 in the human lung, and an internal temperature of 37°C, calculate the solubility of quartz and amorphous silica in the human lung. Give your answers in ppm.

58. Given the following data (from Norton and Gunter, 1999), calculate the SMRs and SE(SMR)s for farmers and nonfarmers for lung cancer, bronchitis, emphysema, and asthma. The reference population is the Entire Population. Note that the populations are not the same size. You should use the Rate % to calculate the SMRs. Discuss the significance of your calculations. See Case Study 7–5.

	Lung cancer	Bronchitis	Emphysema	Asthma
		Entire population		
Number	7892	384	2233	376
Rate %	4.3	0.2	1.2	0.2
		Farmers		
Number	877	68	407	46
Rate %	3.8	0.3	1.8	0.2
SMR				
SE(SMR)				
		Nonfarmers		
Number	7015	316	1826	330
Rate %	4.4	0.2	1.2	0.2
SMR				
SE(SMR)				

The Atmospheric Environment

In this chapter we will consider the physical behavior of the atmosphere, air pollution, rainwater chemistry, greenhouse gases and climate change, and other topics. These topics require a broad understanding of the earth as a whole. For example, the history of climate change requires a reading and interpretation of the climate record as preserved in ice, lake sediments, and ocean sediments. Atmosphere–hydrosphere interactions, in terms of both physical and chemical processes, exert a significant influence on the atmosphere. Consider that the problem of CO_2 addition and greenhouse warming contains as an important component the rate at which CO_2 is sequestered in the oceanic reservoir. The formation and distribution of ozone and volatile organic compounds in the atmosphere requires a knowledge of photochemistry. As a final example, rainwater chemistry can be significantly modified by water–mineral interactions. Thus, many of the processes and topics considered in earlier chapters will play a role in determining the chemistry of the atmospheric environment.

PROPERTIES OF THE EARTH'S ATMOSPHERE

Structure and Composition of the Atmosphere

The atmosphere is conveniently divided into a number of layers on the basis of its thermal structure. These layers and their boundaries are illustrated in Figure 8–1 (p. 244). In the lowermost layer (troposphere) there is a regular decrease in temperature with height. The top of this layer (tropopause) is marked by a reversal in the trend of decreasing temperature. This reversal is largely due to the concentration of ozone in the stratosphere. Above the stratosphere we once again see a layer of decreasing temperature (mesosphere), and then the outermost layer (thermosphere) shows a dramatic increase in temperature with height. This increase in temperature occurs because there are few gas molecules at this distance from the earth and incoming solar photons can interact directly with the gas molecules. The thermosphere neatly illustrates the difference between *temperature* (*average kinetic energy of the gas molecules*) and *heat* (*total kinetic energy of the gas molecules*). Although the temperature of the thermosphere is high because individual gas molecules have a high velocity (kinetic energy $= \frac{1}{2}mv^2$), the total heat energy is very small because there are few molecules. The two atmospheric layers of increasing temperature act as thermal lids on the underlying layers. The troposphere is a convective layer in which the gases are generally well mixed. It is overlain by the stratosphere, which acts as a thermal lid. Transfer of mass between the troposphere and stratosphere is either by diffusion or mass transfer at the tropopause. Because of this, many atmospheric contaminants released at or near the surface are largely confined to the troposphere.

As noted, gases in the troposphere are well mixed. The average composition of the atmosphere reported in many textbooks, and in Table 8–1 (p. 244), is based on the composition of the troposphere plus the stratosphere to a height of 25 km. Permanent gases are those whose relative concentration remains constant on a long (but not geologic) time scale. The concentrations of other gases vary on a shorter time scale (days to years).

Figure 8–1

The U.S. Standard Atmosphere, 1976. Note the various temperature reversals, which act as thermal lids on the lower parts of the atmosphere. In the troposphere, gases are well mixed. From Neiburger et al. (1982).

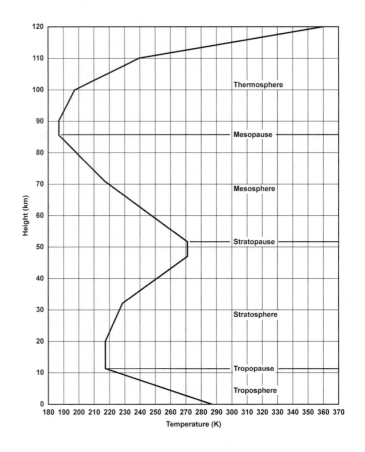

Nitrogen, oxygen, argon, neon, and xenon concentrations are essentially constant on a 1000-year time scale. However, we do know that these gases, particularly O_2, vary on a geologic time scale, and the magnitude and causes of these variations is a current area of research. Helium is a noble gas and does not react with other elements. It also has a very small mass, and given the current temperature of the earth's atmosphere, helium atoms can achieve escape velocity. Water vapor varies on a daily basis depending on rates of evaporation and precipitation. Carbon dioxide, methane, and nitrous oxide have both natural and anthropogenic sources. Carbon dioxide is derived from the burning of fossil fuel as is, indirectly, nitrous oxide. The source of methane is the decay of plant matter under anoxic conditions and release by animals. Modern animal husbandry practices have resulted in increased emission of methane by animals (particularly feedlot cattle). Near-surface ozone (sometimes referred to as *bad ozone* because of its deleterious effect on human health) is

Table 8–1   Average Composition of the Earth's Atmosphere to a Height of 25 km

Gas	Volume %	Source	Variability
N_2	78.08	biologic	permanent
O_2	20.95	biologic	permanent
Ar	0.93	radiogenic	permanent
Ne	0.0018	interior	permanent
He	0.0005	radiogenic	escaping
Xe	0.000009	interior	permanent
H_2O	0 to 4	evaporation	variable
CO_2	0.036	biologic, industrial	increasing
CH_4	0.00017	biologic	increasing
N_2O	0.00003	biologic, industrial	increasing
O_3	0.000004	photochemical	variable

increasing in the atmosphere in response to various hydrocarbon emissions, and stratospheric ozone (sometimes referred to as *good ozone* because it absorbs a significant fraction of the harmful UV radiation coming from the sun) has been decreasing due to chemical reactions with chlorofluorocarbons.

As we move further from the earth's surface, the composition of the atmosphere is no longer homogeneous and concentration gradients are observed that depend on the mass of the different molecules. For example, in the outermost part of the atmosphere the ratio of N_2/CO_2 is relatively greater because nitrogen has a lower molecular weight than carbon dioxide.

Solar and Terrestrial Radiation

A *perfect radiator* is *any substance that emits the maximum amount of electromagnetic energy at all wavelengths*. Solids, liquids, and highly compressed gases (which behave as liquids) all approximate perfect radiators. The shape of the spectrum emitted by a perfect radiator is shown in Figure 8–2. The total amount of energy emitted (i.e., the area under the curve in Figure 8–2) is a function of the temperature of the radiator and is mathematically described by the Stefan–Boltzmann law:

$$E = \sigma T^4 \qquad (8\text{–}1)$$

where E is the energy emitted in W m^{-2}, T is the temperature of the radiator in K, and σ the Stefan–Boltzmann constant ($\sigma = 5.670 \times 10^{-8}$ W m^{-2} K^{-4} = 5.670 $\times$ 10^{-8} J m^{-2} K^{-4} s^{-1}). Note that a doubling of the temperature results in a 16-fold increase in the amount of emitted energy. The wavelength of maximum emitted energy varies inversely with the temperature and is mathematically described by the Wien displacement law:

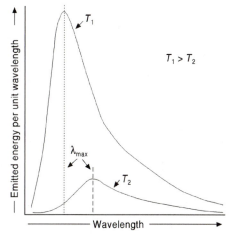

Figure 8–2
Schematic representation of the variation in total energy emitted per unit wavelength as a function of the temperature of the perfect radiator. With increasing temperature the total amount of energy increases and the wavelength of maximum energy decreases.

$$\lambda_M = aT^{-1} \qquad (8\text{–}2)$$

where λ_M is the wavelength of maximum energy, T is the temperature in K, and a is the Wien displacement law constant ($a = 2.898 \times 10^{-3}$ m K = 2898 μm K). The effective temperature of the sun's surface is approximately 5800 K. For this temperature the wavelength of maximum energy is 0.5 μm. The average temperature of the earth is approximately 290 K, for which the wavelength of maximum energy is 10 μm. Given the difference in wavelength maximums for solar and terrestrial radiation, radiation coming from the sun is often referred to as *short wavelength radiation* and radiation coming from the earth is referred to as *long wavelength radiation*.

EXAMPLE 8–1 A distant star has a surface temperature of 8000 K. Calculate the amount of energy emitted by the star per unit surface area and the wavelength of maximum energy.

$$E = \sigma T^4 = (5.670 \times 10^{-8} \text{ W m}^{-2} \text{ K}^{-4})(8000 \text{ K})^4 = 2.322 \times 10^8 \text{ W m}^{-2}$$

$$\lambda_M = aT^{-1} = (2898 \ \mu\text{m K})/(8000 \text{ K}) = 0.36 \ \mu\text{m} \qquad \blacksquare$$

What is the significance of these differences in wavelength maximums in terms of what happens on earth? In Chapter 1 we developed the concept of emission and absorption spectra. This was done in terms of electronic transitions within the atom. Molecules also have emission and absorption spectra. For molecules, the spectra are more complex because in addition to electronic transitions there are also quantized vibration, rotation, nuclear, and electron spin transitions. Also, superposition of lower energy transitions causes electronic molecular absorption bands to be broader than those observed for free atoms. Because a number of these quantized transitions are very close in energy, for any particular molecule we often see regions where there is almost continuous absorption of energy as a function of wavelength. A detailed discussion of this topic is well beyond the scope of this book, but the important point is that gas molecules also absorb photons and that each type of gas molecule absorbs photons of different energy. The absorptivity of various atmospheric gases, and the atmosphere as a whole, as a function of wavelength is shown in Figure 8–3. Note that ozone (O_3) is a significant absorber of photons whose wavelengths are less than 0.3 μm. Carbon dioxide and methane (CH_4) absorb photons of longer wavelengths, in the 3- to 30-μm range. Using the Wien displacement law, we found that the wavelength maximum for solar radiation is at 0.5 μm. Thus, there are a large number of photons in the <0.3-μm range, and ozone is an effective absorber of these wavelengths. A decrease in the amount of stratospheric ozone leads to an increase in the number of photons reaching the earth's surface. This part of the electromagnetic spectrum is in the ultraviolet region, and photons of these energies have been implicated in skin cancer. There is also some evidence that an increase in the number of

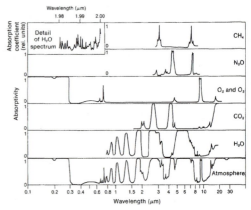

Figure 8–3

Absorptivity of various atmospheric gases, and the atmosphere as a whole, as a function of wavelength. From Fleagle and Businger (1963).

photons in this energy range may have a negative impact on photosynthesis. These are the reasons for our concern about the ozone hole, a topic that will be considered later in this chapter. Energy radiated from the earth back to space has a maximum at about 10 μm. Carbon dioxide and methane are good absorbers in this part of the electromagnetic spectrum, and an increase in these gases might be expected to lead to an increase in the earth's temperature. Carbon dioxide and methane are referred to as *greenhouse gases*, an inexact analogy to what happens in a greenhouse, where the glass inhibits the loss of heat energy leading to an increase in the temperature of the greenhouse. We will consider greenhouse gases later in this chapter. Also note that water vapor would be a good absorber of long wavelength radiation. Water is usually not considered a greenhouse gas, but if global warming sped up the hydrologic cycle so that there was more water vapor in the atmosphere, this increase in water vapor would also contribute to global warming.

The radiant energy reaching the earth from the sun is responsible for essentially all the surface processes that occur on earth (geothermal heat flow from the core is a very distant second). These processes include atmospheric and oceanic circulation and photosynthesis. The total amount of available solar energy at the earth's average distance from the sun is approximately 8.37×10^4 J m^{-2} min^{-1} (2.0 cal cm^{-2} min^{-1}). This is referred to as the solar constant. The **solar constant** is *a measure of the amount of energy passing through a unit surface area perpendicular to the direction of the solar radiation at the average distance of the planet from the sun*. This constant is different for other planets because the radiant flux varies inversely as the square of the distance from the sun, i.e., radiant flux $\propto 1/d^2$. Also note that changes in the sun's temperature would change the energy output of the sun and hence the solar constant.

We can make the simplifying assumption that the amount of solar energy reaching the earth is balanced by the amount of energy returned to space from the earth. If this were not the case, the surface temperature of the earth would change. From both recorded history (e.g., the *Little Ice Age*) and geologic history (the ice ages), we know that the earth's temperature does change with time. The reason for these changes in temperature is an active area of research. On a short-term basis, however, the two radiant fluxes are in balance. What happens to the incoming solar energy? Approximately 30% is reflected directly back to space from the clouds (20%), the air (6%), and the surface (4%). This reflected energy represents the **earth's albedo**, *the total amount of energy reflected back to space by the surface and the atmosphere.* The albedo of the earth will vary depending on the amount of cloud cover and the nature of the surface. For example, snow is a much better reflector of shortwave radiation than a blacktop parking lot. The remaining 70% of the incoming radiation is absorbed by the atmosphere and surface and then reradiated by water vapor in the air (38%), clouds (26%), and the surface (6%).

Insolation (incoming solar radiation) is *the amount of energy that reaches the earth's surface.* This must be less than the value of the solar constant because 30% of the incoming radiation is reflected directly back to space. Additional absorption of shortwave radiation by the atmosphere removes about another 20%, so only 50% of the initial solar energy reaches the earth's surface. If we look at the distribution of solar energy across the earth's surface, we see large variations. At our distance from the sun, the sun's rays are parallel when they approach the earth. At a point on the earth's surface directly under the sun, the rays are perpendicular to the surface and the radiant flux has its maximum density. As we move away from a point directly under the sun, the sun's rays are spread over a greater area and the density of the radiant flux decreases. This is easily understood using a flashlight analogy. If we hold the flashlight perpendicular to a surface we get a round, relatively intense spot. If we tilt the flashlight so the light shines on the surface at an angle, the light is dispersed over a greater area and the intensity decreases. What this means is that there is much more solar energy per unit area near the equator than at high latitudes, resulting in a significant variation in incoming energy as a function of latitude (Figure 8–4). On the other hand, variations in surface and atmospheric temperatures, which determine the amount of energy radiated back to space, are relatively small on a latitudinal basis. Hence, the variations in outgoing energy, as a function of latitude, are much less (Figure 8–4). What we observe is a heat excess at the equator and a heat deficit at high latitudes (Figure 8–4). However, the poles are not getting colder with time, nor is the equator getting warmer, so heat must be transferred from low to high latitudes. It is this transfer of heat that causes atmospheric and oceanic circulation. As a first approximation we can look at the circulation as that of a simple convective system. This circulation pattern is modified by the Coriolis effect, as described in the next section.

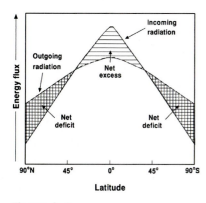

Figure 8–4

Incoming (shortwave) and outgoing (longwave) radiation as a function of latitude. The crossover occurs at ~40°. At lower latitudes there is a heat excess, at higher latitudes a heat deficit.

Atmospheric Circulation

Once large-scale motion begins on the earth, its direction is changed by the Coriolis force. This is not an actual force, but rather a frame of reference problem. Consider the simple case of a missile launched from the North Pole toward the equator. If we observe the path of this missile from space, we will see that it travels in a straight line. However, during the time the missile is in flight, the earth is rotating in a west to east direction, so the direction the missile travels, with respect to the earth's surface, appears to deviate to the west. If we look in the direction the missile is traveling, it will appear to have deflected to the right. If we do the same analysis for the Southern Hemisphere, the missile will appear to deflect to the left. These are referred to as *Coriolis deflections* and occur at all latitudes, except at 0° (the equator). Note that this is a large-scale effect, it cannot be observed on the bathtub scale. Mathematically, the Coriolis force can be written

$$F_c = (2\Omega \sin \varphi)v \qquad (8–3)$$

where Ω is the angular velocity of the earth's rotation in radians (7.29×10^{-5} rad s^{-1}), φ is the latitude, and v is the velocity of the moving mass. Note that when you do this calculation the answer is in m s^{-2}, which is acceleration. What we have actually calculated is the force per unit mass (F_c). This is the standard format for the Coriolis calculation. At the equator the Coriolis force would be zero, and at the poles the Coriolis force would achieve its maximum value.

Figure 8–5

Schematic representation of the general circulation of the earth's atmosphere. L, low pressure; H, high pressure. See text for discussion. Modified from Miller and Thompson (1975).

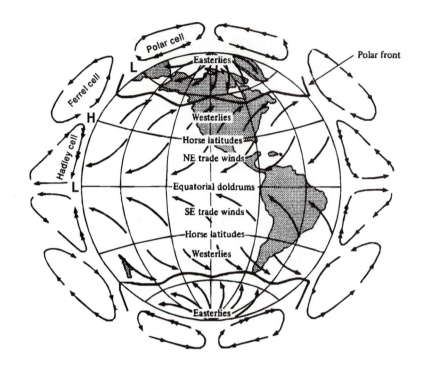

If we now combine the thermal circulation that arises because of the unequal distribution of heat across the earth's surface with the Coriolis effect, we can develop a model for the circulation of the earth's atmosphere and the observed wind patterns (Figure 8–5). We begin with simple thermal convection. Warm air rises at the equator and moves toward the poles, and cold air from the poles moves toward the equator. Because of the rotation of the earth, this simple circulation is broken down into three convective cells, referred to as the Hadley, Ferrel, and Polar cells. At low levels, between 0° and 30° latitude, air is moving toward the equator. Right-handed deflection of this moving air in the Northern Hemisphere (left-handed deflection in the Southern Hemisphere) results in winds that blow from the northeast to the southwest, the NE trade winds. Note that winds are named for the direction from which they come. Between 30° and 60° latitude, air is moving toward the poles. Right-handed deflections in the Northern Hemisphere and left-handed deflections in the Southern Hemisphere set up the westerlies. Between 60° and 90° latitude, air is once again moving toward the equator, which gives rise to the easterlies. Note that at upper levels winds tend to be westerly. This explains why weather patterns move from west to east and why it takes longer to fly a given distance in an east to west direction then in a west to east direction. If we consider the vertical circulation, we find that air is rising at the equator (region of low pressure), descending at 30° latitude (region of high pressure), rising at 60° latitude (region of low pressure), and descending over the poles. This vertical circulation exerts an important influence on precipitation patterns for the earth's surface. Zones of high precipitation are found where air is rising, and zones of low precipitation are found where air is descending. The reason for this will be described in a subsequent section. Lastly, note that the region of rising air along the equator tends to isolate the general circulation patterns of the Northern and Southern hemispheres. Pollutants added to the air in a particular hemisphere are distributed quickly through that hemisphere and more slowly to the other hemisphere.

Temperature and Water Vapor Variations

As a first approximation, we can assume that a parcel of air is isolated from its surroundings both in terms of mass and heat transfer. In a thermodynamic sense, a parcel of air will behave as an adiabatic system. If the air parcel cannot exchange heat with its surroundings, the temperature of the air parcel will change as work is performed on the parcel (compression, which occurs when a parcel descends and pressure increases) or work is done by the parcel (expansion, which occurs when a parcel rises and pressure decreases). Work done on a parcel increases its heat energy and there is a corresponding increase in temperature. When work is done by the parcel, some of the heat energy is converted to the work of expansion and there is a corresponding decrease in temperature.

When both volume and pressure can change, the change in heat energy of a system can be written

$$\Delta H = c_p \Delta T - \alpha \Delta p \qquad (8-4)$$

where c_p is the heat capacity at constant pressure, ΔT is the change in temperature, α is the specific volume (i.e., the reciprocal of density), and Δp is the change in pressure. Because an air parcel acts as an adiabatic system, we set $\Delta H = 0$ and equation 8–4 becomes

$$\Delta T = \frac{\alpha}{c_p} \Delta p \qquad (8-5)$$

A very useful equation for expressing changes in pressure in a fluid is the **hydrostatic equation**, which can be written

$$\Delta p = -\rho g \Delta h \qquad (8-6)$$

where Δp is the change in pressure, ρ is the density of the fluid, g is the acceleration due to gravity, and Δh is the change in height. The equation has a negative sign because increasing height is a positive direction and an increase in height leads to a decrease in pressure. Substituting for Δp in equation 8–5, and remembering that density (ρ) is the reciprocal of specific volume (α), gives

$$\Delta T = -\frac{g}{c_p} \Delta h \qquad (8-7)$$

and rearranging gives

$$\frac{\Delta T}{\Delta h} = -\frac{g}{c_p} \qquad (8-8)$$

Substituting into equation 8–8 $g = 9.8$ m s^{-2} and $c_p = 1003$ J kg^{-1} K^{-1} (for dry air) gives

$$\frac{\Delta T}{\Delta h} = -\frac{9.8 \text{ m s}^{-2}}{1003 \text{ J kg}^{-1} \text{ K}^{-1}} = -9.8 \times 10^{-3} \text{ K m}^{-1} = -9.8 \text{ K km}^{-1} \qquad (8-9)$$

This value is referred to as the **dry adiabatic lapse rate** and is *the rate at which an air parcel cools if lifted in the atmosphere or warms if forced to lower levels, as long as no condensation occurs in the air parcel.*

What happens if condensation occurs in an air parcel? The amount of water vapor that air can hold varies as a function of temperature and pressure. For example, at sea level an air parcel whose temperature is 20°C can hold 15.0×10^{-3} kg of water vapor per kg of dry air. At 10°C this same air parcel can hold 7.76×10^{-3} kg of water vapor per kg of dry air, and at 0°C it can hold 3.84×10^{-3} kg of water vapor per kg of dry air. Meteorologists use two terms in referring to the water content of the atmosphere. **Absolute humidity** is *the amount of water vapor actually present in the air.* **Relative humidity** is *the amount of water vapor in the air divided by the amount of water vapor the air can hold at any particular temperature, expressed in percent.* For example, if at 20°C our air parcel contained 7.5×10^{-3} kg of water vapor per kg of dry air, the relative humidity would be 50%. What would happen if we lifted this air parcel to 1 km? The temperature of the air parcel would decrease to 10°C (dry adiabatic cooling), and at this point the amount of water vapor present would equal the amount the air could hold at this temperature; i.e., the relative humidity is 100% and we say the air is saturated with respect to water vapor. If we lifted the air parcel to higher levels, it would cool further. The air parcel is now supersaturated in water vapor and some of the water vapor would be converted to liquid water. During this process heat energy (latent heat of condensation) is released. We now have an internal source of heat energy for the air parcel and the parcel cools at a slower rate than it did before condensation occurred. The **wet adiabatic lapse rate** is *the rate at which an air parcel cools when condensation occurs*. The wet adiabatic lapse rate varies as a function of temperature and pressure because these variables affect the water-holding capacity of the air. At very low temperatures, the amount of water vapor the air can hold is very small and the amount of heat energy released by condensation is small. Thus, at very low temperatures the value of the wet adiabatic lapse rate approaches that of the dry adiabatic lapse rate. A final point with respect to the vapor-holding capacity of air is the observation that regions where air is rising have significant precipitation and regions where air is descending have little or no precipitation. The reason for these variations is that rising air cools and its vapor-holding capacity decreases, whereas descending air warms and its vapor-holding capacity increases.

A final lapse rate is the **environmental lapse rate**, which is *the observed rate at which temperature changes in a column of air*. This rate is usually not constant with altitude. The environmental lapse rate(s) is determined from a **sounding curve**, which is *the measured change in temperature with altitude in the atmosphere*.

Why do we care about lapse rates? One of the earliest forms of air pollution control was dispersal. The idea was that the total amount of air pollutants emitted was actually small. The problem was that the pollutants were concentrated in a small volume. If the pollutants could be dispersed through a large volume, their concentrations would be reduced to essentially negligible amounts. If air pollutants could not be dispersed, their concentrations would build up to high, and potentially harmful, levels. In the troposphere temperature usually decreases with height. However, under certain conditions the temperature of the atmosphere increases with height. This *reversal of the normal temperature pattern* is referred to as an **inversion**. There are three basic types of inversions:

1. **Radiation inversions**, caused by radiational cooling of the land surface and a decrease in the temperature of the atmosphere at low levels. Inversions of this type are common at high latitudes during the winter.
2. **Subtropical inversions**, caused by the sinking of air in a high pressure center. Remember that when air descends its temperature increases. Referring to Figure 8–5, you will note that there are semipermanent high pressure centers at approximately 30°N and 30°S latitude. The position of the semipermanent highs changes seasonally as the sun moves back and forth across the equator. For example, during Northern Hemisphere summer the semipermanent high located over the ocean along the west coast of the United States moves northward to a position off Los Angeles, California. It is this inversion that is responsible for the extended periods of air pollution that occur in southern California.
3. **Frontal inversions**, caused by the relative movement of warm air over cold air.

The effect of normal and inverted environmental lapse rates on the dispersal of air pollutants is indicated schematically in Figure 8–6. We start with a plant emitting a gaseous plume whose temperature is well above that of the surrounding air. As the plume rises, it first cools at the dry adiabatic lapse rate. If the plume contains water vapor, at some temperature it will become saturated with water vapor and the lapse rate changes to the wet adiabatic lapse rate. These two lapse rates are indicated in Figure 8–6, and the inflection point represents the height and temperature at which the plume becomes saturated with water vapor. First, consider what would happen in the case of a normal lapse rate. As indicated in the diagram, the plume would be warmer than the surrounding air and would continue to rise. The result is that the contaminants in the plume would be dispersed through a significant atmospheric thickness. Second, consider what would happen if a subsidence-type inversion occurs (indicated by the dashed-dotted line). In this case, at some level the temperature begins to increase with height. As the rising plume cools, a point is reached where the temperature of the plume becomes the same as the surrounding air and the plume stops rising. The air pollutants contained in the plume are now trapped below the inversion and the concentration of these pollutants increases in the lowermost part of the atmosphere. If the environmental lapse rate returns to the more normal pattern, the plume will once again rise and the atmosphere will be ventilated.

Figure 8–6

Schematic representation of a subsidence inversion and its effect on the dispersal of air pollutants. In the case of the normal atmospheric lapse rate, the air pollutants will continue to rise and be dispersed, but when an inversion occurs, the air pollutants are trapped below the inversion and can build up in concentration.

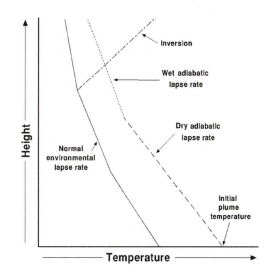

During the winter at high latitudes, one can often observe the buildup and dispersal of pollutants on a daily basis. On a clear, cold night the air near the ground is cooled significantly and a radiation inversion develops. During the early morning hours, air pollutants are trapped near the ground and a ground haze is observed with clear sky above. During the daylight hours, radiation heating occurs and the inversion disappears by midafternoon, as indicated by the disappearance of the haze layer. If an emitter is located in an area where radiation inversions are common, one strategy for pollutant dispersal is to build tall smoke stacks that extend above the height of the inversion. The pollutants are emitted above the inversion and are dispersed throughout the atmosphere.

Is dispersal an effective pollution abatement strategy? It depends on the pollutant and the total amount of pollutant emitted. In general, the strategy has not proved effective. In some cases, such as chlorofluorocarbons, even relatively small amounts of contaminant can have a significant environmental impact. In other cases, such as sulfur dioxide, the total amount emitted is so large that dispersal is not effective in mitigating the adverse effects caused by the contaminant. Thus, in terms of atmospheric contaminants, dispersal is no longer regarded as an effective air pollution control strategy.

AIR POLLUTION

In this section we will briefly look at the question of air pollution, prior to a more detailed discussion of specific aspects of atmospheric chemistry and environmental problems. *Air pollution* can be defined as *the presence in the atmosphere of substances that are toxic, irritant, or otherwise harmful to humans or damaging to vegetation, animals, or property.*

Types of Air Pollutants

Air pollutants can be classified into three broad classes: inorganic gases, organic gases, and particulates. These groups, and examples of each, are listed in Table 8–2 (p. 252). A distinction is usually made between primary and secondary air pollutants. *Primary pollutants* are *the direct products of combustion and evaporation.* *Secondary pollutants* are *formed in the atmosphere by reaction with the primary pollutants.* The primary pollutants are particulates, volatile organic compounds (VOCs), CO, CO_2, SO_x, and NO_x. Secondary pollutants are ozone and other compounds produced by photochemical oxidation. NO_x and VOCs play key roles in photochemical oxidation reactions. One can also consider SO_4 and NO_2 to be secondary pollutants because they are formed by oxidation in the atmosphere of less oxidized forms of these compounds. We will consider these reactions when we look at the problem of acid deposition.

Table 8–2 Classification of Air Pollutants

Major class	Subclass	Examples
Inorganic gases	Oxides of nitrogen	N_2O, NO, NO_2
	Oxides of sulfur	SO_2, SO_3
	Oxides of carbon	CO, CO_2
	Other inorganics	O_3, H_2S, HF, NH_3, Cl_2, Rn
Organic gases	Hydrocarbons	Methane (CH_4), butane (C_4H_{10}), octane (C_8H_{18}), benzene (C_6H_6), acetylene (C_2H_2), ethylene (C_2H_4)
	Aldehydes and ketones	Formaldehyde, acetone
	Other organics	Chlorofluorocarbons, PAHs, alcohols, organic acids
Particulates	Solids	Fume, dust, smoke, ash, carbon soot, lead, asbestos
	Liquids	Mist, spray, oil, grease, acids

Aerosols

Aerosols are *solid particles or liquid droplets ranging in size from clusters of a few molecules to 20 µm in radius.* Aerosols can also result from interactions between gaseous pollutants and small water droplets. For example, consider the following reaction:

$$SO_{3\,(g)} + H_2O_{(l)} \rightarrow H_2SO_{4\,(l)}$$

which produces a sulfuric acid aerosol. This type of aerosol is important both in terms of acid deposition and reflection of shortwave radiation. During major volcanic eruptions, one of the gases emitted is SO_2. This gas is oxidized in the atmosphere and eventually interacts with water to produce sulfuric acid aerosols, which can be transported to the lower stratosphere. It usually takes several years to remove these aerosols from the upper troposphere–lower stratosphere, and during this time the albedo of the earth increases because of the efficiency of the aerosols in reflecting shortwave radiation. The recent eruption of Pinatubo in the Philippines essentially halted, for several years, a trend of rising atmospheric temperature. Table 8–3 lists the various types of aerosols and their sources.

Table 8–3 Sources of Aerosols and Contributions of Natural Versus Anthropogenic Sources*

Source	Natural $(10^{12}\,\text{g y}^{-1})$	Anthropogenic $(10^{12}\,\text{g y}^{-1})$
Soil and rock dust	3000–4000	?
Sea salt	1700–4700	
Biogenic	100–500	
Biomass burning (soot)	6–11	36–154
Volcanic	15–90	
Direct emissions—fuel, incinerators, industry		15–90
Gaseous emissions		
Sulfate from biogenic DMS	51	
Sulfate from volcanic SO_2	18–27	
Sulfate from fossil fuel		105
Nitrate from NO_x	62	128
Ammonium from NH_3	28	37
Biogenic hydrocarbons	20–150	
Anthropogenic hydrocarbons		100
Total	5000–9619	421–614

*Modified from Berner and Berner (1996).

Smogs

Smog is a combination of the words *smoke* and *fog* and is characterized by hazy and irritating air that forms in response to the emission of pollutants and an inversion, which prevents the dispersal of the pollutants. Since the start of the industrial age there have been a number of *famous* air pollution events in which the formation of a smog led to significant health effects and deaths. These extraordinary events have acted as a catalyst for air pollution control efforts. Table 8–4 lists the two major types of smogs and their characteristics. The time of the worst events (Table 8–4) is related to the type of inversion that causes the buildup of the pollutants and the climatic conditions. Industrial smogs occur during the winter because this is the time when radiation inversions are common, coupled with increased burning of fossil fuels for heat. During the course of the day, the radiation inversion often dissipates. Thus, the effects are most significant during the early morning hours. Photochemical smogs are often related to the presence of a semipermanent high. The time of maximum pollutant buildup is midday in the summer because the active agent in the formation of this type of smog is sunlight.

Table 8–4 Types of Smogs and Their Characteristics

Characteristic	Industrial	Photochemical
First occurrence	London	Los Angeles
Principal pollutants	SO_x	O_3, NO_x, HC, CO, free radicals
Principal sources	Industrial and household fuel combustion	Motor vehicle fuel combustion
Effect on humans	Lung and throat irritation	Eye and respiratory irritation
Effect on compounds	Reducing	Oxidizing
Time of worst events	Winter months in the early morning	Summer months around midday

The major impact of industrial smog is the formation of acid aerosols. These acid aerosols corrode building materials. The combination of the acid aerosols and particulates has a major impact on the human respiratory system. In London, the type of locality for industrial smog, the impact is readily apparent in the appearance of major historical buildings. Many of these buildings are faced with limestone, which is readily attacked by acid precipitation. This attack is accompanied by the embedding of particulates in the partially dissolved limestone, and the limestone takes on a dark color rather than its normal white color. Pollution abatement programs over the past several decades, and restoration efforts, have returned many of these buildings to their pre-industrial-age appearance.

The major impact of photochemical smog is the formation of ozone, NO_x, and other compounds that cause significant respiratory distress, and the buildup of a near-surface haze. In the case of Los Angeles, air pollution episodes can persist for a number of days with concomitant buildup of pollutants. The haze can become sufficiently thick that aircraft landing on a bright sunny day may be required to make instrument landings. Individuals who have respiratory problems are at particular risk during times of major air pollutant buildup. Because of the severity of the problem, southern California has the strictest automobile emission standards in the United States.

GREENHOUSE GASES AND CLIMATE CHANGE

Greenhouse gases, and the possibility that these gases may cause climate change, have been one of the "hottest" environmental and political issues in recent years. Greenhouse gases are gases that absorb long wavelength radiation. The most important gases are CO_2, CH_4, N_2O, and the CFCs. Increases in the amounts of these gases lead to an increase in the absorption of outgoing radiation and a potential increase in global temperature. The reasons for the temperature increase are that the increased vibrational or rotational energy of the molecules is transferred by collisions to other molecules in the atmosphere and reradiation of absorbed energy, leading to a heat flux back to the earth from the atmosphere. The size of this potential temperature increase, and its effect on world climate, is a complex problem due to a number of confounding variables.

1. An increase in temperature would be expected to increase the amount of water vapor in the atmosphere. Water vapor is a good absorber of long wavelength radiation, and an increase in atmospheric $H_2O_{(g)}$ would lead to increased absorption of longwave radiation, a positive feedback.
2. The increase in water vapor would lead to an increase in cloud cover. If the increase in cloud cover occurs at high levels, more long wavelength radiation is reflected back to the earth's surface, a positive feedback, but if the increase in cloud cover occurs at low levels, more short wavelength radiation is reflected back to space, a negative feedback.
3. An increase in global temperature should lead to a decrease in snow and ice cover, thus reducing the albedo of the earth, a positive feedback.
4. An increase in soil temperatures, particularly at high latitudes, could lead to increased microbial activity and an increase in CO_2 emitted to the atmosphere because of increased decay, a positive feedback.
5. An increase in CO_2, along with increasing temperature, may lead to an increase in biological activity and the storage of CO_2 in biomass, a negative feedback.

Complex computer models, called *general circulation models* (GCMs), attempt to take all these factors into consideration in order to predict the long-term trends due to the addition of greenhouse gases to the atmosphere. Case Study 8–1 describes one approach to assessing the impact of greenhouse gas additions.

The potential effects of greenhouse warming are also subject to a significant amount of uncertainty. We can divide the potential effects into two types: changes to the hydrologic cycle and sea-level changes. An increase in global temperature would lead to an increase in the rate of evaporation of water and an increase in precipitation—a *speeding up* of the hydrologic cycle. Computer models, however, predict that this increase will be

---

CASE STUDY 8–1
Climate Variations for the Last 1000 Years and Possible Causes for Climate Variations

Crowley (2000) investigated the reasons for temperature changes in the Northern Hemisphere over the past 1000 years. The construction of the past 1000-year temperature record involved a combination of more recent instrumental measurements and various proxy indices such as tree rings, corals, and ice cores. The temperature variations, on an annual basis, were determined by calibrating the proxy indices to the instrumental measurements during the period 1860–1965. This temperature reconstruction showed a warm period from 1000 to 1300, a transition interval from 1300 to 1580, a cold period in the 1600s, a temperature recovery in the 1700s, a cold period in the early 1800s, another cold period in the early 1900s, and a temperature increase of 0.8°C from this early 20th-century temperature minimum to the present day. Between the 17th century and the mid-20th century warm period there was an approximately 0.4°C temperature increase.

Three possible reasons for these temperature variations were considered: volcanic emissions, variations in solar radiation, and changes in the amount of greenhouse gases. Ice cores played a key role in the analysis. Variations in sulfate concentrations in ice cores were used to reconstruct the intensity of volcanic activity. Solar variability was reconstructed using ice core measurements of ^{10}Be, ^{14}C from tree rings, and an estimate of ^{14}C from ^{10}Be fluctuations. Each of these isotopes is sensitive to the intensity of the incoming solar radiation. Concentrations of CO_2 in the atmosphere were determined from the concentration of CO_2 in ice cores.

Variations in temperature in response to changes in volcanic activity, solar radiation, and greenhouse gases were investigated using a linear upwelling/diffusion energy balance model. For the interval 1000–1850, 41 to 64% of the temperature variation could be explained by the variations in volcanic activity and solar radiation. Extending the model into the 20th century, the author found that only about 25% of the temperature increase noted in the second half of the 20th century could be explained by these two factors. The author concluded that the bulk of the 20th-century warming is consistent with that predicted from greenhouse gas increases. Extension of the model into the 21st century, using the current rate of greenhouse gas addition, gave a temperature increase of approximately 2°C by the end of the 21st century.

Source: Crowley (2000).

uneven; in some areas there will be an increase in precipitation while in other areas there will be a decrease in precipitation. In terms of sea-level changes, a global temperature increase will lead to an increase in sea level due to (1) the net transfer of water from the continents to the oceans by the melting of glacial ice and (2) expansion of the water column caused by the increase in seawater temperature. The biggest uncertainty in sea-level change is the possibility that the West Antarctic ice sheet could disintegrate and add a substantial amount of water to the oceans. To once again indicate the complexity of these processes, it is also possible that an increase in ocean temperature around Antarctica would lead to increased evaporation of water from the ocean and increased precipitation over Antarctica, a process that would transfer water from the ocean to the Antarctic ice sheet and thus lower sea level. On a longer time scale, the geologic record, as reflected in ice cores and sediments, has provided insights into past climatic conditions and variations. The characteristic of climate that emerges from the geologic record is that of variability, with times of cold and dry climates contrasting with times of warm and humid climates.

Greenhouse Gases

The major greenhouse gases are CO_2, CH_4, N_2O, and CFCs. These gases, particularly CO_2, are central to most discussions of greenhouse warming. Water vapor is also a strong absorber of infrared radiation, but this gas is usually not considered to be a greenhouse gas. We are all familiar with the effect of water vapor on temperature. The warm, humid days of summer are followed by warm nights because the high water vapor content leads to significant absorption of outgoing radiation. Conversely, the cold, dry days of winter are followed by cold nights because the absence of water vapor means that most of the outgoing radiation is not absorbed by the lower atmosphere. Besides the major greenhouse gases, there are a number of minor gases, of anthropogenic origin, that are exceptionally good absorbers of infrared radiation. Among these are the chlorofluorocarbons, hydrofluorocarbons, perfluorocarbons, SF_6 (sulfur hexafluoride), and the newly discovered SF_5CF_3 (Sturges et al., 2000). Although these gases are present in the part-per-trillion range, on a molecule-to-molecule basis they are thousands of times more efficient absorbers of infrared radiation than CO_2. All of these gases have shown a dramatic increase (Figure 8–7) since the start of the industrial age, and particularly in the 20th century. The chlorofluorocarbons, a product of modern chemistry, did not become significant in the atmosphere until the 1950s.

The significance of greenhouse gases in the atmosphere is often measured in terms of **radiative forcing**, which is the *absorption of longwave radiation (in watts per square meter) due to a particular gas*. Two indices have been developed to quantify the relative

Figure 8–7
Increase in greenhouse gases since 1750. CFCs (as represented by CFC_{11}) are entirely of anthropogenic origin and don't become significant until after 1950. Besides their role as greenhouse gases, CFCs are also important in the breakdown of ozone. From Berner and Berner (1996).

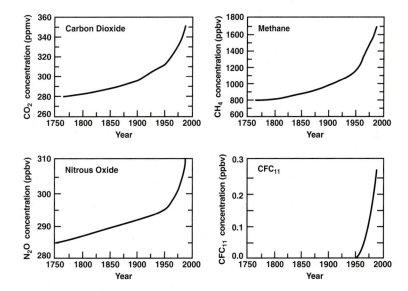

contribution of greenhouse gas additions to radiative forcing: relative instantaneous radiative forcing (RIRF) and global warming potential (GWP). For both indices, the value for carbon dioxide is arbitrarily set at 1 and the other greenhouse gases are compared to carbon dioxide. The ***RIRF*** measures *the effect an incremental addition of gas would have on the infrared absorption for the present atmosphere.* The ***GWP*** of a greenhouse gas is *the ratio of global warming from one unit mass of a greenhouse gas compared to one unit mass of carbon dioxide over a particular time period.* The calculation is generally done by comparing the effect of a 1-kg addition of CO_2 to that of a 1-kg addition of the gas of interest. The calculation of GWP is done for different time intervals depending on the climatic effect of interest. For example, a 100-year interval is appropriate when investigating the maximum temperature change, whereas for maximum sea-level change the calculation should be done for a several-hundred-year interval. The IPCC (Intergovernmental Panel on Climate Change) recommends that GWP be done on a 100-year basis.

Table 8–5 gives summary data for the most important greenhouse gases. In the present atmosphere, CO_2 and CH_4 are the most important gases in terms of the increase in radiative forcing. Of significance is the high GWP for a number of trace atmospheric constituents. Because of their deleterious effect on stratospheric ozone, the production of CFCs is being phased out and other organic molecules are being introduced as substitutes. These substitute compounds, hydrofluorocarbons (HFCs) and perfluorocarbons (PFCs) presumably do not affect ozone levels in the stratosphere. The HFCs and PFCs, do, however, have high GWPs and could have a significant impact on radiative transfer through the atmosphere. The PFCs have long atmospheric lifetimes. Thus, their concentration would be expected to significantly increase over time.

The data shown in Table 8–5 are useful in developing strategies for controlling the increase in radiative forcing. Given the relative contribution of CO_2 to the total radiative forcing, this would seem to be the most important greenhouse gas. Indeed, international meetings have focused on the control of the emission of CO_2, not an easy task because it is the product of the combustion of organic matter (coal and petroleum) on which modern industrialized societies depend for their energy needs. Various schemes have been proposed for the disposal of CO_2 in deep aquifers (Case Study 3–4) or in the ocean. Methane is next in importance, although recently there has been a decrease in the rate of methane

Table 8–5 Data for Greenhouse Gases*

Gas	Concentration 1990 (ppmv)	Positive radiative forcing (W m^{-2})	% Total radiative forcing	Lifetime (y)	Relative instantaneous radiative forcing (molecular basis)	Global warming potential (100 y)
CO_2	354	1.5	61	50–200	1	1
CH_4	1.72	0.42	17	12	43	21
H_2O strat	—	0.14	6			
N_2O	0.310	0.1	4	120	250	310
CFC-11	0.00028	0.062	2.5	65	15,000	3,400
CFC-12	0.000484	0.14	6	130	19,000	7,100
Other CFCs		0.085	3.5			
Total		2.45	100.0			
CFC substitutes						
HFC-23				264		650
HFC-152a				1.5		140
CF_4				50,000		6,500
C_6F_{14}				3,200		7,400

*From Berner and Berner (1996), IPCC (1996), vanLoon and Duffy (2000).

increase. The CFCs are being phased out because of their effect on stratospheric ozone, but some of the substitutes that are being developed for the CFCs are potent greenhouse gases. The long-term impact of these substitutes on atmospheric warming needs to be considered.

Carbon Dioxide (CO_2)   The most widely discussed of the greenhouse gases is carbon dioxide. Starting in 1958, the atmospheric concentration of CO_2 has been measured at Mauna Loa observatory in Hawaii (Figure 8–8). This location, far removed from any region of significant industrialization, is believed to give a reasonable average concentration of CO_2 for the troposphere. The observations clearly show the dramatic increase in CO_2 since the initiation of measurements in 1958. The sawtooth nature of the curve (the annual oscillation) is due to biological processes in the Northern Hemisphere. During the spring and summer, the photosynthetic activity of plants leads to a net depletion of CO_2 in the atmosphere. During the fall and winter, respiration is the dominant process and there is a net addition of CO_2.

The major anthropogenic source of CO_2 is the combustion of fossil fuel (coal and petroleum). Cement production also releases some CO_2 to the atmosphere. The most common type of cement, Portland cement, consists of a mixture of Ca, Si, Al, and Fe. The manufacture of this product involves the heating of a mixture of high-Ca limestone ($CaCO_3$) and the other components to high temperatures in a kiln. During this process, CO_2 is released according to the reaction

$$CaCO_{3 \text{ calcite}} \rightarrow CaO + CO_2$$

An additional source of anthropogenic CO_2 is the clearing of land for agriculture (deforestation), which releases carbon stored in the biomass to the atmosphere.

Anthropogenic carbon added to the atmosphere is removed by transfers to the oceans and terrestrial biosphere. The annual balance for anthropogenic carbon can be represented by the following simple equation:

$$I = F - A - O + B \tag{8–10}$$

where I represents an imbalance in the release and uptake of carbon, F is the fossil fuel addition per year, A is the annual increase in atmospheric CO_2, O is the annual oceanic storage, and B is the net addition due to deforestation (i.e., deforestation − reforestation). Estimates of carbon addition and removal for the period 1980–1989 are given in

Figure 8–8
Mean monthly concentrations of CO_2 at Mauna Loa, Hawaii. From Berner and Berner (1996).

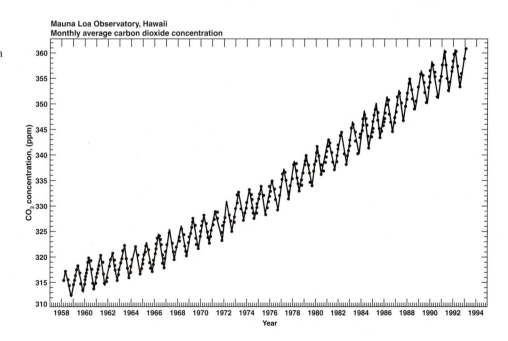

Table 8–6 Annual Additions and Removal of Anthropogenic Carbon*

Sources or sinks	Flux (10^9 tons = 1 Gt)
Fossil fuel (F)	5.4 ± 0.5
Atmosphere (A)	3.2 ± 0.2
Ocean (O)	2.0 ± 0.6
Net deforestation (B)	1.6 ± 1.0

*Data from Siegenthaler and Sarmiento (1993).

Table 8–6. Solution of equation 8–10, using the values in Table 8–6, gives a carbon imbalance of 1.8 ± 1.3 Gt y^{-1}. This result implies that our estimates of the various fluxes are incorrect and/or there is a missing sink for anthropogenic carbon dioxide.

A simplified version of the carbon cycle is shown in Figure 8–9. There are a number of sinks and sources for carbon, and it is the interplay of these various carbon reservoirs that ultimately determines the carbon content of the atmosphere. The most important exchanges are those that occur between the terrestrial biosphere plus soil and the atmosphere and the surface ocean and the atmosphere.

As a first approximation, the ocean can be divided into two reservoirs: a surface reservoir and a deep reservoir separated by the main thermocline (a zone of rapidly changing temperature). The uptake of CO_2 by the surface ocean involves the carbonate system, which we discussed at some length in Chapter 3. This uptake happens on a time scale of tens of years. Removal of carbon from the surface reservoir to the deep ocean occurs on a much longer time scale, on the order of hundreds to thousands of years, and does not play an immediate role in controlling the CO_2 content of the atmosphere. Given the average pH of the surface ocean (8.2 to 8.3), carbon is present in the ocean predominantly as HCO_3^- and in lesser amounts as CO_3^{2-}. CO_2 is taken up by the surface waters according to the following reaction:

$$CO_{2\,(g)} + CO_3^{2-} + H_2O \rightarrow 2HCO_3^-$$

The rate of uptake of anthropogenic CO_2 by the surface ocean cannot be directly measured, and estimates of this uptake are done by indirect means using various tracers of gas exchange (see Case Study 8–2).

Figure 8–9
The carbon cycle. Reservoir concentrations are in 10^{15} g (Gt) carbon. Fluxes are in Gt C y^{-1}. From Berner and Berner (1996).

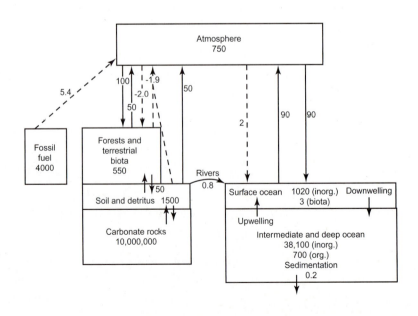

CASE STUDY 8-2
Oceanic Uptake of Fossil Fuel CO_2

Quay et al. (1992) used changes in the $^{13}C/^{12}C$ ratio ($\delta^{13}C$) of atmospheric CO_2 and dissolved organic carbon (DIC) in the surface ocean to estimate the storage of fossil fuel carbon in the surface ocean. For atmospheric CO_2, $\delta^{13}C \approx -8‰$, and for fossil fuel and plant carbon, $\delta^{13}C \approx -27‰$. As discussed in Chapter 6, carbon isotopes are fractionated during photosynthetic fixation of CO_2, and the degree of fractionation depends on whether the plant uses a three-carbon (C_3) or four-carbon (C_4) intermediate to fix the CO_2. $\delta^{13}C$ values for plants range from $-27‰$ (C_3) to $-12‰$ (C_4) cycle. The terrestrial biomass is $> 95\%$ C_3 plants, and thus $\delta^{13}C \approx -27‰$ for the terrestrial biomass. Coal is also derived from C_3 plants, and for coal $\delta^{13}C \approx -27‰$. The addition of both fossil fuel and terrestrial biomass carbon to the atmosphere will lead to a decrease in the $\delta^{13}C$ value for atmospheric CO_2. $\delta^{13}C$ values for the DIC in the surface ocean will also decrease as a function of the amount of CO_2 transferred from the atmosphere to the surface ocean. Over the period 1970–1990, it was estimated that atmospheric $\delta^{13}C$ values decreased by 0.4‰. The $\delta^{13}C$ value for DIC in Pacific Ocean surface waters declined by a similar amount.

The authors proposed the following mass balance equation for anthropogenic CO_2:

$$V(C_t - C_0) = (S_{ff} + S_{br} - S_{bu} - S_{oc})\Delta t$$

where V is the volume of the atmosphere (3.9×10^{18} m^3), C_t is the atmospheric concentration (g C m^{-3}) of CO_2 in 1990, and C_0 is the concentration in 1970, S_{ff} is the time-integrated release of CO_2 from fossil fuel combustion and cement production, S_{br} is the release of CO_2 by the biosphere and S_{bu} is the uptake of CO_2 by the biosphere, S_{oc} is the net uptake by the ocean, and Δt is 20 years. The amount of atmospheric CO_2 stored in the ocean over this time period was determined from a depth-integrated measurement of DIC concentration and $\delta^{13}C$ values. Over the period 1970–1990, 101.6 Gt of anthropogenic carbon was released to the atmosphere. The atmospheric concentration of carbon increased by 58 Gt. During this same time period, using the changes in the $\delta^{13}C$ values, the authors estimated that 42 Gt of carbon were taken up by the ocean, an oceanic uptake of 2.1 Gt y^{-1} of carbon. The difference of 1.6 Gt presumably represents carbon stored in the biosphere.

Source: Quay et al. (1992).

A second oceanic process that will lead to the removal of anthropogenic CO_2 is the so-called *biological pump*. In this process CO_2 is taken up by photosynthetic organisms, i.e., phytoplankton. When the plankton die they sink into deeper waters and decay, thus transferring carbon from the surface waters to the deep waters. The CO_2 released to the deep waters during decay is not rapidly exchanged with the surface waters and is effectively isolated from the atmosphere. Ice core records indicate that periods of lower atmospheric CO_2 can be correlated with increased atmospheric dust (Watson et al., 2000). The interpretation is that the dust provides iron, a limiting nutrient for Southern Ocean phytoplankton growth, which leads to an increase in phytoplankton and an increase in the efficiency of the biological pump. A number of recent studies (Abraham et al., 2000; Boyd et al., 2000; and many others) have found that seeding the Southern Ocean with iron particles does lead to an increase in the phytoplankton population.

Another sink (or source) for atmospheric CO_2 is the terrestrial biosphere. When we solved equation 8–10 we found that there was a positive imbalance in anthropogenic carbon; i.e., some of the released anthropogenic carbon is missing. One possible carbon reservoir is the terrestrial biosphere. Whether or not the terrestrial biosphere has been a net source or a net sink for atmospheric CO_2 has been a contentious issue. Changes in land use will affect the total amount of biomass (see Case Study 8–3, p. 260). Also, the addition of CO_2 to the atmosphere may lead to an increase in plant growth, the so-called *CO_2 fertilization effect*.

Methane (CH_4)   In terms of the present atmosphere, methane is the second most important greenhouse gas. Since 1840, the methane concentration has more than doubled, from 806 ppbv in 1841 to 1730 ppbv in 1996. During the 1990s there was a decrease in the rate of methane increase (Figure 8–10, p. 260). Methane is chemically active and interacts with other species present in the atmosphere. These interactions not only determine the lifetime of methane but also play a role in determining the atmospheric concentrations of tropospheric ozone and stratospheric water, both of which are greenhouse gases. Both natural and anthropogenic sources are important in determining the atmospheric concentration of methane. There are three significant sinks for methane: (1) oxidation by chemical

CASE STUDY 8–3
The U.S. Carbon Budget

Houghton et al. (1999) investigated the accumulation and release of carbon by U.S. ecosystems. The authors used a *bookkeeping model* to calculate annual changes in carbon for each type of ecosystem, land use, and region. The amount of carbon stored in each type of ecosystem was obtained from the ecological literature. The authors used changing land use patterns, i.e., changing geographical distribution of ecosystems, to construct yearly variations in total carbon stored in U.S. terrestrial ecosystems for the period 1770 to 1990. From 1945 onwards, the model shows a net increase in carbon stored in the terrestrial ecosystems. From 1770 to 1945, changes in land use released a total of $27 \pm 6 \times 10^{15}$ g C. From 1945 to 1990, there was a net uptake of $2 \pm 2 \times 10^{15}$ g C. This uptake was largely due to the suppression of forest fires and the growth of forests on abandoned agricultural land. A comparison of the change in carbon storage based on changes in land use versus the direct measurement of wood volumes in forests reveals that the land use model underestimates total carbon storage. The authors suggest that this increased carbon storage may be due to anthropogenic-induced factors, i.e., CO_2 fertilization, climate change (increasing temperatures), and nitrogen deposition. The results of these computations are shown in Figure 8–C3–1, which shows a net uptake (0.037×10^{15} g y^{-1} = 0.037 Gt y^{-1}) of carbon by the U.S. terrestrial biosphere. This amount represents only a small fraction of the total missing anthropogenic carbon (Case Study 8–2).

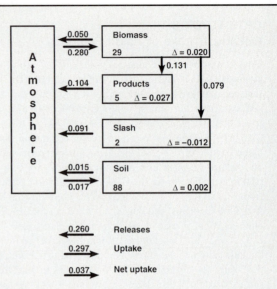

Figure 8–C3–1

Carbon in terrestrial U.S. carbon reservoirs (10^{15} g C), fluxes between carbon reservoirs (10^{15} g y^{-1}), and exchanges between these reservoirs and the atmosphere (Δ) during the 1980s (10^{15} g y^{-1}). From "The U.S. carbon budget: contributions from land use change" by R. A. Houghton, J. L. Hackler and K. T. Lawrence in SCIENCE, 1999, #285, pp. 574–578. Copyright © 1999 American Association for the Advancement of Science. Reprinted with permission.

Source: Houghton et al. (1999).

reaction with tropospheric hydroxyl (OH), (2) stratospheric oxidation, and (3) uptake by soil microbes. The various sources and sinks for methane are listed in Table 8–7. For the present-day atmosphere, anthropogenic sources of methane are approximately twice as important as natural sources.

In Chapter 6 we discussed the carbon isotopic system. The $^{14}C/C$ ratios (or ^{14}C activity) of plants should represent equilibrium with atmospheric CO_2. If all the methane in the present-day atmosphere was derived from modern organic matter, its $^{14}C/C$ ratio should reflect that of the current atmosphere. However, methane is depleted in ^{14}C relative to the atmosphere, which indicates that there are "old" sources of methane. These sources are coal mining, natural gas, and the petroleum industry, and methane trapped in permafrost at high northern latitudes. The methane trapped in the permafrost was produced by anaerobic decay of organic matter during warm interludes in the glacial cycle.

There is the potential, with increasing temperature, for significant positive feedbacks in the methane cycle. Positive feedbacks are those that reinforce the observed change—in this case, the temperature increase. These positive feedbacks include (1) an increase in the rate of anaerobic decay with increasing temperature, (2) the release of methane currently frozen in the permafrost at high northern latitudes, and (3) the release of methane stored

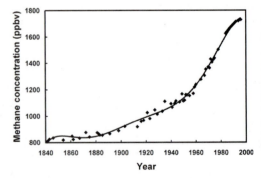

Figure 8–10

Variation in methane abundance from 1841 to 1996. The fitted curve is a sixth-order polynomial. Data from Etheridge et al. (1994) and IPCC (1996).

Table 8–7 *Sources and Sinks for Atmospheric Methane**

Source or sink	CH_4 $(10^{12}$ g C $y^{-1})$	% Total
Sources		
Natural		
Wetlands	86	22.5
Termites	15	3.9
Oceans	8	2.1
Lakes	4	1.0
Methane hydrates	4	1.0
Total natural	117	30.5
Anthropogenic		
Energy production/use	69	18.0
Enteric fermentation	63	16.4
Rice	45	11.8
Animal wastes	20	5.2
Landfills	29	7.6
Biomass burning	21	5.5
Domestic sewage	19	5.0
Total anthropogenic	266	69.5
Total for sources	383	
Sinks		
Atmospheric removal	353	88.2
Removal by soils	23	5.8
Atmosphere	24	6.0
Total for sinks	400	

*Data from Berner and Berner (1996), IPCC (1992).

as methane hydrates in seafloor sediments. Ocean-floor methane hydrates will be discussed in Chapter 10. Because of these positive feedbacks, methane is potentially a very important greenhouse gas.

The most significant sink for methane is removal by atmospheric processes, and the bulk of this removal (~90%) occurs in the troposphere. In the troposphere, methane is converted to carbon dioxide. The net reaction is

$$CH_{4\,(g)} + 2O_{2\,(g)} \rightarrow CO_{2\,(g)} + 2H_2O_{\,(g)}$$

The net reaction is misleading in that the oxidation of methane involves a number of steps. The first step is the reaction of methane with a hydroxyl free radical:

$$CH_4 + OH^{\cdot} \rightarrow CH_3^{\cdot} + H_2O_{\,(g)}$$

A *free radical* is *a molecule that contains an unpaired electron and is highly reactive*. The hydroxyl radical is the key molecule in the oxidation of gases released to the troposphere. A hydroxyl radical can be created through the following reactions:

$$O_3 + \text{UV-photon} \rightarrow O_2 + O^*$$

$$O^* + H_2O_{\,(g)} \rightarrow 2OH^{\cdot}$$

This is an example of a photochemical reaction because the initial step involves the splitting of a molecule by a high-energy photon. Photochemical reactions are key to understanding the chemistry of the troposphere and play a major role in the formation of secondary air pollutants. The absolute abundance of hydroxyl radicals in the troposphere is very small because they are highly reactive. Hence, a balance exists between their rate of formation and destruction. Concerns have been raised that increased emission of methane would significantly deplete the troposphere of hydroxyl radicals, which would not only diminish the rate at which methane is oxidized but would also affect the oxidation of other gaseous emissions. In the stratosphere, methane is broken down by reactions with hydroxyl radicals, chlorine or bromine atoms, or excited atomic oxygen. The reaction with atomic oxygen proceeds as follows:

$$O^* + CH_{4\,(g)} \rightarrow OH^{\cdot} + CH_3^{\cdot}$$

$$OH^{\cdot} + CH_{4\,(g)} \rightarrow H_2O_{\,(g)} + CH_3^{\cdot}$$

Note that H_2O, a significant absorber of infrared radiation, is produced during this reaction. Thus, the decomposition of methane leads to the addition of $H_2O_{\,(g)}$ to the stratosphere, an indirect greenhouse effect. It is estimated that about 25% of the total warming attributed to methane emissions is due to this secondary process of water-vapor formation.

Nitrous Oxide (N_2O)   Because of its long lifetime in the atmosphere and high RIRF, nitrous oxide is considered an important greenhouse gas. The major source of this gas is as a byproduct of biological processes: (1) denitrification under aerobic conditions and (2) nitrification under anaerobic conditions. N_2O also forms when fuels (coal and biomass) that contain nitrogen are burned, in catalytic converters and during the production of nitric acid. There are no sinks for N_2O in the troposphere. The gas diffuses upwards to the stratosphere, where it is photochemically decomposed, usually to N_2 and O^*.

Climate Change and the Geologic Record

Using first principles, climate modelers attempt to predict the impact of greenhouse gases on the earth's climate. The exceedingly complex models used in these calculations have a number of uncertainties, which lead to a great deal of variability in the predicted outcome of greenhouse gas additions. Planet earth has already conducted a number of climate experiments, and the results of these experiments are embedded in the geologic record. Hence, a very active field of research involves the interpretation of the climate data contained in the geologic record. In Case Study 8–1, a brief description was given of climate proxies that were used to determine the last 1000 years of earth's climate history and how these proxies were used to calibrate a greenhouse model. The types of records that are used to determine past climates include ice cores and continental and marine sediments. In Chapter 6 we discussed several methods that use stable isotopes to extract paleotemperature (or temperature variations) from ice and carbonate fossils. In this section we will look at several ways in which the geologic record has been used to investigate past climates and variations in greenhouse gases. There is a vast literature on this topic, and the following is in no way an exhaustive treatment.

Ice Cores   Ice cores have proven invaluable for reconstructing climatic conditions during the past several hundred thousand years. Long cores have been drilled in Greenland and Antarctica. The deepest of these cores is at the Russian station, Vostok, on the Antarctic plateau. Upon completion, this core may yield a record extending back to 500,000 years before the present. Seasonal climatic variations cause layering in the ice, and this layering can be used, much as are tree rings and varved lake sediments, to determine the age of deposition. Each layer represents a one-year period of snow deposition, and the snow is subsequently converted to ice. As long as these layers are undisturbed, the age of each layer can be determined by simply counting the number of layers deposited above the layer of interest. At high pressures, ice behaves plastically. Hence, there is a depth below which the stratigraphic record is no longer readable because the ice is plastically deformed. Oxygen and hydrogen isotopes, as described in Chapter 6, are used to determine the temperature when each layer was deposited. Gas bubbles trapped in the ice are used to determine the atmospheric concentration of the various gases at the time of deposition. Dust layers found in the ice are also of interest because they give information on the extent of the surface exposed to weathering and the nature of the material that was being weathered. The record for the last 240,000 years obtained from the Vostok ice core is shown in Figure 8–11. There is a good correlation between the atmospheric temperatures determined from stable isotope measurements and the concentrations of carbon dioxide and methane. Periods of higher temperature are characterized by increased concentrations of both gases. During this 240,000-year period, concentrations of methane remain low, at preindustrial levels and below, and carbon dioxide concentrations are also at preindustrial levels. During this time interval, the maximum temperature variation is on the order of 7°C. Of interest is the apparent rapidity with which the temperature variations can occur. Note the temperature spike at about 140,000 years and another spike at about 10,000 years, marking the end of the last period of glacial advance. The cause of these rapid temperature variations is not yet understood, but such observations suggest the possibility that the climate can very rapidly change from cool to warm conditions (see Case Study 8–4, p. 264).

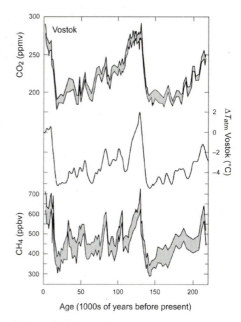

Figure 8–11
Variations in temperature, CO_2, and CH_4 concentrations in Antarctica during the past 240,000 years. From Lorius et al. (1993).

Sediment Record Extensive use has been made of stable isotopes to infer paleotemperatures and other climate parameters. In Chapter 6 the use of oxygen and carbon isotopes in paleotemperature determinations, and the limitations of these methods, was discussed. A record of ocean surface temperatures, based on oxygen isotope measurements from a sediment core from the western Pacific Ocean, is shown in Figure 8–12. This record shows that there was a decrease in seawater surface temperatures from 49 million years ago to 35 million years ago. Glaciers then began to develop in Antarctica. A subsequent period of moderate warming was followed by another period of cooling and the formation of the Antarctic ice sheet. Starting about 2 million years ago, ice sheets appeared in the Northern hemisphere. A more detailed examination of the present glacial cycle would reveal that there have been a number of glacial advances and retreats, with concomitant changes in temperature. Note that present-day Pacific Ocean surface temperatures are still significantly lower than surface temperatures at the beginning of the Eocene.

Oxygen and carbon isotopes have been used for many years to determine paleotemperatures. More recently, other isotopic systems have been developed that yield other types of information. The following case studies illustrate several ways in which climate data is extracted from the sediment record.

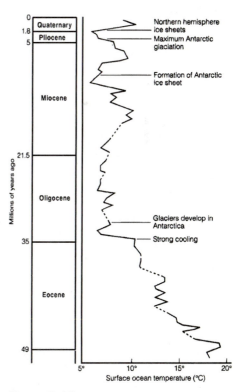

Figure 8–12
Surface temperature of the Pacific Ocean based on oxygen isotope ratios. From THE BLUE PLANET, 2nd Edition by B. J. Skinner, S. C. Porter and D. B. Botkin. Copyright © 1999. This material is used by permission of John Wiley & Sons, Inc.

CASE STUDY 8–4
Variations in Atmospheric N₂O Concentration During Abrupt Climatic Changes

Flückiger et al. (1999) report ice core data for atmospheric N_2O variations for the past 1000 years and for two periods of rapid climate change, from 16,500 to 10,500 years before present and from 37,000 to 32,500 years before present. Samples were taken from cores drilled at Summit (central Greenland) as part of the EUROCORE project and the Greenland Ice Core Project (GRIP). The air trapped in the ice was extracted using the melt-refreezing method and the N_2O contents were measured by gas chromatography. The results for the past 1000 years are shown in Figure 8–C4–1. The data indicate that the preindustrial concentrations of N_2O were relatively stable, with an average value of 270 ± 5 ppbv between 1400 and 1750 A.D. Starting in the late 1700s, there was an increase in atmospheric N_2O, and the rate of increase becomes substantially higher in the 1900s. The ice core N_2O data for the two periods of rapid climate change are shown in Figure 8–C4–2 (16,500 to 10,500 years before present) and Figure 8–C4–3 (37,000 to 32,500 years before present). Also shown on the figures are the variations in methane concentration and $\delta^{18}O$, a measure of temperature variation. During both events, the N_2O and CH_4 concentrations vary directly with changes in $\delta^{18}O$. For the older event, N_2O data obtained from the Camp Byrd, Antarctica, ice core show the same pattern of N_2O variation as the GRIP core. Hence, the authors concluded that the variations observed in the Greenland and Antarctic cores were global events. During both of the natural climate variations there is a significant and rapid change in

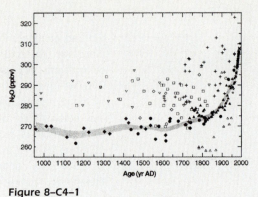
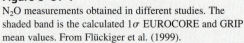

Figure 8–C4–1
N_2O measurements obtained in different studies. The shaded band is the calculated 1σ EUROCORE and GRIP mean values. From Flückiger et al. (1999).

the concentration of N_2O. The two main candidates for the changing sources of N_2O are the oceans and terrestrial soils. Both of these sources would be influenced by climate change. The relative importance of these sources during climate change is still unknown.

Source: Flückiger, et al. (1999). Figures from "Variations in atmospheric N_2O concentration during abrupt climatic changes" by J. Flückiger, A. Dallenbach, T. Blunier, B. Stauffer, T. F. Stocker, D. Raynaud and J. M. Barnola in SCIENCE, 1999, #285, pp. 227–230. Copyright © 1999 American Association for the Advancement of Science. Reprinted with permission.

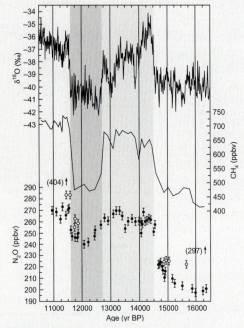

Figure 8–C4–2
GRIP N_2O, CH_4, and $\delta^{18}O$ for the last glacial-interglacial transition. The dark shading represents the Younger Dryas (from 12,700 to 11,600 years before present). During the Younger Dryas there is a significant decrease in N_2O and CH_4 concentrations corresponding to a sharp temperature drop. From Flückiger et al. (1999).

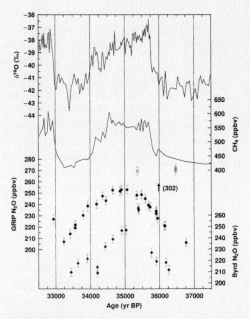

Figure 8–C4–3
GRIP N_2O, CH_4, and $\delta^{18}O$ variations for the Dansgaard–Oeschger event 8 (36,500 to 33,500 years before present), a time of rapid climate change. The solid diamonds show N_2O measurements for the Byrd core from Antarctica. From Flückiger et al. (1999).

CASE STUDY 8–5
Mechanism of Climate Warming at the End of the Paleocene

An abrupt period of warming marked the end of the Paleocene epoch (55.5 Ma). This change is marked by a strong negative (~ -2.5‰) excursion in the $\delta^{13}C$ values of carbonate and organic matter, indicating a large input of isotopically negative carbon to the ocean and atmosphere. Dickens et al. (1997) suggested that this warming, and the source of the isotopically negative carbon, was due to the release of a significant amount of methane from methane hydrates buried on the continental shelves. Bains et al. (1999) obtained oxygen and carbon isotopic values for two deep-sea sediment cores, ODPB 1051B from the Atlantic Ocean and ODP 690B from the Pacific Ocean, which sampled this time interval (Figure 8–C5–1). Both cores show the same pattern of carbon isotope variation. In the case of oxygen isotopes, the temperature excursion is clearly seen in the Pacific core (ODP 690B), but the pattern is not as clearly developed in the core from the Atlantic Ocean (ODP 1051B). The authors conclude that there were three stages of methane release, each marked by sharp drops in $\delta^{13}C$ values. Between these pulses the ocean attempted to return to equilib-

rium, as represented by the plateaus in the carbon isotopic record. In the isotopic records shown in Figure 8–C5–1, the periods of methane release are indicated by the letters c, e, and g, and the intervening periods of cessation of methane release are denoted by the letters d and f. The authors suggest that the original release of methane may have been triggered by submarine seismicity, volcanism, or gravitational slumping that released methane hydrates stored on the continental shelves. Subsequent periods of release may be due to the heating of sediment pore waters by the warmer waters produced by the first methane release. The authors estimate that ~ 600 Gt of carbon were released during the first event (in less than 1000 years), ~ 500 Gt during the second event, and ~ 300 Gt during the last event, giving a total release of ~ 1500 Gt of carbon. In Figure 8–C5–1, the portion of the figure denoted by the letter i indicates the return of the oceanic system to equilibrium conditions. Of note is the rapidity with which these various excursions occurred, suggesting that climate change can occur very quickly. This study also demonstrates how detailed isotopic analysis can be used to unravel past climate changes.

Source: Bains et al. (1999).

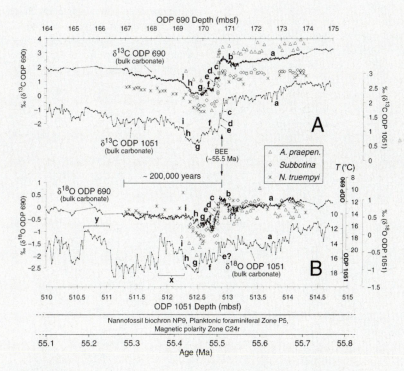

Figure 8–C5–1

Carbon isotope (A) and oxygen isotope (B) data for two deep-sea sediment cores that cross the boundary representing an abrupt episode of global warming. *Acarinina praepentacamerata* and *Subbotina* are planktonic species of foraminifera and *Nuttallides turempyi* is a benthic foraminifera. The similarity in the carbon isotope values for all three species indicates that carbon was well mixed in the oceans. Small letters denote various events during the tempertaure excursion (see text). From "Mechanisms of climate warming at the end of the Paleocene" by S. Bains, R. M. Corfield and R. D. Norris in SCIENCE, 1999, #285, pp. 724–727. Copyright © 1999 American Association for the Advancement of Science. Reprinted with permission.

Case Study 8–5 is an example of the climate data that can be extracted by sampling small intervals across a climate-change boundary. In this case, carbon and oxygen isotopic ratios are determined for two deep-sea sediment cores that span a boundary of rapid climate warming. The small sampling interval made it possible for the authors to present a detailed analysis of what happened during this period of climate warming.

Case Study 8–6 takes the level of sample analysis to an even finer scale. In this case study, bulk oxygen isotope analyses of fish otoliths (ear stones) show little temperature change across a mass extinction boundary. However, microsampling of individual otoliths demonstrates that there is a significant change in the yearly temperature variation across the boundary.

CASE STUDY 8–6
Cooler Winters as a Possible Cause of Mass Extinctions at the Eocene/Oligocene Boundary

One of the largest Cenozoic extinctions of marine invertebrates occurs at the Eocene/Oligocene boundary, ~33.7 Ma. One possible cause for this extinction was a change in mean temperature, from a warm Eocene climate to a cooler Oligocene climate. Ivany et al. (2000) investigated this transition using otoliths (ear stones) from fish in the Congridae and Ophidiidae families collected from sediments deposited on an open marine shelf along the Gulf Coast of the United States. Modern members of these families live in relatively deep water and are non-migratory. Hence, these species should be useful in investigating temperature differences in marine waters over the continental shelves. Bulk analysis of otoliths showed little variation in temperature across the Eocene/Oligocene boundary (Figure 8–C6–1). Hence, the authors concluded that a significant change in mean temperature was not responsible for the mass extinctions. The authors next investigated seasonal varia-

tions by microsampling the otoliths. Otoliths grow throughout the two- to three-year lifespan of the organisms. A set of serial samples from individual otoliths, therefore, will yield the history of seasonal temperature variation during the lifetime of the organism. This set of samples revealed significant differences in the seasonal temperature variations for the Late Eocene versus the Oligocene (Figure 8–C6–2). Statistical tests showed that summer temperatures remained reasonably constant across the boundary, but winter temperatures were about 4°C colder in the Oligocene than the Eocene. The calculated temperatures depend on the oxygen isotope values for ocean water, and there is some uncertainty in this number. However, changes in the sea-water oxygen isotope value will not change the observed relative temperature variations. The authors suggest that the colder winter temperatures found for the Oligocene may be responsible for the observed mass extinctions.

Source: Ivany et al. (2000).

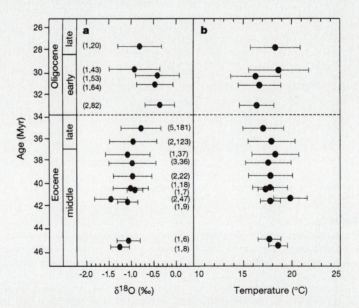

Figure 8–C6–1
Mean $\delta^{18}O$ values and calculated temperatures for otoliths from the middle Eocene to the Oligocene. Error bars are 1 standard deviation around the mean. From Ivany et al. (2000).

CASE STUDY 8–6 (continued)

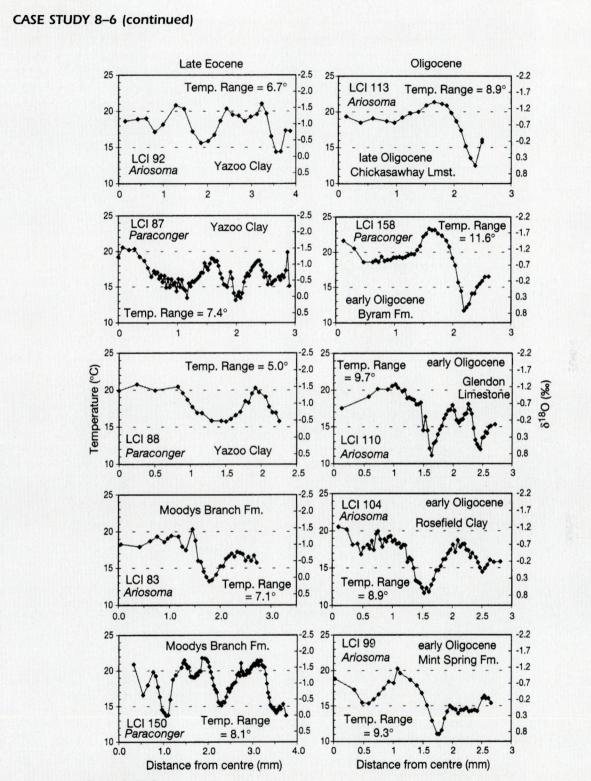

Figure 8–C6–2
Individual temperature records from microsampled congrid otoliths. Larger seasonal temperature variations are found for the early Oligocene compared to the late Eocene. From Ivany et al. (2000).

In Case Study 8–7, the authors use the boron isotopic composition of foraminifera to infer the pH of ancient seawater, and from this pH they calculate the P_{CO_2} of the ancient atmosphere. Boron exists as two species in seawater, $B(OH)_3$ and $B(OH)_4^-$, and the relative abundance of these species is largely controlled by pH. There is a strong isotopic fractionation between the two species, which can be used to determine the relative abundance of the two species and, hence, the pH. Assuming that the seawater was in equilibrium with atmospheric CO_2, it is then possible to calculate the P_{CO_2} of the atmosphere. The appropriate equations for these calculations are given in Case Study 8–7.

OZONE

Environmental scientists often refer to ozone as *good ozone* and *bad ozone*. Good ozone is found in the stratosphere, where it plays an important role in absorbing ultraviolet radiation. Bad ozone is found in the troposphere and is a significant air pollutant.

Stratospheric Ozone

Stratospheric ozone forms as a result of interactions between high-energy photons and oxygen atoms. The ozone-forming reactions are

$$O_2 + h\nu \text{ (UV photon)} \rightarrow O^* + O^* \qquad (\Delta H_R^0 = +498.4 \text{ kJ mol}^{-1})$$

and then

$$O^* + O_2 + M \rightarrow O_3 + M \qquad (\Delta H_R^0 = -106.5 \text{ kJ mol}^{-1})$$

where M is a catalytic particle (usually oxygen or nitrogen) and O^* is oxygen in an excited state. The first reaction is endothermic (energy is taken up by the molecule), and the energy required for the reaction to occur is provided by the UV photon. The second reaction is exothermic, and the energy released during the process is transferred to the catalytic particle. The net reaction for ozone formation is

$$3O_2 + h\nu \rightarrow 2O_3$$

Ozone destruction occurs as follows

$$O_3 + h\nu \text{ (UV photon)} \rightarrow O_2 + O^* \qquad (\Delta H_R^0 = +386.5 \text{ kJ mol}^{-1})$$

and

$$O^* + O_3 \rightarrow O_2 + O_2 \qquad (\Delta H_R^0 = -391.1 \text{ kJ mol}^{-1})$$

The net reaction for ozone destruction is

$$2O_3 + h\nu \rightarrow 3O_2$$

The minimum energy required for a photochemical reaction, i.e., the photon energy, can be calculated from Einstein's equation (1–11) for the photoelectric effect. A sample calculation is given in Example 8–2 for the photodissociation of oxygen.

EXAMPLE 8–2 Calculate the minimum energy a photon must have in order to split an oxygen molecule. For this reaction, $\Delta H_R^0 = +498.4 \text{ kJ mol}^{-1}$. This is the energy required to split 1 mole of oxygen molecules. The number of oxygen molecules in a mole is represented by Avogadro's number, 6.022×10^{23} molecules mol^{-1}.

The energy required per molecule is

$$E = \frac{498.4 \text{ kJ mol}^{-1}}{6.022 \times 10^{23} \text{ molecules mol}^{-1}} = 8.276 \times 10^{-19} \text{ J molecule}^{-1}$$

CASE STUDY 8–7
Middle Eocene Seawater pH and Atmospheric Carbon Dioxide Concentrations

The dissociation of boric acid can be written

$$B(OH)_{3\,(aq)} + H_2O \rightarrow B(OH)_4^- + H^+$$

At 25°C and species concentrations in molality, for seawater the stoichiometric ionization constant is $pK^* = 8.830$. (Hershey et al., 1986). Because this is a stoichiometric equation, we use measured concentrations; i.e., the measured concentrations are not converted to activity. Recalling our discussion of acid–base equilibria in Chapter 3, at pH = 8.83 (and 25°C) the two boron species occur in equal abundance. At lower pH values, $B(OH)_{3\,(aq)}$ is the dominant species, and at higher pH values $B(OH)_4^-$ is the dominant species. The isotope exchange reaction between these species can be written

$$^{10}B(OH)_3 + {}^{11}B(OH)_4^- \rightleftharpoons {}^{11}B(OH)_3 + {}^{10}B(OH)_4^-$$

The equilibrium equation for the reaction is

$$K = \alpha = \frac{[^{11}B(OH)_3]/[^{10}B(OH)_3]}{[^{11}B(OH)_4^-]/[^{10}B(OH)_4^-]}$$

and at 25°C, $K = 1.0194$ (Kakihana et al., 1977). As we did in Chapter 6, writing this equation in delta notation gives

$$\alpha_{B(OH)_4}^{B(OH)_3} = 1.0194 = \frac{\delta_{B(OH)_3} + 1000}{\delta_{B(OH)_4} + 1000}$$

Hemming and Hanson (1992) used these relationships to investigate the boron isotopic chemistry of marine carbonates. The boric acid dissociation and isotopic equilibrium equations can be solved simultaneously for various relative proportions of the two boric acid species. These proportions vary as a function of pH, and hence the isotopic composition of boron in each species also varies as a function of pH. The solutions to these equations, using seawater $\delta^{11}B = +39.5‰$, are shown in Figure 8–C7–1. Hemming and Hanson (1992) analyzed a number of modern marine carbonates and came to the conclusion that boron was incorporated into the carbonates by preferential absorption of $B(OH)_4^-$. Hence, isotopic analysis of carbonates will yield the isotopic composition of the $B(OH)_4^-$ species in seawater at the time the carbonate was formed. The relative proportions of the two boric acid species determine the boron isotopic composition, and the relative abundances of these species are controlled by seawater pH, so boron isotopic measurements of carbonates can yield past seawater pHs.

Pearson and Palmer (1999) used boron isotopic compositions of planktonic foraminifera formed during the Middle Eocene (~45 Ma) to determine oceanic pH. Once the pH of the surface ocean has been determined, it is possible to calculate the carbon dioxide concentration for the atmosphere in equilibrium with the surface ocean. Uncertainties in the calculation are the $\delta^{11}B$ value for the Middle Eocene ocean (this difference is expected to be small) and the total carbonate content of the ocean. The results of these calculations, assuming three different seawater $\delta^{11}B$ values, are shown in Figure 8–C7–2. The authors' best pH estimate gives an atmospheric CO_2 concentra-

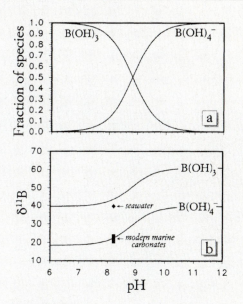

Figure 8–C7–1

(a) Calculated distribution of aqueous boron species versus pH. (b) $\delta^{11}B$ values for the two aqueous boron species versus pH. Note that modern carbonates plot on the $B(OH)_4^-$ curve at pH = 8.2 (the modern ocean), indicating that this species is selectively incorporated into the carbonates. From Hemming and Hanson (1992).

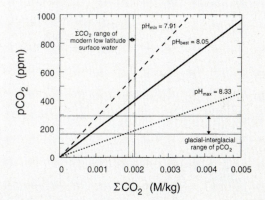

Figure 8–C7–2

Surface seawater pCO_2 versus ΣCO_2 based on various estimates of sea-surface pH. From "Middle Eocene seawater pH and atmospheric carbon dioxide concentrations" by P. N. Pearson and M. R. Palmer in SCIENCE, 1999, #284, pp. 1824–1826. Copyright © 1999 American Association for the Advancement of Science. Reprinted with permission.

tion of 370 to 400 ppm, only slightly higher than today's value. The authors conclude that the mid-Eocene CO_2 content of the earth's atmosphere was very similar to today's atmosphere and that the decline in temperature from the mid-Eocene to the present is probably not due to changes in the atmospheric concentration of CO_2, but rather to changes in oceanic circulation.

Source: Hemming and Hanson (1992), Pearson and Palmer (1999).

Solving equation 1–11 for wavelength gives

$$\lambda = \frac{hc}{E} = \frac{(6.626 \times 10^{-34} \text{ J s})(3.0 \times 10^8 \text{ m s}^{-1})}{(8.276 \times 10^{-19} \text{ J molecule}^{-1})}$$

$$= 2.402 \times 10^{-7} \text{ m} = 240.2 \text{ nm}$$

In order for a photon to split an oxygen molecule it must have a wavelength of $\sim$240 nm or less. ∎

The calculation done in Example 8–2 gives the minimum energy, or wavelength, needed to split an oxygen molecule. Any photon of higher energy (or shorter wavelength) will split an oxygen molecule. The most efficient wavelength is a function of the absorption characteristics for the particular molecule. The absorption spectra for diatomic oxygen and ozone, in the 100 to 300 nm wavelength region, are shown in Figure 8–13. Note that the maximum absorption for diatomic oxygen occurs at $\sim$150 nm, and photons of this wavelength would be most efficient in splitting the oxygen molecule. For ozone, the optimum wavelength would be $\sim$250 nm, assuming that photons of this wavelength had sufficient energy (see problem set).

Figure 8–13

Absorption cross sections for oxygen and ozone in the 100 to 300 nm wavelengths. Also shown is the solar flux density and the wavelengths of biologically harmful radiation (UV-B and UV-C). From vanLoon and Duffy (2000).

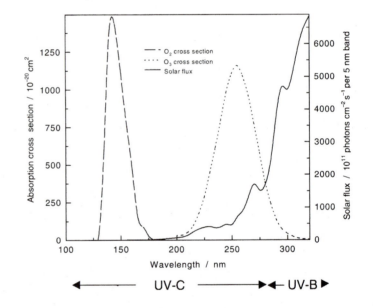

Also shown in Figure 8–13 are the solar flux density and the regions of biologically harmful radiation (UV-B and UV-C). UV-B is harmful to plant and animal life, and UV-C damages plant life. UV-B radiation has also been implicated as a cause of melanoma. A reduction in the amount of stratospheric ozone would lead to a significant increase in the UV-B radiation reaching the earth's surface.

From these reactions, it is apparent that the formation and destruction of ozone is a continuous process and there is a balance between the rate of ozone production and destruction. The two factors that control the rates of production and destruction are the availability of high-energy (UV) photons and the density of the atmosphere (number of gas molecules per unit volume). As we move toward the surface of the earth, more of the incoming photons are absorbed by atmospheric gases and the photon flux decreases. At the same time, there is an increase in the density of the atmosphere. Note that atmospheric pressure, and therefore density, varies exponentially as a function of height. Hence, as a function of altitude, one of the factors favoring ozone production is decreasing while the other is increasing (Figure 8–14). The optimum value for these two variables occurs at approximately 25 km (at the equator), and the density of stratospheric ozone decreases exponentially both above and below this altitude.

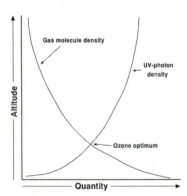

Figure 8–14

Altitude versus variations in photon and molecular densities. The optimum altitude for ozone formation occurs where these curves cross.

The concentration of ozone varies both with latitude and season. Ozone concentrations are measured in a number of ways, but one commonly used measurement is the Dobson unit (DU). One DU is equivalent to a 0.001 mm layer of pure ozone at 1 atm pressure. The measurements are made from the ground up by comparing the intensity of two wavelengths of UV radiation—one absorbed by ozone, the other not absorbed by ozone. The ozone concentration varies from about 250 DU at the equator to 350 DU at temperate latitudes to 450 DU in the polar regions. The reason for this variation is the transport of ozone, by winds in the stratosphere, from equatorial latitudes to polar latitudes. Seasonally, ozone tends to decline during the winter months and increase during the summer months (Figure 8–15).

Figure 8–15

Seasonal variation of ozone concentrations (in Dobson units) at Halley Bay, Antarctica, for two different time periods. From Solomon (1990).

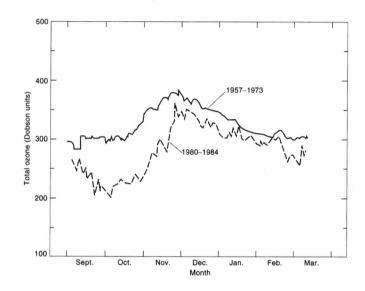

Observations made over a number of years in Antarctica indicated that there was a significant decrease in ozone (Figure 8–15) with time. This led to the discovery of the ozone hole over Antarctica and an intense investigation into the reasons for the ozone depletion. During the Antarctic winter, a strong polar vortex develops and the ozone hole deepens. With the onset of spring, the polar vortex breaks down, spilling ozone-depleted air across the Southern Hemisphere. Concerns were raised over the impact increased UV radiation (due to the reduction of ozone) might have on biological systems, particularly oceanic systems at high southern latitudes, which are areas of major oceanic upwelling and biological productivity (see Chapter 10). Subsequent investigations revealed that anthropogenic additions of N_2O, chlorofluorocarbons, and bromine compounds were responsible for the ozone decline. N_2O and bromine compounds have both anthropogenic and natural sources, and the ozone layer would presumably be in equilibrium with the natural sources. Thus, it is the anthropogenic additions that are of interest.

Ozone-Destroying Reactions   Many ozone-destroying reactions involve atomic oxygen and a species that acts as a catalyst. Reactions of this type can be represented, using X as the catalytic species, as follows:

$$X + O_3 \rightarrow XO + O_2$$
$$\underline{XO + O \rightarrow X + O_2}$$
$$O + O_3 \rightarrow 2O_2$$

Two of the most important reactions of this type involve NO and Cl. NO is formed from N_2O as follows:

$$N_2O + O^* \rightarrow 2NO$$

The subsequent reactions are

$$NO + O_3 \rightarrow NO_2 + O_2$$
$$\underline{NO_2 + O \rightarrow NO + O_2}$$
$$O + O_3 \rightarrow 2O_2$$

Chlorine is released by the photochemical decomposition of methyl chloride or chlorofluorocarbons. A typical reaction is

$$CFCl_3 + h\nu \rightarrow CFCl_2^{\cdot} + Cl^{\cdot}$$

The chlorine radical then reacts with ozone as follows:

$$Cl^{\cdot} + O_3 \rightarrow ClO^{\cdot} + O_2$$
$$\underline{ClO^{\cdot} + O \rightarrow Cl^{\cdot} + O_2}$$
$$O + O_3 \rightarrow 2O_2$$

In these cycles, NO and $Cl^{\cdot}$ act as catalysts; they participate in the reactions but are not consumed by the reactions.

Kinetic Considerations   Calculation of reaction rates for the various ozone-destroying cycles are done to determine the relative importance of the different cycles. For any particular cycle, the slowest step is the one that determines the overall rate of the cycle. The reactions are second order (see Chapter 2). A sample calculation is done in Example 8–3, using the data from Table 8–8.

Table 8–8   Kinetic Data for Various Reactants in the Catalytic Destruction of Ozone at 235 K*

		$X + O_3$		
X	Concentration (molecules cm^{-3})	A (cm^3 molecules^{-1} s^{-1})	E_a (kJ mol^{-1})	k_{235} (cm^3 molecules^{-1} s^{-1})
O	1.0×10^9			
H	2.0×10^{15}	1.4×10^{-10}	3.9	1.9×10^{-11}
OH	1.0×10^6	1.6×10^{-12}	7.8	3.0×10^{-14}
NO	5.0×10^8	1.8×10^{-12}	11.4	5.3×10^{-15}
Cl	Very small	2.8×10^{-12}	21	6.0×10^{-17}

		$XO + O$		
XO	Concentration (molecules cm^{-3})	A (cm^3 molecules^{-1} s^{-1})	E_a (kJ mol^{-1})	k_{235} (cm^3 molecules^{-1} s^{-1})
O_2	5.0×10^{16}	8.0×10^{-12}	17.1	1.3×10^{-15}
HO	1.0×10^6	2.3×10^{-11}	0	2.3×10^{-11}
HO_2	2.5×10^7	2.2×10^{-11}	-0.1	2.3×10^{-11}
NO_2	5.0×10^{19}	9.3×10^{-12}	0	9.3×10^{-12}
ClO	2.0×10^7	4.7×10^{-11}	0.4	3.8×10^{-11}

*The concentrations of the species are for an altitude of 30 km with the exception of ClO, which is for an altitude of 35 km. The concentration of ozone at 30 km is 2.0×10^{12} molecules cm^{-3}. From vanLoon and Duffy (2000).

EXAMPLE 8–3   Calculate the rate for the reaction, $Cl^{\cdot} + O_3 \rightarrow ClO^{\cdot} + O_2$ at 30 km and 235 K. Assume that the concentration of $Cl^{\cdot} = 5.0 \times 10^{11}$ molecules cm^{-3}. For this reaction,

$$Rate = k[Cl][O_3]$$

First we need to calculate the rate constant (k) using the Arrhenius equation.

$$k = Ae^{-E_a/RT} = (2.8 \times 10^{-12})e^{-(21 \text{ kJ mol}^{-1})/(8.314 \times 10^{-3} \text{ kJ mol}^{-1} \text{ K}^{-1})(235 \text{ K})}$$
$$= 6.0 \times 10^{-17} \text{ cm}^3 \text{ molecules}^{-1} \text{ s}^{-1}$$

Next, we calculate the rate for the reaction.

$$\text{Rate} = (6.0 \times 10^{-17} \text{ cm}^3 \text{ molecules}^{-1} \text{ s}^{-1})[5.0 \times 10^{11} \text{ molecules cm}^{-3}]$$
$$\times [2.0 \times 10^{12} \text{ molecules cm}^{-3}]$$
$$= 6.0 \times 10^7 \text{ molecules cm}^{-3} \text{ s}^{-1} \qquad \blacksquare$$

The Antarctic Ozone Hole The special climatic conditions that are present in Antarctica are responsible for the rapidity of the ozone depletion that occurs in the spring. During the winter months, a strong vortex develops over the South Pole and gases are trapped in the vortex. Also formed are clouds that contain nitric acid and water ices. These particles take part in heterogeneous reactions that release chlorine molecules (Cl_2) and hydrogen hypochlorite (HOCl). When the sun reappears in the polar region, photochemical reactions occur that form $Cl^\cdot$ and $OH^\cdot$ radicals from the chlorine-containing molecules. Interactions between these radicals and ozone leads to a rapid decrease in the ozone content of the polar stratosphere. As illustrated here, not all these cycles involve atomic oxygen.

$$
\begin{aligned}
2Cl^\cdot + 2O_3 &\rightarrow 2ClO^\cdot + 2O_2 \\
ClO^\cdot + ClO^\cdot &\rightarrow ClOOCl \\
ClOOCl + h\nu &\rightarrow ClOO^\cdot + Cl^\cdot \\
ClOO^\cdot &\rightarrow Cl^\cdot + O_2 \\
\hline
2O_3 + h\nu &\rightarrow 3O_2
\end{aligned}
$$

The subsequent breakup of the polar vortex spills this ozone-depleted air across the Southern Hemisphere. Of concern is the increase in surface UV radiation due to the reduction in stratospheric ozone. Recent international meetings (e.g., the Montreal Protocol) have addressed this problem of ozone depletion, and the use of ozone-destroying chemicals is being phased out.

Tropospheric Ozone

Tropospheric ozone is the so-called bad ozone because it is an important constituent of photochemical smog and elevated levels of ozone have a negative impact on plant growth and exert stress on mammalian respiratory systems. A variety of volatile organic compounds (VOCs), NO_x, and CO are produced by anthropogenic activities, e.g., combustion processes in automobiles and biomass burning. VOCs are also released to the atmosphere by natural biological processes. There are a number of reaction pathways involved in the production of the components of photochemical smog. Many of these involve the OH radical and NO, and ozone is produced by these reactions.

In the internal combustion engine, nitric oxide (NO) is produced from atmospheric nitrogen and oxygen. The emitted NO molecule is then converted to NO_2 by one of the following reactions:

$$(1) \ 2NO + O_2 \rightarrow 2NO_2$$
$$(2) \ NO + O_3 \rightarrow NO_2 + O_2$$
$$(3) \ RO_2^\cdot + NO \rightarrow RO^\cdot + NO_2$$

In reaction (3), the radicals are referred to as peroxyl radicals and R represents the root organic molecule. The NO_2 then undergoes photochemical decomposition.

$$NO_2 + h\nu \ (\lambda < 400 \text{ nm}) \rightarrow NO + O$$

The atomic oxygen now reacts with atmospheric oxygen molecules to produce ozone.

$$O + O_2 + M \rightarrow O_3 + M$$

where M is the catalytic particle (usually N_2 or O_2) that absorbs the excess energy released during the formation of the ozone molecule. As was discussed previously in the section on stratospheric ozone, the ozone molecule is photochemically decomposed according to the following reaction:

$$O_3 + h\nu \rightarrow O_2 + O*$$

The excited atomic oxygen now reacts with water vapor to form two hydroxyl radicals.

$$O* + H_2O \rightarrow 2OH^{\cdot}$$

Hydroxyl radicals play an important role in the oxidation of VOCs. As an example, consider the oxidation of a generic aliphatic hydrocarbon (from vanLoon and Duffy, 2000). The reaction starts with the hydroxyl radical, and a number of other radicals, including peroxyl radicals, are produced and consumed through the course of the reaction pathway.

$$
\begin{aligned}
OH^{\cdot} + RCH_3 &\rightarrow RCH_2 + H_2O \\
RCH_2 + O_2 + M &\rightarrow RCH_2OO^{\cdot} + M \\
RCH_2OO^{\cdot} + NO &\rightarrow RCH_2O^{\cdot} + NO_2 \\
RCH_2O^{\cdot} + O_2 &\rightarrow RCHO + HOO^{\cdot} \\
\underline{HOO^{\cdot} + NO} &\underline{\rightarrow NO_2 + OH^{\cdot}} \\
RCH_3 + 2O_2 + 2NO &\rightarrow RCHO + 2NO_2 + H_2O
\end{aligned}
$$

These illustrative reactions, plus many others, have been written to describe the chemical reactions that take place during the development of a photochemical smog. An idealized plot of the species produced during the development of a photochemical smog, as a function of time, is shown in Figure 8–16. These reactions are ultimately driven by photochemistry and hence have a diurnal cycle. Because of the short half-life of many of the compounds, the photochemical products dissipate during the nighttime hours. In the early morning hours, the concentration of NO, due to automobile combustion, increases. As we

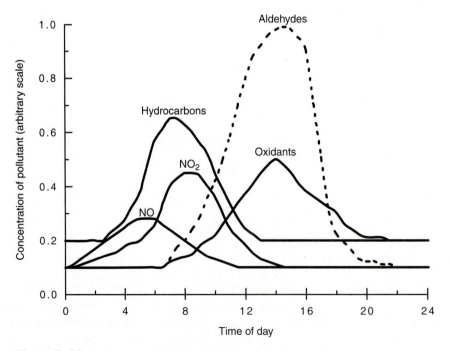

Figure 8–16

Variation in abundances of various species, on a 24-hour cycle, produced during a photochemical smog event. From vanLoon and Duffy (2000).

might expect, because all of the reactions consume NO, once photochemical reactions become important (i.e., sunlight becomes significant), the concentration of NO will decrease. NO_2 is a product and its concentration will increase. Ozone and hydroxyl radicals are essential species in the oxidation of hydrocarbons. As these oxidants increase in abundance, there is a decrease in the abundance of hydrocarbons. Ultimately, aldehydes, which are important products of several of the oxidation pathways, become significant.

RADON

There are a number of so-called *indoor air pollutants*. These pollutants are due to emissions within buildings, such as formaldehyde from insulation. In recent years, these pollutants have become more significant because increased insulation of buildings to reduce heating and cooling costs have diminished ventilation of building air with external air. Thus, the concentrations of these pollutants in the indoor air can build up. One of these pollutants, radon, has a geological source. For a steady-state situation, the net rate of production of an indoor air pollutant is

$$R_i = k_e C_i - k_e C_o \tag{8-11}$$

where R_i is the net rate of production (concentration per unit time), k_e is the first-order rate constant (time^{-1}) for the exchange of air between the indoor and outdoor environment, C_i is the indoor concentration, and C_o is the outdoor concentration. Solving for the steady-state indoor air concentration of the pollutant gives

$$C_i = C_o + R_i/k_e \tag{8-12}$$

In Chapter 6 we discussed the various uranium and thorium decay chains. Radon is produced in all three decay chains, but only in the ^{238}U chain (Table 6–1) does the radon isotope (^{222}Rn) have a significant half-life (3.8235 days). The health hazard associated with radon is believed to be due to the two short-lived polonium progeny, ^{218}Po and ^{214}Po. These progeny are solids and attach to particles in the air. The particles are then inhaled and lodge in the lung. The subsequent alpha decay of the polonium progeny can damage the lining of the lung. Homes constructed on bedrock containing high abundances of uranium can have elevated radon concentrations, particularly in their basements, and elevated radon activity often shows a relationship to the subsurface geology. If radon activities are found above the EPA recommended limit of 4 pCi L^{-1} of air, remedial action is suggested. To conform to SI units, the following conversion is required: 1 pCi = 0.037 Bq and 1 pCi L^{-1} = 0.037 Bq L^{-1} = 37 Bq m^{-3}.

Radon is a radioactive gas, so we need to modify equation 8–11 as follows:

$$R_i = k_e A_i + k_d A_i - k_e A_o \tag{8-13}$$

where A_i is the indoor radon activity, k_d is the radioactive decay constant, A_o is the outdoor radon activity, and the other terms are as defined earlier. Solving for the indoor radon activity gives

$$A_i = (R_i + k_e A_o)/(k_d + k_e) \tag{8-14}$$

Recall from Chapter 6 that

$$k_d = \ln 2/t_{1/2} = \ln 2/3.8235 \text{ d} = 0.1813 \text{ d}^{-1} = 7.554 \times 10^{-3} \text{ h}^{-1}$$

which gives the decay constant for ^{222}Rn.

EXAMPLE 8–4 Radon released from soils surrounding a basement enters the basement at a rate of 0.01 Bq L^{-1} h^{-1}. The outdoor air has a radon activity of 4×10^{-3} Bq L^{-1} (the global average). Assuming a steady state and an air exchange rate constant of $10\ h^{-1}$, calculate the steady-state activity of radon in the basement. Is remedial action required?

$$A_i = (R_i + k_e A_o)/(k_d + k_e)$$
$$= [(0.01\ Bq\ L^{-1}\ h^{-1}) + (10\ h^{-1})(4 \times 10^{-3}\ Bq\ L^{-1})]/[7.554 \times 10^{-3}\ h^{-1} + 10\ h^{-1}]$$
$$= 5.0 \times 10^{-3}\ Bq\ L^{-1}$$

The calculated radon activity is well below the EPA recommended limit for remedial action. ∎

Indoor radon concentrations are determined by a number of factors. Radon enters structures by diffusion from soils and/or bedrock and from groundwater used for water supply. Thus, both the type of bedrock and the radon content of groundwaters are important in determining indoor radon concentrations. Radon transport in soil occurs by convective flow and diffusion. In soils of low permeability, diffusion is the dominant radon transport process. Thus, the permeability of the soil is an important factor and needs to be considered when evaluating the radon potential of an area. Soil radon concentrations also vary in response to meteorological and climatic conditions. The following meteorological factors seem to be most important (Schumann et al., 1992):

1. Precipitation, which determines soil moisture content. Significant soil moisture impedes the diffusion of radon, which is important given the short half-life for radon. For example, during its mean lifetime a radon atom can travel 1 to 2 meters by diffusion through dry soil but only several centimeters in saturated soil.
2. Temperature, which affects the partitioning of radon between the gas and water phases in soil pore spaces. Decreasing temperature favors the water phase.
3. Diurnal freezing and thawing cycles. The frozen soil acts as a cap trapping the radon gas below the cap. This leads to an increase in soil radon concentration.
4. Changes in barometric pressure. Decreasing pressure favors transfer of soil gases to the atmosphere (increasing the radon flux across the soil/air interface), and increasing pressure inhibits transfer of radon to the atmosphere.

The type of structure also plays a role in determining indoor radon concentration. For example, Lanctot et al. (1992) found that the probability of a structure having elevated indoor radon concentrations increased in the order: mobile homes < woodframe homes < brick or stone homes. Lastly, there is a seasonal factor, with increased indoor radon concentrations found during the winter when home heating systems tend to reduce indoor air pressures and structures are tightly sealed.

Gundersen et al. (1992) developed a radon index (Table 8–9) to evaluate the potential for elevated indoor radon values in a particular region. The indoor radon value for a particular area is based on a simple average of the measurements made in the area. Aerial radioactivity is calculated from airborne gamma-ray surveys. The equivalent uranium (eU) is based on the number of counts recorded by a gamma-ray detector using the ^{214}Bi 1.76 MeV gamma ray. It is assumed that the uranium decay chain is in secular equilibrium. The

Table 8–9 *Radon Index Matrix**

Point value	Avg. indoor radon ($Bq\ m^{-3}$)	Aerial radioactivity (ppm eU)	Geology	Soil permeability	Type of architecture
1	<75	<1.5	Negative	Low	Mostly slab
2	75–150	1.5–2.5	Variable	Moderate	Mixed
3	>150	>2.5	Positive	High	Mostly basement

*Modified from Gundersen et al. (1992).

total number of assigned points are summed to give the probability of high indoor radon values: RI = 3–8, low; RI = 9–11, moderate; and RI = 12–17, high probability. The complexities that can exist at the local level are illustrated by Case Study 8–8, describing indoor radon variations in a small village in Hungary.

CASE STUDY 8–8
Radon Variations in a Hungarian Village

Farm houses in the village of Mátraderecske, Hungary, had high indoor radon values, up to 10,000 Bq m^{-3}, with an arithmetic mean of 300 Bq m^{-3}. These high values seemed anomalous because the underlying geologic units did not have particularly high uranium concentrations (< 4 ppm for all units). Tóth et al. (1997) investigated the geological and geochemical setting of the area in an attempt to explain the measured indoor radon values. The village is built on tuff and tuffsite overlying Eocene andesite. Deeper units are Triassic shales and limestones. Surface concentrations of uranium in the tuffsite reach 30 ppm, and there is a rapid decrease in uranium concentration with depth to background values on the order of 4 ppm. Geothermal heating of the limestone at depth releases carbon dioxide, which percolates upward. Locally, carbon dioxide concentrations were sufficiently high that homes had to be insulated from carbon dioxide. The authors suggest that ura-

nium mobilized in the surface environment moves downward until it reaches the rising CO_2-rich water. At this horizon an insoluble uranyl carbonate is precipitated. These horizons tend to be localized in the vicinities of faults and the edges of clay pillows. The amount of radon gas released is sensitive to groundwater levels. If the uranium-enriched horizons are saturated, the diffusion of radon is inhibited. A second factor is the relative abundance of clay (formed from the weathering of feldspar minerals in the andesitic tuffs) in the soil horizon. The clays are largely montmorillonite, and when wet expand and reduce the porosity and permeability of the soils. Thus, the authors found that the radon content of individual homes varied seasonally (lower when conditions were wet) and with the siting of the home relative to the geologic features that concentrated uranium. A generalized approach to assessing potential radon hazards would not be effective in this case.

Source: Tóth et al. (997).

As mentioned previously, the other major source of indoor radon is from groundwater used for water supply. Assuming secular equilibrium for the ^{238}U chain, Wanty et al. (1992) derived a mathematical model for the activity of radon in groundwater in equilibrium with U-bearing rocks. At secular equilibrium,

$$\lambda_U C_U = \lambda_{Rn} C_{Rn} \tag{8–15}$$

where λ represents the decay constant, C the number of atoms present in the water–rock system, and the subscripts U and Rn represent the number of ^{238}U and ^{222}Rn atoms, respectively. Equation 8–15 is rearranged to give

$$\left(\frac{\lambda_U}{\lambda_{Rn}}\right) \times C_U = (C_{Rn})_{rock} + (C_{Rn})_{water} \tag{8–16}$$

Assuming that $[^{238}U]_{water} << [^{238}U]_{rock}$, a reasonable assumption because rock uranium abundances are normally on the order of parts per million and groundwater uranium abundances are normally on the order of parts per billion, rearranging equation 8–16 and substituting gives

$$\left(\frac{\lambda_U}{\lambda_{Rn}}\right) \times \rho \times V_t(1 - \varphi) \times 4.2 \times 10^{-6} \times [U]_{rock} = \frac{(C_{Rn})_{rock} + (C_{Rn})_{water}}{A} \tag{8–17}$$

Where ρ is the density in kilograms per cubic decimeter, V_t is the unit volume in cubic decimeters, φ is the fractional porosity, 4.2×10^{-6} converts uranium concentration in ppm to uranium concentration in moles, and A is Avogadro's number. The rock uranium concentration is given in ppm. An additional factor, the emanating efficiency of the rock with respect to the release of radon (E) is added to the equation. This factor takes into account the fraction of the radon atoms that move from the rock to the water. The diffusion of radon through solid rock is slow, hence only those radon atoms near the surface of a void have an opportunity to move into the void. If we assume that this diffusion is randomly oriented in space, then even for surface uranium atoms only 50% of the radon atoms would diffuse into the void; the other atoms would diffuse into the solid rock. Thus, even in the best case, where all the uranium is at the surface, only 50% of the radon atoms would diffuse into the void space. Mathematically, the emanation factor is written

$$E \equiv \frac{(^{222}Rn)_{water}}{(^{222}Rn)_{water} + (^{222}Rn)_{rock}} \tag{8-18}$$

Substituting 8–18 into 8–17 gives

$$\left(\frac{\lambda_U}{\lambda_{Rn}}\right) \times \rho \times V_t(1 - \varphi) \times 4.2 \times 10^{-6} \times [U]_{rock} = \frac{(C_{Rn})_{water}}{E \times A} \tag{8-19}$$

Substituting the appropriate constants into equation 8–19, and solving for radon concentration in activity per unit volume, gives

$$(Rn)_{water} = 1.247 \times 10^{-2} \times \rho V_t\left(\frac{1 - \varphi}{\varphi}\right) \times E \times [U]_{rock} \tag{8-20}$$

where $(Rn)_{water}$ is in Bq L^{-1}, density is in kg m^{-3}, V_t is a unit volume (i.e., 1 m^3), φ is the fractional porosity (0 to 0.99), E is the emanation efficiency (effectively 0 to 0.5), and $[U]_{rock}$ is the uranium concentration in the rock in ppm.

EXAMPLE 8–5 The Conway granite of the White Mountain batholith has an average uranium content of 100 ppm. A water well is drilled in a fracture zone in the Conway granite. The porosity is 10% and the emanation efficiency is 0.1. Given a rock density of 2700 kg m^{-3}, calculate the radon activity for groundwater in equilibrium with the granite.

$$(Rn)_{water} = 1.247 \times 10^{-2} \times \rho V_t\left(\frac{1 - \varphi}{\varphi}\right) \times E \times [U]_{rock}$$

$$= (1.247 \times 10^{-2})(2700 \text{ kg m}^{-3})(1 \text{ m}^3)\left(\frac{1 - 0.1}{0.1}\right)(0.1)(100) = 3030 \text{ Bq L}^{-1}$$

This is a very high radon activity, but similar values have been reported for other groundwaters in fractured granites (e.g., Banks et al., 1995). ∎

RAINWATER CHEMISTRY

Berner and Berner (1996) give a detailed discussion of the chemistry of rainwater. Much of this section represents a summary of the major inferences of these authors. The interested student should consult Berner and Berner (1996, Chapter 3) for a more detailed discussion.

Ionic Species

As a first approximation, rainwater can be viewed as consisting of two components: marine and continental (terrestrial and pollution). The major ions in rainwater, and their sources, are listed in Table 8–10.

From Table 8–10, we note that the major source of Cl^- is seawater. In most cases, the major source of Na^+ would also be seawater. If these two ions have only a seawater source, then we would expect the weight ratio of Na^+/Cl^- to be the same in the rainwater as it is in seawater. There are anthropogenic sources for Cl^-, but in most cases these are minor. If we make the assumption that Cl^- has only a seawater source, then we can subtract the marine component from the chemical analysis for any rainwater sample to obtain what are called "excess ion" concentrations (excess with respect to a marine input). These excess ion concentrations can be related to terrestrial and pollution sources. In order to do this calculation, we need to know the relative proportions of the various ions in seawater to Cl^- (Table 8–11). The amount of excess ion is calculated as follows:

Excess ion X = Total amount of ion X − (ratio of ion X in seawater to Cl^-)

$$\times \ Cl^- \text{ concentration} \tag{8-21}$$

Table 8–10 *Sources of Individual Ions in Rainwater**

| | Origin | | |
Ion	Marine inputs	Terrestrial inputs	Pollution inputs
Na^+	Sea salt	Soil dust	Biomass burning
Mg^{2+}	Sea salt	Soil dust	Biomass burning
K^+	Sea salt	Biogenic aerosols	Biomass burning
		Soil dust	Fertilizer
Ca^{2+}	Sea salt	Soil dust	Cement manufacture
			Fuel burning
			Biomass burning
H^+	Gas reaction	Gas reaction	Fuel burning
Cl^-	Sea salt	None	Industrial HCl
SO_4^{2-}	Sea salt	DMS, H_2S, etc. from	Biomass burning
	DMS from	biological decay	
	biological decay	Volcanoes	
		Soil dust	
NO_3^-	N_2 plus lightning	NO_2 from biological	Auto emissions
		decay	
		N_2 plus lightning	Fossil fuels
			Biomass burning
			Fertilizer
NH_4^+	NH_3 from	NH_3 from bacterial	NH_3 fertilizers
	biological activity	decay	Human, animal waste
			decomposition
			(Combustion)
PO_4^{3-}	Biogenic aerosols	Soil dust	Biomass burning
	adsorbed on sea salt		Fertilizer
HCO_3^-	CO_2 in air	CO_2 in air	None
		Soil dust	
SiO_2, Al, Fe	None	Soil dust	Land clearing

*From Berner and Berner (1996).

Table 8–11 *Weight Ratios of Major Ions in Seawater Relative to Cl^- or Na^+**

Ion	Weight ratio to Cl^-	Weight ratio to Na^+
Cl^-	1.00	1.80
Na^+	0.56	1.00
Mg^{2+}	0.07	0.12
SO_4^{2-}	0.14	0.25
Ca^{2+}	0.02	0.04
K^+	0.02	0.04

*Source of data for ratio calculations, Wilson (1975).

If there is a reason to suspect an anthropogenic source for some of the Cl^-, then Na^+ can be used instead of Cl^-. The relative proportion of the various ions in seawater to Na^+ are also given in Table 8–11. A sample calculation is given in Example 8–6.

In terms of their sources, the various ions can also be grouped into associations (Table 8–12, p. 280). This is a useful concept that we will revisit when we discuss how we can determine the source(s) of various atmospheric pollutants. Note that various species may occur in several associations. What we are interested in is which species vary together. For example, if Al, Fe, Si, and Ca were all high relative to other species, we might conclude that there is a soil dust input. There is a certain amount of subjective judgment in identifying

Table 8–12 Primary Associations for Rainwater*

Origin	Association
Marine	Cl–Na–Mg–SO_4
Soil	Al–Fe–Si–Ca–(K, Mg, Na)
Biological	NO_3–NH_4–SO_4–K
Biomass burning	NO_3–NH_4–P–K–SO_4–(Ca, Na, Mg)
Industrial pollution	SO_4–NO_3–Cl
Fertilizers	K–PO_4–NH_4–NO_3

*From Berner and Berner (1996).

the source(s) for the various species. In a later section we will briefly describe a statistical technique that can be used to group the species into associations.

EXAMPLE 8–6 The following data were obtained for rainwater from coastal Washington, on the west coast of the United States, and from Penn State, Pennsylvania, in the interior of the North American continent. The calculated excess ions are listed in the table and a sample calculation is given following the table.

Rainwater Composition and Excess Ions

	Coastal Washington		Penn State, PA	
	Composition (mg L^{-1})	Excess ion (mg L^{-1})	Composition (mg L^{-1})	Excess ion (mg L^{-1})
Na^+	1.81	−0.14	0.07	−0.06
K^+	0.12	0.05	0.05	0.05
Mg^{2+}	0.22	−0.02	0.02	0
Ca^{2+}	0.08	0.01	0.14	0.14
Cl^-	3.49	0	0.24	0
SO_4^{2-}	0.73	0.24	3.14	3.11
NO_3^-	0.14	—	1.93	—
NH_4^+	0.04	—	0.32	—
pH	5.1		4.15	

For coastal Washington, the excess sulfate ion was calculated as follows:

$$\text{Excess } SO_4^{2-} = \text{Measured } SO_4^{2-} - (\text{seawater } SO_4^{2-}/Cl^-) \text{ ratio} \times \text{measured } Cl^-$$
$$= 0.73 - 0.14 \times 3.49 = 0.24 \text{ mg } L^{-1}$$

The calculations for the other ionic species are done in the same way.

In the case of coastal Washington, the only significant excess ions are K^+ and SO_4^{2-}. The most likely sources for these excess ions are fertilizers (for K^+) and the burning of fossil fuel (for SO_4^{2-}). In both cases, these excess amounts are not large, and the bulk of the rainwater chemistry is due to marine inputs. Penn State has an inland location, which is reflected in the low Cl^- content for the rainwater. K^+ and Ca^{2+} are in excess, and there are high absolute amounts of NO_3^- and NH_4^+. From the associations given in Table 8–12, these excess ions are most likely due to fertilizer and some contribution from soil dust. This is an agricultural region, so this is a reasonable conclusion. The excess SO_4^{2-} is large and is most likely due to the anthropogenic addition of SO_4^{2-} by the burning of fossil fuel. This part of Pennsylvania is downwind from a number of coal-fired electrical generating stations. Also note the very low pH for the rainwater. This is a very acidic rain, a topic that will be considered in a later section. ∎

Figure 8–17
Average Cl^- concentration (mg L^{-1}) of rainwater for the United States from July 1955 to June 1956. From Berner and Berner (1996).

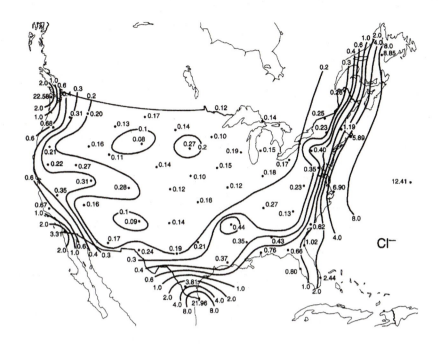

The marine influence on rainwater chemistry is geographically restricted. This is illustrated by Figure 8–17, which shows the variation in Cl^- in rainwater over the United States. The highest values are found in narrow strips along each coast, and there is a rapid decrease in Cl^- away from the coasts. Because the prevailing wind direction is from west to east, the region of marine influence is wider along the west coast of the United States than along the east coast. Along parts of the west coast, inland movement of air is restricted by the presence of mountains (topographic barriers). Where these barriers are minimal, there is a substantial transport of ocean-derived species toward the interior of the continent. Other landmasses will show different patterns depending on prevailing wind directions and the presence of topographic barriers.

The Cl^-/Na^+ weight ratio is sensitive to the inputs of anthropogenic Cl^- and Na^+ from soil dust. The former input would lead to an increase in the Cl^-/Na^+ weight ratio; the latter input would lead to a decrease in the Cl^-/Na^+ weight ratio. Variations in the Cl^-/Na^+ weight ratio for the continental United States are shown in Figure 8–18 (p. 282). Near the coasts the ratio is close to 1.8, the value for marine-derived Na^+ and Cl^-. However, away from the coasts there is a rapid decrease in this ratio, presumably due to the input of Na^+ derived from soil dust. The lowest Cl^-/Na^+ weight ratios are found in the western interior of the United States. This is an arid region underlain by rocks relatively rich in sodium and with abundant saline (Na-rich) dry lake beds. These two sources provide significant inputs of Na to the atmosphere and lead to sodium-rich rainwaters.

Gaseous Species

A variety of gases, both of natural and anthropogenic origin, occur in the atmosphere. These include CO_2, SO_2, NO_2, HCl, and NH_3. The solubility of these gases in rainwater is a function of the partial pressure of the gases and the temperature of the water. As you will recall from previous chapters, the solubility of a gas in water increases with decreasing temperature. All these gases, with the exception of NH_3, form acids when they dissolve in water.

We have already dealt at some length, in Chapter 3, with the carbonic acid system. For SO_2, the relevant reactions are

$$SO_{2\,(g)} + 2OH_{(g)} \rightarrow H_2SO_{4\,(aq)} \rightarrow 2H^+ + SO_4^{2-} \qquad \text{(in the gas phase)}$$

$$SO_{2\,(g)} + H_2O_{2\,(aq)} \rightarrow H_2SO_{4\,(aq)} \rightarrow 2H^+ + SO_4^{2-} \qquad \text{(in liquid cloud droplets)}$$

Figure 8–18
Average Cl^-/Na^+ weight ratio of rainwater for the United States from July 1955 to June 1956. From Berner and Berner (1996).

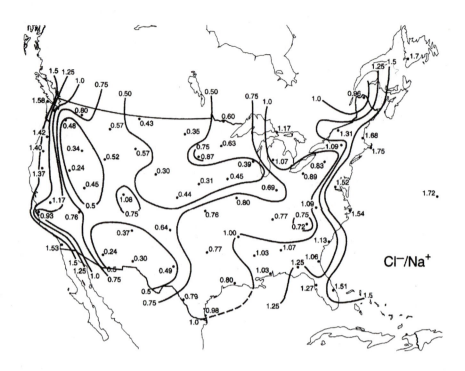

For NO_2, the relevant reaction is

$$NO_{2\,(g)} + OH_{(g)} \rightarrow HNO_{3\,(aq)} \rightarrow H^+ + NO_3^-$$

$HCl_{(g)}$ directly dissolves in water to form hydrochloric acid. With the exception of CO_2, the acidic gases form strong acids. The remaining gas, NH_3, forms a weak base via the following reaction:

$$NH_{3\,(g)} + H_2O \rightarrow NH_4OH_{(aq)} \rightarrow NH_4^+ + OH^-$$

Sulfur dioxide has a number of both natural and anthropogenic sources. Although estimates vary, approximately 75% of the atmospheric sulfur dioxide has an anthropogenic source. Natural sources for sulfur dioxide are volcanic emissions, biogenic reduced sulfur (which is subsequently oxidized in the atmosphere to sulfur dioxide), soil dust, and sea salt. The major anthropogenic source is fossil fuel combustion, with a lesser contribution from biomass burning. The increased use of fossil fuel on a global scale has led to a significant increase in the amount of sulfur dioxide released to the atmosphere (Figure 8–19).

The nitrogen cycle is much more complicated because there are a number of possible pathways. Here we are concerned with the production of NO_x. There are four major

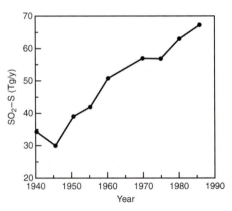

Figure 8–19
Global SO_2 produced by the burning of fossil fuel, 1940 to 1986, in Tg SO_2–S y^{-1} (1 Tg = 10^6 metric tons = 10^{12} g). From Berner and Berner (1996).

natural sources of NO_x: (1) NO produced when atmospheric lightning heats up the air, causing nitrogen and oxygen to combine to form NO; (2) NO and NO_2 produced by photochemical oxidation of N_2O in the stratosphere; (3) oxidation of ammonia in the atmosphere to NO_x; and (4) the production of NO in soil by microbial processes. The two major anthropogenic sources are fossil fuel combustion and biomass burning. Approximately 65% of the NO_x is of anthropogenic origin, and fossil fuel combustion is the dominant pro-

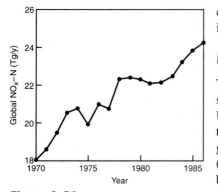

Figure 8–20
Global NO_x produced by the burning of fossil fuel, 1970 to 1986, in Tg NO_x–N y^{-1}. From Berner and Berner (1996).

cess. As was the case for sulfur dioxide, increased use of fossil fuel has led to a significant increase in the global emissions of NO_x (Figure 8–20).

National Atmospheric Deposition Program (NADP)

The U.S. National Atmospheric Deposition Program started in 1978 with 22 measurement sites. Since 1978 the number of sites has increased to over 200, spanning the continental United States, Alaska, Puerto Rico, and the Virgin Islands. The purpose of the program is to collect data on the chemistry of precipitation and to use these data to monitor long-term geographical and temporal trends. Currently pH, sulfate (SO_4), nitrate (NO_3), ammonium (NH_4), chloride (Cl), and cation concentrations (Ca, Mg, Na, K) are measured on a weekly basis. The program was expanded in 1995 to include the acquisition of mercury data (Mercury Data Network), and there are currently more than 35 sites where weekly mercury values are obtained. Another network (Atmospheric Integrated Research Monitoring Network) was established to analyze precipitation on a daily basis, thus allowing a finer resolution of temporal trends. The data collected by the NADP can be accessed via the web *(nadp.sws.uiuc.edu)*. This is a very useful compilation of rainwater chemistry for the United States.

pH of Rainwater and Acid Rain

In Chapter 3 we calculated the pH for water in equilibrium with atmospheric CO_2. This calculation gave a value of pH = 5.7. By definition, acid rain is any rainwater that has a pH of less than 5.7. The two most important components of acid rain are sulfuric and nitric acid. Hence, there is a direct relationship between the SO_4 and NO_x content of rainwater and rainwater acidity. Geographically distributed values for SO_4^{2-} and NO_3^- in rainwater are shown in Figures 8–21 and 8–22, respectively. These data come from the NADP. Note the relatively high abundance of sulfate in rainwater over the midwestern part of the United States, an area with numerous coal-fired electrical generating stations, and the relatively high nitrate concentrations in the same region. Both the absolute and relative abundances of SO_4^{2-} and NO_3^- provide information about the source of the acid rain components. As an illustration, Table 8–13 (p. 284) lists partial chemical analyses for rain over the oceans and the land. In all cases, the rainwater has a pH of less than 5.7. Over the

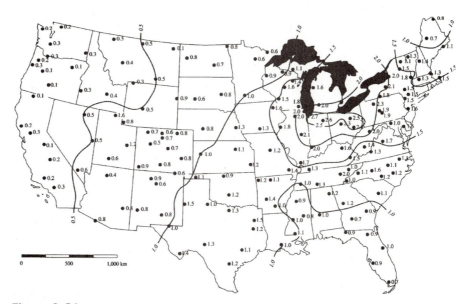

Figure 8–21
Generalized isoconcentration contours for SO_4^{2-} (in mg L^{-1}) for atmospheric precipitation over the contiguous United States in 1995. Source of data is the NADP. From Langmuir (1997).

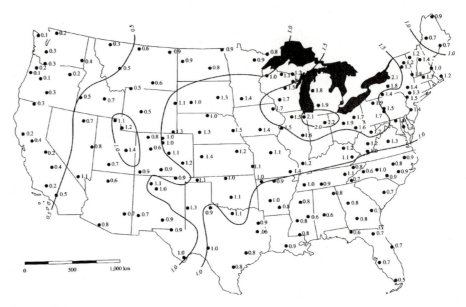

Figure 8–22

Generalized isoconcentration contours for NO_3^- (in mg L^{-1}) for atmospheric precipitation over the contiguous United States in 1995. Source of data is the NADP. From Langmuir (1997).

Table 8–13 Concentration (mg L^{-1}) of Selected Species in Rainwater

Location	Cl^-	SO_4^{2-}	Excess SO_4^{2-}	NO_3^-	SO_4^{2-}/NO_3^-	Reference
Hawaii	9.63	1.92	0.57	0.2	2.9	1
N. Atlantic	5.3	1.2	0.61	0.2	3.1	2
Bermuda	6.2	1.74	0.88	0.34	2.6	3
Poker Flats, Alaska	0.09	0.25	0.24	0.12	2.8	3
New Jersey Pine Barrens	2.82	5.09	4.70	0.39	12.0	4
Hubbard Brook, New Hampshire	0.22	2.08	2.05	1.46	1.4	5
Bishop, California	0.64	2.26	2.17	1.03	2.1	6

Data from: [1]Eriksson (1957), [2]Gambell and Fisher (1964), [3]Galloway et al. (1982), [4]Means et al. (1981), [5]Butler and Likens (1991), [6]Miller (1974).

oceans the absolute abundance of both species is low and $SO_4^{2-}/NO_3^- = 2.6$ to 3.1. The source of the excess sulfur is thought to be either long-range transport from continental areas or locally derived biologically reduced sulfur. For Poker Flats, Alaska, a continental region with minor anthropogenic inputs of these species, the absolute abundances are very low and the $SO_4^{2-}/NO_3^- = 2.8$. In all of these cases, there is apparently no, or only a small, anthropogenic input of nitrate. For the New Jersey Pine Barrens, absolute abundances are high, and the very high SO_4^{2-}/NO_3^- ratio indicates that anthropogenic input of sulfur is the most important factor in determining rainwater pH. Lower ratios for Hubbard Brook, New Hampshire, and Bishop, California, indicate that anthropogenic inputs of NO_x are also important. This type of information is useful when developing pollution control strategies.

The two most important species in determining the pH of acid rain are sulfate and nitrate, and the pH of contaminated rain can be calculated directly from the concentration of these species (Example 8–7).

EXAMPLE 8–7 Calculate the pH for the Hubbard Brook, New Hampshire, acid rain. First calculate the number of moles of sulfate and nitrate.

$$\text{Moles of sulfate} = 2.08 \times 10^{-3} \text{ g L}^{-1}/96.062 \text{ g mol}^{-1} = 2.165 \times 10^{-5} \text{ mol L}^{-1}$$

$$\text{Moles of nitrate} = 1.46 \times 10^{-3} \text{ g L}^{-1}/62.005 \text{ g mol}^{-1} = 2.355 \times 10^{-5} \text{ mol L}^{-1}$$

For every sulfate molecule there will be two hydrogen ions, and for every nitrate molecule there will be one hydrogen ion. The concentration of hydrogen ions is

$$\text{Moles of hydrogen ions} = (2)(2.165 \times 10^{-5} \text{ mol L}^{-1}) + (1)(2.355 \times 10^{-5} \text{ mol L}^{-1})$$
$$= 6.685 \times 10^{-5} \text{ mol L}^{-1}$$

$$\text{pH} = -\log[\text{H}^+] = -\log[6.685 \times 10^{-5} \text{ mol L}^{-1}] = 4.17$$

The measured pH of the Hubbard Brook rainwater is 4.24, in good agreement with the calculated value, indicating that the concentration of the sulfate and nitrate species determines the pH of the rainwater. ∎

Rainwater pH measurements have been made over a number of years at a number of localities around the world. In general, with time there has been a decrease in the pH of rainwater. This is presumably due to the increased emissions of sulfates and nitrates (as shown in Figures 8–19 and 8–20). Rainwater pH for the eastern United States, at various times in the past, is shown in Figure 8–23, along with the pH of rainwater in 1980 for the

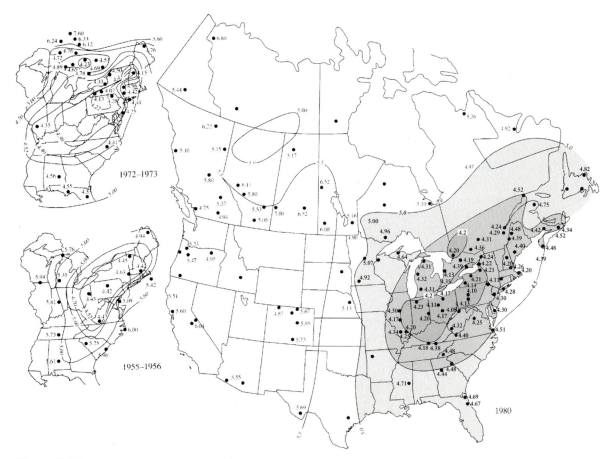

Figure 8–23
Average pH for precipitation in 1955–1956 and 1972–1973 for the northeastern United States and Canada and in 1980 for the contiguous United States and Canada. From Langmuir (1997).

contiguous United States and Canada. Note the increase in the area of acid precipitation from 1955–1956 to 1972–1973. By 1980, the lowest recorded pH value had declined to 4.08. The region of acid rain is elongated in a SW to NE direction because the prevailing winds are from the west. Acid rain is most significant in the eastern half of the United States because this region has the largest number of coal-fired generating stations. Because of the prevailing wind direction, acid rain is exported downwind from its source and areas that are not direct producers of the emissions are also affected by acid rain. This problem becomes particularly significant when the acid rain crosses international borders (in this case, into Canada).

Inspection of Figure 8–23 reveals that there are areas where the pH of the precipitation is greater than 5.7. Hence, the rain has a higher pH than would be expected for rainwater in equilibrium with atmospheric CO_2. These elevated pH values occur because calcite soil dust is entrained by the water droplets and the calcite goes into solution, releasing Ca^{2+} ions. You will recall from Chapter 3 that a weak acid and the salt of a weak acid form a buffer system, and we discussed the carbonate buffer system at some length in Chapter 3. When calcite soil dust is present, the amount of calcite dissolved in the rainwater will determine the pH.

EXAMPLE 8–8 The average pH of precipitation in a rural area is 6.0. Assuming that the pH is controlled only by the carbonate system, i.e., no anthropogenic acid gases are present in the atmosphere, calculate the amount of calcite dissolved per liter of rainwater. In doing this calculation we will ignore ion activity coefficients and ion pairs. Assume that the temperature is 25°C.

At pH = 6.0, the dominant ionic carbonate species in solution is HCO_3^-. For rainwater in equilibrium with atmospheric CO_2,

$$[HCO_3^-] = \frac{[H_2CO_3]K_{a_1}}{[H^+]} = \frac{K_{CO_2}P_{CO_2}K_{a_1}}{[H^+]} = \frac{10^{-1.47}10^{-3.5}10^{-6.33}}{[H^+]} = \frac{10^{-11.30}}{[H^+]}$$

Given that the pH = 6.0, $[H^+] = 10^{-6}$ mol L^{-1}. Solving the preceding equation, $[HCO_3^-] = 10^{-5.30}$. If Ca^{2+} is the only cation, other than H^+, in the rainwater, the charge balance equation is

$$H^+ + 2Ca^{2+} = HCO_3^-$$

Solving for the concentration of Ca^{2+},

$$Ca^{2+} = (HCO_3^- - H^+)/2 = (10^{-5.30} - 10^{-6.0})/2 = 10^{-5.70} = 2.01 \times 10^{-6} \text{ mol L}^{-1}$$

The amount of dissolved $CaCO_3$ per liter of rainwater is

$$CaCO_3 = (100.09 \text{ g mol}^{-1})(2.01 \times 10^{-6} \text{ mol L}^{-1})$$
$$= 2.012 \times 10^{-4} \text{ g L}^{-1} = 0.201 \text{ mg L}^{-1} \quad \blacksquare$$

Systematic measurement of rainwater pH only began in the last half of the 20th century. What were the trends for rainwater pH and the major species (SO_4 and NO_x) responsible for acid rain prior to this time? These questions can be addressed using ice cores. Care must be taken in interpreting the results because variations in regional circulation (both seasonally and over a period of years) will lead to changes in the source of the rainwater components. In addition, intermittent periods of melting can wash out acid cations. Despite these potential difficulties, a number of studies have provided useful information regarding pH variation and its causes for the past several hundred years (Case Study 8–9).

CASE STUDY 8–9
Variations in H⁺ and SO₄²⁻ in Ice Cores as a Record of Acid Deposition

Precipitation over Svalbard, a cluster of islands in the Barents Sea between Greenland and Scandinavia, shows seasonal variations in atmospheric pollution due to changes in the general circulation pattern. During the summer, the Arctic front is well north and polluted air from industrial areas of the northeastern United States is cleansed by precipitation over the open ocean before arriving in Svalbard. In winter, the Arctic front moves further south and cyclones move across the industrialized areas of central Europe, Scandinavia, and northwestern Russia. Svalbard is the first landmass that this pollutant-laden air encounters. Simões and Zagorodnov (2001) investigated temporal variations in pH and sulfate loading using several ice cores collected in Svalbard (Figure 8–C9–1). The ice core retrieved from Skobreen encompassed the time span from 1930 to 1985. The core was analyzed for both acidity (expressed as $\mu Eq\ L^{-1}\ H^{+}$) and $[SO_4^{2-}]^{*}$ (Figure 8–C9–1a), where $[SO_4^{2-}]^{*} = [SO_4^{2-}]_{meas} - 0.103[Cl^{-}]$, measured sulfate corrected for the sea-salt contribution. The sulfate concentrations tend to follow the acidity, although there are exceptions, presumably due to the sampling interval used for the sulfates. The decrease in

acidity from the mid-1950s to the mid-1960s is ascribed to changes in the pattern of precipitation. This was a warm period with relatively high precipitation, which would lead to both greater elution of H⁺ and dilution. The Austfonna core covers the time period 1700 to 1987 (Figure 8–C9–1b). The significant observation from this core is the dramatic increase in H⁺ post-1900, indicating the impact of anthropogenic pollution sources. Also of note is the high acidity point, indicated by the arrow, which corresponds to the Tambora (1816) volcanic eruption. This eruption was marked by the emission of significant amounts of sulfur dioxide, which led to acid precipitation. In the preindustrial period, the acidity oscillated between 2.3 and 2.8 $\mu Eq\ L^{-1}$, with a mean value of 2.6 $\mu Eq\ L^{-1}$. Since the mid-19th century there has been a trend toward higher values, which is reasonably well correlated with increasing sulfate, due to anthropogenic inputs. Present-day values exceed 4.0 $\mu Eq\ L^{-1}$, with sporadic high concentrations of H⁺ exceeding 10 $\mu Eq\ L^{-1}$, associated with high concentrations of sulfate. Thus, the ice cores reflect the anthropogenic inputs to the atmosphere.

Source: Simões and Zagorodnov (2001).

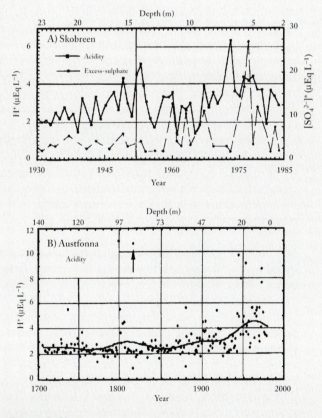

Figure 8–C9–1
Variations in acidity and excess sulfate in ice cores. From Simões and Zagorodnov, 2001.

CHEMISTRY AND SOURCES OF ATMOSPHERIC PARTICULATES (AEROSOLS)

As defined earlier, aerosols are solid or liquid particles ranging in size from clusters of a few molecules to 20 μm in radius. The term *particulate* is more inclusive and applies to any particulate matter found in the atmosphere. Particulates come from a variety of sources (Figure 8–24), both natural and anthropogenic. These particulates are mostly confined to the troposphere, although very fine dust particles and acid aerosols are sometimes transported to the stratosphere. For example, the 1815 eruption of Tambora in the East Indies injected a large amount of sulfur dioxide into the stratosphere, where it was converted to sulfuric acid aerosols. These aerosols caused an increase in the earth's albedo, reducing the shortwave radiation reaching the earth's surface. The result was global cooling and the so-called "year without summer" of 1816.

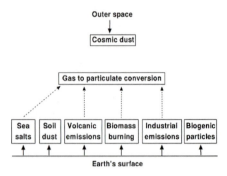

Figure 8–24
Sources of atmospheric particulates. Arrows with dashed lines indicate that there is a gaseous emission associated with the source.

Types of Particulates

Mineral Dust   Mineral dust constitutes the largest fraction of atmospheric particulates. Dust particles are largely derived from semiarid to arid areas. The wind blowing across the surface liberates soil particles, which are then transported by the wind. The distance of transport is a direct function of the size of the particles. Small particles can be transported for great distances (more than 2000 km). These dust particles form the so-called natural background for atmospheric particulates, although there is clearly an anthropogenic aspect related to land-use practices. Soil dust is composed of the common rock-forming minerals: quartz, feldspars, mica, various clay minerals, and calcite. In general, clay minerals are enriched, because of their relatively smaller grain size, in dust that has undergone long-distance transport. A common observation is that there is a decrease in mean grain size of the dust as the transport distance increases.

During transport, various reactions can occur between the dust particles and other aerosols that modify the chemistry of the dust particles. The common reactions are between minerals and the two major acid species, sulfur dioxide and nitrous oxides. As was discussed in Chapter 3, silicate minerals and calcite act as buffers that can modify the acidity of the acid aerosols. Reactions between calcite and sulfur dioxide can lead to the formation of gypsum and hence the sequestering of sulfate.

Sea Salt   Sea salts are the dominant particulates in the marine atmosphere. They are produced by the bursting of air bubbles at the ocean's surface. The number of bubbles at the ocean surface is approximately exponentially related to the wind speed. Hence, with increasing wind speed more particles are released. The initial chemistry of the particle is essentially that of the ocean plus some organic compounds that were scavenged by the rising bubble. Once the particles enter the atmosphere, reactions can occur between the sea-salt particles and the gases in the atmosphere. The result is that Cl is lost and additional sulfate and nitrate are formed on the particles. Sea-salt particles can be identified on the basis of cation ratios involving Na, Mg, K, and Ca, species that are neither deposited nor removed during atmospheric reactions. Conversely, variations in the Cl/Na, S/Na, and N/Na ratios can be used to study the reactions and atmospheric histories of the sea-salt particles. For example, reactions with nitric acid aerosols can lead to the complete conversion of a sea-salt aerosol to $NaNO_3$.

Sulfates   Because NH_3 is abundant in the atmosphere over the continents, continental aerosol sulfate particles are typically composed of $(NH_4)_2SO_4$. In the marine aerosol, sulfate particle compositions vary from H_2SO_4 to $(NH_4)_2SO_4$. The precursor gases for the sulfate aerosols are released by anthropogenic (largely fossil fuel combustion), biogenic (dimethyl sulfide and carbonyl sulfide), and volcanic activity. In the case of the biogenic emissions, the reduced sulfur gases are oxidized through intermediaries to SO_2. The formation of sulfate aerosols often involves another particle (heterogeneous nucleation). Given the association of carbon particles (soot) and SO_2—both are released during the combustion of coal—the particle commonly involved is soot.

Carbonaceous Particles   Carbonaceous particles are composed of carbon and can be divided into three groups: black carbon (soot), organic aerosols, and bioaerosols. *Soot is particles produced by the incomplete combustion of fossil fuels or vegetation*. *Black carbon (BC)* refers to *carbonaceous particles that are good absorbers of solar radiation* (in fact, the strongest among all atmospheric aerosol types). These terms are often used interchangeably, although their definition is based on different properties. Elemental chemistry may be useful in determining the source(s) of soot. Soot derived from biomass burning invariably contains significant K. Soot generated by the combustion of coal is associated with fly ash and contains the usual fly ash elements (such as Fe, Mn, and Zn). Soot generated by oil combustion tends to contain high V. Finally, soot produced by the burning of highly refined petroleum products (gasoline and jet fuel) tends to contain only C and O. *Organic aerosols* show *negligible absorption of solar radiation*. Sources are both anthropogenic (fossil fuel combustion) and natural (organic gases emitted by plants). For example, the major component of photochemical smog is organic aerosols. *Bioaerosols* are *primary biological particles released into the atmosphere from water surfaces, vegetation, soil, and anthropogenic activities*. Examples are spores, pollen, bacteria, fungi, and viruses.

Removal of Particulates from the Atmosphere

Dry Deposition   The *removal of particles by impaction or sedimentation* is referred to as *dry deposition*. Impaction occurs when a particle sticks to a surface, such as a tree leaf or tree trunk. Sedimentation occurs when the settling velocity of a particle exceeds the upward velocity of the air. The settling velocity of a spherical particle is determined by Stokes' law:

$$v_s = \frac{g(\rho_s - \rho_f)d^2}{18\eta}$$

(8–22)

where v_s is the settling velocity, g is the acceleration of gravity (9.8 m s^{-1}), ρ_s is the density of the particle, ρ_f is the density of air, d is the diameter of the particle, and η is the dynamic viscosity of air. At sea level, for the standard atmosphere, density = 1.225 kg m^{-3} and viscosity = 1.789×10^{-5} Pa s. Settling velocities become significant at diameters greater than 10 μm. For smaller particles, turbulent diffusion is the dominant mechanism and vertical transport is proportional to the concentration gradient. In this case, particles move from regions of high concentration to regions of low concentration. The actual removal of this size particle occurs across a thin laminar air layer that separates the atmosphere from the surface. Thus, the removal process is sensitive to the type of surface.

Wet Deposition   The *removal of particles by rainout or washout* is referred to as *wet deposition*. Once the atmosphere becomes saturated in water vapor (100% humidity), condensation should occur. *Homogeneous condensation* is the *conversion of water vapor to water liquid or water solid without the presence of another particle*. In practice, spontaneous nucleation requires significant oversaturation (up to 700%) of the air with water vapor. *Heterogeneous condensation involves a particle that acts as a condensation nucleus*. *Cloud condensation nuclei* are *aerosols that serve as sites of condensation*. Some aerosols are *hygroscopic*, which means that *they attract water vapor*. Such aerosols serve as very effective condensation nuclei. In terms of atmospheric chemistry, a number of reactions can occur between the aerosol and its surrounding water droplet that change the chemistry of the aerosol. Once cloud droplets achieve a significant size (generally greater than 200 μm in diameter), they are removed from the atmosphere by sedimentation. This component of wet deposition is referred to as *rainout*. The falling droplets may encounter other particulates and these particulates may be taken up by the droplet. This component of wet deposition is referred to as *washout*. It is possible to distinguish between rainout and washout by sampling incrementally during a precipitation event. If a particular constituent is removed by rainout, its concentration will remain constant throughout the precipitation event (the constituent is serving as a condensation nucleus). If the constituent is removed by washout, its concentration will decrease with time because the bulk of the aerosols are removed during the early stages of precipitation.

Particulate Sampling

Real-time measurements are made in a variety of ways. High-volume and dicot samplers draw air through a set of filters. The filters are size graded so that the size distribution of the particulates can be determined. Separation into particles less than 2.5 μm in diameter and between 2.5 μm and 10 μm in diameter is frequently used. Particles in these size ranges are readily taken into the human lung. Trace metal concentrations may vary between the two size fractions, and the inference is sometimes made that the larger particles are locally derived whereas the smaller particles come from more distant sources. Although this type of measurement gives us information as to the chemistry of aerosols suspended in the atmosphere, it does not give us the chemistry of aerosols that are removed from the atmosphere by dry and/or wet deposition. For this reason, other types of sampling are done. One type involves the collection of atmospheric precipitation (wet deposition). Various collecting devices are used, and the nature of the collectors is in part controlled by the species that are being measured, inorganic or organic. A second type involves the deposition of airborne particulates onto surfaces (dry deposition). This can be done in a variety of ways: using surfaces coated with sticky material or buckets partially filled with water to approximate deposition onto the surface of lakes or oceans. Historical particulate concentrations and variations are determined from the stratigraphic record. Commonly used for this purpose are ice cores, peat layers, and sediment layers in lakes. Various biological systems are also used, such as trees and other plants that show seasonal growth patterns.

Air Mass Trajectories

This topic is beyond the scope of the book, but it is an important idea in terms of identifying the source of atmospheric pollutants and will be briefly discussed. In the next section we will consider a variety of methods used to fingerprint the source(s) of atmospheric contaminants. A complete analysis of the source of the contaminants requires a knowledge of the path the air mass followed. While we can say, in general, that air mass transport is from west to east, the exact path is important in determining the specific sources for a contaminant. In the case of dust, we can often visually see the path using satellite imagery. In other cases, computer models are used to back-calculate air mass trajectories. At any particular site, we may find that the type and concentration of atmospheric pollutants vary as a function of wind direction. For example, a site located along the western margin of an ocean basin might be expected to show significant marine inputs when winds are from a NE or SE direction. On the other hand, winds from the west might bring large amounts of sulfate deposition. However, we might find that there was considerable variation in the sulfate deposition as a function of time, even given a generally westerly wind pattern. Calculation of air mass trajectories might indicate that they varied with time. Sometimes the air was traveling across agricultural land, and at other times the air was traveling across an industrialized region, hence explaining the differences in sulfate deposition. Air trajectory calculations are best done by specialists conversant with atmospheric circulation models, one of many reasons why multidisciplinary teams are necessary when studying environmental problems.

Source Identification and Apportionment

Of fundamental importance is the source of various atmospheric aerosols, natural or anthropogenic. The nature and relative importance of various sources for trace metals and other species of environmental interest are determined in a number of ways. The most common discrimination techniques use elemental, molecular, or isotopic signatures to characterize anthropogenic, crustal, and marine sources.

Enrichment Factor   The enrichment factor represents the amount of a particular element in excess of that expected from a natural (rock or soil) source. It is often assumed that the aluminum content of a particulate is due solely to crustal sources. Iron, titanium, silicon, and scandium may also be reasonable choices for elements of totally crustal origin. Iron and scandium, but not aluminum, titanium, or silicon, are routinely determined by neutron activation analysis, a common technique used to determine the chemistry of particulates. Other elements are then referenced to Al (or Fe). The crustal source can either be average crust (Table 8–14) or local rock or soil. The aerosol–crust enrichment factor (EF_{crust}) of element X is

$$EF_{crust} = \frac{\left(\dfrac{X}{RE}\right)_{particulate}}{\left(\dfrac{X}{RE}\right)_{crust}} \tag{8–23}$$

where RE is the reference element. Enrichment factors near unity (some authors consider a factor of 3 to be near unity) indicate that the element has come from the crust, either directly as windblown soil or indirectly as coal fly ash. Fly ash is a potential source because

Table 8–14   Elemental Composition of the Continental Crust*

Element	Concentration (ppm) Upper crust	Bulk crust	Element	Concentration (ppm) Upper crust	Bulk crust
Al	80,400	84,100	Se	0.05	0.05
Fe	35,000	70,700	Mo	1.5	1.0
Sc	11	30	Ag	0.050	0.080
Ti	3000	5400	Cd	0.098	0.098
V	60	230	Sn	5.5	2.5
Cr	35	185	Sb	0.2	0.2
Mn	600	1400	W	2.0	1.0
Co	10	29	Au	0.0018	0.003
Ni	20	105	Pb	20	8.0
Cu	25	75	Th	10.7	3.5
Zn	71	80	U	2.8	0.91
As	1.5	1.0			

*Data from Taylor and McLennan (1985). Both bulk continental crust and upper continental crust have been used to calculate crust ratios. In some cases, the results may differ significantly. For example, the Pb/Al ratio for the upper crust is 2.5×10^{-4}, but for the bulk crust the ratio is 9.5×10^{-5}. Given a sample that has a Pb/Al ratio of 2.5×10^{-4}, the upper crust gives an EF of 1 while the bulk crust gives an EF of 2.6. The difference is great enough that different conclusions might be drawn regarding the source of the Pb (natural for upper crust normalization and anthropogenic for bulk crust normalization). Cr would show an even greater difference, but in the opposite sense. The Cr/Al ratio for the upper crust is 4.4×10^{-4}, but for the bulk crust the ratio is 2.2×10^{-3}. Given a sample with a Cr/Al ratio of 2.2×10^{-3}, the bulk crust normalization gives EF = 1 while the upper crust normalization gives EF = 5, suggesting that there is an anthropogenic contribution to the Cr content of the sample.

it is the noncombustible mineral component of coal; i.e., it represents sedimentary material. The noncrustal component ($X_{noncrustal}$) is calculated as follows:

$$X_{noncrustal} = X_{total} - RE_p\left(\frac{X}{RE}\right)_{crust} \tag{8-24}$$

where X_{total} is the concentration of element X in the particulate, RE_p is the concentration of the reference element in the particulate, X is the concentration of element X in the crust (or soil), and RE is the concentration of the reference element in the crust (or soil).

EXAMPLE 8–9 An aerosol sample contains 1000 ppm Al and 7 ppm Cr. Calculate the enrichment factor and noncrustal concentration of Cr using bulk crust for the normalization.

$$EF_{crustal} = \frac{\left(\dfrac{X}{RE}\right)_{particulate}}{\left(\dfrac{X}{RE}\right)_{crust}} = \frac{\left(\dfrac{7}{1{,}000}\right)_{particulate}}{\left(\dfrac{185}{84{,}100}\right)_{crust}} = 3.2$$

$$X_{noncrustal} = X_{total} - RE_p\left(\frac{X}{RE}\right)_{crust} = 7 - 1000\left(\frac{185}{84{,}100}\right) = 7 - 2.2 = 4.8 \quad \blacksquare$$

Elemental Signatures   The chemical composition of an aerosol is a function of its source(s). A great deal of effort has been expended to try and characterize these sources using elemental signatures. The signatures can be either source- or region-specific. In a general way, we can consider three basic types of sources: crustal, marine, and anthropogenic (pollution). Any particular aerosol sample can represent one or more of these sources. Because our intent is to distinguish various anthropogenic sources, as a first step the natural source component of the aerosol chemistry must be identified. Various techniques have been used, including factor analysis and chemical mass balance. Rahn (1999) recently proposed a graphical approach in which the chemistry of a set of aerosol samples is used to delineate the crustal, marine, and anthropogenic components. For any element X, we can write a mass balance equation.

$$X = X_{crust} + X_{marine} + X_{pollution} \tag{8-25}$$

Using Al, Na, and Se as reference elements for crust, marine (note that this is exactly the same approach used when calculating the marine component of rainwater) and pollution inputs, and assuming that these elements are only derived from one specific source, equation 8–25 becomes

$$X = Al\left(\frac{X}{Al}\right)_{crust} + Na\left(\frac{X}{Na}\right)_{marine} + Se\left(\frac{X}{Se}\right)_{pollution} \tag{8-26}$$

To derive an equation for a particular component, divide both sides of equation 8–26 by the reference element for that component. For the crustal component, equation 8–26 becomes

$$\frac{X}{Al} = \left(\frac{X}{Al}\right)_{crust} + \frac{Na}{Al}\left(\frac{X}{Na}\right)_{marine} + \frac{Se}{Al}\left(\frac{X}{Se}\right)_{pollution} \tag{8-27}$$

Similar equations can be written for the marine component or the pollution component. Note that the crust, marine, and pollution ratios are constants for any particular set of aerosols. In order to solve this problem in terms of two axes, we select subsets of equation 8–27. Thus, a plot of X/Al versus Na/Al yields the crustal and marine components

$$\frac{X}{Al} = \left(\frac{X}{Al}\right)_{crust} + \frac{Na}{Al}\left(\frac{X}{Na}\right)_{marine} \tag{8-28}$$

and a plot of X/Al versus Se/Al yields the crustal and pollution components.

$$\frac{X}{Al} = \left(\frac{X}{Al}\right)_{crust} + \frac{Se}{Al}\left(\frac{X}{Se}\right)_{pollution} \tag{8–29}$$

Because linear plots are highly skewed, with most of the points near the origin, the distribution is more clearly shown in logarithmic plots. Theoretical plots are shown in Figure 8–25. Figure 8–25a shows a theoretical plot of the crustal versus marine components given a crustal ratio of X/Al = 0.5 and a marine ratio of X/Na = 0.2, where X is the element of interest. The left-hand, horizontal part of the graph gives the crustal ratio; the right-hand part of the graph, the region of constant slope, gives the marine ratio. For any particular aerosol sample, the vertical distance between the horizontal line and the sloping portion of the curve gives the noncrustal (marine) fraction of X. Figure 8–25b shows a similar plot, but in this case using the crustal and pollution components for X. Plots of real data, given a sufficient number of individual aerosol analyses, would be expected to yield plots of similar shape. If the only significant component is crustal in origin, then the data points will fall on a horizontal line denoting the crustal ratio. If the only significant component is marine or pollution, then the data points will fall on a sloping curve and the distribution will not show a horizontal tail. If the data are not adequately represented by a two-component model, then the distribution will broaden; i.e., the data points will not plot

Figure 8–25

Theoretical two-component logarithmic plots for (a) crustal/marine and (b) crustal/pollution components. From Rahn (1999).

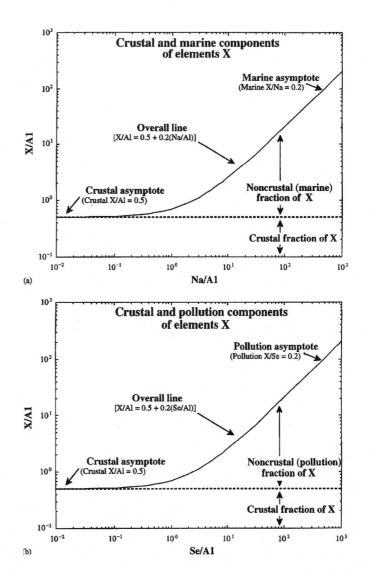

in a tight trend. Examples of the various types of distributions, using aerosol data collected from Narragansett, RI, and Mace Head, Ireland, can be found in Rahn (1999) and Huang et al. (2001), respectively. Once the crustal, marine, and pollution enrichment factors have been determined from the graphical distributions, the crustal (or marine or pollution) component can be calculated for each element. These values can then be used in a discriminant analysis to determine the source of the aerosols (see Case Study 8–11).

Another approach to delineating sources is factor analysis, a standard statistical technique for identifying variables that covary. Using elemental data, we can identify groups of elements and then attempt to relate them to sources such as crustal, marine, or pollution. If an element has more than one significant source, it will usually fall into several groups. An example of the use of factor analysis is given in Case Study 8–10.

Once the potential sources and their elemental characteristics are identified, the contribution of each source to the observed aerosol chemistry can be determined by a least-squares multiple regression of the total data set. In essence, we write a series of linear equations, one for each element, and then solve the complete set of equations for a best-fit solution. A number of standard statistical packages will carry out these sorts of calculations. An example of this type of calculation is given in Case Study 8–11. One widely used computational method is the CMB (chemical mass balance) receptor model (Watson, 1984; Watson et al., 1984, 1990). A set of mass balance equations are written, one linear equation for each element or tracer, and a least-squares, best-fit solution is calculated.

Molecular Signatures   The same types of analysis as those just described can be carried out for organic contaminants, using molecular signatures. Among the most important organic contaminants are VOCs (volatile organic compounds) and PAHs. Source

---

CASE STUDY 8–10
Source of Aerosols Deposited on Massachusetts Bay

Underhill (1995) and Golomb et al. (1997) investigated the deposition of various metals (and organics) onto Massachusetts Bay. The major goal of the study was to determine the importance of atmospheric aerosols relative to the total amount of metals found in the surface waters. As a part of this study, source identification was carried out using factor analysis.

Airborne particulates were collected using a dichot sampler, and wet and dry particle deposition was determined using sampling buckets. Two sample sites were maintained, one at Nahant, near Boston, and the other at Truro on Cape Cod. The data were collected biweekly over a one-year period, and total deposition was calculated for the various metals and organics. Factor analysis was performed to identify groups of elements that covaried. As an example, for dry deposition at Nahant four elemental groups were identified (following table). These groups were then related to the elemental signatures of various anthropogenic sources. F1 was attributed to automotive and crustal sources, F2 to oil combustion, F3 to soils, and F4 to coal combustion and incineration. The elements that made up F2 were derived from several oil-fired power plants in the immediate vicinity. The elements comprising F4 were in part derived from the nearby Saugus incinerator. It was suggested that the primary pathway for Se and Sb in this last group might be long-range transport from distant coal combustion sources.

Rotated Factor Loadings for Nahant Dry Deposition*

Element	F1	F2	F3	F4	Communality
As	0.84				75.3
Ce	0.91				87.8
Fe	0.88				89.1
La	0.86				86.9
Sc	0.94				93.9
Sm	0.96				94.7
Cd		0.83			80.5
Co		0.80			70.0
Cr		0.82			86.2
Cu		0.76			72.4
Ni		0.50	0.49		51.5
Al			0.60		46.3
Mn			0.46		46.8
Pb			0.72		52.1
Sb				0.70	71.3
Se				0.66	46.9
Zn				0.40	44.7
% Variance	46.6	18.9	11.0	6.1	82.7

*From Underhill (1995).

Source: Underhill (1995), Golomb et al. (1997).

CASE STUDY 8–11
Elemental Tracers of Distant Regional Pollution Aerosols

Rahn and Lowenthal (1984) developed a seven-element tracer system that could be used to identify the regional sources of pollution aerosols. Their criteria for selecting this set of elements was that they were pollution derived, associated with fine particles, and their abundances could be accurately determined. The analytical method of choice was neutron activation analysis, a method commonly used because of its low detection limit for many elements. The elements selected were As, Sb, Se, noncrustal V, Zn, noncrustal Mn, and In. Element ratios, relative to Se, were developed for a number of source areas. These ratios were then used to characterize the source for aerosols found in remote areas. A subset of this data set (for individual regional sources) is given in the table opposite.

Source	As/Se	Sb/Se	Noncrustal V/Se	Zn/Se	Noncrustal Mn/Se	In/Se ($\times$ 1000)
NE	0.13	0.45	11.3	32	9.2	9.8
BOS	0.68	0.82	35	37	4.1	5.3
NYC	1.10	1.63	11.1	40	6.5	9.6
WASH	1.46	0.82	9.9	22	4.0	7.1
INT	0.92	0.28	1.96	10.8	2.6	3.9
SONT	8.0	0.75	1.77	57	13.9	46

NE—New England, BOS—Boston, NYC—New York City, WASH—Washington, DC, INT—Interior U.S., SONT—Southern Ontario Smelter.

Individual samples are classified (using discriminant analysis) by comparison to the elemental ratios determined for each source region. For example, 101 aerosol samples were collected at Narragansett, RI. On the basis of the classification scheme, the sources were NE—17 samples, BOS—17 samples, NYC—8 samples, WASH—37 samples, and INT—22 samples. As expected, coastal sources dominated. The major determinant of the source would presumably be the wind direction at the time the sample was collected.

In most cases, an aerosol sample will be a mixture of source areas. It is possible to determine the relative importance of the various sources (source apportionment) by doing a least-squares analysis involving the entire set of elemental tracers. For the sources here (excluding SONT), the average elemental abundances are given in the table opposite.

Element	Concentration (ng m^{-3})				
	NE	BOS	NYC	WASH	INT
As	0.060	0.49	2.0	3.2	1.54
Sb	0.143	0.83	3.1	2.1	0.55
Se	0.37	1.00	1.88	2.4	1.78
Noncrustal V	4.0	35	20	23	3.4
Zn	11.1	37	70	60	18.2
Noncrustal Mn	2.2	4.2	13.0	9.2	4.3
In	0.0028	0.0050	0.0160	0.020	0.0064

The problem is solved by writing a linear equation for each element and then simultaneously solving the set of linear equations using a least-squares technique. Most statistics software packages have a routine to solve this sort of problem. For example, let us suppose that an aerosol collected at Narragansett has an As concentration of 0.67 ng m^{-3} (the observed value). We can write a linear equation for As as follows:

$$\text{As (observed)} = X_1 \cdot \text{NE} + X_2 \cdot \text{BOS} + X_3 \cdot \text{NYC} + X_4 \cdot \text{WASH} + X_5 \cdot \text{INT}$$

where X_1 to X_5 represent the fraction of each source contributing to the concentration of As in the aerosol. Similar equations are written for the other six elements. These equations are then solved simultaneously. Weightings are often used because of the difference in absolute magnitude of the elemental abundances and the relative importance of a particular element. Elements that are not as precisely determined are given a lower weighting. The result of such a calculation, for a Narragansett aerosol, is given next.

Source: Rahn and Lowenthal (1984).

Element	Weighting factor	Concentration (ng m^{-3})				Total predicted	Observed
		NE	BOS	WASH	INT		
As	300	0.03	0.06	0.20	0.36	0.65	0.67
Sb	30	0.08	0.10	0.13	0.13	0.44	0.55
Se	100	0.20	0.13	0.15	0.42	0.90	0.90
Noncrustal V	20	2.20	4.42	1.44	0.8	8.9	9.0
Zn	4	6.11	4.68	3.76	4.31	18.9	18.4
Noncrustal Mn	0.4	1.21	0.53	0.58	1.02	3.34	2.00
In	100	0.0015	0.0006	0.00	0.0015	0.0036	0.0040

From Rahn and Lowenthal (1984).

characteristics for the VOCs can be found in Watson et al. (2001). Case Study 8–12 deals with the identification of the sources of PAHs.

Isotopic Signatures—Pb   Because of its importance to human health (some suggest that empires may have fallen due to lead poisoning), a significant amount of effort has been devoted to identifying the sources of atmospheric lead. Pb isotopes have proven to be particularly useful for source identification. In Chapter 6 we discussed at some length various isotopic systems, including the U and Th decay chains. With reference to Tables 6–1 and 6–2, the ^{238}U chain ends with ^{206}Pb, the ^{235}U chain with ^{207}Pb, and the ^{232}Th chain with ^{208}Pb. These are the so-called radiogenic progeny and can be produced only by the decay of their respective radioactive parents. The isotopic abundances of the various radiogenic progeny are often expressed in terms of ratios to ^{204}Pb, the only stable isotope of lead. This gives rise to the atomic ratios of the Pb isotopes: $^{206}Pb/^{204}Pb$, $^{207}Pb/^{204}Pb$, and $^{208}Pb/^{204}Pb$. Lead isotopic compositions are determined using mass spectrometers and the isotopic ratios are reported directly in terms of ^{204}Pb. These ratios can be combined in various ways to give other lead isotopic ratios. For example, the $^{206}Pb/^{207}Pb$ ratio can be calculated from

$$(^{206}Pb/^{204}Pb)/(^{207}Pb/^{204}Pb) = \frac{^{206}Pb}{^{204}Pb} \times \frac{^{204}Pb}{^{207}Pb} = \frac{^{206}Pb}{^{207}Pb}$$

EXAMPLE 8–10   Pb isotopic measurements of an aerosol sample give $^{206}Pb/^{204}Pb = 18.004$ and $^{208}Pb/^{204}Pb = 38.08$. Calculate the $^{208}Pb/^{206}Pb$ ratio for this sample.

$$(^{208}Pb/^{204}Pb)/(^{206}Pb/^{204}Pb) = \frac{38.08}{18.004} = 2.115$$

■

CASE STUDY 8–12
Sources of PAHs in the Coastal Atmosphere of Chicago and Lake Michigan

High burdens of PAHs have been reported in the atmosphere over Lake Michigan. Simcik et al. (1999) used multivariate statistical techniques to determine the sources of these PAHs and the relative contribution of each source. Both factor analysis and multiple regression were used in the analysis. The diagnostic PAH ratios used in the analysis are listed here.

The factor analysis indicated that the most important sources for the PAHs were coal combustion, natural gas combustion, vehicle emissions (gasoline + diesel), and coke ovens. Using these sources, and their appropriate tracers, multiple regression analysis apportioned the sources for the PAHs as follows: coal combustion—48%, natural gas combustion—26%, coke ovens—14%, and vehicle emissions—9%.

Source: Simcik et al. (1999).

Source	BaP/BghiP	Phen/Anthr	45MePh/MePh	BaA/Chry	BeP/BaP
Chicago/Lake Michigan	0.84	7.6	0.43	0.66	1.1
Vehicles	0.3–0.78	2.7		0.63	
Street dust		8	0.3		1.4
Gasoline exhaust	0.3–0.4	3.4–8		0.28–1.2	1.1–1.3
Firewood fire		3	1	0.93	0.44
Coal	0.9–6.6	3	0.3	1.0–1.2	0.84–1.6
Diesel	0.46–0.81	7.6–8.8		0.17–0.36	2–2.5
Coke oven	5.1	0.79		0.70	2.6
Incinerators	0.14–0.6				
Oil-burning power plants	>2				
Petroleum refineries	0.65–1.7				

BaP—benzo[a]pyrene, BghiP—benzo[ghi]perylene, Phen—phenanthrene, Anthr—anthracene, 45MePh—4,5-methylene phenanthrene, MePh—total methyl phenanthrenes, BaA—benz[a]anthracene, Chry—chrysene, BeP—benzo[e]pyrene, and BaP—benzo[a]pyrene.

Anthropogenic inputs are the dominant source of Pb in atmospheric aerosols. These anthropogenic sources include combustion of leaded gasoline, production of steel and base metals, mining and smelting of lead, and combustion of coal. The natural sources include windblown dust, plant exudates, forest fires, volcanic eruptions, radioactive decay, and sea-salt spray.

Pb isotopic ratios are determined by the U/Pb and Th/Pb ratios and the age of the sample (or system). The three radioactive parents that give rise to the radiogenic lead progeny have distinctly different half-lives: $^{235}U = 0.7038 \times 10^9$ y, $^{238}U = 4.468 \times 10^9$ y, and $^{232}Th = 14.010 \times 10^9$ y. Thus, not only do the radiogenic lead progeny to stable lead ratios increase with time but the rate of increase also varies with time. For example, increases in the $^{207}Pb/^{204}Pb$ ratio were much more rapid earlier in geologic history than they are today. This is due to the relatively shorter half-life of ^{235}U compared to the other radioactive parents. The Pb isotopic ratios are also a function of the time a particular reservoir becomes isolated from the bulk earth system. For example, lead ore deposits are formed at various times in geologic history. The lead ore mineral, galena (PbS), does not have a structural site that can accommodate U or Th. When the ore mineral is formed, it "freezes in" the Pb isotopic ratio that existed at the time of formation. Hence, the isotopic ratios of Pb in galena from the Proterozoic-age Broken Hill ore deposit ($^{206}Pb/^{204}Pb = 16.0$) are significantly lower than those of the younger Paleozoic-age Woodlawn ore deposit ($^{206}Pb/^{204}Pb = 18.1$). The earth system as a whole continues to evolve, giving even higher ratios for modern lead. The result is that a wide range of Pb isotopic ratios are found for natural materials. Thus, one must determine the Pb isotopic ratios of the various sources that may contribute to a particular collection of aerosol samples. For example, one source of atmospheric Pb is leaded gasoline. Lead is added to the gasoline to increase its octane number. The lead is obtained from different lead ore deposits that have different Pb isotopic ratios. Hence, Pb isotopic ratios must be measured for the different gasolines (petrols) used in a region. To complicate the problem, there may be a time-dependent variation in the Pb isotopic characteristics of atmospheric Pb emitted by gasoline combustion due to variations in the Pb sources and relative proportions of the sources used for the lead additive (Chiaradia and Cupelin, 2000).

In Chapter 6 we developed the concept of simple mixing and described a graphical method (the lever rule) to determine the relative proportions of the different components in a particular sample. An example of a simple two-end-member system is given in Case Study 8–13 (p. 298). If there are more than two end members, the data will fall in a field on a graphical plot, and the relative proportions of the various components can be determined by a graphical solution. We can also solve this type of problem by writing a series of linear equations. In the case of the Pb isotopic system, we have two independent Pb ratios: $^{207}Pb/^{206}Pb$ and $^{208}Pb/^{206}Pb$. A three-end-member system can be described by writing the following equations:

$$\left(\frac{^{207}Pb}{^{206}Pb}\right)_{meas} = \left(\frac{^{207}Pb}{^{206}Pb}\right)_A \cdot X_A + \left(\frac{^{207}Pb}{^{206}Pb}\right)_B \cdot X_B + \left(\frac{^{207}Pb}{^{206}Pb}\right)_C \cdot X_C \quad (8\text{–}30)$$

$$\left(\frac{^{208}Pb}{^{206}Pb}\right)_{meas} = \left(\frac{^{208}Pb}{^{206}Pb}\right)_A \cdot X_A + \left(\frac{^{208}Pb}{^{206}Pb}\right)_B \cdot X_B + \left(\frac{^{208}Pb}{^{206}Pb}\right)_C \cdot X_C \quad (8\text{–}31)$$

$$X_A + X_B + X_C = 1 \quad (8\text{–}32)$$

where $(^{207}Pb/^{206}Pb)_{meas}$ and $(^{208}Pb/^{206}Pb)_{meas}$ are the measured Pb isotopic ratios for the aerosol sample; the ratios denoted by the subscripts A, B, and C are the ratios for the three end members; and X_A, X_B, and X_C are the fractions of end members A, B, and C, respectively.

CASE STUDY 8–13
The Origin of Atmospheric Lead in Oslo, Norway

Åberg et al. (1999) used Pb isotopes to identify the sources of atmospheric lead in Oslo, Norway, aerosols. Historically, the major source of Pb has been the combustion of leaded gasoline. Over time, the amount of Pb in gasoline has been reduced and today all gasolines are unleaded. Low-leaded gasolines (0.15 g Pb L^{-1} of gasoline) marketed in Norway had ^{206}Pb/^{204}Pb $\approx$ 17. The phaseout of leaded gasoline in Norway started in the early 1990s. The reduction in the use of leaded gasoline was marked both by a decrease in the total Pb content of atmospheric aerosols and an increase in the ^{206}Pb/^{204}Pb to 17.5 to 17.7. Unleaded gasoline does contain some lead ($\sim$0.01 g L^{-1} of gasoline derived from crude oil), but the general increase in the ^{206}Pb/^{204}Pb was interpreted as indicating that other sources of lead were becoming relatively more important. In Stockholm, Sweden, where all Pb additives in gasoline were phased out in 1995, ^{206}Pb/^{204}Pb ratios rose to, and stabilized at, 18.0. In Oslo, similar ratios were achieved in 1998. Another potential source of atmospheric lead is the combustion of wood. During the winter season wood is widely used as a fuel in Norway. Using tree rings, variations in lead isotopic ratios were investigated for the period 1970 to 1994. During this period, ^{206}Pb/^{204}Pb ratios varied between 18.1 and 18.3, with a slight decline in the ratios noted for the more recent tree rings. Trees take up lead either by absorption from the air or through the root system from the soil. The tree ring data, combined with rainwater pH measurements, revealed that there was a relationship between rainwater pH and the ^{206}Pb/^{204}Pb ratios, higher acidity being correlated with higher ^{206}Pb/^{204}Pb ratios. Lead emitted during the burning of coal in England and Central Europe has ^{206}Pb/^{204}Pb ratios of about 18.5. The authors suggested that the lead from these sources is transported with acid rain components, explaining the correlation between pH and ^{206}Pb/^{204}Pb ratios. Combustion of wood, which has ^{206}Pb/^{204}Pb ratios of 18.0 to 18.2, would account for the increase in the ^{206}Pb/^{204}Pb ratios after the phaseout of leaded gasoline.

The authors further investigated the question of the sources, and their relative importance, using an isotope diagram (Figure 8–C13–1). The isotopic ratios of the various samples fall on an approximately straight line, with gasoline Pb and coal Pb as the apparent end members. From the positions of the bark and soot samples on the diagram, the authors concluded that the Pb in the bark largely came from atmospheric deposition of gasoline Pb and Pb derived from coal combustion was more important in the case of the wood. The authors also concluded that the combustion of wood was an important source for the lead observed in the Oslo aerosols.

Source: Åberg et al. (1999).

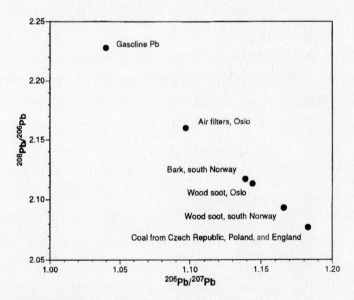

Figure 8–C13–1
Plot of ^{208}Pb/^{206}Pb versus ^{206}Pb/^{207}Pb ratios for various materials. The samples fall on a straight line, suggesting that the various materials represent simple mixtures of the gasoline and coal end members. From Åberg et al. (1999).

EXAMPLE 8–11 The Pb isotopic ratios for an aerosol sample, and the Pb isotopic ratios for the three sources that can contribute lead to this aerosol, are given next.

	$^{206}Pb/^{204}Pb$	$^{207}Pb/^{206}Pb$	$^{208}Pb/^{206}Pb$
Aerosol	17.625	0.8849	2.1301
Coal	18.525	0.8468	2.0895
Gasoline	16.764	0.9217	2.1697
Soil	18.126	0.8606	2.0995

For $^{207}Pb/^{206}Pb$, we can write the following equation:

$$\left(\frac{^{207}Pb}{^{206}Pb}\right)_{meas} = \left(\frac{^{207}Pb}{^{206}Pb}\right)_A \cdot X_A + \left(\frac{^{207}Pb}{^{206}Pb}\right)_B \cdot X_B + \left(\frac{^{207}Pb}{^{206}Pb}\right)_C \cdot X_C$$

where X_A is the coal component, X_B is the gasoline component, and X_C is the soil dust component. Substituting the appropriate values gives

$$(0.8849)_{meas} = (0.8468)_A \cdot X_A + (0.9217)_B \cdot X_B + (0.8606)_C \cdot X_C$$

A similar equation can be written for $^{208}Pb/^{206}Pb$. The third equation is

$$X_A + X_B + X_C = 1$$

Solving this set of linear equations gives $X_A = 0.45$, $X_B = 0.50$, and $X_C = 0.05$, the relative contributions of each source to the aerosol lead. ∎

Isotopic Signatures—C   Carbon isotopes can be used in several ways to characterize the source(s) of organic species. In Chapter 6 we developed the radiocarbon (^{14}C) system of dating. Geologically old samples, such as fossil fuels, do not contain any radiocarbon. The carbon in modern plants and animals is in equilibrium with the present atmosphere and contains ^{14}C. The bomb tests of 30 to 40 years ago added significant ^{14}C to the atmosphere. Hence, the current atmosphere is not in equilibrium with respect to ^{14}C. With time, this excess ^{14}C decays and the activity of atmospheric ^{14}C decreases. The ^{14}C activity of the atmosphere at the time a sample is collected must be known in order to determine the fraction of modern carbon in the sample. ^{14}C concentrations are determined with respect to the ^{14}C content of the NIST Standard Reference Material 4990C (oxalic acid). The *percent modern carbon (pmc)* is *equal to 0.95 times the ^{14}C content of the NIST standard*. To determine the *percent carbon of biogenic origin* in a sample, the *measured ^{14}C content (in pmc) is divided by the ^{14}C content of the atmosphere at that particular time (in pmc) and the result is multiplied by 100.*

EXAMPLE 8–12 Glasius et al. (2000) measured the ^{14}C content of formic acid in rural aerosols in Denmark. The sample collection was done in 1997, at which time pmc equalled 110.5 for the atmosphere. For sample AAR-3678, the measured ^{14}C content was 88.7 pmc. Calculate the percent biogenic carbon in the formic acid.

$$\% \text{ Biogenic} = \frac{(^{14}C)_{measured}}{(^{14}C)_{modern}} \times 100 = \frac{88.7}{110.5} \times 100 = 80 \qquad ∎$$

With modern chromatography techniques it is possible to separate specific organic molecules or a group of related molecules. This allows us to fingerprint the source(s) of a particular species. For example, VOCs (volatile organic compounds) can have both anthropogenic and natural sources. Plant exudates are a major source of VOCs. Using the ^{14}C activity of a specific VOC, it is possible to determine the relative amount of the VOC derived from natural sources.

As was discussed in Chapter 6, stable isotopes of carbon have different masses depending on the biological pathway—C_3, C_4, or CAM—that was responsible for the synthesis of the compound. For example, for *n*-alkanes, $\delta^{13}C = -31‰$ to $-39‰$ for C_3

plants, $\delta^{13}C = -18‰$ to $-25‰$ for the C_4 plants, and $\delta^{13}C = -25‰$ to $-27‰$ for the CAM plants (Collister et al., 1994). The C_4 plants are found in a savannah; the C_3 and CAM plants are found in a wet climate with mixed vegetation. The carbon isotopic composition of petroleum-derived n-alkanes is similar to that of the C_3 plants and, hence, is isotopically indistinguishable from natural sources.

The distinction between petroleum and natural n-alkanes can be made using the **carbon preference index (CPI)**, which is the *sum of the odd carbon number homologs divided by the sum of the even carbon number homologs*. A **homologous series** *consists of a series of carbon compounds in which each member differs from the next member by a constant amount*. The alkanes form such a series because the difference between each member is CH_2 (CH_4, C_2H_6, C_3H_8, etc.). A **homolog** is a *member of a homologous series*. Carbon compounds derived from petroleum have a CPI of approximately 1; natural compounds have a CPI significantly greater than 1. As an example of how this approach can be applied, Pio et al. (2001) determined the CPI for n-alkanes (1.8), n-ketones (2.05), n-alkanols (7.9), and n-alkanoic acids (3.0) from lipids in aerosols collected in rural Portugal. The low CPI for the n-alkanes and n-ketones indicated a significant anthropogenic input. The other species had a largely biological origin.

Temporal Variations in Particulate Compositions

Temporal variations are determined in a number of ways: tree rings (e.g., variations in Pb isotopes, Tommasini et al., 2000), plants of different age (e.g., variations in Pb deposition using sphagnum mosses, Weiss et al., 1999), snow and ice cores, and lake (e.g., trace metals in lake sediments, Gélinas et al., 2000) and ocean sediments (e.g., Hg and As in ocean sediments, Siegel et al., 2001). The most intensely studied ice cores are those from Antarctica and Greenland. We have previously discussed the reading of the climate record using stable isotope and gas composition data from ice cores. These same cores have also been analyzed for a variety of trace metals. Because we are mainly concerned with anthropogenic effects, only the past several thousand years are of real interest.

The components of these types of studies are broadly similar. A chronologic record must be established, the metals of interest extracted and analyzed (because in many cases we are dealing with very low absolute abundances, very sensitive analytical techniques must be used), questions of element mobility must be addressed, and the data interpreted. Data interpretation is done using the same techniques previously developed for interpreting atmospheric aerosol data. Several problems at the end of the chapter deal with these types of data analysis. Depending on the monitoring system that is used, the chronologic record can be developed from historical information, counting of tree rings, counting of sediment layers and ice layers, and the various radiometric techniques developed in Chapter 6. Of concern with studies of this type is the question of element mobility in the natural environment. For example, if we are looking at elemental variations in a sediment core, the pore spaces are filled with water and there is the possibility that an element will be distributed over some distance. Some elements are considered to be mobile (such as S, Mn, Fe, As, Mo, and Ba), whereas others are considered to be immobile (V, Cr, Co, Ni, Cu, Zn, Cd, Sn, Sb, Hg, Pb, and Bi). In practice, element mobility should be determined for the system of interest because factors such as pH (Chapter 3) and oxidation–reduction potential (Chapter 4) play a role in element mobility.

Finally, in terms of data interpretation, we must also consider possible variations in the general circulation of the atmosphere. For example, if we are investigating the change in Pb content in an ice core, variations in the prevailing wind directions may have a significant influence on Pb deposition. If the winds are from a region with low anthropogenic inputs, the Pb deposition will be low, but if the prevailing wind direction changes and the trajectories are now across an industrialized area, there will be a significant increase in Pb deposition. The ice core records the variations in Pb deposition with time, but if we do not take into account changes in prevailing wind direction, we may interpret these data to indicate periods of lesser and greater Pb emissions—in this case, an incorrect inference.

QUESTIONS AND PROBLEMS

1. Distinguish between *temperature* and *heat*.

2. Define *tropopause*, *stratopause*, and *mesopause*.

3. What are the distinguishing characteristics of the *troposphere, stratosphere, mesosphere*, and *thermosphere*?

4. What is a *perfect radiator*?

5. Distinguish between *longwave* and *shortwave* radiation.

6. What is the *solar constant*?

7. Define *albedo*.

8. What is *insolation*?

9. Draw a graph showing the fluxes of incoming and outgoing radiation as a function of latitude. What is the significance of the observed relationship in terms of atmospheric circulation?

10. What is the *Coriolis force*?

11. Draw the general circulation pattern for the earth, both horizontal and vertical, and explain the reasons for the observed pattern.

12. Distinguish between *absolute humidity* and *relative humidity*.

13. Distinguish between *dry adiabatic, wet adiabatic*, and *environmental lapse rates*.

14. What is an *inversion*? How do inversions form, and what is their significance in terms of air pollution?

15. Define *air pollution*.

16. Distinguish between *primary* and *secondary pollutants*.

17. Define *aerosol*.

18. Define *smog*. Name the two major types of smog and list their characteristics and environmental impact.

19. Case Study 8–1 presented one author's view of greenhouse warming. This article, which appeared in *Science*, should be readily accessible. Read the article and answer the following questions. You may need to refer to other sections of this book and, perhaps, articles referenced by the author.

    a. What assumptions and simplifications did the author make in the construction of the time–temperature variation curve? What steps did he take to validate his time–temperature curve?

    b. Are there other variables, besides volcanic emissions, variations in solar radiation, and greenhouse gases, that could affect atmospheric temperature? Did the author satisfactorily account for these other possible variables?

    c. How could ^{14}C data from tree rings be used to reconstruct variations in solar radiation? You might want to refer to Chapter 6 for help in answering this question.

    d. How did the author test his model for temperature variations, and what steps did he take to convince the reader that his model provided an appropriate representation of future temperature variations?

20. Define *radiative forcing*.

21. Define and distinguish between *RIRF* and *GWP*.

22. Why does the CO_2 curve shown in Figure 8–8 have a sawtooth shape?

23. What are the major anthropogenic sources of CO_2?

24. With reference to Figure 8–9, what are the major short- and intermediate-term sinks for CO_2? What is the ultimate repository for CO_2?

25. With reference to Case Study 8–2, describe how carbon isotopes were used to determine the amount of anthropogenic CO_2 transferred from the atmospheric reservoir to the ocean reservoir.

26. What is the *biological pump*, and how does it work?

27. What is a *bookkeeping model*? Refer to the paper on which Case Study 8–3 is based to answer this question.

28. What are the major sources and sinks for atmospheric methane?

29. What is a *free radical*?

30. What are the major conclusions of Case Study 8–4, and what line of reasoning did the authors use to arrive at these conclusions?

31. With reference to Case Study 8–5, describe and discuss how the authors used carbon and oxygen isotopes to develop a model for the warming trend at the end of the Paleocene.

32. How does *good ozone* differ from *bad ozone*?

33. Discuss the formation and destruction of stratospheric ozone, and write the appropriate equations that describe these processes.

34. Explain why the ozone maximum, over the equator, occurs at an altitude of approximately 25 km.

35. Define *Dobson unit*.

36. Write equations for reactions involving NO and Cl that lead to the destruction of stratospheric ozone.

37. What is the *Antarctic ozone hole*, and how does it form?

38. List and describe the factors that determine indoor radon concentrations.

39. With reference to Case Study 8–8, why did the radon content of individual homes vary seasonally?

40. What are the natural and anthropogenic sources for atmospheric SO_2 and NO_x?

41. What is *NADP*? Go to the NADP website and look at isopleth maps for SO_4 from 1994 to the present. Describe and explain the spatial and temporal changes that have occurred over this time.

42. How can the sulfate/nitrate ratio be used to identify the source(s) of acid components in rainwater?

43. In some geographic locations rainwater has a pH > 5.7. Suggest a possible cause for these relatively high pH values.

44. With reference to Case Study 8–9, why is a knowledge of past climatic conditions important when interpreting acid deposition data from a single locality?

45. Name and briefly describe the four major types of particulates.

46. Define *soot* and *black carbon*. How do these definitions differ?

47. Distinguish between *organic aerosols* and *bioaerosols*.

48. Distinguish between *homogeneous condensation* and *heterogeneous condensation*.

49. What are *cloud condensation nuclei*?

50. What is meant by the term *hygroscopic*?

51. How are aerosols removed from the atmosphere?

52. Distinguish between *dry* and *wet deposition*.

53. Soot particles were collected at four different locations. Partial chemical data for the particles is reported in the following table. Assuming only a single source for each sample, classify the samples as to origin: jet engine, automobile, coal-fired power plant, or forest fire. You may want to refer to Chapter 5 for data concerning the trace element content of coal and oil.

Sample	Concentrations (ppm)					
	K	V	Cr	Ni	Zn	Pb
1	0	0.2	0.1	0.1	0.1	0.01
2	50	0.5	0.3	0.1	1.0	0.01
3	1.0	50	1.0	100	2.0	0.05
4	1.0	6.0	10	2.0	30	5.0

54. Why is the size of an aerosol particle important in determining the source of metals in the aerosol?

55. Why are air mass trajectories important in source identification?

56. Define *enrichment factor*.

57. What is *factor analysis*, and how is it used in air pollution studies (see Case Study 8–10)?

58. What is *percent modern carbon*, and how is it measured?

59. Lipids extracted from an aerosol sample have an average $\delta^{13}C = -21.2‰ \pm 1.1‰$. What can you conclude about the source of the lipids found in the aerosol?

60. Define *carbon preference index*.

61. Define *homologous series* and *homolog*.

62. In a far-off galaxy, a planet with an earth-like atmosphere revolves about a star similar in size to the sun but with a surface temperature of 9500 K. Remarkably, the average distance of this planet from the star is the same as the average distance of the earth from the sun.
    a. For both our sun (surface $T = 5770$ K) and the distant star, calculate the energy emitted per unit surface area and the wavelength of maximum radiation.
    b. Predict the value of the solar constant for the planet in this far-off galaxy. How did you arrive at this value? Do you think this distant planet would be warmer or colder than the earth? Explain your answer.
    c. Refer to Figure 8–3, which shows the absorption spectra for the various atmospheric gases. Would ozone loss from the atmosphere of the distant planet be more or less significant than in the case of the earth? Explain your answer.

63. The surface temperature of Venus is estimated to be 673 K. Calculate the wavelength of maximum energy for Venus. Only in terms of carbon dioxide, will the Venusian atmosphere be a better absorber of back radiation than the earth's atmosphere? Explain your answer.

64. An electric utility is planning to build a coal-fired electric generating plant in Your Town. During the winter, subsidence inversions often occur in this area. During a typical inversion, the air temperature found at 2000 m, at the top of the inversion, is 10°C higher than that found at the surface. From the top of the inversion upwards, the atmosphere usually cools, with a lapse rate of 7°C km^{-1}. The company plans to build a smokestack 100 m high and the gases will emerge from the smokestack at a temperature of 30°C. Assume that the gases emerging from the smokestack cool at the dry adiabatic lapse rate. On a day when the surface temperature is 5°C, will air pollution be a problem? Explain your answer. This problem is best solved graphically using an altitude–temperature plot.

65. In Case Study 8–6, Ivany et al. (2000) used ^{18}O isotopic variations in aragonite otoliths to determine paleotemperatures across the Eocene–Oligocene boundary. For a Late Eocene otolith, $\delta^{18}O$ varies from -0.5 to $-1.5‰$. For an Early Oligocene otolith, $\delta^{18}O$ varies from 0.8 to $-2.0‰$. Using the aragonite geothermometer of Horibe and Oba (1972), calculate the temperature range for both otoliths.

$$T\,(°C) = 13.85 - 4.54(\delta_c - \delta_w) + 0.04(\delta_c - \delta_w)^2$$

For Late Eocene seawater, $\delta^{18}O = -0.8‰$, and for Early Oligocene seawater, $\delta^{18}O = -0.5‰$. Comment on the significance of the results in terms of climatic change across the Eocene–Oligocene boundary.

66. In Case Study 8–7, the boron isotopic system was used to estimate the pH and partial pressure of atmospheric CO_2 in the geologic past. An ancient marine carbonate sample gives $\delta^{11}B = +24.7$. Ancient seawater had $\delta^{11}B = +40.0$. Assume that the seawater temperature was 25°C and that the total carbonate content of the ocean was 2×10^{-3} mol L^{-1}. Calculate the partial pressure of CO_2, in atm, for the ancient atmosphere. You will first need to calculate $\delta_{B(OH)_3}$. Remember that the $\delta^{11}B$ of the carbonate represents $\delta_{B(OH)_4^-}$ because this species is selectively incorporated in the carbonates. Next calculate the relative proportions of $B(OH)_3$ and $B(OH)_4^-$ using the isotopic value for seawater. Then calculate the pH of the seawater. You can now determine P_{CO_2} using the carbonate equilibrium relationships discussed in Chapter 3. Is the calculated P_{CO_2} greater or less than that of the present-day atmosphere?

67. Calculate the minimum energy (and wavelength) that a photon must have to split an ozone molecule. For this reaction, $\Delta H_R^0 = +386.5$ kJ mol^{-1}.

68. One set of ozone-destroying reactions involves NO. The reactions are

$$NO + O_3 \rightarrow NO_2 + O_2$$
$$NO_2 + O \rightarrow NO + O_2$$
$$\overline{O + O_3 \rightarrow 2O_2}$$

Calculate the reaction rate for each step at an altitude of 30 km and a temperature of 235 K. Which step controls the overall rate of the cycle?

69. Radon released from soil surrounding a basement enters the basement at a rate of 0.1 Bq L^{-1} h^{-1}. The outdoor air has a radon activity of 4×10^{-3} Bq L^{-1}. Assume a steady-state situation. During the summer the air exchange rate constant is 10 h^{-1}. During the winter, when the house is tightly sealed, the air exchange rate constant is 0.5 h^{-1}. Calculate the steady-state radon activity during the summer and during the winter. Is remedial action required in either case?

70. A radon activity of 200 Bq L^{-1} is measured in soil overlying a mylonite zone. A house with a basement is built on this mylonite zone. Assuming that the radon enters the basement at a rate of 2 Bq L^{-1} h^{-1}, an outdoor radon activity of 4×10^{-3} Bq L^{-1}, and an air exchange rate constant of 5 h^{-1}, calculate the steady-state radon activity. Is remedial action required?

71. A conglomerate has eU = 5.4 ppm and a well-cemented sandstone has eU = 0.9 ppm. A street runs perpendicular to the contact between the sandstone and the conglomerate. At one end of the street, a house with a basement is built on the sandstone. At the other end of the street, a slab house is built on the conglomerate. For the neighborhood, the average indoor radon activity is 80 Bq m^{-3}. The geology factor for the sandstone is rated as negative because it has a low eU and, because it is well cemented, little permeability. The geology factor for the conglomerate is rated positive because it has a high eU, and field observation has indicated that the conglomerate is very permeable (in fact, an aquifer). Calculate the radon index for each home. What is the probability of the home with the basement having a high indoor radon content? What is the probability of the slab home having a high indoor radon content?

72. A sandstone aquifer has an average uranium concentration of 10 ppm. The porosity is 30% and the bulk density is 2160 kg m^{-3}. Given an emanation efficiency of 0.2, calculate the radon activity for groundwater in equilibrium with the granite. Should homeowners who derive their water from wells drilled in this aquifer be concerned about radon activity?

73. The average composition for precipitation at Hubbard Brook, New Hampshire, is given in the accompanying table.

Species	Concentration (mg L^{-1})	Species	Concentration (mg L^{-1})
Cl$^-$	0.51	Mg^{2+}	0.05
Na$^+$	0.12	Ca^{2+}	0.17
SO$_4^{2-}$	2.87	K$^+$	0.07

a. Relative to ocean-derived rainwater, calculate the excess or deficiency of each of these species.

b. What assumption did you have to make in order to do these calculations?

c. Given the calculated excesses and deficiencies, what is the source of each of the species in the rainwater?

74. The average composition for precipitation at Menlo Park, California, is given in the accompanying table.

Species	Concentration (mg L^{-1})	Species	Concentration (mg L^{-1})
Cl$^-$	3.43	Mg^{2+}	0.37
Na$^+$	2.00	Ca^{2+}	0.79
SO$_4^{2-}$	1.39	K$^+$	0.25

a. Relative to ocean-derived rainwater, calculate the excess or deficiency of each of these species.

b. Given the calculated excesses and deficiencies, what is the source of each of the species in the rainwater?

75. Precipitation at Katherine, Northern Territory, Australia, has SO$_4^{2-}$ = 0.19 mg L^{-1} and NO$_3^-$ = 0.25 mg L^{-1}. The measured pH = 4.74. Calculate the pH for the Katherine precipitation. How does the calculated value compare to the measured value?

76. Precipitation at Torres del Paine, Chile, has SO$_4^{2-}$ = 0.211 mg L^{-1} and NO$_3^-$ = 0.031 mg L^{-1}. The measured pH = 5.31. Calculate the pH for the Torres del Paine precipitation. How does the calculated value compare to the measured value?

77. The following data were collected for precipitation at Beijing, China (Zhao and Sun, 1986):

Species	Concentration (mg L^{-1})	Species	Concentration (mg L^{-1})
Na$^+$	3.24	SO$_4^{2-}$	13.11
K$^+$	1.57	NO$_3^-$	3.11
Ca^{2+}	3.68	NH$_4^+$	2.54
Cl$^-$	5.59	pH	6.8

a. Calculate the pH for the Beijing precipitation. Remember sulfate and nitrate are the major contributors to acidity.

b. Does the calculated pH agree with the measured pH?

c. Using the measured pH, calculate the bicarbonate ion concentration. Assume that the temperature is 25°C.

d. Write a charge balance equation using all the positive and negative species. Is there a charge balance?

e. Taking into account the results from part (d), explain any difference that exists between the observed pH and the calculated pH.

78. Samples of aerosols were collected at a particular location on the open ocean on three consecutive days. The accompanying table gives the atomic elemental ratios for these three samples and for seawater. A polluted air mass with a high burden of sulfur dioxide moved into the area during the time the samples were collected.

	Cl/Na	S/Na	Mg/Na	K/Na	Ca/Na
Seawater	1.16	0.06	0.11	0.022	0.022
Sample 1	1.16	0.06	0.11	0.022	0.022
Sample 2	0.64	0.48	0.11	0.022	0.022
Sample 3	0	1.0	0.11	0.022	0.022

a. Interpret sample 1 and sample 3 in terms of source and aerosol modification.
b. Interpret sample 2. If you believe this is a mixed sample, identify and calculate the percents of the end members.

79. A sulfuric acid aerosol particle has a density of 1100 kg m^{-3} and a diameter of 30 μm. A dust particle has a density of 2300 kg m^{-3} and a diameter of 30 μm.
a. Calculate the settling velocity for both particles. How do they compare?
b. Calculate the diameter of a sulfuric acid aerosol particle settling at the same velocity as the dust particle.

80. The accompanying elemental concentrations were determined for an atmospheric aerosol sample.

Element	Concentration (μg m^{-3})	Element	Concentration (μg m^{-3})	Element	Concentration (μg m^{-3})
Al	2130	Cr	1.0	Sb	0.25
Fe	932	Zn	1.2	W	1.55
Sc	0.3	As	1.8	Pb	3.2

a. Using Al as the crust reference element, calculate the crust enrichment factor (relative to upper crust, Table 8–14) for each of the other elements. Which elements are significantly enriched over crust abundances? Given the uncertainty in these types of calculations, if the EF is less than 3, the element is considered to have a crust source.
b. Calculate the noncrustal amount of each element in the aerosol sample.
c. From the crust enrichment factors, what inferences can you make about the source(s) for the aerosol?

81. An air sampler is used to collect particles less than 10 μm in diameter. The sampler is located downwind from a steel plant. The accompanying data are for particulate samples collected on two different days.

	Sample 1	Sample 2
Element	Concentration (ppm)	Concentration (ppm)
Al	2130	1940
Fe	1024	1520
Mn	16.8	20.6
Cr	1.2	2.9
Ni	0.7	1.8

a. For both samples, calculate the noncrustal abundance of Fe, Mn, Cr, and Ni. Use Al as the reference element. Use the upper crust values from Table 8–14.

b. For particles emitted by the steel plant, $Mn/Fe = 0.009$, $Cr/Fe = 0.003$, and $Ni/Fe = 0.002$. Using the noncrustal abundances of Fe, Mn, Cr, and Ni, do emissions from the steel plant contribute to the aerosol chemistry? Explain.

c. If your answer to part (b) is yes, what percent of aerosol sample 1 and aerosol sample 2 is due to emissions from the steel plant?

82. A dichot sampler was used at Nahant, Massachusetts, to separate aerosol particles into two size ranges, 10 μm to 2.5 μm and < 2.5μm (Case Study 8–10). The accompanying data table lists concentrations for some of the elements found in the Nahant particles (data from Underhill, 1995).

	< 2.5μm (ppm)				10 μm to 2.5μm (ppm)			
Sample	Fe	Na	Cr	Sb	Fe	Na	Cr	Sb
1	20415	20881	1078	84	36966	179592	145	71
2	34247	32576	1689	175	67192	144368	566	97
3	5583	15065	369	55	17017	32103	112	36
4	1357	18479	102	19	4494	114859	54	8
5	1998	8747	76	15	3980	23263	48	5
6	160907	44640	374	nd	418201	28913	422	nd
7	4850	50762	229	9	3889	21990	71	25
8	4640	43741	74	4	1905	18561	60	22
9	171259	15310	418	57	6286	26847	173	20
10	7977	96528	271	31	5698	75389	28	17
11	13088	136982	231	45	15681	253655	231	nd
12	6227	67051	324	16	10747	34339	59	22
13	11025	36870	166	108	26503	113185	192	70
14	8548	93871	533	44	20267	149843	215	29
15	26879	149837	157	24	11779	103022	351	73
16	11785	107719	124	15	8028	52197	313	36

a. For both the fine and coarse particulates, construct the following logarithmic plots: Cr/Fe vs. Na/Fe, Cr/Fe vs. Se/Fe, Sb/Fe vs. Na/Fe, and Sb/Fe vs. Se/Fe. In these plots Fe is used as the crust reference element. The plots are easily done using a spreadsheet.

b. Interpret the plots and estimate the crust, marine, and anthropogenic (pollution) ratios for Cr and Sb. You may want to refer to Rahn (1999) for additional information.

c. Are there any differences between the fine and coarse particulates in terms of the source(s) of Cr and Sb?

83. The accompanying table lists concentrations for some of the elements found in the dry deposition at Nahant, Massachusetts (see Case Study 8–10, data from Underhill, 1995).

Sample	Elemental concentrations (ppm)					
	Fe	Na	Se	Cr	Sb	Sc
600	196648	61926	18.0	814	99.5	45.6
601	331519	42293	62.4	1506	45.5	92.0
604	71470	16439	20.1	358	11.7	10.8
606	136202	142618	25.9	1220	20.5	34.2
607	315507	132056	71.9	1687	30.5	91.8
609	395531	179382	17.8	2152	106.5	99.6
611	93773	36937	nd	426	15.4	24.9
614	83121	34011	7.5	849	12.2	17.3
618	214130	84661	nd	901	97.8	51.0
623	155502	50738	15.0	2006	35.7	35.8
627	113669	46882	nd	444	28.4	26.2
631	152369	52858	25.3	968	51.8	43.2
635	182099	79145	32.1	798	100.3	43.3
643	445281	187117	46.8	1525	113.1	121.0
647	219412	99236	nd	793	75.3	54.0
651	161082	41526	85.5	577	97.8	33.3
655	292078	131253	20.1	1162	42.3	72.9
659	244506	101157	31.4	1081	131.1	64.7
663	272539	117891	52.5	1133	159.2	68.1
667	113128	45424	13.6	592	31.7	26.7
671	99260	47942	nd	783	30.4	30.6
675	468623	208053	52.3	1703	116.0	123.5
679	541651	235780	12.8	1539	91.8	137.8
683	725924	312133	29.5	2291	115.6	208.4
687	503814	240355	25.9	1480	73.5	147.4
691	733165	318919	49.8	2632	123.8	195.7
695	388366	168338	nd	1122	135.7	100.2
699	685132	304651	nd	1878	105.6	190.0
701	211676	112937	nd	945	69.7	51.0
703	140205	61701	nd	703	26.5	37.6
707	527247	199747	11.7	1689	114.0	133.7
711	163972	65609	nd	869	31.3	39.3

a. Construct the following logarithmic plots: Cr/Fe vs. Na/Fe, Cr/Fe vs. Se/Fe, Sb/Fe vs. Na/Fe, Sb/Fe vs. Se/Fe, Sc/Fe vs. Na/Fe, and Sc/Fe vs. Se/Fe. In these plots Fe is used as the crust reference element. The plots are easily done using a spreadsheet.

b. Interpret the plots and estimate the crust, marine, and anthropogenic (pollution) ratios for Cr, Sb, and Sc. You may want to refer to Rahn (1999) for additional information.

c. How do the source(s) for the various elements compare to those suggested in Case Study 8–10? Comment on any differences.

84. Despite its distant location from significant sources of air pollutants, Bar Harbor, Maine (site of Acadia National Park) does have a significant air pollution problem, particularly during the summer months. The accompanying table gives the concentration of As, V, and Se in a Bar Harbor aerosol sample and average values for the three possible sources of the Bar Harbor contaminants. The data are the noncrustal concentrations for these elements.

Element	Boston	Concentrations (ppm) New York City	Interior U.S.	Bar Harbor
As	0.49	2.0	1.54	1.15
V	35	20	3.4	24.18
Se	1.0	1.88	1.78	1.42

Assuming that Boston, New York City, and the interior United States are the only possible sources, calculate the percent contribution of each source to the Bar Harbor aerosol. There are three unknowns and three equations that relate these unknowns, so the problem can be solved by solution of simultaneous equations.

85. Pb isotopic measurements of an aerosol sample give $^{206}Pb/^{204}Pb = 18.673$, $^{207}Pb/^{204}Pb = 15.667$, and $^{208}Pb/^{204}Pb = 38.857$. Calculate the $^{206}Pb/^{207}Pb$ and $^{208}Pb/^{206}Pb$ ratios for this sample.

86. The Pb isotopic ratios for an aerosol sample and the Pb isotopic ratios for the three sources that can contribute lead to this aerosol follow:

	$^{206}Pb/^{204}Pb$	$^{207}Pb/^{206}Pb$	$^{208}Pb/^{206}Pb$
Aerosol	17.252	0.9006	2.1466
Coal	18.525	0.8468	2.0895
Gasoline	16.764	0.9217	2.1697
Soil	18.126	0.8606	2.0995

Calculate the relative contribution of each source to the aerosol lead.

87. With reference to Case Study 8–13, graphically calculate the relative proportions of gasoline-derived lead and coal-derived lead in the Oslo air filters, the south Norway bark, and the south Norway wood soot.

88. Véron et al. (1999) used Pb isotopes to identify the sources of lead pollution in northwestern France. The data in the following table were collected in 1994 and 1995. The potential anthropogenic sources are gasoline, a steel metallurgy plant, a Fe-Mn metallurgy plant, and a lead smelter. Wimereux is located on the coast. The Mercantour samples are from the French Alps. The North Atlantic sediments are considered to represent natural inputs. The Mercantour and North Atlantic Pb ratios were estimated from Figure 5 in Véron et al. (1999).

$^{206}Pb/^{207}Pb$	$^{208}Pb/^{206}Pb$	$^{206}Pb/^{207}Pb$	$^{208}Pb/^{206}Pb$	$^{206}Pb/^{207}Pb$	$^{208}Pb/^{206}Pb$
Gasoline		Lead smelter		Paris	
1.1314	2.1223	1.1336	2.1263	1.1306	2.1284
1.1366	2.1193	1.1323	2.1275	1.133	2.1262
1.1371	2.1178	North Atlantic sediments		1.1483	2.1147
Steel metallurgy plant		1.2215	2.0525	Mercantour	
1.2226	2.0035	1.2143	2.0562	1.1822	2.0824
1.2064	2.0189	1.2115	2.0594	1.1654	2.0956
1.1944	2.0315	1.2084	2.0618	1.1574	2.1019
1.1829	2.0466	1.2044	2.0639	1.1521	2.1060
1.1786	2.0537	Wimereux		1.1451	2.1116
1.1897	2.0390	1.1512	2.1035		
Fe-Mn metallurgy plant		1.1468	2.1072		
1.1176	2.1406	1.1461	2.1057		
1.1064	2.1575	1.1516	2.1003		
1.1099	2.1537	1.1534	2.0981		

    a. Plot the data on a $^{208}Pb/^{206}Pb$ vs. $^{206}Pb/^{207}Pb$ diagram.

    b. Interpret the data in terms of the sources of Pb in the aerosol samples collected at Wimereux and Paris. Using the graph, estimate the percent contribution of the various sources.

    c. Interpret the Pb isotopic data for the Mercantour samples. Remember that these samples were collected in the Alps, away from any immediate source of anthropogenic Pb.

89. Formic acid collected in an aerosol sample at Anholt, Denmark, had ^{14}C = 98.6 pmc (Glasius et al., 2000). At the time this sample was collected, ^{14}C = 110.5 pmc for the atmosphere. Calculate the percent biogenic carbon in the formic acid sample.

90. Plant wax alkanes were extracted from an atmospheric aerosol. The composition of the alkanes was determined by GC-MS analysis. The results are reported in the accompanying table.

Carbon number	Concentration (ng m^{-3})	Carbon number	Concentration (ng m^{-3})
20	5	29	33
21	10	30	22
22	10	31	17
23	20	32	8
24	13	33	7
25	18	34	4
26	13	35	5
27	21	36	3
28	20		

    a. Plot the concentration of each alkane versus the carbon number.

    b. Calculate the CPI for the sample. What is the source of the lipid? Why?

    c. For petroleum-derived n-alkanes, CPI = 1. Subtracting the petroleum contribution gives the biological component. This is done by subtracting the average concentration of even-numbered alkanes from the concentration of the intervening odd-numbered alkane. Negative values are taken as zero. If you need an equation: $C_{odd} = C_{odd} - (C_{odd-1} + C_{odd+1})/2$. Calculate the biological component for this lipid sample.

91. The following table lists Zn, Pb, and Al concentrations for ice samples from the 3028.8-m European Greenland Ice Core Project (GRIP) deep ice core drilled in 1990–1992 at Summit in central Greenland. Also listed are the depths and estimated age for each of the samples. The Zn and Al data are from Hong et al. (1997). The lead concentrations are estimated from Figure 1 in Hong et al. (1994).

    a. Calculate the enrichment factor for Zn and Pb. Use *bulk crust* for the normalization.

    b. Plot the enrichment factor for each element versus age. Do this on the same graph so you can compare the results. The two oldest samples are included to give the prehuman abundances for these elements. You may want to use only the period 400 to 3000 years for your plot so you can see the details. All of this is easily done using a spreadsheet program.

    c. Interpret the data in terms of human history for the past 8000 years. You may find a history book or the papers on the which the problem is based useful in answering this question.

    d. Recalculate the enrichment factor for Zn and Pb using *upper crust* for the normalization. Replot the data using the recalculated enrichment factors. If you had done this plot first, would you have arrived at the same conclusions regarding the natural and anthropogenic inputs of Zn and Pb? Explain.

Depth (m)	Age (y)	Measured concentrations (pg g^{-1})		
		Zn	Pb	Al
129.3	471	18	3.9	7600
184.2	723	50	4.1	14000
239.3	985	27	2.4	11000
294.3	1248	32	1.25	9400
349.3	1521	26	0.66	6900
399.3	1774	19	0.8	6200
401.0	1783	17	1.55	8200
404.3	1801	19	1.05	7200
416.9	1866	14	1.05	5800
422.4	1894	25	1.2	6600
430.7	1936	28	1.5	7700
435.1	1958	18	1.2	5900
446.1	2015	19	1.4	6300
457.1	2073	34	3.0	16000
459.3	2084	27	1.3	5500
469.2	2137	20	1.3	6200
492.3	2260	43	2.1	14000
509.3	2351	23	1.75	9700
511.0	2360	18	1.64	8300
569.3	2674	11	0.65	3900
619.3	2956	19	0.35	5100
1230.4	7290	23	0.80	11000
1286.5	7760	23	0.55	6900

92. Weiss et al. (1999) determined the lead isotopic ratios for herbarium *Sphagnum* mosses in Switzerland. The accompanying data were used to reconstruct past atmospheric lead deposition in Switzerland.

Sample ID	Year	$^{206}Pb/^{204}Pb$	$^{207}Pb/^{204}Pb$	$^{208}Pb/^{204}Pb$
16519	1992	17.77	15.57	37.57
13596	1988	17.49	15.54	37.31
12793	1987	17.44	15.55	37.26
11166	1985	17.46	15.54	37.25
10046	1984	17.47	15.54	37.25
1755	1975	17.78	15.60	37.71
19449	1969	17.93	15.57	37.83
2293/3	1951	18.13	15.59	38.02
2293/6	1896	18.29	15.61	38.26
2293/1	1895	18.35	15.61	38.34
2293/5	1894	18.36	15.62	38.36
2293/2	1867	18.39	15.66	38.48

a. Calculate the $^{206}Pb/^{207}Pb$ ratios for the samples.

b. Plot $^{206}Pb/^{204}Pb$ and $^{206}Pb/^{207}Pb$ ratios versus age (two graphs).

c. Leaded gasoline was introduced in Switzerland in 1947 and unleaded gasoline was introduced in 1985. Interpret the lead isotopic data in the context of these two events.

d. Plot $^{206}Pb/^{207}Pb$ vs. $^{206}Pb/^{204}Pb$ ratios. Interpret the graph in terms of what you know about isotopes in general and lead in particular. You may want to refer to the paper of Weiss et al. (1999).

93. Velde et al. (2000) measured Ag, Au, Pt, Pd, and Rh concentrations at various depths in a 140-m snow/ice core from the French-Italian Alps. A subset of their data follows:

Age	Measured concentrations (pg g^{-1})				
	Ag	Au	Pt	Pd	Rh
1991	0.3	0.07	0.08	0.5	0.01
1987	4.3	0.11	0.24	4.9	0.21
1981	3.2	0.35	0.26	8.0	0.18
1978	0.3	0.08	0.16	0.6	0.03
1974	1.4	0.23	0.22	1.3	0.08
1970	0.9	0.23	0.08	0.1	0.07
1965	0.4	0.22	0.21	0.1	0.06
1960	1.7	0.35	0.38	2.1	0.10
1957	2.9	0.18	0.20	4.8	0.18
1954	0.5	0.07	0.14	0.7	0.03
1945	0.4	0.19	0.28	1.7	0.08
1940	0.2	0.18	0.17	2.2	0.05
1918	0.2	0.17	0.28	2.7	0.05
1916	0.8	0.20	0.26	3.7	0.06
1911	0.4	0.12	0.13	1.7	0.06
1887	0.2	0.14	0.17	1.0	0.02
1855	0.3	0.17	0.31	2.3	0.05
1803	0.3	0.24	0.38	2.1	0.07
1782	0.2	0.20	0.33	1.3	0.06
1778	0.4	0.20	0.34	1.3	0.04

   a. Calculate the Au/Pt and Pd/Rh ratios.
   b. Plot the elemental concentrations and the Au/Pt and Pd/Rh ratios versus time.
   c. What inferences can you draw about the variations in these elements with time? Are there any discernable trends?
   d. What can you conclude about the source(s) of these metals in atmospheric particulates? The following table lists Au/Pt and Pd/Rh ratios for potential sources (from Velde et al., 2000):

Source	Au/Pt	Pd/Rh
Upper continental crust	6.2	6.7
Sea-salt spray	0.25	0.05
Cosmic dust	4.2	1.4
Russian nickel ores	0.6–0.7	12.3–15.5

The Continental Environment

In this chapter we will consider the geochemistry of lakes, rivers, and groundwater and processes that occur in the surface environment. It is this environment that is most susceptible to pollution. We will start by investigating the natural processes that determine the geochemistry of waters and soils. We will then investigate anthropogenic inputs and their effect on these systems and the modeling of the transport of metals and other species.

THE HYDROLOGIC CYCLE

The various reservoirs for surface water, and their water content, are listed in Table 9–1. The major reservoir for water, by a significant margin, is the oceans. The second major reservoir is the polar ice caps and glaciers. These two reservoirs account for 98.93% of the water at the earth's surface. Note that most of the rest of the surface water (1.09%) is contained in the groundwater reservoir. Thus, that portion of the water cycle that we are most familiar with (or aware of), the lakes and rivers, accounts for only a very small part of the total surface waters. However, this is the part of the hydrologic cycle that has the shortest residence times and is most susceptible to change. In Chapter 8 we discussed the atmospheric reservoir and in Chapter 10 we will consider the oceanic reservoir. In this chapter we will discuss rivers, groundwater, and lakes.

Table 9–1 Inventory of Water at the Earth's Surface*

Reservoir	Volume 10^6 km^3	% of Total
Oceans	1400	95.96
Mixed layer	50	
Thermocline	460	
Abyssal	890	
Ice caps and glaciers	43.4	2.97
Groundwater	15.3	1.05
Lakes	0.125	0.009
Rivers	0.0017	0.0001
Soil moisture	0.065	0.0045
Atmosphere	0.0155	0.001
Terrestrial	0.0045	
Oceanic	0.0110	
Biosphere	0.002	0.0001
Total	1459	

*From Berner and Berner (1996).

The *hydrologic cycle* (Figure 9–1) *describes the transfer of water among the various water reservoirs.* Water evaporated from the ocean or rivers and lakes is transported through the atmosphere as water vapor. When this vapor condenses it reacts with gases and particulate matter in the atmosphere (see Chapter 8). These reactions lead to either acidic or basic precipitation, which contains various amounts of trace constituents. The precipitation may fall on either the ocean or the continents. Here we are concerned with the continental (land) portion of the cycle.

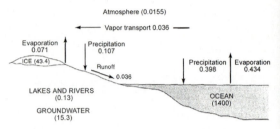

Figure 9–1
Schematic representation of the hydrologic cycle. Numbers in parentheses are the volume of water (10^6 km^3) in each reservoir. Fluxes are given in 10^6 km^3 y^{-1}. From Berner and Berner (1996).

If vegetation is present at the surface, the precipitation interacts with the vegetation and is modified by this interaction. This modification often consists of the addition of various organic molecules and interaction with dust particles attached to the vegetation. The precipitation subsequently flows to rivers and lakes by a series of processes including groundwater flow. A variety of rock (mineral)–water interactions occur during this transfer. These interactions, and their effect on water chemistry, are described in the following sections.

WEATHERING

Weathering can be defined as *the various physical and chemical processes that lead to the decomposition of minerals and the breakdown of rocks to form soil.* This weathering is of two types: physical and chemical. Physical weathering (which is the mechanical breakdown of material) increases the surface-to-volume ratio so that subsequent chemical weathering, which is a surface process, is more effective. Chemical weathering takes place through interactions between water and gases and solid particles. For example, rainwater in equilibrium with atmospheric CO_2 has a pH of 5.7. In areas of acid input the rainwater can be even more acidic. These acid waters react with the minerals in rocks and soils, breaking down individual minerals and in the process releasing various species to the water, thus modifying its chemical composition. Chemical weathering is the major contributor of dissolved species to surface and ground waters.

There are four major types of chemical weathering: oxidation, congruent dissolution by water, congruent dissolution by acids, and incongruent dissolution by acids.

1. Oxidation occurs when oxygen reacts with minerals that contain reduced forms of various elements, usually Fe and S, and oxidizes these elements. As an example of this type of reaction, we can write the following equation for the oxidation of pyrite:

$$4FeS_{2 \text{ pyrite}} + 15O_2 + 8H_2O \rightarrow 2Fe_2O_{3 \text{ hematite}} + 8H_2SO_4$$

Both Fe and S have been oxidized in this reaction. Hematite represents the final stable form that would be produced during this type of oxidation reaction, but intermediate species serve as precursors to the final product.

2. Congruent dissolution by water involves the simple dissolution of minerals, such as halite (NaCl) and anhydrite ($CaSO_4$), which are soluble in water. During congruent dissolution the ions go directly into solution, hence the dissolution of halite in water could be written

$$NaCl_{\text{ halite}} \rightarrow Na^+ + Cl^-$$

3. Congruent dissolution by acids occurs when a mineral is dissolved by acid attack. In the natural environment this acid is often a soil acid formed by organic processes. However, for simplicity, we usually use carbonic acid when writing

equations for acid attack. The silicate minerals, olivine, pyroxene, and amphibole, and the carbonate minerals normally dissolve congruently. As examples, consider the dissolution of olivine by acid attack, which can be written

$$Mg_2SiO_{4\ olivine} + 4H_2CO_{3\ (aq)} \rightarrow 2Mg^2 + 4HCO_3^- + H_4SiO_{4\ (aq)}$$

and the dissolution of calcite by acid attack, which can be written

$$CaCO_{3\ calcite} + H_2CO_{3\ (aq)} \rightarrow Ca^{2+} + 2HCO_3^-$$

In these equations we have used the aqueous form of carbonic acid and the bicarbonate ion. As we know from previous discussions, the carbonate species actually present in solution depend on the pH. For the silicate weathering reaction, one of the products is aqueous silicic acid. Except at very high pH values, virtually all the silicic acid will exist in the nondissociated form.

4. Incongruent dissolution by acids occurs when a mineral breaks down to ions in solution and a solid of different composition. Most silicate minerals undergo incongruent dissolution. A typical reaction is the chemical decomposition of feldspar, which can be written

$$2NaAlSi_3O_{8\ Na\text{-}plagioclase} + 2H_2CO_{3\ (aq)} + 9H_2O \rightarrow Al_2Si_2O_5(OH)_{4\ kaolinite}$$
$$+ 2Na^+ + 2HCO_3^- + 4H_4SiO_{4\ (aq)}$$

The Na-plagioclase is converted to kaolinite with the concomitant release of Na^+ and Si to solution.

Table 9–2 lists the common rock-forming minerals and the usual chemical decomposition process(es). The *minerals that are decomposed by the chemical weathering process*

Table 9–2 Decomposition Reactions for Common Primary Minerals*

Mineral	Composition	Commonly occur in rock type(s)	Reaction
Olivine	$(Mg,Fe)SiO_4$	Igneous	Oxidation of Fe Congruent dissolution by acids
Pyroxenes	$(Mg,Fe)SiO_3$ $Ca(Mg,Fe)Si_2O_6$	Igneous	Oxidation of Fe Congruent dissolution by acids
Amphiboles	$Ca_2(Mg,Fe)_5Si_8O_{22}(OH)_2$ (also some Na and Al)	Igneous Metamorphic	Oxidation of Fe Congruent dissolution by acids
Plagioclase feldspar	$NaAlSi_3O_8$ to $CaAl_2Si_2O_8$	Igneous Metamorphic	Incongruent dissolution by acids
K-feldspar	$KAlSi_3O_8$	Igneous Metamorphic Sedimentary	Incongruent dissolution by acids
Biotite	$K(Mg,Fe)_3(AlSi_3O_{10})(OH)_2$	Metamorphic Igneous	Incongruent dissolution by acids Oxidation of Fe
Muscovite	$KAl_3Si_3O_{10}(OH)_2$	Metamorphic	Incongruent dissolution by acids
Volcanic glass (not a mineral)	Ca,Mg,Na,K,Al,Fe-silicate	Igneous	Incongruent dissolution by acids Incongruent dissolution by water
Quartz	SiO_2	Igneous Metamorphic Sedimentary	Resistant to dissolution
Calcite	$CaCO_3$	Sedimentary	Congruent dissolution by acids
Dolomite	$CaMg(CO_3)_2$	Sedimentary	Congruent dissolution by acids
Pyrite	FeS_2	Sedimentary	Oxidation of Fe and S
Gypsum	$CaSO_4 \cdot 2H_2O$	Sedimentary	Congruent dissolution by water
Anhydrite	$CaSO_4$	Sedimentary	Congruent dissolution by water
Halite	$NaCl$	Sedimentary	Congruent dissolution by water

*Adapted from Berner and Berner (1996).

are referred to as ***primary minerals***. The *minerals that are produced by the weathering process*, such as hematite, gibbsite, kaolinite, etc., are referred to as ***secondary minerals***.

In an old but classic study, Goldich (1938) investigated the breakdown of minerals in the weathering environment. Observational evidence suggested that there was a regular pattern to the order in which minerals in igneous and metamorphic rocks decomposed during chemical weathering. Other studies have essentially confirmed the original observations of Goldich. Based on these previous studies, Berner and Berner (1996) prepared a list of mineral weatherability (Table 9–3).

Table 9–3 Mineral Weatherability Listed in Order of Increasing Resistance to Weathering*

Halite
Gypsum, anhydrite
Pyrite
Calcite
Dolomite
Volcanic glass
Olivine
Ca-plagioclase
Pyroxenes
Ca-Na plagioclase
Amphiboles
Na-plagioclase
Biotite
K-feldspar
Muscovite
Vermiculite, smectite
Quartz
Kaolinite
Gibbsite, hematite, goethite

---

*Modified from Berner and Berner (1996).

The list is headed by the nonsilicate minerals halite, gypsum-anhydrite, pyrite, calcite, and dolomite. Halite is readily soluble in water. Gypsum and anhydrite are relatively insoluble in water, but calcite and dolomite are relatively soluble in acid waters. Thus, a thermodynamic explanation of the weatherability of these minerals is not straightforward. A free-energy calculation for the oxidation of pyrite yields a large negative free energy, indicating that the reaction should go spontaneously to the right; i.e., the oxidation of pyrite is thermodynamically favored.

The situation is different for the silicate minerals, and Curtis (1976) was able to show that the weatherability of silicate minerals did follow thermodynamic principles. The weathering reactions investigated by Curtis and their corresponding free energies are listed in Table 9–4. Also included in this table is the free energy for each reaction on a gram atom^{-1} basis. This last value is calculated by dividing the free energy of the reaction by the number of product atoms. This calculation is required because the free energy determined for the reaction is a function of the chemical equation used to describe the reaction, while we want to make our comparison on the basis of the number of product atoms. The order of weatherability determined by this approach is in good agreement with observational evidence. The order of weatherability determined from the thermodynamic calculations is olivine (fayalite and forsterite) $\rightarrow$ pyroxene (clinoenstatite and diopside) $\rightarrow$ amphibole (anthophyllite and tremolite) $\rightarrow$ anorthite (Ca-plagioclase) $\rightarrow$ albite (Na-plagioclase) $\rightarrow$ K-feldspar (microcline) $\rightarrow$ muscovite. With the exception of anorthite (Ca-plagioclase), this is the same order shown in Table 9–3, which was based on the observed breakdown of various silicate minerals in the weathering environment.

Table 9–4   *Weathering Reactions and Free Energies per Gram Atom**

Mineral	Weathering reaction	ΔG_R^0 kJ mol^{-1}	ΔG_R^0 kJ (g atom)$^{-1}$
Fayalite	$Fe_2SiO_4 + \frac{1}{2}O_2 \rightarrow Fe_2O_3 + SiO_2$	−220.5	−27.53
Forsterite	$Mg_2SiO_4 + 4H^+ \rightarrow 2Mg^{2+} + 2H_2O + SiO_2$	−184.1	−16.74
Clinoenstatite	$MgSiO_3 + 2H^+ \rightarrow Mg^{2+} + H_2O + SiO_2$	−87.4	−12.47
Diopside	$CaMgSi_2O_6 + 4H^+ \rightarrow Mg^{2+} + Ca^{2+} + 2H_2O + 2SiO_2$	−159.4	−11.38
Anthophyllite	$Mg_7Si_8O_{22}(OH)_2 + 14H^+ \rightarrow 7Mg^{2+} + 8H_2O + 8SiO_2$	−574.0	−10.42
Tremolite	$Ca_2Mg_5Si_8O_{22}(OH)_2 + 14H^+ \rightarrow 5Mg^{2+} + 2Ca^{2+} + 8H_2O + 8SiO_2$	−515.5	−9.37
Anorthite	$CaAl_2Si_2O_8 + 2H^+ + H_2O \rightarrow Al_2Si_2O_5(OH)_4 + Ca^{2+}$	−100.0	−5.52
Albite	$2NaAlSi_3O_8 + 2H^+ + H_2O \rightarrow Al_2Si_2O_5(OH)_4 + 4SiO_2 + 2Na^+$	−96.7	−3.14
Microcline	$2KAlSi_3O_8 + 2H^+ + H_2O \rightarrow Al_2Si_2O_5(OH)_4 + 4SiO_2 + 2K^+$	−72.4	−1.34
Muscovite	$2KAl_3Si_3O_{10}(OH)_2 + 2H^+ + 3H_2O \rightarrow 2K^+ + 3Al_2Si_2O_5(OH)_4$	−72.4	−1.34

*Data from Curtis (1976).

Dissolution and Precipitation of Silica

After feldspar, the crystalline silica phases are the most abundant minerals in the rocks of the earth's crust. In most environments, quartz is very resistant to weathering. Amorphous silica occurs in deep-sea sediments and in volcanic rocks. Hence, an understanding of the behavior of silica compounds in the surface environment is necessary in order to describe the chemistry of surface waters. The dissolution of silica can be described by the following equation:

$$SiO_{2\,(s)} + 2H_2O \rightarrow H_4SiO_{4\,(aq)}$$

Rimstidt and Barnes (1980) developed the following equations for the solubility of silica compounds up to temperatures of about 200°C. For quartz,

$$\log K_{sp} = 1.8814 - 2.028 \times 10^{-3}T - \frac{1560.46}{T} \tag{9–1}$$

and for amorphous silica,

$$\log K_{sp} = 0.338037 - 7.8896 \times 10^{-4}T - \frac{840.075}{T} \tag{9–2}$$

where T is in Kelvin. Silicic acid ($H_4SiO_{4\,(aq)}$) is a weak acid, and its dissociation has been discussed in some detail in Chapter 3. At pH < 9.9, the dominant species is $H_4SiO_{4\,(aq)}$. At higher pHs, other species are dominant. The total silica in solution, as a function of pH, can be represented by the following equation (see Chapter 3):

$$Si_T = [H_4SiO_4]\left(1 + \frac{K_{a_1}}{[H^+]} + \frac{K_{a_1}K_{a_2}}{[H^+]^2}\right) \tag{9–3}$$

where K_{a_1} and K_{a_2} are the dissociation constants for the first and second dissociation steps. Solution of equations 9–1 and 9–2 for various temperatures gives us the concentration of [H_4SiO_4] in solution. Equation 9–3 can then be used to determine the total silica in solution as a function of pH. At 25°C, solution of equations 9–1 and 9–2 gives $K_{sp} = 10^{-3.96}$ for quartz and $K_{sp} = 10^{-2.71}$ for amorphous silica. Assuming activity equals concentration, these solubility products are used to determine the total solubility of quartz and amorphous silica as a function of pH (Figure 9–2). Note that the solubility of both quartz and amorphous silica increases significantly at high pHs. This is because the dissociated forms of silicic acid become important contributors to the total amount of dissolved silica. In most natural environments, silica solubility will be low, but in environments such as alkaline lakes, relatively large amounts of silica will go into solution. Note that the behavior of silica is opposite to that of the carbonates, for which increasing pH leads to reduced

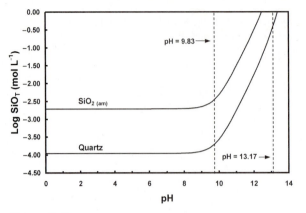

Figure 9–2
Solubility, at 25°C, of quartz and amorphous silica as a function of pH. pH = 9.83 and pH = 13.17 correspond to the first and second dissociation constants, respectively, of silicic acid.

solubility. These differences in behavior give rise to the common observation in the geologic record of silica replacing carbonates or vice versa, such replacements being controlled by the pH of the solutions.

Dissolution and Precipitation of Aluminum and Iron Hydroxides

The solubility of aluminum and iron hydroxides in the weathering environment exerts a significant influence on the transport of these elements in solution. In many weathering studies these two elements are considered to be immobile; i.e., they are believed to stay in the weathered material. Later in this chapter we will look at the effect of chelating agents on the transport of these, and other, species. Here we are concerned with the variation in solubility of the aluminum and iron hydroxides as a function of pH. The approach is similar to that already used for silica.

The dissolution of aluminum hydroxide can be written

$$Al(OH)_3 \rightarrow Al^{3+} + 3OH^-$$

For this reaction,

$$K_{sp} = [Al^{3+}][OH^-]^3 \tag{9-4}$$

Recall that $K_w = [H^+][OH^-]$. We can use this expression to substitute for OH^- in equation 9–4. After this substitution we have the following:

$$K_{sp} = [Al^{3+}]\left(\frac{K_w^3}{[H^+]^3}\right) = \frac{[Al^{3+}]K_w^3}{[H^+]^3} \tag{9-5}$$

Solving for Al^{3+} in solution,

$$[Al^{3+}] = \frac{K_{sp}[H^+]^3}{K_w^3} \tag{9-6}$$

The total solubility of aluminum in solution is the sum of the solubility of Al^{3+} plus all the Al-OH complexes.

$$\Sigma\, Al_{(aq)} = [Al^{3+}] + [AlOH^{2+}] + [Al(OH)_2^+] + [Al(OH)_{3\,(aq)}] + [Al(OH)_4^-] \tag{9-7}$$

For each of the complexes we can write a complexation reaction, and each of the complexation reactions has a corresponding equilibrium constant.

$$Al^{3+} + H_2O \rightleftharpoons AlOH^{2+} + H^+$$

and

$$K_{\beta_1} = \frac{[AlOH^{2+}][H^+]}{[Al^{3+}]} \qquad (9\text{--}8)$$

$$Al^{3+} + 2H_2O \rightleftharpoons Al(OH)_2^+ + 2H^+$$

and

$$K_{\beta_2} = \frac{[Al(OH)_2^+][H^+]^2}{[Al^{3+}]} \qquad (9\text{--}9)$$

$$Al^{3+} + 3H_2O \rightleftharpoons Al(OH)_{3\ (aq)} + 3H^+$$

and

$$K_{\beta_3} = \frac{[Al(OH)_{3\ (aq)}][H^+]^3}{[Al^{3+}]} \qquad (9\text{--}10)$$

$$Al^{3+} + 4H_2O \rightleftharpoons Al(OH)_4^- + 4H^+$$

and

$$K_{\beta_4} = \frac{[Al(OH)_4^-][H^+]^4}{[Al^{3+}]} \qquad (9\text{--}11)$$

Rearranging equations 9–8 through 9–11 and substituting into equation 9–7 gives

$$\Sigma\ Al_{\ (aq)} = [Al^{3+}] + \frac{K_{\beta_1}[Al^{3+}]}{[H^+]} + \frac{K_{\beta_2}[Al^{3+}]}{[H^+]^2} + \frac{K_{\beta_3}[Al^{3+}]}{[H^+]^3} + \frac{K_{\beta_4}[Al^{3+}]}{[H^+]^4} \qquad (9\text{--}12)$$

Next we factor out $[Al^{3+}]$ and substitute equation 9–6 for $[Al^{3+}]$. The final equation is

$$\Sigma\ Al_{\ (aq)} = \frac{K_{sp}}{K_w^3}\left([H^+]^3 + K_{\beta_1}[H^+]^2 + K_{\beta_2}[H^+] + K_{\beta_3} + \frac{K_{\beta_4}}{[H^+]}\right) \qquad (9\text{--}13)$$

The constants required to solve this equation are found in Table 9–5. The total aluminum in solution, as a function of pH, for a solution in equilibrium with gibbsite is shown in Figure 9–3 (p. 320). Note that the solubility is strongly pH dependent. The minimum solubility is found at pH $\approx$ 6.4, within the range of natural waters. Acidic and alkaline waters in equilibrium with gibbsite will contain large amounts of Al. This is a significant observation in terms of the aluminum content of acid streams and lakes and alkaline lakes. We will consider this topic further in a later section.

Ferric iron behaves similarly to Al^{3+}. The equation for total ferric iron in solution is identical to equation 9–13, except that the complexation and solubility constants are different (Table 9–5).

Table 9–5   Constants for Al and Fe^{3+} Solubility Calculations at 25°C*

Phase	pK_{sp}	pK_{β_1}	pK_{β_2}	pK_{β_3}	pK_{β_4}
$Al(OH)_{3\ (am)}$	31.2	5.00	10.1	16.9	22.7
$Al(OH)_{3\ gibbsite}$	33.9	5.00	10.1	16.9	22.7
$Fe(OH)_{3\ (am)}$	37.1	2.19	5.67	12.56	21.6
$Fe(OH)_{3\ goethite}$	44.2	2.19	5.67	12.56	21.6
	pK_0	pK_1	pK_2	pK_3	pK_4
Kaolinite	-3.72	1.28	6.38	13.18	18.98

*Data from Nordstrom et al. (1990) and Macalady et al. (1990).

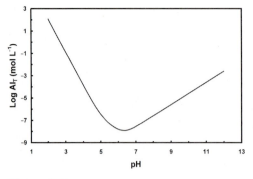

Figure 9–3
Total concentration of aluminum in solution, as a function of pH, for a solution in equilibrium with gibbsite.

In the next section, dealing with stability diagrams, we will find that at $[H_4SiO_{4\,(aq)}]$ $\approx 10^{-4}$ mol L^{-1}, kaolinite replaces gibbsite [Al(OH)$_3$] as the stable aluminum-containing phase. When this happens the aluminum solubility will be controlled by reactions involving kaolinite. Langmuir (1997) develops this equilibrium in detail. The resulting equation is given here and the appropriate constants are listed in Table 9–5.

$$\Sigma \, Al_{(aq)} = \frac{1}{[H_4SiO_{4\,(aq)}]}\left(K_0[H^+]^3 + K_1[H^+]^2 + K_2[H^+] + K_3 + \frac{K_4}{[H^+]}\right) \quad (9\text{–}14)$$

Stability Diagrams

Stability diagrams are *graphical representations of equilibria between minerals and aqueous solutions.* Such diagrams are very useful in inferring what will happen when waters of various composition interact with solid phases. In this section we will go through the process of constructing one of these diagrams from thermodynamic data and then use the diagram to make inferences about what will happen during the weathering of silicate minerals. In order to simplify our calculations, we will assume that the solid phases are pure and fixed in composition and that aluminum is insoluble and remains in the solid phases. At low and high pH values, aluminum will enter solution, and the latter assumption will not be valid. As is always the case when making equilibrium thermodynamic calculations, the usefulness of these calculations in understanding natural processes is dependent on how close the reactions come to achieving equilibrium and the similarity between the solid phases used in the calculations and the solid phases in the natural environment.

From Table 9–2 we know that plagioclase and K-feldspar both break down by incongruent dissolution. Plagioclase and K-feldspar, along with quartz, are among the most abundant minerals in the earth's crust. We will construct a stability diagram that can be used to understand the chemical breakdown of Na-plagioclase to a variety of weathering products. We will also include on this diagram the stability limits for quartz and amorphous silica. The other stability diagrams of interest involve the chemical weathering of Ca-plagioclase and K-feldspar. The construction of these diagrams appears in the problem set. These stability diagrams were derived by Bricker and Garrels (1965) and Garrels (1984). In order to maintain consistency with the literature, Table 9–6 lists the thermodynamic values to be used in these calculations (including the problem set). All the calculations are done at 25°C.

We start by determining the equilibrium constants for the dissolution of quartz and amorphous silica. For quartz the reaction is

$$SiO_{2\,quartz} + 2H_2O \rightleftharpoons H_4SiO_{4\,(aq)}$$

This is an example of a hydrolysis reaction, another type of chemical weathering in which the solid reacts with H$_2$O. For this reaction, $\Delta G_R^0 = 22.6$ kJ mol^{-1} and $K_{eq} = 10^{-3.96}$.

Table 9–6   Thermodynamic Data for Stability Diagram Calculations

Species or mineral	ΔG_R^0 kJ mol^{-1}	Source
Na$^+$	−261.9	Robie et al. (1978)
K$^+$	−282.5	Robie et al. (1978)
Ca^{2+}	−553.5	Robie et al. (1978)
H$_2$O	−237.15	Robie et al. (1978)
H$_4$SiO$_{4 \text{ (aq)}}$	−1308.0	Robie et al. (1978)
Quartz [SiO$_2$]	−856.3	Robie et al. (1978)
Amorphous SiO$_2$	−849.1	Drever (1997)
Albite [NaAlSi$_3$O$_8$]	−3711.7	Robie et al. (1978)
Microcline [KAlSi$_3$O$_8$]	−3742.3	Robie et al. (1978)
Anorthite [CaAl$_2$Si$_2$O$_8$]	−4017.3	Robie et al. (1978)
Muscovite [KAl$_3$Si$_3$O$_{10}$(OH)$_2$]	−5600.7	Robie et al. (1978)
Kaolinite [Al$_2$Si$_2$O$_5$(OH)$_4$]	−3799.4	Robie et al. (1978)
Pyrophyllite [Al$_2$Si$_4$O$_{10}$(OH)$_2$]	−5269.3	Garrels (1984)
Gibbsite [Al(OH)$_3$]	−1159.0	Garrels (1984)
Illite [K$_{0.8}$Al$_{1.9}$(Al$_{0.5}$Si$_{3.5}$)O$_{10}$(OH)$_2$]	−5471.8	Garrels (1984)
Na-beidellite [Na$_{0.33}$Al$_{2.33}$Si$_{3.67}$O$_{10}$(OH)$_2$]	−5368.1	Langmuir (1997)
Ca-beidellite [Ca$_{0.167}$Al$_{2.33}$Si$_{3.67}$O$_{10}$(OH)$_2$]	−5371.6	Calculated
Montmorillonite [K$_{0.3}$Al$_{1.9}$Si$_4$O$_{10}$(OH)$_2$]	−5303.2	Garrels (1984)

Remember that for consistency we are using the data in Table 9–6, not equation 9–1 or 9–2. The equilibrium equation is

$$K_{eq} = 10^{-3.96} = [\text{H}_4\text{SiO}_{4 \text{ (aq)}}] \tag{9–15}$$

At equilibrium, $[\text{H}_4\text{SiO}_{4 \text{ (aq)}}] = 10^{-3.96}$ mol L^{-1}. A similar calculation for amorphous silica gives $\Delta G_R^0 = 15.4$ kJ mol^{-1}, $K_{eq} = 10^{-2.70}$, and at equilibrium $[\text{H}_4\text{SiO}_{4 \text{ (aq)}}] = 10^{-2.70}$ mol L^{-1}.

We can now write a set of equations that describe the interaction between solutions and various minerals during the incongruent dissolution of Na-plagioclase (albite). The phases (minerals) of interest are gibbsite, kaolinite, Na-beidellite (a montmorillonitic clay), and albite. For the reaction kaolinite → gibbsite,

$$\text{Al}_2\text{Si}_2\text{O}_5(\text{OH})_{4 \text{ kaolinite}} + 5\text{H}_2\text{O} \rightleftharpoons 2\text{Al(OH)}_{3 \text{ gibbsite}} + 2\text{H}_4\text{SiO}_{4 \text{ (aq)}}$$

For this reaction, $\Delta G_R^0 = 51.15$ kJ mol^{-1} and $K_{eq} = 10^{-8.96}$. The equilibrium equation is

$$K_{eq} = 10^{-8.96} = [\text{H}_4\text{SiO}_{4 \text{ (aq)}}]^2 \tag{9–16}$$

and at equilibrium $[\text{H}_4\text{SiO}_{4 \text{ (aq)}}] = 10^{-4.48}$ mol L^{-1}. For the reaction albite → gibbsite,

$$\text{NaAlSi}_3\text{O}_{8 \text{ albite}} + \text{H}^+ + 7\text{H}_2\text{O} \rightleftharpoons \text{Al(OH)}_{3 \text{ gibbsite}} + 3\text{H}_4\text{SiO}_{4 \text{ (aq)}} + \text{Na}^+$$

For this reaction, $\Delta G_R^0 = 26.85$ kJ mol^{-1} and $K_{eq} = 10^{-4.70}$. The equilibrium equation is

$$K_{eq} = 10^{-4.70} = \frac{[\text{Na}^+][\text{H}_4\text{SiO}_{4 \text{ (aq)}}]^3}{[\text{H}^+]} \tag{9–17}$$

Note that there are three aqueous species, Na$^+$, H$^+$, and H$_4$SiO$_{4 \text{ (aq)}}$. Subsequent reactions will have the same three aqueous species. For the reaction albite → kaolinite,

$$2\text{NaAlSi}_3\text{O}_{8 \text{ albite}} + 2\text{H}^+ + 9\text{H}_2\text{O} \rightleftharpoons \text{Al}_2\text{Si}_2\text{O}_5(\text{OH})_{4 \text{ kaolinite}} + 4\text{H}_4\text{SiO}_{4 \text{ (aq)}} + 2\text{Na}^+$$

For this reaction, $\Delta G_R^0 = 2.55$ kJ mol^{-1} and $K_{eq} = 10^{-0.45}$. The equilibrium equation is

$$K_{eq} = 10^{-0.45} = \frac{[\text{Na}^+]^2[\text{H}_4\text{SiO}_{4 \text{ (aq)}}]^4}{[\text{H}^+]^2} \tag{9–18}$$

For the reaction Na-beidellite → kaolinite,

$$6Na_{0.33}Al_{2.33}Si_{3.67}O_{10}(OH)_2 \text{ Na-beidellite} + 2H^+ + 23H_2O \rightleftharpoons 7Al_2Si_2O_5(OH)_4 \text{ kaolinite}$$
$$+ 8H_4SiO_{4\,(aq)} + 2Na^+$$

For this reaction, $\Delta G_R^0 = 79.45$ kJ mol^{-1} and $K_{eq} = 10^{-13.92}$. The equilibrium equation is

$$K_{eq} = 10^{-13.92} = \frac{[Na^+]^2[H_4SiO_{4\,(aq)}]^8}{[H^+]^2} \tag{9–19}$$

For the reaction albite → Na-beidellite,

$$7NaAlSi_3O_8 \text{ albite} + 6H^+ + 20H_2O \rightleftharpoons 3Na_{0.33}Al_{2.33}Si_{3.67}O_{10}(OH)_2 \text{ Na-beidellite}$$
$$+ 10H_4SiO_{4\,(aq)} + 6Na^+$$

For this reaction, $\Delta G_R^0 = -30.8$ kJ mol^{-1} and $K_{eq} = 10^{5.40}$. The equilibrium equation is

$$K_{eq} = 10^{5.40} = \frac{[Na^+]^6[H_4SiO_{4\,(aq)}]^{10}}{[H^+]^6} \tag{9–20}$$

We have now written equations for all the reactions of interest in the weathering of albite. Given the aqueous species, on our diagram we will plot log([Na$^+$]/[H$^+$] versus log[H$_4$SiO$_{4\,(aq)}$]. For the kaolinite → gibbsite reaction, the only aqueous species is H$_4$SiO$_{4\,(aq)}$. For the other reactions, all three aqueous species are present. The easiest way to plot these reactions is to select two Na$^+$/H$^+$ ratios and then calculate the corresponding activity of H$_4$SiO$_{4\,(aq)}$ from the equilibrium equations. The curves are plotted in Figure 9–4a. Inspection of the crossover points leads to delineation of the stability fields for each mineral. These stability fields are plotted in Figure 9–4b, along with the saturation curves for quartz and amorphous silica.

Stability diagrams can be used to understand chemical changes that occur during water–rock (mineral) interactions. With reference to Figure 9–4b, we will consider two cases—closed system and open system. The curve labeled A through F in Figure 9–4b shows the path that will be taken during the weathering of albite in a closed system. We start with albite in equilibrium with a solution of composition A. This is a situation that

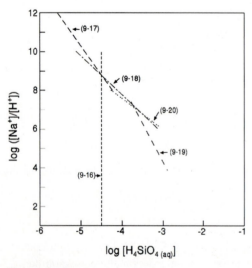

Figure 9–4a
Equilibrium equations 9–16 to 9–20 plotted on a log([Na$^+$]/[H$^+$]) versus log[H$_4$SiO$_{4\,(aq)}$] diagram. Numbers on the diagram indicate corresponding equations in the text.

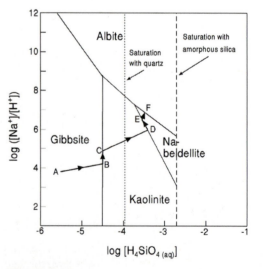

Figure 9–4b
Mineral stability fields as delineated by equilibrium equations plotted in Figure 9–4a. The labeled curve indicates the changes in chemistry of a solution in equilibrium with albite during weathering in a closed system. See text for discussion.

would exist during the start of weathering when a dilute solution is in contact with albite. We are in the gibbsite stability field and the weathering reaction can be written

$$NaAlSi_3O_{8 \text{ albite}} + H^+ + 7H_2O \rightarrow Al(OH)_{3 \text{ gibbsite}} + 3H_4SiO_{4 \text{ (aq)}} + Na^+$$

The conversion of albite to gibbsite involves the consumption of hydrogen ions and the release of silicic acid and sodium ions. Thus, both silicic acid activity and the sodium/hydrogen ratio of the solution increase and the solution follows the path A $\rightarrow$ B. When the gibbsite–kaolinite boundary is encountered, gibbsite is converted to kaolinite. The reaction can be written

$$2NaAlSi_3O_{8 \text{ albite}} + 4Al(OH)_{3 \text{ gibbsite}} + 2H^+ \rightarrow 3Al_2Si_2O_5(OH)_{4 \text{ kaolinite}} + 2Na^+ + H_2O$$

As long as gibbsite and kaolinite are present, the reaction will occur at constant silicic acid activity; i.e., the system is buffered in terms of silicic acid activity. This process is represented by segment B $\rightarrow$ C. After all the gibbsite is converted to kaolinite, the solution follows the path C $\rightarrow$ D and the reaction is

$$2NaAlSi_3O_{8 \text{ albite}} + 2H^+ + 9H_2O \rightarrow Al_2Si_2O_5(OH)_{4 \text{ kaolinite}} + 4H_4SiO_{4 \text{ (aq)}} + 2Na^+$$

During this reaction hydrogen ions are consumed and silicic acid and sodium ions are released to solution. The result is an increase in the Na^+/H^+ ratio and an increase in the activity of silicic acid. When the kaolinite–Na-beidellite boundary is encountered, kaolinite is converted to Na-beidellite. The reaction can be written

$$3NaAlSi_3O_{8 \text{ albite}} + 2Al_2Si_2O_5(OH)_{4 \text{ kaolinite}} + 2H^+ + 2H_2O \rightarrow$$
$$3Na_{0.33}Al_{2.33}Si_{3.67}O_{10}(OH)_{2 \text{ Na-beidellite}} + 2H_4SiO_{4 \text{ (aq)}} + 2Na^+$$

As long as kaolinite and Na-beidellite are present, the reaction will move along the boundary separating the stability field of these two minerals (D $\rightarrow$ E). After the kaolinite has been converted to Na-beidellite, the solution chemistry changes along the curve E $\rightarrow$ F until the Na-beidellite–albite stability boundary is reached.

For the open system, the important variable is the rate at which water moves through the weathering environment. This is sometimes referred to as the *flushing rate*. If we have high rainfall and good infiltration, the concentration of silicic acid and various ions in solution will be low. Under these conditions albite will weather to gibbsite. This is the situation observed in tropical settings where there is deep weathering and the weathered material largely consists of aluminum (gibbsite) and iron hydroxides. In regions of lower rainfall and less rapid infiltration, the concentration of ions in solution is greater and albite will weather to kaolinite or montmorillonitic clays. This is the situation usually observed in temperate settings. Hence, returning to our discussion of the formation of soils in Chapter 5, there is a relationship between soil type and climatic conditions (Case Study 9–1).

CASE STUDY 9–1
Variations in Clay Mineralogy as an Indicator of Paleoclimatic Conditions

Price et al. (2000) determined the oxygen isotope compositions for belemnite genera from the Speeton Clay Formation, Filey Bay, England. The Speeton Clay Formation, consisting of claystones and calcareous mudrocks, was deposited in an epicontinental sea during the Early Cretaceous, approximately 142 to 133 million years ago. The oxygen isotope data revealed that paleotemperatures had varied from approximately 9 to 15°C (see problem 56 in Chapter 6). The variations in paleotemperature were accompanied by variations in clay mineral content. Kaolinite was the dominant clay mineral when paleotemperatures were high, and smectite (e.g., Na-beidellite) was the dom-

inant clay mineral when paleotemperatures were low. This variation in clay mineral content was related to climatic conditions in the continental source region, the high smectite content representing arid conditions and the high kaolinite content representing humid conditions. This inference is related to the idea of flushing rates discussed earlier. For an arid climate the flushing rate would be low and smectite would be the stable clay mineral (Figure 9–4b), whereas during humid conditions the flushing rate would be higher and kaolinite would be the stable clay mineral. Thus, the authors were able to use variations in clay mineral content to make inferences about climatic conditions in the Early Cretaceous.

Source: Price et al. (2000).

GEOCHEMISTRY OF SURFACE AND GROUND WATERS

We begin this section with a brief discussion of the convention used to distinguish between dissolved and suspended matter and the variability in water chemistry and the ways in which we can represent this variability. In standard water-chemistry analysis, water samples are first filtered to remove suspended particles. By convention, this filtration is done using a 0.45-μm filter. Anything that passes through this filter is considered to be in solution. Thus, the distinction between dissolved and suspended material is somewhat arbitrary. The important point, however, is that *by convention substances less than 0.45 μm in size are considered to be in solution; i.e., they are dissolved species*. The significance of this criteria will be considered further later in the chapter.

Variations in River and Groundwater Chemistry

Chemical analysis of surface and ground waters reveals a wide variation in relative proportions and total concentrations of dissolved species (Table 9–7). Several methods are commonly used to graphically portray these variations in water chemistry and to classify water types.

Table 9–7 Compositions of Selected Surface and Ground Waters

| | Concentration (mg L^{-1}) | | | | | | | | | |
	Ca^{2+}	Mg^{2+}	Na^+	K^+	Cl^-	SO_4^{2-}	HCO_3^-	SiO_2	TDS	Ref.
River										
Colorado	83	24	95	5.0	82	270	135	9.3	703	1
Columbia	19	5.1	6.2	1.6	3.5	17.1	76	10.5	139	1
Mississippi	39	10.7	17	2.8	19.3	50.3	117	7.6	265	1
Rio Grande	109	24	117	6.7	171	238	183	30	881	2
U. Rhine	41	7.2	1.4	1.2	1.1	36	114	3.7	307	3
U. Amazon	19	2.3	6.4	1.1	6.5	7.0	68	11.1	122	4
L. Amazon	5.2	1.0	1.5	0.8	1.1	1.7	20	7.2	38	4
L. Negro	0.2	0.1	0.4	0.3	0.3	0.2	0.7	4.1	6	4
Zambeze	9.7	2.2	4.0	1.2	1	3	25	12	58	1
Nile	25	7.0	17	4.0	7.7	9	134	21	225	1
Ganges	25.4	6.9	10.1	2.7	5	8.5	127	8.2	194	5
Yellow	42	17.7	55.6	2.9	46.9	71.7	182	5.1	424	6
Groundwater										
(dominant rock)										
Central Florida (carbonate)	34	5.6	3.2	0.5	4.5	2.4	124	12	—	7
Central Pennsylvania (carbonate)	83	17	8.5	6.3	17	27	279	—	—	8
Montana (sandstone)	3.0	7.4	857	2.4	71	1.6	2080	16	3098	9
New Mexico (gypsum)	636	43	17	—	24	1570	143	29	2480	9
California (serpentine)	34	242	184	18	265	6.6	1300	175	2226	9
Rhode Island (granite)	6.5	2.6	5.9	0.8	5	0.9	38	20	82	9
Maryland (gabbro)	5.1	2.3	6.2	3.2	1.0	9.2	37	39	109	9
Hawaii (basalt)	17	42	38	3.1	63	15	84	18	251	9
New Mexico (rhyolite)	6.5	1.1	38	2	17	15	77	103	222	10
North Carolina (mica schist)	17	1.7	6.4	1	1.1	6.9	69	29	98	10
West Virginia (sand and gravel)	58	13	23	2.8	39	116	101	10	338	10
Alabama (limestone)	46	4.2	1.5	0.8	3.5	4.0	146	8.4	222	10

References: [1]Meybeck (1979), [2]Livingstone (1963), [3]Zobrist and Stumm (1980), [4]Stallard (1980), [5]Sarin et al. (1989), [6]Gordeev and Siderov (1993), [7]Back and Hanshaw (1970), [8]Langmuir (1971), [9]Matthess (1982), [10]Hem (1970).

Graphical Representations of Water Chemistry Two types of diagrams are commonly used to portray water chemistry, Stiff and Piper. Stiff diagrams show the concentrations (in milliequivalents) of the major ions (both cations and anions) as a shape that gives both the relative abundance of the various species and the total abundance. Piper diagrams are trilinear representations of cation, anion, and combined cation and anion proportions. Piper diagrams are often used to classify water types. The following two examples illustrate the plotting of Stiff and Piper diagrams.

EXAMPLE 9–1 In this example we will plot the chemical data for the Columbia and Rio Grande rivers and the Pennsylvania groundwater (Table 9–7) on a Stiff diagram. The following table illustrates the calculation for the Columbia River water. We start by determining the concentration of each ion in millimoles per liter and then calculate the concentration of each ion in milliequivalents. The concentrations in milliequivalents are plotted on the Stiff diagram, as shown in Figure 9–5, and the various points are connected by straight lines. Note that the concentrations of the Na^+ and K^+ ions have been combined.

	Columbia River						
	Ca^{2+}	Mg^{2+}	Na^+	K^+	Cl^-	SO_4^{2-}	HCO_3^-
mg L^{-1}	19	5.1	6.2	1.6	3.5	17.1	76
mmol L^{-1}	0.47	0.21	0.27	0.04	0.1	0.18	1.25
meq L^{-1}	0.95	0.42	0.27	0.04	0.1	0.36	1.25

The shape of the field is a representation of the relative proportions of the various ions, and the size of the field represents the total ionic concentration. The Rio Grande river and Central Pennsylvania ground waters are also plotted in Figure 9–5. From the shape of the Stiff diagrams we can easily infer that the Rio Grande has a much greater concentration of all ionic species than the Columbia River and that the bicarbonate ion, relative to the sulfate ion, is much less important in the Rio Grande than it is in the Columbia River. Inspection of the Central Pennsylvania groundwater diagram reveals that the dominant cation is Ca^{2+} and the dominant anion is HCO_3^-, so this is a Ca-carbonate water. Stiff diagrams are particularly useful when plotted on a map because they give a graphical representation of regional variations in water chemistry.

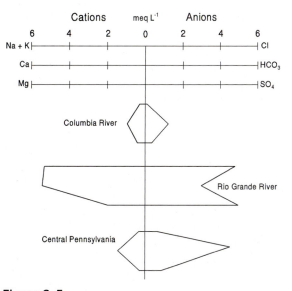

Figure 9–5
Stiff diagram for Columbia and Rio Grande river waters and
Central Pennsylvania groundwater. See text for discussion.

EXAMPLE 9–2 In this example we plot the Columbia River data from Example 9–1 on a Piper diagram. As in Example 9–1, we start by calculating the concentration of each ion in milliequivalents. We now normalize the cations and the anions to 100% (as shown in the following table) so that we can plot them on the Piper diagram.

	Cation concentrations			Anion concentrations	
	meq L^{-1}	Normalized		meq L^{-1}	Normalized
Ca^{2+}	0.95	56.5	Cl$^-$	0.10	5.8
Mg^{2+}	0.42	25.0	SO$_4^{2-}$	0.36	21.1
Na$^+$ + K$^+$	0.31	18.5	HCO$_3^-$	1.25	73.1
Total	1.68	100.0		1.71	100.0

The normalized cation and anion concentrations are plotted in their appropriate triangle (Figure 9–6). The data plotted on each triangle are then projected into the quadrilateral by drawing a line from the point on the cation triangle parallel to the Mg axis into the quadrilateral and by drawing a line from the point in the anion triangle parallel to the SO$_4$ axis into the quadrilateral. The intersection of these two lines marks the location of the point to be plotted on the quadrilateral (Figure 9–6). Symbols of different size are sometimes used to indicate differences in total dissolved species. This convention is used in the example. The data for the Rio Grande river and the Central Pennsylvania ground waters are also plotted in Figure 9–6. ■

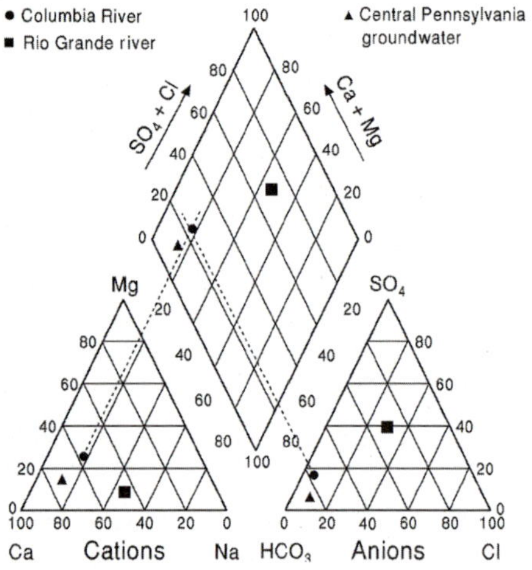

Figure 9–6
Piper diagram for Columbia and Rio Grande river waters and Central Pennsylvania groundwater. See text for discussion.

Piper diagrams have two main uses. The first is the graphical representation of water chemistry for the purpose of water classification (see the next section). The second is to determine if a series of water compositions represent the mixing of two end members. If the samples are the result of two-end-member mixing, they will plot along straight lines in each of the fields of the diagram. If they do not plot along straight lines, then their compositions are not controlled by simple two-end-member mixing (Case Study 9–2).

Hydrofacies Back (1966) divided the ion triangles of the Piper diagram into various fields that correspond to water type or *chemical facies* (Figure 9–7). This is a convenient way to classify water types on the basis of their major ion chemistry. Examples of chemical facies names are calcium bicarbonate and sodium chloride. If the water plots in the center of an ion triangle, it is referred to as *mixed-cation* or *mixed-anion facies*. With reference to the waters plotted in Figure 9–6, the Columbia River water belongs to the calcium-bicarbonate facies, the Rio Grande river water to the mixed-cation–mixed-anion facies, and the Central Pennsylvania groundwater to the calcium-bicarbonate facies. The hydrochemical facies in which the water sample plots potentially reveals information about the factors controlling the water chemistry. For example, the observation that the Columbia River water plots in the calcium-bicarbonate facies suggests that rock weathering is the major factor controlling water chemistry (see next section). The observation that the Central Pennsylvania groundwater plots in the calcium-bicarbonate facies suggests that dissolution of limestone, which would provide the Ca and bicarbonate ions, is important

CASE STUDY 9–2
Mixing of Acid Mine Drainage and Lake Water

In Case Study 2–1, Foos (1997) investigated the mixing of acid mine drainage from an abandoned coal mine with water discharged from a lake. The chemical compositions of the acid mine drainage, lake water, and the stream water downstream from the point of mixing were plotted on a Piper diagram (Figure 9–C2–1). This plot indicated that simple mixing would explain the concentrations of the major ions in the stream (mixed water). Further testing of the mixing model revealed that Ca^{2+}, Mg^{2+}, Na^+, K^+, and Cl^- concentrations could be explained by simple mixing, but not HCO_3^- and SO_4^{2-} concentrations. For HCO_3^- and SO_4^{2-}, the deviation from simple mixing was about 14%, so as a first approximation simple mixing is a reasonable assumption. The deviations from simple mixing for Fe^{2+} (1800%) and Mn^{2+} (39%) were significant. The reasons for the nonconservative behavior of these two species was discussed in Case Study 2–1. In this study, the Piper diagram was used to make an initial evaluation of the conservative versus nonconservative behavior of various species in solution.

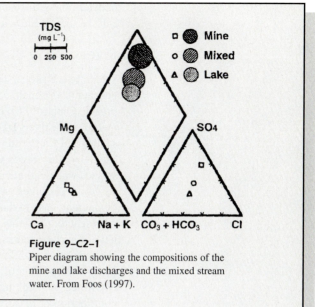

Figure 9–C2–1
Piper diagram showing the compositions of the mine and lake discharges and the mixed stream water. From Foos (1997).

Source: Foos (1997).

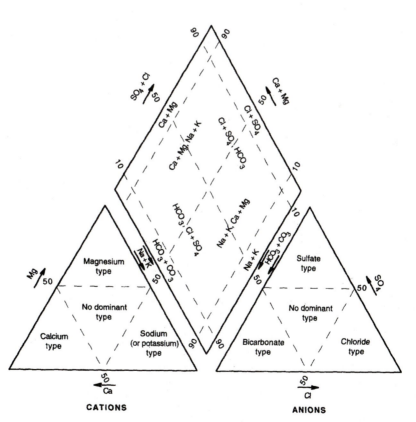

Figure 9–7
Hydrochemical facies. After Back (1966).

in determining the chemistry of the groundwater. The processes that determine the chemistry of surface and ground waters are considered further in the following sections.

Rivers

What are the processes that determine the chemical composition of rivers, and how can geochemical principles be used to answer this question? In a somewhat controversial approach, Gibbs (1970) investigated this question by summarizing the chemical data for numerous rain, river, lake, and ocean samples. In terms of cations, he considered Na (high saline) and Ca (freshwater) to be the best representatives of surface water end members. Similarly, the anions Cl (high saline) and HCO_3 (freshwater) were selected to represent surface water end members. Plots of the relative abundances of the cations and anions versus total dissolved solids (Figures 9–8a and 9–8b) gave two diagonal fields (and a boomerang-shaped diagram) anchored by the two end members. Based on these graphical representations, Gibbs suggested that there are three mechanisms that control the chemistry of surface waters. The first is atmospheric precipitation, which is represented by the surface waters with low total dissolved solids and relatively high Na and Cl that plot in the lower right portion of each diagram. The chemistry of most tropical rivers is presumably controlled by precipitation because these areas are well leached and there should be little contribution of dissolved materials from the rocks. The second mechanism is essentially weathering (referred to as *rock dominance*), in which the chemical breakdown of rocks (minerals) in the drainage basin provides the dissolved components. The positions in the

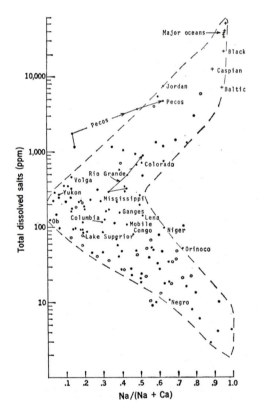

Figure 9–8a

Plot of total dissolved solids versus relative cation abundances for surface waters. Filled circles are river, unfilled circles are lake, and pluses are ocean waters. From "Mechanisms controlling world water chemistry by R. J. Gibbs in SCIENCE, 1970, #170, pp. 1088–1090. Copyright © 1970 American Association for the Advancement of Science. Reprinted with permission.

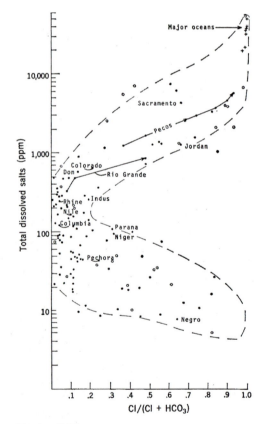

Figure 9–8b

Plot of total dissolved solids versus relative anion abundances for surface waters. Filled circles are river, unfilled circles are lake, and pluses are ocean waters. From "Mechanisms controlling world water chemistry by R. J. Gibbs in SCIENCE, 1970, #170, pp. 1088–1090. Copyright © 1970 American Association for the Advancement of Science. Reprinted with permission.

diagrams of waters that are controlled by this mechanism are determined by the climate, relief, and the rock types exposed in each basin. The third mechanism is evaporation–fractional crystallization. During this process the concentration of total dissolved solids increases due to evaporation and the concentrations of various ionic species are controlled by the precipitation of solids from solution as the total ionic concentration increases. The effect of these three mechanisms on surface water chemistry is shown graphically in Figure 9–9.

A simple way to approach the Gibbs model of surface-water chemistry is to start with waters that are in partial equilibrium with the weathering products of their basin, the *rock dominance* field in Figure 9–9. We can envision two surface-water chemistry sequences. In the one, represented by the lower right part of the boomerang, the surface-water chemistry is a reflection of the relative importance of precipitation to groundwater + overland flow. If most of the ions are derived from precipitation, the chemistry of the surface waters will approximate that of dilute seawater. Remember from our discussion in Chapter 8 that one rainwater component can be viewed as dilute seawater. For the upper right arm of the boomerang, the controlling factor is the amount of evaporation and precipitation. For example, a river flowing through an arid landscape would be expected to show an increase in total dissolved solids as evaporation occurs and a decrease in Ca relative to Na due to the precipitation of calcium carbonate.

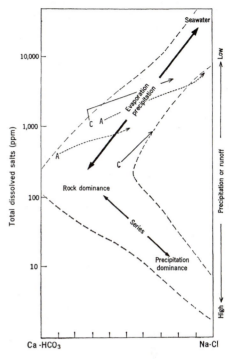

Figure 9–9

Graphical representation of the processes that control the chemistry of surface waters. See text for discussion. From "Mechanisms controlling world water chemistry" by R. J. Gibbs in SCIENCE, 1970, #170, pp. 1088–1090. Copyright © 1970 American Association for the Advancement of Science. Reprinted with permission.

A number of authors have challenged the Gibbs model. The crux of the debate is the relative importance of weathering in determining the chemistry of rivers. This debate is particularly focused on the two ends of the boomerang. For example, for some rivers that plot in the lower right portion of the boomerang, it can be shown that the marine (i.e., precipitation) component is very small, leading to the conclusion that the ions are largely derived from weathering of rocks within the basin. For example, if carbonate and mafic igneous rocks are absent or of minor abundance, compared to rocks such as granite, one would expect Na to dominate over Ca. On the other hand, a number of the surface waters that plot in the upper right portion of the boomerang flow through regions that contain evaporite deposits, and it has been argued that it is the weathering of these deposits that is responsible for the high Na, Cl, and total dissolved solids content of these rivers. The Gibbs model also does not consider ion-exchange reactions that may occur between ions in solution and clay minerals. These reactions may be of significance in some cases. In summary, the objections to the Gibbs model are largely related to the significance of rock weathering, which many authors consider to be the most important process in determining the chemistry of surface waters. There are cases, however, where the Gibbs model provides a reasonable explanation for the surface-water chemistry.

Stallard and Edmond (1983) proposed a classification of river water chemistry that emphasizes the importance of rock weathering. Although this classification was developed specifically for the Amazon basin, it seems to have general applicability. The classification scheme is summarized in Table 9–8.

Table 9–8 Stallard and Edmond (1983) River Classification*

Type	Total cationic charge (μeq L^{-1})	~TDS (mg L^{-1})	Predominant source–rock type	Characteristic water chemistry (mole ratios)	Examples	Gibbs category
1	< 200	< 20	Intensely weathered (cation-poor) siliceous rocks and soils (thick regolith)	Si-enriched; low pH; Si/(Na + K) = 2; high Na/Na + Ca)	Amazon tributaries	Atmosphere-precipitation controlled
2	200–450	20–40	Siliceous (cation-rich); igneous rocks and shales (sedimentary silicates)	Si-enriched; (higher Si from igneous and metamorphic rocks); Si/(Na + K) = 2; intermediate Na/(Na + Ca)	L. Amazon, Orinoco, Zaire	Between atmosphere-precipitation controlled and rock-dominated
3	450–3000	40–250	Marine sediments; carbonates, pyrite; minor evaporites	Na/Cl = 1; (Ca +Mg)/(0.5HCO$_3$ + SO$_4$) = 1; low Na/(Na + Ca)	Most major rivers	Rock-weathering dominated
4	> 3000	> 250	Evaporites; CaSO$_4$ and NaCl	Na/Cl = 1; (Ca + Mg)/(0.5 HCO$_3$ + SO$_4$) = 1; high Na/(Na + Ca)	Rio Grande	Evaporation–crystallization

*From Berner and Berner (1996).

The characteristic water chemistry of the various river types provides information about the material that is being weathered. For the first two river types, the rock that is weathered contains variable amounts of feldspar (K-feldspar and Na-plagioclase). We can write a weathering reaction for K-feldspar as follows:

$$2KAlSi_3O_8 \text{ K-feldspar} + 2H^+ + 9H_2O \rightarrow Al_2Si_2O_5(OH)_4 \text{ kaolinite} + 2K^+ + 4H_4SiO_4 \text{ (aq)}$$

Note that in this weathering reaction Si and K are released in a 2:1 ratio. A similar weathering reaction can be written for Na-plagioclase. For type 1 rivers, the Na/(Na + Ca) ratio is high because there is little Ca in the plagioclase and other Ca-bearing ferromagnesian silicates are rare. For type 2 rivers, there is a greater amount of Ca relative to Na due to an increase in the Ca content of the plagioclase in the rocks plus the presence of Ca-containing ferromagnesian silicates. For the type 3 rivers, Ca and Mg are significant because of the presence of carbonate-containing sediments. For the type 4 rivers, the high Na relative to Ca reflects the presence of evaporites. For the type 4 rivers, note that in basins that contain a significant amount of gypsum and/or anhydrite, the Ca content would be much more significant than the Na content. For the type 3 and type 4 rivers, representative evaporite and carbonate weathering reactions are

$$CaSO_4 \rightarrow Ca^{2+} + SO_4^{2-}$$

and

$$CaCO_3 + H_2CO_3 \rightarrow Ca^{2+} + 2HCO_3^-$$

For both reactions the $(Ca^{2+} + Mg^{2+})/(0.5HCO_3^- + SO_4^{2-})$ is 1:1. Hence, we would conclude that for the last two water types in Table 9–8 the weathering of carbonate rocks and evaporites is a major factor in determining their water chemistry.

Table 9–9 summarizes the relative importance of various sources for the ionic species found in river water. Note that contributions from pollution are included in the table and that for certain species pollution is a significant source. This is particularly notable for SO_4^{2-}, which has a predominant pollution source—mostly as acid deposition. Other sources of sulfate are natural biogenic emissions, volcanism, and the weathering of pyrite and other sulfide minerals.

Table 9–9   Sources of Major Elements in River Water (%)*

Species	Atmosphere Cyclic salt	Weathering Carbonates	Weathering Silicates	Weathering Evaporites	Pollution
Ca^{2+}	0.1	65	18	8	9
HCO_3^-	<< 1	61	37	0	2
Na^+	8	0	22	42	28
Cl^-	13	0	0	57	30
SO_4^{2-}	2	0	0	22	54
Mg^{2+}	2	36	54	<< 1	8
Na^+	1	0	87	5	7
H_4SiO_4	<< 1	0	> 99	0	0

*From Berner and Berner (1996).

In addition to the processes we have discussed, relief and climate also play a role in the chemistry of surface waters. Chemical weathering occurs at a maximum rate in wet and warm climates and at a minimum rate in dry and cold climates. The degree of relief determines the rapidity of the erosional process. For example, an area of low relief subjected to wet and warm climatic conditions would undergo extensive chemical weathering and most of the mobile components would be leached from the rocks and soil. This situation is typified by the type 1 river waters in Table 9–8. Streams draining the area have low total cation abundances. In an area of high relief, rapid erosion occurs and the time available for chemical weathering is reduced. Under these conditions we would expect chemical weathering processes, such as dissolution of evaporites, that proceed rapidly to contribute a substantially greater proportion of dissolved species to the streams than processes that occur much more slowly, such as the chemical breakdown of feldspar to kaolinite and dissolved species. For example, in an area of high relief where arkosic (feldspar-rich) sandstones are the major rock type, but minor amounts of evaporites or carbonate-bearing rocks exist, the latter rock types would disproportionately contribute to the dissolved load for the streams draining the area.

Groundwaters

A simplified groundwater system is shown in Figure 9–10. Precipitation infiltrates the subsurface materials and moves through the vadose zone to the zone of saturation. The boundary between the **vadose zone** (in which *the pores are not completely filled with water*) and the **saturated zone** (in which *the pores are completely filled with water*) is the water table. In the saturated zone the groundwater moves along curved flow paths to the point of discharge—in this case, a stream. The water entering the system (1) has essentially the same chemical composition as the precipitation. As the water moves through the vadose zone (2), its chemical composition changes due to additions of organic material and

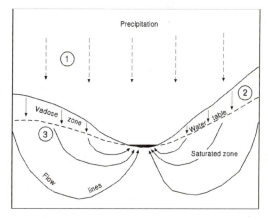

Figure 9–10

Simplified groundwater system showing the movement of water through the system. See text for discussion.

soluble salts. Below the water table (3), the water is isolated from the atmosphere and a variety of chemical changes can occur. As the water moves through the groundwater system, oxygen is consumed by the oxidation of dissolved organic carbon. This changes the oxidation–reduction potential of the system. A variety of rock (or sediment)–water interactions occur, which are controlled by the minerals present in the rocks (or sediments). If carbonates are present, their dissolution leads to changes in pH and P_{CO_2}. Soluble minerals, such as halite and gypsum, go into solution, adding various anions, such as Cl^- and SO_4^{2-}, to the groundwater. A number of important reactions occur between silicate minerals and the groundwater. These will be discussed more fully later. Because there is a kinetic factor involved in these reactions, the chemistry of the groundwater is modified as it moves through the groundwater system (Case Study 9–3).

Interactions between silicate minerals and groundwater are significant for aquifers consisting of sandstone, conglomerate, or fractured (or porous) igneous rock (and sediments of similar mineralogical composition). In what is now considered a classic paper, Garrels (1967) investigated the chemistry of groundwaters from igneous rocks and developed a model to explain it. The following assumptions are relevant to the construction of the model.

CASE STUDY 9–3
Chemical Evolution of Groundwater in the Floridian Aquifer System

We previously encountered the Floridian aquifer in Case Study 3–1. In that case study we were concerned with changes in pH, HCO_3^-, and P_{CO_2}. Here we will look at the variety of chemical changes that occur as groundwaters move through this aquifer. The Floridian aquifer occurs in a sequence of Tertiary carbonate rocks that underlie Florida and extend northward into Alabama, Georgia, and South Carolina. This is one of the most productive aquifers in the world and is a major source of drinking water for residents in the southeastern United States. One of the major recharge areas for the Floridian aquifer is in central Florida, and groundwater moves away from this recharge area (potentiometric high) toward the Gulf of Mexico and Atlantic Ocean. Along the coastal margins the groundwater mixes with seawater.

The chemical evolution of the waters in the Floridian aquifer, as they move away from the potentiometric high, is shown in Figure 9–C3–1. In the recharge area the waters are

dominantly Ca^{2+} and HCO_3^-. As the water moves away from the recharge area, there are relative increases in the concentrations of Mg^{2+} and SO_4^{2-}. Finally, an increase in $Na^+ + K^+$ and Cl^- is observed in the discharge areas.

Why do these changes occur? In the recharge area the dominant process is calcite dissolution, which contributes the Ca^{2+} and HCO_3^- to the water. As the groundwater moves away from the recharge area, minor amounts of gypsum in the carbonate rock dissolve. The dissolution of gypsum releases Ca^{2+}, and the increase in Ca^{2+} concentration leads to oversaturation (due to the common ion effect) of the water in calcite. Dolomite $[CaMg(CO_3)_2]$ is also present in the carbonate rocks, and as the calcite precipitates, dolomite dissolves. The combination of these three reactions—gypsum dissolution, calcite precipitation, and dolomite dissolution—leads to the observed increases in Mg^{2+} and SO_4^{2-}. Ion-exchange reactions also add Na^+ to the water. In the coastal discharge areas the groundwaters encounter modern or ancient seawater. Mixing with the seawater is responsible for the observed increases in Na^+ and Cl^- in the discharge areas.

The common ion effect was previously mentioned in Chapter 3 without explanation. Let us consider the dissolution of calcite. We can write the chemical reaction as follows:

$$CaCO_3 + H_2CO_3 \rightleftharpoons Ca^{2+} + 2HCO_3^-$$

For this reaction the equilibrium equation is

$$K_{eq} = [Ca^{2+}][HCO_3^-]^2$$

The dissolution of gypsum can be written

$$CaSO_4 \cdot 2H_2O \rightarrow Ca^{2+} + SO_4^{2-} + 2H_2O$$

During gypsum dissolution Ca^{2+} ions are released to solution (a common ion with respect to both gypsum and calcite solubility). The increase in Ca^{2+} ions in solution can lead to saturation of the solution in calcite and calcite precipitation. The other ion, SO_4^{2-}, released during gypsum dissolution does not participate in the calcite equilibrium reaction.

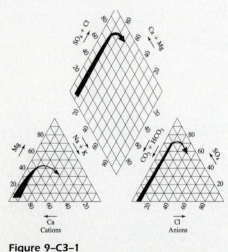

Figure 9–C3–1
Piper diagram showing chemical evolution of groundwater in the Floridian aquifer from recharge areas to discharge areas. From Back and Hanshaw (1970).

Source: Back and Hanshaw (1970).

1. Silicate minerals are decomposed through aggressive attack by waters that contain CO_2, i.e., acid attack by carbonic acid. Organic acids in soils can be a major contributor to the total acidity of soil waters, but treating all acid as carbonic is a reasonable simplification.
2. Plagioclase feldspars and the ferromagnesian silicate minerals (olivine, pyroxene, amphibole, and biotite) are attacked much more rapidly than K-feldspar or quartz.
3. K^+ and Mg^{2+} come chiefly from biotite and/or amphibole.
4. After correction for the rainwater Na^+, the Na^+ and Ca^{2+} in the groundwater are solely due to plagioclase dissolution. The correction for rainwater Na^+ is done by subtracting a molar amount of Na^+ from the analysis equal to the molar amount of Cl^-. An implicit assumption is that calcite, dolomite, or gypsum, all of which could contribute Ca^{2+} ions, are not present in the aquifer. Although it might be possible to correct for Ca^{2+} ions derived from the dissolution of gypsum, using the amount of SO_4^{2-}, this is an uncertain correction because the oxidation of sulfide minerals in the aquifer will also contribute SO_4^{2-}.

From the earlier section dealing with stability diagrams, we know that the common weathering products for plagioclase would be gibbsite, kaolinite, and smectites. Given the total ionic concentration of groundwaters, gibbsite is an unlikely phase. Hence, the two most important weathering products would be kaolinite and the smectite minerals (Garrels used montmorillonite, one of the smectites, in his model). Examples of weathering reactions for plagioclase feldspar to kaolinite are

$$8Na_{0.75}Ca_{0.25}Al_{1.25}Si_{2.75}O_{8 \text{ plagioclase}} + 10CO_2 + 15H_2O \rightarrow$$
$$5Al_2Si_2O_5(OH)_{4 \text{ kaolinite}} + 12SiO_{2 \text{ (aq)}} + 6Na^+ + 2Ca^{2+} + 10HCO_3^-$$

and

$$4Na_{0.5}Ca_{0.5}Al_{1.5}Si_{2.5}O_{8 \text{ plagioclase}} + 6CO_2 + 9H_2O \rightarrow$$
$$3Al_2Si_2O_5(OH)_{4 \text{ kaolinite}} + 4SiO_{2 \text{ (aq)}} + 2Na^+ + Ca^{2+} + 6HCO_3^-$$

The relative proportions of the soluble weathering products vary as a function of the composition of the plagioclase feldspar. For the first reaction, $Na^+/Ca^{2+} = 3$ and $HCO_3^-/SiO_{2 \text{ (aq)}} = 0.83$; for the second reaction, $Na^+/Ca^{2+} = 1$ and $HCO_3^-/SiO_{2 \text{ (aq)}} = 1.5$. By writing a number of these reactions, Garrels was able to construct the plagioclase–montmorillonite, plagioclase–kaolinite, and plagioclase–gibbsite curves shown in Figure 9–11.

The chemical compositions for waters from a number of igneous rocks tend to plot between the plagioclase–montmorillonite and plagioclase–kaolinite curves (Figure 9–11), suggesting that these waters are in equilibrium with mixed kaolinite–montmorillonite weathering products. Note that the agreement between theoretical and measured water compositions is best for volcanic (extrusive) rocks, with waters from rhyolites in equilibrium with oligoclase, waters from andesites in equilibrium with andesine, and waters from basalts in equilibrium with labroadorite, the appropriate plagioclase compositions for these rock types.

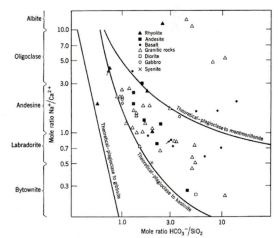

Figure 9–11

Mole ratio of Na^+/Ca^{2+} versus mole ratio of $HCO_3^-/SiO_{2 \text{ (aq)}}$ for waters from various types of igneous rocks. Most of the waters plot between the theoretical curves for the incongruent dissolution of plagioclase to montmorillonite and the incongruent dissolution of plagioclase to kaolinite. See text for discussion. From Garrels (1967).

The intrusive igneous rocks—granites, diorites, gabbros, and syenite—show a greater amount of scatter. One possibility is that the much larger grain size for the intrusive igneous rocks, and hence the much smaller surface-to-volume ratio, inhibits the approach to equilibrium. Although not a perfect match, the results of Garrels (1967) indicate that the chemical compositions of groundwaters are strongly affected by rock (mineral)–water interactions.

Garrels (1967) also investigated the question of whether or not the groundwaters were behaving as open or closed systems with respect to atmospheric CO_2. Figure 9–12 is a schematic diagram showing the approximate compositions of waters in equilibrium with some of the phases in the system $CaO–Al_2O_3–SiO_2–H_2O$. Calculated reaction paths for open and closed systems, starting with a plagioclase of composition $Na_{0.66}Ca_{0.34}Al_{1.34}Si_{2.66}O_8$, are shown in Figure 9–12. For the open system, $P_{CO_2} = 10^{-3.5}$ atm, and this pressure is maintained throughout the reaction. For the closed system, initial $P_{CO_2} = 10^{-1.5}$ atm (corresponding to dissolved $CO_2 = 0.001$ mol L^{-1}). As the reaction between H_2CO_3 and plagioclase proceeds, the composition of the waters changes, as shown by the arrows in Figure 9–12. Eventually, the boundary between the kaolinite and montmorillonite fields is reached, and at this point the waters are in equilibrium with both kaolinite and montmorillonite. The composition of the waters will then move along this stability boundary until all the kaolinite has been converted to montmorillonite. An important observation is that natural groundwaters tend to plot along the closed system curve and in the region above this curve. Virtually all the samples plot below the open system curve. This suggests that, in most cases, the chemical evolution of groundwater occurs in a closed system.

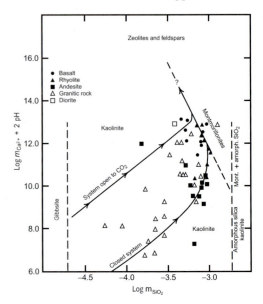

Figure 9–12

Plot of Ca^{2+} + pH versus SiO_2 for various groundwater samples. Stability boundaries for various phases are shown in the diagram. Solid curves with arrows indicate evolution of groundwater chemistry for systems open or closed with respect to atmospheric CO_2. See text for discussion. From Garrels (1967).

In summary, the preceding discussion suggests that there are several sources for the aqueous species found in groundwater. These are rainwater, dissolution of soluble minerals, and rock–water interactions. Table 9–10 lists the origin of the major aqueous species.

Lakes

Because of their complexity, in terms of physical, chemical, and biological processes, lakes have long attracted scientific interest. **Limnology** is *the study of lakes*. Lakes are of environmental interest because they can be readily affected by anthropogenic activities. For example, the addition of phosphorus to a lake by sewage effluent can lead to significantly enhanced primary biological production. Subsequent decay of the plant matter exhausts the oxygen from the lake waters, leading to anoxic conditions. This is an example of **eutrophication**, which is *overnourishment of the lake*. As a second example, trace metals entering lakes can be removed by particles in the lake and sequestered in the bottom sediments, leading to a buildup of these trace metals.

Lakes can be viewed as temporary water storage reservoirs. Water input to lakes occurs by precipitation onto the lake surface, streams and rivers flowing into the lake, springs discharging into the lake, and groundwater flow into the lake. Water output occurs by evaporation, streams flowing out of the lake, and seepage through the lake floor. If the lake has been modified by anthropogenic activities, additional inputs and outputs can be via artificial channels. Most lakes have outlets, but in the case of interior drainage there may not

Table 9–10 *Origin of Major Aqueous Species in Groundwater**

Aqueous species	Origin
Na^+	NaCl dissolution (some pollution)
	Plagioclase weathering
	Rainwater addition
K^+	Biotite weathering
	K-feldspar weathering
Mg^{2+}	Amphibole and pyroxene weathering
	Biotite (and chlorite) weathering
	Dolomite weathering
	Olivine weathering
	Rainwater addition
Ca^{2+}	Calcite weathering
	Plagioclase weathering
	Dolomite weathering
HCO_3^-	Calcite and dolomite weathering
	Silicate weathering
SO_4^{2-}	Pyrite weathering (some pollution)
	$CaSO_4$ dissolution
	Rainwater addition
Cl^-	NaCl dissolution (some pollution)
	Rainwater addition
$H_4SiO_{4\,(aq)}$	Silicate weathering

*From Berner and Berner (1996).

be an outlet. In this case, water is lost via evaporation or seepage. This is a significant difference in that the concentration of dissolved species will increase in the lake waters, leading to high salt concentrations. Such lakes, sometimes referred to as *alkaline lakes*, are relatively common in arid areas.

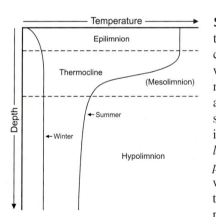

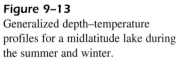

Figure 9–13
Generalized depth–temperature profiles for a midlatitude lake during the summer and winter.

Structure and Mixing of Lake Waters Mixing and overturn of lake waters are controlled by the water depth and climatic conditions. Mixing and overturn are important processes because they redistribute substances in the lake waters. For example, if overturn of water in the lake does not occur, or only occurs sporadically, deep waters may become enriched in some substances, such as organic matter, and depleted in other substances, such as dissolved oxygen. As a starting point for our discussion, we will look at the thermal structure of a midlatitude lake during the summer (Figure 9–13, summer curve). The lake is divided into three vertical layers: (1) the **epilimnion**, which is the *well-mixed surface layer*; (2) the **thermocline (or mesolimnion)**, which is a *zone of rapidly decreasing temperature*; and (3) the **hypolimnion**, which is the *well-mixed deep layer*. With the onset of winter, the surface layer begins to cool and eventually reaches a temperature of 4°C, the temperature of maximum density for freshwater (Chapter 1). The denser surface waters now sink and the deeper, less dense waters are brought to the surface. As winter cooling continues, stratification returns to the lake with lower-density, colder waters overlying higher-density, warmer waters (winter curve in Figure 9–13). With the onset of summer the surface waters warm to 4°C and a second overturn will occur. Finally, the cycle is completed with a return to the summer temperature profile shown in Figure 9–13. In this case, two convective overturns have occurred. At lower latitudes (Mediterranean-type climates), winter temperatures do not reach 4°C. In this case, there is continuous mixing of water throughout the winter until stratification returns in the spring. At high latitudes, temperatures never get above 4°C. In this case, there is continuous mixing throughout the summer and then stratification during the winter when the lake is ice covered. In the tropics, air temperature is relatively constant throughout the year and there is no seasonally well-defined overturn. Other factors determine the overturn of the lakes under these conditions. In certain unusual settings the absence of a seasonal overturn can lead to a natural disaster (Case Study 9–4). Shallow lakes are constantly stirred by wind action and a hypolimnion does not develop. Conversely, in the case of deep lakes a well-developed hypolimnion forms that is effectively isolated from the atmosphere.

Lake (Box) Models   Substances can behave either conservatively or nonconservatively in any particular system. *Conservative behavior* occurs when the *variations in concentration of a substance can be described by simple mixing*. *Nonconservative behavior* occurs

---

CASE STUDY 9–4
"Killer Lakes" of the Cameroon Volcanic Line

The Cameroon volcanic line consists of a linear chain of Tertiary to Recent volcanoes extending from the Atlantic island of Pagalu to the interior of the African continent. Mount Cameroon is currently active, and there is ample evidence of recent volcanic activity throughout the continental portion of the province. Many of the volcanic craters are occupied by lakes. The Cameroon volcanic line occurs very close to the equator and the crater lakes do not have a seasonal overturn. On August 15, 1984, a lethal gas burst at Lake Monoun resulted in the death of 37 people. On August 21, 1986, a similar burst from Lake Nyos resulted in the death of more than 1700 people. Because of the loss of human life, and the unique character of the events, both lakes were subjected to intense scientific scrutiny.

Sigurdsson et al. (1987) reports the results of an extensive study of Lake Monoun, and the following description is derived from this paper. Lake Monoun is the result of a lava flow that formed a volcanic barrier lake. Subsequent volcanic activity formed two craters on the lake bottom. The larger crater in the eastern part of the lake extends to a depth of 96 m. Temperature measurements made in March of 1985 showed very little variation in surface temperatures (22.9 to 23.8°C) and deep-water temperatures of a similar value, around 22°C. The bottom sediments were predominantly fine-grained reddish brown silt composed of siderite, quartz, and kaolinite with minor muscovite, biotite, pyrite, and gibbsite. The siderite was considered to be an endogenic mineral (derived from within the lake), and the other minerals were considered to be allogenic (derived from outside the lake). Lake waters were anoxic below 50 m and the dominant species in solution were Fe^{2+} (~ 600 mg L^{-1}) and HCO_3^- (≥ 1900 mg L^{-1}). When water samples were brought to the surface, large amounts of gas escaped from the samples. This gas was dominantly CO_2 with minor CH_4. Sulfur compounds were below detection in both the water and the gas. Oxygen and hydrogen isotopic analyses showed that the waters were stratified below about 15 m. The water in the hypolimnion plotted on the meteoric water line, indicating that magmatic waters are not present in the lake. Carbon isotopic data for both the dissolved bicarbonate and the CO_2 gas indicated a magmatic origin. Carbon isotopic data for methane suggested that this gas was produced by anaerobic bacterial reduction of organic matter. ^{14}C dating of the lake water revealed that only about 10% of the total carbon in the lake water was modern carbon. The authors concluded that the most likely scenario for the Lake Monoun disaster was the gradual buildup of CO_2 in the hy-

polimnion by degassing of a deeper magmatic source. Hydrothermal solutions were not involved, as indicated by the meteoric character of the water isotopic chemistry, the virtual absence of sulfur and chlorine (associated with hydrothermal activity), and the temperature of the lake water. Over a period of years, the release of CO_2 and the buildup of dissolved ferrous iron led to a siderite-saturated, higher-density hypolimnion within the larger lake bottom crater, overlain by a well-mixed, lower-density epilimnion. The lake was now density stratified and stable. This equilibrium was disturbed when a landslide occurred on the eastern rim of the crater lake, leading to the release of large volumes of CO_2 gas.

Lake Nyos occupies a 400-year-old (radiocarbon age) maar crater sited in Precambrian granite (Lockwood and Rubin, 1989). The lake is partly surrounded by poorly consolidated volcanic surge deposits. Field evidence suggests that carbon dioxide was the principal volatile involved in the formation of the maar. The maximum water depth is 220 m and, despite an increase in temperature with depth, the lake is density stratified due to the increase in dissolved species in the deeper waters. Like Lake Monoun, the deep waters are characterized by high concentrations of Fe^{2+} and HCO_3^-. Stable isotope ratios indicate a meteoric source for the lake water, but a mantle source for the carbon (Kusakabe et al., 1989). Kling et al. (1989) attributed an increase in bottom water alkalinity with time to the addition of thermal spring water. The mechanism responsible for the gas release is still uncertain. However, once deep water starts to ascend toward the surface, the process is self-sustaining. As the water rises, the hydrostatic pressure decreases and eventually equals the dissolved gas pressure. When this occurs, gas bubbles begin to form and the resulting decrease in density leads to a rapid rise of the deep waters to the surface. Of major concern is the integrity of the spillway at the northwest edge of the lake. At this location a 40-m-thick layer of pyroclastic material resting on granitic basement acts as a plug for the lake. The failure of this plug would result in a 40-m drop in water level and another catastrophic release of CO_2 from the deep waters. International efforts are now focused on decreasing the CO_2 concentration of the deeper waters so the lake level can be safely lowered. This degassing project involves the use of pipes that bring deep waters to the surface. The process is the same as described for the rise of the deep CO_2-rich waters, and, once initiated, the energy released during degassing is sufficient to drive the pumping operation.

Source: Sigurdsson et al. (1987), Lockwood and Rubin (1989), Kusakabe et al. (1989), Kling et al. (1989).

when *the concentration of a substance can be affected by processes other than simple mixing, i.e., by chemical and biological processes that occur within the system.* For example, a stream carrying iron in solution enters a lake. If the Eh and pH of the lake water is different from that of the stream, an iron–hydroxide flocculate may form and the iron is removed from solution. In this case, because it is removed, iron behaves as a nonconservative substance.

We can develop a generalized box model (Figure 9–14) for a lake that describes the behavior of both conservative and nonconservative substances (a similar model is applicable to the oceanic system). The lake is divided into two compartments, the epilimnion and hypolimnion. Additions to the epilimnion are by streams and transfer from the hypolimnion. Precipitation, airborne particles, and groundwater inputs are ignored. There may be cases where such inputs are significant and cannot be ignored in the calculation. For example, in Case Study 9–4, Sigurdsson et al. (1987) concluded that soil dust was the major source of the iron in Lake Monoun and Kling et al. (1989) concluded that thermal spring waters entering Lake

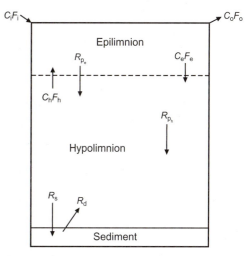

Figure 9–14
Two-compartment box model for a lake. See text for details.

Nyos were an important source of alkaline species. A substance is removed from the epilimnion by outflow, transfer to the hypolimnion, and as particles. In the case of chemical species, they may be removed as primary precipitates, by incorporation into the hard or soft parts of organisms, and by adsorption onto pre-existing particles. The removal of a species by particles can occur in the epilimnion and/or hypolimnion. Once formed, particles settle deeper into the lake, where changing chemical conditions may lead to re-solution of a species. For example, changes in Eh and pH may lead to the release of Fe and other Eh-sensitive elements to solution. For the epilimnion, we can write the following equation, which describes the change in amount of a substance with time:

$$\frac{\Delta M_e}{t} = C_i F_i + C_h F_h - C_o F_o - C_e F_e - R_{p_e} \qquad (9\text{--}21)$$

where ΔM_e is the change, with time, in the amount of the substance in the epilimnion, t is time, C_i is the concentration of the substance in the inlet water, F_i is the flux of water into the lake, C_h is the concentration of the substance in the hypolimnion, F_h is the flux of water from the hypolimnion to the epilimnion, C_o is the concentration of the substance in the outlet water, F_o is the flux of water out of the lake, C_e is the concentration of the substance in the epilimnion, F_e is the flux of water from the epilimnion to the hypolimnion, and R_{p_e} is the rate of removal of the substance by particles in the epilimnion. If the system is in a steady state, then $\Delta M_e/t = 0$ and equation 9–21 becomes

$$C_i F_i + C_h F_h = C_o F_o + C_e F_e + R_{p_e} \qquad (9\text{--}22)$$

For the hypolimnion, we can write the following equation for the change in the amount of a substance with time:

$$\frac{\Delta M_h}{t} = C_e F_e + R_d - C_h F_h - R_{p_h} \qquad (9\text{--}23)$$

where ΔM_h is the change, with time, in the amount of the substance in the hypolimnion, R_d is the rate of re-solution of the substance, and R_{p_h} is the rate of removal of the substance by particles in the hypolimnion. If the system is in a steady state, equation 9–23 becomes

$$C_e F_e + R_d = C_h F_h + R_{p_h} \qquad (9\text{--}24)$$

The amount of the substance that is ultimately stored in the sediment is

$$R_s = R_{P_e} + R_{P_h} - R_d \tag{9–25}$$

where R_s is the rate at which the substance is sequestered in the sediment.

Conceptually, it is important to understand the significance of each term in equation 9–25. In most cases, a certain amount of recycling will occur in the epilimnion. For example, phosphorus is a nutrient element and will be incorporated in organic matter. Some of this organic matter will break down in the epilimnion and the phosphorus will be returned to the water column. Therefore, R_{P_e} represents the net amount of the substance (in this case, phosphorus) removed from the epilimnion by particles. Similarly, R_{P_h} represents the net amount of a substance removed by particles from the hypolimnion. R_d is the amount of the substance returned to solution by interaction between the bottom waters and the sediments plus the release of the substance from particles originating in the epilimnion.

EXAMPLE 9–3 The concentration of Fe^{2+} in the epilimnion remains constant. Stream water inflow is $100 \text{ m}^3 \text{ s}^{-1}$ and the stream waters have an Fe^{2+} concentration of 3 mg L^{-1}. The lake is located in a temperate region and there is a net addition of water by precipitation (precipitation − evaporation) of $20 \text{ m}^3 \text{ s}^{-1}$. In order for the lake to maintain a constant volume, the outflow must be $120 \text{ m}^3 \text{ s}^{-1}$. The outflow has a mean Fe^{2+} concentration of 1.5 mg L^{-1}. Calculate the rate at which Fe^{2+} is removed from the epilimnion by particles. Assume that the Fe^{2+} concentration is the same for the epilimnion and the hypolimnion so the transfer of Fe^{2+} between the two layers can be ignored. Remember that $1 \text{ m}^3 = 1000$ L.

Rearranging equation 9–22 and solving gives

$$R_{P_e} = C_i F_i - C_o F_o = (3000 \text{ mg m}^{-3})(100 \text{ m}^3 \text{ s}^{-1}) - (1500 \text{ mg m}^{-3})(120 \text{ m}^3 \text{ s}^{-1})$$
$$= 120{,}000 \text{ mg s}^{-1} = 120 \text{ g s}^{-1}$$

Suppose that Fe^{2+} was not homogeneously distributed and that the concentration of Fe^{2+} was 2 mg L^{-1} in the epilimnion and 3 mg L^{-1} in the hypolimnion. If the volume of the epilimnion and hypolimnion remains constant, $F_e = F_h$. If $F_e = F_h = 500 \text{ m}^3 \text{ s}^{-1}$, calculate the rate at which Fe^{2+} is removed from the epilimnion as particles.

Rearranging equation 9–22 and solving gives

$$R_{P_e} = C_i F_i + C_h F_h - C_o F_o - C_e F_e$$
$$R_{P_e} = (3000 \text{ mg m}^{-3})(100 \text{ m}^3 \text{ s}^{-1}) + (3000 \text{ mg m}^{-3})(500 \text{ m}^3 \text{ s}^{-1})$$
$$- (1500 \text{ mg m}^{-3})(120 \text{ m}^3 \text{ s}^{-1}) - (2000 \text{ mg m}^{-3})(500 \text{ m}^3 \text{ s}^{-1})$$
$$= 620{,}000 \text{ mg s}^{-1} = 620 \text{ g s}^{-1} \qquad ■$$

For a lake in a steady state, the total input of a substance to the lake must equal its total output. With reference to Figure 9–14, we see that there is only one input, rivers and streams flowing into the lake (but don't forget that there are other possible inputs such as precipitation and soil dust), and two outputs, water flowing out of the lake plus accumulation of the substance in the sediments. In this case, we can represent the rate of accumulation of the substance in the sediments as

$$R_s = C_i F_i - C_o F_o \tag{9–26}$$

For a system that is in a steady state, we can calculate both a residence time and a relative residence time for any substance. *The length of time, on average, that a substance remains in a lake* is referred to as the **residence time**. As you will recall from Chapter 1, if a system is in a steady state (inputs = outputs) with respect to a particular substance, the residence time can be calculated from the following relationship:

$$\text{Residence time} = \frac{\text{Total amount of substance}}{\text{Rate of input (output)}} \tag{9–27}$$

We can also calculate the **relative residence time**, which is the *residence time of a particular substance compared to the residence time of water*. If the lake volume remains constant, water acts as a conservative constituent. Comparison of other substances to water allows us to differentiate conservative from nonconservative behaviors. If a substance has a relative residence time of 1, then it shows conservative behavior. Values other than 1 indicate nonconservative behavior. If the relative residence time is greater than 1, the substance accumulates in the lake. If the relative residence time is less than 1, the element is removed from the lake water to the sediment. We can derive a very simple relationship for relative residence time.

$$\tau_{rel} = \frac{\tau_s}{\tau_w} = \frac{M/C_iF_i}{V/F_i} = \frac{C_lV/C_iF_i}{V/F_i} = \frac{C_l}{C_i} \tag{9–28}$$

where τ_s is the residence time for the substance, τ_w is the residence time for the water, V is the volume of water in the lake, C_l is the concentration of the substance in the lake water, and M is the total mass of the substance in the lake ($M = C_lV$).

EXAMPLE 9–4 A lake has achieved steady-state conditions with respect to Fe, Cl, and P. For the incoming water, Fe = 4 mg L^{-1}, Cl = 14 mg L^{-1}, and P = 6 μg L^{-1}. For the lake water, Fe = 3.5 mg L^{-1}, Cl = 14 mg L^{-1}, and P = 10 μg L^{-1}. Calculate the relative residence time for each substance.

For Fe,

$$\tau_{rel} = \frac{C_l}{C_i} = \frac{3.5 \text{ mg L}^{-1}}{4.0 \text{ mg L}^{-1}} = 0.88$$

For Cl,

$$\tau_{rel} = \frac{C_l}{C_i} = \frac{14 \text{ mg L}^{-1}}{14 \text{ mg L}^{-1}} = 1.0$$

For P,

$$\tau_{rel} = \frac{C_l}{C_i} = \frac{10 \ \mu\text{g L}^{-1}}{6 \ \mu\text{g L}^{-1}} = 1.67$$

The calculations indicate that Cl acts as a conservative species, Fe acts as a nonconservative species (removed by sequestration in the lake sediments), and P acts as a nonconservative species (added because of recycling by biological processes). ∎

pH of Surface and Ground Waters

In Chapter 3 we explored acid–base equilibria in some detail. Here we will look at specific processes that affect the pH of surface and ground waters and consider the impact of acid deposition on pH, an environmentally important topic. Recall from Chapter 3 that the carbonate system exerts the major control on the pH of natural waters. On a longer time frame, water–rock interactions can also exert a significant control on pH. We will consider both of these systems, plus other factors, in the following discussion.

There are three processes that determine the pH of surface and ground waters:

1. Ion-exchange reactions that occur between silicate minerals (particularly clay minerals in the soil horizon) and acid waters in which H$^+$ ions in solution are adsorbed by the clay minerals. We will discuss ion exchange in a subsequent section.
2. Buffering reactions involving the carbonic acid system. In this case, the important parameter is the availability of bicarbonate (HCO$_3^-$).
3. Water–rock interactions involving carbonate and silicate minerals.

Carbonic Acid System We discussed this buffer system at some length in Chapter 3. Figure 9–15 (p. 340) shows the distribution of the various species as a function of pH. The calculations have been done for a total carbonate content of 1×10^{-3} mol L^{-1}, a typical value for surface waters, and a temperature of 25°C. Except in very alkaline waters, the buffering reaction is

$$H^+ + HCO_3^- \rightleftharpoons H_2CO_{3 \text{ (aq)}}$$

The effectiveness of this buffer is determined by the availability of HCO_3^-, which is determined by the pH of the system. When $H^+ = HCO_3^-$, there is no longer any buffering capacity. For the case shown here, this equality occurs at pH = 4.65. For waters with lower amounts of total carbonate, the buffering capacity is lost at even higher pH values (see problem 87). Consider the impact of acid rain or acid mine runoff on this system. If the pH of the acid rain or acid mine runoff is less that 4.65, continued addition of these acid waters will lower the pH of the system until all buffering capacity is lost. As the pH decreases, the buffering capacity decreases (see Chapter 3, Figure 3–7) and, thus, the rate of pH decline increases; i.e., the acidification of surface waters is an accelerating process.

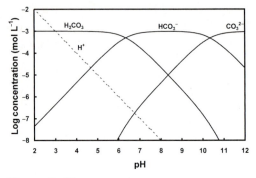

Figure 9–15
Variations in the concentration of the various carbonate species as a function of pH, in freshwater at a temperature of 25°C, given a total carbonate concentration of 1×10^{-3} mol L^{-1}. For these conditions the carbonate system loses its buffering capacity at pH = 4.65.

Water–Mineral Interactions   Waters in contact with carbonate minerals will tend to have basic pH values. In Chapter 3 we determined that water saturated with calcite and in equilibrium with atmospheric CO_2 would have a pH of 8.26. We also determined the buffering capacity as a function of pH for this system (Figure 3–8). Waters in equilibrium with carbonate minerals have a significant capacity to resist pH changes. Groundwaters in equilibrium with carbonate minerals will also have basic pH values. If we add acid to the calcite-carbonate system, the following reaction occurs:

$$H^+ + CaCO_{3\ calcite} \rightarrow Ca^{2+} + HCO_3^-$$

This reaction replenishes bicarbonate ion that has been consumed in the buffering reaction. It also releases Ca^{2+} (and if dolomite is involved, Mg^{2+}) to the water. This is one reason why lakes that have been subjected to acid inputs tend to have higher concentrations of Ca and Mg than nonacidic lakes.

The other important set of pH-controlling reactions involves the silicate minerals. We considered several of these reactions in Chapter 3. Other silicate mineral–water reactions that may be important in consuming hydrogen ions are

$$Al_2Si_2O_5(OH)_{4\ kaolinite} + 6H^+ \rightarrow 2Al^{3+} + 2H_4SiO_{4\ (aq)} + H_2O$$
$$NaAlSi_3O_{8\ albite} + 4H_2O + 4H^+ \rightarrow Na^+ + Al^{3+} + 3H_4SiO_{4\ (aq)}$$

and

$$KMg_{1.5}Fe_{1.5}(AlSi_3O_{10})(OH_2)_{biotite} + 10H^+ \rightarrow$$
$$K^+ + 1.5Mg^{2+} + 1.5Fe^{2+} + Al^{3+} + 3H_4SiO_{4\ (aq)}$$

Note that these reactions release Al^{3+} and other ions. One characteristic of acid lakes is an increase in dissolved aluminum, which is often correlated with a decline in fish populations. Recall from our discussion earlier in the chapter of aluminum solubility that decreasing pH leads to a greater solubility for aluminum. The other cations released to solution can be used to fingerprint the hydrogen-consuming mineral reactions. For example, the albite reaction leads to an increase in Na^+ and the biotite reaction releases K^+, Mg^{2+}, and Fe^{2+}. There is a stoichiometric relationship between these various ionic species, and in the case of our biotite example, the ratio of the ions should be K^+: $1.5Mg^{2+}$: $1.5Fe^{2+}$. In regions where mafic and ultramafic igneous rocks (in which olivine and pyroxene are common minerals) are abundant, the following silicate mineral–water reactions can lead to strongly alkaline waters:

$$Mg_2SiO_{4\ forsterite} + 4H^+ \rightarrow 2Mg^{2+} + H_4SiO_{4\ (aq)}$$

and

$$MgSiO_{3 \text{ enstatite}} + 2H^+ + 2H_2O \rightarrow Mg^{2+} + H_4SiO_{4 \text{ (aq)}}$$

These waters also tend to have high concentrations of bicarbonate ion and, thus, high buffering capacities.

ADSORPTION–DESORPTION PROCESSES

The adsorption (and desorption) of inorganic and organic species by particles plays an important role in the distribution of these species in the environment. As discussed in Chapter 7, the smaller the size of a particle, the greater the surface-to-volume ratio. The surface area of the particle is a critical factor in adsorption–desorption processes because broken bonds are found at the surface and this is where the particle has a net electrostatic charge exposed to the surrounding solution. This surface charge arises because of (1) substitutions of cations of lesser charge for cations of higher charge in the crystal structure or (2) reactions involving functional groups on the mineral surface and ions in solution. The charge due to ion substitution is considered to be fixed, whereas the charge due to surface complexation is variable and depends on the pH of the solution. The binding of the charged species to the surface is a function of several factors, including the charge density at the surface, the total charge, and the competing charged species in solution. If we look at this process in terms of the solution, there are several important factors: (1) pH, (2) oxidation–reduction potential, and (3) ionic strength of the solution. What this means is that changing environmental conditions influence whether or not a particular species is adsorbed or released by particles.

For particles that have a fixed charge (i.e., the 2:1 clay minerals), the net negative charge is balanced in several ways. (1) *Cations may attach directly to oxygen ions (or more precisely, silanol functional groups, Chapter 7) exposed at the particle surface.* This type of complex is called an ***inner-sphere complex*** and the bonds are relatively strong. (2) *Cations that are surrounded by water molecules (sometimes called solvated cations) are attached to the particle surface by bonds between the water molecules and exposed surface functional groups.* This type of complex is called an ***outer-sphere complex***. These are much weaker bonds and outer-sphere complex ions are readily exchanged with ions in solution. (3) The remainder of the negative charge is balanced by a diffuse layer of ***counter ions***, *cations that surround the mineral surface.* These cations are not attached to the mineral surface. In the so-called *diffuse double layer*, the positive counter ions are more abundant than the negative co-ions, which offsets the remaining negative surface charge.

For particles that do not have a fixed charge (e.g., 1:1 clays, oxyhydroxides), the surface charge is variable and varies as a function of pH. In this case, the charge is due to complexation reactions involving surface hydroxyls. At low pH values, the surface hydroxyls combine with H^+ in solution to form positively charged surface complexes:

$$S–OH + H^+ \rightarrow S–OH_2^+$$

where S represents a surface metal cation (Al^{3+}, Mg^{2+}, Fe^{3+}, etc.). The end result of this reaction is a positively charged surface. With increasing pH the hydrogen ions will return to solution, as represented by the following reaction:

$$S–OH + OH^- \rightarrow S–O^- + H_2O$$

The result of this reaction is a negatively charged surface. There are also other possible types of surface complexation reactions that involve metal cations and negatively charged ligands.

In summary, the surface charge of a particle is a function of a number of different processes. We can write the following equation to represent these relationships:

$$NC = FC + NP + ISC + OSC \tag{9–29}$$

where NC is the net charge, FC is the fixed charge due to substitutions in the crystal structure,

Table 9–11 Point of Zero Net Proton Charge*

Material	pH_{pznpc}	Material	pH_{pznpc}
α-Al(OH)$_3$	5.0	δ-MnO$_2$	2.8
γ-AlOOH	8.2	SiO$_2$	2.0
Fe$_3$O$_4$	6.5	Feldspars	2–2.4
α-FeOOH	7.8	Kaolinite	4.6
α-Fe$_2$O$_3$	8.5	Montmorillonite	2.5
Fe(OH)$_{3\ (am)}$	8.5	Albite	2.0

*From Kehew (2001).

NP is the net proton charge due to the binding or release of hydrogen ions from the surface, ISC is the charge due to the presence of inner-sphere complexes, and OSC is the charge due to the presence of outer-sphere complexes. If the net charge is not zero, it is balanced by the ions in the diffuse double layer adjacent to the surface. In equation 9–29, with the exception of the fixed charge, all the charge components on the right-hand side vary as a function of pH. Because charge varies as a function of pH, the surface may be either negatively or positively charged. It follows that there is some *pH at which the surface has no charge (**point of zero charge, PZC**). We can also define a **point of zero net proton charge**, PZNPC* (Table 9–11), which occurs *when the charge due to the binding and release of protons (NP) is zero* (see Chapter 7 for further details). At pH values below these points of zero charge, the particle has a positively charged surface; at pH values above the zero point charge, the particle has a negatively charged surface. Hence, the relationship between pH and surface charge is an important factor in determining the adsorptive characteristics of particles.

Adsorption of Metal Cations

The adsorption of metal cations varies as a function of their ionic potential. ***Ionic potential** is the ratio of the charge of the cation divided by its radius*—in effect, the charge density of the surface of the cation. Ions that have low ionic potential more readily lose their attached water molecules and, hence, form inner-sphere complexes. As noted, inner-sphere complexes are more tightly bound to the surface of the particle. For ions of the same charge, the selectivity series (order in which the ions are preferentially adsorbed, from most strongly adsorbed to least strongly adsorbed) is simply related to ionic radius. Increasing ionic radius leads to lower ionic potential. For example, consider the first column of elements in the periodic table, all of which form $+1$ ions. With reference to Appendix III, the ionic radii for these ions (in VI-fold coordination) are Li (0.82), Na (1.10), K (1.46), Rb (1.57), and Cs (1.78). The selectivity series for this group of ions is

$$Cs^+ > Rb^+ > K^+ > Na^+ > Li^+$$

For the elements in the second column, the order would be

$$Ba^{2+} > Sr^{2+} > Ca^{2+} > Mg^{2+}$$

For metals in the transition series, the electron configuration becomes more important than ionic radius in determining selectivity. For example, the following selectivity series has been established for a subset of the transition metals:

$$Cu^{2+} > Ni^{2+} > Co^{2+} > Fe^{2+} > Mn^{2+}$$

As noted, when the adsorbing particles have variable surface charge, pH becomes an important factor. At low pH, the surface is positively charged, and at higher pH, above the PZNPC, the surface becomes negatively charged. Once the surface becomes negatively charged, metals are rapidly adsorbed by the surface over a relatively narrow pH range (Figure 9–16). The degree of adsorption increases with increasing pH because the negative surface charge is increasing. Within any particular group of charged ions, those with higher selectivity are adsorbed at lower pH values. Of particular importance here is to note that changing pH exerts a significant influence on the adsorption of various metal cations. Hence, changes in pH can lead to either adsorption of cations from solution or desorption of cations from particles.

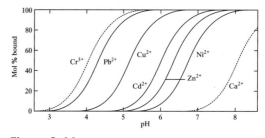

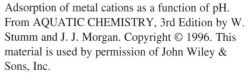

Figure 9–16

Adsorption of metal cations as a function of pH. From AQUATIC CHEMISTRY, 3rd Edition by W. Stumm and J. J. Morgan. Copyright © 1996. This material is used by permission of John Wiley & Sons, Inc.

Colloids

The types of particles involved in adsorption–desorption reactions can be mineral particles (Chapter 7), organic particles (Chapter 5), and colloids. As mentioned previously, by definition a species is in solution if it passes through a 0.45-μm filter. Molecular-size species pass through this filter, as expected, but a portion of the colloidal size range is less than 0.45 μm and colloids will also pass through the filter. *Colloids* are *particles that occupy the size range between true solutions and suspensions. This size range is not precisely defined but is generally considered to encompass particles between 10 μm and 0.01 μm in diameter. In this size range Brownian motion plays an important role in maintaining particle suspension.* In natural waters colloids are usually clay minerals, silica, oxyhydroxides, organic matter, and bacteria. The colloidal size range and amount vary as a function of the hydrologic environment (Table 9–12).

Colloidal sizes vary as a function of the dynamics of the hydrologic environment. Rapidly moving water carries much larger colloidal particles than slow-moving water. For example, most groundwater, which has a very low velocity, only transports very small colloidal particles. But note the much larger colloidal particles found in the Noiraigue spring, which is located in a karst topography where high groundwater flow velocities would be expected. In general, the greatest absolute amounts of colloidal particles are found in rivers, largely because of the high input of dissolved materials. Even in the case of systems with low absolute and relative amounts of colloidal particles, colloidal particles can be major transporters of metals because of their small size and correspondingly large surface area. This is particularly true when the colloidal particles are clay minerals, which have large specific surface areas (Chapter 7).

Earlier in this chapter we developed equations that describe the solubility of $Al(OH)_3$ in water. Calculation of the solubility of $Al(OH)_3$ as a function of pH showed that minimum

Table 9–12 *Size Ranges for Colloidal Particles in Natural Waters**

Sample	Amount (mg L^{-1})	Size observed (μm)	Size peaks (μm)
Rainwater	0.006	0.08	
Northern Pacific Ocean		0.38–1	
Gulf of Mexico		0.02–8	
Biscayne Bay		0.5–150	
Lake		0.04–0.4	0.1
Chuckawa Creek	50	0.3–1.3	0.2
Mississippi River	350		0.3
Yarra River	1–10	0.1–0.5	0.22
Noiraigue spring	0.5–10	0.5–60	0.8–1
Groundwater (Gorleben)			0.005–0.01
Groundwater (Grimsel)	0.1	0.04–1	

*Data from Atteia et al. (1998).

CASE STUDY 9–5
Removal of Dissolved Zinc by Oxyhydroxide Colloids

Kwong (1996) investigated the chemistry of a naturally acidic lake in the central Yukon, Canada. The lake had a uniform pH of 2.8. The acid pH was due to subsurface sulfide oxidation of an adjacent massive sulfide lens. The lake had dissolved zinc and sulfate concentrations of 2.7 and 1210 mg L^{-1}, respectively. The lake is surrounded by carbonate rocks. Because of these carbonate rocks neutralization reactions take place in streams flowing from the lake. Within 1 km of the outlet from

the lake the acidic drainage has been neutralized. This is accompanied by the removal of zinc, iron, and manganese. Zinc content in the stream sediment increases to a maximum of 0.13% 2 km downstream from the outlet. During aqueous transport the dissolved zinc is removed to the sediments by adsorption onto the precipitating iron and manganese oxyhydroxides. These colloidal-sized particles increase in size because of coagulation and are subsequently removed from the water column.

Source: Kwong (1996).

solubility occurs around pH = 6.4 (Figure 9–3). An analysis of the solubility of Fe(OH)$_3$ would yield a similar result. _Compounds such as Al(OH)$_3$ and Fe(OH)$_3$ are referred to as_ **oxyhydroxides**. At the intermediate pH values typical of natural waters, the oxyhydroxides are very insoluble and exist in solution in colloidal form. Depending on the pH, the surface of these particles is either negatively or positively charged (Chapter 7). At some particular pH, the surface is uncharged (the PZNPC, Chapter 7). At pH values far from this ZNPC, the surfaces are highly charged and mutual repulsion keeps the colloidal particles in suspension. The high surface charge also means that various ionic species can be adsorbed by the colloidal particle. This adsorption varies as a function of pH. Increases in the ionic strength of the solution, which would tend to neutralize the charge, or changes in pH, which can reduce the net surface charge, lead to flocculation and settling out of the colloidal particles. Thus, substantial quantities of various metals can be carried in natural waters in colloidal form, and these metals will be removed from solution in response to changing pH or increasing ionic strength (Case Study 9–5). A common place where this type of process occurs is where a river enters the ocean. Changes in pH and total ionic strength of the solution lead to flocculation of the colloidal particles and the accumulation of various metals in the bottom sediments of coastal marine environments (Chapter 10).

Isotherms and Partitioning of Species

The partitioning of species between solutions and particles is described in a number of ways (see Chapters 5 and 7 for a more extensive discussion). In the simplest case, where the partitioning is independent of concentration, we can represent the partitioning by the linear partition coefficient, K_d:

$$K_d = \frac{C_{ads}}{C_{soln}} \tag{9–30}$$

where C_{ads} is the concentration of the substance of interest adsorbed on the solid phase and C_{soln} is the concentration of the substance of interest in solution. We can also write an equation for a related quantity called the _activity_ K_d. In this case, we use the activity of the substance in solution, rather than the concentration, for the K_d calculation

$$K_d = \frac{C_{ads}}{a_{soln}} \tag{9–31}$$

where a_{soln} is the activity of the substance of interest. Linear partitioning best describes the behavior of trace molecular organic compounds, which are nonionized and, thus, weakly adsorbed. In the case of soils and sediments, it has been found that the adsorption of organic substances is essentially dependent on the amount of solid organic matter in the soils and sediments. In these cases we can determine an organic carbon distribution coefficient, K_{oc}, as follows:

$$K_{oc} = \frac{\mu g \text{ Adsorbed/g organic carbon}}{\mu g/g \text{ Solution}} \tag{9–32}$$

For more complex situations in which the adsorption characteristics vary as a function of the concentration of the substance in solution, Freundlich or Langmuir isotherms are commonly used to describe the adsorption behavior. For the Freundlich isotherm, the equation is

$$C_{ads} = KC_{soln}^n \qquad (9\text{--}33)$$

where K is the adsorption coefficient and n is an exponent. For the Langmuir isotherm,

$$C_{ads} = \frac{S_{max}KC_{soln}}{1 + KC_{soln}} \qquad (9\text{--}34)$$

where S_{max} is the maximum sorptive capacity for the surface (i.e., the number of adsorption sites available per unit surface area). The linearization of these equations and the calculation of the various parameters from experimental data were discussed in Chapter 7.

EXAMPLE 9–5 An experimental study of the partitioning of Cd (concentrations in μg) between soil water and soil revealed that the adsorption of Cd by the soil obeyed a Freundlich isotherm with $K = 0.034$ L g^{-1} and $n = 0.82$. If the concentration of Cd in the soil water is 4×10^{-6} mol L^{-1}, calculate the concentration of Cd adsorbed per gram of soil.

$$C_{ads} = KC_{soln}^n = (0.034 \text{ L g}^{-1})(4 \text{ } \mu g \text{ L}^{-1})^{0.82} = 0.106 \text{ } \mu g \text{ g}^{-1}$$

Suppose we started with 10 g of soil, which we place in a 1-L solution containing $4 \text{ } \mu g \text{ L}^{-1}$ Cd. This is not the same problem we've just done, because none of the Cd is initially adsorbed on the soil. Calculate the final concentrations of Cd on the soil and in solution. This type of problem is most easily done by successive approximations.

As a starting point, we know from our previous calculation that soil in equilibrium with a $4 \text{ } \mu g \text{ L}^{-1}$ Cd solution has $0.106 \text{ } \mu g \text{ g}^{-1}$ adsorbed Cd. Starting with this initial input, and recalling that we have 10 g of soil, we will assume that $1.06 \text{ } \mu g$ of Cd is adsorbed on the soil.

$$C_{ads} = KC_{soln}^n = (0.034 \text{ L g}^{-1})(4.0 - 1.06 \text{ } \mu g \text{ L}^{-1})^{0.82} = 0.082 \text{ } \mu g \text{ g}^{-1}$$

Redo the calculation using the new value for the total amount of adsorbed Cd ($0.082 \text{ } \mu g$ $g^{-1} \times 10$ g $= 0.82 \text{ } \mu g$):

$$C_{ads} = KC_{soln}^n = (0.034 \text{ L g}^{-1})(4.0 - 0.82 \text{ } \mu g \text{ L}^{-1})^{0.82} = 0.088 \text{ } \mu g \text{ g}^{-1}$$

Subsequent iterations (actually only one more is required) give the final answer of $0.087 \text{ } \mu g \text{ g}^{-1}$. For our system, at equilibrium, $0.87 \text{ } \mu g$ of Cd is adsorbed on the soil and $3.13 \text{ } \mu g$ of Cd remains in solution. Calculations involving the Langmuir isotherm are done in an analogous manner. ∎

Because a variety of factors, such as pH and the ionic strength of the solution, can affect the adsorption coefficient, it is often best to determine in situ K values. Assuming that the substance of interest is already present in the system, this is done by measuring the concentration of the substance in solution and adsorbed to particles. For example, if one were interested in the adsorption of Cd by a particular soil, one would analyze the pore water and the soil for their Cd content and then calculate an empirical (in situ) K value.

Ion Exchange

Adsorption processes involve ion exchange; i.e., an ion attached to the surface of the particle must be exchanged with an ion in solution when adsorption occurs. A number of models are used to describe the ion-exchange process (Appelo and Postma, 1996). For a simple binary exchange—i.e., only two ionic species are involved in the exchange—we can write the following general reaction:

$$aA^y + bBX \rightleftharpoons bB^z + aAX$$

where A and B are ionic species, a and b are the number of moles of each species, y and z are the valence of each species, and X is the solid phase involved in the ion-exchange process. The expression for the ion-exchange constant is

$$K_{ex} = \frac{[B^z]^b[AX]^a}{[A^y]^a[BX]^b} \tag{9-35}$$

You will note that this equation has been written in terms of activity. The activity of the ionic species in solution can be easily calculated using one of the activity-coefficient models. However, calculation of the activity of the solid, or perhaps, more specifically, the activity of the sorbing site, is a much more difficult problem. For this reason, the ion-exchange expression is often written in terms of concentrations and what is calculated is the selectivity coefficient, K':

$$K' = \frac{(B^z)^b(AX)^a}{(A^y)^a(BX)^b} \tag{9-36}$$

The concentration of the ions in the solid phase is usually written in terms of equivalent fractions (*Gaines–Thomas* convention) or molar fractions (*Vanselow* convention). For homovalent exchanges (same ionic charge for the ions involved in the exchange reaction) either convention gives the same results. However, for heterovalent exchanges (ions involved in the exchange reaction have different ionic charges) the two conventions do not give the same result.

EXAMPLE 9-6 Ten grams of sodium-saturated smectite (CEC = 100 meq/100 g) are mixed with 1 L of water containing 1×10^{-3} mol L^{-1} sodium and 1×10^{-3} mol L^{-1} potassium. The ion-exchange reaction is

$$K^+ + Na_{clay} \rightleftharpoons Na^+ + K_{clay} \quad \text{and} \quad K_{K/Na} = \frac{(K_{clay})(Na)}{(Na_{clay})(K)} = 2$$

Note that we have used a subscript for the selectivity coefficient that shows the order in which the solute ions appear in the reaction.

For 10 g of smectite there are 10 meq of exchangeable cation sites. Therefore, the total meq for $K_{clay} + Na_{clay}$ must equal 10. When K^+ is adsorbed from solution, an equal amount of Na^+ must be released from the clay. Using x to represent the meq of K^+ adsorbed by the clay, we can write the preceding equation as follows:

$$x = (K_{K/Na})\left[\frac{(K - x)}{(Na + x)}\right](Na_{clay} - x)$$

Substituting the appropriate numerical values, and remembering that initially $Na_{clay} = 10$ meq, gives

$$x = (2)\left[\frac{(1 \text{ meq } L^{-1} - x)}{(1 \text{ meq } L^{-1} + x)}\right](10 \text{ meq} - x)$$

The quadratic form of this equation is

$$x^2 - 23x + 20 = 0$$

Solving the equation using the quadratic formula gives $x = 0.905$. At equilibrium, $K_{clay} = 0.905$ meq, $Na_{clay} = 9.095$ meq, $K^+ = 0.095$ meq L^{-1}, and $Na^+ = 1.905$ meq L^{-1}. ∎

In the case of homovalent ion-exchange reactions, the strength of the solution does not affect the exchange reaction. This is not the case for heterovalent ion-exchange reactions. Consider the following reaction:

$$2Na^+ + Ca_{solid} \rightleftharpoons Ca^{2+} + 2Na_{solid}$$

The ion-exchange reaction is written

$$K_{\text{Na/Ca}} = \frac{(\text{Na}_{\text{solid}})^2(\text{Ca})}{(\text{Ca}_{\text{solid}})(\text{Na})^2}$$

EXAMPLE 9–7 For the ion-exchange reaction

$$2\text{Na}^+ + \text{Ca}_{\text{solid}} \rightleftharpoons \text{Ca}^{2+} + 2\text{Na}_{\text{solid}}$$

$K_{\text{Na/Ca}} = 0.5$. When $\text{Ca}^{2+} = \text{Na}^+ = 0.1 \text{ mol L}^{-1}$, calculate the distribution of these two ions on the solid.

$$\frac{(\text{Na}_{\text{solid}})^2}{(\text{Ca}_{\text{solid}})} = K_{\text{Na/Ca}} \frac{(\text{Na})^2}{(\text{Ca})} = (0.5) \frac{(0.1)^2}{(0.1)} = 0.05$$

$\text{Na}_{\text{solid}} + \text{Ca}_{\text{solid}} = 1$. Setting $\text{Na}_{\text{solid}} = x$ and $\text{Ca}_{\text{solid}} = 1 - x$ gives the following quadratic equation:

$$x^2 + 0.05x - 0.05 = 0$$

Solving for x using the quadratic formula gives $x = 0.22$, the mole fraction of Na on the solid. ■

Because of the exponential terms in the heterovalent cation-exchange equation, the exchange process is sensitive to the total ionic strength of the solution. The student can investigate this (and, needless to say, we will provide a problem) by solving the equation in Example 9–7 for order of magnitude differences in cation concentrations. The effect of total ionic strength is illustrated by doing a range of calculations for Ca and Na partitioning onto a solid using $K_{\text{Na/Ca}} = 0.5$ (Figure 9–17). In this example, the total Ca + Na in solution is maintained at a constant value, 1.0N or 0.01N. Note that we are using normality; hence, what is being kept constant is the total equivalents of Ca + Na. The important observation is that the selectivity of the solid

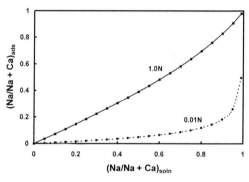

Figure 9–17

Partitioning of monovalent and divalent cations between solution and adsorber. Concentrations are given as an equivalent fraction of Na. $K_{\text{Na/Ca}} = 0.5$. The divalent cation is much more strongly adsorbed in the low ionic strength (low normality) solution.

for the divalent cations increases substantially at lower total ionic strength. This has a real-world significance in that cation exchangers in freshwater will relatively concentrate the divalent cation, whereas in seawater the monovalent cation will be favored. Thus, the cation-exchange properties of the same solid will differ between a freshwater and a marine environment. Case Study 9–6 (p. 348) describes the partitioning of monovalent and divalent cations in a hypersaline environment (oil field brine).

Adsorption and Contaminant Transport

We have considered the question of partitioning of metals between solutions and solids in some detail. What is the significance of this exercise? From the standpoint of environmental problems, one of the things we wish to know is the behavior of metals and other deleterious species in natural waters. Will they be transported, adsorbed, or liberated in response to changing conditions? Some of these questions have been dealt with earlier. Here we will specifically look at the question of contaminant transport in a groundwater system as an example of the use of distribution coefficients. A number of standard chemical hydrology and aqueous geochemistry books treat this topic in more detail. The interested student should consult one of these sources for further information.

CASE STUDY 9–6
Impact of Ion-Exchange Processes on the Chemistry of Subsurface Brines

In petroleum-producing regions a major potential groundwater contaminant is oil field brine. These brines are associated with the extraction of liquid hydrocarbons and present a disposal problem. They are often held in surface storage ponds prior to ultimate disposal by injection into deep stratigraphic units. Leakage of these brines to the groundwater system can occur both during storage and stratigraphic injection. Total dissolved solids content in these brines can exceed 100,000 mg L^{-1} (significantly in excess of that of seawater). The composition of a typical oil field brine is shown in the table.

Cations	Concentration (mg L^{-1})	Anions	Concentration (mg L^{-1})
Ca^{2+}	11,980	HCO$_3^-$	13.2
Mg^{2+}	2,180	CO$_3^{2-}$	0
Na$^+$	68,500	SO$_4^{2-}$	2.1
K$^+$	10	Cl$^-$	120,000

By far the dominant anion is chloride. The major cations are sodium, which is monovalent, and calcium and magnesium, which are divalent. Given the high ionic strength of the brine, and our earlier discussion concerning the partitioning of heterovalent cations, we would expect sodium to behave differently from calcium and magnesium during brine transport.

Cates et al. (1996) investigated the ion-exchange properties of the brines with soils. The significant observation from these experiments is that the transport of sodium is retarded with respect to calcium and magnesium (as expected). In their column experiments they found that the peak concentrations for calcium and magnesium exceeded the concentration of these two elements in the brine (a *hardness halo*). Two explanatory points are required. (1) The way column experiments are done is to take a measured amount of sorbent, in this case soil, and pass a fixed volume of solution through the column. The solution is collected in increments as it exits the column, and each increment is analyzed for the species of interest. The result is a concentration curve for each species as a function of the volume of solution passed through the column. In these experiments calcium and magnesium appeared in the solutions before sodium, and the maximum amount of calcium and magnesium (on a meq L^{-1} basis) exceeded that of the initial brine. (2) In water chemistry, hardness has a very specific meaning: **Hardness** is *the concentration of ions in water that will react with a sodium soap to precipitate an insoluble residue.* It is usually reported in terms of equivalent CaCO$_3$ per liter, which can be calculated from measured concentrations of calcium and magnesium as follows:

$$\text{Equivalent CaCO}_3 = 2.5(\text{mg Ca L}^{-1}) + 4.1(\text{mg Mg L}^{-1}) \qquad (9\text{–}37)$$

Cates et al. then investigated the utility of Piper diagrams for identifying brine contamination episodes. The dominant anion in the brine is chloride, and this species behaves conservatively. On the anion portion of a Piper diagram, the mixing between groundwater and brine is represented as a simple mixing process. Because of the very large content of Cl in the brine, relative to groundwater, even a small amount of brine added to a groundwater sample will move the anion composition of the water to the chloride corner. For the cations, the relationship varies from simple mixing because of the retardation of the sodium ion relative to calcium and magnesium. The result is that the compositions plot off the simple mixing line. This discrepancy is best seen by projection into the diamond-shaped portion of the Piper diagram (Appendix IV). The sample that shows the greatest displacement from the simple mixing line is inferred to represent the position of the hardness halo.

Source: Cates et al. (1996).

The most commonly used adsorption model in contaminant transport calculations is the distribution coefficient model, K_d. In doing these calculations the assumption is made that the distribution coefficient remains constant. As long as there are no significant changes in the bulk chemistry of the solution, nor in the chemistry of the adsorbing particles, this is probably a reasonable assumption. The other issue is the determination of an appropriate K_d, and, as noted, if possible this is best done using water and solid samples from the system of interest. The standard way to do this type of problem (Freeze and Cherry, 1979) is to measure the transport distance of the adsorbed species relative to a tracer species that is not adsorbed. Let us assume that both the tracer and the species of interest were introduced simultaneously into the aquifer. As the species move through the groundwater system, dispersion occurs; i.e., the introduced species spread out over a horizontal distance. The lines for the adsorbed and nonadsorbed species in Figure 9–18 indicate the dispersal of the species. The calculations are done by measuring the distance traveled by a slug of water having exactly one-half the concentration that the species had when they were introduced into the system. These midpoint distances are indicated by lines a (for the adsorbed species) and b (for the nonadsorbed species). For the nonadsorbed species, the distance traveled is equal to the average groundwater velocity times the

Figure 9–18
Relative transport distances in a groundwater system for a nonadsorbed tracer species and an absorbed species. C/C_o is the measured concentration of the species relative to its original concentration at distance x. Note that the concentration of the species varies with distance because of dispersion.

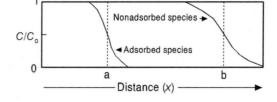

elapsed time ($x_{\text{non-adsorbed}} = \bar{v}t$). For the adsorbed species the distance traveled (x_{adsorbed}) in the same time is less because of adsorption. The ratio $x_{\text{nonadsorbed}}/x_{\text{adsorbed}}$ is the retardation factor. The retardation factor can be calculated as follows:

$$R_{\text{d}} = 1 + \left(\frac{\rho}{\varphi}\right)K_{\text{d}} \tag{9–38}$$

where ρ is the bulk density of the sediment (or rock), φ is the porosity, and K_{d} is the distribution coefficient. The velocity at which the contaminant moves through the groundwater system (note that we are using the midpoint of the concentration) can be found using the following relationship:

$$v_{\text{contaminant}} = \frac{\bar{v}}{R_{\text{d}}} \tag{9–39}$$

where $\bar{v}$ is the average velocity of groundwater flow.

EXAMPLE 9–8 Cd^{2+} ions are introduced into a groundwater system. For this particular system, $K_{\text{d}} = 2.5\ \text{cm}^3\ \text{g}^{-1}$ for Cd^{2+}. The average density of the grains that comprise the sediment is 2.6 g cm^{-3} and the porosity is 30%. The measured groundwater flow rate is 200 m y^{-1}. Determine the distance that Cd^{2+} will travel in one year in this aquifer.

First calculate the bulk density of the aquifer material. Assume that the pore spaces are filled with water with a density of 1.0 g cm^{-3}. The bulk density of the aquifer material is

$$\rho = (2.6\ \text{g cm}^{-3})(0.7) + (1.0\ \text{g cm}^{-3})(0.3) = 2.12\ \text{g cm}^{-3}$$

The retardation factor is

$$R_{\text{d}} = 1 + \left(\frac{\rho}{\varphi}\right)K_{\text{d}} = 1 + \left(\frac{2.12\ \text{g cm}^{-3}}{0.3}\right)(2.5\ \text{cm}^3\ \text{g}^{-1}) = 18.7$$

The velocity of Cd^{2+} transport in the groundwater system is

$$v_{\text{contaminant}} = \frac{\bar{v}}{R_{\text{d}}} = \frac{200\ \text{m y}^{-1}}{18.7} = 10.7\ \text{m y}^{-1}$$

In one year the Cd^{2+} ion will travel 10.7 m. Remember this is the point in the aquifer where the Cd^{2+} concentration is 50% of the Cd^{2+} concentration at the time of injection. Cd^{2+} will be found in the aquifer at a greater distance than 10.7 m, but its concentration will be less than 50% of that at the time of injection. ∎

COMPLEX FORMATION AND CHELATION

We previously discussed complex formation and chelation (Chapter 5). Here we briefly return to this topic because the formation of chelates plays an important role in the retention (or transport) of metals. Consider the following reaction:

$$M^{2n^+}{}_{\text{(aq)}} + nC_2O_4^{2-}{}_{\text{(aq)}} \rightleftharpoons M(C_2O_4)_{n\ \text{(aq)}}$$

where M is a metal and $C_2O_4^{2-}{}_{\text{(aq)}}$ is a ligand (in this case, a bidentate ligand). A *ligand* is defined as *an anion or molecule that forms a coordination compound with a metal*. A unidentate ligand contains one ligand atom; a multidentate ligand contains more than one ligand atom. In the preceding reaction, the ligand atom is oxygen and there are two oxygens in the oxalate; hence, it is a bidentate ligand. **Chelation** occurs when *complexes are formed with multidentate ligands*, and *the complex is referred to as a **chelate***. Recall our discussion of coordination number in Chapter 7. A similar concept applies to chelates, in which the ligands arrange themselves around a central metal atom in two-, four-, six-, and, occasionally, eightfold coordination. In the preceding reaction we used the oxalate species as the ligand. This is a common ligand found in the natural environment and results from the dissociation of oxalic acid ($H_2C_2O_4$), a common organic acid produced by biological processes.

Precipitation of the metal oxalates can be represented by the following equation:

$$M^{2n+}_{(aq)} + nC_2O_4^{2-}_{(aq)} + xH_2O \rightarrow M(C_2O_4)_n \cdot xH_2O_{(aq)}$$

With the exception of the alkali metals (Li, Na, K), ammonium, and Fe^{3+}, most of the simple metal oxalates have a low solubility in water (Table 9–13). This low solubility means that metal oxalates will precipitate from solution in response to biological processes.

Table 9–13 Solubility Products of Some Metal Oxalates*

Metal oxalate	T (°C)	pK_{sp}
$Cd_2C_2O_4 \cdot 3H_2O$	25	7.85
$CaC_2O_4 \cdot H_2O$	25	8.59
CuC_2O_4	25	7.54
PbC_2O_4	25	9.07
$MgC_2O_4 \cdot 2H_2O$	25	5.32
$MnC_2O_4 \cdot 2H_2O$	25	6.77
$Ag_2C_2O_4$	25	11.27
$Hg_2C_2O_4$	25	12.76
$SrC_2O_4 \cdot H_2O$	18	7.25
$ZnC_2O_4 \cdot 2H_2O$	25	8.86

*From Gadd (2000).

METALS

As you will recall from earlier discussions, elements that occur on the left-hand side of the periodic table and that normally form cations in solution are referred to as metals. **Heavy metals**, which are of environmental concern, are generally defined as *metals with atomic number 20 or greater* (i.e., calcium and higher atomic number metals). **Metalloids** are *elements that show both metallic and nonmetallic properties; i.e., depending on physicochemical conditions they can occur as either cations or anions*. The metalloids—As, Sb, Se, and Te—are found in columns VA and VIA of the periodic table. The metalloids will be included in the following discussion.

Metal Cycles

Natural   Revisiting earlier themes, in terms of metals (and nonmetals) we can view the earth as consisting of four reservoirs. These reservoirs are the *geosphere* (the solid earth), the *hydrosphere* (streams, rivers, lakes, groundwater, oceans), the *atmosphere* (the gaseous envelope), and the *biosphere* (living organisms). The interaction among these reservoirs (Figure 9–19) determines the transport and fate of the various metals. With the exception of a very minor cosmic flux, the source of all metals is the geosphere and, in the context of geologic time, the geosphere is also the sink for the metals. The atmosphere, biosphere, and hydrosphere serve as temporary reservoirs, in which important chemical changes can occur, for the metals.

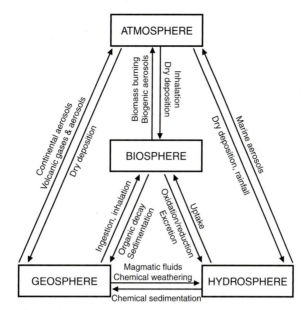

Figure 9–19
Interactions between various earth reservoirs.
Adapted from Larocque and Rasmussen (1998).

Metals are transferred from the geosphere to the atmosphere by volcanic processes (as gases and aerosols) and as dust particles (aerosols) and to the hydrosphere by chemical weathering and magmatic fluids. As we noted in Chapter 8, volcanic emanations and continent-derived dust particles can have a significant impact on atmospheric chemistry. As discussed earlier in this chapter, weathering processes can release various species to solution. A process we have not yet considered is the interaction between magmatic fluids and water chemistry. This is particularly significant for the ocean system (Chapter 10), where the cycling of seawater through fractures in warm seafloor basalts and the input of metals from ocean ridge hydrothermal vent systems control the long-term chemistry of the oceans.

The atmosphere serves as a very short-term reservoir for most metals, for which oxidation is the most important atmospheric process. The atmosphere plays a role in the long-distance transport of metals either as fine particles or as gaseous aerosols. Metals are ultimately removed from the atmosphere by dry and wet deposition (Chapter 8) or inhalation.

Plants and animals acquire metals by inhalation (gaseous form), ingestion (eating), and uptake of elements during plant growth. Many metals (such as Se and Co) play an important role in biological processes and are considered essential for the proper functioning of biological systems. A casual perusal of the ingredients in a multivitamin tablet will make this point. One of the interesting problems of modern agriculture is that enhanced crop production, due to the use of fertilizers and irrigation, has led to the depletion of soils in trace metals and the occurrence of diseases associated with trace metal deficiencies. As previously discussed (Chapters 4 and 7), biological processes play an important role in mediating a variety of chemical reactions, particularly oxidation–reduction reactions. Metals are removed from the biosphere by organic decay, sedimentation (also an important process in the ocean, Chapter 10), and excretion.

The processes that occur in the hydrosphere have been a major focus of this chapter. Metals are added to the hydrosphere by magmatic fluids and weathering, excretion by plants and animals, and dry and wet deposition. The transport and residence time of metals in the hydrosphere is a function of pH, oxidation–reduction potential, and the presence of sorbates such as clay minerals and oxyhydroxides. As we will see in Chapter 10, some metals can have very long residence times in the ocean. The major removal processes for metals are uptake by plants and chemical sedimentation.

Anthropogenic All the metals in the periodic table occur at some level in the natural environment. The impact of anthropogenic inputs is to significantly increase the concentrations of these elements. Hence, an understanding of the functioning of an uncontaminated (natural) system allows us to make inferences regarding the impact of anthropogenic inputs. It is worth noting here that this does not apply to many of the manufactured carbon compounds that were not previously found in the natural environment. The impact of these chemicals is much more problematic, and unanticipated results are not uncommon (DDT is a good example of an unanticipated result). In addition, an understanding of the "natural" levels of a particular metal is important in determining the impact of anthropogenic releases and remediation methods. For example, arsenic in drinking water has both natural and anthropogenic sources. There are some groundwaters that have naturally high arsenic concentrations due to the presence of bedrock containing various arsenic (and/or arsenic-containing) minerals. Without an understanding of this natural level, it is impossible to make informed inferences about the anthropogenic input of arsenic to the groundwater system.

The human role in the "metal cycle" is illustrated in Figure 9–20. Metals are extracted from the lithosphere by mining. One can also envision other, unintentional, extraction methods such as agriculture and disturbance of the surface for construction of roads and structures. The metals are returned to the geosphere through remediation or to the hydrosphere by waste disposal. For example, remediation of acid mine drainage, which usually takes the form of neutralizing the acidic pH, leads to the precipitation of various metal-containing compounds. Discharge of acid mine drainage directly to a water course is an example of (in this case) negative waste disposal to the hydrosphere. The combustion of fossil fuels, and other anthropogenic activities, releases metals to the atmosphere as aerosols or gases. Humans acquire these metals through inhalation and ingestion, and it is the occurrence of metals at levels hazardous to human (and, to a lesser extent, other organisms') health that motivates remediation of metal-contaminated air, soil, and water.

Acid Drainage

Acid mine drainage is a common problem associated with coal and metal sulfide mining. Acidic waters occur when metal sulfide minerals are oxidized. We have previously visited this topic in Chapters 3 and 4. As an example, consider the oxidation of pyrite, a complex cycle that can be represented by the following reactions:

$$2FeS_{2\ pyrite} + 2H_2O + 7O_2 \rightarrow 2Fe^{2+} + 4SO_4^{2-} + 4H^+$$

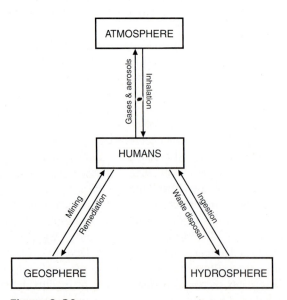

Figure 9–20
Human impact on the "metal cycle." Adapted from Larocque and Rasmussen (1998).

The ferrous iron is subsequently oxidized to ferric iron by the following biologically mediated reaction that consumes protons:

$$4Fe^{2+} + O_2 + 4H^+ \rightarrow 4Fe^{3+} + 2H_2O$$

The ferric iron now acts as an electron acceptor (oxidant) for further pyrite oxidation, during which protons are released to solution:

$$FeS_{2\,pyrite} + 14Fe^{3+} + 8H_2O \rightarrow 15Fe^{2+} + 2SO_4^{2-} + 16H^+$$

A hydrolysis reaction may also occur, and this reaction releases additional protons to solution:

$$Fe^{3+} + 3H_2O \rightarrow Fe(OH)_{3\,(aq)} + 3H^+$$

The overall sequence of reactions can be summarized as

$$4FeS_{2\,pyrite} + 14H_2O + 15O_2 \rightarrow 4Fe(OH)_{3\,(aq)} + 8SO_4^{2-} + 16H^+$$

Because most groundwaters do not contain significant amounts of oxygen, oxidation of sulfide minerals is generally confined to the surface weathering environment (however, note that there are exceptions; see Case Study 9–5).

Pyrite is a common accessory mineral in coal, and the preceding reaction is the major source of hydrogen ions for the acid waters associated with coal mining. Besides pyrite, base metal deposits contain other metal sulfide minerals, such as chalcopyrite ($CuFeS_2$), chalcocite (Cu_2S), covellite (CuS), galena (PbS), and sphalerite (ZnS). There is an important difference between the MS_2 (such as pyrite) and MS sulfide minerals during oxidation. As shown, the oxidation of an MS_2 sulfide mineral ($FeS_{2\,pyrite}$) leads to the release of H^+ ions to solution. For MS sulfide minerals, this is not the case. For example, the oxidation of sphalerite does not directly release H^+ ions to solution:

$$ZnS_{\,sphalerite} + 2O_2 \rightarrow Zn^{2+} + SO_4^{2-}$$

Subsequent hydrolysis reactions may release H^+ ions to solution, but the impact on overall acidity is much less than that due to the oxidation of MS_2 sulfides.

As we know from earlier discussions in this chapter, the solubility of iron in solution increases with decreasing pH, and under acidic conditions most metals are not adsorbed by oxyhydroxides and mineral particles. Thus, these metals stay in solution and acid mine drainage can contain significant amounts of heavy metals. A subsequent increase in pH leads to a decrease in the solubility of iron and the formation of oxyhydroxides. At the same time, the sorptive capacity of the particles in the water for the metal ions increases. The result is the removal of metals to the sediments and a different kind of potential environmental problem, metal-rich sediments.

There are a number of possible sources for acid mine drainage. (1) An active mine functions as a large well and groundwater flows into the mine. Thus, the mine must be continually pumped. Depending on the oxygen content of the waters, and the reaction times, these waters may be acidic. Because this is a controlled situation, the acid waters can be neutralized before they are returned to the natural environment. (2) After a mine is closed it floods to the water table. Water percolating down through the mine to the water table can become acidic, and discharge of this water leads to acid drainage. (3) During the mining operation a large amount of material is extracted from the mine and is piled at the surface (spoils). If they contain a significant amount of pyrite, these spoils are a potential source of acid drainage. (4) During the processing of metal ores the waste material (called *tailings*) is pumped to holding ponds. The result is a tailings pile that may contain significant amounts of sulfide minerals. During, or after cessation of, mining operations rainwater percolating through the tailings can become acidified. In abandoned mining districts spoils and tailings are significant sources of acid drainage. The impact, treatment, and control of acid drainage from these abandoned sites is in part a function of the site history (Case Study 9–7, p. 354).

In the previous discussion we have considered acid drainage due to mining activities or exposed sulfide deposits. Sulfide minerals are also precipitated in reducing marine en-

CASE STUDY 9–7
Sulfide Mine Tailings at Two Sites in New Mexico, U.S.A.

Boulet and Larocque (1998) investigated the characteristics of two sulfide mine tailings sites. For both deposits the common ore minerals were chalcopyrite, sphalerite, and galena. At one site (CPA) the tailings were deposited in a polyethylene-lined bermed impoundment. Mining at this site ended in 1995. At the other site (CM) the tailings were deposited in a valley at the headwaters of an ephemeral stream. Mining ceased at this site in 1950.

At the CPA site the dominant tailings minerals are calcite, quartz, and pyrite. A lime slurry was added to the tailings before they were discharged to the tailings pond. The tailings pond water is alkaline (pH = 7 to 8.3) and mineralogical investigation showed no evidence of dissolution of the primary calcite, indicating that the added lime slurry is still providing sufficient buffering capacity. Under the present conditions, the metals in the CPA tailings are immobile and acid drainage is not a problem at this site.

At the CMT site significant weathering and oxidation has occurred. Besides the sulfide minerals, the other primary minerals in this tailing pile were quartz, calcite, and several other silicate minerals. Because of weathering and oxidation, the current assemblage of primary minerals are quartz and the sulfide minerals pyrite, sphalerite, and chalcopyrite. Essentially all the primary calcite has been consumed in buffering reactions. Water draining from the site is strongly acidic (pH = 2.3) and has high metal concentrations. There has been a significant redistribution of metals within the tailings pile. Abundant secondary minerals (the result of the weathering of the primary minerals) have formed throughout the tailings pile. These secondary minerals consist of various iron oxides, hydroxides, and sulfates plus gypsum. In places, these secondary minerals have cemented the tailings to form *hardpans*. The hardpan represents a local zone of neutralization and precipitation, and the highest metal concentrations are found in the hardpan. Elements such as Ag, Au, Pb, and Sb were found at highest concentrations in the layer above the hardpan. These elements form highly insoluble complexes with sulfate, and the slight increase in pH as the waters percolated down through the tailings pile was sufficient to lead to the precipitation of the metal sulfates. The trapping of metals (Ag, Au, As, Pb, and Sb) in the hardpan layers was an important process that inhibited movement of these metals into the local environment. Metals (Cd, Cu, Fe, and Zn) not trapped in the hardpan layers were readily transported by the acid waters draining the site.

Source: Boulet and Larocque (1998).

---

vironments, such as deltas. Subsequent exposure of this material to subaerial weathering leads to acid drainage and the release of metals. This exposure can be due to either uplift or artificial drainage for land recovery (Case Study 9–8).

Heavy Metals

As defined earlier, heavy metals are those metals of atomic number 20 and greater. Of the heavy metals, the transition elements, As and Se (strictly speaking, metalloids), Cd, Hg, and Pb have attracted the greatest environmental interest. Other metals specifically involved in issues of nuclear waste disposal and radioactive fallout will be considered in the next section.

In a strongly reducing environment, and in the presence of sulfur, all these elements form insoluble sulfides. This, of course, is the reason why coals, which form in strongly reducing environments, contain significant amounts of heavy metals. As long as the environment remains reducing, the elements are immobile and are retained in the sediment. Changes in oxidation–reduction conditions can lead to a release of these elements. For less strongly reducing conditions, or environments that lack sulfur, the heavy metals can become variably mobile. The factors that control element mobility are pH, Eh, and the presence of organic compounds. The most important factors differ depending on the element. Thus, it is convenient to form three metal groups: (1) transition metals, Zn, Cd, and Pb; (2) As and Se; and (3) Hg.

Transition Metals, Zn, Cd, and Pb Under most oxidation–reduction conditions, these elements occur in solution as divalent or trivalent cationic species. Brookins (1988), a very useful reference, gives a number of calculated Eh–pH diagrams for virtually all systems of geochemical interest, and the student should consult this source for Eh–pH diagrams for specific metals. At high pH, many of these elements form insoluble oxyhydroxides or, in the presence of carbonate, insoluble carbonates. As we know from

CASE STUDY 9–8
The Effect of Acid Soil Leaching on Metal Concentrations in Streams

Sulfide-bearing marine and lacustrine sediments are common in the coastal regions of Finland. Artificial drainage of these areas during the past few decades has exposed these sediments to atmospheric O_2 and a high oxidation rate. The result has been the development of acid soils (pH = 2.5 to 5.0). Water-chemistry measurements were made for a stream draining this region under various flow conditions (Åström, 2001). The headwaters originate in an unpopulated area underlain by glacial till, peat, and glaciofluvial deposits. Water-chemistry measurements made at this point constitute the metal baseline concentrations. The concentrations of the metals in the outlet waters (after the stream had traversed the sulfide-bearing sediments) were compared to the baseline values. The outlet-to-baseline ratios varied from 0.3 to 55 as a function of the metal, the magnitude of the stream flow, and the season. The metals that showed downstream depletion were As, Pb, Sb, Ti, and V. The limited leaching of these metals from the acid sulfate soils was believed to be due to low absolute abundances, low concentrations in easily weathered minerals, or low mobility in the soils due to complexation with oxyhydroxides and humic substances. For the elements that showed downstream enrichment, the order, from most enriched to least enriched, was Ni > Co ≈ Cd ≈ Zn > U ≈ Al ≈ Cu ≈ Mn. Åström noted that the downstream variations for all the elements were closely similar to that for SO_4^{2-}, suggesting that the metal and sulfate had a similar hydrogeochemical behavior.

Figure 9–C8–1 shows a typical pH–depth profile for acid sulfate soils and the various redox environments. Seasonal changes in this profile can be directly related to variations in metal release to the stream. Åström proposed the following model to explain the seasonal variations in metal and sulfate concentrations in the stream:

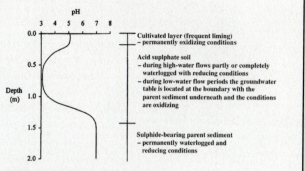

Figure 9–C8–1

pH, redox, and hydrological characteristics of a typical acid sulfate soil. After Åström (2001).

1. During the summer, higher temperatures and a lowered groundwater table favor aerobic bacterial oxidation of sulfide minerals and the release of metals and sulfate to the soils. Substantial alteration of soil minerals also occurs.
2. During periods of heavy rainfall in the autumn, the soluble weathering products are washed from the soils, and sulfate, metal ions, and hydrogen ions are discharged to the stream. As winter approaches, temperatures fall and the water table rises, leading to a cessation in the formation of soluble weathering products.
3. Metal and sulfate inputs to the stream during the early summer are small because most of the soluble weathering products were removed during the fall rains and there has been little formation of new material.
4. Under low flow conditions (summer and early autumn), the contribution from the acid soils is negligible compared to that of the areas covered with the other sediments.

Source: Åström (2001).

---

earlier discussions, adsorption is an important process. Thus, another important factor is the presence of suitable sorbates, among which are Mn and Fe oxyhydroxides. At low pH, sorbtion is negligible for these metals, but with increasing pH, adsorption becomes an important process (Figure 9–16), effectively removing the metals from solution by adsorption to particles and ultimately to the sediments. These elements also form complexes with dissolved organic matter (humic acids). Tipping and Hurley (1992) determined the following binding order for various metals and humic substances: $Mn^{2+} < Cd^{2+} < Co^{2+} < Ni^{2+} \sim Zn^{2+} < Pb^{2+} < Cu^{2+} < VO^{2+}$.

One notable exception to the behavior just described is vanadium. Under reducing conditions, V occurs as the relatively immobile V^{3+}. In oxic environments, the solubility of V increases dramatically because of conversion to higher oxidation states, V^{4+} and V^{5+}, and the formation of vanadyl cations, $VO_2(OH)_3^{2-}$, H_2VO^{4-}, and HVO_4^{2-}.

Selective Extraction   In any given environment, the actual distribution of a metal among various phases may differ substantially from that predicted from first principles. One way to address this question is by a technique known as *selective extraction*, in which various leaching experiments are done to remove the metals from specific phases. The leaching agents are selected so that they remove metals that occur in different ways, i.e., as exchangeable cations, bound to carbonates, bound to amorphous Fe and Mn hydroxides, bound to organic matter and sulfides, and as essential components of soil minerals. The experiment is done in a stepwise fashion, starting with a leaching agent that removes only the least tightly held cations. The last step is often a strong acid leach that breaks down the soil minerals. From these types of experiments, estimates can be made of the mobility of various metals in the specific environment. Case Study 9–9 describes a typical selective extraction experiment.

Arsenic and Selenium   Arsenic and selenium differ from the transition metals in that they occur in solution as neutral and negatively charged species. Under oxidizing conditions, arsenic is in the +5 state. Given the range of pH values found for the natural environment, As^{5+} exists in solution as $H_2AsO_4^-$ and $HAsO_4^{2-}$. Under reducing conditions, arsenic is in the +3 state and exists in solution as $H_3AsO_{3\,(aq)}$ and $H_2AsO_3^-$. Selenium also shows multiple oxidation states. Under strongly oxidizing conditions, it is in the +6 state (SeO_4^{2-}), under intermediate oxidizing conditions in the +4 state ($HSeO_3^-$ and SeO_3^{2-}), and under reducing conditions, it occurs in its elemental form or in the −2 state (HSe^-). Under reducing conditions, if sulfur is present, both As and Se are incorporated in sulfide minerals. Because As and Se exist in solution as negatively charged species, adsorption increases with declining pH (the surface becomes less negatively charged or positively charged). The arsenate (+5) species form inner-sphere complexes at the iron oxyhydroxide surface and are strongly adsorbed near neutral pH (note from Table 9–11 that for the oxyhydroxides $pH_{pznpc} = 7.8\text{–}8.5$), whereas the arsenite (+3) species are only weakly adsorbed at all pH values. For selenium, the situation is reversed and the selenite (+4) species are strongly adsorbed whereas the selenate (+6) species are weakly adsorbed.

The difference in adsorption behavior for the various species of arsenic and selenium is important in terms of their transport in the surface environment. In the case of arsenic, a strongly oxidizing environment that favors the formation of arsenate species will lead to adsorption by oxyhydroxides and removal from the aqueous environment. If conditions become reducing, the bound arsenate will be converted to arsenite and can be liberated from the particles to the solution. The reverse behavior would be noted for the various selenium species; i.e., selenium tends to remain in solution under oxidizing conditions but is removed under reducing conditions. The important point is that changes in the redox conditions of the environment can lead to either removal or liberation of arsenic and selenium. The other important point is that the kinetics of the redox reactions is slow, so

---

CASE STUDY 9–9
Geochemical Behavior and Mobility of Metals in a Sediment Retention Pond

Lee et al. (1997) studied the mobility of metals in a sediment retention pond located along a motorway in France. The purpose of these retention ponds is to accumulate particulates contaminated by heavy metals, thus preventing their discharge to surrounding streams. Eight sediment cores were driven to depths of 16 to 20 cm. The interstitial fluids were extracted and analyzed for their metal content. The metals associated with the solid phases were then analyzed using sequential chemical extraction. The authors used a five-step process, as follows:

1. Fraction I: *Exchangeable*, 1 M $MgCl_2$ at pH 7
2. Fraction II: *Bound to carbonate*, 1 M sodium acetate adjusted to pH 5.0 with acetic acid for 5 h
3. Fraction III: *Bound to amorphous Fe and Mn hydroxides*, 0.04 M hydroxylamine hydrochloric acid in 25% (v/v) acetic acid at 96°C for 6 h
4. Fraction IV: *Bound to organic matter and sulfides*, 30% H_2O_2 and 0.02 M nitric acid at 85°C for 5 h
5. Fraction V: *Residual*, concentrated HNO_3 and $HClO_4$

For the four metals of interest, Pb was concentrated in Fraction V (35–60%) and Fraction III (20–44%), Zn in Fraction V (30–65%) and Fraction III (13–47%), and Cd in Fraction II (25–68%). Mn showed a more complex behavior, with the bulk of the Mn in the surface layers in the exchangeable fractions (I, II, and III) and a notable increase with depth of Mn in the nonexchangeable Fraction V. The authors ascribed this increase in the nonexchangeable fraction to either the precipitation of Mn^{4+} oxides or the incorporation of Mn in Fe oxyhydroxides. Metals concentrated in Fraction V would be essentially immobile in the surface environment.

Source: Lee et al. (1997).

disequilibrium is common. Thus, predicting the mobility of arsenic and selenium under a particular set of environmental conditions is difficult.

Both elements are of environmental interest. Arsenic in very high concentrations (not those encountered in surface and ground waters, but perhaps due to human intervention) is a poison. Low-level, long-term exposure to arsenic may lead to an increased risk of cancer (National Academy of Sciences, 1999). For this reason, the USEPA (U.S. Environmental Protection Agency) has recommended lowering the allowable amount of arsenic in drinking water. The major anthropogenic source of arsenic is pesticides applied to orchards and other crop lands. Selenium is an essential metal for human health, but at slightly higher levels it becomes a toxic element. Certain plants can accumulate selenium, causing toxicity problems for ruminants that graze on the plants. The major anthropogenic source for selenium is deposition of atmospheric aerosols derived from the burning of coal.

Mercury   Under a wide range of Eh–pH conditions mercury exists as mercury metal (remember that at room temperature mercury metal is a liquid). Liquid mercury is essentially insoluble in water, but mercury is transferred to the atmosphere by volatilization. Under strongly oxidizing conditions mercury exists either as the Hg^{2+} ion or its hydrolysis product $Hg(OH)_{2\ (aq)}$. In a reducing environment, if sulfur is present, mercury can form the insoluble sulfide HgS (cinnabar).

The global cycle of mercury is dominated by the vapor phase transport of Hg^0 (Figure 9–21). The residence time of mercury in the atmosphere is 1 year. However, because of the equilibrium between the atmosphere and the surface ocean, the rate constant for mercury removal from the atmosphere is 0.29 y^{-1}. Because of the low removal rate of mercury by the surface ocean, the bulk of the mercury is deposited on land, where it is sequestered in the soils.

In its metallic form mercury is essentially inert. While this means it does not represent an immediate environmental hazard, it also means that mercury can persist for an extended period of time in the natural environment. Certain microorganisms, such as anaerobic bacteria, can mediate the methylation of mercury (Figure 9–21) to monomethyl mercury (CH_3Hg) or dimethyl mercury [$(CH_3)_2Hg$], which are highly soluble species that are readily taken up by aquatic organisms, particularly fish and shellfish. The mercury is concentrated up the food chain and can reach concentrations of several thousand ppm in organisms residing at the highest trophic levels. At these concentration levels, mercury is

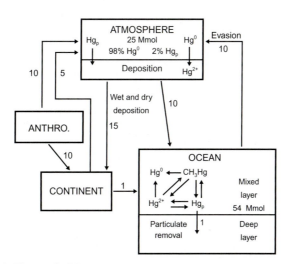

Figure 9–21
Present-day mercury cycle. The fluxes are given in Mmol L^{-1}. Hg_p represents mercury adsorbed to particles. Anthropogenic inputs are approximately 20 Mmol y^{-1}; half is returned to the surface close to the source and the other half is transferred to the atmosphere as volatile mercury. Adapted from Mason et al. (1994).

hazardous to human health. Because methylation of mercury is mediated by anaerobic bacteria, some of the mercury deposited in sediments in a reducing environment, even if sulfur is present, may be converted to soluble methyl species rather than insoluble mercury sulfide.

Metal Variations in Space and Time A variety of approaches are used to determine both natural and anthropogenic distributions of metals in the surface environment. One such approach is a regional geochemical survey in which the metal content of soils, streams, and groundwater is determined (Reimann et al., 2001; Sewell, 1999). The purpose of such studies is to gather information on both the background concentrations of the metals and areas of abnormally high metal concentrations. The bulk of these techniques were developed in the 1950s and 1960s, when they were used in geochemical exploration programs for base metal ore deposits. The sampling of different types of material can be used to infer the source of the metals. For example, Reimann et al. (2001) used four different sample materials: terrestrial moss, which predominantly reflects the atmospheric input of metals; the O-soil horizon, which reflects the interplay among atmosphere, biosphere, and lithosphere; the B-soil horizon, which reflects weathering processes; and the C-soil horizon, which represents the bedrock geochemistry.

The geochemical patterns can be compared to bedrock geology, surficial geology, land use, etc. Such studies have been greatly facilitated by the advent of Geographical Information Systems (GIS), which enables a multidimensional correlation of various parameters. For example, in the case of high levels of arsenic in groundwater, which can have both natural (bedrock) and anthropogenic sources (pesticides), a comparison of bedrock geology, land use, and arsenic well water concentrations can be used to distinguish between natural and anthropogenic sources of arsenic (Ayotte et al., 1999).

We are also interested in changes in metal concentrations with time. In Chapter 8 we discussed how snow, ice, lake sediments, and biological indicators can be used to map changes in atmospheric deposition of metals. In the case of the continental environment, the most useful information is contained in lake sediments. In general, lake sediments are the result of vertical sedimentation. As we have discussed in previous chapters, it is possible to determine the age of lake sediments as a function of depth. Chemical or isotopic analysis of the various layers can then be used to trace changes in anthropogenic input with time.

Metal Contamination of Soils and Sediments A number of processes sequester metals in sediments. One way to look at the anthropogenic enrichment of metals in sediments is to compare the abundance of the metals to background (or reference) values. One of the commonly used references is average shale. The calculation is analogous to that for enrichment factors (discussed in Chapter 8); i.e.,

$$\text{Enrichment factor} = \frac{\text{Concentration in sediment}}{\text{Concentration in reference material}} \quad (9\text{--}40)$$

The geoaccumulation index (Müller, 1979) can be used to assess the degree of contamination. The index is calculated as follows

$$I_{\text{geo}} = \log_2\left(\frac{C_n}{1.5B_n}\right) \quad (9\text{--}41)$$

where I_{geo} is the geoaccumulation index, $\log_2$ is log base 2, C_n is the concentration in the sediment, and B_n is the background or reference concentration. After the geoaccumulation index has been calculated, it can be used to classify the sediment in terms of quality (Table 9–14).

Radioactive Isotopes and Radioactive Waste Disposal

Radioactive isotopes in the environment come from both natural sources (K, Th, and U in rocks and minerals) and anthropogenic sources (bomb testing, nuclear medicine, nuclear accidents, and the nuclear fuel cycle). Radioactivity and radioactive decay were discussed

Table 9–14 I_{geo} Classes and Sediment Quality

I_{geo} Class	Sediment quality
0	Unpolluted
1	Unpolluted to moderately polluted
2	Moderately polluted
3	Moderately to highly polluted
4	Highly polluted
5	Highly to very highly polluted
6	Very highly polluted

*From Müller (1979).

in Chapter 6, and the student should refer to that chapter for a review of the basic principles. In Chapter 8 we considered the radioactive gas radon, which is produced in the U and Th decay chains. In this section we will be concerned with radioactive isotopes associated with the operation of nuclear reactors and the transport and sequestration of these radioactive isotopes in the natural environment.

Nuclear Fuel Cycle   The nuclear fuel cycle (Figure 9–22) starts with the extraction of uranium ore. The ore is then processed and the ^{235}U isotope is enriched relative to the ^{238}U isotope. This step is necessary because the fissionable form of uranium is ^{235}U. Natural uranium consists of 0.0057% ^{234}U, 0.719% ^{235}U, and 99.275% ^{238}U. Although some reactors are designed to operate with unenriched uranium, most require enriched uranium. In enriched fuel, ^{235}U constitutes 1.8 to 3.7% of the total uranium. The uranium is embedded in fuel elements for use in a nuclear reactor. During the operation of a nuclear reactor, the ^{235}U in the fuel elements undergoes fission (see the next section) and fission products build up in the fuel elements. The fuel elements eventually reach the end of their useful lifetime, not because all the ^{235}U has been used, but rather because of the buildup of fission products, which effectively "poison" the fuel element. After on-site storage, the fuel elements can be reprocessed (not currently done in the United States) to separate the uranium from the fission products. The recovered uranium can be recycled into new fuel elements and the radioactive waste products are sequestered in a radioactive waste repository. Currently in the United States there is no operating repository for spent fuel elements, and the elements are stored on-site at nuclear power plants. Excluding reactor accidents (Chernobyl, see the section on radioactive wastes), radioactive materials can be released at two points in the nuclear fuel cycle: during the mining of the ore and during the disposal of the radioactive waste (Figure 9–22). It is important to distinguish between low-level and high-level wastes. ***Low-level radioactive wastes*** have *relatively low radioactivity and typically contain radioactive isotopes with short half-lives.* Examples of

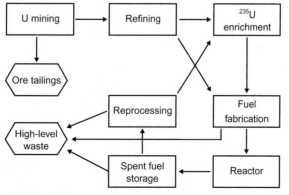

Figure 9–22
The nuclear fuel cycle. Modified from Faure (1998).

this type of waste are gloves, wipes, protective clothing, etc. used when handling radioactive materials for medical or research purposes. The volume of waste can be substantial, but the total radioactivity is low. These types of wastes are sequestered in specially designed landfills. ***High-level radioactive wastes comprise the used fuel elements and the waste generated during the processing of these fuel elements.*** Although the total volume of material is small, the total radioactivity is high. These materials present a major waste disposal problem.

Nuclear Fission   Thermal energy is released in a conventional power reactor through the fission of ^{235}U. The reaction can be written

$$^{235}U + {}^1_0n \rightarrow \text{Fission fragments} + 2 \text{ or } 3 \text{ neutrons} + \text{energy}$$

In this reaction a neutron is captured by the ^{235}U nucleus, which causes the splitting of the ^{235}U nucleus. The mass of the products is less than the mass of the reactants—i.e., mass is lost—and the missing mass appears as energy according to Einstein's famous relationship, $E = mc^2$. The neutrons produced in the fission reaction have an energy of 2 MeV and are high-energy neutrons. The probability of a high-energy neutron interacting with a ^{235}U nucleus is exceedingly small. In order for this reaction to occur, the neutron must lose energy. The role of water in a conventional water reactor is to moderate the neutron energy (i.e., slow down the neutron) so that it will interact with a ^{235}U nucleus and to remove the thermal energy produced by the fission process. In a graphite reactor, the graphite acts as a moderator. The rate of the nuclear reaction is controlled by regulating the number of neutrons. This is done by inserting control rods, which are made of materials (such as Ag/In/Cd or B_4C) that are good neutron absorbers, into the reactor. Changes in power level are achieved by varying the amount of neutron-absorbing material in the reactor core and thus regulating the number of neutrons. A fundamental difference between the operation of a water-moderated and a graphite-moderated (Chernobyl-type) reactor is the response of the moderator to increasing temperature. In the case of water, increasing temperature leads to a decrease in the ability of the water to slow down neutrons. Hence, water-moderated reactors tend to be self-quenching if there is a temperature increase. Graphite behaves in the opposite manner and becomes a better moderator at higher temperature. This difference in moderator behavior was a contributing factor (besides, most importantly, incorrect operating procedures) in the Chernobyl nuclear reactor disaster.

Radioactive Wastes Due to Nuclear Fission   Several different groups of radioactive isotopes are produced during the operation of a nuclear reactor. Fission fragments (or fission products) are essentially confined to two atomic mass ranges, ~85 to 105 and ~130 to 145. In the first mass range, radioactive isotopes of Kr, Sr, Y, Zr, Nb, and Ru (Table 9–15) are among the most important; and in the second mass range, radioactive

Table 9–15   Representative Radioactive Isotopes for Nuclear Wastes

Isotope	Half-life	Decay mode	Isotope	Half-life	Decay mode
Fission products			Fission products		
^{85}Kr	10.8 y	β	^{137}Cs	30 y	β
^{89}Sr	51 d	β	^{141}Ce	33 d	β
^{90}Sr	28 y	β	^{147}Pm	2.6 y	β
^{95}Zr	64 d	β	Transuranics		
^{95}Nb	35 d	β			
^{99}Tc	2.1×10^5 y	β	^{237}Np	2.1×10^6 y	α
^{106}Ru	1 y	β	^{239}Pu	2.4×10^4 y	α
^{131}I	8 d	β	^{240}Pu	6.6×10^3 y	α
^{133}Xe	5.2 d	β	^{241}Am	433 y	α

isotopes of I, Xe, Cs, Ba, and the rare earth elements (REE) are important (Table 9–15). Besides the fission products, other radioactive isotopes of higher atomic number than uranium are produced by neutron capture and the subsequent decay of the resulting radioactive nucleus. As an example, consider the following reaction:

$$^{238}U + {}^{1}_{0}n \rightarrow {}^{239}U \rightarrow {}^{239}Np + \beta^{-} \rightarrow {}^{239}Pu + \beta^{-}$$

The ^{239}Pu isotope is a fissionable isotope (by fast neutrons, i.e., neutrons with an energy greater than 1 MeV). This reaction is the basis of the breeder reactor cycle, in which ^{238}U is converted to fissionable ^{239}Pu, thus utilizing the other 99.3% of the uranium that is not useable in conventional uranium reactors. There are a number of possible pathways, involving neutron capture and radioactive decay, that lead to the formation of transuranic (atomic number greater than 92) radioisotopes (Table 9–15).

The level of radioactivity is a function of the power level and length of time the fuel was irradiated in the reactor. Once the radiation time (operation time) becomes long relative to the half-life of a particular radioactive isotope, secular equilibrium is achieved and there is no increase in radioactivity for that isotope. For example, after six months of operation, the ^{95}Nb would have reached its maximum activity but the activity of ^{137}Cs would still be increasing. After a fuel element is removed from a reactor, the radioactive isotopes decay at a rate determined by their half-life (or radioactive decay constant). Thus, with time, the short-lived isotopes become less important in determining the total radioactivity of the spent fuel and the total radioactivity declines (Figure 9–23a). At the same time, the amount of heat released by the decay processes also declines (Figure 9–23b). For this reason, spent fuel elements are stored on-site for a period of time (five years is often used as an optimum time) before disposal.

In terms of disposal of waste from the nuclear reactor fuel cycle, a useful distinction can be made between high-level waste (HLW) and spent unreprocessed fuel (SURF). If fuel elements are reprocessed, the uranium is recycled and the radioactive waste (HLW) consists of the fission products and transuranic elements separated during the reprocessing. If the fuel elements are not reprocessed, the waste consists of the total inventory of radioactive isotopes, i.e., the fission products, uranium, and transuranic elements (SURF). The disposal of SURF is a greater problem than the disposal of HLW because the total amount of material and the total radioactivity are significantly greater (Figure 9–23a).

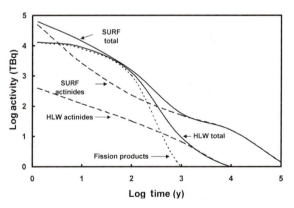

Figure 9–23a

Radioactivity as a function of time after removal of a fuel element from a power reactor. Curves are drawn for both SURF and HLW (after reprocessing) and the two types of radioactive components, fission products and actinides (Th, Pa, U, Np, Pu, Am, and Cm). The data are for a pressurized water reactor (PWR) with a fuel burn-up of 33 GW-day tonne^{-1} normalized to 1 metric ton of heavy metal in the original fuel element. TBq = 1×10^{12} Bq. Plotted from the data of Roxburgh (1987).

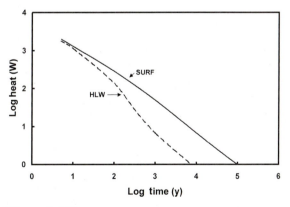

Figure 9–23b

Heat production as a function of time, after a five-year cooling period, for SURF and HLW. The data are for a pressurized water reactor (PWR) with a fuel burn-up of 33 GW-day tonne^{-1} normalized to 1 metric ton of heavy metal in the original fuel element. Plotted from the data of Roxburgh (1987).

The total amount of heat generated, as a function of time, is also greater for SURF (Figure 9–23b). Hence, repositories for SURF disposal have to be designed to deal with higher activities, higher heat flows, and longer decay times. The total volume of material is also significantly greater for SURF compared to HLW.

The waste disposal cycle represents the controlled release of radioactive materials to the environment; nuclear accidents represent uncontrolled releases. The difference is of fundamental importance. During the Chernobyl nuclear accident large amounts of radioactive material were released, with the majority of this radioactivity due to relatively short half-life radioactive isotopes. Among these short half-life radioactive isotopes is ^{131}I. Although this isotope has a short half-life, its release represents a serious health risk because iodine is concentrated in the thyroid gland. Note that all the fission products listed in Table 9–15 are beta emitters. Beta particles are of low energy and do not penetrate to great depths in most materials. For example, beta rays emitted external to the human body are totally absorbed by the skin. The major hazard with beta emitters is ingestion. Ingested ^{131}I will be concentrated in the thyroid gland and the radioactive dose is delivered directly to the thyroid. It has been proposed that individuals who may be exposed to ^{131}I take large amounts of KI. The iodine in this compound would saturate the thyroid and prevent the uptake of radioactive iodine. The long-term radioactive exposures from the Chernobyl accident will be due to ^{90}Sr and ^{137}Cs.

Mobility and Sequestering of Radioactive Isotopes in the Surface Environment   From the preceding discussion it becomes apparent that the elements of greatest interest in terms of radioactivity are Sr, Cs, Ba, Tc, I, the REE, and the actinides (U, Np, and Pu). After a long period of time, Pb and Ra, which are produced in various decay chains, also become important. There is no fundamental difference between the chemical behavior of an element and its radioactive isotopes. Thus, our goal is to understand the behavior of these particular elements in the environment using the same methodologies previously developed in this chapter (and other chapters). From the standpoint of sequestration of SURF or HLW in a waste repository, additional factors are the high levels of radioactivity, which can degrade the structural integrity of the material encasing the radioactive isotopes, and the significant heat production, which leads to a significant temperature increase in the immediate environment. Questions regarding the encapsulation of the radioactive isotopes and the design of the repository are beyond the scope of this book. Here we will briefly consider what happens to these elements if they are released into the natural environment.

As has been the case for all of the other systems we have investigated, the two major factors controlling the mobility of these elements in the surface environment are the formation of insoluble phases and ion adsorption. Many of the elements of interest can exist in several oxidation states in natural waters. Thus, the oxidation–reduction potential of the environment is important in determining their solubility. Many of these species (particularly in the case of the actinides) exist in solution as hydoxyl complexes. At pH values above 6 to 7, various carbonate complexes become important in determining the solubility of the actinides. Assuming that the wastes in a repository exist as oxides, Table 9–16 gives estimated solubilities for some of the more important radioactive isotopes. Note that the solubility of Sr, Cs, Tc, and I is high under most conditions. The mobility of these elements in the natural environment will largely be determined by adsorption properties. Ra and Pb are also generally soluble. The actinides have significantly lower solubilities under most conditions, but note the exception for uranium. Under oxidizing conditions, uranium exists in the +6 state and forms highly soluble uranocarbonate complexes. When solutions transporting uranium encounter a reducing environment, the uranium is reduced to U^{4+} and under these conditions precipitates as an insoluble uranium oxide, uraninite [UO_2], or silicate, coffinite [$U(SiO_4)_{1-x}(OH)_4$]. A detailed discussion of actinide chemistry in the natural environment can be found in Langmuir (1997).

As we know from our earlier discussion of adsorption, this is a complex process affected by the pH, the nature of the adsorbing material, the redox conditions, and the overall ionic strength of the solution. Given measured adsorption coefficients, and the porosity

Table 9–16 Estimates of Solubilities (mg L^{-1}) of Important Radioisotopes at 25°C and 1 atm*

Element	Reducing conditions (Eh = −0.2 V)		Oxidizing conditions (Eh = +0.2 V)	
	pH 9	pH 6	pH 9	pH 6
Sr	0.6	high	0.6	high
Cs	high	high	high	high
Tc	10^{-10}	high	high	high
I	high	high	high	high
U	10^{-3}	10^{-6}	high	high
Np	10^{-4}	10^{-4}	10^{-2}	10^{-1}
Pu	10^{-5}	10^{-4}	10^{-5}	10^{-3}
Am	10^{-8}	10^{-5}	10^{-8}	10^{-5}
Ra	10^{-3}	10^{-1}	10^{-3}	10^{-1}
Pb	10^{-1}	1	10^{-1}	1

*From Krauskopf (1986).

and density of the geomedia through which a solution passes, it is possible to calculate a retardation factor (equation 9–38) for the transport of a particular species. Table 9–17 gives a range of retardation factors for the elements of interest in various geomedia that might host a radioactive waste disposal repository. The range of values indicates least favorable to most favorable conditions for adsorption of a particular species. The low end of the range is believed to be a very conservative estimate; i.e., in a real-world situation we would expect higher retardation factors.

The Oklo Fossil Fission Reactors

As is often the case, nature has already done an experiment on the long-term sequestration of nuclear wastes. In 1970 uranium mining was initiated at Oklo, Gabon. The ores were processed and the uranium was enriched for use as reactor fuel. Yields of ^{235}U during the enrichment process were less than those normally obtained. Subsequent isotopic analysis revealed that the ores were depleted in ^{235}U, and some ores contained as little as 0.440% ^{235}U (compared to the expected 0.719%). It was concluded that the ores were depleted in ^{235}U because of prior fission reactions.

The uranium mineralization is confined to a 5 to 7-meter-thick layer at the top of a stream-deposited sandstone and conglomerate sequence. The original uranium minerals were dispersed throughout the sandstone–conglomerate sequence but were subsequently

Table 9–17 Retardation Factors for Radioactive Isotopes in Various Geomedia*

Element	Granite	Basalt	Volcanic ash	Shale (or clay)
Sr	20–4,000	50–3,000	100–100,000	100–100,000
Cs	200–100,000	200–100,000	500–100,000	200–100,000
Tc	1–40	1–100	1–100	1–40
I	1	1	1	1
U	40–500	100–500	40–400	100–2,000
Np	20–500	20–200	20–200	50–1,000
Pu	20–2,000	20–10,000	20–5,000	50–100,000
Am	500–10,000	100–1,000	100–1,000	500–100,000
Ra	50–500	50–500	100–1,000	100–200
Pb	20–50	20–100	20–100	20–100

*From Krauskopf (1986).

dissolved and the uranium redeposited as uraninite in lenses and pods containing up to 50 to 70% UO_2. Radiometric dating gave an age of 2.05×10^9 years for this ore-forming event (Gancarz, 1978). At this time, ^{235}U would have constituted approximately 3.8% of the total uranium (remember that the half-life of ^{235}U is much less than that of ^{238}U, see Table 6–1), similar to the amount of ^{235}U in reactor fuel elements. Interstitial water in the sandstone acted as a moderator. Once the fission reaction started, the temperature of the groundwater increased and groundwater was thermally expelled from the immediate vicinity. This led to a cessation of the fission reactions. When conditions returned to normal, the fission reaction was reinitiated. This process apparently continued for about 500,000 years (Hagemann et al., 1975), and approximately 1×10^4 kg of fission products and 4×10^4 kg of ^{239}Pu were generated by the reactors (Cowan et al., 1975). Today, the reactor zones are surrounded by clay minerals; i.e., the rock is no longer a sandstone, because the quartz was removed by hydrothermal (hot-water) solutions emanating from the reactor zones.

A number of studies have been done on the mobility of the fission products and actinides in the Oklo reactors (Brookins, 1978; Curtis et al., 1989; Loss et al., 1989; Hidaka et al., 1994). These studies have distinguished between primary and secondary retention. Primary retention occurs when radioactive isotopes are retained at their site of generation, i.e., within the UO_2 ore horizons. Secondary retention occurs when radioactive isotopes migrate from the site of generation but are then retained in the surrounding rock. The sum total of the studies to date indicate that the transuranics, REEs, PGEs (platinum group elements), Y, Zr, Nb, Th, and Bi were largely retained at their site of generation. Mo, Ag, I, Kr, and Xe migrated from the site of generation and were only partially contained in the surrounding rocks. Rb, Cs, Sr, Cd, and Pb underwent extensive migration and were largely lost from the reactor zones. Of particular interest is the essentially 100% retention of the long half-life transuranic elements over a 2-billion-year period in a geologic horizon that is less than optimum for radioactive waste disposal. A number of the fission products also show a high degree of retention, suggesting that it is possible to design a repository for nuclear wastes that can sequester these materials for hundreds of thousands, if not millions, of years.

NONMETALS

Carbon

Of the nonmetals, carbon and its many compounds form the most widely distributed and varied group of chemicals in the environment. Garrison et al. (1977) proposed a classification scheme for organic compounds in natural waters that included 24 major classes and more than 100 subclasses. Domenico and Schwartz (1990) condensed and reordered this scheme into 16 classes (Table 9–18). Organic compounds are classified in a hierarchical order starting with class 1. Going in the order 1 to 16, an organic compound is placed into the first class that describes its chemistry.

The student should refer to Chapter 5 for a much more complete discussion of carbon chemistry. Briefly, as with the metal species, we are concerned with the mobility and persistence of carbon compounds in the natural environment. Because carbon compounds are largely covalently bonded, they tend to form uncharged species that are only weakly adsorbed by soil particles and oxyhydroxides. Hence, they tend to be relatively mobile. Carbon compounds are degraded by microbially mediated oxidation reactions. These reactions can occur under aerobic or anaerobic conditions. For example, the following reaction, representing the degradation of toluene, occurs under aerobic conditions:

$$C_6H_5-CH_3 + 9O_2 \rightarrow 7CO_2 + 4H_2O$$

Under anaerobic conditions, nitrate or ferric iron can act as an oxidant. For example, consider the following toluene degradation reactions:

$$C_6H_5-CH_3 + 7.2H^+ + 7.2NO_3^- \rightarrow 7CO_2 + 7.6H_2O + 3.6N_2$$

Table 9–18   Classification of Organic Compounds in Natural Waters*

		Examples	
Class	Name	Naturally occurring	Anthropogenic
1	Miscellaneous nonvolatile compounds	Fulvic acid, humic acid, chlorophyll, xanthophylls, enzymes	Tannic acids, dyes, optical brighteners
2	Halogenated hydrocarbons		Solvents (methylene chloride, chloroform, carbon tetrachloride), pesticides (aldrin, DDT, dieldrin), industrial chemicals (vinyl chloride, methyl chloride, PCBs)
3	Amino acids	Glycine, alanine, aspartic acid	
4	Phosphorous compounds		Pesticides (diazinon, malathion, parathion)
5	Organometallic compounds		Tetraethyllead, diethylmercury, copper phthalocyanine
6	Carboxylic acid	Acetic acid, benzoic acid, butyric acid, formic acid	Acetic acid, benzoic acid, butyric acid, formic acid, phenoxy acetic pesticides (2,4-D, silvex, 2,4,5-T)
7	Phenols	Phenol, cresol, p-hydroxybenzoic acids	Cresol, phenol, pyrocatechol, napthol, pesticides (dinitrocresol, 2,4-dinitrophenol)
8	Amines		Diethylamine, dimethylamine, benzidine, pyridine
9	Ketones		Acetone, 2-butane, methyl propenyl
10	Aldehydes		Formaldehyde
11	Alcohols		Methanol, glycerol, terpinol, ethyleneglycol
12	Esters		Dimethrin, omite, vinyl acetate
13	Ethers	Diethyl ether, diphenyl ether	Tetrahydrofuran, 1,4-dioxane
14	PAHs		Anthracene, benzo[a]pyrene
15	Aromatic hydrocarbons	Benzene, ethylbenzene, toluene	Benzene, ethylbenzene, toluene
16	Alkanes, alkenes, and alkynes	Methane, propane, propene, 2-hexene, 2-hexyne	Methane, propane, propene, 2-hexene, 2-hexyne

*From Domenico and Schwartz (1990).

and

$$C_6H_5\text{–}CH_3 + 36Fe(OH)_3 + 72H^+ \rightarrow 7CO_2 + 36Fe^{2+} + 94H_2O$$

Note that during the latter reaction dissolved Fe^{2+} is added to the solution. In a contaminated aquifer in which benzene, toluene, and xylenes were removed by this process, the degradation of the organic chemicals would be accompanied by an increase in Fe^{2+} in solution and a decrease in Fe^{3+} oxides in the contaminated sediment.

Halogens

As was the case for the metals, solubility and adsorption are the major controls on the distribution of negatively charged species. For the normal range of pH values encountered in natural waters, most surfaces tend to have a negative charge (Table 9–11). A notable exception is the iron oxyhydroxides, which have positively charged surfaces at intermediate pH values. Because the role of adsorption is more limited in the case of negatively charged species, solubility becomes more important in controlling the abundance of anions in solution.

Fluorine   Fluoride (F^-) is often added, in small quantities, to drinking water to improve dental health. However, at higher concentrations (> 3 mg L^{-1}) fluoride is considered a health hazard. At these higher concentrations fluoride can cause dental fluorosis (tooth mottling) and skeletal fluorosis (bone deformation and painful brittle joints). Natural sources of fluorine are volcanic emissions and the weathering of fluoride-containing minerals. Anthropogenic sources are mining and industrial emissions.

The concentration of fluoride in natural waters is largely controlled by dissociation reactions involving fluoride-containing minerals, of which the most important is fluorite. The dissociation reaction is

$$CaF_{2 \text{ fluorite}} \rightarrow Ca^{2+} + 2F^-$$

From the data in Appendix II, source 3, the equilibrium constant for this reaction, at 25°C, is

$$K_{\text{fluorite}} = [Ca^{2+}][F^-]^2 = 10^{-10.51}$$

Writing this equation in logarithmic form gives

$$\log K_{\text{fluorite}} = \log[Ca^{2+}] + 2[\log F^-] = -10.51$$

On a logarithmic plot, the relationship between the concentrations of the two ionic species in solution and fluorite solubility plots as a straight line (Figure 9–24). This is a very useful plot because it allows us to determine if a water is saturated with respect to fluorite, the major mineral controlling F^- concentrations. Consider a water sample of composition A. The water sample is undersaturated with respect to fluorite, and fluoride can be added to the water until it becomes saturated (A → B). At this point the concentration of fluoride will remain constant unless we reduce the Ca^{2+} concentration, in which case the F^- concentration can increase (B → C). If we increase the Ca^{2+} concentra-

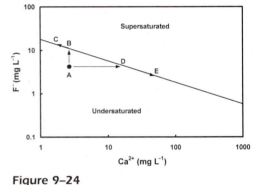

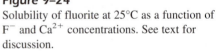

Figure 9–24
Solubility of fluorite at 25°C as a function of F^- and Ca^{2+} concentrations. See text for discussion.

tion (A → D), the solution will eventually become saturated in fluorite. Continued addition of Ca^{2+} will lead to a decrease in the F^- concentration in solution (D → E). It has been suggested that Ca^{2+} could be added (in the form of gypsum) to high F^- waters, leading to the precipitation of fluorite and a reduction in the F^- concentration. Most natural waters are undersaturated or just saturated with respect to fluorite, suggesting that fluorite solubility controls the F^- concentration of natural waters (Case Study 9–10).

Cl^- and Br^- Chlorine and bromine have both natural and anthropogenic sources. Natural sources include seawater, subsurface brines, and evaporite deposits. Anthropogenic sources for bromine include pesticides, medicines, industrial solvents, gasoline additives, and water purification. For chlorine, the list is even more extensive because it is one of the most widely used elements in modern industrial chemistry. Degradation of these anthropogenic compounds releases Cl^- and Br^- to solution, apparently largely in the form of simple ions. Hydrochemical characteristics of both compounds are (1) low concentration in most rock-forming minerals, (2) high solubility of most nonsilicate compounds, (3) virtually no adsorption by particles, (4) lack of volatile compounds, and (5) generally low bioconcentrations in aqueous systems. Because of these characteristics, both elements behave as conservative species and they have been widely used as tracers in hydrological systems. Bromine and chlorine, however, do differ in several important ways: (1) Bromide compounds are even more soluble than chloride compounds. During the formation of evaporite deposits Br is concentrated with respect to Cl in the late-stage brines, leading to a decrease in the Cl/Br ratio. (2) Br does show some adsorption characteristics, particularly at low pH, on kaolinite and iron oxide surfaces.

CASE STUDY 9–10
F⁻ Concentrations in Streams and Groundwaters from Nalgonda District, India

Ground and surface waters in the Nalgonda District of India contain elevated levels (0.4 to 20 mg L^{-1} for groundwater and 0.4 to 6.6 mg L^{-1} for surface waters) of fluoride, which has led to endemic fluorosis (Ramamohana Rao et al., 1993). There are no fluorine-based industries in the area (nor volcanoes), so the fluorine must originate from the minerals in the rocks and soils. The area is underlain by several varieties of granites that contain the fluorine-bearing minerals fluorite (0 to 3.3%), biotite (0.1 to 1.7%), and hornblende (0.1 to 1.1%). Whole-rock F content ranges from 325 ppm to 3200 ppm, with a mean of 1440 ppm. Acid-soluble fluoride content (fluoride contained in fluorite and fluoroapatite, which can be removed by acid leaching) ranges from 40 to 1150 ppm, with a mean of 410 ppm. The overlying soils are also relatively high in fluoride, have relatively low Ca and Mg concentrations and high Na concentrations. There is an inverse relationship between hardness (Ca + Mg) and fluoride concentration and a direct relationship

between alkalinity (due to the presence of Na in the soils) and fluoride concentration. All the soils have basic pH values. In a series of experiments the authors found that the solubility of fluorite was significantly increased in the presence of sodium carbonate and sodium bicarbonate, due to the following reactions:

$$CaF_{2\ fluorite} + 2Na^+ + CO_3^{2-} \rightarrow CaCO_{3\ calcite} + 2Na^+ + 2F^-$$

and

$$CaF_{2\ fluorite} + 2Na^+ + 2HCO_3^- \rightarrow$$
$$CaCO_{3\ calcite} + 2Na^+ + 2F^- + H_2O + CO_2$$

The authors concluded that these reactions led to the precipitation of calcite and a reduction in the amount of Ca^{2+} available for fluorite precipitation, thus increasing the amount of F^- in solution.

Source: Ramamohana Rao et al. (1993).

(3) There is some evidence that marine plants and surface plants do concentrate Br (Davis et al., 1998).

Because there are natural processes that will change the Cl/Br ratio, plus anthropogenic inputs that may lead to relative increases in Br or Cl concentrations and changes in the Cl/Br ratio, Davis et al. (1998) suggested that Cl/Br ratios could be used to fingerprint the sources of various waters in an aquifer. Once Cl and Br have entered a groundwater system, they tend to behave conservatively. Thus, variations in the Br/Cl ratio reflect variable degrees of end-member mixing. In Chapter 6 we developed mixing equations for isotopic systems. Analogous mixing equations can be written for conservative species. For a simple binary mixture involving two end members, the mathematics are straightforward if we use the abundances of the conservative species rather than ratios. For a mixture involving one conservative species and two end members, we can write

$$X_M = X_A f_A + X_B(1 - f_A) \tag{9–42}$$

where X_M is the concentration of species X in the mixture, X_A is the concentration of species X in end member A, X_B is the concentration of species X in end member B, and f_A (the mixing parameter) is the fraction of end member A. For a second conservative species in a mixture, we can write

$$Y_M = Y_A f_A + Y_B(1 - f_A) \tag{9–43}$$

where Y is the second conservative species. For a binary mixture involving conservative species, f_A must be the same for all species. We can solve equations 9–42 and 9–43 for f_A and equate the results:

$$f_A = \frac{X_M - X_B}{X_A - X_B} = \frac{Y_M - Y_B}{Y_A - Y_B} \tag{9–44}$$

Cross multiplying equation 9–44 and solving for Y_M gives

$$Y_M = X_M \frac{Y_A - Y_B}{X_A - X_B} + \frac{X_A Y_B - X_B Y_A}{X_A - X_B} \tag{9–45}$$

which is the equation for a straight line (Figure 9–25, p. 368). For any simple binary mixture of two conservative species, the data will fall on a straight line. The purpose of this

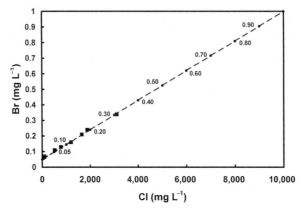

Figure 9–25
Binary mixture of groundwater (0.05 mg L^{-1} Br and 5 mg L^{-1} Cl) and brine (1 mg L^{-1} Br and 10,000 mg L^{-1} Cl). Labeled points are fraction of brine in the mixture. Groundwater samples from an aquifer (filled squares) fall along this line, suggesting they represent mixtures of uncontaminated groundwater and brine. The maximum amount of brine in the groundwater is approximately 31%.

exercise is to determine if a set of data obeys a simple binary mixing model. If the data do obey a simple binary mixing model, and if we know the composition of the end members, either equation 9–42 or 9–43 can be solved for the fraction of each end member in the individual samples. For example, groundwater samples were collected from a number of wells in an aquifer (filled squares in Figure 9–25). It is believed that subsurface brines are mixing with the groundwater. The well data plot on a straight line. The end-member compositions, based on measurements of groundwater entering the aquifer and the brine, are 0.05 mg L^{-1} of Br and 5 mg L^{-1} of Cl for the uncontaminated groundwater and 1 mg L^{-1} of Br and 10,000 mg L^{-1} of Cl for the brine. The labeled points on the diagram represent the fraction of brine. Inspection of Figure 9–25 reveals that the groundwater samples represent mixtures containing from 1% to about 31% brine.

Suppose our water samples consist of a mixture of three end members. As was discussed in Chapter 6, the data will fall within a triangular field. We can determine the fraction of the three end members in any particular sample using a graphical solution (Figure 6–14). If we know the end-member compositions, we can determine the fraction of each end member by solving a set of linear equations. For a three-end-member (A, B, and C) mixture, the concentration of X in the mixture is

$$X_M = X_A f_A + X_B f_B + X_C f_C \qquad (9\text{--}46)$$

and the concentration of Y in the mixture is

$$Y_M = Y_A f_A + Y_B f_B + Y_C f_C \qquad (9\text{--}47)$$

The symbols are as defined earlier. For our third equation,

$$f_A + f_B + f_C = 1 \qquad (9\text{--}48)$$

because the total end-member fractions must add up to 1. Simultaneous solution of these three equations yields the end-member fractions.

Nitrogen

Nitrogen shows a complex and varied behavior in the natural environment (Figure 9–26). In Chapter 4 we constructed the Eh–pH diagram (Figure 4–9) for the various nitrogen species found in water. Depending on redox conditions, nitrogen can occur in natural wa-

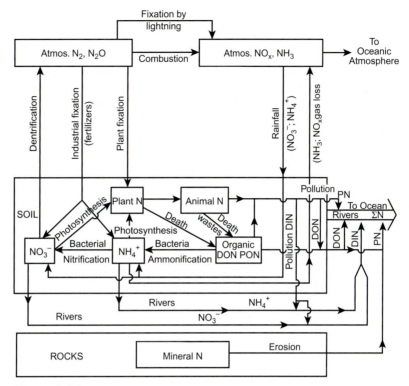

Figure 9–26

The terrestrial nitrogen cycle. DON = dissolved organic nitrogen; DIN = dissolved inorganic nitrogen; PN = particulate nitrogen; PON = particulate organic nitrogen: ΣN = total nitrogen. From Berner and Berner (1996).

ters as NO_3^-, NO_2^-, N_2, NH_4^+, and NH_3. In Chapter 4 we also discussed the nitrification and denitrification reactions mediated by microorganisms. Nitrogen species play an important role in various biological processes and nitrogen is often a limiting nutrient. In Chapter 6 we discussed at some length the use of stable isotopes to identify the source of nitrogen in surface and ground waters. In Chapter 8 we considered the role of nitrogen in the atmosphere and in acid deposition.

Here we are concerned with the mobility of the various nitrogen species in surface and ground waters. Under reducing conditions, the dominant nitrogen species is NH_4^+. Only at relatively high pH values, under reducing conditions, is NH_3 the dominant species. Under oxidizing conditions, nitrogen occurs as NO_3^- or NO_2^-. Except at very acid pH values, the surface charge of most particles (with the exception of the oxyhydroxides) is negative. Thus, NH_4^+ can be geochemically immobilized by adsorption onto clays and organic material. In contrast, the negatively charged nitrogen species that exist under oxidizing conditions are not adsorbed. Additionally, nitrates and nitrites usually form only weak complexes with most metals. The result is that the nitrates and nitrites behave as geochemically mobile species. The most important control on the mobility of the nitrates and nitrites is uptake by plants. The important point here is that the oxidation–reduction potential of natural waters plays a key role in the mobility of nitrogen species. Under oxidizing conditions, nitrogen species are mobile; but under reducing conditions, nitrogen species are relatively immobile.

Phosphorus

Phosphorus has received a great deal of attention because it is a limiting nutrient and is often implicated in the process of eutrophication. The natural source of phosphorus is the weathering of phosphorus-containing minerals in rocks, the most common of which is apatite [$Ca_5(PO_4)_3(OH,F,Cl)$]. At sufficiently high concentrations, apatite, and similar

CASE STUDY 9–11
Phosphate Pollution in Groundwater

Srinivasa Rao and Rajendra Prasad (1997) investigated the distribution of phosphorus in groundwater (113 wells) from the lower Vamsadhara, India, river basin. The area is underlain by garnet-bearing granite gneisses and granite gneisses. Apatite occurs as an accessory mineral in the gneisses. This is a typically rural area in which, starting in 1990, various phosphate-containing fertilizers came into widespread use. Many of the wells contain elevated phosphate levels. What is the source of the phosphate in these wells?

The authors found that the phosphate levels in the well water could be subdivided into three groups: < 2 mg L^{-1}, 3–4 mg L^{-1}, and > 4 mg L^{-1}. The authors compared the geographical distribution of the three phosphate concentration groups with the surface and subsurface geology. The regions of > 2 mg L^{-1} P coincided either with regions of thick sandy formations or red sandy soils. They ascribed the low phosphorus abundance for waters associated with these units to the very low adsorption characteristics of sandy material. The regions of 3–4 mg L^{-1} P coincided with clayey soils. The clayey soils showed significant adsorption of phosphate. The groundwater table fluctuated seasonally between the clayey soils and the underlying sandy material. The soils were believed to be enriched in phosphorus due to the application of fertilizer in the region, and this phosphorus was subsequently transferred to the groundwater. The regions of > 4 mg L^{-1} P coincided with thick, weathered gneiss horizons. The authors concluded that the gneisses were the source of the high phosphorus content in these wells. Because fertilizer application was uniformly distributed throughout the region, the authors concluded that both the nature of the surface material and the availability of phosphorus from the weathering of the gneisses played a key role in determining the phosphorus levels of the groundwater. Water-chemistry modeling revealed that the high-P groundwaters were saturated with respect to apatite, suggesting that the solubility of this mineral was controlling the phosphorus content of these waters.

Source: Srinivasa Rao and Rajendra Prasad (1997).

---

phosphate minerals, forms ore deposits, which are mined for the phosphorus used in fertilizer. Anthropogenic releases of phosphorus are associated with fertilizers and sewage effluent (the residues of phosphate-containing detergents and human wastes).

The abundance of phosphorus in natural waters is controlled both by the solubility of phosphorus-containing minerals and adsorption (see Case Study 9–11). At low pH values, the least soluble phosphorus phases are those that contain iron [e.g., vivianite, $Fe_3(PO_4)_2 \cdot 8H_2O$, and strengite, $FePO_4 \cdot 2H_2O$] and aluminum (e.g., variscite, $AlPO_4 \cdot 2H_2O$). At high pH values, the least soluble phosphorus phases are the various varieties of the mineral apatite. Phosphorus solubility is greatest at intermediate pH values, i.e., those typically found in natural waters. You may recall from Chapter 3 that phosphorus forms a weak acid, H_3PO_4. Thus, the distribution of phosphorus species in solution is pH dependent (problem 9–107). In the pH region of greatest phosphorus solubility, $H_2PO_4^-$ is the most abundant species. Phosphate is also strongly adsorbed by ferric (Fe^{3+}) oxides and oxyhydroxides (Table 9–11).

Eutrophication occurs when a water body, for which phosphorus is the limiting nutrient, receives excess amounts of phosphorus. The result is a rapid increase in plant biomass. The subsequent decomposition of this organic material, in the hypolimnion, depletes the deeper waters in oxygen, leading to anoxic conditions. At the same time, the depletion of oxidizing agents used to break down the organic matter leads to strongly reducing conditions in the deep waters. Lakes can be classified in terms of their trophic status (Table 9–19) from those that are poorly nourished (ultraoligotrophic) to those that are extremely overnourished (hypertrophic).

Once a lake becomes eutrophic, it turns out to be very difficult to reverse the process. The immediate action would be to limit the input of phosphorus into the lake. However, a great deal of the phosphorus previously deposited in the lake remains in available form. Although some of the phosphorus is removed in the organic matter, anaerobic bacteria in the bottom sediments that break down the organic matter concentrate the phosphorus. This phosphorus can be released when the lake becomes aerobic. In addition, phosphorus that has been adsorbed on ferric oxide and oxyhydroxides in the surface waters is released in the deep waters. This occurs because reducing conditions in the hypolimnion cause the

Table 9–19 Trophic Status of Lakes*

	Total P (μg L^{-1})	Chlorophyll *a* (μg L^{-1})
Ultraoligotrophic	<4	<1.0
Oligotrophic	4–10	1.0–2.5
Mesotrophic	10–35	2.5–8
Eutrophic	35–100	8–25
Hypertrophic	>100	>25

*From Rast and Holland (1988).

reduction of Fe^{3+} to Fe^{2+}, leading to the breakdown of these oxides and oxyhydroxides and the formation of Fe^{2+}, FeS_2, and/or $FeCO_3$.

Variations in nitrogen and phosphorus concentrations in rivers are illustrated in Figure 9–27. In rivers unaffected by anthropogenic activity, both total phosphorus and nitrogen are low, less than 1 μmol L^{-1} PO_4^{3-} (0.025 mg L^{-1} P) and less than 30 μmol L^{-1} NO_3^- (0.42 mg L^{-1} N). As we can see from Figure 9–27, both the total nitrogen (as nitrate) and phosphorus (as phosphate) and the N/P ratio vary over a relatively large range. The N/P ratio shown in this diagram is the atomic (or mol/mol) ratio, not the weight ratio, for these two elements. Of particular significance is N/P = 16. This is the so-called Redfield ratio, which represents the proportion in which marine phytoplankton utilize these two nutrient elements during photosynthesis. In general, rivers that have significant indus-

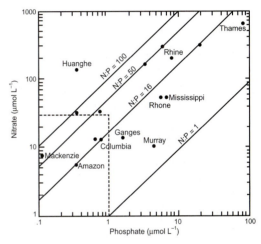

Figure 9–27

Total nitrogen (as nitrate) and phosphorus (as phosphate) for selected rivers. Area enclosed by dashed lines represents unpolluted rivers. The diagonal lines represent constant N/P ratios (atomic). N/P = 16 is the Redfield ratio. From Berner and Berner (1996).

trial/urban inputs have low N/P ratios, and rivers that have significant agricultural inputs tend to have high N/P ratios. These relationships reflect the high amounts of phosphorus in sewage effluent and certain industrial wastes relative to the high amounts of nitrogen in fertilizers and manure. Thus, rivers entering the marginal marine environment that have significant industrial/urban inputs will be relatively depleted in nitrogen, and the limiting nutrient in this case will be nitrogen.

Sulfur

In natural waters sulfur exists as four aqueous species: HSO_4^-, SO_4^{2-}, $H_2S_{(aq)}$, and HS^-. The latter two species only exist under strongly reducing conditions (problem 4–45). $H_2S_{(aq)}$ is the dominant species under acidic conditions and HS^- is the dominant species under alkaline conditions. With reference to Figure 4–8, note that under strongly reducing conditions iron sulfide minerals can precipitate. The same is true for a number of other metal sulfide minerals. Thus, the abundance in solution of the reduced sulfur species is largely controlled by solubility reactions. Over the rest of the Eh–pH range normally encountered in natural waters, SO_4^{2-} is the dominant sulfur species.

Sulfate has a number of sources, both natural and anthropogenic. Natural sources include the weathering of evaporite deposits that contain sulfate minerals, e.g., gypsum

($CaSO_4 \cdot 2H_2O$) and anhydrite ($CaSO_4$), and the weathering of rocks that contain sulfide minerals, e.g., pyrite (FeS_2), sphalerite (ZnS), and galena (PbS). Anthropogenic sources are acid deposition associated with the burning of fossil fuel and acid mine drainage. These anthropogenic sources release hydrogen ions to the environment, thus raising the overall acidity. In general, sulfate mobility is determined by anion-exchange processes and biological uptake (sulfate is an essential nutrient). As was the case for phosphate, sulfate anions are adsorbed by aluminum and iron oxides at low pH. Phosphate is even more strongly adsorbed than sulfate, but it is usually present at such low levels that it does not compete with sulfate for adsorption sites. Also, as is the case for cations, divalent anions are preferentially adsorbed with respect to monovalent anions. Hence, sulfate is preferentially adsorbed with respect to nitrate and chloride (and the other halides).

QUESTIONS AND PROBLEMS

1. What is the *hydrologic cycle*?

2. Define *weathering*.

3. Name and briefly describe the four major types of chemical weathering.

4. How do *congruent dissolution* and *incongruent dissolution* differ?

5. In terms of weathering, what is the difference between a *primary* and a *secondary* mineral? Give several examples of both types of minerals.

6. Table 9–3 lists a number of minerals in order of their "weatherability." From a thermodynamic perspective, explain why minerals weather in this order. Note any discrepancies that exist between thermodynamic predictions and the observed weathering sequence.

7. Why is there a significant increase in the solubility of quartz and amorphous silica at pH greater than 10?

8. What are *stability diagrams*?

9. Explain why albite weathers to gibbsite in tropical climates and to kaolinite or smectite in temperate climates.

10. With reference to Case Study 9–1, explain how the authors were able to determine the climate from the clay minerals present in the Speeton Clay Formation.

11. With reference to Case Study 9–2, how was a Piper diagram used to distinguish between conservative and nonconservative species?

12. What is the Gibbs model for river water chemistry? In terms of the Gibbs model, how are the range of water compositions shown in Figures 9–8a and 9–8b generated?

13. How does the Stallard and Edmond model of river water chemistry differ from the Gibbs model?

14. With reference to the Stallard and Edmond model of river water chemistry, how can the chemistry of river water be used to infer the type of rock that is being weathered? Give specific examples.

15. Briefly describe the types of chemical changes that can take place as water enters and flows through a groundwater system.

16. What reactions are responsible for the observed increases in Mg^{2+} and SO_4^{2-} in the waters of the Floridian aquifer (Case Study 9–3)?

17. Explain the *common ion effect*.

18. Describe the Garrels model for groundwater chemistry. List and discuss the major assumptions used in constructing the model. Are these assumptions reasonable?

19. Define *limnology*.

20. What happens to a lake during *eutrophication*?

21. Define *epilimnion*, *thermocline*, and *hypolimnion*.

22. List and briefly discuss the factors that control the mixing and overturn of lake waters.

23. Distinguish between *conservative behavior* and *nonconservative behavior*.

24. What is *relative residence time*?

25. Briefly discuss the three processes that determine the pH of surface and ground waters.

26. Why does the dissolved Al content increase in an acid lake?

27. Distinguish between fixed and variable charge on a particle surface.

28. Distinguish between *inner-sphere* and *outer-sphere* complexes.

29. What role do *counter ions* play in balancing the charge at the surface of a particle?

30. Distinguish between the *PZC* and *PZNPC*.

31. Explain why particles have positively charged surfaces at pH values below the PZC and negatively charged surfaces at pH values above the PZC.

32. Discuss the relationship between ionic potential and the adsorption of cations.

33. What is a *colloid*?

34. What are *oxyhydroxides*?

35. What factors affect the size and abundance of colloidal particles in natural waters?

36. What factors favor the precipitation of colloidal particles?

37. How do *homovalent* ion-exchange processes differ from *heterovalent* ion-exchange processes?

38. Define *ligand*.

39. Define *chelate* and *chelation*.

40. Define *heavy metals* and *metalloids*.

41. Briefly discuss the natural (Figure 9–19) and anthropogenic (Figure 9–20) metal cycles.

42. Explain why, during weathering, the MS_2 sulfide minerals are much more important contributors of H^+ ions than the MS sulfide minerals.

43. List and briefly describe the sources for acid mine drainage.

44. With reference to Case Study 9–7, explain why the *hardpan layer* trapped certain metals.

45. With reference to Case Study 9–8, explain the seasonal variations in metal concentrations in the stream.

46. What are the two major factors that control the mobility of the transition metals, Zn, Cd, and Pb in the natural environment?

47. With reference to Case Study 9–9, what do we mean by *sequential extraction*? What type of information can one expect to get from this analytical technique?

48. Why is arsenic relatively immobile under oxidizing conditions? Compare the behavior of arsenic under oxidizing conditions to that of selenium.

49. Briefly discuss the mercury cycle (Figure 9–21).

50. Briefly discuss the nuclear fuel cycle (Figure 9–22). At what points in the cycle can radioactive material be released to the environment?

51. Distinguish between *low-level radioactive wastes* and *high-level radioactive wastes*.

52. In a nuclear reactor, what is the role of water?

53. Distinguish between *fission products* and *transuranic elements*.

54. Distinguish between *HLW* and *SURF*.

55. Why are the redox conditions important when considering the transport of uranium in solution?

56. What happened at Oklo, Gabon, and why is this important?

57. In a contaminated aquifer, after oxygen has been depleted, Fe^{3+} becomes the oxidizer. What evidence might we find in the water and aquifer sediments to indicate that this process has occurred?

58. With reference to Case Study 9–10, why was there a negative correlation between hardness and the concentration of fluoride in ground and surface waters?

59. List the hydrochemical characteristics of Cl and Br. How do these elements behave in a groundwater system?

60. Why is NO_3^- much more mobile in the natural environment than NH_4^+?

61. With reference to Case Study 9–11, why did the authors conclude that the gneisses were the source of the high phosphate levels in the water wells?

62. Why is it difficult to reverse the process of eutrophication?

63. What are the reasons for the variations in N/P ratios shown in Figure 9–27?

64. What processes control the mobility of sulfate in the natural environment?

65. For $T = 10°C$, calculate the solubility of quartz and amorphous silica at pH = 5.7 and pH = 12.2. Your final answers should be in ppm. Remember that not only must you calculate the K_{sp} for each phase at this temperature but you must also correct the dissociation constants (see Chapters 2 and 3) to this temperature. Use the thermodynamic values from Appendix II, source 2. Comment on the changes in solubility with pH.

66. For a hyperalkaline lake in East Africa, $T = 32°C$ and pH = 12.5. Calculate the solubility, in ppm, of quartz in this lake. Assume that activity equals concentration. Remember that not only must you calculate the K_{sp} for this temperature but you must also correct the dissociation constants (see Chapters 2 and 3) to this temperature. Use the thermodynamic values from Appendix II, source 2.

67. Construct a plot, similar to that shown in Figure 9–2, of the solubility of calcite, at $T = 25°C$, as a function of pH. The solution is in equilibrium with the atmosphere, i.e., $P_{CO_2} = 10^{-3.5}$ atm. This is most easily done using a spreadsheet. The necessary information is found in Chapters 2 and 3. Assume that activity equals concentration. With reference to Figure 9–2, comment on the difference between the solubility of calcite as a function of pH and the solubility of quartz as a function of pH.

68. Calculate the total concentration of Al, as a function of pH, for a solution in equilibrium with aluminum hydroxide at 25°C. Assume that activity equals concentration. With reference to Figure 9–3, compare the concentration of Al in a solution in equilibrium with aluminum hydroxide relative to a solution in equilibrium with gibbsite. Comment on the solubility of Al in the weathering environment. Is the assumption that Al is immobile during weathering justified? If not, are there conditions under which the assumption would be justified?

69. In the weathering environment kaolinite is forming as the result of the breakdown of feldspars. Assume that $T = 25°C$ and that activity equals concentration.
    a. Water passing through the soil is in equilibrium with atmospheric CO_2 and has a pH of 5.7. Given that $[H_4SiO_{4\ (aq)}] = 10^{-3.5}$ mol L^{-1}, calculate the concentration of Al in the soilwater solution.
    b. Because of the construction of coal-fired power plants, the region is impacted by acid rain. Now the water passing through the soil has a pH of 4.0. Assume that the amount of $[H_4SiO_{4\ (aq)}]$ is the same as in part (a). Calculate the concentration of Al in the soilwater solution. What impact has acid rain had on Al mobility? Give a quantitative answer.

70. Calculate the total concentration of Al, as a function of pH, for a solution in equilibrium with aluminum hydroxide at 10°C. You will need to adjust the various equilibrium constants to this temperature using the approach given in Chapter 2 and the appropriate thermodynamic values from Appendix II, source 2. Assume that activity equals concentration.

71. **a.** Calculate the total concentration of Fe^{3+}, as a function of pH, for a solution in equilibrium with ferric hydroxide at 25°C. Assume that activity equals concentration.
    b. Calculate the total concentration of Fe^{3+}, as a function of pH, for a solution in equilibrium with goethite at 25°C. Assume that activity equals concentration.
    c. Compare the solubility of Fe^{3+} for solutions in equilibrium with ferric hydroxide and goethite. Comment on the solubility of Fe^{3+} in the weathering environment. Is the assumption that Fe is immobile during weathering justified? If not, are there conditions under which the assumption would be justified?

72. A river is saturated with respect to $Fe(OH)_{3 (aq)}$. The pH of the stream is 5.1. Assume that activity equals concentration and that $T = 25°C$.
    a. Calculate the concentration of Fe^{3+} in the river.
    b. The river enters the ocean, which has a pH of 8.2. Calculate the solubility of Fe^{3+} in the ocean. What happens to the dissolved Fe^{3+} in the river when the river water mixes with ocean water?

73. Calculate the total concentration of Fe^{3+}, as a function of pH, for a solution in equilibrium with goethite at 30°C. You will need to adjust the various equilibrium constants to this temperature using the approach given in Chapter 2 and the appropriate thermodynamic values from Appendix II, source 2. Assume that activity equals concentration.

74. **a.** Construct a stability diagram, at $T = 25°C$, that represents the relationships for K-feldspar and its various weathering products. Plot $\log([K^+]/[H^+])$ on the y-axis and $\log[H_4SiO_{4 (aq)}]$ on the x-axis. The various phase boundaries for which you will have to derive thermodynamic equations are kaolinite–gibbsite, K-feldspar–kaolinite, K-feldspar–muscovite, muscovite–kaolinite, and muscovite–gibbsite. Use the thermodynamic data from Table 9–6.
    b. On the diagram, plot the path that will be followed during the weathering of K-feldspar in a closed system. Explain the path.
    c. Compared to the closed system case, discuss what will happen in an open system with low rainfall compared to an open system with high rainfall.

75. **a.** Construct a stability diagram, at $T = 25°C$, that represents the relationships for anorthite and its various weathering products. Plot $\log([Ca^{2+}]/[H^+]^2)$ on the y-axis and $\log[H_4SiO_{4 (aq)}]$ on the x-axis. The various phase boundaries for which you will have to derive thermodynamic equations are kaolinite–gibbsite, anorthite–gibbsite, anorthite–kaolinite, anorthite–Ca-montmorillonite (Ca-beidellite), and Ca-montmorillonite (Ca-beidellite)–kaolinite. Use the thermodynamic data from Table 9–6.
    b. On the diagram, plot the path that will be followed during the weathering of anorthite in a closed system. Explain the path.
    c. Compared to the closed system case, discuss what will happen in an open system with low rainfall compared to an open system with high rainfall.

76. On the same Stiff diagram, plot the cation and anion concentrations for the Colorado and Nile rivers (Table 9–7). Comment on the similarities and differences between the chemistries of these two rivers.

77. Plot the Colorado and Nile river data (Table 9–7) on a Piper diagram (Appendix IV). Classify these two rivers in terms of the hydrochemical facies of Back (1966).

78. A stream draining an abandoned coal mine is a tributary of a major river. When the stream enters the river, the waters mix. The chemical compositions of the tributary water and the river water upstream from the point where the tributary enters the river follow.

	Concentration (mg L^{-1})								
	Ca^{2+}	Mg^{2+}	Na^+	K^+	Cl^-	SO_4^{2-}	HCO_3^-	SiO_2	TDS
Tributary	44	21	4	1	10	300	0	7	387
River	33	7.7	15	3.6	19	69	63	7.9	221

  a. Plot the data for the Tributary and the River on a Piper diagram (Appendix IV).
  b. Using the Piper diagram, predict the cation and anion ratios for a 50:50 mixture of tributary and river water.

79. The following data were obtained for three rivers:

	Concentration (mg L^{-1})								
River	Ca^{2+}	Mg^{2+}	Na^+	K^+	Cl^-	SO_4^{2-}	HCO_3^-	SiO_2	TDS
1	207	55	310	15	401	544	409	69	2015
2	33	10.4	7.0	1.1	8.9	36.1	111	3.0	211
3	0.2	0.1	0.4	0.3	0.3	0.2	0.7	4.1	6.3

  a. Plot the data for the three rivers on the Gibbs diagrams (Figures 9–7a and 9–7b).
  b. Classify these three rivers according to the Gibbs scheme, i.e., evaporation–precipitation, rock, or precipitation dominance.
  c. Classify these three rivers according to the Stallard and Edmond's scheme (Table 9–8).

80. Plot the various groundwater samples listed in Table 9–7 on a Piper diagram (Appendix IV). Determine the hydrochemical facies of each sample (Figure 9–6).

81. Plot the chemical analyses for groundwater from serpentine, granite, gabbro, basalt, rhyolite, and mica schist (Table 9–7) on Figure 9–11. Are these groundwaters in equilibrium with their host rocks? Explain.

82. The following questions refer to Case Study 9–4, the lethal gas releases at Lake Monoun and Lake Nyos.
  a. Use the hydrostatic equation, $p = \rho g h$ (Chapter 8), to calculate the pressure at the bottom of Lake Monoun (96 m) and Lake Nyos (220 m). The density (ρ) of water is 1000 kg m^{-3}. Express the final answers in atmospheres.
  b. Plot the saturation concentration of CO_2 (in mmol L^{-1}) for a depth range of 0 to 200 m. You will first need to calculate the pressure as a function of depth using the hydrostatic equation. Then calculate the number of mmol L^{-1} of CO_2, at saturation, using the ideal gas law (Chapter 1). Do the calculations for a temperature of 22°C. This problem is most easily done using a spreadsheet program.
  c. The following table lists the concentration of dissolved CO_2 at various depths for Lake Nyos (data from Kling et al., 1989). One set of measurements was made on January 23, 1987 (Set 1) and the other set was made on May 18, 1987 (Set 2). Plot these data on the graph you made in part (b). Assume that mmol kg^{-1} is equivalent to mmol L^{-1}. Comment on the CO_2 saturation of Lake Nyos at the times these measurements were made. At what depth would the 198-m sample of May 18, 1987, become saturated in CO_2?

Depth (m)	$CO_{2\,(aq)}$ (mmol kg^{-1})	
	Set 1	Set 2
17	22.5	
35	48.6	
61		62.8
76	70.0	
99		92.26
107	110.0	
131	108.0	
143		129.5
171		131.8
198	159.0	193.2

d. Calculate the solubility constant for siderite ($FeCO_3$). Use the appropriate thermodynamic values from Appendix II, source 3. Do the calculation at 25°C.

e. Given the following data for Lake Nyos bottom waters in May 1989, are the bottom waters saturated with respect to siderite? All values in mg L^{-1}. $HCO_3^- = 964$, $Ca^{2+} = 66$, $Mg^{2+} = 78$, $Na^+ = 27$, $K^+ = 8$, $Fe^{2+} = 95$, and $Mn^{2+} = 2$ (Data from Kling et al., 1989).

In order to do these calculations, assume pH = 5.2. You will first have to calculate the activity of CO_3^{2-}. Report your final result in terms of the saturation index. You will need to take the ionic strength of the solution and activity into consideration when you do this calculation (Chapter 2).

f. When deep waters in Lake Nyos come to the surface, the dissolved Fe^{2+} is oxidized to Fe^{3+}. One possible reaction is

$$2Fe^{2+} + 4HCO_3^- + H_2O + 0.5O_{2\,(g)} \rightarrow 2Fe(OH)_{3\,\text{ferrihydrite}} + 4CO_{2\,(g)}$$

Calculate the equilibrium constant for the reaction at 25°C. Use the appropriate thermodynamic values from Appendix II, source 2. Using the concentration for HCO_3^- reported for the Lake Nyos deep waters in part (e), $P_{O_2} = 0.21$ atm and $P_{CO_2} = 10^{-3.5}$ atm, calculate the solubility of Fe^{2+} in the surface waters. Calculate the amount of ferrihydrite precipitated in mg L^{-1}. During the Lake Nyos disaster several observers reported seeing streaks of reddish-brown material at the lake's surface. Based on your calculations, does the precipitation of ferrihydrite seem a reasonable explanation for this observation?

g. The ^{14}C activity of modern carbon is 13.56 dpm g^{-1}. Lake Monoun water has a carbon activity of 1.48 dpm g^{-1}. Calculate the apparent age of the Lake Monoun water. Estimate the amount of modern carbon in the Lake Monoun water. If necessary, refer to Chapter 6.

h. CO_2 emanating directly from the mantle would be expected to have $\delta^{13}C = -2.5‰$. Biogenic carbon in the Lake Nyos and Lake Monoun region would have $\delta^{13}C = -30‰$. Lake Nyos water has an average $\delta^{13}C = -2.7‰$ and Lake Monoun has an average $\delta^{13}C = -5.8‰$. Comment on the relative importance of biogenic carbon versus mantle carbon for both lakes. In the case of Lake Monoun, calculate the percent biogenic carbon. How does this agree with the amount of modern carbon estimated in part (g) using ^{14}C activity? If necessary, refer to Chapter 6.

i. The following table lists the stable isotope data for Lake Monoun and Lake Nyos (from Kusakabe et al., 1989). Plot the data for both lakes on a δD versus $\delta^{18}O$ diagram. Plot the meteoric water line on this diagram (see Chapter 6). Given that for magmatic waters $\delta D = -40$ to $-80‰$ and $\delta^{18}O = +5.5$ to $+9.5‰$, what can you conclude about the source of the lake waters?

Lake Monoun			Lake Nyos		
Depth (m)	δD	$\delta^{18}O$	Depth (m)	δD	$\delta^{18}O$
0	−16.3	−3.8	0	−9.7	−3.0
15	−20.4	−4.4	7	−7.7	−2.9
25	−20.8	−4.2	40	−7.5	−2.6
50	−22.3	−4.1	80	−7.7	−2.9
75	−21.7	−4.5	130	−8.1	−2.9
95	−22.6	−4.5	200	−11.9	−2.8

83. Water entering a lake contains 2.0 mg Ni L^{-1}. The total inflow is 200 $m^3 \, s^{-1}$. The total outflow from the lake is 170 $m^3 \, s^{-1}$ and the outflow water contains 1.5 mg Ni L^{-1}. The lake is located in an arid area and evaporation exceeds precipitation. Hence, water inflow is greater than water outflow. The lake is in a steady state with respect to both its water volume and nickel content. Calculate the rate at which nickel is stored in the lake sediments.

84. Acid mine drainage with a pH of 4.5 enters a lake whose pH is 7.0. The acid mine drainage has an Al content of 10^{-5} mol L^{-1}. The water flowing out of the lake has an Al content of 10^{-7} mol L^{-1}. Total inflow = total outflow = 175 $m^3 \, s^{-1}$. The lake is in a steady-state condition with respect to Al. Calculate the rate of Al removal in mg s^{-1}.

85. Acid rain lowers the pH of a lake to 4.8. Water (pH = 4.2) flowing into the lake contains 90 μg Al L^{-1}. Water flowing out of the lake contains 11 μg Al L^{-1}. Total inflow = total outflow = 150 $m^3 \, s^{-1}$. Because of the lake's acid condition, Al remains in solution and the lake is not in a steady-state condition. Assume that the Al is uniformly distributed throughout the lake (outflow Al concentration equals lake Al concentration), that the pH of the lake water remains constant, and that concentration equals activity.
    a. Calculate the Al concentration, at 25°C, when the lake water is saturated with respect to gibbsite.
    b. Calculate the rate at which the Al content of the lake is increasing. Will this rate remain constant? Explain.
    c. When the lake achieves a steady state, what is the concentration of Al in the water flowing out of the lake? Assuming no change in the Al concentration of the water entering the lake, under the steady-state conditions calculate the rate of removal of Al from the lake.

86. A lake has achieved steady state with respect to Al, Si, and N. For the incoming water, Al = 30 $\mu g \, L^{-1}$, Si = 2.8 mg L^{-1}, and N = 41 $\mu g \, L^{-1}$. For the lake water, Al = 5 $\mu g \, L^{-1}$, Si = 2.8 mg L^{-1}, and N = 68 $\mu g \, L^{-1}$.
    a. Calculate the relative residence times for each of these elements.
    b. Which are behaving conservatively? Which are behaving nonconservatively?
    c. For the nonconservative elements, what processes might be affecting their concentrations? Explain.

87. Using a spreadsheet program, calculate the variations in the concentration of the various carbonate species as a function of pH given a temperature of 25°C and a total carbonate concentration of 1×10^{-4} mol L^{-1}. You will need to refer to Chapter 3. At what pH does this system lose its buffering capacity?

88. An experimental study was done on the partitioning of Zn^{2+} between groundwater in an aquifer and the aquifer material. The adsorption of Zn by the aquifer material obeyed a Freundlich isotherm, with $K = 0.12$ L g^{-1} and $n = 0.9$.

a. Calculate the amount of zinc adsorbed per gram of aquifer material when the concentration of zinc in the pore waters is 2×10^{-6} mol L^{-1}.

b. A surface spill introduces 4 mg L^{-1} of Zn to the aquifer. Assuming complete and instantaneous mixing, calculate the concentration of Zn in the groundwater and adsorbed per gram of aquifer material after equilibrium has been established. Note that you will first have to convert the concentration of Zn to moles per liter. Add this to the zinc already in solution and then determine the new equilibrium values.

89. Schlüter (1997) experimentally determined the adsorption of mercury by organic-rich soil at pH = 3–3.3 (acid) and pH = 5.5–5.7 (normal) conditions. For both cases, the sorption of mercury could be described by a Freundlich isotherm. For acid conditions, $K = 0.89$ L μg^{-1} and $n = 1.52$. For normal conditions, $K = 0.101$ L μg^{-1} and $n = 2.076$.

a. An organic-rich soil has a total mercury content of 0.23 μg g^{-1}. For the pore waters, pH = 3.1. Assume that the average density of the soil particles is 2.1 g cm^{-3} and the average porosity is 40%. Calculate the fraction of the total mercury bound to the soil particles.

b. An organic-rich soil has a total mercury content of 0.23 μg g^{-1}. For the pore waters, pH = 5.7. Assume that the average density of the soil particles is 2.1 g cm^{-3} and the average porosity is 40%. Calculate the fraction of the total mercury that is in solution in the pore water.

90. The distribution of phosphorus between lake sediment and pore water obeys a Langmuir isotherm. $S_{max} = 4 \times 10^{-7}$ mol g^{-1} and $K = 2.7 \times 10^{6}$ L mol^{-1}.

a. In its pristine state the lake water contains 4 μg L^{-1} phosphorus. Assume that the concentration in the pore waters is the same. Calculate the amount of phosphorus adsorbed per gram of lake bottom sediment.

b. Summer cottages are built around the lake. Effluent from septic-tank drain fields enters the lake. This effluent is rich in phosphorus (the summer residents like bright whites). Assume that all the phosphorus was added instantaneously. The total phosphorus added is equivalent to 40 μg L^{-1}. The same amount of phosphorus is added to the pore waters. Immediately after the addition of the phosphorus, the pore waters will have a phosphorus concentration of 44 μg L^{-1}. Calculate the concentration of phosphorus in the pore waters and adsorbed per gram of lake bottom sediment after equilibrium is achieved between the pore waters and the sediments.

91. Ten grams of calcium-saturated smectite (CEC = 100 meq/100 g) is mixed with 1 L of water containing 1×10^{-3} mol L^{-1} calcium and 1×10^{-3} mol L^{-1} cadmium. The exchange reaction is

$$Cd^{2+} + Ca_{clay} = Ca^{2+} + Cd_{clay}$$

and the selectivity coefficient is $K = 0.9$. Calculate the amount of Cd^{2+} in solution after equilibrium is achieved.

92. Ten grams of calcium-saturated smectite (CEC = 100 meq/100 g) is mixed with 1 L of water containing 1×10^{-2} mol L^{-1} calcium and 1×10^{-3} mol L^{-1} lead. The exchange reaction is

$$Pb^{2+} + Ca_{clay} = Ca^{2+} + Pb_{clay}$$

and the selectivity coefficient is $K = 0.6$. Calculate the amount of Pb^{2+} in solution after equilibrium is achieved.

93. A midnight dumper is disposing of Cd-contaminated waste. The underlying soil has an average CEC = 50 meq/100 g. The porosity of the soil is 40%. The average density of the minerals in the soil is 2.65 gm cm^{-3}. If the waste contains 250 mg L^{-1} Cd, calculate the amount of cadmium per liter of wastewater after it equilibrates with the soil. The exchange reaction is

$$Cd^{2+} + Ca_{clay} = Ca^{2+} + Cd_{clay}$$

for which the measured selectivity coefficient is $K = 1.4$.

94. Zinc-contaminated wastewaters (150 mg L^{-1} Zn) are released into a pond. The pond is underlain by a silty sandstone that contains 10 wt% Ca-montmorillonite (CEC = 100 meq/100 g), 70% quartz, and 20% K-feldspar. The rock porosity is 30%. The average density of the minerals is 2.65 g cm^{-3}. As a simplifying assumption, assume that the other minerals have no ion-exchange capacity. Calculate the amount of zinc per liter of pore water after the water has equilibrated with the Ca-montmorillonite. The exchange reaction is

$$Zn^{2+} + Ca_{clay} = Ca^{2+} + Zn_{clay}$$

and the measured selectivity coefficient is $K = 0.4$.

95. For a Na-montmorillonite we can write the following exchange reaction for Cd^{2+}:

$$Cd^{2+} + Na_{clay} = Na^{+} + Cd_{clay}$$

and the measured selectivity coefficient is $K = 0.3$. Calculate the distribution of Na^{+} and Cd^{2+} on the montmorillonite, as a function of the equivalent fraction of Na^{+} in solution, for a total normality of 1.0N (similar to that of seawater) and 0.001N (similar to that of freshwater). (*Hint:* Use a spreadsheet and refer to Example 9–7 and Figure 9–17 for guidance on how to do this problem.)

96. Refer to Case Study 9–5 in order to answer this question. Figure 9–P96 shows the locations of various brine storage areas, monitoring wells (S for shallow and D for deep), and a water well drilled for a cattle corral. The emergency pit was opened in 1959 and closed in 1987. The lined pit was installed in 1964 and abandoned in 1971. In October of 1989, cattle first refused to drink from the corral well. The site overlies a portion of the recharge area for the Ogallala aquifer. There are two saturated zones separated by a continuous clay layer. The corral well was drilled to a depth of 34 m and penetrates both zones. The groundwater flow direction is to the north. Given this, well MW5S is updip from the contaminant input, and water from this well represents the initial groundwater composition for the shallow aquifer. Sample MW6D is updip from the contaminant input, and water from this well represents the initial groundwater composition for the deep aquifer. The following table gives the water chemistry for the various monitoring wells.

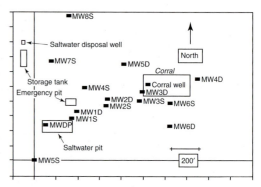

Figure 9–P96
Brine contamination site showing location of shallow (S) and deep (D) monitoring wells and the corral well. From "The Impact of Ion Exchange Processes on Subsurface Brine Transport as Observed on Piper Diagrams" by D. A. Cates, R. C. Knox and D. A. Sabatini, GROUND WATER; Vol. 34, No. 3, p. 540 (Figure 6). May/June 1996. Reprinted from Ground Water with permission of the National Ground Water Association. Copyright © 1996.

Chemical Composition of Monitoring Well Samples and Average Brine*

Well sample	Concentration (mg L^{-1})					
	Ca	Mg	Na	HCO$_3$	SO$_4$	Cl
MW1D	57	10	41	238	17	62
MW2D	59	9	15	213	14	42
MW3D	730	101	954	253	17	3,040
MW4D	53	10	64	209	18	132
MW5D	699	82	160	122	17	1,820
MW6D	48	9	18	217	13	38
MWDP	71	13	273	292	98	406
MW1S	668	154	427	191	17	2,550
MW2S	702	110	736	226	11	2,880
MW3S	364	38	242	260	25	1,030
MW4S	776	108	1,362	261	19	3,250
MW5S	67	11	15	261	16	23
MW6S	250	24	34	204	15	476
MW7S	291	41	904	253	19	2,190
MW8S	573	65	531	198	14	2,310
BRINE	11,980	2,180	68,500	13.2	2.1	120,000

*Data from Cates et al. (1996).

a. Plot the data for the brine and the deep water samples on a Piper diagram. Use the water chemistry for sample MW5S for the uncontaminated groundwater composition. Draw simple mixing lines between the end-member compositions. Which sample represents the *hardness halo* and why?

b. On another Piper diagram plot the brine and shallow water samples. Draw simple mixing lines between the end-member compositions. Which sample represents the *hardness halo* and why?

c. Interpret the two Piper diagrams in terms of the mixing of brine and uncontaminated groundwater. Is the deeper aquifer also contaminated by brine? If so, how might this have happened when there is an impermeable layer between the two aquifers?

d. Calculate the hardness for sample MW5S.

e. Calculate the hardness for sample MW4S.

97. In groundwater, Cl$^-$ is a conservative species. A slug of contaminated water containing Cd^{2+} and Cl$^-$ is injected into a groundwater system. After 1 year, the Cl$^-$ has traveled downgradient from the well a distance of 120 m. For this particular system, $K_d = 2.0$ cm^3 g^{-1} for Cd. The average density of the grains that comprise the sediment is 2.6 g cm^{-3} and the porosity is 25%. Assume that the pore spaces are filled with water. How far will the Cd^{2+} travel in 1 year?

98. A radioactive tracer, which behaves as a conservative species, is injected into a groundwater system. Measurements over a period of several years give an average groundwater velocity of 85 m y^{-1}. This aquifer serves as a water source for a municipality. Along the edge of the aquifer are several plants, one of which is dumping mercury into the sewer system. Unfortunately, there is a leak in the pipe that connects to the sewer system and some of this water is leaking into the aquifer. A water supply well is located 500 m from the plant. During routine water quality monitoring, mercury is found in the water. The aquifer has a porosity of 30% and the minerals that

constitute the sediment have an average density of 2.60 g cm^{-3}. An experiment is done to measure the partitioning of mercury between the aquifer material and water. The experimental value is $K_d = 1.3$ cm^3 g^{-1}. If the plant is the source of the mercury in the well water, how long ago did this mercury enter the groundwater system? The plant was built 30 years ago. Given this additional information, is the plant the source of the mercury contamination in the groundwater system? Explain.

99. In Case Study 5–6 the authors investigated the impact of discharges from a manufactured-gas plant on shallow, anaerobic groundwater. As part of this study, the authors investigated the partitioning of napthalene and toluene between the groundwater and the aquifer material. For napthalene, $K_d = 137$ L kg$^{-1} \approx 137$ cm^3 g^{-1} and for toluene, $K_d = 10$ L kg$^{-1} \approx 10$ cm^3 g^{-1}. The average density of the grains that comprise the aquifer sediment is 2.6 g cm^{-3} and the porosity is 20%. Assume that the pore spaces are filled with water and the average groundwater velocity is 100 m y^{-1}.

   a. How far will the napthalene and toluene travel in 1 year?
   b. A water supply well is located 100 m from the manufactured-gas plant. How long will it take for each chemical to reach the well?
   c. Under anaerobic conditions, the following first-order rate constants were determined for the breakdown of these organic compounds: napthalene, $K = -0.000046$ d^{-1}; and toluene, $K = -0.002$ d^{-1}. Given that the breakdown of the organic chemicals is a first-order reaction, calculate the percent napthalene and toluene from an injection event remaining in the groundwater when the water reaches the water supply well. If you don't remember how to do a kinetics problem, refer to Chapter 2.
   d. Should the users of this water supply well be concerned about contaminants released by the manufactured-gas plant? Explain.

100. Singh et al. (1997) analyzed sediment samples from the Gomati River, a tributary of the Ganga River, India. This river passes through New Hyderabad. A series of sediment samples were collected along the river. Sediment metal concentrations and downstream distances are given for these samples in the following table. G1 is upstream from the industrialized area and is considered to represent background concentrations. Also included in the data table are values for a reference shale.

Sample	Distance (km)	Concentration (mg kg^{-1})							
		Cd	Co	Cr	Cu	Mn	Ni	Pb	Zn
G1	0	0.26	44.6	24.3	65.7	663	58.5	25	90.3
G2	1.5	0.30	54.5	30.0	80.1	842	78.9	27	105.5
G3	4.0	0.45	54.9	32.5	99.9	604	80.9	29	120.7
G4	5.0	0.76	51.8	38.8	108.0	672	86.4	27	138.7
G5	6.5	2.34	46.4	32.9	173.7	476	44.9	47	240.4
G6	8.5	0.42	61.7	38.8	109.8	638	53.0	29	134.9
G7	10.5	1.65	44.1	30.8	129.6	502	48.3	49	167.2
G8	11.7	0.30	46.4	31.7	96.3	714	59.2	30	114.0
G9	13.0	3.62	48.6	43.1	294.3	442	63.9	77	389.6
G10	14.5	2.92	47.3	33.8	235.8	442	72.1	67	314.5
Shale		0.30	45	19.0	90	850	68	20	95

The easiest way to answer the following questions is to enter the data from the table into a spreadsheet. You can then make the required plots and do the calculations directly on the spreadsheet. Parts (c) and (d) are best answered in table form.

   a. Plot elemental concentrations versus downstream distance. Interpret the plots.

b. Calculate the enrichment factors, and plot the enrichment factors versus downstream distance. Interpret the plots.
c. Calculate the geoaccumulation index for each metal and each sample.
d. Classify each sample in terms of sediment quality.
e. Are there any other types of plots that might be useful for interpreting the data?
f. From the preceding analysis, what do you conclude about heavy metal contamination in this river? You may want to refer to the original paper.

101. Plutonium-containing radioactive waste from a fuel-reprocessing facility is accidentally discharged into volcanic ash. Using the minimum and maximum retardation factors given in Table 9–17 and a flow velocity of 200 m y^{-1} through the volcanic ash, calculate the minimum and maximum distance the plutonium will migrate in 100 years.

102. Calculate the solubility of pyrrhotite (FeS) and pyrite (FeS$_2$) as a function of pH (from 2 to 9) when $P_{H_2S} = 10^{-4}$ atm and $T = 25°C$. Use the appropriate thermodynamic values from Appendix II, source 4. The relevant reactions are

$$FeS_{\text{pyrrhotite}} + 2H^+ \rightarrow Fe^{2+} + H_2S_{(g)}$$

and

$$FeS_{2\ \text{pyrite}} + 4H^+ \rightarrow Fe^{2+} + 2H_2S_{(g)}$$

You may want to do this problem on a spreadsheet. Comment on the relative solubility of the two minerals as a function of pH.

103. Monna et al. (2000) determined the Pb-isotopic chemistry of sediment cores collected from a lake in southern France. The sampled sediment record encompassed the last several hundred years of earth history. The data for one of these cores follow. You may want to refer to the section in Chapter 6 that discusses radioactive isotopes and the section in Chapter 8 dealing with lead isotopes before you try this problem.

Lead Isotopic Data for a Sediment Core from Thau Lake, France

Depth (cm)	$^{206}Pb/^{204}Pb$	$^{207}Pb/^{204}Pb$	$^{208}Pb/^{204}Pb$	$^{206}Pb/^{207}Pb$
0.5	18.290	15.612	38.23	1.172
1.6	18.362	15.673	38.50	1.172
2.8	18.411	15.679	38.53	1.174
4.0	18.417	15.637	38.44	1.178
5.2	18.409	15.630	38.45	1.178
6.4	18.433	15.668	38.60	1.176
7.8	18.395	15.635	38.44	1.177
9.0	18.421	15.675	38.62	1.175
10.3	18.388	15.625	38.42	1.177
11.8	18.386	15.638	38.53	1.176
13.2	18.431	15.633	38.53	1.179
14.4	18.488	15.659	38.53	1.181
15.6	18.474	15.645	38.61	1.181
16.8	18.585	15.677	38.77	1.185
18.0	18.606	15.649	38.63	1.189
19.3	18.624	15.687	38.77	1.187
20.8	18.550	15.625	38.56	1.187
22.3	18.535	15.634	38.52	1.186
23.5	18.538	15.659	38.67	1.184
26.5	18.484	15.641	38.60	1.182
31.5	18.455	15.614	38.60	1.182
36.5	18.693	15.667	38.70	1.193

Possible lead end members, estimated from the graphs of Monna et al. (2000), are listed in the accompanying table.

Lead Isotopic Data for Pb Sources for the Thau Lake Sediments

End member	$^{206}Pb/^{204}Pb$	$^{208}Pb/^{204}Pb$
Gasoline	16.60	36.50
Gasoline	16.85	36.75
Gasoline	17.05	37.00
Industry	17.80	37.75
Industry	18.00	38.00
Industry	18.15	38.25
Miocene sediments	18.70	38.85
Jurassic sediments	20.00	38.35
Jurassic sediments	20.00	39.10

a. Plot sediment depth versus the $^{206}Pb/^{207}Pb$ ratio. Is there any pattern? If so, what might it mean?

b. Plot $^{208}Pb/^{204}Pb$ versus $^{206}Pb/^{204}Pb$. Include in this diagram both the core data and the isotopic ratios for the possible end members. From this diagram what can you conclude about the source(s) of Pb in the lake sediments? What does this tell us about the surrounding environment?

104. The following chemical data were obtained for shallow groundwaters at Owens Lake, California (Levy et al., 1999). A spreadsheet will prove very useful for answering the following questions. Assume that the groundwater temperature is 25°C.

Groundwater sample	Concentration (mg L^{-1})						
	Ca	Na	K	CO_3^{2-}	SO_4^{2-}	Cl	F
Keeler	33	160	29	290	110	98	1.2
Well AW	1.7	570	36	630	47	210	2.2
Sulfate well	1.2	700	21	880	1	260	1.1
SP-1	1.12	114,000	11,300	43,000	23,600	133,000	45
MP-1	0.92	30,500	3,690	6,000	3,280	48,900	12.2
MP-2	0.66	9,225	995	2,200	634	13,450	6.38
SP-2	0.82	124,000	11,000	55,000	23,800	135,000	37
MP-3	1.68	43,700	4,450	9,600	6,600	57,950	20.5
MP-4	0.92	9,330	1,240	1,900	533	15,900	7.58

a. Plot the Ca and F data on a graph on which you've drawn the fluorite solubility curve (Figure 9–24). Are the groundwaters saturated with respect to fluorite?

b. Calculate the ionic strength of each of the solutions. Given these ionic strengths, should activity be taken into account?

c. Calculate the activity coefficient for Ca and F in each water sample and the resulting activity of Ca and F. Use the Truesdell–Jones activity coefficient model (see Chapter 2). Now what do you conclude about the saturation of the groundwaters with respect to fluorite?

d. With reference to Chapter 2, what is the ionic strength limit for the Truesdell–Jones model? Identify those points on your diagram that exceed this limit. Now what do you conclude about the saturation of the groundwaters with fluorite?

105. Groundwater samples were collected from a number of wells in Oklahoma. The Cl and Br data for 12 of the well water samples are listed in the following table. Remember that these two elements usually behave as conservative species.

Well	Cl (mg L^{-1})	Br (mg L^{-1})
1	5.5	0.030
2	13.8	0.031
3	27	0.033
4	58	0.036
5	114	0.428
6	229	0.054
7	450	0.078
8	837	0.120
9	1,763	0.219
10	4,009	0.460
11	7,168	0.799
12	30,110	3.258

   a. Using a spreadsheet program, plot the data for the 12 wells. Fit a linear regression equation to the data. Can the data be explained by simple mixing? Explain.

   b. An oil field brine found in this region has Cl = 32,600 mg L^{-1} and Br = 3.525 mg L^{-1}. Would this be an appropriate mixing end member for our well water samples? Explain.

   c. The well water sample with the lowest concentration of these two anions has Cl = 2 mg L^{-1} and Br = 0.030 mg L^{-1}. Would this be a suitable end member for our well water samples? Explain.

   d. If well water sample #7 was a mixture of oil field brine and uncontaminated groundwater (c), calculate the amount of oil field brine in this sample.

106. Ethylene dibromide ($C_2H_4Br_2$) is a gasoline additive. Decomposition of this compound releases Br. Leakage from an underground storage tank at a gasoline (petrol) station introduced significant quantities of this compound into the subsurface. Measurements of groundwater chemistry at the edge of the property gave Cl = 3.0 mg L^{-1} and Br = 2.5 mg L^{-1}. Uncontaminated groundwater in the region has Cl = 2.0 mg L^{-1} and Br = 0.030 mg L^{-1}. In order to assess the extent of groundwater contamination, a series of wells were drilled from the gas station in the direction of groundwater flow. The data are listed in the following table.

Distance from station (m)	Cl (mg L^{-1})	Br (mg L^{-1})
50	3.00	2.49
100	2.85	2.12
150	2.70	1.77
200	2.55	1.38
250	2.40	1.015
300	2.25	0.651
350	2.10	0.274
400	1.98	0.029
450	2.01	0.030
500	2.00	0.029

   a. Plot the Cl/Br ratio as a function of distance. What can you conclude from this plot?

  b. Plot Br versus Cl. Can the data be explained by simple mixing? You may have to think about this and refer to the graph and your answer for part (a).

  c. Given the two possible end members, calculate the amount of contaminated groundwater in the sample collected at a distance of 250 m from the gas station.

107. Using a spreadsheet program, calculate the distribution of the phosphorus species, as a function of pH, given a temperature of 25°C and a total phosphorus concentration of 1×10^{-6} mol L^{-1}. Dissociation constants for phosphoric acid at $T = 25$°C are given in Table 3–1. Which are the major phosphorus species for the pH range normally encountered in natural waters?

The Marine Environment

10

The oceans cover approximately 71% of the earth's surface and constitute the major water reservoir (~96% of the total surface water) for the earth. The oceans are the final sink for material eroded from the continents. While ocean chemistry is reasonably constant, long-term ocean chemistry is controlled by a number of processes, including various water–sediment and water–rock interactions. The oceans play a major role in the sequestration of carbon dioxide (Chapter 8), and we have already discussed the importance of seafloor sediments in paleoclimate studies (Chapter 8).

The marginal marine environment is a critical area for a number of reasons. This environment plays a major role in the biological productivity of the ocean. Freshwater streams entering the ocean interact with the often vastly different marine environment (in terms of total ionic strength, pH, and oxidation–reduction potential), leading to a number of precipitation and dissolution reactions. The marginal marine environments are also regions in which there are significant anthropogenic interactions, and, hence, these tend to be highly stressed environments.

In this chapter we will investigate the basic properties of the marine environment, both the open ocean and the marginal environment. It is convenient to consider these two environments separately because the chemistry of the open ocean is much more uniform than that of the marginal marine environment. We will discuss the basic physical and chemical properties and the various processes that control the chemistry of these environments, and we will consider the impact of anthropogenic activities on the marine system.

PHYSICAL ASPECTS OF THE OPEN OCEAN

As a first approximation, we can look at the earth as consisting of two topographic levels, the ocean basins and the continents (Figure 10–1). *The submerged edge of a continent is referred to as the* **continental shelf**, *which is essentially flat-lying. The point where there is a change in slope (from ~0° to 2–4°) is referred to as the* **shelf break**. The continental slope then

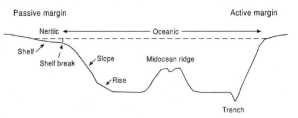

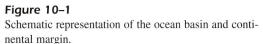

Figure 10–1

Schematic representation of the ocean basin and continental margin.

descends to the deep ocean basin. The bottom of the continental slope is often marked by the continental rise, which represents sedimentary material transported from the continents, by turbidity flows and gravity slides, to the deep ocean. Along active continental margins, the continental slope may be separated from the ocean basin by an oceanic trench. These trenches mark the location of subduction zones, where seafloor is returned (subducted) to the mantle. Midocean ridges represent regions of extension and seafloor spreading where material (basalt) is added to the ocean floor. In these regions seawater interacts with the hot rocks of the seafloor in a complex series of reactions that change both the chemistry of the seafloor basalts and the ocean. We will return to this topic later in the chapter. It is sometimes convenient to divide the ocean into two environments: neritic and oceanic. The **neritic environment** is that part of the *ocean that overlies the continental shelves*; the **oceanic environment** is that part of the *ocean that overlies the ocean basin*.

Vertical Structure of the Ocean

The ocean can be divided into three layers; a surface mixed layer, a *zone of changing temperature* (the **main thermocline**), and a deep layer (Figure 10–2). In subsequent sections, when we discuss ocean chemistry, we will often consider the ocean to consist of two separate reservoirs, the surface and deep ocean. These two reservoirs are separated by the main thermocline. At high latitudes the surface mixed layer disappears. In effect, the deep waters "crop out" at the surface. At these latitudes ocean temperature would be relatively constant with depth, and the water column is isothermal.

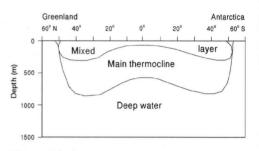

Figure 10–2
Schematic cross section showing the three-layer ocean.

Oceanic Circulation

There are essentially two separate, but related, oceanic circulation patterns: the surface currents, which are driven by the winds, and the vertical circulation (thermohaline), which is driven by differences in density. Both are important in terms of understanding the transport of heat and matter in the oceans. As you may recall from Chapter 8, there is a thermal imbalance between the earth's poles and the equator. There is a net heat loss at the poles and a net heat gain at the equator. It is this differential in thermal energy that is responsible for the general atmospheric circulation pattern. Atmospheric flows redistribute approximately two-thirds of the thermal excess. The rest is redistributed by the ocean currents. For this reason, oceanic circulation plays an important role in climate, and changes in oceanic circulation can lead to climate change.

Surface Currents   The surface currents are established by the interaction between the winds and the ocean surface. We can gain a first-order understanding of this process by considering a simple rectangular ocean basin (Figure 10–3). The general circulation pattern for the earth's atmosphere is shown in Figure 8–5. The winds interact with the ocean and transfer energy to the ocean surface. This energy (or momentum) transfer sets the surface waters in motion at roughly 45° to the wind direction. In effect, each prevailing wind direction acts as one-half of a force couple. It is the interaction of these couples that causes the surface currents. Consider the couple that consists of the westerly winds and the trade winds (in the Northern Hemisphere). The trade winds cause the surface water to move to the west while the westerly winds cause the surface waters to flow to the east. The result is a large gyre with a clockwise circulation. It is not necessary that we have two different prevailing wind directions. Simple changes in velocity are also adequate to set up gyral circulation patterns. This is shown by the couple that forms the north tropical gyral, which consists of the trade winds to the north and a zone of no horizontal air motion (the doldrums) to the south. The other feature to note from this schematic diagram is the displacement of the center of the gyres to the west. This occurs because the earth rotates from west to east, and as the earth rotates it slips from under the water above, thus shifting the whole current system toward the west. Because the volume of water flowing through any particular gyral must remain constant, the current along the western margin of an ocean basin is narrower, deeper, and faster than the current along the eastern margin of an ocean basin. The observed surface currents are shown in Figure 10–4. Note the currents

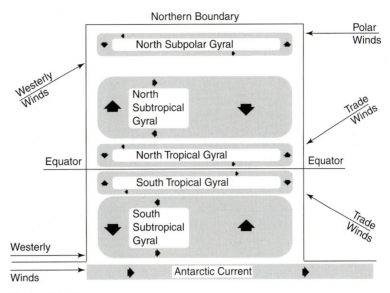

Figure 10–3
Schematic representation of the interaction between winds and the ocean surface. From Williams (1962).

flowing toward the equator from high latitudes, which bring cold waters to low latitudes, and the reverse situation with currents flowing from the equator to high latitudes. For example, the Labrador Current brings cold water down along the east coast of Canada (Labrador), causing a year-round cold and moist climate. Conversely, the Gulf Stream brings warm water to the British Isles, leading to a very temperate climate (at the same latitude as Labrador). In fact, palm trees grow on the southwest coast of England.

Upwelling and Downwelling   There is another important consequence of the ocean surface circulation. Where ocean surface currents meet, convergence or divergence can occur. For example, at 60°S the surface currents diverge. The same is true at the equator.

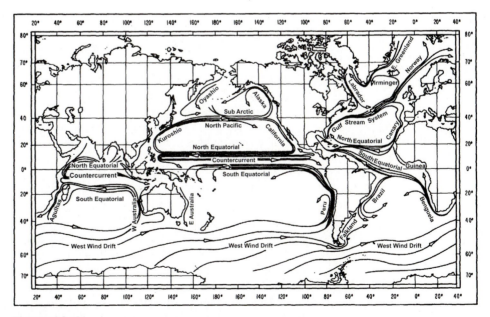

Figure 10–4
Average surface currents of the world's oceans. From Williams (1962).

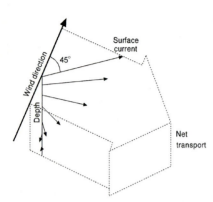

Figure 10–5
Schematic representation of the Ekman spiral in the Northern Hemisphere. Surface water moves at 45° to the right of the prevailing wind direction. With increasing depth, the water velocity decreases and the Coriolis deflection increases. The net movement of water in the water column is at 90° to the prevailing wind direction.

Conversely, at 30°S and 30°N the surface currents converge. In regions where currents diverge, surface waters move away. The resulting mass deficiency is compensated by the upward movement of deep waters (upwelling). Conversely, where currents converge, water accumulates and the mass excess is compensated by downward movement of the surface waters (downwelling). As we shall subsequently see, this has a significant impact on the availability of nutrient elements in the surface waters and thus biological productivity.

Another kind, and equally important type, of upwelling and downwelling is driven by the Ekman spiral. When the wind acts on the water surface, the water motion is initially at approximately 45° to the wind. In the Northern Hemisphere the deflection would be to the right of the wind direction (Figure 10–5). Momentum is transferred downward in the water column. With increasing depth, the water velocity decreases and the Coriolis deflection increases. At some depth the water is moving exactly opposite to the surface waters, but with a much smaller velocity. This change in velocity and direction of water motion with depth forms what is called the *Ekman spiral*. If one sums up all the water directions and current magnitudes, the net movement of water in the water column is at 90° to the surface wind direction. Continuing with our example, let us assume that there is a landmass along the right-hand edge of the diagram. Because of Ekman transport, the water will pile up against this landmass. Mass continuity requires that this excess mass be compensated. This leads to the downward displacement of the excess water, i.e., downwelling. Conversely, if there was a landmass to the left-hand side of the diagram, the water would be moving away from the landmass, resulting in a mass deficit. Mass continuity requires that this water be replaced by water from depth, i.e., upwelling. Ekman-driven upwelling and downwelling is a very important process in coastal waters. Areas of coastal upwelling are areas of high productivity because nutrient elements are brought back into the surface waters.

Thermohaline Circulation   The thermohaline circulation is driven by density differences that are due to differences in temperature and salinity. Vertical currents have a much lower velocity than surface currents, on the order of several kilometers per day versus several or more kilometers per hour for surface currents. The ocean deep waters originate at the surface where they acquire temperatures and salinities characteristic of their source area. For example, deep water that originates off the coast of Norway is characterized by extremely low temperatures, near 0°C, and a slightly lower salinity than the ocean as a whole because of the addition of freshwater from rivers and precipitation. In contrast, water that originates in the Mediterranean Sea is characterized by high temperatures and salinities. When this water enters the Atlantic Ocean at the Straits of Gibraltar, it forms a characteristic body of water that can be traced at intermediate depths for some distance into the eastern Atlantic Ocean. These *distinctive bodies of water* are referred to as **water masses** and can be *characterized by their temperature and salinity*. Both temperature and salinity (to be discussed later) are conservative properties. When water masses interact, the temperature and salinity of the mixed waters are linearly related to the relative proportions of the contributing water masses.

As an example of the vertical circulation of the oceans, we will look at the thermohaline circulation and water masses of the eastern Atlantic Ocean (Figure 10–6). The individual water masses are named on the basis of their position in the ocean. For example, North Atlantic Deep Waters (NADW) comprise much of the deep water of the North Atlantic Ocean and extend well into the South Atlantic. This water mass has its origin at high latitudes off the coast of Norway. If we follow the path of this water mass, we see that it moves into the South Atlantic, where it encounters even colder (and denser) waters (AADW and AABW) that originate off the coast of Antarctica, and it is forced to rise above these waters. The deep water (AABW) that originates off the coast of Antarctica extends some distance into the North Atlantic until it meets the North Atlantic Bottom Waters (NABW). Also note the region occupied by the Mediterranean Intermediate Waters, which originate in the Mediterranean. Other water masses are formed in regions of convergence, such as the Antarctic Intermediate Water (AAIW), which originates at the Antarctic Convergence (AC), the area in the South Atlantic where surface currents converge.

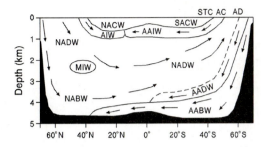

Figure 10–6
Generalized pattern of vertical circulation in the Atlantic Ocean. The water masses are NACW (North Atlantic Central Water), SACW (South Atlantic Cental Water), AIW (Arctic Intermediate Water), AAIW (Antarctic Intermediate Water), NADW (North Atlantic Deep Water), NABW (North Atlantic Bottom Water), AADW (Antarctic Deep Water), AABW (Antarctic Bottom Water), and MIW (Mediterranean Intermediate Water). The labels at the top of the diagram identify the positions of the Subtropical Convergence (STC), Antarctic Convergence (AC), and the Antarctic Divergence (AD).

As noted, water masses are identified in terms of their temperature and salinity, both conservative properties. As might be expected, the different water masses often do not have a unique temperature and salinity, but rather a range of characteristic temperatures and salinities. The data for various water masses are generally plotted on temperature–salinity (T-S) diagrams, and these diagrams can then be used to identify water masses in the ocean basins. An example of such a diagram, for the Atlantic Ocean, is shown in Figure 10–7. T-S diagrams can also be used to predict what will happen during the mixing of water masses. The analytical approach is exactly the same as that used in previous chapters when we looked at the percent of various end members in a mixture.

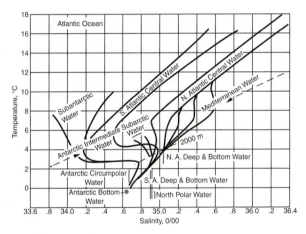

Figure 10–7
T-S characteristics for various water masses in the Atlantic Ocean. From von Arx (1962).

EXAMPLE 10–1 Water samples were collected in the North Atlantic in the depth range 850 m to 1500 m. The sample collected at 850 m had temperature and salinity values typical of MIW, whereas the sample collected at 1500 m had temperature and salinity characteristics typical of NADW. The temperature and salinity values for the five samples are given in the table.

Depth (m)	T (°C)	S (‰)
850	13	37.3
1200	11	36.7
1300	9.4	36.2
1350	8.4	35.9
1500	4.8	34.8

When plotted on a T-S diagram (Figure 10–8), the data lie on a straight line, with the MIW and NADW as end members. Thus, the samples in the depth range 1200 m to 1350 m represent simple mixtures of MIW and NADW. The numbers in parentheses indicate the fraction of MIW in each sample.

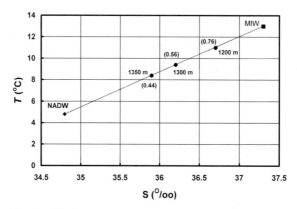

Figure 10–8
T-S diagram for water samples in Example 10–1. The intermediate waters are a mixture of MIW and NADW. The numbers in parentheses are the fraction of MIW in each sample.

Constancy of Composition and Salinity

The beginning of chemical oceanography, and oceanography in general, is often ascribed to the voyage of HMS *Challenger* (1872–1876). Although naturalists were often found on voyages of exploration (Darwin and Banks being particularly notable examples), the voyage of HMS *Challenger* was the first for the sole purpose of oceanographic research. The British Admiralty was persuaded to outfit and fund this expedition. During its five-year mission (echoes of *Star Trek*) the HMS *Challenger* covered a distance in excess of 68,000 nautical miles and collected a variety of oceanographic data. The interested student should consult a standard oceanography textbook for the details of this voyage. Of interest here are the 77 water samples from the *Challenger* expedition that were analyzed by Dittmar. These analyses revealed that the *major species in seawater exist in constant proportion*, an observation that came to be known as **constancy of composition**. Although the total concentration of the individual species varied from place to place in the ocean, their relative proportions remained constant. The only variable that affected the total concentration of dissolved species was dilution due to the addition of freshwater or concentration due to evaporation.

In previous sections we used salinity to characterize water masses and have assumed that salinity is a conservative property. What is salinity? In a sense it is a measure of the total salt content of seawater, but this measurement is done in a very particular way. Given that the major ionic species in seawater remain in constant proportion to each other, one can determine the total concentration of dissolved material by determining the concentration of one of these species. Because it is easily determined by titration, Cl^- is the species usually used to determine salinity. The titration is straightforward and was already mentioned in Chapter 1. A silver nitrate solution is added to seawater. The Ag^+ ions combine with the halide ions in solution to form an insoluble precipitate (silver halide). Once all the halide ions have been precipitated from solution, the next drop of titrant reacts with an indicator, signaling the end of the titration. Note that we have referred to the halides. The titration does not distinguish between Cl^-, Br^-, and I^-. The combination of these three ions is often referred to as the *chlorinity*. Given the way salinity is determined, the accepted definition for **salinity** is *the total amount in grams of solid material dissolved in 1 kg of seawater when all the carbonate has been converted to oxide, all the iodine and* *bromine have been replaced by chlorine, and all organic matter has been completely oxidized.* Mathematically, salinity is expressed as

$$S‰ = 1.80655 \, Cl‰ \qquad (10-1)$$

where Cl‰ is the chlorinity (determined by titration) in parts-per-thousand. Note that salinity is reported in parts-per-thousand, i.e., grams of dissolved material per kg of seawater. More recently, salinity measurements have been made using electrical conductivity because these measurements are inherently more precise.

CHEMICAL COMPOSITION OF THE OPEN OCEAN

The average composition of seawater, along with the world average for rivers, is listed in Table 10–1. Because the rivers empty into the oceans, one might expect that ocean chemistry would mirror river chemistry. Soluble material weathered from the continents is added to the oceans throughout geologic time, thus increasing the abundances of the various soluble species. In fact, Joly (1901) used this model to estimate the age of the earth. He started with the assumption that the oceans were originally freshwater, that all the Na and Cl produced by weathering was carried to the oceans, where it remained in solution, and that the present rate of Na and Cl addition is a reasonable average for all of geologic time. Using this approach, Joly arrived at an age of 100 million years for the earth, a not unreasonable estimate at the time. The present estimate of the age of the earth, ~4.5 billion years, and the observation that the oceans have existed for much of earth history lead us to the conclusion that this simple model does not explain ocean chemistry. Additionally, an inspection of Table 10–1 reveals that the dissolved species in river water exist in very different molecular proportions than they do in the oceans. Thus, we are led to conclude that processes must occur in the ocean that not only change the relative proportions of the species in solution but also remove or add dissolved species.

Table 10–1 Average Chemical Composition of Rivers and Seawater[1]

	Average river water			Average seawater		
	$mg \, kg^{-1}$	$mmol \, kg^{-1}$	Species/Cl^-	$g \, kg^{-1}$	$mol \, kg^{-1}$	Species/Cl^-
Cl^-	8.3	0.23		19.344	0.546	
Na^+	7.2	0.31	1.35	10.773	0.467	0.855
SO_4^{2-}	11.5	0.12	0.52	2.717	0.028	0.051
Mg^{2+}	3.7	0.15	0.65	1.294	0.053	0.097
Ca^{2+}	14.7	0.37	1.61	0.412	0.010	0.018
K^+	1.4	0.036	0.16	0.399	0.010	0.018
[2]HCO_3^-	53.0	0.87	3.78	0.142	0.002	0.004
Br^-				0.0674		
Sr^{2+}				0.0079		
B				0.00445		
F^-				0.00128		
Total	99.8	2.086		35.162	1.116	

[1]Data for average river water from Berner and Berner (1996), and for average seawater from Libes (1992).
[2]HCO_3^- is a nonconservative species.

Residence Time of Seawater Species

We first developed the idea of residence time in Chapter 1. In the case of the ocean, we start by making the assumption that the ocean is in a steady state. The rate of addition of species to the ocean is equal to the rate of removal. Residence times in the oceanic system

Table 10–2   Elemental Oceanic Residence Times (in years)
Based on River Input*

	Log (y)		Log (y)		Log (y)		Log (y)
Li	6.3	S	6.9	Cu	4.0	Sb	4.0
B	7.0	Cl	7.9	Zn	4.0	I	6.0
C	4.9	K	6.7	As	5.0	Cs	5.8
N	6.3	Ca	5.9	Se	4.0	Ba	4.5
O	4.5	Sc	4.6	Br	8.0	La	6.3
F	5.7	Ti	4.0	Rb	6.4	Au	5.0
Na	7.7	V	5.0	Sr	6.6	Hg	5.0
Mg	7.0	Cr	3.0	Zr	5.0	Pb	2.6
Al	2.0	Mn	4.0	Mo	5.0	Ra	6.6
Si	3.8	Fe	2.0	Ag	5.0	Th	2.0
P	4.0	Co	4.5	Cd	4.7	U	6.4

*Data source, Holland (1978).

for various elements are listed in Table 10–2. The residence times have been calculated by dividing the concentration of the particular element in seawater by the yearly rate of addition by the rivers. Note that this is a very specific calculation that ignores other possible inputs, of which the most likely and important would be due to interactions between seawater and the seafloor basalts. However, this type of calculation gives us some idea of the reactivity of various elements in seawater.

We can somewhat arbitrarily, in terms of residence time, divide the elements into three groups: short (100–1,000 years), medium (1,000–1,000,000 years), and long (> 1,000,000 years). Using this division, elements with short residence times are Al, Cr, Fe, Pb, and Th, and elements with long residence times are Li, B, Na, Mg, S, Cl, K, Br, Rb, Sr, I, La, Ra, and U. Elements with short residence times are highly reactive—i.e., they are quickly removed from seawater—whereas elements with long residence times have a low reactivity in seawater.

The Global Chemical Cycle for Seawater Species

The sources of oceanic species are dissolved gases from the atmosphere, inputs from the weathering of continents, and seawater–seafloor interactions. Species are removed from seawater by transfer to the atmosphere, precipitation or adsorption and removal to the seafloor sediments, and interaction with the seafloor basalts. This global cycle is shown in Figure 10–9.

Let us look at the various components of the cycle. The major elements in seawater are supplied by the uplift and erosion of the continents, and for most of the minor elements, erosion of the continents (supplemented by anthropogenic inputs) is also the major source. The dissolved gases, N_2, CO_2, O_2, and SO_2, are derived from the atmosphere, and at the surface the ocean waters are approximately in equilibrium with the atmosphere. Here we need to mention biological processes because photosynthesis and respiration can add or remove O_2 and CO_2 from the top 100–200 m of the water column, leading to short-term disequilibrium with the atmosphere. At deeper levels, the ocean water is effectively isolated from the atmosphere and the O_2 and CO_2 are not in equilibrium with the atmosphere. Evaporation at the ocean's surface can lead to saturation in halite (NaCl) and anhydrite ($CaSO_4$). These small particles are transferred to the atmosphere, where they act as aerosols. Most are returned directly to the ocean, but some are transported over the continents and the constituents return to the oceans via runoff. These comprise what are known as the *cyclical salts*. Within the water column elements are removed by primary precipitation (actually, relatively rare), by biological processes, and by adsorption onto pre-existing particles, of which the clay minerals and various oxyhydroxides are most important. Given the average ocean pH of 8.1 to 8.3, and recalling from Chapter 9 that the clay minerals and the oxyhydroxides in this pH range will have negatively charged surfaces, the

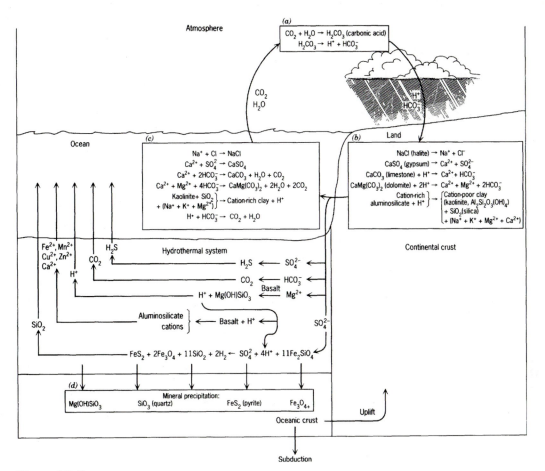

Figure 10-9
Chemical balance between the (a) atmosphere, (b) continental crust, (c) ocean, and (d) oceanic crust. From Edmond and von Damm (1983).

particles will adsorb positively charged species. The interactions between seawater and the oceanic basalts (hydrothermal system in Figure 10–9) are complex and lead to both the addition and removal of various species from seawater. This system (Figure 10–10, p. 396) has received a good deal of attention because of the deep sea smokers and hydrothermal vents on midocean ridges and their associated biological communities. The energy source for the biological communities is chemoautotrophs (bacteria), which utilize various oxidation–reduction reactions for their energy source. Once the species dissolved in the hydrothermal (hot water) solutions reach the seafloor, they are precipitated by a variety of reactions, as indicated in Figure 10–10. We will return to some of these processes in later sections of this chapter.

Processes Controlling Seawater Composition

Biological Controls on Seawater Composition   In previous chapters we have used a simple equation to represent photosynthesis and the production of organic matter from CO_2 and H_2O. We can write a more detailed equation for this process, which has the following form:

$$106CO_2 + 16NO_3^- + HPO_4^{2-} + 122H_2O + 18H^+ \rightarrow$$

$$(C_{106}H_{263}O_{110}N_{16}P_1)_{\text{algal protoplasm}} + 138O_2$$

According to this equation, during photosynthesis phosphorus, nitrogen, and carbon are utilized in the ratio 1P:16N:106C. This ratio is often referred to as the **Redfield ratio** (Redfield, 1958), which is the *proportions in which the various elements are found in marine algae*. During respiration and breakdown of the organic matter, these elements are

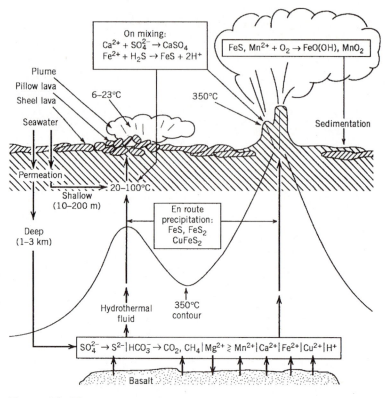

Figure 10–10
Schematic diagram showing inorganic chemical processes that occur at a hydrothermal vent. From "Geomicrobiology of the deep sea hydrothermal vents," by H. W. Jannasch and M. J. Mottl in SCIENCE, #229, pp. 717–725. Copyright © 1985 American Association for the Advancement of Science. Reprinted with permission.

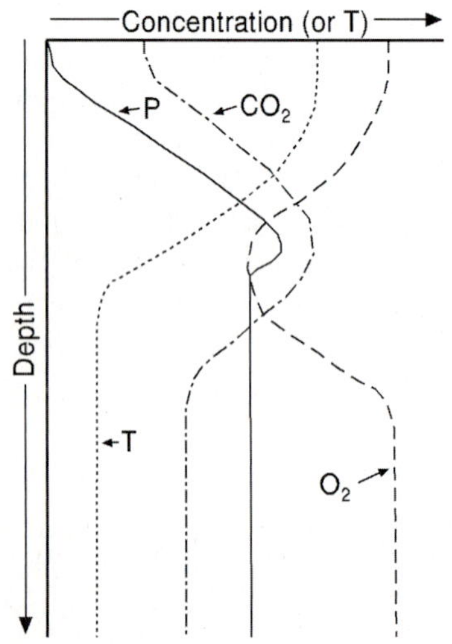

Figure 10–11
Schematic representation of depth versus the variations in temperature (T), phosphorus, carbon dioxide, and oxygen in a region with high biological activity.

released in the same proportion. The concept of the Redfield ratio can also be extended to other elements (such as trace metals) that are used during photosynthesis. For example, a recent study by Kuss and Kremling (1999) determined the abundances of a number of elements in near-surface organic particles from the North Atlantic. For this data set, the approximate Redfield-type composition was 180C:23N:**1P**:0.005Fe:0.002Zn,Mn:0.001 Ni:0.0005Cd:0.0004Cu:0.0002Co:0.00004Pb. Because CO_2 and H_2O are present in abundance in surface waters, the limiting factor in primary production is the availability of phosphorus and nitrogen. There are rare (?), but interesting, cases in which another element can become the limiting nutrient. The most notable case is iron, which acts as a limiting nutrient for phytoplankton growth in the waters around Antarctica (Martin et al., 1990). Light is also a seasonal limiting factor at mid-to-high latitudes, and this results in one or two seasonal plankton blooms (periods of high plankton production).

The activity of autotrophic and heterotrophic organisms plays a major role in the distribution of biologically utilized elements in the upper part of the ocean (Figure 10–11). Photosynthesis is restricted to the topmost layer of the ocean (referred to as the ***euphotic zone***), in which there is adequate light. Because the intensity of light decreases exponentially with depth, phytoplankton (floating plants) tend to concentrate near the very top of the ocean. In an area of high biological productivity, the density of organisms in the surface waters shades the deeper waters significantly, reducing the depth range through which net primary production can occur. In the surface waters, oxygen is produced and carbon dioxide is consumed. Oxygen concentrations may, on a daily basis, exceed atmospheric equilibrium concentrations. Nutrient elements are consumed, and if the availability of nutrient elements is the limiting factor in phytoplankton growth, N and P will be depleted in the surface waters. Heterotrophs consume the organic matter produced by the phytoplankton and release carbon dioxide. Below the zone of photosynthesis, oxygen is consumed and carbon dioxide is released. The organic matter produced in the surface layers is ultimately transported to depth (either directly or as the waste products of heterotrophic consumption), and subsequent decay releases the elements to the seawater. The resulting vertical distribution of the elements is shown schematically in Figure 10–11. Note that oxygen and carbon dioxide concentrations are mirror images. Nitrogen would show the same pattern as phosphorus. It is depleted in the surface waters and then released to seawater throughout the region of the main thermocline. Below the main thermocline, we encounter the deep waters of the ocean. These waters have compositions similar to their source regions. The net result of these processes is that nutrient elements are transferred from the surface reservoir to the deep reservoir of the ocean. In regions of upwelling, the nutrient-rich deep waters are brought to the surface, leading to areas of high biological productivity.

Broecker (1971) developed a two-box model (Figure 10–12) to describe the cycling of biologically utilized elements. In this model the only source of the biologically utilized elements is river inflow. The elements are extracted from seawater to make both organic matter and hard parts [skeletal material: $CaCO_3$ (calcite and aragonite) and SiO_2 (opaline silica)]. These materials are then transported to the deep ocean as particles. Some of the particulate material decomposes and releases its constituents to the seawater, while the remainder is removed, at least temporarily, from the ocean and sequestered in the bottom sediments. Because the ocean is in a steady state, the total river inflow on a yearly basis must be offset by evaporation (the volume of the surface reservoir remains constant) and the exchange of water between the two reservoirs (by upwelling and downwelling) must be equal. Broecker (1971) determined that the flow between the surface and deep-water reservoirs was 20 times the river inflow and also determined an average residence time of 1600 years for water in the deep reservoir (which is also the time required for one complete oceanic overturn or cycle). The derivation of the following equations can be found in Broecker (1971). The fraction of a biogenic element that is transferred to the deep-water reservoir in particle form and ultimately sequestered in the bottom sediment can be calculated as follows:

$$f = \frac{1}{1 + 20\left(\dfrac{[D]}{[R]} - \dfrac{[S]}{[R]}\right)} \tag{10–2}$$

where f is the fraction of the biogenic element sequestered in the bottom sediments, D is the concentration of the element in the deep-water reservoir, S is the concentration of the element in the surface-water reservoir, and R is the concentration of the element in river water. The number 20 is the ratio of upwelling to the surface water to riverwater inflow to the surface reservoir. The fraction of the element removed as biogenic particles can be calculated as follows:

$$g = 1 - \left(\frac{20[S]/[R]}{1 + 20[D]/[R]}\right) \tag{10–3}$$

where g is the fraction of the element removed as biogenic particles. We can calculate a third parameter, the residence time for a biologically utilized element, using the following equation:

$$\tau = \frac{1600 \text{ y}}{f \times g} \tag{10–4}$$

where τ is the residence time in years and 1600 y is the time required for one complete oceanic cycle.

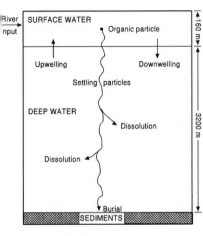

Figure 10–12

Two-box model for the cycling of biologically utilized elements. After Broecker (1971).

EXAMPLE 10–2 The following average concentrations, in μmol L^{-1}, were determined for P in the Pacific ocean: surface water = 0.2 and deep water = 2.5. For average river water, P = 0.7 μmol L^{-1}. Calculate f, g, and τ for P.

$$f = \frac{1}{20\left(\dfrac{[D] - [S]}{[R]}\right) + 1} = \frac{1}{20\left(\dfrac{[2.5] - [0.2]}{[0.7]}\right) + 1} = 0.015$$

$$g = 1 - \left(\frac{20[S]/[R]}{20[D]/[R] + 1}\right) = 1 - \left(\frac{20[0.2]/[0.7]}{20[2.5]/[0.7] + 1}\right) = 0.92$$

$$\tau = \frac{1600 \text{ y}}{f \times g} = \frac{1600 \text{ y}}{0.015 \times 0.92} = 115{,}942 \text{ y} \quad\blacksquare$$

What is the significance of the results in Example 10–2? The value calculated for g, 0.92, tells us that 92% of the phosphorus is removed from the surface waters as biogenic particles. The value calculated for f, 0.015, tells us that over 98% of the P contained in the organic particles is returned to the water column by decomposition and dissolution. This is not a surprising result given that phosphorus is contained in the easily decomposed organic portion of the biogenic particle flux. The calculated residence time for P in the ocean is relatively long because it is continually recycled through the oceanic system. Based on these types of calculations, Broecker (1974) identified three groups of elements: *biolimiting elements, which are almost totally depleted in the surface waters; biointermediate elements, which are only partly depleted in the surface waters*; and *biounlimited elements, which show no measurable depletion in the surface waters*. According to this classification scheme, N, P, and Si are biolimiting elements, and Ca, C, Ba, and Ra are biointermediate elements.

Seawater–Sediment Interactions In a later section we will look in some detail at the particle flux in the ocean. For the moment it is sufficient to know that it consists of particulate organic carbon, carbonate and silica shell material, and a variety of silicate minerals (dominantly quartz and clay minerals) derived from the continents. Some of these particles may have oxyhydroxide coatings. Given the pH of seawater, these particles will have negatively charged surfaces and will, to various degrees, adsorb cations. Given the much higher ionic strength of seawater, differences in ion-exchange processes are also important. As you may recall from Chapter 9, in high-ionic-strength solutions monovalent cations preferentially exchange for divalent cations. Thus, in the ocean Na$^+$ will exchange for divalent cations on clay minerals.

Ocean waters at all depths are well oxygenated (Figure 10–11). Thus, the environment at the seawater–sediment interface, and throughout the water column, is oxidizing. However, below the seawater–sediment interface the redox conditions of the environment will change. This is largely determined by the availability of organic matter. In regions of high organic input to the sediments, such as deltas, the pore waters quickly become anoxic. When this occurs, other oxidation reactions become important (Chapter 4). The common oxidation reactions that would occur in a marine sediment are listed in Table 10–3. As each oxidant is exhausted, the next reaction in the list becomes the oxidation reaction. Because these reactions are microbially mediated, they depend on the presence of the appropriate bacteria; i.e., reactions that are mediated by anaerobic bacteria will not occur in the presence of free oxygen.

An important reaction is the reduction of sulfate, an abundant species in seawater, to hydrogen sulfide. If iron or other metals are present in the sediment pore waters, metal–sulfide precipitates are formed. This reaction not only sequesters the metals in the sulfide minerals but also removes sulfate from the ocean.

Seawater–Basalt Interactions New seafloor is created at the midocean ridges by the intrusion and extrusion of basalt magmas derived from the asthenosphere. At the location

Table 10–3 Oxidation of Organic Matter in Marine Sediments*

Oxygenation (oxic): $CH_2O + O_2 \rightarrow CO_2 + H_2O$
Nitrate reduction (mainly anoxic): $5CH_2O + 4NO_3^- \rightarrow 2N_2 + CO_2 + 4HCO_3^- + 3H_2O$
Manganese oxide reduction (mainly anoxic): $CH_2O + 2MnO_2 + 3CO_2 + H_2O \rightarrow 2Mn^{2+} + 4HCO_3^-$
Ferric oxide (hydroxide) reduction (anoxic): $CH_2O + 4Fe(OH)_{3\ (s)} + 7CO_2 \rightarrow 4Fe^{2+} + 8HCO_3^- + 3H_2O$
Sulfate reduction (anoxic): $2CH_2O + SO_4^{2-} \rightarrow H_2S + 2HCO_3^-$
Methane formation (anoxic): $2CH_2O \rightarrow CH_4 + CO_2$

*From Berner and Berner (1996).

of the active basalt magmatism, seafloor temperatures are on the order of 400°C. As the newly formed basalt is rifted away from the midocean ridge, the seafloor cools and ultimately reaches ambient temperatures. In the context of active magmatism and subsequent cooling of the seafloor, two hydrothermal (hot water–rock interaction) regimes can be identified: high temperature (200–400°C) and low temperature (50–200°C). The nature of the seawater–basalt interaction varies as a function of temperature.

The major products of basalt magmatism are the minerals olivine [$(Fe,Mg)_2SiO_4$], pyroxene [$Ca(Mg,Fe)Si_2O_6$ and $(Mg,Fe)SiO_3$], Ca-plagioclase [$CaAl_2Si_2O_8$], and volcanic glass, all of which are chemically unstable in seawater. Figures 10–9 and 10–10 illustrate the types of reactions that occur in high-temperature hydrothermal systems. We have a reasonably good understanding of what happens in these systems both through measurements made on active systems and high-temperature laboratory experiments. In these reactions, Mg^{2+} and SO_4^{2-} are removed from seawater and Ca^{2+}, $H_4SiO_{4\ (aq)}$, and K^+ are added. High-temperature experiments indicate that SO_4^{2-} is originally removed as $CaSO_4$ and subsequently reduced to H_2S. The calcium required for the precipitation of sulfate comes from the basalt. The net result of these various reactions is summarized in Table 10–4. Note that the total Ca^{2+} released from the basalt is the amount of Ca^{2+} in the hot spring waters plus the Ca^{2+} consumed in the precipitation of $CaSO_4$.

Interactions that occur at low temperature are much less certain. This is due both to the difficulty of conducting low-temperature experiments and the absence of seafloor hot springs associated with the low-temperature regime. In total, existing studies suggest that K^+ and Mg^{2+} are taken up during low-temperature reactions and Ca^{2+} and $H_4SiO_{4\ (aq)}$ are released. In sediment sequences containing volcanic glass, with increasing depth K^+ and Mg^{2+} decrease in the pore waters and Ca^{2+} increases. These changes are presumably due to reactions between the interstitial seawater and volcanic glass that form smectite (Gieskes and Lawrence, 1981).

Table 10–4 Concentration Changes for Seawater Species During High-Temperature Seawater–Basalt Interactions*

	Concentrations (mmol L^{-1})		Δ (mmol L^{-1})
	Seawater	Hot springs	
Mg^{2+}	54	0	−54
Ca^{2+}	10	36	26
K^+	10	26	16
SO_4^{2-}	28	0	−18
$H_4SiO_{4\ (aq)}$	~0	20	~20
$\Delta Ca^{2+} - \Delta SO_4^{2-}$			54

*From Berner and Berner (1996).

Summary of Processes Controlling Seawater Composition   The processes that control the concentrations of the major species in seawater are summarized in Table 10–5.

Table 10–5   Processes Affecting the Concentration of Various Species in Seawater in Order of Importance*

Species	Input	Output
Cl^-	River water	Evaporative NaCl deposition
		Net sea–air transfer
		Pore water burial
Na^+	River water	Evaporative NaCl deposition
		Net sea–air transfer
		Cation exchange
		Basalt–seawater reaction
		Pore water burial
SO_4^{2-}	River water	Evaporative $CaSO_4$ deposition
	Polluted rain and dry deposition	Biogenic pyrite formation
		Net sea–air transfer
Mg^{2+}	River water	Volcanic–seawater reaction
		Biogenic Mg–calcite deposition
		Net sea–air transfer
K^+	River water	Low-T volcanic–seawater reaction
	High-T volcanic–seawater reaction	Fixation on clays near river mouths
		Net sea–air transfer
Ca^{2+}	River water	Biogenic $CaCO_3$ deposition
	Volcanic–seawater reaction	Evaporative $CaSO_4$ deposition
	Cation exchange	
HCO_3^-	River water	$CaCO_3$ deposition
	Biogenic pyrite formation	
$H_4SiO_{4\,(aq)}$	River water addition	Biogenic silica deposition
	Basalt–seawater reaction	
P species	River water	Burial of organic P
	Rain and dry deposition	$CaCO_3$ deposition
		Adsorption on volcanogenic ferric oxides
		Phosphorite formation
N species	N_2 fixation	Denitrification
	River water	Burial of organic N
	Rain and dry deposition	

*From Berner and Berner (1996).

SEAWATER CHEMISTRY

The concepts that we have developed in the previous nine chapters are applicable to understanding the chemical processes that take place in seawater. The major difference is the much higher ionic strength of seawater compared to the freshwater environment that we considered in Chapter 9. We can calculate the ionic strength of seawater, as a function of salinity, from the following relationship (Millero, 1982):

$$I = \frac{19.92 \times S\text{‰}}{(1000 - 1.005 \times S\text{‰})} \tag{10–5}$$

where S‰ is the salinity in parts-per-thousand. For a seawater salinity of 35‰, the ionic strength is 0.723 moles of charge kg^{-1}. When we are dealing with high-ionic-strength solutions, a distinction is made between the stoichiometric ionic strength and the effective ionic strength. What we have calculated using equation 10–5 is the stoichiometric ionic strength. Effective ionic strength is less due to the formation of aqueous species that remove charged species from solution. In the case of freshwater, the difference between the two ionic strength calculations is less than 1%. In seawater, at 35‰, the difference is on the order of 7%.

In terms of themodynamic calculations, this higher ionic strength necessitates the use of appropriate activity-coefficient models (Truesdell–Jones, Pitzer) or the use of stoichiometric equilibrium constants (experimentally determined as a function of salinity) in which we directly use the measured concentrations of the species in solution. Ion-exchange processes are also affected by the higher ionic strength, as noted previously. Other differences involve the equilibria that control ocean pH, alkalinity, adsorption processes, and the abundance of both inorganic and organic ligands that form metal complexes.

Seawater Alkalinity and Dissolved Inorganic Carbon (DIC)

We initially developed the idea of alkalinity and pH in Chapter 3. Up to now, we have essentially dealt with alkalinity in the form developed in Chapter 3 because we were dealing with dilute solutions. In seawater, however, we have many more ionic species that are contributors to alkalinity. Unfortunately, as we shall see in the ensuing paragraphs, different investigators have defined alkalinity (and pH) in different ways. This leads to a certain amount of confusion in the literature. We will not attempt to sort out all these differences here. Because of our concern with the buildup of CO_2 in the atmosphere, and the observation that the oceans are a major sink for CO_2, a great deal of effort has been devoted to understanding the relationships among the fugacity of CO_2, pH, dissolved inorganic carbon (DIC), and total alkalinity (TA).

The following definitions have been used for total seawater alkalinity. Dickson (1981):

$$TA = [HCO_3^-] + 2[CO_3^{2-}] + [B(OH)_4^-] + [OH^-] + 2[HPO_4^{2-}] + 3[PO_4^{3-}]$$
$$+ [H_3SiO_4^-] + [HS^-] + 2[S^{2-}] + [NH_3] - [H^+] - [HSO_4^-]$$
$$- [HF] - [H_3PO_4] \tag{10–6}$$

Takahashi et al. (1982):

$$TA = [HCO_3^-] + 2[CO_3^{2-}] + [H_2BO_3^-] \tag{10–7}$$

and Peng et al. (1987):

$$TA = [HCO_3^-] + 2[CO_3^{2-}] + [H_2BO_3^-] + [H_3SiO_4^-] + [H_2PO_4^-]$$
$$+ 2[HPO_4^{2-}] + 3[PO_4^{3-}] + [OH^-] \tag{10–8}$$

Of the species included in the alkalinity calculations, boron, sulfate, and fluoride are conservative species; i.e., they exist in constant ratio to the other major species in seawater. Hence, their concentrations can be determined directly from the measured seawater salinity using the following relationships (from Millero, 1982). The calculated concentrations are in moles kg^{-1}.

$$B_T = 0.000416 \ (S‰/35) \tag{10–9}$$

$$SO_4^{2-} = 0.0293 \ (S‰/35) \tag{10–10}$$

$$F^- = 0.00007 \ (S‰/35) \tag{10–11}$$

The concentrations of $B(OH)_4^-$, HSO_4^-, and HF can then be calculated from the appropriate equilibria relationships. The other species must be measured directly or calculated from various equilibria equations.

An important point emphasized by equation 10–6 is that in seawater a number of species contribute to the total alkalinity. We can, however, gain a first-order understanding of changes in alkalinity and DIC by limiting our analysis to HCO_3^- and CO_3^{2-}. This is because variations in the concentrations of these two species exert the major control on variations in ocean alkalinity. The boron species are also important, but ignoring them does not introduce a significant error into our analysis.

Recall that a fundamental tenet of solution chemistry is charge balance. The major cations and anions of the ocean exist in constant proportion to each other. If we sum up the total positive and negative charges (per kg of seawater) for the major species, we will find that there is a slight positive charge imbalance. This must be balanced by negative charges derived essentially from the dissociation of carbonic acid. The other negatively charged species, with the exception of the boron species, are present at concentrations orders of magnitude less than those of the carbonate species. The key here is the relative abundance of the HCO_3^- species (with one negative charge) and the CO_3^{2-} species (with two negative charges). The excess positive charge is balanced by changes in the relative abundance of the two carbonate species. At seawater pHs, boron can also exist as two species, H_3BO_3 and $B(OH)_4^-$, which is why a complete analysis of alkalinity variations requires the inclusion of boron.

Let us write several simple equations defining the alkalinity of the ocean in terms of carbonate species. The total dissolved carbon in seawater is

$$\Sigma CO_2 = HCO_3^- + CO_3^{2-} = DIC \qquad (10\text{-}12)$$

Recall from Chapter 3 that for the range of pH values encountered in the ocean, the abundance of $H_2CO_3{}_{(aq)}$ is orders of magnitude less than that of the other carbonate species. We can write carbonate alkalinity as

$$CA = HCO_3^- + 2CO_3^{2-} \qquad (10\text{-}13)$$

If the carbonate alkalinity (CA) remains constant, changes in total dissolved CO_2 (DIC) lead to changes in the relative abundances of the bicarbonate and carbonate species (Figure 10–13). Remember, charge balance must be preserved. Changes in DIC are caused by exchange with the atmosphere and photosynthesis and respiration. Photosynthesis lowers the total DIC of surface waters. Because CA remains the same, this decline in total DIC is matched by an increase in the proportion of carbonate relative to bicarbonate species. Respiration (and organic decay) releases carbon to the water. In order to maintain charge balance, there must be a decrease in the proportion of carbonate to bicarbonate species.

The precipitation and dissolution of carbonate shell material changes both the calcium ion and total DIC of seawater (Figure 10–13). During precipitation of carbonate shell material, calcium and carbon are removed in a 1:1 molar ratio ($CaCO_3$). But the removal of calcium, which has two positive charges, changes carbonate alkalinity (CA) by a factor of 2 with respect to DIC; i.e., for every mole of carbon removed from the ocean, 2 moles of positive charge are removed. During dissolution, CA increases at twice the rate of DIC.

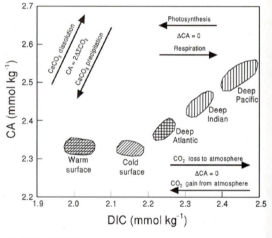

Figure 10–13
Carbonate alkalinity (CA) versus dissolved inorganic carbon (DIC). Arrows indicate changes in CA and DIC in response to various biological and chemical processes. Modified from Broecker (1974).

Various water masses have different carbonate alkalinity and dissolved inorganic carbon concentrations (Figure 10–13), which reflect the history of the water mass. Cold surface waters have a higher DIC than warm surface waters because the solubility of CO_2 in water increases with decreasing temperature. However, the alkalinity of the surface waters is essentially the same. What this means is that cold surface waters must contain relatively less carbonate ion than warm surface waters. Earlier we discussed the general vertical circulation pattern of the ocean. A simplified version of this circulation has cold waters formed in the North Atlantic Ocean moving through the Atlantic Ocean into the Indian Ocean and then into the Pacific Ocean, with surface returns occurring at sites of regional upwelling. Organic material and carbonate shell material from the surface break down in the deep waters, releasing both carbon and calcium ions to solution. This release leads to an increase in both carbonate alkalinity and total dissolved carbon. The deep Pacific waters have the longest travel path in the oceans, so they have the highest CA and DIC values. Also note that the carbonate ion concentration decreases as we move from warm surface waters to deep Pacific waters. We will return to this point in a later section when we discuss the dissolution of calcium carbonate shell material in the ocean.

EXAMPLE 10–3 Calculate the amount of carbonate ion in warm surface water and Pacific deep water.

From Figure 10–13, we estimate for average surface water, CA = 2.35 mmol kg^{-1} and DIC = 2.0 mmol kg^{-1}, and for Pacific deep water, CA = 2.55 mmol kg^{-1} and DIC = 2.45 mmol kg^{-1}. We do this problem by writing two equations, one for CA and the other for DIC, and then simultaneously solving the equations. For the surface waters,

$$CA = HCO_3^- + 2CO_3^{2-} = 2.35 \text{ mmol kg}^{-1}$$

and

$$DIC = HCO_3^- + CO_3^{2-} = 2.0 \text{ mmol kg}^{-1}$$

Subtracting DIC from CA gives CO_3^{2-} = 0.35 mmol kg^{-1}. Doing the same calculation for the Pacific deep waters gives CO_3^{2-} = 0.10 mmol kg^{-1}. The Pacific deep water has a higher total DIC than the surface waters but less carbonate ion. ∎

Seawater pH

Several different pH scales have been used in marine chemistry. The two commonly used are the total pH scale and the seawater pH scale. For total pH,

$$pH_T = -\log[H^+]_T = -\log\{[H^+] + [HSO_4^-]\} \tag{10–14}$$

and for seawater pH,

$$pH_{sws} = -\log[H^+]_{sws} = -\log\{[H^+] + [HSO_4^-] + [F^-]\} \tag{10–15}$$

The student will find the literature confusing when it comes to ocean pH scales because authors do not always specify which scale they are using. The actual differences in pH values are relatively small; e.g., at 20°C and salinity of 35‰ the pH values on the total scale are 0.01 units higher than the pH values on the seawater scale. Because most of the equilibrium constants are determined by experiment using stoichiometric concentrations, the student does need to be aware of which pH definition was used when the equilibrium constants were calculated from the experimental results.

pH changes with depth are shown for the typical ocean in Figure 10–14. Carbon dioxide is depleted in the surface waters because of photosynthesis. Consumption of CO_2 leads to an increase in pH (see Chapter 3). Below the euphotic zone, respiration dominates and CO_2 is returned to the water column, thus lowering the pH. The pH decrease continues through the region of the main thermocline, where particulate organic matter derived from the surface waters is oxidized by microbial activity, releasing carbon and nutrient elements

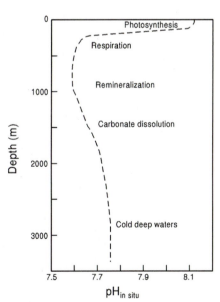

Figure 10–14
Depth versus pH for the average open ocean. See text for discussion.

to the water column (a process referred to as ***remineralization***). Subsequent reaction with calcium carbonate (aragonite or calcite) particles in the water column consumes some of the hydrogen ions, with a concomitant rise in pH and DIC. In the deep waters the pH is similar to that of surface waters at high latitudes. This pH is less than that found in the surface waters because the solubility of CO_2 is greater in the cold high-latitude waters than in the warm surface waters (see Figure 10–13).

Speciation and Stoichiometric Equilibrium Constants

Consider the following speciation reaction for boron:

$$H_3BO_{3\,(aq)} + H_2O \rightleftharpoons H^+ + B(OH)_4^-$$

The equilibrium equation is

$$K_B = \frac{[H^+][B(OH)_4^-]}{[H_3BO_{3\,(aq)}]} \qquad (10\text{–}16)$$

Here we are interested in the concentration of the various species in seawater as a function of pH. The equation for total boron concentration is

$$B_T = H_3BO_{3\,(aq)} + B(OH)_4^- \qquad (10\text{–}17)$$

Solving equation 10–16 for $B(OH)_4^-$ and substituting into equation 10–17 gives

$$B_T = [H_3BO_{3\,(aq)}]\left(1 + \frac{K_B}{[H^+]}\right) \qquad (10\text{–}18)$$

As we did in Chapter 3, we define a variable called α_H as

$$\alpha_H = \left(1 + \frac{K_B}{[H^+]}\right) \qquad (10\text{–}19)$$

We can then solve for both H_3BO_3 and $B(OH)_4^-$ in terms of total boron and pH. The relevant equations are

$$[H_3BO_{3\,(aq)}] = \frac{B_T}{\alpha_H} \qquad (10\text{–}20)$$

and

$$[B(OH)_4^-] = \frac{B_T K_B}{[H^+]\alpha_H} \qquad (10\text{–}21)$$

For any given salinity, B_T can be calculated from equation 10–9.

As noted in previous sections, we can deal with equilibrium reactions in the ocean either by calculating activity coefficients using appropriate models for high-ionic-strength solutions or by experimentally determining equilibrium constants as a function of salinity, temperature, and pressure. The latter approach is commonly used for seawater thermodynamic calculations. For the dissociation of boric acid, using the total pH scale, the equilibrium constant is calculated from the following equation (Millero, 1995):

$$\begin{aligned}\ln K_B = &(-8966.90 - 2890.51S^{0.5} - 77.942S + 1.726S^{1.5} - 0.0993S^2)/T \\ &+ (148.0248 + 137.194S^{0.5} + 1.62247S) \\ &+ (-24.4344 - 25.085S^{0.5} - 0.2474S)\ln T + 0.053105S^{0.5}T \qquad (10\text{–}22)\end{aligned}$$

(don't you just love empirical equations) where S is salinity in parts-per-thousand and T is temperature in K. Note that the result is $\ln K_B$. We need to divide by 2.303 to convert the natural log result to log base 10. Solving equation 10–22 for salinity = 35‰ and $T = 20°C$ gives $\log K_B = -8.639$. At this salinity, $B_T = 0.000416$ mol L^{-1} (equation 10–9). Solving equations 10–20 and 10–21 over a range of pH values gives the speciation of boron as a function of pH (Figure 10–15). The concentration of the two species is equal at $pH_T = 8.64$. Within the normal range of seawater pH values, $H_3BO_{3 (aq)}$ will be the dominant species. In subsequent sections, and the problem set, we will investigate other equilibrium reactions that are of interest in marine chemistry.

As a rough rule-of-thumb, for every 10 m increase in depth in the ocean, pressure increases by 1 bar. Given that ocean depths can exceed 5000 m, stoichiometric equilibrium constants may also need to be corrected for pressure when dealing with the marine system. This is done empirically by fitting experimental results

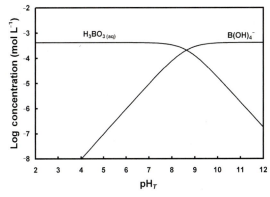

Figure 10–15

Concentration of boron species as a function of pH at 20°C and salinity = 35‰.

to appropriate equations and extracting the requisite constants. The effect of pressure is represented by the following equation:

$$\ln\left(\frac{K_i^P}{K_i^0}\right) = -\left(\frac{\Delta V_i}{RT}\right)P + \left(\frac{0.5\Delta k_i}{RT}\right)P^2 \tag{10–23}$$

where K_i^0 is the equilibrium constant determined on the basis of temperature and salinity, K_i^P is the equilibrium constant corrected for pressure, ΔV_i is the molal volume change, Δk_i is the compressibility change, R is the gas constant (83.145 bar-cm³ mol⁻¹ K⁻¹), P is the pressure in bars, and T is the temperature in K. The values for ΔV_i and Δk_i are calculated as follows:

$$\Delta V_i = a_0 + a_1 t + a_2 t^2 \tag{10–24}$$

and

$$\Delta k_i = b_0 + b_1 t + b_2 t^2 \tag{10–25}$$

where t is in °C and the appropriate constants are taken from Table 10–6. Values for b_2 are normally zero and are not shown in the table.

EXAMPLE 10–4 Calculate the boric acid dissociation constant at a depth of 1000 m ($P = 100$ bar). Seawater temperature is 5°C and the salinity is 35‰. At this temperature and salinity, $\log K_B = -8.827$.

Using the data from Table 10–6 on page 406,

$$\Delta V_i = a_0 + a_1 t + a_2 t^2 = -29.48 + (0.1622)(5) + (0.002608)(5)^2 = -28.604$$

$$\Delta k_i = b_0 + b_1 t + b_2 t^2 = -0.00284 + (0)(5) + (0)(5)^2 = -0.00284$$

$$\ln\left(\frac{K_i^P}{K_i^0}\right) = -\left(\frac{\Delta V_i}{RT}\right)P + \left(\frac{0.5\Delta k_i}{RT}\right)P^2$$

$$= -\left(\frac{-28.604}{(83.145)(278.15)}\right)(100) + \left(\frac{(0.5)(-0.00284)}{(83.145)(278.15)}\right)(100)^2 = 0.1231$$

$$\log\left(\frac{K_i^P}{K_i^0}\right) = \left(\frac{0.1231}{2.303}\right) = 0.0535$$

$$\log K_i^P = \log K_i^0 + 0.0535 = -8.827 + 0.0535 = -8.774$$

The dissociation constant has increased by approximately 13%. ∎

Table 10–6   Coefficients Used to Correct for the Effect of Pressure on Dissociation Constants of Acids and the Solubility Products of Carbonate Solids*

Compound	a_0	a_1	$a_2 (10^{-3})$	$b_0 (10^{-3})$	$b_1 (10^{-3})$
$H_2CO_{3 (aq)}$	−25.50	0.1271		−3.08	0.0877
HCO_3^-	−15.82	−0.0219		1.13	−0.1475
$B(OH)_{3 (aq)}$	−29.48	0.1622	2.608	−2.84	
H_2O	−25.60	0.2324	−3.6246	−5.13	0.0794
HSO_4^-	−18.03	0.0466	0.316	−4.53	0.0900
$HF_{(aq)}$	−9.78	−0.0090	−0.942	−3.91	0.054
$H_2S_{(aq)}$	−14.80	0.0020	−0.400	2.89	0.054
NH_4^+	−26.43	0.0889	−0.905	−5.03	0.0814
$H_3PO_{4 (aq)}$	−14.51	0.1211	−0.321	−2.67	0.0427
$H_2PO_4^-$	−23.12	0.1758	−2.647	−5.15	0.09
HPO_4^{2-}	−26.57	0.2020	−3.042	−4.08	0.0714
$CaCO_3$ calcite	−48.76	0.5304		−11.76	0.3692
$CaCO_3$ aragonite	−45.96	0.5304		−11.76	0.3692

*From Millero (1995) as modified by Lewis and Wallace (1998).

Seawater Buffers and Buffering Capacity

The concept of buffers and buffering capacity was developed in Chapter 3. Included in our earlier discussion was the carbonate system, which plays a key role in controlling freshwater pH. Here we will look specifically at the marine system, which is similar to the freshwater system with some modifications. In the ocean we have two important buffer systems: carbonate and borate. Analogous to what we did in Chapter 3, for the buffer system water–boric acid–carbonic acid, we can write the following buffering equation:

$$B = 2.3\left[\frac{K_B B_T(H^+)}{(K_B + H^+)^2} + \frac{K_{C_1} C_T(H^+)}{(K_{C_1} + H^+)^2} + \frac{K_{C_2} C_T(H^+)}{(K_{C_2} + H^+)^2} + \frac{K_w}{H^+} + H^+\right] \quad (10\text{–}26)$$

The first term describes the boric acid system, the second and third terms the carbonate system, and the last two terms the water system. Here we are using stoichiometric equilibrium constants so they must be calculated as described earlier. We have already written the equation (10–22) for the dissociation of boric acid. For the dissociation of carbonic acid, using the seawater pH scale, the equations (Millero, 1995) are

$$\ln K_{C_1} = 2.83655 - 2307.1266/T - 1.5529413 \ln T + (-0.20760841 - 4.0484/T)S^{0.5}$$
$$+ 0.0846834S - 0.00654208S^{1.5} \quad (10\text{–}27)$$

and

$$\ln K_{C_2} = -9.226508 - 3351.6106/T - 0.2005743 \ln T$$
$$+ (-0.106901773 - 23.9722/T)S^{0.5} + 0.1130822S$$
$$- 0.00846934S^{1.5} \quad (10\text{–}28)$$

The variation in buffer capacity as a function of pH for the various seawater buffers is shown in Figure 10–16. The calculations were done at 20°C and S‰ = 35, appropriate values for surface waters. For warm surface waters total carbonate is approximately 2 mmol kg^{-1}. For these conditions, pK_{C_1} = 5.89 and pK_{C_2} = 8.99. You might want to compare these dissociation constants with those calculated in Chapter 3 for the freshwater system. Note that the total pH scale was used for the boric acid equilibrium calculation and the seawater pH scale was used for the carbonic acid equilibrium calculations. Given the conditions at which these calculations were done, the difference between the two scales is insignificant, and the results are plotted without correction in Figure 10–16.

The maximum buffering capacity for both the boric acid and carbonic acid systems falls outside the normal range, from about 7.5 to 8.2, of seawater pH values. Within the

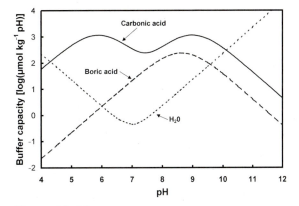

Figure 10–16
Variations in buffer capacity as a function of
pH for boric acid and carbonic acid in seawater.

normal range of pH values, buffering capacity decreases as the pH decreases
(Figure 10–17). Thus, on a short-term
basis seawater is not a well-buffered solution. This is why changes in CO_2 content due to photosynthesis and respiration
(plus decay, which releases carbon to the
water column) can lead to fairly significant pH variations. However, as we noted
in Chapter 3, on a long-term basis seawater pH is maintained within a relatively
narrow range due to reactions between
carbonate solids (calcite and aragonite)
and seawater (see problem set).

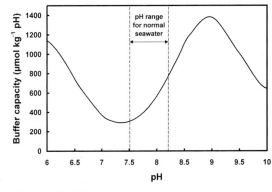

Figure 10–17
Variation in total buffering capacity for ocean
surface water as a function of pH.

Because of the interrelationships
among dissolved CO_2, pH, dissolved inorganic carbon, and total alkalinity, it is possible
to calculate the amount of dissolved CO_2 using any of the other two parameters (Kitack et
al., 1996; Lueker et al., 2000). Zhang (2000) has recently suggested that pH and buffer intensity can be used to quantify the carbon cycle in the ocean. The interested student should
consult these papers for further information. Lewis and Wallace (1998) have developed a
computer code (DOS-based, not a warm and cuddly GUI, but it's good to broaden one's
horizons) to do these sorts of calculations. The code can be downloaded from
http://cdiac.esd.ornl.gov/oceans/co2rprt.html. The student should also print, and study,
the accompanying documentation before using the code. The documentation is very informative and describes some of the pitfalls in doing these types of calculations.

Case Study 10–1 (p. 408) describes how changes in nutrient element concentrations,
DIC, and pH can be used to quantify the remineralization of organic matter.

Inorganic and Organic Ligands in Seawater

Both inorganic and organic ligands occur in the ocean. Depending on the availability of
the different types of ligands, metals will be bound to either or both types of ligands. In
general, organic ligands preferentially bind metals with respect to inorganic ligands.
Organic ligands, in particular, may control the concentration of free ions of bioactive metals. Some studies have suggested that plants secrete these ligands in order to control their
chemical microenvironment. In addition, as was the case for inorganic complexes discussed in Chapter 2, both inorganic and organic ligands may increase the solubility of various metals in the ocean.

CASE STUDY 10–1
Remineralization of Organic Matter in Antarctic Intermediate Waters

Zhang (2000) used changes in nutrient element concentrations, DIC, and pH to quantify the remineralization of organic matter in the Antarctic Intermediate Waters. Organic matter produced in the euphotic zone is oxidzed as it sinks through the water column, utilizing oxygen and releasing carbon and nutrients to the water column. The remineralization reaction, normalized to phosphorus, can be written

$$(CH_2O)_b(NH_3)_c(H_3PO_4) + aO_2 \rightarrow bHCO_3^-$$
$$+ H_2PO_4^- + cNO_3^- + (b + c + 1)H^+ \quad (C10–1a)$$

where the coefficients a, b, and c are, respectively, the remineralization ratios for oxygen, organic carbon, and nitrogen with respect to phosphorus. During the remineralization process, inorganic carbon, phosphorus, nitrogen, and hydrogen ions are released to solution. The release of hydrogen ions leads to a decrease in pH. Some of the hydrogen ions react with particulate calcium carbonate, a reaction that consumes hydrogen ions and releases calcium ion and DIC to seawater.

$$CaCO_{3\ carbonate} + H^+ \rightarrow Ca^{2+} + HCO_3^- \quad (C10–1b)$$

From the stoichiometry of reactions C10–1a and C10–1b, we can deduce the following relationships. The change in phosphate concentration is directly proportional to the amount of organic matter that has been oxidized and

$$\Delta H_2PO_4^- = x \quad (C10–1c)$$

where x is the moles of phosphate released to the seawater. The amount of dissolved oxygen consumed during the process is

$$-\Delta O_2 = ax \quad (C10–1d)$$

where a is a stoichiometry coefficient that represents the number of moles of oxygen needed to oxidize 1 mole of organic matter. For nitrate,

$$\Delta NO_3^- = cx \quad (C10–1e)$$

where c is the stoichiometry coefficient for nitrate. Plots of abundance of phosphate versus apparent oxygen utilization (the difference between the saturation oxygen content and the measured oxygen content) and nitrate yield the coefficients a and c (slopes of straight lines), respectively. Changes in bicarbonate ion due to oxidation of organic matter and dissolution of particulate carbonate material can be written

$$\Delta HCO_3^- = bx + y \quad (C10–1f)$$

where y is the number of moles of particulate carbonate dissolved during the remineralization process. We can also write the following equation, which relates changes in hydrogen and bicarbonate ions:

$$\Delta(HCO_3^- + C_H) = (2b + c + 1)x \quad (C10–1g)$$

The change in hydrogen ion concentration is determined from the change in buffer capacity. A plot of $\Delta(HCO_3^- + C_H)$ versus $\Delta H_2PO_4^-$ gives a straight line with the slope $(2b + c + 1)$. Given a value for c from the nitrate equation (C10–1e), it is possible to calculate a value for b. Finally, one can obtain y from equation C10–1f from the change in DIC (HCO_3^-) during the remineralization process. The relative contribution of organic matter oxidation (OC) versus particulate carbonate dissolution (IC) to the observed DIC change can be estimated from the following equation:

$$IC/OC = y/bx \quad (C10–1h)$$

For Antarctic Intermediate Waters, in a south to north direction, dissolved oxygen decreased from 272 to 80 $\mu mol\ kg^{-1}$, phosphate increased from 1.6 to 2.7 $\mu mol\ kg^{-1}$, and nitrate increased from 23.4 to 39.1 $\mu mol\ kg^{-1}$. The total change in DIC was 150 $\mu mol\ kg^{-1}$. From these data, $a = 173$, $c = 14.3$, b (calculated from C10–1g) = 107, and y (calculated from C10–1f) = 32.3. Solving equation C10–1h, we find that the dissolution of particulate carbonate accounted for 21.5% of the total DIC released during the remineralization process.

Source: Zhang (2000).

Inorganic Ligands For metals, the most important inorganic ligand is HS^-, which can form complexes with metal ions. The dissociation of H_2S is described by the following two reactions:

$$H_2S_{(aq)} \rightleftharpoons H^+ + HS^- \quad and \quad HS^- \rightleftharpoons H^+ + S^{2-}$$

For the first dissociation step, at $T = 25°C$ and $S = 35‰$, $\log K_1 = -6.50$ (calculated from the equation in Millero, 1995; see problem set). The second dissociation constant is approximately $\log K_2 = -13.6$, and hence the abundance of S^{2-} will be orders of magnitude less than that of the other sulfur species. For total sulfur as H_2S, the relative amount of HS^- can be calculated from the following equation:

$$[HS^-] = [H_2S]_T / \left(1 + \frac{[H^+]}{K_1}\right) \quad (10–29)$$

At pH = 8, 96.9% of the total sulfur occurs as HS^-, and this is the important ligand for the complexation of metal ions.

We can write the complex formation reaction as

$$M^{n+} + HS^- \rightleftharpoons MHS^{(n-1)+}$$

where M^{n+} is the metal ion. For this reaction, the conditional stability constant (uncorrected for side reactions) is

$$K' = \frac{[MHS^{(n-1)+}]}{[M^{n+}][HS^-]} \tag{10–30}$$

Experimental values determined for various complexation reactions, at $T = 25°C$ and $S = 35‰$, are tabulated in Table 10–7.

Note that these are very large positive stability constants. What this means is that the complex is strongly favored over the ionic species in solution. Oxygenated seawater contains sulfide concentrations in the range 0.1 to 2 nM kg^{-1}. If we added 1 nM of Cu^{2+} to 1 L of seawater containing 1 nM of HS^-, essentially all the HS^- would be combined with Cu^{2+} in the $CuHS^+$ complex. If we had more HS^- than Cu^{2+}, then some HS^- would remain in solution. Conversely, if $HS^- < Cu^{2+}$, then some of the copper would remain in solution as Cu^{2+} (free metal). With reference to Table 10–7, we see that the $CuHS^+$ complex has the largest stability constant. If we had a multielement mixture, the HS^- would first combine with Cu^{2+}, and any excess HS^- would form a complex with Cd^{2+}. As long as HS^- remained in the solution, complexes with metal ions would be formed in the order indicated in the table. While the total concentration of HS^- in oxygenated seawater is low, at the oxic/anoxic boundary in seafloor sediments (or in a stratified water column that is anoxic at depth) we would expect a significant increase in HS^-, and metal sulfide complexation would be significant. If organic ligands are present, they will compete with HS^- ligands for metal ions, thus reducing the importance of the HS^- ligands.

Organic Ligands   Organic ligands are associated with both dissolved organic carbon (DOC) and colloidal organic carbon (COC). As we noted in Chapter 9, the distinction between colloids and dissolved material is not sharp. Different authors use different criteria, but one commonly used criterion is that COC is $<0.2\mu m$ in diameter and has a mass >1000 Da (recall from Chapter 5 that the Dalton is a unit of mass defined as $^{12}C = 12.0000$ Da). It follows that DOC has a mass <1000 Da. Note that this distinction is different from that previously used, i.e., that DOC was any material that passed through a 0.45-μm filter. The role of organic ligands, in terms of bioactive metals (those used by organisms), is still uncertain. One possibility is that the ligands act as buffers, maintaining the amount of free metal in solution (not bound to a ligand) at a constant value. This regulation is presumably mediated by biological processes because the bioactive metals are utilized by organisms. The other possibility is that the formation of ligands may lead to the removal of bioactive metals from the ocean by coagulation into larger particles that are removed by sedimentation.

Table 10–7   Stability Constants for Various Metal Sulfide Complexes*

Metal ion	log K'
Cu^{2+}	11.50
Cd^{2+}	6.83
Pb^{2+}	6.44
Co^{2+}	6.13
Ni^{2+}	4.69
Zn^{2+}	4.69

*From Al-Farawati and van den Berg (1999).

The number of ligands and their stability constants are determined by experiment. For any particular metal, one or more organic ligands may form complexes. The experiments tell us the number of ligands and their conditional stability constants and abundances. The ligands are normally arranged in order of decreasing stability. As an example, Muller (1999) found that copper in a water sample from the Arran Deep, Scotland, was bound to three types of ligands. The stability constant and abundance for each ligand were L1 ($\log K_1 = 13.92$, $C_{L1} = 3.35$ nmol kg^{-1}), L2 ($\log K_2 = 11.28$, $C_{L2} = 7.9$ nmol kg^{-1}), and L3 ($\log K_3 = 9.03$, $C_{L3} = 60.60$ nmol kg^{-1}). Note that the stability constant for the copper–L1 ligand complex is over two orders of magnitude greater than that for the CuHS$^-$ complex (Table 10–7). Hence, copper would preferentially bind with the organic ligand. Other elements, such as Cd, usually bind with only one organic ligand. Experiments also indicate that organic ligands are metal-specific. This poses a potentially significant problem when using radioactive tracers to measure the rate of removal of material from the surface reservoir. If the radioactive tracer does not behave in the same way as the metal of interest (which it most likely will not), then removal rates based on the radioactive tracer will not be applicable to the metal of interest.

The organic ligands are distributed between DOC and COC, and the relative distribution of the ligands is important in answering the question of removal of bioactive metals. If the ligands are concentrated in the COC, then metal–ligand complexes may represent an effective way to remove bioactive elements from surface waters. In a study of Cu, Zn, Cd, and Pb complexing ligands in Narragansett Bay, Rhode Island, Wells et al. (1998) found that the Pb-complexing organic ligands were approximately equally distributed between the DOC and COC phases, while for the other metals the ligands were concentrated in the DOC. Case Study 10–2 describes the role of organic ligands in controlling the abundance of free Zn in the Atlantic Ocean.

Trace Metals in Marine Colloids

Significant, but variable, amounts of dissolved trace metals in seawater are associated with colloidal organic material (COM). Hence, COM may play a significant role in the distribution and removal of trace metals. What are the source(s) of trace metals in colloids? We can think of several possibilities—terrestrial microparticles, marine plants, and humic substances. Similar to what we did in Chapter 8, we can compare elemental abundance ratios in possible sources (Table 10–8) to observed colloidal abundance ratios. A commonly used element for this type of comparison is aluminum, which is assumed to have only a terrestrial source. In most cases, these comparisons reveal that colloids have largely, or exclusively, a humic substance or marine plant origin.

Table 10–8 *Average Metal/Aluminum (Me/Al) Atomic Ratios for Various Colloidal Sources**

Element	Average soils	Upper crust	Humic substance	Marine plants
Cu	0.00018	0.000078	0.06–0.29	0.078
Pb	0.000033	0.000029	0.001–0.003	0.018
Zn	0.00052	0.00028	0.014–0.14	1.04
Cd	0.000001	< 0.000001	0.00023–0.0004	0.008
Co	0.000058	0.000069	—	0.005
Ni	0.00032	0.00011	0.019	0.023
Cr	0.00051	0.00023	0.03	0.009
Fe	0.27	0.19	7.1	5.6
Mn	0.0038	0.0033	—	0.43
V	0.00067	0.00036	—	0.017
Ba	0.00138	0.0016	0.00012	0.098
Ti	0.039	0.022	0.033	0.43

*From Guo et al. (2000).

CASE STUDY 10–2
Zinc Speciation in the Northeastern Atlantic Ocean

Ellwood and Van den Berg (2000) used cathodic stripping voltammetry with ligand competition to determine the speciation of zinc in the northeastern part of the Atlantic Ocean. Their analytical results for water samples from *Challenger* station 13 are given in the following table and shown graphically in Figure 10–C2–1. Also listed in the table are concentrations for Si, a biologically utilized element.

Zn_T is the total amount of zinc in the water sample, and Zn^{2+} represents the free zinc, e.g., zinc not bound to natural organic ligands. The log of the conditional stability constant varies between 10.1 and 10.6, indicating that the Zn–organic-ligand complex is strongly favored. There is no systematic change in the conditional stability constant with depth, and the mean stability constant is 10.3 ± 0.2. Comparison of total zinc to Zn^{2+} (free zinc) reveals that in the surface waters 96 to 99% of the zinc is bound to an organic ligand and thus not biologically available. However, the authors concluded that the free zinc concentrations were not low enough to inhibit the growth of typical oceanic species of phytoplankton. In the surface waters the ligand concentration was significantly greater than the Zn_T concentration, and in the deep waters the two concentrations were approximately equal. Thus, the availability of ligands does not limit the formation of zinc complexes.

The authors considered the source of the zinc-binding organic ligands. Some phytoplankton produce ligands in order to reduce the concentration of toxic free-metal ions in their environment. The authors concluded that this was not the source of the Zn-binding ligands because there was no correlation between phytoplankton abundance and ligand abundance. In part

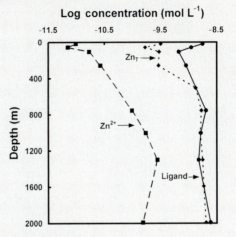

Figure 10–C2–1
Depth versus variations in the abundance of Zn^{2+} (free zinc), Zn_T, and ligands in the northeastern Atlantic Ocean. Note that in the surface waters the abundance of the ligands significantly exceeds that of total zinc and in the deep waters the abundances of ligands and total zinc are approximately equal.

because of the relatively constant concentration of zinc-binding ligands in the deep waters, the authors suggested that the ligands may be produced by the breakdown of organisms sinking out of the euphotic zone. One possibility is polypeptide molecules produced during the decay of phytoplankton.

Source: Ellwood and Van den Berg (2000).

Depth (m)	Si (μmol L^{-1})	Zn_T (nmol L^{-1})	Zn^{2+} (pmol L^{-1})	Ligand (nmol L^{-1})	Log $K_{Zn\,L}$
18	0.5	0.32	9.8	1.79	10.3
58	2.3	0.17	7.0	1.13	10.4
107	2.2	0.30	17	0.68	10.6
258	4.7	0.29	27	0.91	10.1
506	7.5	1.34			
754	8.5	1.68	97	2.04	10.4
1003	9.0	1.61	173	1.66	10.2
1299	9.8	1.77	285	1.53	10.1
1592	11.5	1.87			
1991	11.6	2.03	154	2.44	10.2

Because organic carbon (OC) is a significant component of marine colloids, we are interested in the relationship between the abundance of metals and organic carbon. If the metals are bound to organic ligands, we would expect to see a correlation between metal concentration and OC. On the other hand, if organic ligands do not play an important role in the metal concentration of the colloid (adsorption is another possibility), then we would expect to see poor or little correlation between metal and OC abundances. We do this analysis in two steps. First, we determine the correlation between the metal concentration and carbon content. If the metal concentration and carbon content are highly correlated, then we assume that organic ligands are important. Second, we calculate the metal/carbon (Me/C) ratio. For elements whose concentrations are controlled by binding to organic ligands, there is a direct correlation between this ratio and the affinity of the metal for the ligand (often referred to as the Irving–Williams order). For humic substances, the Irving–Williams order is Mg < Ca < Cd, Mn < Co < Zn, Ni < Cu < Hg (Mantoura et al., 1978).

EXAMPLE 10–5 Guo et al. (2000) determined the following Me/C atomic ratios ($\times 10^5$) for colloids from Galveston Bay, Texas:

Element	Cu	Pb	Zn	Cd	Co	Ni	Cr	Mn
Ratio	4.9	0.016	2.0	0.014	0.28	2.9	2.7	0.62

From these data the Irving–Williams order for the metals in the Galveston Bay colloids is Cd, Pb < Co < Mn < Zn < Cr, Ni < Cu. Note that this is similar to the order found for humic substances. ∎

There are also significant differences in Me/C ratios for various colloidal sources. These differences may be useful in distinguishing COM composed of humic substances from COM composed of the degradation products of marine plants. The approach is the same as that used for the terrestrial component of a colloid, except that here we are looking at the organic component. Representative ratios are given in Table 10–9.

Trace Metals in Seawater

Virtually every element in the periodic table has been detected in seawater, although some of these elements occur at exceptionally low concentrations. Those elements that occur at concentrations of less than 0.05 μmol kg^{-1} are referred to as *trace elements* (most of which are metals). Many of the trace metals have significant (besides natural) anthropogenic sources and, thus, are of environmental interest.

Sources of Trace Metals   There are a number of sources, both natural and anthropogenic, for trace metals.

Table 10–9   Me/C Atomic Ratios ($\times 10^5$) for Continental Crust, Humic Substances, and Marine Plankton*

Element	Continental crust	Humic substances	Marine plants
Cu	13	10	0.61
Pb	4.7	0.19	0.14
Zn	45	2.9	8.1
Cd	0.05	0.034	0.063
Co	11	—	0.042
Ni	18	0.29	0.18
Cr	38	0.46	0.068
Fe	3.2×10^4	107	44
Al	1.6×10^5	145	7.8
Mn	1.4×10^4	—	3.4
V	59	—	0.14
Ba	2.8×10^2	0.0017	0.77
Ti	3.7×10^3	0.50	3.5

*From Guo et al. (2000).

1. Hydrothermal systems. In an earlier section we described seawater–seafloor basalt and midocean ridge hydrothermal systems, which both add and remove various elements from seawater. These hydrothermal solutions are rich in metal ions. When the hydrothermal solutions enter seawater, many of the metal ions are precipitated as sulfide minerals. Others, such as Mn, Fe, Ba, Li, and Rb, enter the seawater column, leading to a net addition of these elements to seawater.

2. Rivers. Rivers are a major source of both dissolved and particulate metal. When the streams enter the ocean, a number of changes occur (which will be discussed more completely in a later section) that lead to both precipitation and mobilization of the metals. For metals adsorbed to clay minerals, the higher ionic strength of seawater leads to the release of metals to seawater. Changes in Eh and pH lead to the precipitation of iron and manganese oxyhydroxides. For particulates that are deposited on the seafloor, changes in pore water chemistry in response to oxidation of organic matter can lead to the release of these metals to the pore waters and ultimately to the ocean. The metals in rivers can have both natural (from the weathering of rocks) and anthropogenic sources. For some metals, such as Hg, As, Se, Zn, Cu, Cr, Mo, and Pb, the anthropogenic source is often dominant.

3. Atmosphere. As you will recall from Chapter 8, aerosols can be transported for great distances. Hence, they can be an important source of metals, particularly for oceanic regions that are far from land. The aerosols can have both natural (such as soil dust) and anthropogenic (such as fly ash from coal-fired power plants) sources.

Removal of Trace Metals from the Ocean Trace metals are removed from the ocean by particles. The metals can be either incorporated into the particles or adsorbed on the surface. There are two basic types of particles, lithogenic and biogenic. The lithogenic particles consist of various silicate minerals derived from river runoff and atmospheric deposition. The biogenic particles are of three types: organic matter produced by organisms, calcium carbonate shell material, and opaline shell material. Any given particulate sample may consist of all these components. The mixture of components in a particulate sample may be determined from chemical analysis for calcium carbonate and silica (shell material), particulate organic carbon (organic matter), and aluminum (crustal component). In the open ocean, suspended particulate matter (SPM) is dominantly biogenic. For example, in a recent study of SPM in the Atlantic Ocean, Kuss and Kremling (1999) found that the SPM (on average) consisted of 75% organic matter, 14.3% carbonate shell material, 4.4% opal shell material, and 6.3% lithogenic material.

1. Removal by biogenic materials. Many of the trace metals are micronutrients that are essential for phytoplankton growth. In the previous section on organic ligands, we briefly discussed how organisms may attempt to regulate the concentration of trace metals in their environment by secreting organic ligands. The Redfield ratio, discussed earlier, can be extended to include the micronutrient elements. For example, Kuss and Kremling (1999) determined the following Redfield ratio for open-ocean plankton:

C:N:**P**:Fe:Zn,Mn:Ni:Cd:Cu:Co:Pb

$$= 180:23:\mathbf{1}:0.005:0.002:0.001:0.0005:0.0004:0.0002:0.00004$$

The degree to which a metal is concentrated in a marine organism relative to seawater is given by the enrichment factor (EF), which is defined as

$$\mathrm{EF} = \frac{\text{Metal concentration in biogenic material}}{\text{Metal concentration in seawater}} \qquad (10\text{--}31)$$

Enrichment factors for metals in the organic tissue of phytoplankton and brown algae are given in Table 10–10 (p. 414). Metals are also enriched, relative to seawater, in shell material. Among these metals are Ba, Sr, Cu, Ag, Zn, Pb, Ti, Cr, Mn, Fe, and Ni.

Table 10–10 Enrichment Factors for Plankton and Brown Algae*

Element	Plankton	Brown algae
Al	25,000	1,550
Cd	910	890
Co	4,600	650
Cr	17,000	6,500
Cu	17,000	920
Fe	87,000	17,000
Mn	9,400	6,500
Mo	25	11
Ni	1,700	140
Pb	41,000	70,000
Sn	2,900	92
V	620	250
Zn	65,000	3,400

*Data from Libes (1992).

2. Removal by adsorption and precipitation. Most metals exist in the oceans at concentrations far below those required to precipitate a solid phase. The one exception is oxyhydroxides, which can precipitate from seawater. Thus, dissolved metals are mostly removed by adsorption onto sinking particles. These particles are largely clay minerals, oxyhydroxides, particulate organic matter, and shell material, all of which have a small net negative surface charge at the pH of seawater. In addition, metals can be adsorbed or bound to ligands associated with organic coatings on inorganic particles and colloids. The *removal of metals from the surface waters by particles* is referred to as **scavenging**. The rate at which metals are scavenged from the surface waters is a function of their speciation, the abundance of particulate matter, and the concentrations of other solutes. Metals can exist in seawater as a variety of species, including various types of complexes, and in multiple oxidation states. Multiple oxidation states apparently occur because of kinetic inhibitions, because the more oxidized form of a metal is generally thermodynamically favored. Given the complexity of the seawater system, the sorption behavior of metals is often determined by experiment.

Types of Metal Distributions

Measurements of metal concentrations as a function of horizontal distance or depth reveal that metal concentrations vary across the ocean. These distribution patterns have been divided into a number of types. The most important are conservative distributions and nutrient-type distributions. Other types of distributions are surface-water enrichments and mid-depth maxima or minima. Each distribution type reflects the processes that control the trace metal concentrations.

Conservative-Type Distributions We have considered the difference between conservative and nonconservative behavior earlier in this chapter. You will recall that certain oceanic properties are conservative; i.e., their variability is solely due to mixing. The most commonly used conservative properties are temperature and salinity. If the variation in the concentration of a trace metal is due solely to mixing, a plot of concentration of the trace metal versus salinity (or temperature) will yield a straight line (Figure 10–18). If the end members can be identified, it is possible to estimate the contribution of each end member to a particular water sample. Rb and Cs are examples of trace metals that show conservative behavior. Nonconservative behavior is demonstrated by deviations from the simple mixing line. If the species concentration is greater than that expected by simple mixing

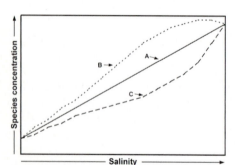

Figure 10–18
Schematic plot of concentration versus salinity. Species A shows conservative behavior. Species B and C show nonconservative behavior. Internal processes are adding species B to the seawater. Internal processes are removing species C from the seawater.

(Figure 10–18), an internal process (for example, the breakdown of organic matter) is adding the species to the water column. If the species concentration is less than that expected by simple mixing (Figure 10–18), an internal process (for example, adsorption onto particles) is removing the species from the water column.

Nutrient-Type Distributions   Vertical profiles that show surface depletion and deep-water enrichment in a trace metal are classified as nutrient-type if the trace metal shows a good correlation with one of the nutrient elements (P, N, Si). This will be the case for bi-olimiting trace elements. For biointermediate trace elements, the deep-water enrichment will not be as marked and there will be a relatively poor correlation with the nutrient elements. If the trace metal is associated with the organic carbon component, it will be returned to the water column at intermediate depths, resulting in a mid-depth maximum. In this case, the trace metal concentration will be strongly correlated with nitrogen and phosphorus concentrations. If the trace metal is associated with shell material, there is a kinetic impediment and the release of the trace metal will be delayed. In this case, there will be a deep-water maximum. For a trace metal that is associated with opaline shell material, there will be a correlation between the concentration of dissolved silica and the trace metal. A similar analysis cannot be done for a trace metal associated with carbonate shell material because the calcium concentration is essentially constant at all depths. If the trace metal is associated with both the organic carbon fraction and shell material, two maxima—mid-depth and deep-ocean—may be observed.

Surface-Water Enrichments   These distributions are represented by much higher surface-water concentrations of the trace metal relative to the water column. This type of distribution generally indicates a significant atmospheric source for the trace metal. The classic example is the deposition of Pb by Pb-containing aerosols that resulted from the combustion of leaded gasoline. With the phaseout of leaded gasoline, lead concentrations in the surface ocean have declined. Surface maxima can also be caused when trace metals carried by rivers or trace metals released from continental shelf sediments are dispersed by surface currents.

Mid-Depth Maxima and Minima   There are several possible reasons for mid-depth maxima and minima: (1) horizontal transport (advection) of waters depleted or enriched in trace metals, (2) scavenging of the metal in the mid-depth region, and (3) suboxic regions in which reduced metals, which are the more soluble form, are released to the water column.

Determination of Scavenging Rates

How rapidly a species is removed from the surface ocean is clearly an environmentally important question. Here we will look at two ways in which this rate can be measured. What we are calculating is the rate of removal of the species from the surface ocean. Recall from previous discussions that a number of elements, particularly those associated with organic material, are returned to the ocean at depth during the breakdown of the organic material (remineralization). Hence, the scavenging rate is not the same as the residence time of an element in the ocean. Residence time is a measure of how long the element (species) remains in the ocean before it is sequestered in the seafloor sediments.

Sediment Traps   A direct measurement of the rate of removal of an element can be made by trapping particles descending through the water column. The trap is positioned at some depth in the water column and measurements of both total particle mass and metal concentrations are made on a periodic basis. From these measurements a removal rate can be determined. For example, Cutter (1993) measured sediment trap fluxes at a depth of 150 m in the Sargasso Sea and determined a flux of 1 μmol m^{-2} y^{-1} for arsenic and 0.18 μmol m^{-2} y^{-1} for antimony.

Radioactivity Measurements   It is possible to measure the removal rate of radioactive species from the ocean, and this removal rate has been used to estimate the rate at which various trace constituents are scavenged from the surface waters. An implicit assumption is that the trace constituent behaves in the same way as the radioactive species, an assumption that may not be valid. In Chapter 6 we dealt at some length with radioactivity and various radioactive decay systems. The student may want to review this material. The most widely used systems utilize the radioactive elements in the uranium decay chains (Table 6–1). Recall from earlier discussions that uranium has a very long residence time in the ocean (~1 million years). Hence, it is generally assumed that the ocean is in a steady state with respect to uranium concentration. Secular equilibrium is achieved when the rate of formation of a radioactive progeny is the same as its rate of decay. When secular equilibrium is achieved, the activity ratio of the radioactive parent to the radioactive progeny is 1. In the ocean, for a number of the radioactive progeny, particularly the various radioactive Th isotopes, the ratio is much less than 1, indicating that these radioactive progeny are removed from the ocean by particle scavenging.

For a system at steady state, the activity of the parent equals the activity of the radioactive progeny plus the activity lost when the progeny is removed from the system by particle scavenging. To describe this process we can write the following equation:

$$A_N = A_P + k_P[P] \tag{10–32}$$

where A_N is the activity of the radioactive parent, A_P is the activity of the radioactive progeny, P is the concentration of the radioactive progeny, and k_P is the removal rate for the radioactive progeny. Because $A_P = \lambda_P[P]$, P can be eliminated from equation 10–32, giving

$$A_N = A_P + \left[\frac{k_P A_P}{\lambda_P} \right] \tag{10–33}$$

where λ_P is the decay constant for the radioactive progeny. Finally, solving for k_P gives

$$k_P = \left[\frac{1 - (A_P/A_N)}{A_P/A_N} \right] \lambda_P \tag{10–34}$$

The **scavenging turnover time** is *equal to* $1/k_P$ and the **chemical half-life** (or half-scavenging time) is *equal to* $0.693/k_P$.

EXAMPLE 10–6 ^{226}Ra and ^{210}Pb both occur in the ^{238}U decay chain. At mid-depths in the ocean the ^{210}Pb/^{226}Ra activity ratio is less than 1, suggesting that lead is removed faster than radium from seawater. Given an activity ratio of 0.8, calculate the removal rate for ^{210}Pb and the scavenging turnover time and chemical half-life. For ^{210}Pb, $\lambda_P = 0.0311$ y^{-1}.

$$k_P = \left[\frac{1 - (A_P/A_N)}{A_P/A_N} \right] \lambda_P = \left[\frac{1 - 0.8}{0.8} \right] (0.0311 \text{ y}^{-1}) = 0.0078 \text{ y}^{-1}$$

The scavenging turnover time = $1/k_P$ = $1/0.0078$ y^{-1} = 128 y and the chemical half-life = $0.693/k_P$ = $0.693/0.0078$ y^{-1} = 89 y. ∎

Case Study 10–3 describes how ^{234}Th disequilibria can be used to estimate the rate of removal of PAHs from the surface ocean.

GEOCHEMISTRY OF MARINE SEDIMENTS

The components of the sediments on the continental margins and in the deep ocean are derived from outside the ocean basins and from processes that occur within the ocean basin. We can identify several types of sedimentary components.

CASE STUDY 10–3
Using ^{234}Th Disequilibria to Estimate Vertical Removal Rates for PAHs from the Surface Ocean

Gustafsson et al. (1997) used ^{234}Th disequilibria to estimate the rate of removal of PAHs from the surface ocean. Like other hydrophobic organic contaminants (HOCs), PAHs exhibit a strong affinity for marine particulates (as does thorium). Assuming that ^{238}U is nonreactive in seawater, the vertical flux of ^{234}Th can be determined from the following relationship:

$$F_{Th} = \lambda_{Th}([^{238}U]_{tot} - [^{234}Th]_{tot}) \qquad (C10\text{–}3a)$$

where F_{Th} is the vertical flux, λ_{Th} is the decay constant (0.0288 d^{-1}), $[^{238}U]_{tot}$ is the measured activity for ^{238}U, and $[^{234}Th]_{tot}$ is the measured activity for ^{234}Th. Combining the vertical flux of ^{234}Th with measurements of particle-associated HOC concentrations gives

$$F_{HOC} = \frac{z_{mix}\lambda_{Th}([^{238}U]_{tot} - [^{234}Th]_{tot})[HOC]_{part}}{[^{234}Th]_{part}} \qquad (C10\text{–}3b)$$

where F_{HOC} is the hydrophobic organic contaminant flux, z_{mix} is the thickness of the surface mixing layer, $[HOC]_{part}$ is the concentration of the particulate hydrophobic organic contaminant, and $[^{234}Th]_{part}$ is the activity associated with the particulate phase. The authors found that there was a high degree of correlation between the partitioning of Th between the aqueous and particulate phase and the particulate organic carbon (POC). The authors calculated an apparent mean K_{oc}^{Th} of 3×10^6 L kg$_{oc}^{-1}$, which indicates that Th is strongly bound to the particulate organic carbon phase. Thus, it is anticipated that removal rates for Th and HOCs would be similar. The following data were obtained for PAHs in Portland Harbor, Maine. Given $z_{mix} = 8$ m, using equation C10–3b, the authors calculated the HOC fluxes given in the table.

The residence time for an HOC can be calculated from the following relationship:

$$\tau = \frac{[HOC]_{tot}}{F_{HOC}/z_{mix}} \qquad (C10\text{–}3c)$$

For the HOCs listed in the table, the residence time is ~8 days.

Source: Gustafsson et al. (1997).

Particulate PAH	Concentration (pmol m^{-3})	HOC flux (pmol m^{-2} d^{-1})	^{234}Th (dpm m^{-3}) Particulate	^{234}Th (dpm m^{-3}) Total	^{238}U (dpm m^{-3})
Pyrene	6300	6200	334	504	1920
Benzo[a]pyrene	1100	1100			
Dibenz[a,h]anthracene	94	92			

1. ***Detrital*** components are the product of rock weathering and are carried to the ocean basins by rivers. We discussed rock weathering at some length in Chapter 9. In brief, the weathered material consists of quartz, a variety of clay minerals whose mineralogy is determined by the physical environment during weathering, and poorly crystalline material such as degraded aluminosilicates, various oxyhydroxides, opal, and organic matter. The relatively unstable (under surface conditions) minerals, such as the olivines, pyroxenes, amphiboles, and Ca-plagioclases found in igneous and metamorphic rocks, are rarely found in oceanic sediments. There are, however, notable exceptions. For example, the black sand beaches of Hawaii consist largely of olivine, pyroxene, and Ca-plagioclase, because these are the major minerals of the rocks that comprise Hawaii. The clay mineralogy is largely determined by the climatic conditions, which control the degree of chemical weathering. In dry, cold climates there is little chemical weathering and the clay minerals are those present in the original rock. With increasing temperatures and rainfall, chemical weathering progresses along the path outlined in Figure 9–4b, with smectite, kaolinite, and ultimately gibbsite becoming the stable phase. Because of its size, most of the sediment carried by the rivers is deposited in the marginal marine environment. Only the finest-grained material is transported across the continental shelf and into the deep ocean. If we looked at the mineralogy of the bottom sediments as we moved from the edge of the continent to the deep-ocean basin, we would find that there is a variation in the relative abundance of the minerals as a function of their grain size. The finest-grained material, which is essentially composed of the clay minerals, is the most widely distributed.

2. **Biogenic** components are the product of biological processes that take place in the water column. These components may be either soft parts (organic material) or hard parts (skeletal material). The skeletal material is usually calcium carbonate (calcite or aragonite) or silica (opal). During the descent of particles to the seafloor much of the organic matter is decomposed and its constituents returned to the water column (remineralization). The degree of decomposition of the hard parts depends on a number of factors, a topic that will be considered in a later section.

3. **Hydrogenous** components are formed within the water column. The most important hydrogenous materials are the iron and manganese oxyhydroxides. These form where metal-rich hydrothermal solutions enter the ocean or at the water–sediment interface where dissolved iron and manganese have diffused from the underlying pore waters. Perhaps the most interesting example of iron and manganese oxyhydroxides is the manganese nodule fields found in the deepest parts of the Pacific Ocean. These nodules range up to several centimeters in size, are essentially at or within 1 m of the seafloor, have concentric growth rings, and apparently grow at extremely slow rates (on the order of several mm per million years). In the 1970s and early 1980s both American and Japanese companies investigated the feasibility of extracting economically important metals from these fields. The mineralogy and geochemistry of the various Fe-Mn oxide deposits reflect the source of the metals: directly from the water column, from hydrothermal solutions, or by recycling from the underlying sediments (Table 10–11).

Table 10–11 Chemical Composition of Oceanic Fe-Mn Deposits*

Element	Element abundance ($\mu g\ g^{-1}$)			
	Hydrogenous crust	Oxic nodule	Suboxic nodule	Hydrothermal crust
Mn	222,000	316,500	480,000	550,000
Fe	190,000	44,500	4,900	2,000
Co	1,300	280	35	39
Ni	5,500	10,100	4,400	180
Cu	1,480	4,400	2,000	50
Zn	750	2,500	2,000	2,020
Mn:Fe	1.2	7.1	98	275
Mineralogy of Mn phase	δ-MnO_2	todorokite δ-MnO_2	todorokite	birnessite todorokite
Growth rate (mm/10^6 y)	1–0.2	10–50	100–200	1000–2000

*From Aplin (2000).

4. **Authigenic** components form within the sediment by interactions between the pore waters and the least stable components of the sediment. These interactions also change the chemistry of the pore waters and may lead to the liberation of various species to solution or the sequestration of various species in the sediments (for example, Se and Hg, see Case Study 4–5). This topic will be considered in more detail in a subsequent section.

Distribution of Deep-Sea Sediments

The distribution of the major marine sediment types is shown in Figure 10–19. The ocean margin and terrigenous (land-derived) sediments are those sediments that accumulate on the continental shelf, slope, and rise. This is a diverse group of sediment types, ranging from coral reefs to the turbidite deposits of the continental rise. Glacial marine sediments are found off glaciated areas where blocks of ice calved off the continental glaciers float out to sea and melt, dropping their sediment load on the seafloor. As you may know, glaciers can carry material of virtually any size (from clay to house-size boulders) and the

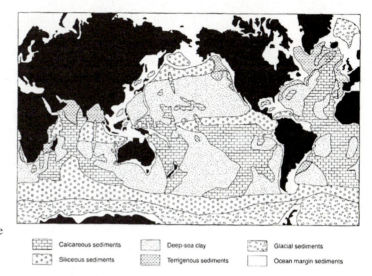

Figure 10–19
Distribution of major marine sediment types. From Aplin (2000).

Calcareous sediments	Deep-sea clay	Glacial sediments
Siliceous sediments	Terrigenous sediments	Ocean margin sediments

sediments deposited on the seafloor by the melting of the ice reflect this size range. Hence, glacial marine sediments are a very distinctive seafloor sediment type.

A simple, but informative, way to look at the distribution of deep-ocean sediments is to view their deposition as a particle rain coming from the surface ocean. These particles consist of detrital minerals derived from the continents (mostly clays, i.e., the deep-sea clay of Figure 10–19) and biogenic material (hard parts) formed as the result of biological activity in the surface waters. The detrital (clay–mineral) portion of the particle rain is essentially of the same intensity everywhere in the ocean basins, and the ocean floor is everywhere covered by clay minerals. The calcareous and siliceous sediments occur where there is a significant input of biogenic material to the particle rain. By definition, bottom sediments are mapped as calcareous or siliceous if they contain more than 30% biogenic material. Therefore, clay minerals may also be a significant component of the calcareous and siliceous sediments. The question is what controls the distribution of the siliceous and calcareous material.

Siliceous Sediments   The oceans are undersaturated with respect to both quartz and opal at all depths. Thus, silica shell material formed in the surface waters will dissolve as it descends through the water column. We can think of several factors that might favor the preservation of silica shells: (1) size of the shells—the greater the size, the less the surface-to-volume ratio; (2) thickness of the shell walls; (3) the presence of organic coatings that isolate the shells from seawater; and (4) the production of shell material in the surface ocean. Of these four factors, production is by far the most important. Hence, what we find is that siliceous sediments occur under regions of significant upwelling and biological productivity. Earlier in the chapter we noted that in areas where surface currents diverge (the equator and 60°) there is oceanwide upwelling. Referring to Figure 10–19, you will note that siliceous sediments are found at the equator and 60°N and S latitudes. Because of seafloor spreading, the position of the seafloor changes with time. When the seafloor moves under a region of oceanwide upwelling, it will accumulate siliceous sediments. When it moves away from the region of upwelling, siliceous sediment accumulation ceases. If we look at the vertical record of sediment deposition in a particular location, the presence or absence of siliceous sediment in the vertical record, and the age of the siliceous sediment, gives us information about past positions of the seafloor.

Calcareous Sediments   Calcareous shell material consists of both aragonite and calcite (the two polymorphs of $CaCO_3$). The surface ocean is saturated (in fact, supersaturated) with respect to both of these minerals, but direct precipitation of calcium carbonate from seawater does not occur in the open ocean. At deep levels, the ocean is undersaturated with respect to calcium carbonate and the shell material will dissolve. The ***saturation horizon*** is the *level at which the ocean becomes saturated (or undersaturated) with respect to calcite or aragonite.* The ***carbonate compensation depth (CCD)*** is the *depth below which calcareous sediments are not found on the seafloor.* The ***lysocline*** is the *depth zone in which the waters become increasingly undersaturated with respect to calcite (or aragonite).*

These terms are sometimes loosely used in the scientific literature. Consider that there is a kinetic component to the dissolution of carbonate shell material. Once the shells enter the undersaturated portion of the water column, they begin to dissolve. However, this dissolution is not instantaneous, so shells still accumulate on the seafloor at depths below the saturation horizon. A rough rule-of-thumb is that the seawater must be > 50% undersaturated with the carbonate phase before complete dissolution occurs. Referring to the preceding definitions, in effect the top of the lysocline is defined by the saturation horizon and the bottom of the lysocline by the CCD.

The concentration of Ca^{2+} is reasonably constant in the open ocean (it varies as a function of salinity, but salinity variations are small). Thus, the position of the saturation horizon is essentially determined by the concentration of CO_3^{2-} and the solubility product of calcite and aragonite. The aragonite stoichiometric solubility product (concentrations in $mol\ kg^{-1}$) can be calculated from the following equation (Mucci, 1983):

$$\log K_{arg} = -171.945 - 0.077993T + 2903.293/T + 71.595 \log T$$
$$+ (-0.068393 + 0.0017276T + 88.135/T)S^{0.5} - 0.10018S + 0.0059415S^{1.5}$$
$$(10\text{–}35)$$

and for calcite (Mucci, 1983),

$$\log K_{cal} = -171.9065 - 0.077993T + 2839.319/T + 71.595 \log T$$
$$+ (-0.77712 + 0.0028426T + 178.34/T)S^{0.5} - 0.07711S + 0.0041249S^{1.5}$$
$$(10\text{–}36)$$

where T is in Kelvin and S is the salinity in parts-per-thousand. The concentration for Ca^{2+} is calculated from the salinity using the following equation (Millero, 1982):

$$(Ca^{2+}) = 0.01028(S/35) \qquad (10\text{–}37)$$

where S is the salinity in parts-per-thousand. Finally, we can calculate the saturation state of calcium carbonate in seawater from the following equations:

$$\Omega_{arg} = (Ca^{2+})(CO_3^{2-})/K_{arg} \qquad (10\text{–}38)$$

and

$$\Omega_{cal} = (Ca^{2+})(CO_3^{2-})/K_{cal} \qquad (10\text{–}39)$$

EXAMPLE 10–7 Calculate the saturation state for calcite in surface waters having a temperature of 23°C and a salinity of 34.5‰. The carbonate ion concentration is $0.3\ mmol\ kg^{-1}$.

$$\log K_{cal} = -171.9065 - (0.077993)(296) + 2839.319/(296) + 71.595 \log(296)$$
$$+ (-0.77712 + (0.0028426)(296) + 178.34/296)(34.5)^{0.5}$$
$$- (0.07711)(34.5) + (0.0041249)(34.5)^{1.5}$$

$$\log K_{cal} = -6.38 \quad \text{and} \quad K_{cal} = 10^{-6.38} = 4.17 \times 10^{-7}$$

$$(Ca^{2+}) = (0.01028)(34.5/35) = 0.0101\ mol\ kg^{-1}$$

$$\Omega_{cal} = (Ca^{2+})(CO_3^{2-})/K_{cal} = (0.0101)(0.3 \times 10^{-3})/(10^{-6.38}) = 7.27$$

The surface waters are strongly oversaturated with respect to calcite.

For the same temperature and salinity, $K_{arg} = 6.43 \times 10^{-7}$ and $\Omega_{arg} = 4.7$. Aragonite is approximately 50% more soluble than calcite. ∎

Using the relationships in equations 10–35 and 10–39, and the carbonate ion concentration, it is possible to calculate the saturation state of aragonite and calcite as a function of depth. As expected, given the larger solubility constant for aragonite, seawater becomes undersaturated with respect to aragonite at much shallower depths than for calcite. Because all the oceans have essentially the same temperature structure and salinity, the depth at which a particular ocean becomes undersaturated with respect to aragonite and calcite is essentially a function of the carbonate ion concentration. Referring to Figure 10–13, we note that the carbonate ion abundance in the deep waters decreases in going from the Atlantic to Indian to Pacific oceans. Therefore, we would expect the saturation horizon, and the carbonate compensation depth, to be much greater in the Atlantic than the Pacific. This is in fact observed. Approximately 15% of the Atlantic Ocean seafloor is below the CCD, while approximately 60% of the Pacific Ocean seafloor is below the CCD. If we superimposed a map of ocean depths (a bathymetric map) on Figure 10–19, we would find that the areas of deep-sea clay lie below the CCD.

Let us return to our analogy that the deep-ocean sediments represent the accumulation of a particle rain from the surface waters. At all locations, the particle rain contains fine-grained clays. Under regions of high biological productivity an additional component of the particle rain is silica shell material. In these regions the bottom sediments are siliceous. At depths above the CCD, the particle rain also contains carbonate shell material, leading to the carbonate sediments. Additionally, at high latitudes off glaciated areas melting icebergs add a glacial component to the seafloor sediments. But remember that all of these latter sediments are superimposed on the fine-grained clays that are found everywhere in the ocean basins.

Authigenesis in Marine Sediments

Seafloor sediments, in part depending on the location at which the sediments are accumulating, are a complex mixture of alumino-silicate minerals, carbonates, opaline silica, and organic matter. As the sediment pile accumulates, a variety of reactions occur between the pore waters and the various components of the sediment. The waters of the open ocean contain measurable dissolved oxygen at all depths. Thus, the marine sediments are initially deposited under oxic conditions. The oxidation of organic matter in the sediments leads to a depletion in oxygen and ultimately the utilization of other oxidizing agents (refer to Chapter 4 if you need to review this topic). Recall that seawater (the initial pore water in marine sediments) is undersaturated at all depths with respect to silica. If silica shell material is present in the sediments, it will begin to dissolve and the silica will go into solution in the pore waters. Also, at great depths, seawater is undersaturated in aragonite and calcite, so if these minerals are present in the sediment, they will also begin to dissolve. The pore waters also react with the alumino-silicate minerals. Hence, over a period of time, the chemistry of the pore waters changes in response to the solid phases in the sediment. If adequate organic matter is present, the deeper pore waters become anoxic and the precipitation of pyrite and other metal sulfide minerals becomes possible. The result of these reactions is the formation of new minerals in the sediment (authigenesis). The sum total of these reactions is referred to as *diagenesis*. Figure 10–20 (p. 422) illustrates a number of the reactions that can occur during the early stages of diagenesis. From an environmental point of view, the important observation is that changing conditions in the sediment column can lead to either sequestration (e.g., as insoluble sulfides) or release of various elements to the water column. Deep-ocean disposal schemes—e.g., for radioactive elements—need to consider the chemical changes that will occur in the sedimentary column and the effect these may have on the mobility of the species of interest.

Age of Seafloor Sediments

As was noted in Chapter 8, the seafloor sediments contain an extensive record of climatic conditions as a function of time (paleoclimates). The seafloor record is structurally simple because the sediments are the product of vertical sedimentation; i.e., they are in a simple stratigraphic order, horizontally layered younger sediments overlying horizontally layered

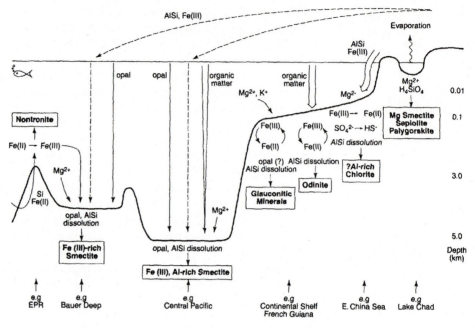

Figure 10–20
Schematic representation of the reactions that can occur between reactive minerals (opal, degraded alumino-silicates, Fe and Al oxyhydroxides, organic matter) and the clay minerals in marine sediments. From Aplin (2000).

older sediments. Sedimentation rates in the open ocean are low, and sediments accumulate at a rate of 0.1 to 1 cm per thousand years. Hence, a significant amount of earth history is contained in a relatively short vertical section of seafloor sediment. In order to use this sedimentary record to reconstruct, for example, climate change, we need to be able to correlate the sediment from ocean basin to ocean basin (i.e., we need to be sure we are looking at sediment layers deposited at the same time) and we need to determine the age of the sediments. Ages are determined in a number of ways, including using the geomagnetic time scale (based on the reversal of the earth's geomagnetic field) and radiometric dating systems. These radiometric systems were discussed at some length in Chapter 6.

Methane Hydrates

Over the past several decades, methane hydrates (clathrates) have attracted a great deal of attention both as a potential energy source and as an unexpected, and perhaps catastrophic, potential contributor to greenhouse warming. Hydrates are cages of water molecules that surround and trap gas molecules and form crystalline solids. Three hydrate structures have been identified, each with a different number of water and gas molecules. Structure I hydrates contain 46 water molecules per 8 gas molecules. The voids in the structure are small and can only accommodate small gas molecules, such as methane and ethane. Structure II hydrates contain 136 water molecules per 24 gas molecules. These hydrates contain a set of slightly larger voids that can accommodate larger gas molecules, such as propane and butane. Structure H hydrates contain 34 water molecules per 6 gas molecules. The voids in this structure are large enough to contain pentane.

Methane hydrates can form at moderate pressures and low temperatures (Figure 10–21), close to but greater than the freezing point of water. Methane hydrates are found underlying regions of permafrost and in marine sediments. The methane is either derived by bacterially mediated decomposition of organic matter under anaerobic conditions or by venting of methane from deeper levels in the crust along faults. Most of the methane hydrates are formed from biogenic methane and are, therefore, most abundant in regions of rapid accumulation of both dead organic matter and sediments (which protect the organic

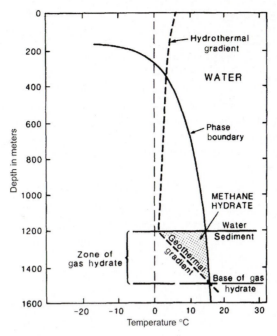

Figure 10–21
Methane hydrate stability in ocean sediment. From
Kvenvolden (1988).

matter from oxidation). The actual amount of methane that exists as methane hydrates is
still uncertain, but these hydrates are potentially a large source of methane. A warming of
the ocean, or a release of pressure due to landslides, could result in the breakdown of the
methane hydrates, releasing methane to the atmosphere and contributing to greenhouse
warming. You may recall that in Case Study 8–5 the authors suggested such a causal link
between the breakdown of methane hydrates and a global temperature increase.

THE MARGINAL MARINE ENVIRONMENT

The marginal marine environment encompasses all bodies of seawater that have salinities
significantly different from that of the open ocean. These environments include estuaries
and semi-isolated basins. Salinity can range from close to that of freshwater to significantly greater than seawater. Within any particular environment one may find a significant
range in salinity, temperature, pH, and Eh. Consider a river entering the ocean. The river
has essentially zero salinity, in most cases an acidic pH, and a temperature that may be different from that of the open ocean. When the river encounters the ocean, the zone of mixing (the estuary) will show a range of pH values, salinity, and, perhaps, temperature. As
we know, adsorption properties and solubilities vary with changes in ionic strength and
pH. Thus, when river and ocean waters meet some species may be removed from the system while others are added. Estuaries and other marginal marine environments are important in terms of biological productivity (they have sometimes been referred to as the
"nurseries of the sea") and, hence, are of environmental interest. Anthropogenic activities
may add contaminants to marginal marine systems, and these additions may prove deleterious for marine ecosystems. Lagoons and semi-isolated basins represent another type of
marginal marine environment. The salinity of these environments is determined by the
freshwater input and degree of interaction with the open ocean. For example, a basin located in an arid environment that has a restricted circulation with the open ocean may, because of evaporation, become much more saline (hypersaline) than the open ocean. In
addition, restricted circulation and a high organic input may lead to anoxic conditions in
the deep waters of the basin, a redox change that can have significant implications for the
mobility of certain species.

Estuaries

An *estuary* is a *semi-enclosed coastal body of water that has a free connection with the open ocean and within which seawater is measurably diluted with freshwater derived from land drainage*. From this definition we see that not all semi-isolated basins are estuaries. For example, lagoons that are largely isolated from the open ocean by barrier islands with only small tidal inlets are not estuaries. Similarly, basins in which seawater is not measurably diluted by freshwater (such as the Mediterranean Sea) are not estuaries. Estuaries are divided into structural types on the basis of their mode of formation and dynamic types on the basis of the interactions between the freshwater flow of the river and the saltwater wedge.

There are five structural (morphological) types of estuaries:

1. Bar-built estuaries occur when offshore bars restrict circulation with the open ocean. These estuaries do not need to be embayments in the coastline. Bar-built estuaries are usually shallow, elongate, narrow water bodies running parallel to the coast.
2. Drowned-river-valley estuaries occur in regions where relative sea-level rise has inundated the continental margin. These estuaries tend to have a dendritic pattern.
3. Fjord-type estuaries are found where continental glaciation has carved deep valleys marginal to the continent that are subsequently inundated by the ocean. Fjord-type estuaries tend to be narrow, deep, and of significant length.
4. Tectonic estuaries are formed when movement along faults depresses the land surface.
5. Stream-cut estuaries are found where rivers cut channels for some distance into the continental margin.

The dynamical classification of estuaries is based on the relative significance of the freshwater versus saltwater (tidal) flow. There are three types of dynamic estuaries, cleverly named Type I, II, and III salt-wedge estuaries.

1. Type I salt-wedge estuaries are characterized by a large freshwater flow relative to tidal volume. The interface between the impinging saltwater wedge and the freshwater is steep, and current shear near the wedge produces turbulence that helps mix the freshwater and saltwater. In these systems, there is limited intrusion of the tidal saltwater wedge into the freshwater flow.

2. Type II salt-wedge estuaries are characterized by a tidal flow greater than the freshwater flow. The predominant flow is tidal, and this flow provides the energy to mix the inflowing and outflowing waters. The boundary between the fresh and salt waters is diffuse. In these types of estuaries, the Coriolis force becomes significant and the saltwater–freshwater interface is inclined. For example, in the Northern Hemisphere saltwater flowing into an estuary would be deflected to the right and the interface will slope from right to left.

3. Type III salt-wedge estuaries are characterized by an insignificant freshwater flow relative to the saltwater tidal flow. The interface between fresh and saltwater may approach vertical. In the Northern Hemisphere (looking up-channel) the saltwater would be on the right side of the channel. In such a system, the change in salinity with depth would be small but there would be a significant salinity change across the channel.

Water Circulation in Semi-Isolated Basins

A semi-isolated basin is a depression filled with seawater that is partially separated by land or submarine barriers from the open ocean. Somewhere in the barrier there is a depression that forms an entrance to the basin. The *sill depth* is the *maximum depth of the entrance in the barrier that separates the basin from the open ocean*. Of considerable interest is how water circulates in a semi-isolated basin. This determines a number of properties of the water column, including the variations of chemical properties in the vertical and the rate at which a contaminant can be removed from the system. Below the sill depth there are no horizontal exchanges between the basin and the ocean, and any exchange at greater depths will take place by vertical currents. If the basin is much deeper than the sill depth, exchange with the open ocean is determined by the difference between evaporation and precipitation. If the sill depth is great relative to the depth of the basin, the flow is into the basin at sill depth and the main exchange takes place in the upper layers.

For a semi-isolated basin, we can write two mass balance equations that describe the exchange of water between the basin and the open ocean. If the total volume of water in the basin remains constant, the volume of water entering the basin must equal the volume of water leaving the basin.

$$T_i + P = T_o + E \tag{10-40}$$

where T_i is the amount of water per unit time flowing into the basin from the open ocean, P is the amount of water added to the basin per unit time by precipitation and runoff, T_o is the amount of water per unit time flowing from the basin into the open ocean, and E is the amount of water per unit time lost from the basin by evaporation. If the salinity of the basin remains constant, there must be a mass balance between the total amount of salt added to the basin and the total amount of salt leaving the basin.

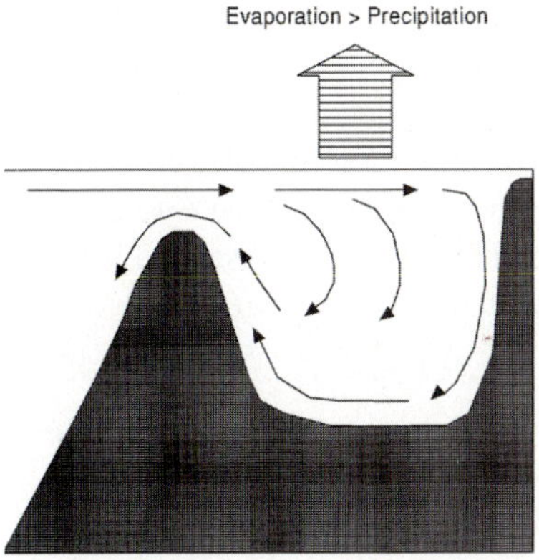

$$T_i S_i = T_o S_o \tag{10-41}$$

where S_i is the average salinity of the water entering the basin and S_o is the average salinity of the water leaving the basin. Substituting equation 10–41 into equation 10–40, and solving for either T_i or T_o, gives

$$T_i = \frac{S_o(E - P)}{S_o - S_i} \tag{10-42}$$

and

$$T_o = \frac{S_i(E - P)}{S_o - S_i} \tag{10-43}$$

Evaporation Greater Than Precipitation   Let us first consider a basin for which evaporation exceeds precipitation. These types of basins would occur in regions of descending air, around 30° (and perhaps 60°) latitude. A classic example of this type of semi-isolated basin is the Mediterranean Sea. The basin is isolated from the open ocean by the sill at Gibralter, and evaporation exceeds precipitation. We know water flows both into and out of the basin, so equations 10–42 and 10–43 both must yield positive numbers. Consider equation 10–42. Because $E > P$ for T_i to be a positive value, $S_o > S_i$. The salinity of the inflowing water is less than that of the outflowing water, and it follows that the density of the inflowing water is less than that of the outflowing water. The result is that surface water moves from the ocean to the basin while deeper waters move from the basin to the open ocean (Figure 10–22). During the summer, evaporation of the surface water leads to an increase in salinity. During the winter, the cooling of this water increases its density and it sinks, creating large vertical currents. This water ultimately exits the basin across the sill. Hence, oxygen-rich waters from the surface regularly replace the deeper waters and the deep waters are well oxygenated.

Precipitation Greater Than Evaporation   When precipitation exceeds evaporation, as it would at middle and high latitudes, $E < P$. In order for T_i (equation 10–42) to be a positive value, $S_i > S_o$. In this case, surface waters move across the sill from the basin to the open ocean and ocean water enters the basin at depth. The result is that a lower-density surface water overlies a denser deep water and there is little vertical circulation (Figure 10–23). Given relatively small salinity differences, we also see that the flow across the sill is much less for a basin in which precipitation exceeds evaporation compared to a basin for which evaporation exceeds precipitation. Because of the relatively small amount of

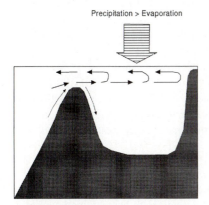

Evaporation > Precipitation

Figure 10–22

Water circulation in a basin where evaporation exceeds precipitation. Note that the circulation extends throughout the entire depth of the basin.

Precipitation > Evaporation

Figure 10–23

Water circulation in a basin where precipitation exceeds evaporation. Note that there is very little vertical circulation. The water flow is largely confined to the surface layers above the sill depth.

exchange occurring between the basin and the open ocean, and the flow of surface water across the denser deeper water, the deep waters below sill depth tend to become stagnant. If there is a significant organic input into the deep waters, anoxic conditions may result. Examples of this type of basin are the Black Sea, Baltic Sea, many Norwegian fjords, and the Arctic Ocean.

Flushing Time

If a contaminant is introduced into an estuary or a semi-isolated basin, one of the things we would like to know is how long it will take to remove the contaminant. The time it takes to exchange the contaminated water with uncontaminated water, either river water or seawater from the open ocean, is known as the *flushing time*. In this case, we are considering a conservative species, one that is neither added nor removed from the water column by processes occurring within the water column. We can measure flushing times in a variety of ways. In each of these calculations we are assuming that there is no further addition of the contaminant and that it is uniformly distributed throughout the layer of interest. First let us consider a case where there is a well-developed surface layer and water is only added by river runoff. In this case,

$$\text{Flushing time} = F/R \qquad (10\text{--}44)$$

where F is the volume of the surface layer and R is the rate of addition of water to this layer. As a second case, let us consider a basin where there is significant vertical circulation, the contaminant is mixed throughout the basin, and the inflow of water from the ocean greatly exceeds the freshwater flow. Such a situation would exist in the case of an estuary where tidal flow was significantly greater than river flow or a semi-isolated basin where evaporation exceeded precipitation. For this situation,

$$\text{Flushing time} = V/T_o \qquad (10\text{--}45)$$

where V is the volume of contaminated water and T_o is the flow from the basin into the open ocean. Note that in both of these cases we are assuming no mixing between the contaminated waters in the basin and the additions of clean water. Thus, the calculated flushing times represent minimum flushing times.

A more realistic model is one in which there is mixing between the uncontaminated water and the contaminated water. We can solve this type of problem relatively easily by using an iterative approach. We assume that each addition of uncontaminated water mixes completely with the contaminated water. This results in a dilution of the concentration of the contaminant equal to the volume of the contaminated water divided by the volume of the contaminated water plus the uncontaminated water addition. With each successive addition, the concentration of the contaminant decreases. The relative decrease is constant, so with time (or number of additions) the concentration of the contaminant asymptotically approaches zero. Mathematically, one can never reach zero, but, practically, low-enough concentrations are achieved so that the contaminant is no longer significant. If the contaminant is continually added to the system, its concentration will increase until (if the rate of addition is constant) a steady-state concentration is achieved. The value of doing flushing-time calculations is that they tell us how long it will take the system to return to a noncontaminated state after the contaminant is no longer discharged to the system.

EXAMPLE 10–8 The discharge of a heavy metal by a plant located along the edge of an estuary has led to the buildup of the heavy metal in the estuary. The current concentration of the heavy metal is 25 μmol kg^{-1}. An environmental action group has obtained a court order prohibiting any further release of this contaminant to the estuary. The mixed layer has a volume of 1,000,000 m^3, and 100,000 m^3 d^{-1} of freshwater is added to the estuary. Assuming complete mixing between the uncontaminated water and the contaminated water in the estuary, plot the change in concentration of the heavy metal over a 50-day period. How long will it take to reduce the heavy metal concentration to 10% of its present-day value? During each daily mixing cycle there is an approximately 10% dilution in the concentration of the heavy metal.

$$\text{Fraction remaining after each cycle} = 1{,}000{,}000/1{,}100{,}000 = 0.909$$

$$\text{Metal concentration after an addition} = 0.909 \times \text{metal concentration before addition}$$

Solving for multiple cycles using a spreadsheet gives the graph shown in Figure 10–24. From this figure we see that after 17 days the heavy metal concentration has been reduced to 10% of its initial value. Note that the concentration asymptotically approaches the x-axis. Only at infinity will the concentration of the contaminant be reduced to zero.

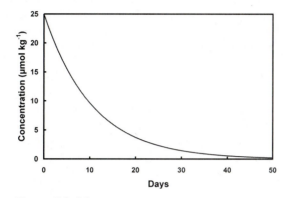

Figure 10–24
Change in concentration of heavy metal contaminant with time.   ■

For a conservative contaminant in an estuarine system, the most realistic flushing time is obtained by dividing the estuary into two layers, a surface layer (S) of mixed freshwater and seawater and a deeper layer (D) of unmixed seawater. The model assumes that a pollutant introduced to the estuary will rapidly mix with the deep layer, will gradually mix upward, and will then exit with the surface waters. The model is solved on the basis of a tidal cycle. For the surface layer, the net outflow is

$$Q_S = Q_D + R \tag{10–46}$$

where Q_S is the volume of surface outflow, Q_D is the volume of water transferred from the deep layer to the surface layer, and R is the volume of river water added per tidal cycle. The exchange ratios for the two layers can be calculated from the following equations:

$$r_S = Q_S/S \quad \text{and} \quad r_D = Q_D/D \tag{10–47}$$

where S is the volume of the surface layer, Q_D is the inflow per tidal cycle, and D is the volume of the deep layer. The mean exchange ratio within the estuarine system is

$$r = (r_S + r_D)/2 \tag{10–48}$$

The equation for the volume of contaminant (C) remaining in a tidal estuary after t tidal cycles is

$$C_t = S(1 - r_S)^t + \frac{tr}{(1 - r)} D(1 - r)^t \tag{10–49}$$

Note that what we are actually calculating is the removal of the water that was initially present in the mixed layer. The assumption is that the contaminant is behaving conservatively. The fraction of contaminant remaining after each tidal cycle can be determined by reference to the total volume at the start of the process ($S + D$). This is an exponential equation, and the decrease in concentration of the contaminant with time approaches the x-axis asymptotically. One would achieve zero concentration only after an infinite number of tidal cycles.

Water Column Chemistry

The chemical interactions that occur in an estuarine environment, where fresh and salt waters mix, are among the most complex interactions in natural waters. Freshwater entering the ocean contains a variety of inorganic and organic particles. Various charged species are adsorbed to these particles. When mixing occurs between fresh and salt waters, there is a significant increase in ionic strength and adsorbed ions are often released to solution. As we know from earlier sections, changes in salinity affect equilibrium constants, and changes in these constants play a role in the precipitation or re-solution of metal-containing solids. In addition, increases in ionic strength can lead to the flocculation of colloidal particles. For example, when freshwater enters the marine environment, there is often a rapid decrease in the amount of iron in solution. This decrease is due to either the flocculation of colloids or precipitation. For acid streams, the increase in pH that occurs when the freshwater mixes with seawater leads to the precipitation of Fe (and Al) oxyhydroxides (see the discussion of aluminum solubility in Chapter 9). Changes in oxidation–reduction potential can also have a significant effect. For example, consider a situation where well-oxygenated freshwater enters a marine environment of significantly lower oxidation potential (such as might be the case in an estuary where large inputs of organic matter have led to anoxic conditions in the deeper waters). Fe^{3+} oxyhydroxides, with adsorbed phosphate ions, are reduced, releasing Fe^{2+} and PO_4^{2-} to solution. If the dissolved constituents are then transferred into a region where the water is aerobic, Fe^{2+} will be oxidized to Fe^{3+}, and Fe^{3+} hydroxide will reprecipitate and the phosphate ions will be readsorbed.

Given all the possible variables, how does one go about analyzing the chemical processes that occur in an estuary? One starting point is to separate species that are behaving conservatively from those that are not. This is done by plotting the species of interest against a conservative parameter such as salinity or chloride ion. If the variation in the species can be explained by simple mixing (between two end members), then the data will fall on a straight line. However, in a complex estuarine environment it is possible that there are more than two simple (freshwater and seawater) end members. As you will recall from earlier discussions, under these conditions the data will plot within a field. If the species distribution cannot be explained by simple mixing, the observed concentrations will either fall above (net addition) or below (net removal) the mixing line. We have already dealt with these situations in the open ocean. Here we will look specifically at the marginal marine environment.

Equilibrium Constants Both in the preceding sections and in the problem set, equations are given to calculate stoichiometric equilibrium constants for a number of reactions of interest in seawater. These equations are determined by experiments done over a range of salinities and temperatures. However, most, but not all, experiments do not encompass waters of very low salinity (5‰ is a typical limit at the low-salinity end of the experiments) or waters of very high salinity (40‰ is a typical limit at the high-salinity end of the experiments). We are not going to consider hypersaline waters here; the interested student should consult other sources. However, we will consider estuarine and brackish waters, where salinities in the range of 0 to 5‰ are encountered. We can deal with this salinity range either by applying the stoichiometric equilibrium equations or by using the relationships developed for freshwater and correcting for ionic strength (activity). Either option may be acceptable. If the stoichiometric equilibrium constant equations give results in good agreement, at zero salinity, with the thermodynamic equilibrium constants, then one might have some confidence that the stoichiometric equations could be extended to very low salinities.

Adsorption–Desorption In the estuarine environment we find suspended particulate matter (both organic and inorganic particles), organic ligands (inorganic ligands are generally less important), and truly dissolved species. The partitioning of species among these various phases is a complex process. As an example, for the partitioning of a metal we can write the following equilibria:

$$M + SPM \rightleftharpoons MSPM$$

$$M + DOM \rightleftharpoons MDOM$$

and

$$MDOM + SPM \rightleftharpoons MDOMSPM$$

where M is the metal, SPM is the suspended particulate matter, and DOM is the dissolved organic matter. Because the MDOM complex generally has a much lower affinity for suspended particulate matter than the species in true solution, the presence of dissolved organic matter increases the amount of the metal in solution and decreases the amount adsorbed to the particles relative to a system without DOM. Hence, variations in total DOM will affect the adsorption of a metal. Experimental studies have shown that there is an increase in the adsorption of metals when dissolved organic matter is removed from the system (Turner et al., 1998). Other variables are the nature of the suspended particulate matter and the pH.

An additional factor is that the metal may be occluded in the suspended particle. Consider the precipitation of oxyhydroxides, which may trap metals. These trapped metals are not exchangeable. If one analyzes the bulk particulate phase, the exchangeable metals will be overestimated. For this reason, experimental protocols have been developed using a variety of leachates so that only the adsorbed species are included in the analysis.

Metal distribution adsorption coefficients are determined by batch or column experiments using both radiotracers and measurements of the concentration of the element of interest. The experimental data are then fitted to appropriate isotherms (see Chapters 7 and 9). Experiments are done to determine both the sorption and desorption coefficients, because not all sorption reactions are reversible (or are not reversible on a short time scale). For metals, there is often a relationship between salinity and the distribution coefficient, which can be expressed by the following equation (Bale, 1987; Turner and Millward, 1994):

$$\ln K_D = b \ln(S + 1) + \ln K_D^0 \tag{10–50}$$

or

$$K_D = K_D^0 (S + 1)^b \tag{10–51}$$

where K_D is the adsorption coefficient at some salinity S, b is a constant, and K_D^0 is a constant that essentially represents the adsorption coefficient at zero salinity. A plot of $\ln K_D$ versus $\ln(S + 1)$ yields a straight line (not always) with a slope b and a y-intercept $\ln K_D^0$.

The adsorption of hydrophobic organic contaminants (HOCs) occurs because of a weak solute–solvent interaction rather than a large, specific sorbate–sorbent interaction. What this means is that HOCs are only sparingly soluble in water and, hence, adsorb to particles because they are excluded from the aqueous phase. In Chapter 5 we defined the octanol/water partition coefficient (K_{ow}), which is a measure of the partitioning of an organic chemical between a nonaqueous phase and an aqueous phase. The greater the K_{ow}, the more strongly the HOC is partitioned into the nonaqueous phase (the less soluble it is in water). Thus, the greater the K_{ow}, the greater is the affinity of the HOC for particles. Because HOCs have a strong affinity for organic carbon, the adsorption coefficient is normalized to the amount of organic carbon in the particle (or sediment).

$$K_{OC} = K_D/f_{OC} \tag{10–52}$$

where K_{OC} is the organic carbon normalized adsorption coefficient and f_{OC} is the fraction of organic carbon in the particle. The type of organic carbon in the particles (sediment) also plays a role in the adsorption of the HOC (e.g., the adsorption of PAHs by sediments, Mitra et al., 1999). You may recall from Chapter 2 that for uncharged species, increasing ionic strength leads to activity coefficients greater than 1; i.e., the solubility of the uncharged species decreases. This is sometimes referred to as a *salting out* effect. We would expect that with increasing salinity the adsorption coefficient of uncharged species would increase, and this increase has been observed for aromatic hydrocarbons (Means, 1995).

From this, we see that inorganic and organic contaminants will behave differently in estuarine systems. If salinity is the only variable, increasing salinity will lead to the release of adsorbed metals to solution. Conversely, increasing salinity will lead to the removal of organic contaminants. But remember, there are other variables that affect the solubility of species in solution, and all of these need to be considered in explaining the behavior of species in an estuarine system.

Redox Reactions   In marginal marine environments where there is limited overturn of the deep waters, the deep waters can become anoxic because of the consumption of oxygen during the decomposition of organic matter. Various names have been used to describe the transition zone between the oxic and anoxic waters; **oxycline**, which denotes a *zone of rapidly changing oxygen content*; or **redoxcline**, which denotes a *zone of rapidly changing oxidation–reduction potential*. Although anoxic conditions are unusual in estuarine systems, during the summer months, when there is generally a low flow of freshwater into the estuary and high biological productivity, estuarine waters can show significant oxygen depletion. The waters still contain measurable dissolved oxygen, but the concentrations are much less than saturation values. Changes in redox conditions are particularly important for metals such as iron and manganese, which form oxyhydroxide precipitates under oxidizing conditions but which remain in solution under reducing conditions. These oxyhydroxides also trap other metals (occlusion, as noted earlier), and changes in redox conditions that lead to the dissolution of the oxyhydroxides will release the occluded metals to solution.

At the oxic (O_2)/anoxic (H_2S) interface, there is often an increase in dissolved Mn, Fe, and Co and a decrease in Cd, Cu, and Zn. The increase in Mn, Fe, and Co is presumably due to the reduction of these elements to their soluble form and the concomitant breakdown of the oxyhydroxides. The decrease in dissolved Cd, Cu, and Zn is due to either the precipitation of insoluble sulfides onto the surface of particles or the formation of metal sulfide complexes that attach to the surface of silicate particles. The introduction of oxygenated waters into the anoxic portions of basins will lead to the formation of oxyhydroxides and the removal of metals by occlusion and adsorption to the oxyhydroxides. With the return of anoxic conditions, these metals will once again be released to the water column. Thus, changing redox conditions, due to changes in circulation, can have a significant impact on the concentration of dissolved metal species (see Case Study 10–4).

Suspended Particulate Matter (SPM)   The suspended particles in the water column are derived from a number of sources and are of variable composition. Sources include atmospheric aerosols, river inputs, and biological processes. There is a significant potential range of particle types: organic material at various stages of decomposition (including organic colloidal particles), calcium carbonate and silica shell material, silicate and other minerals derived from erosion of the continents, oxyhydroxide precipitates, and various minerals that precipitate in the water column (e.g., insoluble metal sulfides) and anthropogenic particles (e.g., fly ash from coal-fired power plants). Because we are interested in the significance of anthropogenic sources for various metals, we will briefly consider one approach to answering this question, enrichment factors, which is already familiar from previous discussions in Chapter 8 and this chapter.

We will simplify the analysis by assuming that in the natural environment metals are associated with organic particles or silicate and other minerals. This is not a new idea. We have already discussed the Redfield ratio, which gives the proportions of various metals to carbon in biological systems. Thus, in order to determine the biogenic association we will normalize metal concentrations to carbon and use the Redfield ratio to identify variations from natural biogenic concentrations. For the silicate and other minerals, we will normalize the metals to the concentration of an element that is not believed to have an anthro-

CASE STUDY 10–4
Effect of Redox Processes on the Partitioning of Cd, Pb, Cu, and Mn Between Dissolved and Particulate Phases in the Baltic Sea

Because the shallow Belt Sea permits only a slow exchange of water between the Baltic Sea and the North Sea, the deep waters of the Baltic Sea can become anoxic. A stagnation period of approximately 17 years was terminated in 1993 by an inflow of dense, oxygen-rich water into the Baltic Sea. This resulted in a change in the redox conditions of the deep waters from anoxic to oxic. Prior to the inflow, the redoxcline occurred at approximately 125 m depth and H_2S increased from this depth toward the bottom. Dissolved Fe, Mn, and Co increased dramatically in the anoxic waters, which also had high phosphate. Mn increased continuously toward the bottom, but Co and Fe displayed maximum concentrations at 150 m and 200 m, respectively (Brügmann et al., 1998). The decreases in Fe and Co concentrations may have been due to the formation of insoluble sulfides. Cd, Cu, Zn, and Pb showed a more complex behavior, first decreasing with depth and then increasing near the bottom. Brügmann et al. suggested that this might be due to the formation of metal complexes with sulfide and/or colloids. After the influx, there was a dramatic drop in the deep waters in the concentrations of dissolved phosphate, Fe, Mn, and Co and an increase in Zn, Cu, and Cd concentrations. The decreases were explained by the precipitation of oxyhydroxides, which incorporated phosphate and Co. The increase in the concentrations of the other metals may be due to the oxidation of sulfide species and the breakdown of metal-sulfide complexes.

Over a period of four years, the deep waters slowly returned to anoxic conditions. Pohl and Hennings (1999) investigated the changes that took place in the distribution of various metals between solution and particles in response to these changes in redox condition. Under anoxic conditions, the Mn was largely found in solution because the reduced form of Mn is soluble. The change from anoxic to oxic conditions was accompanied by a significant increase in the amount of Mn found in particulate form. Presumably, this is due to the direct precipitation of oxyhydroxides. From 1995 to 1997 both the relative and absolute amounts of dissolved Mn increased in the deep waters. This increase was related to diffusion of Mn from the water–sediment interface, where Mn is concentrated, to higher levels in the water column. Because the sediment pore waters are anoxic, Mn diffuses to the sediment–water interface and then into the overlying water. Under anoxic conditions, Cd is strongly associated with particles, presumably due to the precipitation of CdS onto the surface of particles. After the addition of oxic waters, Cd adsorption coefficients decrease, presumably because of the breakdown of CdS, but then increase as Cd is adsorbed onto oxyhydroxide precipitates formed as a result of the oxidation of Mn. Oxic conditions also favored the adsorption of Pb on particles. Before the addition of the oxic water, 40% of the Cu was associated with particles. After the waters became oxic, Cu adsorption decreased, and after several years only 5% of the Cu was associated with particles. The authors suggested that the decrease in Cu adsorption might be due to the formation of soluble hydroxy-, chloro-, and/or organo-Cu complexes after the inflow event. As indicated by this study, the distribution of various metals between particles and solution is strongly influenced by redox conditions, and changing redox conditions can either remove or liberate metals to the aqueous environment.

Source: Brügmann et al. (1998); Pohl and Hennings (1999).

pogenic source. We have already used this approach in Chapter 8 as a way of identifying the sources of atmospheric aerosols. Appropriate reference elements are Al, in many cases Fe, and Sc. Average crustal abundances are often used as the reference for the nonbiogenic association. The enrichment factor is defined as follows:

$$EF = \frac{(X/RE)_{particle}}{(X/RE)_{reference}} \quad (10\text{–}53)$$

where X is the element of interest, RE is the reference element, and the subscripts particle and reference refer to the sediment particle and the reference material, respectively. We can also calculate enrichment factors for potential sources for the SPM and use these to identify the source of the particles.

EXAMPLE 10–9 Atmospheric aerosols deposited onto the surface of a bay have Cd = 79 ppb and Sc = 14.2 ppm. The river flowing into the bay has Cd = 13 ppb and Sc = 17.5 ppm. For suspended particles collected in the top 10 m of the bay, Cd = 4 ppb and Sc = 5.2 ppm. Calculate the enrichment factor for each component using average upper crust as the reference (Table 8–14).

For the aerosol,

$$EF = \frac{(X/RE)_{particle}}{(X/RE)_{reference}} = \frac{(0.079/14.2)_{particle}}{(0.098/11)_{reference}} = 0.62$$

Similarly, EF = 0.083 for the river and EF = 0.086 for the suspended particles. The EF for the particles is very close to that of the river, suggesting that the river is the source of the Cd in the particles. If the results indicated that the particles were a mixture of aerosol and river input, assuming simple linear mixing, we could calculate the percent of each component using the mixing equations developed elsewhere (e.g., equations 9–42 to 9–45). ∎

Pore Water Chemistry

In the preceding sections, we looked at processes that occur in the water column. A number of metals and organic contaminants are associated with suspended particulate matter and are eventually removed by sedimentation of the particles. The particulate matter may consist of various types of biogenic material (e.g., decomposed plant remains at all stages of degradation and shell material), silicate particles, and oxyhydroxides and other precipitates. In the sediment column, a number of reactions occur between the particles and the pore waters. Many of these interactions, such as the oxidation of organic matter and the dissolution of various phases that are undersaturated in the pore waters, are already familiar processes. Added to these are the possible redox changes in the pore waters that can lead to the release or sequestration of redox-sensitive species. Because we are generally dealing with relatively shallow waters, the bottom sediments can support a diverse and productive biological community, and the impact of contaminants on this community is of environmental concern. Also, changing conditions may release these contaminants to the water column, thus making them available to other systems.

Release of Metals to Pore Waters and the Water Column   In most cases, given the high organic input associated with marginal marine environments, we can expect the pore waters to become anoxic at a relatively shallow depth. Hence, except within a few centimeters of the surface, the redox-sensitive species will exist in their reduced state. The change in redox conditions can lead to the release of various species, such as Mn and Fe, to the pore waters. The release of Mn and Fe is due to the dissolution of oxyhydroxides. Metals that coprecipitated with or were adsorbed to the oxyhydroxides will also be released to the pore waters during dissolution of the oxyhydroxides. Other reactions also occur in the sediment column that lead to the degradation of organic carbon and a change in the nature of the DOC in the pore waters. One of the consequences of these reactions is the release of metals to the water column that were associated with the organic material. These various changes have several important consequences: (1) metals associated with particles that were initially sequestered in the sediments can be released to the pore waters, and (2) the release of these metals can establish concentration gradients in the pore waters that lead to the diffusion of the metals back into the overlying water.

Transport of Metals in Pore Waters   The changes in soluble species in water moving through a porous material, in one dimension, can be described by the following equation (the *advection–diffusion equation*):

$$\frac{\partial C_i}{\partial t} = D_i \underbrace{\frac{\partial^2 C_i}{\partial x^2}}_{\substack{\text{diffsion} \\ \text{term}}} - v \underbrace{\frac{\partial C_i}{\partial x}}_{\substack{\text{advection} \\ \text{term}}} \pm (\text{reaction terms}) \qquad (10\text{--}54)$$

where C_i is the concentration of the species, t is time, x is distance, D is the diffusion coefficient, and v is the velocity of the fluid. The diffusion term represents the transport of the species by diffusion, and the advection term represents the transport of the species by fluid movement (advection). An additional factor (*reaction terms*) involves processes that take place between the water and the solid material that can add or remove the species or processes in the water (such as precipitation) that can remove the species. The sum of all these processes is the change in concentration of the species, as a function of time, at a specific point in the system. This equation has general applicability and is the fundamental equation used to model chemical changes in a hydrodynamic system. Hence, this equation would be applicable to groundwater flow, the movement of fluids through seafloor hydrothermal systems, etc. Here we are concerned with the diffusive part of the process.

Diffusion along a concentration gradient can be represented by *Fick's first law*:

$$J = -D\frac{\partial C}{\partial x} \tag{10–55}$$

where J is the flux (mass per unit area per unit time), D is the diffusion coefficient, C is the change in concentration of the species, and x is the distance over which this change in concentration occurs. The negative sign indicates that the flux is in the opposite direction to the concentration gradient (i.e., the species diffuse from higher concentrations to lower concentrations). D can have a variety of meanings: ***diffusion coefficient*** when *species transport is by simple diffusion (essentially a static fluid)*, ***dispersion coefficient*** when the *transport is due to hydrodynamic dispersion (laminar flow)*, and ***eddy diffusivity*** when the *transport is by turbulent flow*. The change in concentration of a species in a specific volume is the difference between the flux into the volume and the flux out of the volume. For a one-dimensional case,

$$\Delta C\Delta x = (J_{\text{in}} - J_{\text{out}})\Delta t \tag{10–56}$$

where $\Delta C\Delta x$ is the change in concentration over distance x, $J_{\text{in}} = -D(\partial C/\partial x)_{x=x}$, $J_{\text{out}} = -D(\partial C/\partial x)_{x=x+\Delta x}$, and t = time. Equation 10–56 is equivalent to the diffusion term (*Fick's second law*) in equation 10–54. Diffusion through a sediment is slower than diffusion through a liquid because only part of the cross-sectional area is occupied by the liquid and the path followed by the diffusing species is not a straight line. Hence, we must modify the diffusion coefficient to take these two factors into account. This is done in a variety of ways, most of which are empirical. We will not pursue these topics any further here. For the interested student, Drever (1997) gives an excellent introductory description of transport and reaction modeling.

EXAMPLE 10–10 For a particular marginal basin, Mn concentrations are measured in sediment pore waters 5 cm below the sediment–water interface (2000 nmol L^{-1}), in seawater at the sediment–water interface (1200 nmol L^{-1}), and at 10 m above the seafloor (20 nmol L^{-1}). For the seawater column, assume an eddy diffusion coefficient of 1×10^{-1} cm^2 s^{-1}, and for seawater, a diffusion coefficient for Mn of 5×10^{-5} cm^2 s^{-1}. The diffusion coefficient for the pore water in the sediment is $D_S = \varphi^2 D_0$, where φ is the porosity of the sediment and D_0 is the diffusion coefficient for Mn in seawater. Calculate the flux of Mn from the sediment to the basin and the Mn flux in the bottom 10 m of the water column. The sediment has a porosity of 80%.

In order to maintain dimensional consistency, we must convert concentration in nmol L^{-1} to concentration in nmol cm^{-3} (the diffusion coefficients are in cgs units). There are 1000 cm^3 per liter. Therefore, Mn in pore waters = 2.0 nmol cm^{-3}, Mn at the sediment–water interface = 1.2 nmol cm^{-3}, and Mn at 10 m above the seafloor = 0.02 nmol cm^{-3}. For the Mn flux from the sediment to the seafloor,

$$J = -\varphi^2 D_0 \frac{\partial C}{\partial x} = -(0.80)^2(5 \times 10^{-5} \text{ cm}^2 \text{ s}^{-1})\left[\frac{(2.0 - 1.2) \text{ nmol cm}^{-3}}{5 \text{ cm}}\right]$$

$$= -5.12 \times 10^{-6} \text{ nmol cm}^{-2} \text{ s}^{-1} = -1162 \text{ nmol cm}^{-2} \text{ y}^{-1}$$

For the Mn flux in the bottom 10 m of the seawater column,

$$J = -D\frac{\partial C}{\partial x} = -(1 \times 10^{-1} \text{ cm}^2 \text{ s}^{-1})\left[\frac{(1.2 - 0.02) \text{ nmol cm}^{-3}}{1000 \text{ cm}}\right]$$

$$= -1.18 \times 10^{-4} \text{ nmol cm}^{-2} \text{ s}^{-1} = -3724 \text{ nmol cm}^{-2} \text{ y}^{-1} \qquad \blacksquare$$

Removal of Metals from Pore Waters There are several processes that can remove metals from the pore waters. (1) If reduced sulfur is present, metal-sulfide minerals may precipitate, thus removing (or decreasing) the concentration of certain metals in the pore waters. (2) The presence of an oxic layer at the seawater–sediment interface may inhibit the transport of metal ions into the overlying seawater. This occurs because oxyhydroxides can form within the oxic sediments and these oxyhydroxides remove Mn and Fe and other metals by coprecipitation or adsorption.

Sediment Chemistry

As noted earlier, seafloor sediments contain a record of changing surface conditions. Particularly in marginal marine environments, the sediment record has been used to document anthropogenic changes. It is important to recognize that pore water sediment interactions can change the sedimentary record by redistributing metals, and the decomposition of organic matter, as a function of time, changes the molecular characteristics of the organic material. Bearing these caveats in mind, analysis of the sediment record has proven to be a very useful tool for understanding anthropogenic environmental changes.

Sediment Age and Accumulation Rate In order to use the various markers that occur in the sediment, we need to determine the age and sedimentation rate for the sedimentary sequence. In Chapter 6 we developed the basic principles of radiometric dating and discussed in some detail the various U-series disequilibrium relationships that are used to date sedimentary sequences. We also described event dating, which can be used when some well-established event of short duration can be identified in the sediment record. As an example, in terms of radiometric event dating, the period of atmospheric testing of nuclear weapons provides a time line. On a more local basis, changes in land use or specific contamination events, such as a one-time mercury release, can serve as time lines. Once a sedimentation rate has been established, a sediment accumulation rate can be calculated. The basic equation for calculating the sediment accumulation rate (SAR) is

$$\text{SAR} = \text{SR} \times \rho_s(1 - \varphi) \tag{10–57}$$

where SAR is the sediment accumulation rate in $\text{g cm}^{-2} \text{ y}^{-1}$, SR is the sedimentation rate in cm y^{-1}, ρ_s is the dry density of the sedimentary material, and φ is the porosity. The accumulation rate for a component of the sediment (e.g., organic carbon, silica shell material, carbonate shell material) is calculated as follows:

$$\text{Ar}_{\text{comp}} = \text{comp}\% \times \text{SR} \times \rho_s(1 - \varphi)/100 \tag{10–58}$$

where Ar_{comp} is the accumulation rate for the component in $\text{g cm}^{-2} \text{ y}^{-1}$ and comp% is the content of the component in weight percent.

Redox Indicators The most widely used redox indicators involve the oxidation state of iron. The degree of pyritization (DOP) is defined as

$$\text{DOP}(\%) = \text{Fe}_{\text{pyrite}}/(\text{Fe}_{\text{pyrite}} + \text{Fe}_{\text{reactive}}) \tag{10–59}$$

where $\text{Fe}_{\text{pyrite}}$ is that fraction of the iron bound in pyrite and $\text{Fe}_{\text{reactive}}$ is the amount of iron that is available for reactions in the biogeochemical cycle. $\text{Fe}_{\text{pyrite}}$ is often calculated from the sulfur content of the sediment using the ratio of 0.5 mol of S to 1 mol of Fe (based on the formula of pyrite, FeS_2). An increase in DOP indicates more reducing conditions. Roden and Tuttle (1993) proposed a more sensitive measure of redox conditions, the AVS (= FeS) to CRS (= pyrite and S^0) ratio. Increases in the AVS/CRS ratio indicate a more reducing environment because S in FeS is in a lower oxidation state than S in FeS_2 (and S^0).

Metal Indicators Numerous studies have been done in which the variation of metal concentrations with depth have been correlated with anthropogenic inputs (see Case Study 10–5 and problems). The data are often normalized to a reference value, usually the concentration of the metal in sediment layers deposited before anthropogenic activities became important and (a now very familiar concept) enrichment factors are calculated. Some authors (e.g., Skowronek et al., 1994) have used an element that is presumed not to have a significant anthropogenic source (a commonly used element is Al) as a reference element

CASE STUDY 10–5
The Impact of Human Activities on the Sediments of San Francisco Bay, California

San Francisco Bay is surrounded by a highly urbanized area comprising San Francisco, Oakland, and various suburban communities. The Bay Area is also the locus of a number of universities and research institutes. Given the potential of high anthropogenic impact on San Francisco Bay and the abundance of research groups, it is not surprising that San Francisco Bay is one of the most intensely studied estuaries in the world. In 1999, an issue of *Marine Chemistry* was devoted to papers dealing with anthropogenic impacts on San Francisco Bay.

The watershed for San Francisco Bay encompasses about 40% of California. The bay is shallow, with a mean depth of 6 m, and the tidal range at the mouth is 1.7 m. Communication with the open ocean is limited (Figure 10–C5–1) to the narrow inlet of the Golden Gate. Sediment input to the bay has been variable due to human activities. Maximum sediment input occurred between 1852 and 1914 because of hydraulic mining for gold. At the same time the filling and construction of levies reduced the salt marsh surrounding the bay (originally, nearly double the area of the bay) to less than 4 to 8% of its original area. Because of water management projects begun in the 1930s, more than 60% of the original freshwater flow is now diverted before it reaches the bay. At the same time, the construction of dams has led to a significant reduction in the amount of sediment delivered to the bay. Thus, human activities have had a substantial impact on the hydrogeologic characteristics of the bay.

There are numerous potential sources of contamination. These include mining for Hg, Au, and Cu; extensive agricultural use of pesticides; untreated urban runoff; treated wastewater; and a variety of industrial activities (power plants, oil terminals, and chemical, metal, and paper facilities). Surface sediments were sampled and two sediment cores were obtained (Figure 10–C5–1). Various radiometric methods were used to establish the geochronology and sedimentation rate for the cores. It should be noted that there is substantial mixing of the sediment, so each horizon in the cores is a mixture of sediment deposited over a period of years. A number of parameters were measured for each core, and these results are summarized in Figure 10–C5–2. Note that PAHs and Hg are the earliest contaminants to appear in the sediment record. The PAH signature may represent early land clearing by fire. The Hg signature corresponds with the onset of gold mining because Hg was used to amalgamate and extract gold. The subsequent appearance of other metals in the sedimentary sequence is correlated with mining and industrial activities. Pesticides appear with the onset of extensive agricultural activities in the 1930s. Since the 1970 ban on DDT, this pesticide has been decreasing in the sediments. It is still found at higher levels because of sediment mixing, but at much lower absolute concentrations. In total, the sediments of San Francisco Bay record the anthropogenic activities that have occurred around the bay since the arrival of the first colonizers.

Source: Hostettler et al. (1999); van Geen and Luoma (1999).

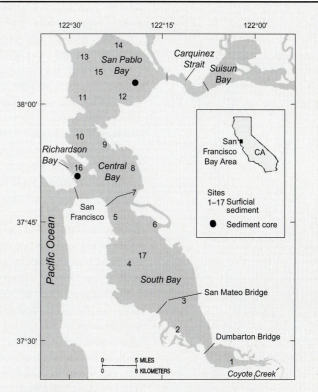

Figure 10–C5–1
Map of San Francisco Bay showing location of surface sediment samples and sediment cores. From Hostettler et al. (1999).

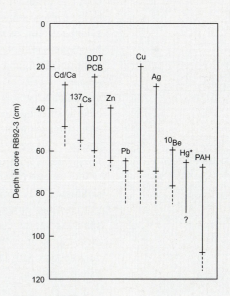

Figure 10–C5–2
Appearance of contaminants in the sediment record. The dashed lines connect the shallowest interval where a contaminant was not detected to the nearest interval where it was detected. The solid line connects the depth interval where a contaminant was first detected to the depth of maximum concentration. From van Geen and Luoma (1999).

against which the concentrations of other metals from uncontaminated layers are compared. The comparison is done by linear regression. What this calculation actually gives us is the concentrations of the various metals relative to the reference metal for eroded material transported to, and deposited in, the basin. Using these equations, for any particular sediment layer, the "natural" concentration of a metal can be calculated from the observed abundance of the reference element. This calculated "natural" concentration is then compared to the measured concentration (an enrichment factor is calculated). For enrichment factors significantly greater than 1, the implication is that the metal has been added to the sediment either by diagenetic processes or anthropogenic inputs. Besides bulk sediments, shells from a particular species have been used to investigate metal variations. By using the same species, we presumably eliminate variations due to differences in biochemical behavior of organisms. Finally, Pb isotopes are often used to identify, and quantify, anthropogenic versus natural sources of lead.

Bioindicators   There are a variety of biological materials that can be used as indicators of changing surface conditions. For example, Zimmerman and Canuel (2000) used variations in the abundance of ambrosia (ragweed) pollen in the sediment record to identify periods of agricultural activity. Increased land clearing was marked by increased amounts of ambrosia pollen in the sedimentary record (and an increase in the ragweed/oak pollen ratio). Using pollen and spores as indicators of past surface conditions is an active field of both biological and geological research (palynology). **Biomarkers** are *organic molecules that originated in a biological system and that retain enough of their original structure in the sedimentary record so that information can be obtained about their source and maturity.* For example, Zimmerman and Canuel used a variety of organic compounds as markers for algal, bacterial, and terrestrial sources of organic carbon in Chesapeake Bay sediments. The data were normalized to total organic carbon (TOC) so that comparisons could be made on a percentage basis. Enrichment factors were calculated for each group of biomarkers relative to the composition of organic matter deposited prior to 1825, presumably before any significant anthropogenic inputs. A plot of enrichment factors versus age (Figure 10–25) shows the impact of anthropogenic activities on the Chesapeake Bay sediments. The organic material becomes enriched with respect to algal and bacteria biomarkers after 1934, which corresponds with the increased use of fertilizer.

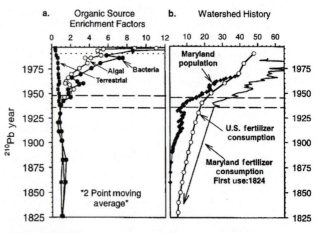

Figure 10–25

Change in bioindicators, as a function of time, in response to changes in land use in the Chesapeake Bay area. From Zimmerman and Canuel (2000).

Organic Contaminants   A variety of organic contaminants are found in the sediment record. Among the more common are PAHs, petroleum derivatives, and various pesticides. PAHs are refractory and persistent in sediments. PAHs have numerous sources, although the most common are high-temperature combustion (pyrolytic) and crude oils, coal, coal tar, or various refinery products (petrogenic). The phenanthrene/anthracene (P/A) and fluoranthene/pyrene (Fluo/Pyr) ratios can be used to distinguish between pyrolytic and petrogenic sources (Gschwend and Hites, 1981). PAHs that have a pyrolytic origin tend to have P/A ratios between 4 and 10, whereas the slow thermal maturation of organic matter in petroleum leads to much higher P/A ratios. Fluo/Pyr ratios greater than

1 are typically associated with coal combustion (but there are exceptions), whereas lower ratios have a petrogenic association. Thus, PAHs that have a pyrolitic origin tend to have lower P/A and higher Fluo/Pyr ratios than PAHs that have a petrogenic origin. The combination of these two ratios can be used to distinguish the source of the PAHs in the sediment (Figure 10–26). The ratio of methylphenanthrenes to phenanthrene (MP/P) has also been used to distinguish PAH sources (Pereira et al., 1999). The MP/P ratio is generally less than 1 for pyrolytic PAHs and between 2 and 6 for petrogenic PAHs. You may recall from Chapter 5 (if not, review Case Study 5–3) that petroleum hydrocarbons can be distinguished from biological hydrocarbons on the basis of their (1) odd/even ratios of approximately 1 for *n*-alkanes and (2) the presence of an unresolved complex mixture (UCM) that cannot be separated by high-resolution gas chromatography. Both of these characteristics can be used to make a semiquantitative assessment of the relative importance of petroleum versus biological hydrocarbons in a sediment. For other organic contaminants that are persistent in the environment, such as DDT and its metabolites and PCBs, direct concentration measurements can be used to monitor changes in the input of these chemicals to the marginal marine environment.

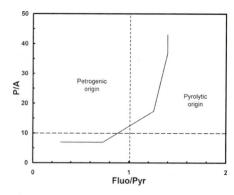

Figure 10–26
Distinction between pyrolytic and petrogenic PAHs based on P/A and Fluo/Pyr ratios. Adapted from Budzinski et al. (1997).

QUESTIONS AND PROBLEMS

1. Distinguish between *active* and *passive* continental margins.
2. Distinguish between the *neritic environment* and the *oceanic environment*.
3. What is the *main thermocline*?
4. Why are currents along the western margins of ocean basins faster than those along the eastern margins?
5. What is *upwelling* and *downwelling*, and why are these processes important?
6. Along the coast of Queensland, Australia, the prevailing wind direction is from south to north. Will upwelling or downwelling occur along this coastline? Explain. A drawing might be useful.
7. Explain *thermohaline circulation*.
8. Define *water mass*.
9. What is meant by *constancy of composition*?
10. Define *salinity*.
11. What are the major inputs and outputs for dissolved species in the ocean?
12. What are *cyclical salts*?
13. What is the *Redfield ratio*?
14. Briefly discuss the factors that affect biological productivity in the ocean.
15. With reference to Figure 10–11, explain why the O_2 and CO_2 curves are mirror images and why phosphorus concentration increases with depth.
16. Distinguish among *biolimiting*, *biointermediate*, and *biounlimited* elements.
17. What is the major factor that determines if pore waters in marine sediments will become anoxic?
18. Compare and contrast the uptake and release of various species from and to seawater during high-temperature versus low-temperature hydrothermal processes.

19. Why do changes in dissolved CO_2, at constant salinity, lead to changes in the relative proportions of bicarbonate ions to carbonate ions?

20. Why are changes in alkalinity twice as great as changes in DIC when calcite is precipitated or dissolved in seawater?

21. With reference to Figure 10–13, explain the variations in carbonate alkalinity and dissolved inorganic carbon between the different water masses.

22. With reference to Figure 10–14, explain the variations in pH as a function of depth.

23. What are *stoichiometric equilibrium constants*, and how do they differ from the equilibrium constants used in previous chapters?

24. What is *remineralization*?

25. Name some possible sources of organic ligands.

26. What are the possible sources of trace metals in colloids, and how can we distinguish among these sources?

27. What is the *Irving–Williams order*?

28. What are the possible sources of trace metals found in the ocean?

29. Name and briefly describe the various components of suspended particulate matter (SPM).

30. Briefly describe the various ways in which trace metals can be removed from the surface ocean.

31. What is *scavenging*?

32. How can trace metals that behave conservatively be distinguished from those that behave nonconservatively?

33. What might a surface-water enrichment suggest about the source of a trace metal?

34. What are possible reasons for mid-depth maxima and minima?

35. Name and briefly describe the major components of sediments deposited on the continental margins and in the deep ocean.

36. Explain the distribution of deep-sea sediments shown in Figure 10–19.

37. What factors favor the preservation of silica shell material?

38. Define *saturation horizon*, *carbonate compensation depth*, and *lysocline*.

39. Briefly describe the authigenic processes that occur in deep-sea sediments.

40. What is a *methane hydrate*?

41. Define *estuary*.

42. Name and briefly describe the five morphological types of estuaries.

43. Name and describe the different types of dynamical estuaries.

44. Discuss the role of sill depth and climate (i.e., precipitation versus evaporation) in determining the vertical circulation within a semi-enclosed basin. Why is the extent of vertical circulation an important parameter in determining the chemistry of waters in a semi-isolated basin?

45. Can the concentration of a contaminant in an estuary ever be reduced to zero by flushing? Explain.

46. In terms of adsorption, in an estuarine environment where freshwater mixes with saltwater, compare and contrast the behavior of metal ions and organic species.

47. What is the *salting out* effect?

48. Define *oxycline* and *redoxcline*.

49. With reference to Case Study 10–4, what happened to Mn, and why, when the environment changed from anoxic to oxic?

50. Distinguish among *diffusion coefficient*, *dispersion coefficient*, and *eddy diffusivity*.

51. What processes can remove metals from pore waters?

52. What are *biomarkers*?

53. With reference to Case Study 10–5, briefly describe the changes in anthropogenic activities around San Francisco Bay as determined from the sediment record.

54. The following temperature and salinity data were collected from a midlatitude station in the North Atlantic:

Depth (m)	Salinity (‰)	Temperature (°C)
0	33.68	12.1
10	33.70	12.1
20	33.72	11.9
50	33.79	11.0
100	33.97	8.9
200	34.09	7.5
300	34.23	6.0
400	34.32	5.0
500	34.36	4.5
600	34.42	4.1
700	34.40	4.0
800	34.38	4.0

   a. Plot the data on a temperature–salinity diagram.
   b. Can the observed pattern be explained by simple mixing? Remember that real data don't necessarily form perfectly straight lines.
   c. Can you identify any potential mixing end members? If so, identify the end members.
   d. If the water sample collected at 400 m is the result of simple mixing, calculate the percent of the end members in this sample.

55. The following data were collected in the central North Pacific Ocean.

Depth (m)	Salinity (‰)	Temperature (°C)
20	34.658	26.53
49	34.663	26.53
99	34.811	25.21
149	35.007	21.32
248	34.403	13.97
544	34.235	6.10
742	34.429	5.27
988	34.513	4.18
1185	34.542	3.57
1534	34.583	2.72
2126	34.622	2.15
2530	34.640	1.82
3130	34.671	1.29
3735	34.680	1.16
4385	34.687	1.07
4953	34.698	0.91

   a. Plot the data on a temperature–salinity diagram.
   b. Interpret the data, including the characteristics of identifiable water masses.
   c. Can any portion of the water column be explained by simple mixing? If so, identify the mixing end members and calculate the percents of these end members in the sample collected at 2126 m.

56. A seawater sample has a chlorinity of 19.10‰. Calculate the salinity of this sample.

57. The following table gives average concentrations of N, Si, C_{inorg}, and Ca in surface and deep-ocean waters and river waters. Calculate f, g, and τ for these elements.

Element	Concentration (μmol L^{-1})		
	Surface water	Deep water	River water
N	3	35	20
Si	2	180	170
C_{inorg}	2,050	2,480	870
Ca	10,000	10,090	367

58. Calculate the ionic strength of a seawater sample with S‰ = 31.25.

59. Calculate the concentration of boron species as a function of pH (from 2 to 12) at $T = 2°C$ and salinity = 34.2‰.

60. Calculate the concentration of boron species as a function of pH (from 2 to 12) at $T = 2°C$, salinity = 34.2‰, and a depth of 3000 m.

61. You will recall from Chapter 3 that for the range of pH values encountered in the natural environment, silica exists in solution as silicic acid ($H_4SiO_{4\ (aq)}$) and $H_3SiO_4^-$. For the equilibria

$$H_4SiO_{4\ (aq)} \rightarrow H^+ + H_3SiO_4^-$$

Millero (1995) gives the following equation for the calculation of the dissociation constant:

$$\ln K_{si} = 117.40 - 8904.2/T - 19.334 \ln T + (3.5913 - 458.79/T)I^{0.5}$$
$$+ (-1.5998 + 188.74/T)I + (0.07871 - 12.1652/T)I^2$$

  a. Calculate the equilibrium constant at $T = 18°C$ and S = 33.8‰.
  b. Given a total silica content of 100 μmol kg^{-1}, calculate the concentration of $H_3SiO_4^-$ at pH = 8.3.
  c. For this temperature and salinity, calculate the concentration of the two silica species as a function of pH (from 2 to 12).

62. Bisulfate dissociates according to the following reaction:

$$HSO_4^- \rightarrow H^+ + SO_4^{2-}$$

  a. Khoo et al. (1977) give the following equation for the calculation of the dissociation constant:

$$\ln K_{HSO_4} = 647.59/T - 6.3451 + 0.019085T - 0.5208I^{0.5}$$

  Calculate the equilibrium constant at $T = 20°C$ and S = 34.0‰.
  b. For a seawater sample with a temperature of 20°C and a salinity of 34.0‰, calculate the concentration of HSO_4^- in solution at pH = 8.0. The sulfur value calculated from equation 10–10 represents the total sulfur as sulfate.
  c. Calculate the concentration of bisulfate and sulfate ion as a function of pH.
  d. Based on the results from parts (b) and (c), what do you conclude about the dominant species of sulfur in an oxic oceanic environment?
  e. Calculate the concentration of HSO_4^- in a seawater sample with $T = 2°C$, S = 34.52‰, and pH = 8.0 at a depth of 3000 m.
  f. For the seawater in part (e), calculate the concentration of bisulfate and sulfate ion as a function of pH.

63. Under anoxic conditions sulfur occurs in the reduced state. Hydrogen sulfide dissociates according to the following reaction:

$$H_2S \rightarrow HS^- + H^+$$

a. Millero (1995) gives the following equation for the calculation of the dissociation constant:

$$\ln K_{H_2S} = 225.838 - 13{,}275.3/T - 34.6435 \ln T + 0.3499S^{0.5} - 0.0274S$$

Calculate the equilibrium constant at $T = 20°C$ and S = 34.0‰.

b. For a seawater sample with a temperature of 20°C and a salinity of 34.0‰, calculate the concentration of HS^- in solution at pH = 8.0. The sulfur value calculated from equation 10–10 represents the total sulfur as sulfate.

c. Calculate the concentration of H_2S and HS^- as a function of pH.

d. Based on the results from parts (b) and (c), what do you conclude about the dominant species of sulfur in an anoxic oceanic environment?

e. Calculate the concentration of HS^- in a seawater sample with $T = 2°C$, S = 35.12‰, and pH = 8.0 at a depth of 3000 m.

f. For the seawater in part (e), calculate the concentration of H_2S and HS^- as a function of pH.

64. Under anoxic conditions nitrogen occurs in the reduced state. Ammonium dissociates according to the following reaction:

$$NH_4^+ \rightarrow NH_3 + H^+$$

a. Millero (1995) gives the following equation for the calculation of the dissociation constant:

$$\ln K_{NH_4^+} = -6285.33/T + 0.0001635T - 0.25444$$
$$+ (0.46532 - 123.7184/T)S^{0.5} + (-0.01992 + 3.17556/T)S$$

Calculate the equilibrium constant at $T = 20°C$ and S = 34.0‰.

b. For a seawater sample with a temperature of 20°C and a salinity of 34.0‰, calculate the concentration of NH_3 in solution at pH = 8.0. Total nitrogen = 40×10^{-6} mol kg^{-1}.

c. Calculate the concentration of NH_4^+ and NH_3 as a function of pH.

d. Based on the results from parts (b) and (c), what do you conclude about the dominant species of nitrogen in an anoxic oceanic environment?

e. Calculate the concentration of NH_3 in a seawater sample with $T = 2°C$, S = 34.6‰, and pH = 8.0 at a depth of 3000 m. Total nitrogen = 40×10^{-6} mol kg^{-1}.

f. For the seawater in part (e), calculate the concentration of NH_4^+ and NH_3 as a function of pH.

65. Phosphoric acid (H_3PO_4) is a triprotic acid and undergoes a three-step dissociation. Millero (1995) gives the following equations for calculating the dissociation constants:

$$\ln K_{P_1} = 115.54 - 4576.752/T - 18.453 \ln T$$
$$+ (0.69171 - 106.736/T)S^{0.5} + (-0.01844 - 0.65643/T)S$$

$$\ln K_{P_2} = 172.1033 - 8814.715/T - 27.927 \ln T$$
$$+ (1.3566 - 160.340/T)S^{0.5} + (-0.05778 + 0.37335/T)S$$

$$\ln K_{P_3} = -18.126 - 3070.75/T + (2.81197 + 17.27039/T)S^{0.5}$$
$$+ (-0.09984 - 44.99486/T)S$$

a. Given a total phosphorus concentration of 0.5 μmol kg^{-1}, calculate the concentration of phosphorus species as a function of pH (from 2 to 12) at $T = 2°C$ and salinity $= 34.2‰$.

b. Given a total phosphorus concentration of 3.5 μmol kg^{-1}, calculate the concentration of phosphorus species as a function of pH (from 2 to 12) at $T = 2°C$, salinity $= 34.2‰$, and a depth of 3000 m.

66. Calculate the buffer capacity, as a function of pH, for the boric acid system at $T = 2°C$, S $= 34.2‰$, and depth $= 3000$ m.

67. Calculate the buffer capacity, as a function of pH, for the carbonic acid system at $T = 2°C$, S $= 34.2‰$, and depth $= 3000$ m. Total carbonate concentration is 2.4 mmol kg^{-1}.

68. Calculate the buffer capacity, as a function of pH (from pH $= 5$ to pH $= 10$), for the calcite–carbonic acid system at $T = 4°C$, S $= 34.2‰$, and depth $= 3000$ m. The total carbonate concentration is 2.4 mmol kg^{-1}. You will find an example of this type of calculation in Chapter 3 (Example 3–15). The equation used to calculate the stoichiometric solubility product for calcite is found in the section dealing with marine sediments.

69. In Case Study 10–1, we investigated the quantification of the remineralization process. The following data were obtained for North Atlantic Deep Waters in the southern Atlantic Ocean. In a north-to-south direction, phosphate concentrations increase from 1.5 to 2.9 μmol kg^{-1} and nitrate concentrations increase from 21.7 to 42.1 μmol kg^{-1}. Over this distance, the DIC concentration increases by 190 μmol kg^{-1}. C_H was determined by measuring the change in buffer capacity of the ocean over this distance. Using the approach developed in Case Study 10–1, and the following data, plot the appropriate graphs required to determine the a, b, and c coefficients, and then determine the relative percentages of inorganic and organic carbon contributing to the observed DIC change.

$H_2PO_4^-$	NO_3^-	AOU	DIC $+$ C_H
1.50	21.7	257	2035
1.60	23.3	270	2052
1.65	23.8	281	2062
1.70	24.8	291	2078
1.80	26.0	304	2098
1.85	26.9	315	2111
1.90	27.4	323	2120
1.95	28.5	328	2125
2.00	28.7	344	2142
2.10	30.8	357	2160
2.15	31.2	366	2184
2.25	32.6	380	2190
2.30	33.7	394	2208
2.40	35.0	402	2235
2.50	36.3	430	2251
2.55	37.0	438	2261
2.70	39.0	461	2301
2.75	40.1	468	2308
2.80	40.6	473	2317
2.90	42.1	493	2338

70. Anoxic waters in a stagnant basin contain 0.001 mol kg^{-1} H$_2$S. Calculate the concentration of HS$^-$ at pH $= 7.7$, $T = 25°C$, and S $= 36.5‰$. See problem 63 for the equation to calculate ln K_{H_2S}.

71. With reference to Case Study 10–2, calculate, in percent, the amount of zinc in the deep water (1991 m) that is bound to organic ligands.

72. With reference to Case Study 10–2, for the northeastern Atlantic waters, plot depth versus the concentration of Si and the Zn_T/Si ratio (μmol mol^{-1}).
    a. Given the variation in silicon concentration with depth, is silicon a biologically utilized element? Explain.
    b. Given the following information, interpret the Zn_T/Si profile. (1) Similar measurements made in the Pacific Ocean show a relatively constant Zn_T/Si ratio of approximately 50 with depth. Surface waters show a somewhat lower ratio (20–40). (2) Mediterranean outflow water contains elevated levels of zinc and other metals and, given its density, Mediterranean water is typically found at depths between 1000 and 1200 m in the northeastern Atlantic Ocean.

73. The following chemical data were obtained for colloidal material in surface waters from the Middle Atlantic Bight (Guo et al., 2000). *The concentration of the colloidal organic matter (OC) is in mg g^{-1} and the concentrations of the metals are in μg g^{-1}.* The following questions are most easily answered by entering the data and doing the required calculations on a spreadsheet.

Station	OC	Cu	Pb	Zn	Cd	Co	Ni	Cr	Fe	Al	Mn
8	161	16.7	0.24	10.8	0.087	0.93	20.6	31.1	160	6.6	1.99
10	216	18.1	3.2	13.6	1.15	0.65	25	70.7	178	6.3	4.04
11	182	8.3	0.62	7.5	0.084	0.32	12.4	53.3	138	3.9	0.8
1	166	16.6	1.8	27.7	0.32	0.24	16.5	26.4	123	11.3	10.7
12	164	10.5	3.1	13.7	1.42	0.75	32.5	130	453	5.6	4.99
13	153	9.9	7.4	21.1	1.86	1.04	98.7	457	2165	13.2	7.86

    a. Calculate the Me/Al atomic ratios for each metal. Report both the average Me/Al ratio and the standard deviation for each ratio. Compare the average ratios to those given in Table 10–8 for possible colloidal sources. What can you conclude about the source of the metals in the colloids?
    b. Calculate the Me/C atomic ratios ($\times 10^5$) for each metal. Report both the average Me/C $\times 10^5$ ratio and the standard deviation for each ratio. Compare the average ratios to those given in Table 10–9 for possible colloidal sources. What can you conclude about the source of the metals in the colloids?
    c. Plot the various metal concentrations versus OC (organic carbon). You may want to do some curve fitting in order to investigate the correlation between organic carbon and metal concentrations. From these graphs, what can you conclude about the binding of the various metals to organic ligands associated with the colloids?

74. Given a surface-water Cd concentration of 1.32×10^{-9} g kg^{-1}, calculate the Cd concentration for the organic tissue of plankton and brown algae.

75. Given a surface-water Cr concentration of 4.2×10^{-7} g kg^{-1}, calculate the Cr concentration for the organic tissue of plankton and brown algae.

76. The data on the following page were collected at a station in the central North Pacific (GEOSECS program). Potential T is the temperature a water sample would have if it was brought adiabatically to the surface. Recall from Chapter 8 that during adiabatic processes changes in pressure cause pressure–volume work to be done that either uses internal heat energy or releases heat energy to the system. Water is not as compressible as air, and the change in temperature is not nearly as great. However, if a water sample was brought to the surface from great depth, it would expand and there would be a drop in temperature. Enter the data into a spreadsheet.

Depth (m)	Potential T (°C)	S (‰)	O_2 (mmol kg^{-1})	CO_2 (mmol kg^{-1})	PO_4 (μmol kg^{-1})	SiO_2 (μmol kg^{-1})
20	26.53	34.658	0.203	—	0.18	1.8
49	26.53	34.663	0.204	1.946	0.18	1.8
99	25.21	34.811	0.209	1.973	0.18	1.6
149	21.32	35.007	0.205	2.041	0.23	1.8
248	13.97	34.403	0.169	2.093	0.85	10.3
544	6.10	34.235	0.053	2.303	2.61	71.7
742	5.27	34.429	0.044	2.299	2.83	86.8
988	4.18	34.513	0.050	2.331	2.88	107.8
1185	3.57	34.542	0.058	2.357	2.90	121.6
1534	2.72	34.583	0.073	2.365	2.90	139.2
2126	2.15	34.622	0.102	2.353	2.78	156.4
2530	1.82	34.640	0.117	2.366	2.70	162.2
3130	1.29	34.671	0.135	2.375	2.61	164.4
3735	1.16	34.680	0.148	2.341	2.48	162.5
4385	1.07	34.687	0.157	2.323	2.44	152.8
4953	0.91	34.698	0.182	2.411	2.33	136.9

a. Plot depth versus CO_2 and O_2. Comment on the relative shape of the two curves as a function of depth. What do you conclude from the variations shown by the two curves about the regions in which photosynthesis and respiration dominate?

b. Plot depth versus PO_4 and SiO_2. You can either make two graphs or change the scale for one of the species so that the data plot in the same general region. What can you conclude about the bioactivity of these two species? Explain the depth–concentration relationships.

c. Plot temperature versus salinity. Can the observed distribution be explained by simple mixing? If not, how can you explain the temperature–salinity pattern? Think about the thermal structure of the ocean when answering this question. Appropriately label regions of the temperature–salinity plot.

d. For depths greater than 900 m, plot temperature versus salinity. How many water masses are present in the ocean below 900 m? Explain your answer.

e. For depths greater than 900 m, plot SiO_2 and O_2 versus salinity. You can either use two graphs or plot both species on the same graph using different scales.

f. From the graphs you constructed for parts (d) and (e), which properties (or species) behave conservatively and which behave nonconservatively? Explain. For the non-conservative species, are they being added or removed from the ocean? Explain.

77. The following hypothetical data were collected in the North Atlantic. Plot depth versus Sb concentrations. How would you classify the metal distribution pattern? What does this tells us about the source of the Sb and its bioactivity?

Depth (m)	Sb (μmol L^{-1})
0	2.98
10	3.01
20	2.52
30	2.08
50	1.00
100	0.98
250	1.02
500	1.00
1000	0.98
1500	1.01
2000	1.02

78. A subset of the data collected in a study of Cr distributions in the Arctic and Atlantic oceans (Sirinawin et al., 2000) is given in the following table. Enter the data into a spreadsheet.

Arctic Ocean				Atlantic Ocean			
Depth (m)	T (°C)	S (‰)	Cr (nmol L^{-1})	Depth (m)	T (°C)	S (‰)	Cr (nmol L^{-1})
0		32.74	3.79	25	21.58	35.78	4.04
10	−1.74	32.81	4.04	104	16.50	35.55	3.94
20	−1.74	32.81	4.19	280	12.93	35.17	4.42
50	−1.79	33.46	3.93	440	9.98	34.80	3.56
90	−1.30	34.19	4.68	620	6.63	34.43	4.52
190	0.84	34.77	5.02	800	4.52	34.33	4.28
290	1.04	34.89	4.40	961	3.64	34.36	4.57
590	0.28	34.90	4.57	1201	3.01	34.48	4.52
990	−0.33	34.91	4.71	1481	2.96	34.68	5.00
1290	−0.55	34.92	4.78	1801	2.87	34.81	4.86
1690	−0.73	34.93	5.05	2551	2.38	34.82	5.29
1990	−0.84	34.93	4.48				
2390	−0.91	34.93	5.03				
2490	−0.91	34.93	4.94				

a. Plot depth versus temperature curves for both sets of data. Comment on the differences between the two plots.

b. Plot temperature versus salinity for both sets of data. How many water masses can you identify in each plot? Are there any regions of simple mixing? Explain your answers.

c. Plot depth versus Cr concentration for both sets of data. How would you characterize these patterns? Are there any differences in the concentration–depth pattern for the Arctic Ocean versus the Atlantic Ocean?

d. Do you think Cr is a bioactive element? Explain your answer.

e. Plot Cr concentrations versus salinity. Does Cr behave conservatively? Explain.

79. The following data were collected in the western equatorial Pacific (Abe, 2001). Enter the data on a spreadsheet.

Depth (m)	T (°C)	S (‰)	PO$_4$ (mmol L^{-1})	Cd (μmol L^{-1})	Depth (m)	T (°C)	S (‰)	PO$_4$ (mmol L^{-1})	Cd (μmol L^{-1})
0		34.295	0.079	0.012	598	7.10	34.558	2.337	0.580
16	29.77	34.282	0.079	0.014	698	6.19	34.530	2.416	0.817
31	29.76	34.298	0.072	0.012	797	5.11	34.531	2.654	0.869
40	29.67	34.366	0.082	0.010	898	4.77	34.545	2.748	0.906
51	29.60	34.384	0.075	0.011	997	4.34	34.558	2.897	0.924
60	29.56	34.386	0.095	0.017	1097	4.11	34.564	2.850	0.850
70	29.47	34.405	0.095	0.016	1196	3.80	34.573	2.847	0.900
80	29.49	34.468	0.102	0.017	1396	3.23	34.590	2.825	0.900
90	29.54	34.530	0.108	0.015	1595	2.81	34.610	2.825	0.872
120	24.60	34.186	0.489	0.049	2592	1.73	34.664	2.689	0.868
151	21.38	35.849	0.723	0.080	2793	1.66	34.668	2.693	0.889
202	16.84	35.320	1.029	0.152	2992	1.61	34.672	2.668	0.866
299	12.06	34.859	1.587	0.301	3192	1.58	34.677	2.651	0.852
350	10.58	34.760	2.122	0.487	3391	1.56	34.679	2.626	0.850
399	9.38	34.684	2.196	0.630	3490	1.56	34.680	2.608	0.893
498	8.25	34.608	2.173	0.689					

a. Plot depth versus temperature. Identify the surface mixed layer, main thermocline, and deep waters.

b. On the same diagram plot depth versus PO_4 and Cd. Comment on the shape of the concentration–depth curves and the differences or similarities between the two curves. Do you think Cd is used by phytoplankton? Explain.

c. Plot temperature versus salinity. From this diagram what can you conclude about the structure of the ocean and the possibility of water-mass mixing?

d. Plot Cd versus PO_4. Do you see any patterns? Remember these are real data so they don't yield perfect results (or perfect straight lines).

e. Plot depth versus Cd/PO_4. Is there any pattern?

f. Given the previous plots, what do you think is happening in this part of the Pacific Ocean both with regard to the physical behavior of the ocean and the chemical behavior of Cd and PO_4? Are there any other plots that might be useful? Sometimes one can get a better understanding of the data by looking at vertical segments of the ocean. You may want to refer to Abe (2001).

80. Zhang et al. (2001) determined dissolved silver concentrations for the North Atlantic Ocean, Sea of Okhotsk, and Japan Sea. Their data for the North Atlantic Ocean are listed in the following table. Enter the data on a spreadsheet.

Depth (m)	Potential T (°C)	S (‰)	SiO_2 (nmol kg^{-1})	Ag (pmol kg^{-1})
0	20.20	34.119	0.7	4.4
49	15.27	34.575	12.8	5.7
99	11.73	34.117	21.3	5.7
198	10.29	34.177	21.3	5.9
495	4.69	33.944	78.7	10.4
794	3.64	34.235	123.5	19.0
991	3.17	34.362	139.5	23.8
1236	2.68	34.442	153.2	28.5
1483	2.33	34.497	163.6	33.7
1974	1.84	34.585	170.9	40.6
2467	1.56	34.631	168.7	44.6
2967	1.37	34.657	165.1	44.5
3445	1.25	34.665	161.5	44.9
3933	1.16	34.682	158.3	42.5
4420	1.11	34.684	156.8	42.0
4905	1.08	34.687	156.6	40.4
5225	1.07	34.688	154.6	41.7
5321	1.06	34.690	154.7	39.0
5418	1.06	34.688	155.1	41.5
5514	1.06	34.689	155.1	41.4
5564	1.06	34.689	155.3	41.6
5593	1.06	34.689	155.2	43.0
5610	1.06	34.691	154.8	41.5

a. Plot depth versus temperature. Identify the surface mixed layer, the main thermocline, and the deep waters.

b. On the same diagram plot depth versus SiO_2 and Ag concentrations. What type of trace metal distribution is shown by this plot? Comment on the similarities and differences for the two elements.

c. Plot temperature versus salinity. Explain the observed pattern. Does any segment of the water column show conservative behavior? If so, which segment?

d. For water depths greater than 200 m, plot Ag versus salinity. Interpret the diagram in terms of conservative versus nonconservative behavior and the removal and addition of Ag to the water column.

e. Plot depth versus Ag/SiO_2. What can you conclude from this diagram about the behavior of Ag relative to Si?

81. Using sediment traps, Cutter and Cutter (1998) determined scavenging times for the metalloids (As, Sb, and Se) in the North Atlantic Ocean. The sediment traps were located at a depth of 150 m in the Sargasso Sea and the measured fluxes are given in the following table. The depth-integrated (top 150 m) concentrations of these metals is also given. Calculate the scavenging time for each of these elements in the surface waters of the North Atlantic Ocean.

Element	Depth-integrated concentration (μmol L^{-1})	Sediment trap flux (μmol m^{-2} y^{-1})	Scavenging time (y)
As	2583	1.0	
Sb	125.4	0.18	
Se	67.7	1.5	

82. In deep waters in the North Pacific the activity ratio for ^{210}Pb/^{226}Ra $= 0.5$. Calculate the removal rate for ^{210}Pb, the scavenging turnover time, and the chemical half-life.

83. In a surface water the ^{228}Th/^{228}Ra activity ratio $= 0.1$. For ^{228}Th, $\lambda = 0.3623$ y^{-1}. Calculate the removal rate for ^{228}Th, the scavenging turnover time, and the chemical half-life.

84. Gustafsson et al. (1997) (Case Study 10–3) determined the following PAH concentrations and Th and U activities for a surface-water sample from the North Atlantic. Given that the depth of the mixed zone is 30 m, calculate the HOC flux for each of the PAHs and the residence time in the surface water for the HOCs.

Particulate PAH	Concentration (pmol m^{-3})	HOC flux (pmol m^{-2} d^{-1})	^{234}Th (dpm m^{-3}) Particulate	^{234}Th (dpm m^{-3}) Total	^{238}U (dpm m^{-3})
Pyrene	30		220	2270	2610
Benzo[a]pyrene	0.58				
Dibenz[a,h]anthracene	0.061				

85. Surface water in the Arctic Ocean has $T = 1.5°C$ and S $= 32.74‰$. Calculate the saturation state for calcite given a carbonate ion concentration of 0.24 mmol kg^{-1}.

86. The following data were obtained for the Pacific Ocean. Calculate the degree of saturation of calcite with depth and plot a depth–saturation curve. You will need to consider both temperature and pressure when making these calculations. Recall that pressure increases approximately 1 bar per 10-m increase in depth. You can calculate the carbonate ion concentration from the difference between the carbonate alkalinity and DIC. Use a spreadsheet to do this problem. At what depth does the Pacific Ocean become undersaturated with respect to calcite?

Depth (m)	T (°C)	S (‰)	CA (mmol kg^{-1})	DIC (mmol kg^{-1})	Depth (m)	T (°C)	S (‰)	CA (mmol kg^{-1})	DIC (mmol kg^{-1})
0	29.80	34.295	2.35	2.00	997	4.34	34.558	2.44	2.31
31	29.76	34.298	2.35	2.01	1196	3.80	34.573	2.45	2.33
51	29.60	34.384	2.34	2.02	1396	3.23	34.590	2.46	2.36
80	29.49	34.468	2.33	2.05	1595	2.81	34.610	2.47	2.39
120	24.60	34.186	2.35	2.07	1892	2.56	34.621	2.50	2.44
151	21.38	35.849	2.36	2.10	2045	2.12	34.645	2.51	2.45
202	16.84	35.320	2.37	2.15	2265	1.98	34.655	2.52	2.46
299	12.06	34.859	2.38	2.20	2592	1.73	34.664	2.52	2.46
350	10.58	34.760	2.38	2.21	2793	1.66	34.668	2.53	2.47
399	9.38	34.684	2.40	2.24	2992	1.61	34.672	2.53	2.47
598	7.10	34.558	2.41	2.26	3490	1.56	34.680	2.54	2.48
797	5.11	34.531	2.42	2.28					

87. For a semi-enclosed basin located at approximately 30°S, the salinity of the inflowing water is 34.51‰ and the salinity of the outgoing water is 35.69‰. On a daily basis, evaporation exceeds precipitation by 100,000 m^3 d^{-1}. Calculate the daily flow of water from the basin to the open ocean.

88. For a semi-enclosed basin located at approximately 45°N, the salinity of the inflowing water is 34.51‰ and the salinity of the outgoing water is 32.14. On a daily basis, precipitation exceeds evaporation by 90,000 m^3 d^{-1}. Calculate the daily flow of water from the ocean into the basin.

89. A contaminant is introduced into the freshwater portion of an estuary. The volume of the freshwater is 24×10^6 m^3. Daily inflow of freshwater is 2×10^6 m^3 d^{-1}. Calculate the minimum flushing time for the estuary.

90. A contaminant is introduced into a semi-enclosed basin. Assume the contaminant is completely mixed throughout the volume of the basin. The total volume of the basin is 1.1×10^8 m^3. The volume of water flowing out of the basin on a daily basis is 292,000 m^3 d^{-1}. Calculate the minimum flushing time for this basin.

91. A trace metal contaminant has reached a concentration of 12 μmol kg^{-1} in an estuary. The volume of the mixed zone is 5×10^8 m^3. The daily flow of freshwater into the estuary is 20×10^6 m^3 d^{-1}. Calculate the change in concentration of the contaminant with time. Extend your calculation until the contaminant is less than 10% of its initial value. Show the calculations graphically.

92. Plot the change in concentration (fraction remaining) of a contaminant in an estuary over 100 tidal cycles. The surface layer has a volume of 1×10^7 m^3 and the deep layer has a volume of 1.4×10^7 m^3. The exchange ratios are $r_S = 0.04$ and $r_D = 0.02$. How many tidal cycles does it take to reduce the contaminant concentration to 10% of its initial value?

93. The following data were obtained by Delgadillo-Hinojosa et al. (2001) for the Gulf of California. The stations extend along an almost linear SE transect down the center of the gulf. The distance between stations is given in kilometers starting at Station 6. This is a comprehensive problem and the data should be entered on a spreadsheet.

Station	Depth (m)	T (°C)	Salinity	Cd (nmol L^{-1})	PO_4 (μmol L^{-1})
6	10	28.46	35.394	0.219	0.70
(0 km)	20	28.22	35.374	0.212	0.80
	50	26.44	35.343	0.202	0.80
	120	17.67	35.140	0.572	1.50
	200	13.40	34.920	0.731	1.60
	300	11.17	34.782	0.702	1.60
8	10	26.20	35.276	0.358	0.70
(120 km)	30	25.72	35.234	0.379	1.20
	60	17.86	35.149	0.448	1.40
	100	16.69	35.023	0.725	2.20
	200	12.45	34.831	0.931	2.30
	300	11.45	34.765	0.909	2.40
	500	8.16	34.600	1.023	2.60
10	10	29.26	35.220	0.117	0.30
(170 km)	20	29.23	35.223	0.139	0.40
	50	26.50	35.133	0.287	1.20
	300	11.27	34.730	0.938	2.20
	500	8.16	34.588	1.084	2.50
	750	5.78	34.529	1.151	2.80
	1000	4.38	34.543	0.951	2.90

(*continued*)

Station	Depth (m)	T (°C)	Salinity	Cd (nmol L^{-1})	PO_4 (μmol L^{-1})
14	10	29.46	35.073	0.160	0.00
(300 km)	20	29.17	35.066	0.134	0.00
	50	23.49	35.181	0.448	1.00
	100	17.41	35.057	0.509	1.70
	250	12.45	34.818	0.750	1.80
	500	7.90	34.576	0.946	2.60
	750	5.72	34.519	1.126	2.60
	1000	4.51	34.539	1.144	2.70
17	10	29.32	34.915	0.088	0.40
(510 km)	20	28.35	34.887	0.331	0.80
	50	19.83	35.002	0.281	1.20
	100	16.07	34.979	0.585	1.30
	250	12.09	34.737	0.570	—
	750	5.71	34.512	1.039	2.70
	1000	4.35	34.537	1.124	2.80
19	10	29.32	34.961	0.160	1.00
(620 km)	20	28.35	34.967	0.285	1.00
	60	19.83	34.885	0.334	1.00
	100	16.07	34.891	0.378	1.10
	250	12.09	34.749	0.587	—
	500	7.07	34.523	0.960	2.30
	750	5.71	34.515	0.930	2.75
	1000	4.35	34.536	1.030	2.90

a. Draw two cross sections, one on which you plot depth versus the temperature for each station and the other on which you plot depth versus salinity for each station. The horizontal axis of the cross section is distance and the vertical axis is depth. You can use a different scale for each axis. Draw isotherms (lines of constant temperature) and isohalines (lines of constant salinity) on the appropriate diagram. For the isotherms use a 2°C contour interval, and for the isohalines use a 0.10‰ contour interval. Interpret the diagrams in terms of the distribution of water masses. After you have answered part (b), indicate the various water masses on the salinity cross section. Use heavy lines to delineate the boundaries of the water masses.

b. On the same diagram, for each station plot temperature versus salinity (T-S diagram). The following table gives the characteristics for the different water masses identified in this region. On the T-S diagram draw the boundaries for the various water masses and label the fields. Use this information to complete your answer for part (a).

Water mass	Temperature (°C)	Salinity (‰)
Gulf of California Water (GCW)	12–32	35.0– >36.0
Equatorial Surface Water (ESW)	18–32	< 34.4–35.0
Subtropical Surface Water (SSW)	9–18	< 34.4–35.0
Pacific Intermediate Water (PIW)	4–9	< 34.4–34.8
Pacific Deep Water (PDW)	0–4	< 34.4–34.8

c. On the same diagram, for each station plot depth versus Cd concentration. What type of trace metal distribution is shown by these curves?

d. Plot the concentration of Cd versus PO_4. Fit a regression line to the data, and determine the degree of correlation. Taking into consideration the plots from parts (c) and (d), what process controls the distribution of Cd in the Gulf of California? Explain.

94. Mason et al. (1999) determined the following mercury mass balance for Chesapeake Bay. Total Hg in Chesapeake Bay = 700 mol. Inputs to Chesapeake Bay (in mol y^{-1}): atmospheric deposition = 1300, dissolved from rivers = 340, particles from rivers = 1070, erosion = 375, groundwater inflow = 340, particles from ocean = 90, and dissolved from ocean = 840. Assuming that Chesapeake Bay is in a steady state with respect to mercury, calculate the residence time for mercury. Refer to Chapter 1 if you don't remember how to do this type of problem.

95. The following data for the western Mediterranean are reported by Takayanagi et al. (1996).

Depth (m)	Sb (μmol L^{-1})	PO$_4$ (μmol L^{-1})
0	1.31	0.01
15	1.34	0.03
30	1.34	0.03
60	1.27	0.04
100	1.30	0.30
150	1.31	0.34
200	1.33	0.35
500	1.31	0.39
750	1.27	0.39
1000	1.35	0.35
1500	1.33	0.35
2000	1.30	0.35
2515	1.33	0.37

a. Plot depth versus Sb and PO$_4$. From the shape of the curves, comment on the factors that control the distribution of these two species.

b. Takayanagi et al. (1996) present a two-layer steady-state model for the cycling of Sb in the western Mediterranean (Figure 10–P95). You may want to refer to the section in Chapter 1 dealing with steady-state models.

• Calculate the flux balance for the surface and deep layer. If the fluxes are not in balance (remember this is a steady-state model; i.e., the concentration of Sb in each layer remains constant), how might you account for the missing or excess Sb?

• Using the input fluxes, calculate the residence time of Sb in the surface and deep layers.

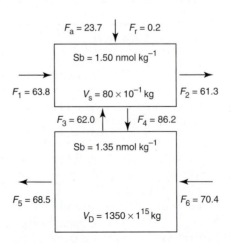

Figure 10–P95
Steady-state model for Sb in the western Mediterranean. Fluxes are in 1×10^6 mol y^{-1}. V_S is the volume of the surface layer and V_D is the volume of the deep layer. Adapted from Takayanagi et al. (1996).

96. The following chemical data were obtained for colloidal material in surface waters from Galveston Bay, Texas (Guo et al., 2000). The samples were collected along a salinity gradient from the Trinity River to the Gulf of Mexico off Galveston Island. *The concentration of the colloidal organic matter (OC) is in mg g^{-1} and the concentrations of the metals are in μg g^{-1}.* The following questions are most easily answered by entering the data and doing the required calculations on a spreadsheet.

Station	Salinity	OC	Cu	Pb	Zn	Cd	Co	Ni	Cr	Fe	Al	Mn
95-TR	0	161	18.1	0.34	7.3	0.266	1.85	28.0	3.7	995	13	3.58
95-02	0.2	313	58.8	0.39	20.8	0.153	5.04	51.3	19.0	475	63.5	13.6
95-03	2.3	264	71.6	1.28	20.4	0.157	2.75	42.6	13.8	294	12.7	8.10
95-04	5	222	64.7	0.31	19.7	0.267	2.34	34.3	9.3	216	3.6	2.49
95-05	7	228	66.8	0.47	27.9	0.145	2.73	43.9	37.0	237	11.3	3.13
95-06	10	203	61.2	0.41	19.9	0.199	2.93	33.9	18.5	221	9.2	3.69
95-07	12	179	50.5	0.50	28.0	0.161	2.60	29.2	19.9	206	22.4	3.63
95-08	15	159	46.3	0.43	21.7	0.176	2.43	22.0	12.4	199	12.2	4.36
95-10	20	171	46.7	0.52	28.5	0.172	3.90	23.3	27.8	218	12.6	8.30
95-12	25	103	28.4	0.41	9.4	0.161	1.32	10.6	17.9	236	31	2.82
95-13	29	107	36.8	0.39	16.2	0.279	1.39	10.0	15.4	200	6.7	6.25
95-14	33	69	19.3	0.23	6.6	0.179	0.73	5.5	20.8	216	4.2	1.21

a. Plot OC and each of the metal concentrations versus salinity. What relationships do you observe between OC and metal concentrations and salinity? Do any of the metals show a rapid decrease at low salinities? If any do, what might be the reason(s) for the rapid decrease?

b. Calculate the Me/Al atomic ratios for each metal. Report both the average Me/Al ratio and the standard deviation for each ratio. Compare the average ratios to those given in Table 10–8 for possible colloidal sources. What can you conclude about the source of the metals in the colloids?

c. Calculate the Me/C atomic ratios ($\times 10^5$) for each metal. Report both the average Me/C $\times 10^5$ ratio and the standard deviation for each ratio. Compare the average ratios to those given in Table 10–9 for possible colloidal sources. What can you conclude about the source of the metals in the colloids?

d. Plot the various metal concentrations versus OC (organic carbon). You may want to do some curve fitting in order to investigate the correlation between organic carbon and metal concentrations. From these graphs, what can you conclude about the binding of the various metals to organic ligands associated with the colloids?

97. The following data were obtained for the partitioning of nickel between water and particles in the Beaulieu estuary (Turner et al., 1998).

Salinity (‰)	K_D (ml g^{-1})
2.0	5360
10.0	5610
17.1	1860
29.1	1680
33.8	633

a. Plot ln K_D versus ln(S + 1). From this plot determine b and K_D^0 by fitting a regression line to the data. (*Note*: Plot the natural logs directly on the graph. Do not plot the data arithmetically and then change the axes to log axes.)

b. Taking into consideration the correlation coefficient, do you think there is a good correlation between K_D and salinity for this estuary? Explain.

c. Calculate the K_D for nickel at S = 15‰.

98. The following data for the partitioning of chromium between water and particles in the Clyde estuary were estimated from Turner (1996).

Salinity (‰)	K_D (ml g^{-1})
0.8	130,000
4.0	85,000
7.0	72,000
15.5	45,000
31.5	30,000

a. Plot ln K_D versus ln(S + 1). From this plot determine b and K_D^0 by fitting a regression line to the data. (*Note*: Plot the natural logs directly on the graph. Do not plot the data arithmetically and then change the axes to log axes.)

b. Taking into consideration the correlation coefficient, do you think there is a good correlation between K_D and salinity for this estuary? Explain.

c. Calculate the K_D for chromium at S = 21.5‰.

99. The following data were obtained for the partitioning of pyrene between water and sediments (Means, 1995).

	Lab Point		
Salinity (‰)	K_D	f_{OC}	K_{OC}
0	2273	0.035	
9	2913	0.035	
18	3337	0.035	
32	4532	0.035	

a. Calculate K_{OC}.

b. Plot K_{OC} versus salinity, and fit a regression line to the data.

c. Calculate the K_{OC} at a salinity of 21.1‰.

100. The following data were obtained for the partitioning of pyrene between water and sediments (Means, 1995).

	West Lake		
Salinity (‰)	K_D	f_{OC}	K_{OC}
0	2091	0.033	
9	4533	0.033	
18	5675	0.033	
32	6233	0.033	

a. Calculate K_{OC}.

b. Plot K_{OC} versus salinity, and fit a regression line to the data.

c. Calculate the K_{OC} at a salinity of 12.5‰.

101. In the text we discussed the distribution of trace metals between colloidal particles and solution. Wen et al. (1999) obtained the following data for Galveston Bay in July, 1993. SPM = suspended particulate matter, OC = organic carbon, D = dissolved (< 0.45 μm in size), and C = colloidal (< 0.45 μm in size and greater than 1 kDa in mass). Recall that the operative definition of colloidal material is that it have a mass greater than 1 kDa. Assuming 100% recovery, the difference between D and C is the amount of material in true solution.

Station	Salinity (‰)	SPM (mg L^{-1})	Phosphate (μmol L^{-1})	Silicate (μmol L^{-1})	OC (D) (μmol L^{-1})	OC (C) (μmol L^{-1})
1	0.1	41.6	1.57	114	649	424
2	1.3	37.6	3.30	127	549	341
3	4.9	23.3	3.54	122	457	283
4	5.9	22.0	4.08	117	466	291
5	9.2	18.4	2.65	91	591	358
37	21.1	9.4	1.25	65	316	183
8	28.4	7.0	0.76	46	291	—
9	30.0	4.9	0.73	46	216	133

Station	Cu (D) (μmol L^{-1})	Cu (C) (μmol L^{-1})	Ni (D) (μmol L^{-1})	Ni (C) (μmol L^{-1})	Pb (D) (μmol L^{-1})	Pb (C) (μmol L^{-1})
1	27.1	8.8	12.3	4.6	0.314	0.125
2	23.3	7.2	11.6	4.3	0.222	0.068
3	21.9	7.2	11.1	4.4	0.101	0.043
4	26.6	2.4	12.4	1.2	0.092	0.048
5	23.4	8.0	7.3	3.1	0.092	0.034
37	14.9	6.1	5.1	2.6	0.077	0.016
8	10.4	2.8	3.6	1.9	0.058	0.007
9	8.2	2.5	3.1	1.7	0.039	0.005

a. On the same graph, plot SPM, phosphate, and silicate versus salinity. Do any of these show conservative behavior? If not, comment on why they do not.

b. Phosphate and silicate are often considered to be nutrient elements. Plot silicate versus phosphate. Assuming that phosphate behaves as a nutrient element, comment on the behavior of silicate.

c. On the same graph plot OC (D) and OC (C) versus salinity. Does organic carbon behave as conservative species?

d. Make a series of graphs on which you plot Cu (D) and Cu (C), Ni (D) and Ni (C), and Pb (D) and Pb (C) versus salinity. Do any of these elements behave conservatively? Comment on the distribution of these elements between colloidal and truly dissolved form.

102. The following data were obtained by Grousset et al. (1995) for aerosols, river particles, and suspended particles in the Mediterranean.

	Sc (ppm)	Pb (ppm)	Zn (ppm)	As (ppm)
Aerosols				
Pic du Midi	14.2	293	695	3.1
City of Lyon	6.1	291	288	0.5
Rivers				
Rhône	17.5	96	252	1.7
Têt	10.3	52	128	4.9
Sediment Traps				
Depth (m)				
50	5.2	136	149	0.4
100	9.9	243	213	0.2
300	10.3	135	109	0.1
600	9.7	100	109	0.1

a. Calculate the enrichment factors for Pb using Sc as the reference element and average upper crust (Table 8–14) as the reference material. What is the most likely source for the Pb in the particles?

    b. What happens to the EF for Pb as a function of depth? What are some possible explanations for the observed changes?

    c. Calculate the enrichment factors for Zn using Sc as the reference element and average upper crust (Table 8–14) as the reference material. Assuming simple linear mixing, calculate the percent aerosol particles in the SPM collected at 50 m.

    d. What happens to the EF for Zn as a function of depth? What are some possible explanations for the observed changes?

103. For a particular basin, pore waters at a depth of 1 cm below the sediment–seawater interface contain 100 nmol L^{-1} Hg. The seawater at the bottom of the basin has a mercury concentration of 4 nmol L^{-1}. Ten meters above the seafloor, the Hg concentration is 0.1 nmol L^{-1}. The seawater diffusion coefficient for mercury is 5×10^{-6} cm^2 s^{-1}, and the eddy diffusion coefficient for the basin is 0.2 cm^2 s^{-1}. The sediment has a porosity of 80%.

    a. Calculate the mercury flux, on a yearly basis, from the bottom sediment to the seafloor.

    b. Calculate the mercury flux, on a yearly basis, at a height of 10 m above the seafloor.

104. For the River Weser estuary, Skowronek et al. (1994) determined the following relationships among Al, Mn, and Cd in uncontaminated estuarine sediments: Mn (g kg^{-1}) = -0.07 + 0.036Al (g kg^{-1}) and Cd (g kg^{-1}) = -0.40 + 0.07Al (g kg^{-1}). The following data, estimated from Skowronek et al. (1994), were obtained for a sediment core.

Depth (cm)	Al (g kg^{-1})	Mn (g kg^{-1})	EF_{Mn}	Cd (mg kg^{-1})	EF_{Cd}
0	30	1.3		1.0	
0.5	29	1.5		1.0	
1.0	34	1.3		1.0	
1.5	35	1.2		1.0	
2.0	35	1.1		1.0	
2.5	30	1.0		1.0	
3.0	18	0.7		0.9	
3.5	21	0.5		0.7	
4.0	30	0.4		0.9	
4.5	32	0.9		1.8	
5.0	38	0.6		0.4	
5.5	23	0.9		2.2	
6.0	21	0.8		2.7	
6.5	12	0.4		0.2	
7.0	26	0.7		2.2	
7.5	36	0.5		1.1	
8.0	31	0.6		1.9	

    a. Calculate the enrichment factors for Mn and Cd.

    b. Plot depth versus the concentrations of Al, Mn, and Cd.

    c. Plot depth versus the enrichment factors.

    d. Interpret the plots. Remember that both diagenetic changes (redox conditions may be very important, and at the time these data were collected the upper 2 cm of the sediment column were oxic) and anthropogenic inputs may both play a role in determining the concentrations of these metals in the sediment.

105. Because mollusks accumulate metals and store these metals in their shells—particularly elements not needed by the organism, such as heavy metals—mollusk shells have been used as bioindicators of environmental metal concentrations. Labonne et al. (1998) determined the lead isotopic composition of mollusk shells from the Thau coastal lagoon in France; from Sete harbor, which has a high density of industries and whose water discharges to Thau lagoon; and for shells from the Mediterranean Sea. Both recent and ancient shells (dating from the time of the Roman Empire) were analyzed. The data for the recent shells are reported in the following table. Also reported in the table are average values (estimated from the figures in Labonne et al., 1998) for possible end members: a road adjacent to the lagoon, cement plants, and Tertiary- and Jurassic-age sedimentary rocks.

Type	Location	$^{206}Pb/^{204}Pb$	$^{207}Pb/^{204}Pb$	$^{208}Pb/^{204}Pb$
Recent shells	Lafarge 1	18.214	15.621	38.171
	Lafarge 2	18.211	15.617	38.203
	Belaruc 1	18.178	15.618	38.156
	Belaruc 2	18.180	15.626	38.190
	Bouzigues 1	18.070	15.618	38.034
	Bouzigues 2	18.072	15.616	38.052
	Marseillan 1	18.131	15.630	38.127
	Onglous 1	18.270	15.644	38.303
	Sea spat	18.367	15.637	38.400
	Sea mussels	18.301	15.588	38.293
Sete harbor	Flesh	18.187	15.615	38.173
	Flesh	18.218	15.638	38.294
	Flesh	18.226	15.619	38.224
	Shell	18.215	15.629	38.239
Road		17.40		37.20
Lafarge cement		18.40		38.28
Tertiary sediments		18.90		38.90
Jurassic sediments		19.80		38.60

   a. Plot $^{208}Pb/^{204}Pb$ versus $^{206}Pb/^{204}Pb$. You may want to make two plots, one that shows all the data and an expanded plot that focuses on the shell data.

   b. What are the most likely sources for the Pb found in the shells? Estimate the percent of the various sources for each group of shells.

106. For a particular basin, the sedimentation rate is 0.22 cm y^{-1}, the sediments have a porosity of 0.85, and the density of the dry sediment is 2.6 g cm^{-3}. Calculate the sediment accumulation rate.

107. For a particular basin, the sedimentation rate is 1.0 cm y^{-1}, the sediments have a porosity of 0.89, and the density of the dry sediment is 2.5 g cm^{-3}. Carbonate shell material makes up 45% of the sediment. Calculate the sediment accumulation rate for the carbonate shell material.

108. The following data were obtained for a 3-m core collected from the mesohaline (intermediate salinity) region of Chesapeake Bay (Zimmerman and Canuel, 2000). Concentrations are on the basis of grams of dry sediment.

^{210}Pb age	Normalized depth (cm)	Organic C (mg g^{-1})	Total N (mg g^{-1})	AVS (μmol g^{-1})	CRS (μmol g^{-1})
1996.2	0.09	34.6	4.7	15.9	234.4
1995.2	1.76	33.8	4.5	39.7	193.6
1994.2	3.35	32.8	4.5	62	204.8
1993.7	4.34	31.4	4.2	32.8	285.7
1991.1	8.57	27.2	3.5	39.3	225.1
1988.7	12.62	26.9	3.4	66.1	238.3
1983.5	21.28	28.5	3.8	43.5	325.6
1975.8	34.11	26.4	3.3	29.2	241
1969.6	44.36	27.7	3.2	17.2	109
1962.2	56.71	25.8	2.8	16.7	108.6
1958.8	62.40	24.1	2.7	14.7	100.8
1954.2	70.11	25.0	2.7	10	107.7
1950.2	76.78	26.0	2.7	14.6	95.8
1946.2	83.36	25.8	2.6	14.4	111.3
1942.3	89.94	24.9	2.5	6.4	126.9
1938.3	96.69	25.1	2.4	11.2	131.1
1929.6	111.17	19.0	2.2	6	402.3
1910	126.51	17.4	2.2	—	—
1891	141.92	17.1	2.1	4.4	423
1872	157.46	17.4	2.2	3.5	365.2
1832	189.61	16.3	2.1	3.3	324.3
1785	226.85	15.2	1.9	1.6	327.1
1738	264.18	15.4	2.0	1.2	300.8
1689	303.34	16.0	1.9	1.7	322
1610	366.88	14.8	1.8	—	—

a. Plot depth versus the concentration of organic carbon. Based on this graph, when did anthropogenic activities begin to have an important impact on Chesapeake Bay?

b. Plot depth versus the molecular C/N ratio. You will have to convert the weight of carbon and nitrogen per gram of sediment to the moles of carbon and nitrogen per gram of sediment. Refer to Figure 5–26 and the accompanying discussion in Chapter 5. What is the significance of this plot in terms of changing C/N ratios with time and the types of organic material that are accumulating in the sediment? What does this tell you about anthropogenic activities around Chesapeake Bay?

c. Plot depth versus the AVS/CRS ratio, and comment on changes in redox condition as a function of time.

109. The following table lists concentrations of PAHs in surface sediments from Hypothetical Bay. Concentrations are in ng g^{-1} of dried sediment. Samples 1–5 are from sediments near the mouth of the river that enters the bay. There is an oil refinery adjacent to the river. Samples 6–10 are from the center of the bay, far removed from the river.

Sample	P	A	Fluo	Pyr
1	12.7	0.5	1.4	1.6
2	21	1	2	2.9
3	31	2	2.4	3.1
4	6.7	0.3	1	1.8
5	41	2.5	2.1	2.5
6	91	23	177	138
7	69	16	126	92
8	252	50	440	387
9	15	3.3	16	12
10	77	16	141	108

a. Calculate the P/A and Fluo/Pyr ratios for each sample.

b. Plot the ratios on a copy of Figure 10–26. Based on this plot, what can you conclude about the source of the PAHs at the various locations in Hypothetical Bay?

110. The following PAH concentrations (Pereira et al., 1999) were determined for a sediment core from Richardson Bay (San Francisco Bay). Concentrations are in ng g^{-1} dry weight of sediment.

Approximate date	Depth (cm)	P	A	Fluo	Pyr	MP
1990	0–10	236	75	537	648	62
	10–20	336	103	693	761	97
	20–30	430	133	773	926	113
1980	30–40	405	144	691	840	111
	40–50	332	90	604	675	82
1950	50–60	200	71	392	484	54
	60–70	418	117	873	980	108
	70–80	311	71	740	915	69
1910	80–90	62	11	228	269	18.3
	90–100	48	9.2	121	136	15.9
	100–110	20	3.6	57	70	8.6
	110–120	12	1.6	25	29	6.1
	120–138	5.3	< 0.5	6.4	7	4.2
	138–148	4.3	< 0.5	2.1	2.2	3.6

a. Calculate the P/A, Fluo/Pyr, and MP/P ratios.

b. Plot the P/A and Fluo/Pyr ratios on a copy of Figure 10–26. From this plot, what do you conclude about the source of the PAHs deposited in Richardson Bay?

c. Plot the MP/P ratio versus mean depth interval (i.e., 0–10 cm = 5 cm). What can you conclude from this plot about changes in anthropogenic activities in the San Francisco Bay region?

References

Abe, K., 2001. Cd in the western equatorial Pacific. *Marine Chemistry* **74**, 197–211.

Åberg, G., Pacyna, J. M., Stray, H., and Skjelkvåle, B. L., 1999. The origin of atmospheric lead in Oslo, Norway, studies with the use of isotopic ratios. *Atmospheric Environment* **33**, 3335–3344.

Abraham, E. R., Law, C. S., Boyd, P. W., Lavender, S. J., Maldonado, M. T., and Bowie, A. R., 2000. Importance of stirring in the development of an iron-fertilized phytoplankton bloom. *Nature* **407**, 727–730.

Aiken, G. R., McKnight, D. M., Wershaw, R. L., and MacCarthy, P., 1985. An introduction to humic substances in soil, sediment, and water. In Aiken, G. R., McKnight, D. M., Wershaw, R. L., and MacCarthy, P. (eds.). *Humic Substances in Soil, Sediment, and Water.* New York: Wiley-Interscience, pp. 1–12.

Al-Asasm, I. S., Clarke, J. D., and Fryer, B. J., 1998. Stable isotopes and heavy metal distribution in *Dreissena polymorpha* (Zebra Mussels) from western basin of Lake Erie, Canada. *Environmental Geology* **33**, 122–129.

Al-Farawati, R., and van den Berg, C. M. G., 1999. Metal-sulfide complexation in seawater. *Marine Chemistry* **63**, 331–352.

Amano, H., Matsunaga, T., Nagao, S., Hanzawa, Y., Watanabe, M., Ueno, T., and Onuma, Y., 1999. The transfer capability of long-lived Chernobyl radionuclides from surface soil to river water in dissolved forms. *Organic Geochemistry* **30**, 437–442.

Anderson, L. D., Kent, D. B., and Davis, J. A., 1994. Batch experiments characterizing the reduction of Cr(VI) using suboxic material from a mildly reducing sand and gravel aquifer. *Environmental Science and Technology* **28**, 178–185.

Anderson, T. F., and Arthur, M. A., 1983. Stable isotopes of oxygen and carbon and their application to sedimentologic and paleoenvironmental problems. In Arthur, M. A., Anderson, T. F., Kaplan, I. R., Veizer, J., and Land, L. S. (eds.). *Stable Isotopes in Sedimentary Geology.* SEPM Short Course 10, pp. 1–151.

Apambire, W. B., Boyle, D. R., and Michel, F. A., 1997. Geochemistry, genesis, and health implications of fluoriferous groundwaters in the upper regions of Ghana. *Environmental Geology* **33**, 13–24.

Aplin, A. C., 2000. Modern marine sediments. In Vaughan, D. J., and Wogelius, R. A. (eds.). *Environmental Mineralogy, European Mineralogical Union Notes in Mineralogy,* v. 2. Budapest: Eötvös University Press, pp. 125–172.

Appelo, C. A. J., and Postma, D., 1996. *Geochemistry, Groundwater and Pollution.* Rotterdam: A. A. Balkema, 536 pp.

Åström, M., 2001. The effect of acid soil leaching on trace element abundance in a medium-sized stream, W. Finland. *Applied Geochemistry* **16**, 387–396.

Atteia, O., Perret, D., Adatte, T., Kozel, R., and Rossi, P., 1998. Characterization of natural colloids from a river and spring in a karstic basin. *Environmental Geology* **34**, 257–269.

Ayliffe, L. K., Chivas, A. R., and Leakey, M. G., 1994. The retention of primary isotope compositions of fossil elephant skeletal phosphates. *Geochimica et Cosmochimica Acta* **58**, 5291–5298.

Ayotte, J. D., Nielsen, M. G., Robinson, G. R., Jr., and Moore, R. B., 1999. Relation of arsenic, iron, and manganese in ground water to aquifer type, bedrock lithogeochemistry, and land use in the New England coastal basins. *Water-Resources Investigations Report* **99–4162**, 61 pp.

Back, W., 1966. Hydrochemical facies and groundwater flow patterns in northern part of Atlantic coastal plain. *United States Geological Survey Professional Paper* **498-A**, 42 pp.

Back, W., and Hanshaw, B. B., 1970. Comparison of chemical hydrology of Florida and Yucatan. *Journal of Hydrology* **10**, 360–368.

Bains, S., Corfield, R. M., and Norris, R. D., 1999. Mechanisms of climate warming at the end of the Paleocene. *Science* **285**, 724–727.

Baird, C., 1995. *Environmental Chemistry.* New York: W. H. Freeman and Company, 484 pp.

Baker, E. W., 1964. Vanadium and nickel in crude petroleum of South America and Middle East origin. *Journal of Chemical and Engineering News* **42**, 307–308.

Bale, A. J., 1987. The characteristics, behaviour and heterogeneous reactivity of estuarine suspended particles. Ph.D. Thesis, Plymouth Polytechnic, 216 pp.

Banfield, J. F., and Nealson, K. H., eds., 1997. *Geomicrobiology: Interactions Between Microbes and Minerals.* Reviews in Mineralogy, v. 35. Washington, DC: Mineralogical Society of America, 448 pp.

Banks, D., Røyset, O., Strand, T., and Skarphagen, H., 1995. Radioelement (U, Th, Rn) concentrations in Norwegian bedrock groundwaters. *Environmental Geology* **25**, 165–180.

Bedient, P. B., Rifai, H. S., and Newell, C. J., 1994. *Ground Water Contamination: Transport and Remediation.* Englewoods Cliffs, NJ: Prentice Hall, Inc., 541 pp.

Ben Rhaiem, H., Tessier, D., and Ben Haj Amara, A., 2000. Mineralogy of the < 2 μm fraction of three mixed-layer clays from southern and central Tunisia. *Clay Minerals* **35**, 375–381.

Berger, A. C., Bethke, C. M., and Krumhansl, J. L., 2000. A process model of natural attenuation in drainage from a historic mining district. *Applied Geochemistry* **15**, 655–666.

Berner, R. A., 1981a. Authigenic mineral formation resulting from organic matter decomposition in modern sediments. *Fortschritte der Mineralogie* **59**, 117–135.

Berner, R. A., 1981b. A new geochemical classification of sedimentary environments. *Journal of Sedimentary Petrology* **51**, 359–365.

Berner, R. A., 1999. A new look at the long-term carbon cycle. *GSA Today* **9, 11**, 1–6.

Berner, E. K., and Berner, R. A., 1996. *Global Environment: Water, Air, and Geochemical Cycles.* Upper Saddle River, NJ: Prentice Hall, Inc., 376 pp.

Berry, L. G., ed., 1974. *Selected Powder Diffraction Data for Minerals.* Swathmore, NY: Joint Committee on Powder Diffraction Standards, 833 pp.

Berthelin, J., Leyval, C., and Mustin, C., 2000. Illustrations of the occurrence and diversity of mineral-microbe interactions involved in weathering of minerals. In Cotter-Howells, J. D., Campbell, L. S., Valsami-Jones, E., and Batchelder, M. (eds.). *Environmental Mineralogy: Microbial Interactions, Anthropogenic Influences, Contaminated Land and Waste Management.* London: The Mineralogical Society of Great Britain and Ireland, pp. 7–25.

Bethke, C. M., 1996. *Geochemical Reaction Modeling.* Oxford, UK: Oxford University Press, 397 pp.

Bish, D. L., and Guthrie, G. D., Jr., 1993. Mineralogy of clay and zeolite dusts (exclusive of 1:1 layer silicates). In Guthrie, G. D., Jr., and Mossman, B. T. (eds.). *Health Effects of Mineral Dusts.* Reviews in Mineralogy Volume 28. Washington, DC: Mineralogical Society of America, pp. 139–184.

Bollhöfer, A., Mangini, A., Lenhard, A., Wessels, M., Giovanoli, F., and Schwarz, B., 1994. High-resolution ^{210}Pb dating of Lake Constance sediments: stable lead in Lake Constance. *Environmental Geology* **24**, 267–274.

Bosma, T. N. P., Ballemans, E. M. W., Hoekstra, N. K., Welscher, A. G., Smeenk, J. G. M. M., Schraa, G., and Zehnder, A. J. B., 1996. Biotransformations of organics in soil columns and an infiltration area. *Ground Water* **34**, 49–56.

Boulet, M. P., and Larocque, A. C., 1998. A comparative mineralogical and geochemical study of sulfide mine tailings at two sites in New Mexico, USA. *Environmental Geology* **33**, 130–142.

Boyd, P. W. and 34 additional authors, 2000. A mesoscale phytoplankton bloom in the polar Southern Ocean stimulated by iron fertilization. *Nature* **407**, 695–702.

Bricker, O. P., and Garrels, R. M., 1965. Mineralogic factors in natural water equilibria. In Faust, S., and Hunter, J. V. (eds.). *Principles and Applications of Natural Water Chemistry.* New York: John Wiley & Sons, pp. 449–469.

Broecker, W. S., 1971. A kinetic model for the chemical composition of seawater. *Quaternary Research* **1**, 188–207.

Broecker, W. S., 1974. *Chemical Oceanography.* New York: Harcourt Brace Jovanovich, Inc., 214 pp.

Broecker, W. S., and Oversby, V. M., 1971. *Chemical Equilibria in the Earth.* New York: McGraw-Hill, 318 pp.

Brookins, D. G., 1978. Retention of transuranic and actinide elements and bismuth at the Oklo natural reactor, Gabon: application of Eh–pH diagrams. *Chemical Geology* **23**, 309–323.

Brookins, D. G., 1988. *Eh–pH Diagrams for Geochemistry.* Berlin: Springer Verlag, 176 pp.

Brownlow, A. H., 1996. *Geochemistry*, 2nd ed. Upper Saddle River, NJ: Prentice Hall, Inc., 580 pp.

Brügmann, L., Hallberg, R., and Larsson, C., 1998. Trace metal speciation in sea and pore water of the Gotland Deep, Baltic Sea, 1994. *Applied Geochemistry* **13**, 359–368.

Budzinski, H., Jones, I., Bellocq, J., Piérard, and Garrigues, P., 1997. Evaluation of sediment contamination by polycyclic aromatic hydrocarbons in the Gironde estuary. *Marine Chemisty* **58**, 85–97.

Budzinski, H., Raymond, N., Nadalig, T., Gilewicz, M., Garrigues, P., Bertrand, J. C., and Caumette, P., 1998. Aerobic biodegradation of alkylated aromatic hydrocarbons by a bacterial community. *Organic Geochemistry* **28**, 337–348.

Butler, T. J., and Likens, G. E., 1991. The impact of changing regional emissions on precipitation chemistry in the eastern United States. *Atmospheric Environment* **25A**, 305–315.

Carson, R., 1962. *Silent Spring.* Boston, MA: Houghton Mifflin, 368 pp.

Cates, D. A., Knox, R. C., and Sabatini, D. A., 1996. The impact of ion exchange processes on subsurface brine transport as observed on Piper diagrams. *Ground Water* **34**, 532–544.

Cember, H., 1983. *Introduction to Health Physics*, 2nd ed. New York: Pergamon Press, 517 pp.

Chiaradia, M., and Cupelin, F., 2000. Behaviour of airborne lead and temporal variations of its source effects in Geneva (Switzerland): comparison of anthropogenic versus natural processes. *Atmospheric Environment* **34**, 959–971.

Christanis, K., Georgakopoulos, A., Fernádex-Turiel, and Bouzinos, A., 1998. Geological factors influencing the concentration of trace elements in the Philippi peatland, eastern Macedonia, Greece. *International Journal of Coal Geology* **36**, 295–313.

Collister, J. W., Rieley, G., Stern, B., Eglinton, G., and Fry, B., 1994. Compound-specific $\delta^{13}C$ analyses of leaf lipids from plants with differing carbon dioxide metabolisms. *Organic Geochemistry* **21**, 265–276.

Coombs, D. S. and others, 1997. Recommended nomenclature for zeolite minerals: report of the subcommittee on zeolites of the International Mineralogical Association, Commission on New Minerals and Mineral Names. *Canadian Mineralogist* **35**, 1571–1606.

Coplen, T. B., 1993. Uses of environmental isotopes. In Alley, W. M., (ed.). *Regional Ground-Water Quality*. New York: Van Nostrand Reinhold, pp. 227–254.

Cowan, G. A., Bryant, E. A., Daniels, W. R., and Maeck, W. J., 1975. Some United States studies of the Oklo phenomenon. *Proceedings of the International Symposium on the Oklo Phenomenon, Libreville*. Vienna: International Atomic Energy Agency, SM-204/5, pp. 341–356.

Cox, J. D., Wagman, D. D., and Medvedev, V. A., 1989. *CODATA Key Values for Thermodynamics*. New York: Hemisphere Publishing Corporation, 271 pp.

Craig, H., 1961. Isotopic variations in meteoric waters. *Science* **133**, 1702–1703.

Craig, H., 1965. The measurement of oxygen isotope paleotemperatures. In *Stable Isotopes in Oceanographic Studies and Paleotemperatures*. Pisa: Consiglio Nazionale delle Ricerche, Laboratorio di Geologica Nucleare, pp. 1–24.

Craig, H., and Gordon, L. I., 1965. Deuterium and oxygen-18 variations in the ocean and the marine atmosphere. In *Stable Isotopes in Oceanographic Studies and Paleotemperatures*. Pisa: Consiglio Nazionale delle Ricerche, Laboratorio di Geologica Nucleare, pp. 25–122.

Crowley, T. J., 2000. Causes of climate change over the past 1000 years. *Science* **289**, 270–277.

Curtis, C. D., 1976. Stability of minerals in surface weathering reactions: a general thermochemical approach. *Earth Surficial Processes* **1**, 63–70.

Curtis, D., Benjamin, T., Gancarz, A., Loss, R., Rosman, K., DeLaeter, J., Delmore, J. E., and Maeck, W. J., 1989. *Applied Geochemistry* **4**, 49–62.

Cutter, G. A., 1993. Metalloids in wet deposition on Bermuda: concentrations, sources, and fluxes. *Journal of Geophysical Research* **98**, 16777–16786.

Cutter, G. A., and Cutter, L. S., 1998. Metalloids in the high latitude North Atlantic Ocean: sources and internal cycling. *Marine Chemistry* **61**, 25–36.

Daniels, F., and Alberty, R. A., 1967. *Physical Chemistry*, 3rd ed. New York: John Wiley & Sons, 767 pp.

Dansgaard, W., 1964. Stable isotopes in precipitation. *Tellus* **16**, 436–468.

Dansgaard, W. S., Johnson, S. J., Moller, J., and Langway, C. C., Jr., 1969. One thousand centuries of climatic record from Camp Century on the Greenland ice sheet. *Science* **166**, 377–381.

Dansgaard, W. S., Johnson, S. J., Clausen, H. B., and Langway, C. C., Jr., 1971. Climate record revealed by the Camp Century ice core. In Turekian, K. K. (ed.). *Late Cenozoic Glacial Ages*. New Haven, CT: Yale University Press, pp. 37–56.

Das, B. K., Singh, M., and Borkar, M. D., 1994. Sediment accumulation rate in the lakes of Kumaun Himalaya, India using ^{210}Pb and ^{226}Ra. *Environmental Geology* **23**, 114–118.

Davies, C. W., 1962. *Ion Association*. Washington, DC: Butterworth, 190 pp.

Davis, S. N., Whittemore, D. O., and Fabryka-Martin, J., 1998. Uses of chloride/bromide ratios in studies of potable water. *Ground Water* **36**, 338–350.

Deer, W. A., Howie, R. A., and Zussman, J., 1992. *An Introduction to the Rock-Forming Minerals*, 2nd ed. Burnt Hill, Harlow, UK: Longman Group Limited, 696 pp.

Deines, P., 1980. The isotopic composition of reduced organic carbon. In Fritz, P., and Fontes, J. Ch. (eds.). *Handbook of Environmental Isotope Geochemistry*, vol. 1. New York: Elsevier, pp. 329–406.

Deines, P., Langmuir, D., and Harmon, R. S., 1974. Stable isotope ratios and the existence of a gas phase in the evolution of carbonate ground water. *Geochimica et Cosmochimica Acta* **38**, 1147–1164.

Delgadillo-Hinojosa, F., Macías-Zamora, J. V., Segovia-Zavala, J. A., and Torres-Valdés, S., 2001. Cadmium enrichment in the Gulf of California. *Marine Chemistry* **75**, 109–123.

Dickens, G. R., Castillo, M. M., and Walker, J. C. G., 1997. A blast of gas in the latest Paleocene: simulating first-order effects of massive dissociation of oceanic methane hydrate. *Geology* **25**, 259–262.

Dickin, A. P., 1997. *Radiogenic Isotope Geology*. Cambridge, UK: Cambridge University Press, 490 pp.

Dickson, A. G., 1981. An exact definition of total alkalinity and a procedure for the estimation of alkalinity and total inorganic carbon from titration data. *Deep-Sea Research* **28A**, 609–623.

Domenico, P. A., and Schwartz, F. W., 1990. *Physical and Chemical Hydrogeology*. New York: John Wiley & Sons, 824 pp.

Drever, J. I., 1997. *The Geochemistry of Natural Waters*, 3rd ed. Englewood Cliffs, NJ: Prentice Hall, Inc., 436 pp.

Duxbury, A. C., 1971. *The Earth and Its Oceans*. Reading MA: Addison-Wesley Publishing Company, 381 pp.

Dyer, A., 2000. Applications of natural zeolites in the treatment of nuclear wastes and fall-out. In Cotter-Howells, J. D., Campbell, L. S., Valsami-Jones, E., and Batchelder, M. (eds.). *Environmental Mineralogy: Microbial*

Interactions, Anthropogenic Influences, Contaminated Land and Waste Management. London: The Mineralogical Society of Great Britain and Ireland, pp. 319–368.

Edmond, J. M., and von Damm, K., 1983. Hot springs on the ocean floor. *Scientific American* **248** 78–93.

Ehrlich, H. L., 1996. *Geomicrobiology*, 3rd ed. New York: Marcel Dekker, Inc., 719 pp.

Ellwood, M. J., and Van den Berg, C. M. G., 2000. Zinc speciation in the Northeastern Atlantic Ocean. *Marine Chemistry* **68**, 295–306.

Epstein, S., Buchsbaum, R., Lowenstram, H. A., and Urey, H. C., 1953. Revised carbonate-water isotopic temperature scale. *Geological Society of America Bulletin* **64**, 1315–1326.

Eriksson, E., 1957. The chemical composition of Hawaiian rainfall. *Tellus* **9**, 509–520.

Etheridge, D. M., Pearman, G. I., and Fraser, P. J., 1994. Historical CH_4 record from the "DE08" ice core at Law Dome. *Trends '93*, 256–260.

Faure, G., 1986. *Principles of Isotope Geology*, 2nd ed. New York: John Wiley & Sons, 589 pp.

Faure, G., 1998. *Principles and Applications of Geochemistry*, 2nd ed. Upper Saddle River, NJ: Prentice Hall, 600 pp.

Feder, G. L., and Lee, R. W., 1981. Water-quality reconnaissance of Cretaceous aquifers in the southeastern coastal plain. *United States Geological Survey Open File Report* **81-696**, 10 pp.

Fleagle, R. G., and Businger, J. A., 1963. *An Introduction to Atmospheric Physics.* New York: Academic Press, 346 pp.

Flückiger, J., Dällenbach, A., Blunier, T., Stauffer, B., Stocker, T. F., Raynaud, D., and Barnola, J.-M., 1999. Variations in atmospheric N_2O concentration during abrupt climatic changes. *Science* **285**, 227–230.

Foos, A., 1997. Geochemical modeling of coal mine drainage, Summit County, Ohio. *Environmental Geology* **31**, 205–210.

Fredrickson, J. K., Zachara, J. M., Kennedy, D. W., Duff, M. C., Gorby, Y. A., Li, S. W., and Krupka, K. M., 2000. Reduction of U(VI) in goethite (α-FeOOH) suspensions by a dissimilatory metal-reducing bacterium. *Geochimica et Cosmochimica Acta* **64**, 3085–3098.

Freeze, R. A., and Cherry, J. A., 1979. *Groundwater.* Englewood Cliffs, NJ: Prentice Hall, Inc., 604 pp.

Fritz, S. J., Bryan, J. D., Harvey, F. E., and Leap, D. I., 1994. A geochemical and isotopic approach to delineate landfill leachates in an RCRA study. *Ground Water* **32**, 743–750.

Gadd, G. M., 2000. Heterotrophic solubilization of metal-bearing minerals by fungi. In Cotter-Howells, J. D., Campbell, L. S., Valsami-Jones, E., and Batchelder, M.

(eds.) *Environmental Mineralogy: Microbial Interactions, Anthropogenic Influences, Contaminated Land and Waste Management.* London: The Mineralogical Society of Great Britain and Ireland, pp. 57–75.

Galloway, J. N., Likens, G. E., Keene, W. C., and Miller, J. M., 1982. The composition of precipitation in remote areas of the world. *Journal of Geophysical Research* **87**, 8771–8786.

Gambell, A. W., and Fisher, D. W., 1964. Occurrence of sulfate and nitrate in rainfall. *Journal of Geophysical Research* **69**, 4203–4210.

Gancarz, A., 1978. U-Pb age (2.05×10^9 years) of the Oklo uranium deposit. *International Atomic Energy Agency, Technical Communication* **119**, 513–520.

Garrels, R. M., 1967. Genesis of some ground waters from igneous rocks. In Abelson, P. H. (ed.). *Researches in Geochemistry*, vol. 2. New York: John Wiley & Sons, pp. 405–420.

Garrels, R. M., 1984. Montmorillonite/illite stability diagrams. *Clays and Clay Minerals* **32**, 161–166.

Garrels, R. M. and Christ, C. L., 1965. *Solutions, Minerals, and Equilibria.* New York: Harper & Row, 450 pp.

Garrels, R. M., Mackenzie, F. T., and Hunt, C., 1975. *Chemical Cycles and the Global Environment: Assessing Human Influences.* Los Altos, CA: William Kaufmann, Inc., 206 pp.

Garrison, A. W., Keith, L. H., and Shackelford, W. H., 1977. Occurrence, registry, and classification of organic pollutants in water, with development of a master scheme for their analysis. In Hutzinger, O., Van Lelyveld, I. H., and Zoeteman, B. C. J. (eds.). *Aquatic Pollutants: Transformation and Biological Effects.* Oxford, UK: Pergamon Press, pp. 39–68.

Gélinas, Y., Lucotte, M., and Schmit, J-P., 2000. History of the atmospheric deposition of major and trace elements in the industrialized St. Lawrence Valley, Quebec, Canada. *Atmospheric Environment* **34**, 1797–1810.

Gibbs, R. J., 1970. Mechanisms controlling world water chemistry. *Science* **170**, 1088–1090.

Gieskes, J. M., and Lawrence, J. R., 1981. Alteration of volcanic matter in deep sea sediments: evidence from the chemical composition of interstitial waters from deep sea drilling cores. *Geochminica et Cosmochimica Acta* **45**, 1687–1703.

Glasius, M., Wessel, S., Christensen, C. S., Jacobsen, J. K., Jørgensen, H. E., Klitgaard, K. C., Petersen, L., Rasmussen, J. K., Hansen, T. S., Lohse, C., Boaretto, E., and Heinemeier, J., 2000. Sources of formic acid studied by carbon isotopic analysis and air mass characterization. *Atmospheric Environment* **34**, 2471–2479.

Goldich, S. S., 1938. A study in rock weathering. *Journal of Geology* **46**, 17–58.

Golomb, D., Ryan, D., Eby, N., Underhill, J., and Zemba, S., 1997. Atmospheric deposition of toxics onto

Massachusetts Bay—I. metals. *Atmospheric Environment* **31**, 1349–1359.

Gordeev, V. V., and Siderov, I. S., 1993. Concentrations of major elements and their outflow into the Laptev Sea by the Lena River. *Marine Chemistry* **43**, 33–45.

Grasby, S. E., Hutcheon, I., and Krouse, H. R., 1997. Application of the stable isotope composition of SO₄ to tracing anomalous TDS in Nose Creek, southern Alberta, Canada. *Applied Geochemistry* **12**, 567–575.

Grim, R. E., Bray, R. H., and Bradley, W. F., 1937. The mica in argillaceous sediments. *American Mineralogist* **22**, 813–829.

Grootes, P. M., Mook, W. G., and Vogel, J. C., 1969. Isotopic fractionation between gaseous and condensed carbon dioxide. *Zeitschrift für Physik* **221**, 257–273.

Gross, M. G., and Gross, E., 1996. *Oceanography, A View of Earth*, 7th ed. Upper Saddle River, NJ: Prentice Hall, 472 pp.

Grousset, F. E., Quetel, C. R., Thomas, B., Donard, O. F. X., Lambert, C. E., Guillard, F., and Monaco, A., 1995. Anthropogenic vs. lithogenic origins of trace elements (As, Cd, Pb, Rb, Sb, Sc, Sn, Zn) in water column particles: northwestern Mediterranean Sea. *Marine Chemistry* **48**, 291–310.

Gschwend, P. M., and Hites, R. A., 1981. Fluxes of the polycyclic aromatic compounds to marine and lacustrine sediments in the northeastern United States. *Geochimica et Cosmochimica Acta* **45**, 2359–2367.

Gundersen, L. C. S., Schumann, R. R., Otton, J. K., Dubiel, D. E. O., and Dickinson, K. A., 1992. Geology of radon in the United States. In Gates, A. E., and Gundersen, L. C. S. (eds.). *Geologic Controls on Radon*. Boulder, CO: Geological Society of America Special Paper 271, pp. 1–16.

Gunter, W. D., Perkins, E. H., and Hutcheon, I., 2000. Aquifer disposal of acid gases: modelling of water–rock reactions for trapping of acid wastes. *Applied Geochemistry* **15**, 1085–1095.

Guo, L., Santschi, P. H., and Warnken, K. W., 2000. Trace metal composition of colloidal organic material in marine environments. *Marine Chemistry* **70**, 257–275.

Gustafsson, O., Gschwent, P. M., and Buesseler, K., 1997. Using ²³⁴Th disequilibria to estimate the vertical removal rates of polycyclic aromatic hydrocarbons from the surface ocean. *Marine Chemistry* **57**, 11–23.

Guthrie, G. D., Jr., and Mossman, B. T., 1993. *Health Effects of Mineral Dusts, Reviews in Mineralogy,* vol. 28. Washington, DC: Mineralogical Society of America, 584 pp.

Hackley, K. C., Liu, C. L., and Coleman, D. D., 1996. Environmental isotope characteristics of landfill leachates and gases. *Ground Water* **34**, 827–836.

Hagemann, R., Devillers, C., Lucas, M., Lecomte, T., and Ruffenach, J-C., 1975. Estimation de la durée de la réaction—Limitations imposées par les données neutroniques. *Proceedings of the International Symposium on the Oklo Phenomenon, Libreville*. Vienna: International Atomic Energy Agency, SM-204/28, pp. 415–423.

Harvey, G. R., and Boran, D. A., 1985. Geochemistry of humic substances in seawater. In Aiken, G. R., McKnight, D. M., Wershaw, R. L., and MacCarthy, P. (eds.). *Humic Substances in Soil, Sediment, and Water*. New York: Wiley-Interscience, pp. 233–248.

Hatcher, P. G., Rowan, R., and Mattingly, M. A., 1980. ¹H and ¹³C BNR of marine humic acids. *Organic Geochemistry* **2**, 77–85.

Hedges, J. I., and Oades, J. M., 1997. Comparative organic geochemistries of soils and marine sediments. *Organic Geochemistry* **27**, 319–361.

Hem, J. D., 1970. *Study and Interpretation of the Chemical Characteristics of Natural Water*, 2nd ed. Geological Survey Water-Supply Paper 1473. Washington, DC: United States Government Printing Office, 363 pp.

Hemming, N. G., and Hanson, G. N., 1992. Boron isotopic composition and concentration in modern marine carbonates. *Geochimica et Cosmochimica Acta* **56**, 537–543.

Henderson, P. J., McMartin, I., Hall, G. E., Percival, J. B., and Walker, D. A., 1998. The chemical and physical characteristics of heavy metals in humus and till in the vicinity of the base metal smelter at Flin Flon, Manitoba, Canada. *Environmental Geology* **34**, 39–57.

Hershey, J. P., Fernandez, M., Milne, P. J., and Millero, F. J., 1986. The ionization of boric acid in NaCl, Na-Ca-Cl and Na-Mg-Cl solutions at 25°C. *Geochimica et Cosmochimica Acta* **50**, 143–148.

Hidaka, H., Sugiyama, T., Ebihara, M., and Holliger, P., 1994. Isotopic evidence for the retention of ⁹⁰Sr inferred from excess ⁹⁰Zr in the Oklo natural fission reactors: implication for geochemical behaviour of fissiogenic Rb, Sr, Cs and Ba. *Earth and Planetary Science Letters* **122**, 173–182.

Hochella, M. F., Jr., 1993. Surface chemistry, structure, and reactivity of hazardous mineral dust. In Guthrie, G. D., Jr., and Mossman, B. T. (eds.). *Health Effects of Mineral Dusts Reviews in Mineralogy*, vol. 28. Washington, DC: Mineralogical Society of America, pp. 275–308.

Hodgson, G. W., 1954. Vanadium, nickel and iron trace metals in crude oils of western Canada. *AAPG Bulletin* **38**, 2537–2554.

Holland, H. D., 1978. *The Chemistry of the Atmosphere and Oceans*. New York: John Wiley & Sons, Inc., 351 pp.

Holser, W. T., and Kaplan, I. R., 1966. Isotope geochemistry of sedimentary sulfates. *Chemical Geology* **1**, 93–135.

Hong, S., Candelone, J.-P., Patterson, C. C., and Boutron, C. F., 1994. Greenland ice evidence of hemispheric lead pollution two millennia ago by Greek and Roman civilizations. *Science* **265**, 1841–1843.

Hong, S., Candelone, J.-P., and Boutron, C. F., 1997. Changes in zinc and cadmium concentrations in

Greenland ice during the past 7760 years. *Atmospheric Environment* **31**, 2235–2242.

Horibe, Y., and Oba, T., 1972. Temperature scales of aragonite-water and calcite-water systems. *Fossils* **23/24**, 69–79.

Hostettler, F. D., Pereira, W. E., Kvenvolden, K. A., van Geen, A., Luoma, S. N., Fuller, C. C., and Anima, R., 1999. A record of hydrocarbon input to San Francisco Bay as traced by biomarker profiles in surface sediment and sediment cores. *Marine Geology* **64**, 115–127.

Houghton, R. A., Hackler, J. L., and Lawrence, K. T., 1999. The U.S. carbon budget: contributions from land-use change. *Science* **285**, 574–578.

Hounslow, A. W., 1995. *Water Quality Data: Analysis and Interpretation*. New York: Lewis Publishers, 397 pp.

Huang, S., Arimoto, R., and Rahn, K. A., 2001. Sources and source variations for aerosol at Mace Head, Ireland. *Atmospheric Environment* **35**, 1421–1437.

Hume, L. A., and Rimstidt, J. D., 1992. The biodurability of chrysotile asbestos. *American Mineralogist* **77**, 1125–1128.

IPCC, 1992. *Climate Change 1992: The Supplementary Report to the IPCC Scientific Assessment*. Cambridge, UK: Cambridge University Press, 212 pp.

IPCC, 1996. *Climate Change 1995: The Science of Climate Change—Contribution of Working Group I to the Second Assessment Report of the IPCC*. Intergovernmental Panel on Climate Change. Cambridge, UK: Cambridge University Press, 584 pp.

Ishiwatari, R., 1985. Geochemistry of humic substances in lake sediments. In Aiken, G. R., McKnight, D. M., Wershaw, R. L., and MacCarthy, P. (eds.). *Humic Substances in Soil, Sediment, and Water*. New York: Wiley-Interscience, pp. 147–180.

IUPAC, 1992. Isotopic compositions of the elements—1989. *Pure and Applied Chemistry* **63**, 991–1002.

Ivany, L. C., Patterson, W. P., and Lohmann, K. C., 2000. Cooler winters as a possible cause of mass extinctions at the Eocene/Oligocene boundary. *Nature* **407**, 887–890.

Jannasch, H. W., and Mottl, M. J., 1985. Geomicrobiology of deep sea hydrothermal vents. *Science* **229**, 717–725.

Jensen, M. L., 1972. Bacteriogenic sulfur in air pollution. *Science* **177**, 1099–1100.

Johnson, N. F., and Mossman, B. T., 2001. Dose, dimension, durability and biopersistence of chrysotile asbestos. In Nolan, R. P., Langer, A. M., Ross, M., Wick, F. J., and Martin, R. F. (eds.). *The Health Effects of Chrysotile Asbestos: Contribution of Science to Risk-Management Decisions*. Canadian Mineralogist, Special Publication 5, pp. 145–154.

Johnson, S. J., Dansgaard, W. S., Clausen, H. B., and Langway, C. C., 1972. Oxygen isotope profiles through the Antarctic and Greenland ice sheets. *Nature* **235**, 429–434.

Joly, J., 1901. An estimate of the geological age of the earth. Washington, DC: *Smithsonian Institution Annual Report for 1899*, pp. 247–288.

Jones, A. M., and Young, E. D., 1998. In-situ UV ablation and IRM-GCMS measurements of subannual oxygen isotope variations in Proboscidean enamel phosphate—prospects for seasonal paleoclimatological and paleoecological studies. *Geological Society of America Abstracts with Programs*, A-373.

Kakihana, H., Kotaka, M., Satoh, S., Nomura, M., and Okamoto, M., 1977. Fundamental studies on the ion exchange separation of boron isotopes. *Bulletin of the Chemical Society of Japan* **50**, 158–163.

Karickhoff, S. W., 1981. Semi-empirical estimation of sorption of hydrophobic pollutants on natural sediments and soils. *Chemosphere* **10**, 833–846.

Kehew, A. E., 2001. *Applied Chemical Hydrogeology*. Upper Saddle River, NJ: Prentice Hall, Inc., 368 pp.

Keeley, D. F., Hoffpaulr, M. A., and Meriwether, J. R., 1988. Solubility of aromatic hydrocarbons in water and sodium chloride solutions of different ionic strengths: benzene and toluene. *Journal of Chemical and Engineering Data* **33**, 87–89.

Keil, R. G., Tsamakis, E., Fuh, C. B., Giddings, J. C., and Hedges, J. I., 1994. Mineralogical and textural controls on the organic composition of coastal marine sediments: hydrodynamic separation using SPLITT-fractionation. *Geochimica et Cosmochimica Acta* **58**, 879–893.

Kenaga, E. E., and Goring, C. A. I., 1980. Relationship between water solubility, soil sorption, octanol-water partitioning and concentration of chemicals in biota. *Proceedings Third Aquatic Toxicology Symposium ASTM STP* **707**, 78–115.

Kielland, J., 1937. Individual activty coefficients of ions in aqueous solutions. *Journal of the American Chemical Society* **59**, 1675–1735.

Khoo, K. H., Ramette, R. W., Culberson, C. H., and Bates, R. G. 1977. Determination of hydrogen ion concentrations in seawater from 5 to 40°C: standard potentials at salinities from 20 to 45‰. *Analytical Chemistry* **49**, 29–34.

Killops, S. D., and Killops, V. J., 1993. *An Introduction to Organic Geochemistry*. New York: John Wiley & Sons, 265 pp.

Kitack, L., Millero, F. J., and Campbell, D. M., 1996. The reliability of the thermodynamic constants for the dissociation of carbonic acid in seawater. *Marine Chemistry* **55**, 233–245.

Klein, C., and Hurlbut, C. S., 1993. *Manual of Mineralogy*, 21st ed. New York: John Wiley & Sons, 681 pp.

Kling, G. W., Tuttle, M. L., and Evans, W. C., 1989. The evolution of thermal structure and water chemistry in Lake Nyos. *Journal of Volcanology and Geothermal Research* **39**, 151–165.

Knauss, K. G., Dibley, M. J., Leif, R. N., Mew, D. A., and Aines, R. D., 1999. Aqueous oxidation of trichloroethene (TCE): a kinetic analysis. *Applied Geochemistry* **14**, 531–541.

Knauss, K. G., Dibley, M. J., Leif, R. N., Mew, D. A., and Aines, R. D., 2000. The aqueous solubility of trichloroethene (TCE) and tetrachloroethene (PCE) as a function of temperature. *Applied Geochemistry* **15**, 501–512.

Komor, S. C., and Anderson, H. W., Jr., 1993. Nitrogen isotopes as indicators of nitrate sources in Minnesota sandplain aquifers. *Ground Water* **31**, 260–270.

Krauskopf, K. B., 1986. Aqueous geochemistry of radioactive waste disposal. *Applied Geochemistry* **1**, 15–23.

Krauskopf, K. B., and Bird, D. K., 1995. *Introduction to Geochemistry*, 3rd ed. New York: McGraw-Hill, 647 pp.

Kusakabe, M., Ohsumi, T., and Aramaki, S., 1989. The Lake Nyos gas disaster: chemical and isotopic evidence in waters and dissolved gases from three Cameroonian crater lakes, Nyos, Monoun and Wum. *Journal of Volcanology and Geothermal Research* **39**, 167–185.

Kuss, J., and Kremling, K., 1999. Spatial variability of particle associated trace elements in near-surface waters of the North Atlantic (30°N/60°W to 60°N/2°W), derived by large volume sampling. *Marine Chemistry* **68**, 71–86.

Kvenvolden, K. A., 1988. Methane hydrate: a major reservoir of organic carbon in the shallow geosphere? *Chemical Geology* **71**, 41–51.

Kwong, Y. T. J., 1996. Metal leaching and immobilization in the vicinity of a naturally acidic lake in central Yukon. *Geological Association of Canada—Mineralogical Association of Canada Program with Abstracts* **21**, A52.

Kyser, T. K., ed., 1987. *Stable Isotope Geochemistry of Low Temperature Fluids*. Mineralogical Association of Canada Short Course Series Volume 13. Toronto: Mineralogical Association of Canada, 452 pp.

Labonne, M., Othman, D. B., and Luck, J.-M., 1998. Recent and past anthropogenic impact on a Mediterranean lagoon: lead isotope constraints from mussel shells. *Applied Geochemistry* **13**, 885–892.

Lambert, S. J., and Balsley, S. D., 1997. Stable-isotopes of groundwaters from the Albuquerque, New Mexico, basin: one decade later. *Environmental Geology* **31**, 199–204.

Lanctot, E. M., Rand, P. W., Lacombe, E. H., Hess, C. T., and Bogdan, G. F., 1992. The influence of season, bedrock, overburden, and house construction on airborne levels of radon in Maine homes. In Gates, A. E., and Gundersen, L. C. S. (eds.). *Geologic Controls on Radon*. Boulder, CO: Geological Society of America Special Paper 271, pp. 79–88.

Landmeyer, J. E., Chapelle, F. H., Petkewich, M. D., and Bradley, P. M., 1998. Assessment of natural attenuation of aromatic hydrocarbons in groundwater near a former manufactured-gas plant, South Carolina, USA. *Environmental Geology* **34**, 279–292.

Langmuir, D., 1971. The geochemistry of some carbonate groundwaters in central Pennsylvania. *Geochimica et Cosmochimica Acta* **42**, 1023–1045.

Langmuir, D., 1997. *Aqueous Environmental Geochemistry*. Upper Saddle River, NJ: Prentice Hall, Inc., 600 pp.

Larocque, A. C. L., and Rasmussen, P. E., 1998. An overview of trace metals in the environment, from mobilization to remediation. *Environmental Geology* **33**, 85–91.

Larsen, D., and Scarbrough, R. A., 2000. Sources of high dissolved manganese concentrations in mildly acidic runoff from a coal mine in Eastern Tennessee. *The Geological Society of America Abstracts with Programs* **32**, 3, A-33.

Lasaga, A. C., Soler, J. M., Ganor, J., Burch, T. E., and Nagy, K. L., 1994. Chemical weathering rate laws and global geochemical cycles. *Geochimica et Cosmochimica Acta* **58**, 2361–2386.

Lawrence, R. W., and Scheske, M., 1997. A method to calculate the neutralization potential of mining waters. *Environmental Geology* **32**, 100–106.

Lee, P-K., Baillif, P., and Touray, J-C., 1997. Geochemical behaviour and relative mobility of metals (Mn, Cd, Zn and Pb) in recent sediments of a retention pond along the A-71 motorway in Sologne, France. *Environmental Geology* **32**, 142–152.

Leenheer, J. A., Malcolm, R. C., McKinley, P. W., and Eccles, L. A., 1974. Occurrence of dissolved organic carbon in selected groundwater samples in the United States. *United States Geological Survey Journal of Research* **2**, 361–369.

Lerman, A., 1990. Transport and kinetics in surficial processes. In Stumm, W. (ed.). *Aquatic Chemical Kinetics*. New York: Wiley-Interscience, pp. 505–534.

Levorsen, A. I., 1967. *Geology of Petroleum*, 2nd ed. San Francisco, CA: W. H. Freeman, 724 pp.

Levy, D. B., Schramke, J. A., Esposito, K. J., Erickson, T. A., and Moore, J. C., 1999. The shallow ground water chemistry of arsenic, fluorine, and major elements: Eastern Owens Lake, California. *Applied Geochemistry* **14**, 53–65.

Lewis, E., and Wallace, D. W. R., 1998. *Program developed for CO₂ system calculations*. Carbon Dioxide Information Analysis Center, Oak Ridge National Laboratory, U.S. Department of Energy, Oak Ridge, Tennessee, ORNL/CDIAC-105, 31 pp.

Li, H., and Misawa, S., 1994. Groundwater chemistry within a plateau neighboring Matsumoto city, Japan. *Environmental Geology* **24**, 166–175.

Libes, S. M., 1992. *An Introduction to Marine Biogeochemistry*. New York: John Wiley & Sons, 734 pp.

Lide, D. R., ed., 2000. *CRC Handbook of Chemistry and Physics*, 81st ed. Boca Raton, FL: CRC Press.

Likens, G. E., Bormann, F. H., Pierce, R. S., Eaton, J. S., and Johnson, N. M., 1977. *Biogeochemistry of a Forested Ecosystem*. New York: Springer-Verlag, 146 pp.

Livingstone, D. A., 1963. Chemical composition of rivers and lakes. *USGS Professional Paper 440G*, 64 pp.

Lockwood, J. P., and Rubin, M., 1989. Origin and age of the Lake Nyos maar, Cameroon. *Journal of Volcanology and Geothermal Research* **39**, 117–124.

Lorius, C., Jouzel, J., and Raynaud, D., 1993. The ice core record: past archive of the climate and signpost to the future. *Antarctica and Environmental Change*. Oxford, UK: Oxford Science Publications, pp. 27–34.

Loss, R. D., Rosman, K. J. R., DeLaeter, J. R., Curtis, D. B., Benjamin, T. M., Gancarz, A. J., Maeck, W. J., and Delmore, J. E., 1989. Fission-product retentivity in peripheral rocks at the Oklo natural fission reactors, Gabon. *Chemical Geology* **76**, 71–84.

Loughnan, F. C., 1969. *Chemical Weathering of Silicate Minerals*. New York: American Elsevier, 154 pp.

Lueker, T. J., Dickson, A. G., and Keeling, C. D., 2000. Ocean pCO$_2$ calculated from dissolved inorganic carbon, alkalinity, and equations for K$_1$ and K$_2$: validation based on laboratory measurements of CO$_2$ in gas and seawater at equilibrium. *Marine Chemistry* **70**, 105–119.

Lyons, P. C., Curtis, A. P., Bostick, N. H., Fletcher, J. D., Dulong, F. T., Brown, F. W., Brown, Z. A., Krasnow, M. R., and Romankiw, L. A., 1989. Chemistry and origin of minor and trace elements in vitrinite concentrates from a rank series from the eastern United States, England, and Australia. *International Journal of Coal Geology* **13**, 481–527.

Macalady, D. L., Langmuir, D., Grundl, T., and Elzerman, A., 1990. Use of model generated Fe^{3+} ion activities to compute Eh and ferric oxyhydroxide solubilities in anaerobic systems. In Melchior, D. C., and Bassett, R. L. (eds.). *Chemical Modeling in Aqueous Systems II*. American Chemical Society Symposium Series 416. Washington, DC: American Chemical Society, pp. 350–367.

Malcolm, R. L., 1985. Geochemistry of stream fulvic and humic substances. In Aiken, G. R., McKnight, D. M., Wershaw, R. L. and MacCarthy, P. (eds.). *Humic Substances in Soil, Sediment, and Water*. New York: Wiley-Interscience, pp. 181–210.

Mangold, D. C., and Tsang, C., 1991. A summary of subsurface hydrological and hydrochemical models. *Reviews of Geophysics* **29**, 51–80.

Mantoura, R. F. C., Dickson, A., and Riley, J. P., 1978. The complexation of metals with humic materials in natural waters. *Estuarine Coastal Marine Sciences* **6**, 387–408.

Marshall, W. L., and Chen, C. A., 1982. Amorphous silica solubilities. V. Prediction of solubility behavior in aqueous mixed electrolyte solutions to 300°C. *Geochimica et Cosmochimica Acta* **46**, 289–291.

Martin, J. H., Fitzwater, S. R., and Gordon, R. M., 1990. Iron deficiency limits phytoplankton growth in Antarctic waters. *Global Biogeochemical Cycles* **4**, 5–12.

Mason, R. P., Fitzgerald, W. F., and Morel, F. M. M., 1994. The biogeochemical cycling of elemental mercury: anthropogenic influences. *Geochimica et Cosmochimica Acta* **58**, 3191–3198.

Mason, R. P., Lawson, N. M., Lawrence, A. L., Leaner, J. J., Lee, J. G., and Sheu, G.-R., 1999. Mercury in the Chesapeake Bay. *Marine Chemistry* **65**, 77–96.

Matthess, G., 1982. *The Properties of Groundwater*. New York: John Wiley & Sons, 406 pp.

Means, J. C., 1995. Influence of salinity upon sediment–water partitioning of aromatic hydrocarbons. *Marine Chemistry* **51**, 3–16.

Means, J. L., Yuretich, R. F., Crerar, D. A., Kinsman, D. J. J., and Borcsik, M. P., 1981. *Hydrogeochemistry of the New Jersey Pine Barrens*. Trenton, NJ: New Jersey Geological Survey Bulletin 76, 107 pp.

Mercone, D., Thomson, J., Croudace, I. W., and Troelstra, S. R., 1999. A coupled natural immobilisation mechanism for mercury and selenium in deep-sea sediments. *Geochimica et Cosmochimica Acta* **63**, 1481–1488.

Merlivat, L., and Jouzel, J., 1979. Global climatic interpretation of the deuterium-oxygen 18 relationship for precipitation. *Journal of Geophysical Research* **84**, 5029–5033.

Meybeck, M., 1979. Concentrations des eaux fluviales en éléments majeurs et apports en solution aux ocèans. *Reviews de Géologique Dynamique et Géographique Physique* **21(3)**, 215–246.

Middelburg, J. J., 1989. A simple rate model for organic matter decomposition in marine sediments. *Geochimica et Cosmochimica Acta* **53**, 1577–1581.

Miller, A., and Thompson, J. C., 1975. *Elements of Meteorology*, 2nd ed. New York: Charles E. Merrill Publishing Company, 362 pp.

Miller, J. M., 1974. A statistical evaluation of U.S. precipitation chemistry. In Semonin, R. G., and Beadle, R. W. (eds.). *Precipitation Scavenging*. ERDA Symposium Series 41, pp. 639–661.

Millero, F. J., 1982. The thermodynamics of seawater: the PVT properties. *Ocean Science and Engineering* **7**, 403–460.

Millero, F. J., 1995. Thermodynamics of the carbon dioxide system in the ocean. *Geochimica et Cosmochimica Acta* **59**, 661–677.

Millero, F. J., and Schreiber, D. R., 1982. Use of the ion pairing model to estimate activity coefficients of the ionic components of natural waters. *American Journal of Science* **282**, 1508–1540.

Mirecki, J. E., and Parks, W. S., 1994. Leachate geochemistry at a municipal landfill, Memphis, Tennessee. *Ground Water* **32**, 390–398.

Mitra, S., Dickhut, R. M., Kuehl, S. A., and Kimbrough, K. L., 1999. Polycyclic aromatic hydrocarbons (PAH) source, sediment deposition patterns, and particle geochemistry as factors influencing PAH distribution coefficients in sediments of the Elizabeth River, VA, USA. *Marine Chemistry* **66**, 113–127.

Monna, F., Clauer, N., Toulkeridis, T., and Lancelot, J. R., 2000. Influence of anthropogenic activity on the lead isotope signature of Thau Lake sediments (southern France): origin and temporal evolution. *Applied Geochemistry* **15**, 1291–1305.

Mook, W. G., Bommerson, J. C., and Staverman, W. H., 1974. Carbon isotope fractionation between dissolved bicarbonate and gaseous carbon dioxide. *Earth and Planetary Science Letters* **22**, 169–176.

Mucci, A., 1983. The solubility of calcite and aragonite in seawater at various salinities, temperatures, and one atmosphere total pressure. *American Journal of Science* **283**, 780–799.

Mukhopadhyay, P. K., Goodarzi, F., Crandlemire, A. L., Gillis, K. S., MacNeil, D. J., and Smith, W. D., 1998. Comparison of coal composition and elemental distribution in selected seams of the Sydney and Stellarton Basins, Nova Scotia, Eastern Canada. *International Journal of Coal Geology* **37**, 113–141.

Muller, F. L. L., 1999. Evaluation of the effects of natural dissolved and colloidal organic ligands on the electrochemical lability of Cu, Pb and Cd in the Arran Deep, Scotland. *Marine Chemistry* **67**, 43–60.

Müller, G., 1979. Schwermetalle in den sedimenten des Rheins–Veränderungen seit 1971. *Umschau* **79**, 778–783.

National Academy of Sciences, 1999. *Arsenic in Drinking Water*. Washington, DC: National Academy of Sciences Press, 273 pp.

Neiburger, M., Edinger, J. G., and Bonner, W. D., 1982. *Understanding Our Atmospheric Environment*, 2nd ed. San Francisco, CA: W. H. Freeman & Co., 453 pp.

Nichols, G., 1999. *Sedimentation and Stratigraphy*. Oxford, UK: Blackwell Science Ltd., 355 pp.

Nordstrom, D. K., and Munoz, J. L., 1986. *Geochemical Thermodynamics*. Palo Alto, CA: Blackwell Scientific, 477 pp.

Nordstrom, D. K., Plummer, L. N., Langmuir, E., Busenberg, E., and May, H. M., 1990. Revised chemical equilibrium data for major water–mineral reactions and their limitations. In Melchior, D. C., and Bassett, R. L. (eds.). *Chemical Modeling in Aqueous Systems II*. American Chemical Society Symposium Series 416. Washington, DC: American Chemical Society, pp. 398–413.

Norton, M. R., and Gunter, M. E., 1999. Relationships between respiratory diseases and quartz-rich dust in Idaho, U.S.A. *American Mineralogist* **84**, 1009–1019.

Oades, J. M., 1989. An introduction to organic matter in mineral soils. In Dixon, J. B., and Weed, S. B. (eds.). *Minerals in Soil Environments*, 2nd ed. Madison, WI: Soil Science Society of America, pp. 89–159.

Ohmoto, H., Yayashi, K-I., and Kajisa, Y., 1994. Experimental study of the solubilities of pyrite in NaCl-bearing aqueous solutions at 250–350°C. *Geochimica et Cosmochimica Acta* **58**, 2169–2185.

O'Neil, J. R., 1986. Terminology and standards. In Valley, J. W., Taylor, H. P., Jr., and O'Neill, J. R. (eds.). *Stable Isotopes in High Temperature Geological Processes. Reviews in Mineralogy,* v. 16, Washington, DC: Mineralogical Society of America, pp. 561–570.

Oremland, R. S., Dowdle, P. R., Hoeft, S., Sharp, J. O., Schaffer, J. K., Miller, L. G., Blum, J. S., Smith, R. L., Bloom, N. S., and Wallschlaeger, D., 2000. Bacterial dissimilatory reduction of arsenate and sulfate in meromictic Mono Lake, California. *Geochimica et Cosmochimica Acta* **64**, 3073–3084.

Pabalan, R. T., and Bertetti, F. P., 2001. Cation-exchange properties of natural zeolites. In Bish, D. L., and Ming, D. W. (eds.). *Natural Zeolites: Occurrence, Properties, and Applications. Reviews in Mineralogy and Geochemistry,* v. 45. Washington: Mineralogical Society of America, pp. 381–446.

Pagenkopf, G. K., 1978. *Introduction to Natural Water Chemistry*. New York: Marcel Dekker, Inc., 272 pp.

Parkhurst, D. L., 1990. Ion-association models and mean activity coefficients of various salts. In Melchior, D. C., and Bassett, R. L. (eds.). *Chemical Modelling of Aqueous Systems II*. American Chemical Society Symposium Series 416, pp. 30–43.

Pauling, L., 1960. *Nature of the Chemical Bond*, 3rd ed. Ithaca, NY: Cornell University Press, 644 pp.

Pearson, P. N., and Palmer, M. R., 1999. Middle Eocene seawater pH and atmospheric carbon dioxide concentrations. *Science* **284**, 1824–1826.

Peng, T. H., Takahashi, T., Broecker, W. S,. and Olafsson, J., 1987. Seasonal variability of carbon dioxide, nutrients and oxygen in the northern North Atlantic surface water: observations and model. *Tellus* **39B**, 439–458.

Pereira, W. E., Hostettler, F. D., Luoma, S. N., van Geen, A., Fuller, C. C., and Anima, R. J., 1999. Sedimentary record of anthropogenic and biogenic polycyclic aromatic hydrocarbons in San Francisco Bay, California. *Marine Chemistry* **64**, 99–113.

Pio, C. A., Alves, C. A., and Duarte, A. C., 2001. Identification, abundance and origin of atmospheric organic particulate matter in a Portuguese rural area. *Atmospheric Environment* **35**, 1365–1375.

Pitzer, K. S., 1973. Thermodynamics of electrolytes I: theoretical basis and general equations. *Journal of Physical Chemistry* **77**, 268–277.

Pitzer, K. S., 1979. Theory: Ion interaction approach. In Pytkowicz, R. M. (ed.). *Activity Coefficients in Electrolyte Solutions, Volume I*. Boca Raton, FL: CRC Press, pp. 157–208.

Pitzer, K. S., 1980. Electrolytes. From dilute solutions to fused salts. *Journal of the American Chemical Society* **102**, 2902–2906.

Plummer, L. N., and Busenberg, E., 1982. The solubilities of calcite, aragonite and vaterite in CO_2-H_2O solutions between 0 and 90°C, and an evaluation of the aqueous model for the system $CaCO_3$-CO_2-H_2O. *Geochimica et Cosmochimica Acta* **46**, 1011–1040.

Plummer, L. N., and MacKenzie, F. T., 1974. Predicting mineral solubility from rate data: application to the dissolution of magnesian calcites. *American Journal of Science* **274**, 61–83.

Plummer, L. N., Michel, R. L., Thurman, E. M., and Glynn, P. D., 1993. Environmental tracers for age dating young ground water. In Alley, W. M. (ed.). *Regional Ground-Water Quality*. New York: V. N. Reinhold, pp. 255–294.

Pohl, C., and Hennings, U., 1999. The effect of redox processes on the partitioning of Cd, Pb, Cu, and Mn between dissolved and particulate phases in the Baltic Sea. *Marine Chemistry* **65**, 41–53.

Price, G. D., Ruffell, A. H., Jones, C. E., Kalin, R. M., and Mutterlose, J., 2000. Isotopic evidence for temperature variation during the early Cretaceous (late Ryazanian–mid-Hauterivian). *Journal of the Geological Society, London* **157**, 335–343.

Quay, P. D., Tilbrook, B., and Wong, C. S., 1992. Oceanic uptake of fossil fuel CO_2: carbon-13 evidence. *Science* **256**, 74–79.

Rahn, K. A., 1999. A graphical technique for determining major components in a mixed aerosol. I. descriptive aspects. *Atmospheric Environment* **33**, 1441–1455.

Rahn, K. A., and Lowenthal, D. H., 1984. Elemental tracers of distant regional pollution aerosols. *Science* **223**, 132–139.

Ramamohana Rao, N. V., Rao, N., Surya Parkash Rao, K., and Schuiling, R. D., 1993. Fluorine distribution in waters of Nalgonda District, Andhra Pradesh, India. *Environmental Geology* **21**, 84–89.

Rast, W., and Holland, M., 1988. Eutrophication of lakes and reservoirs: a framework for making management decisions. *Ambio* **17**, 2–12.

Redfield, A. C., 1958. The biological control of chemical factors in the environment. *American Journal of Science* **46**, 205–222.

Reimann, C., Kashulina, G., de Caritat, P., and Niskavaara, H., 2001. Multi-element, multi-medium regional geochemistry in the European Arctic: element concentration, variation and correlation. *Applied Geochemistry* **16**, 759–780.

Rember, W. C., Erdman, T. W., Hoffmann, M. L., Chamberlain, V. E., and Sprenke, K. F., 1993. Dating of mine waste in lacustrine sediments using cesium-137. *Environmental Geology* **22**, 242–245.

Rimstidt, J. D., and Barnes, H. L., 1980. The kinetics of silica–water reactions. *Geochimica et Cosmochimica Acta* **44**, 1683–1699.

Robertson, W. D., and Blowes, D. W., 1995. Major ion and trace metal geochemistry of an acidic septic-system plume in silt. *Ground Water* **33**, 275–283.

Robie, R. A., Hemingway, B. S., and Fisher, J. R., 1978. *Thermodynamic properties of minerals and related substances at 298.15 K and 1 bar (10^5 pascal) pressure and at higher temperatures*. U.S. Geological Survey Bulletin 1452, 456 pp.

Roden, E. E., and Tuttle, J. H., 1993. Inorganic sulfur cycling in mid and lower Chesapeake Bay sediments. *Marine Ecology Program Series* **93**, 101–118.

Roxburgh, I. S., 1987. *Geology of High-Level Nuclear Waste Disposal*. London: Chapman and Hall, 229 pp.

Rubinson, M., and Clayton, R. N., 1969. Carbon-13 fractionation between aragonite and calcite. *Geochimica et Cosmochimica Acta* **33**, 997–1002.

Saffiotti, U., Daniel, L. N., Mao, Y., Williams, A. O., Kaighn, M. E., Ahmed, N., and Knapton, A. D., 1993. Biological studies on the carcinogenic mechanisms of quartz. In Guthrie, G. D., Jr., and Mossman, B. T. (eds.). *Health Effects of Mineral Dusts Reviews in Mineralogy*, v. 28. Washington, DC: Mineralogical Society of America, pp. 523–544.

Sanchez, A. G., Ayuso, E. A., and de Blas, O. J., 1999. Sorption of heavy metals from industrial waste water by low-cost mineral silicates. *Clay Minerals* **34**, 469–477.

Sargent-Welch Scientific Company, 1980. *Periodic Table of the Elements*. Skokie, Illinois.

Sarin, M. M., Krishnaswami, S., Dilli, K., Somayajulu, B. L. K., and Moore, W. S., 1989. Major ion chemistry of the Ganga-Bramaputra river systems, India. *Geochimicia et Cosmochimica Acta* **53**, 997–1009.

Schidlowski, M., 1988. A 3800-million-year isotopic record of life from carbon in sedimentary rocks. *Nature* **333**, 313–318.

Schlosser, P., Stute, M. Dörr, Sonntag, C., and Münnich, K. O., 1988. Tritium/^{3}He dating of shallow groundwater. *Earth and Planetary Science Letters* **89**, 353–362.

Schlosser, P., Stute, M. Sonntag, C., and Münnich, K. O., 1989. Tritiogenic ^{3}He in shallow groundwater. *Earth and Planetary Science Letters* **94**, 245–256.

Schlüter, K., 1997. Sorption of inorganic mercury and monomethyl mercury in an iron–humus podzol soil of southern Norway studied by batch experiments. *Environmental Geology* **30**, 266–279.

Schnitzer, M., 1978. Humic substances: chemistry and reactions. *Developments in Soil Science* **8**, 1–64.

Schoell, M., 1984. Stable isotopes in petroleum research. In Brooks, J., and Welte, D. (eds.). *Advances in Petroleum Geochemistry*, v. I. New York: Academic Press, pp. 215–245.

Schoell, M., 1988. Multiple origins of methane in the Earth. *Chemical Geology* **71**, 1–10.

Schumann, R. R., Owen, D. E., and Asher-Bolinder, S., 1992. Effects of weather and soil characteristics on temporal variations in soil–gas radon concentrations. In Gates, A. E., and Gundersen, L. C. S. (eds.). *Geologic*

Controls on Radon. Boulder, CO: Geological Society of America Special Paper 271, pp. 65–72.

Scott, M. J., and Morgan, J. J., 1990. Energetics and conservative properties of redox systems. In Melchior, D. C., and Bassett, D. C. (eds.). *Chemical Modeling of Aqueous Systems II*, American Chemical Society Symposium Series 416. Washington, DC: American Chemical Society, pp. 368–378.

Sewell, R., 1999. *Geochemical Atlas of Hong Kong.* Hong Kong: Geotechnical Engineering Office, Civil Engineering Department, The Government of the Hong Kong Special Administrative Region, 110 pp.

Siegel, F. R., Galasso, J. J., and Kravitz, J. H., 2001. Geochemistry of thirteen Voronin Trough cores, Kara Sea, European Arctic: Hg and As contaminants at a 1965 timeline. *Applied Geochemistry* **16**, 19–34.

Siegenthaler, U., and Sarmiento, J. L., 1993. Atmospheric carbon dioxide and the ocean. *Nature* **365**, 119–125.

Sigurdsson, H., Devine, J. D., Tchoua, F. M., Presser, T. S., Pringle, M. K. W., and Evans, W. C., 1987. Origin of the lethal gas burst from Lake Monoun, Cameroun. *Journal of Volcanology and Geothermal Research* **31**, 1–16.

Simcick, M. F., Eisenreich, S. J., and Lioy, P. J., 1999. Source apportionment and source/sink relationships of PAHs in the coastal atmosphere of Chicago and Lake Michigan. *Atmospheric Environment* **33**, 5071–5079.

Simões, J., and Zagorodnov, V. S., 2001. The record of anthropogenic pollution in snow and ice in Svalbard, Norway. *Atmospheric Environment* **35**, 403–413.

Singh, M. Ansari, A. A., Müller, G., and Singh, I. B., 1997. Heavy metals in the freshly deposited sediments of the Gomati River (a tributary of the Ganga River): effects of human activities. *Environmental Geology* **29**, 246–252.

Sirinawin, W., Turner, D. R., and Westerlund, S., 2000. Chromium (VI) distributions in the Arctic and the Atlantic Oceans and a reassessment of the oceanic Cr cycle. *Marine Chemistry* **71**, 265–282.

Sjöström, J., 1993. Ionic composition and mineral equilibria of acidic groundwater on the west coast of Sweden. *Environmental Geology* **21**, 219–226.

Skinner, B. J., Porter, S. C., and Botkin, D. B., 1999. *The Blue Planet*, 2nd ed. New York: John Wiley & Sons, 552 pp.

Skowronek, F., Sagemann, J., Stenzel, F., and Schulz, H. D., 1994. Evolution of heavy-metal profiles in River Weser Estuary sediments, Germany. *Environmental Geology* **24**, 223–232.

Smirnov, A., Abrajano, T. A., Jr., Smirnov, A., and Stark, A., 1998. Distribution and sources of polycyclic aromatic hydrocarbons in the sediments of Lake Erie, Part 1. Spatial distribution, transport, and deposition. *Organic Geochemistry* **29**, 1813–1828.

Solomon, D. K., Poreda, R. J., Cook, P. G., and Hunt, A., 1995. Site characterization using ^{3}H/^{3}He groundwater ages, Cape Cod, MA. *Ground Water* **33**, 988–996.

Solomon, S., 1990. Progress towards a quantitative understanding of Antarctic ozone depletion. *Nature* **347**, 347–354.

Spears, D. A., and Zheng, Y., 1999. Geochemistry and origin of elements in some UK coals. *International Journal of Coal Geology* **38**, 161–179.

Sposito, G., 1989. *The Chemistry of Soils*. New York: Oxford University Press, 277 pp.

Srinivasa Rao, N., and Rajendra Prasad, P., 1997. Phosphate pollution in the groundwater of lower Vamsadhara river basin, India. *Environmental Geology* **31**, 117–122.

Stallard, R. F., 1980. Major element geochemistry of the Amazon River system. Ph.D. dissertation, MIT/Woods Hole Oceanographic Institute, WHOI-80-29, 366 pp.

Stallard, R. F., and Edmond, J. M., 1983. Geochemistry of the Amazon 2: the influence of the geology and weathering environment on the dissolved load. *Journal of Geophysical Research* **88**, 9671–9688.

Stevenson, F. J., 1994. *Humus Chemistry, Genesis, Composition, Reactions*. New York: John Wiley & Sons, 496 pp.

Stoffyn-Egli, P., Buckley, D. E., and Clyburne, J. A. C., 1998. Corrosion of brass in a marine environment: mineral products and their relationship to variable oxidation and reduction conditions. *Applied Geochemistry* **13**, 643–650.

Stumm, W., and Morgan, J. J., 1996. *Aquatic Chemistry*, 3rd ed. New York: John Wiley & Sons, 1022 pp.

Sturges, W. T., Wallington, T. J., Hurley, M. D., Shine, K. P., Sihra, K., Engel, A., Oram, D. E., Penkett, S. A., Mulvaney, R., and Brenninkmeijer, C. A. M., 2000. A potent greenhouse gas identified in the atmosphere: SF_5CF_3. *Science* **289**, 611–613.

Stute, M., Deák, J., Révész, K., Böhlke, J. K., Deseö, É., Weppernig, R., and Schlosser, P., 1997. Tritium/^{3}He dating of river infiltration: an example from the Danube in the Szigetköz area, Hungary. *Ground Water* **35**, 905–911.

Sutton, C., and Calder, J. A., 1975. Solubility of alkybenzenes in distilled and sea water at 25.0°C. *Journal of Chemical and Engineering Data* **20**, 320–322.

Sverdrup, H. U., Johnson, M. W., and Fleming, R. H., 1942. *The Oceans*. Englewood Cliffs, NJ: Prentice Hall, Inc., 1059 pp.

Takahashi, T., Williams, R. T., and Bos, D. L., 1982. Carbonate chemistry. In Broecker, W. S., Spencer, D. W., and Craig, H. (eds.). *GEOSECS Pacific Expedition, Volume 3, Hydrographic Data 1973–1974*. Washington, DC: National Science Foundation, pp. 77–83.

Takayanagi, K., Cossa, D., and Martin, J.-M., 1996. Antimony cycling in the western Mediterranean. *Marine Chemistry* **54**, 303–312.

Tan, K. H., 1996. *Soil Sampling, Preparation, and Analysis*. New York: Marcel Dekker Inc., 408 pp.

Taylor, S. R., and McLennan, S. M. 1985. *The Continental Crust: Its Composition and Evolution.* Oxford, UK: Blackwell Scientific, 312 pp.

Tenzer, G. E., Meyers, P. A., Robbins, J. A., Eadie, B. J., Morehead, N. R., and Lansing, M. B., 1999. Sedimentary organic matter record of recent environmental changes in the St. Marys River ecosystem, Michigan–Ontario border. *Organic Geochemistry* **30**, 133–146.

Thatcher, L. L., 1962. The distribution of tritium fallout in precipitation over North America. *Bulletin of the International Association of Scientific Hydrology* **7**, 48–58.

Thode, H. G., and Monster, J., 1965. Sulfur-isotope geochemistry of petroleum, evaporites and ancient seas. In Young, A., and Galley, J. E. (eds.). *Fluids in Subsurface Environments.* American Association of Petroleum Geologists, Memoir 4, 367–377.

Thurman, E. M., 1985a. *Organic Geochemistry of Natural Waters.* Dordrecht: Martinus Nijhoff/Dr. W. Junk Publishers, 497 pp.

Thurman, E. M., 1985b. Humic substances in groundwater. In Aiken, G. R., McKnight, D. M., Wershaw, R. L., and MacCarthy, P. (eds.). *Humic Substances in Soil, Sediment, and Water.* New York: Wiley-Interscience, pp. 87–104.

Tipping, E., and Hurley, M. A., 1992. A unifying model of cation binding by humic substances. *Geochimica et Cosmochimica Acta* **56**, 3627–3641.

Tissot, B. P., and Welte, D. H., 1984. *Petroleum Formation and Occurrence.* Berlin: Springer-Verlag, 699 pp.

Tommasini, S., Davies, G. R., and Elliott, T., 2000. Lead isotope composition of tree rings as biogeochemical tracers of heavy metal pollution: a reconnaissance study from Firenze, Italy. *Applied Geochemistry* **15**, 891–900.

Tóth, E., Deak, F., Gyurkócza, C. S., Kasztovszky, Z. S., Kuczi, R., Marx, G., Nagy, B., Oberstedt, S., Sajó-Bohus, L., Sükösd, C. S., Toth, G., and Vajda, N., 1997. Radon variations in a Hungarian village. *Environmental Geology* **31**, 123–127.

Truesdell, A. H., and Jones, B. F., 1974. WATEQ, a computer program for calculating chemical equilibria of natural waters. *U.S. Geological Survey Journal of Research* **2**, 233–248.

Tuccillo, M. E., Cozzarelli, I. M., and Herman, J. S., 1999. Iron reduction in the sediments of a hydrocarbon-contaminated aquifer. *Applied Geochemistry* **14**, 655–667.

Turner, A., 1996. Trace-metal partitioning in estuaries: importance of salinity and particle concentration. *Marine Chemistry* **54**, 27–39.

Turner, A., and Millward, G. E., 1994. Partitioning of trace metals in a macrotidal estuary—Implications for contaminant transport models. *Estuarine Coastal Shelf Science* **39**, 45–58.

Turner, A., Nimmo, M., and Thuresson, K. A., 1998. Speciation and sorptive behaviour of nickel in an organic-rich estuary (Beaulieu, UK). *Marine Chemistry* **63**, 105–118.

Turner, J. V., 1982. Kinetic fractionation of carbon-13 during calcium carbonate precipitation. *Geochimica et Cosmochimica Acta* **46**, 1183–1191.

Underhill, J. T., 1995. Atmospheric deposition of metals and organics onto Massachusetts and Cape Cod Bays: a comparison of measurement techniques and source apportionment. Ph.D. Thesis, University of Massachusetts Lowell, 203 pp.

Urey, H. C., 1947. The thermodynamic properties of isotopic substances. *Journal of the Chemical Society* **1947**, 562–581.

Urey H. C., Lowenstam, H. A., Epstein, S., and McKinney, C. R., 1951. Measurement of paleotemperatures and temperatures of the Upper Cretaceous of England, Denmark, and the southeastern United States. *Geological Society of America Bulletin* **62**, 399–416.

van der Heijde, C. H., and Elnawawy, O. A., 1993. *Compilation of Ground-water Models.* U.S. Environmental Protection Agency Report EPA/600R-93/118, 118 pp.

Van Geen, A., and Luoma, S. N., 1999. The impact of human activities on sediments of San Francisco Bay, California: an overview. *Marine Chemistry* **64**, 1–6.

van Krevelen, D. W., 1963. Geochemistry of coal. In Breger, I. A. (ed.). *Organic Geochemistry.* New York: Macmillan Publishing Co., pp. 183–247.

vanLoon, G. W., and Duffy, S. J., 2000. *Environmental Chemistry.* Oxford, UK: Oxford University Press, 492 pp.

Veblen, D. R., and Wylie, A. G., 1993. Mineralogy of amphiboles and 1:1 layer silicates. In Guthrie, G. D., Jr., and Mossman, B. T. (eds.). *Health Effects of Mineral Dusts. Reviews in Mineralogy,* v. 28. Washington, DC: Mineralogical Society of America, pp. 61–138.

Velde, K. V., Barbante, C., Cozzi, G., Moret, I., Bellomi, T., Ferrari, C., and Boutron, C., 2000. Changes in the occurrence of silver, gold, platinum, palladium and rhodium in Mont Blanc ice and snow since the 18th century. *Atmospheric Environment* **34**, 3117–3127.

Véron, A., Flament, P., Bertho, M. L., Alleman, L., Flegal, R., and Hamelin, B., 1999. Isotopic evidence of pollutant lead sources in Northwestern France. *Atmospheric Environment* **33**, 3377–3388.

Von Arx, W. S., 1962. *Introduction to Physical Oceanography.* Reading, MA: Addison-Wesley, 422 pp.

Wagman, D. D., Evans, W. H., Parker, V. B., Schumm, R. H., Halow, I., Bailey, S. M., Churney, K. L., and Nuttall, R. L., 1982. The NBS tables of chemical thermodynamic properties: selected values for inorganic and C_1 and C_2 organic substances in SI units. *Journal of Physical and Chemical Reference Data* **11**, Supplement **2**, 1–392.

Wanty, R. B., Lawrence, E. P., and Gundersen, L. C. S., 1992. A theoretical model for the flux of radon from rock to ground water. In Gates, A. E., and Gundersen, L. C. S. (eds.). *Geologic Controls on Radon.* Boulder, CO: Geological Society of America Special Paper 271, pp. 65–72.

Watson, A. J., Bakker, D. C. E., Ridgwell, A. J., Boyd, P. W., and Law, C. S., 2000. Effect of iron supply on Southern Ocean CO_2 uptake and implications for glacial atmospheric CO_2. *Nature* **407**, 730–733.

Watson, J. G., 1984. Overview of receptor model principles. *Journal of the Air Pollution Control Association* **34**, 619–623.

Watson, J. G., Cooper, J. A., and Huntzicker, J. J., 1984. The effective variance weighting for least squares calculations applied to the mass balance receptor model. *Atmospheric Environment* **18**, 1347–1355.

Watson, J. G., Robinson, N. F., Chow, J. C., Henry, R. C., Kim, B. M., Pace, T. G., Meyer, E. I., and Nguyen, Q., 1990. The USEPA/DRI chemical mass balance receptor model, CMB 7.0. *Environmental Software* **5**, 38–49.

Watson, J. G., Chow, J. C., and Fujita, E. M., 2001. Review of volatile organic compound source apportionment by chemical mass balance. *Atmospheric Environment* **35**, 1567–1584.

Weiss, D., Shotyk, W., Kramers, J. D., and Gloor, M., 1999. *Sphagnum* mosses as archives of recent and past atmospheric lead deposition in Switzerland. *Atmospheric Environment* **33**, 3751–3763.

Weiss, W., Bullacher, J., and Roether, W., 1979. Evidence of pulsed discharges of tritium from nuclear energy installations in central European precipitation. In *Behaviour of Tritium in the Environment.* Vienna, IAEA-SM-232/18, pp. 17–30.

Wells, M. L., Kozelka, P. B., and Bruland, K. W., 1998. The complexation of "dissolved" Cu, An, Cd and Pb by soluble and colloidal organic matter in Narragansett Bay, RI. *Marine Chemistry* **62**, 203–217.

Wen, L.-S., Santschi, P., Gill, G., and Paternostro, C., 1999. Estuarine trace metal distributions in Galveston Bay: importance of colloidal forms in the speciation of the dissolved phase. *Marine Chemistry* **63**, 185–212.

Westerich, J. T., and Berner, R. A., 1984. The role of sedimentary organic matter in bacterial sulfate reduction: the G model tested. *Limnology and Oceanography* **29**, 236–249.

Whittaker, E. J. W., and Muntus, R., 1970. Ionic radii for use in geochemistry. *Geochimica et Cosmochimica Acta* **34**, 945–956.

Williams, J., 1962. *Oceanography.* Boston, MA: Little, Brown and Company, 242 pp.

Williamson, M. A., and Rimstidt, J. D., 1994. The kinetics and electrochemical rate-determining step of aqueous pyrite oxidation. *Geochimica et Cosmochimica Acta* **58**, 5443–5454.

Wilson, T. R. S., 1975. Salinity and the major elements of sea water. In Riley, J. P., and Skirrow, G., (eds.). *Chemical Oceanography,* 2nd ed. London: Academic Press, pp. 365–413.

Zhang, J-Z., 2000. The use of pH and buffer intensity to quantify the carbon cycle in the ocean. *Marine Chemistry* **70**, 121–131.

Zhang, Y., Amakawa, H., and Nozaki, Y., 2001. Oceanic profiles of dissolved silver: precise measurements in the basins of western North Pacific, Sea of Okhotsk, and the Japan Sea. *Marine Chemistry* **75**, 151–163.

Zhao, D., and Sun, B., 1986. Air pollution and acid rain in China. *Ambio* **15**, 2–5.

Zielinski, R. A., Chafin, D. T., Banta, E. R., and Szabo, B. J., 1997. Use of ^{234}U and ^{238}U isotopes to evaluate contamination of near-surface groundwater with uranium-mill effluent: a case study in south-central Colorado, U.S.A. *Environmental Geology* **32**, 124–136.

Zimmerman, A. R., and Canuel, E. A., 2000. A geochemical record of eutrophication and anoxia in Chesapeake Bay sediments: anthropogenic influence on organic matter composition. *Marine Chemistry* **69**, 117–137.

Zorbrist, J., and Stumm, W., 1980. Chemical dynamics of the Rhine catchment area in Switzerland: extrapolation to the "pristine" Rhine river input into the ocean. In Martin, J. M., Burton, J. D., and Eisma, D., (eds.). *River Inputs to Oceans Systems, SCOR/UNEP, UNESCO Review and Workshop.* Rome, FAO, pp. 52–63.

Zuerol, X., Whateley, M. K. G., Fernández-Turiel, and Tuncali, E., 1997. Geological controls on the mineralogy and geochemistry of the Beypazari lignite, central Anatolia, Turkey. *International Journal of Coal Geology* **33**, 255–271.

Zumdahl, S. S., 1989. *Chemistry,* 2nd ed.. Lexington, MA: D. C. Heath & Company, 1091 pp.

APPENDIX I

Electronic Structure and Valences of the Elements*

Period	Z	Element	Electronic structure	Common valences
1	1	H	$1s^1$	+1
	2	He	$1s^2$	0
2	3	Li	$1s^2 2s^1$	+1
	4	Be	$1s^2 2s^2$	+2
	5	B	$1s^2 2s^2 2p^1$	+3
	6	C	$1s^2 2s^2 2p^2$	+4, +2
	7	N	$1s^2 2s^2 2p^3$	+5, +3, −3
	8	O	$1s^2 2s^2 2p^4$	−2
	9	F	$1s^2 2s^2 2p^5$	−1
	10	Ne	$1s^2 2s^2 2p^6$	0
3	11	Na	$[Ne]3s^1$	+1
	12	Mg	$[Ne]3s^2$	+2
	13	Al	$[Ne]3s^2 3p^1$	+3
	14	Si	$[Ne]3s^2 3p^2$	+4
	15	P	$[Ne]3s^2 3p^3$	+5, +3, −3
	16	S	$[Ne]3s^2 3p^4$	+6, +4, +2, −2
	17	Cl	$[Ne]3s^2 3p^5$	−1
	18	Ar	$[Ne]3s^2 3p^6$	0
4	19	K	$[Ar]4s^1$	+1
	20	Ca	$[Ar]4s^2$	+2
	21	Sc	$[Ar]3d^1 4s^2$	+3
	22	Ti	$[Ar]3d^2 4s^2$	+4, +3
	23	V	$[Ar]3d^3 4s^2$	+5, +4, +3, +2
	24	Cr	$[Ar]3d^5 4s^1$	+6, +3, +2
	25	Mn	$[Ar]3d^5 4s^2$	+4, +3, +2
	26	Fe	$[Ar]3d^6 4s^2$	+3, +2
	27	Co	$[Ar]3d^7 4s^2$	+3, +2
	28	Ni	$[Ar]3d^8 4s^2$	+3, +2
	29	Cu	$[Ar]3d^{10} 4s^1$	+2, +1
	30	Zn	$[Ar]3d^{10} 4s^2$	+2
	31	Ga	$[Ar]3d^{10} 4s^2 3p^1$	+3, +1
	32	Ge	$[Ar]3d^{10} 4s^2 3p^2$	+4, +2
	33	As	$[Ar]3d^{10} 4s^2 3p^3$	+5, +3, −3
	34	Se	$[Ar]3d^{10} 4s^2 3p^4$	+6, +4, −2
	35	Br	$[Ar]3d^{10} 4s^2 3p^5$	−1
	36	Kr	$[Ar]3d^{10} 4s^2 3p^6$	0
5	37	Rb	$[Kr]5s^1$	+1
	38	Sr	$[Kr]5s^2$	+2
	39	Y	$[Kr]4d^1 5s^2$	+3
	40	Zr	$[Kr]4d^2 5s^2$	+4
	41	Nb	$[Kr]4d^3 5s^2$	+5, +3

Period	Z	Element	Electronic structure	Common valences
	42	Mo	$[Kr]4d^55s^1$	+6, +4, +3, +2
	43	Tc	$[Kr]4d^55s^2$	+7
	44	Ru	$[Kr]4d^65s^2$	+8, +4, +3, +2
	45	Rh	$[Kr]4d^75s^2$	+4, +3, +2
	46	Pd	$[Kr]4d^85s^2$	+4, +2
	47	Ag	$[Kr]4d^{10}5s^1$	+1
	48	Cd	$[Kr]4d^{10}5s^2$	+2
	49	In	$[Kr]4d^{10}5s^25p^1$	+3, +1
	50	Sn	$[Kr]4d^{10}5s^25p^2$	+4, +2
	51	Sb	$[Kr]4d^{10}5s^25p^3$	+5, +3, −3
	52	Te	$[Kr]4d^{10}5s^25p^4$	+6, +4, −2
	53	I	$[Kr]4d^{10}5s^25p^5$	−1
	54	Xe	$[Kr]4d^{10}5s^25p^6$	0
6	55	Cs	$[Xe]6s^1$	+1
	56	Ba	$[Xe]6s^2$	+2
	57	La	$[Xe]5d^16s^2$	+3
	58	Ce	$[Xe]4f^25d^06s^2$	+4, +3
	59	Pr	$[Xe]4f^35d^06s^2$	+4, +3
	60	Nd	$[Xe]4f^45d^06s^2$	+3
	61	Pm	$[Xe]4f^55d^06s^2$	+3
	62	Sm	$[Xe]4f^65d^06s^2$	+3, +2
	63	Eu	$[Xe]4f^75d^06s^2$	+3, +2
	64	Gd	$[Xe]4f^75d^16s^2$	+3
	65	Tb	$[Xe]4f^95d^06s^2$	+4, +3
	66	Dy	$[Xe]4f^{10}5d^06s^2$	+3
	67	Ho	$[Xe]4f^{11}5d^06s^2$	+3
	68	Er	$[Xe]4f^{12}5d^06s^2$	+3
	69	Tm	$[Xe]4f^{13}5d^06s^2$	+3, +2
	70	Yb	$[Xe]4f^{14}d^06s^2$	+3, +2
	71	Lu	$[Xe]4f^{14}5d^16s^2$	+3
	72	Hf	$[Xe]4f^{14}5d^26s^2$	+4
	73	Ta	$[Xe]4f^{14}5d^36s^2$	+5
	74	W	$[Xe]4f^{14}5d^46s^2$	+6, +4, +3, +2
	75	Rh	$[Xe]4f^{14}5d^56s^2$	+7, +6, +4, +2
	76	Os	$[Xe]4f^{14}5d^66s^2$	+8, +4, +3, +2
	77	Ir	$[Xe]4f^{14}5d^76s^2$	+6, +4, +3, +2
	78	Pt	$[Xe]4f^{14}5d^96s^1$	+4, +2
	79	Au	$[Xe]4f^{14}5d^{10}6s^1$	+3, +1
	80	Hg	$[Xe]4f^{14}5d^{10}6s^2$	+2, +1
	81	Tl	$[Xe]4f^{14}5d^{10}6s^26p^1$	+3, +1
	82	Pb	$[Xe]4f^{14}5d^{10}6s^26p^2$	+4, +2
	83	Bi	$[Xe]4f^{14}5d^{10}6s^26p^3$	+5, +3
	84	Po	$[Xe]4f^{14}5d^{10}6s^26p^4$	+4, +2
	85	At	$[Xe]4f^{14}5d^{10}6s^26p^5$	+7, +5, +3, +1
	86	Rn	$[Xe]4f^{14}5d^{10}6s^26p^6$	0
7	87	Fr	$[Rn]7s^1$	+1
	88	Ra	$[Rn]7s^2$	+2
	89	Ac	$[Rn]6d^17s^2$	+3
	90	Th	$[Rn]5f^06d^27s^2$	+4
	91	Pa	$[Rn]5f^26d^17s^2$	+5, +4
	92	U	$[Rn]5f^36d^17s^2$	+6, +5, +4, +3
	93	Np	$[Rn]5f^46d^17s^2$	+6, +5, +4, +3
	94	Pu	$[Rn]5f^66d^07s^2$	+6, +5, +4, +3
	95	Am	$[Rn]5f^76d^07s^2$	+6, +5, +4, +3
	96	Cm	$[Rn]5f^76d^17s^2$	+3

*From Faure (1998).

Standard State (298.15 K, 10⁵ Pa) Thermodynamic Data

Thermodynamic data are determined by experiment and are thus subject to various types of uncertainties: (1) has the reaction achieved equilibrium, (2) have the proper corrections been applied for the activity of the ions in solution, and (3) are the phases synthesized in the experiments the same as those encountered in the natural environment. For thermodynamic values determined from experimental reactions, changes in the values for one of the reactants or products will change the values for the other species. For example, let us assume that we have determined the solubility product for a particular precipitation reaction. Newer data provide a more accurate (?) free-energy value for one of the products. Because the solubility product doesn't change, the free energies of the other species involved in the reaction must be adjusted to take into account this revised value. This is often not done. When a particular thermodynamic data compilation is published, the author(s) try in ensure that the data are internally consistent. However, different data compilations, while internally consistent, may not agree with other compilations. CODATA values (Cox et al., 1989) represent an attempt to provide a universal reference base for thermodynamic data.

The following compilation of thermodynamic data is intended for classroom exercises. These data should not be used for research purposes without verification. Five sets of thermodynamic values are reported: the CODATA values of Cox et al. (1989); data from Drever (1997), which have been adjusted to conform with the CODATA reference values; the widely used mineralogical database of Robie et al. (1978); the NBS values of Wagman et al. (1982); and free-energy values for the transuranics and other radioactive elements from Brookins (1988). The most consistent results are obtained if all the thermodynamic data used in a calculation are taken from the same compilation.

Species or compound	ΔG_f^0 kJ mol^{-1}	ΔH_f^0 kJ mol^{-1}	S^0 J mol^{-1} K^{-1}	Source
Ag (silver)				
Ag $_{metal}$	0	0	42.6	3, 4
Ag$^+$	—	105.79	73.45	1
	77.1	105.8	73.4	3
	77.1	105.6	72.7	4
Ag^{2+}	269.0	268.6	88	4
AgCl$_2^-$	−215.4	−245.2	231.4	4
Ag(OH)$_2^-$	−260.2	—	—	4
AgCl $_{chlorargyrite}$	−109.8	−127.1	96.2	3, 4
Ag$_2$S $_{acanthite}$	−40.7	−32.6	144.0	4
Ag$_2$CO$_3$ $_{(cr)}$	−436.8	−505.8	167.4	4
Ag$_2$O $_{(cr)}$	−11.2	−31.1	121.3	4
Ag$_2$SO$_4$ $_{(cr)}$	−618.4	−715.9	200.4	4
Ag$_2$SeO$_4$ $_{(cr)}$	−334.2	−420.5	248.5	4
AgNO$_3$ $_{(cr)}$	−33.4	−124.4	140.9	4

Species or compound	ΔG_f^0 kJ mol^{-1}	ΔH_f^0 kJ mol^{-1}	S^0 J mol^{-1} K^{-1}	Source
Al (aluminum)				
Al $_{metal}$	0	0	28.3	3, 4
Al^{3+}	—	−538.4	−325	1
	−487.65	−540	−340	2
	−489.4	−531.0	−308.0	3
	−485	−531	−321.7	4
AlO$_2^-$	−830.9	−930.9	−36.8	4
Al(OH)$^{2+}$	−696.54	−778	−204	2
	−694.1	—	—	4
Al(OH)$_2^+$	−901.7	−1000	−16	2
Al(OH)$_{3 (aq)}$	−1100.6	−1230	108	2
Al(OH)$_4^-$	−1305.8	−1487	160	2
	−1305.3	−1502.5	102.9	4
Al$_2$O$_{3 \ corundum}$	—	−1675.7	50.92	1
	−1582.2	−1675.7	50.9	3
	−1582.3	−1675.7	50.9	4
Al(OH)$_{3 \ gibbsite}$	−1154.86	−1293.1	68.4	2
	−1154.9	−1293.1	68.4	3
	−1155.1	−1293.3	68.4	4
AlO(OH) $_{boehmite}$	−913	−994	—	2
	−918.4	−993.0	48.4	3
	−915.8	−990.4	48.4	4
AlO(OH) $_{diaspore}$	−992.0	−1000.6	35.3	3
	−920.9	−999.4	35.3	4
AlPO$_4$·2H$_2$O $_{variscite}$	−2097.8	—	—	4
Al$_2$Si$_2$O$_5$(OH)$_{4 \ kaolinite}$	−3785.8	−4133	—	2
	−3799.4	−4120.1	203.0	3
	−3799.7	−4119.6	205.0	4
Al$_2$Si$_2$O$_5$(OH)$_{4 \ dickite}$	−3796.3	−4118.8	197.1	3
	−3795.9	−4118.3	197.1	4
Al$_2$Si$_2$O$_5$(OH)$_{4 \ halloysite}$	−3769.4	−4114	—	2
	−3780.7	−4101.5	203.0	3
	−3780.5	−4101.2	203.3	4
Al$_2$Si$_4$O$_{10}$(OH)$_{2 \ pyrophyllite}$	−5273.3	—	—	2
	−5269.4	−5643.3	239.4	3
	−5268.1	−5642.0	239.4	4
Am (americium)				
Am^{3+}	−559.1	—	—	5
Am^{4+}	−373.2	—	—	5
AmOH^{2+}	−793.5	—	—	5
Am(OH)$_5^-$	−1529.6	—	—	5
AmO$_{2 (cr)}$	−880.4	—	—	5
Am$_2$O$_{3 (cr)}$	−1614.9	—	—	5
Am(OH)$_{3 (cr)}$	−1168.0	—	—	5
Am$_2$(CO$_3$)$_{3 (cr)}$	−2996.2	—	—	5
As (arsenic)				
As $_{(cr)}$	0	0	35.7	3
	0	0	35.1	4
AsO$^+$	−163.8	—	—	4
AsO$_2^-$	−350.0	−429.0	40.6	4
AsO$_3^{3-}$	−447.69	—	—	5
AsO$_4^{3-}$	−648.4	−888.1	−162.8	4
HAsO$_4^{2-}$	−714.6	−906.3	−1.7	4
HAsO$_3^{2-}$	−524.30	—	—	5
H$_2$AsO$_3^-$	−587.1	−714.8	110.5	4
H$_2$AsO$_4^-$	−753.2	−909.6	117.0	4

Species or compound	ΔG_f^0 kJ mol^{-1}	ΔH_f^0 kJ mol^{-1}	S^0 J mol^{-1} K^{-1}	Source
H_3AsO_3 (aq)	−639.8	−742.2	195.0	4
H_3AsO_4 (aq)	−766.0	−902.5	184.0	4
As_2O_3 arsenolite	−576.0	−657.0	107.4	3
	−576.2	−657.0	107.1	4
As_2O_3 clauderite	−575.6	−654.8	113.3	3
	−577.0	−654.8	117.0	4
As_2S_2 realgar	−140.6	−142.7	127.0	3
	—	−142.7	—	4
As_2S_3 orpiment	−168.4	−169.0	163.6	3
	−168.6	−169.0	163.6	4
Ba (barium)				
Ba^{2+}	−555.4	−532.5	8.4	2
	−560.7	−537.6	9.6	3
	−560.8	−537.6	9.6	4
$BaCl_2$ (cr)	−810.4	−858.6	123.7	4
BaF_2 (cr)	−1156.8	−1207.1	96.4	4
$BaCO_3$ witherite	−1132.21	−1210.9	112.1	2
	−1132.2	−1210.8	112.1	3
	−1137.6	−1216.3	112.1	4
$BaSO_4$ barite	−1362.2	−1473.2	132.2	2, 3, 4
$Ba(NO_3)_2$ nitrobarite	−796.6	−992.1	213.8	3, 4
Be (beryllium)				
Be metal	0	0	9.50	1, 3, 4
Be^{2+}	−379.7	−383.0	−130.0	3
	−379.7	−382.8	−129.7	4
BeO_2^{2-}	−640.1	−790.8	−159	4
$Be_3(OH)_3^{3+}$	−1801.6	—	—	4
BeO bromellite	—	−609.4	13.77	1
	−580.1	−609.4	13.8	3
	−580.3	−609.6	—	4
$Be(OH)_2$ (cr, α)	−815.0	−902.5	51.9	4
$BeCl_2$ (cr, α)	−445.6	−490.4	82.7	4
$BeSO_4$ (cr, tetragonal)	−1093.8	−1205.2	77.9	4
$BeAl_2O_4$ chrysoberyl	−2178.5	−2300.8	66.3	3, 4
Be_2SiO_4 phenacite	—	—	64.3	3
	−2032.5	−2149.3	64.3	4
C (carbon)				
C graphite	0	0	5.74	1, 2
	0	0	5.7	3, 4
C diamond	2.9	1.9	2.4	3, 4
CCl_4 (l)	−65.3	−135.4	216.4	4
HCN (aq)	119.7	107.1	—	4
CN^-	172.4	150.6	94.1	4
CH_4 (g)	−50.72	−74.81	186.3	2
	−50.7	−74.8	186.3	3, 4
CH_4 (aq)	−34.3	−89.0	83.7	4
CH_2O (aq)	−129.7	—	—	4
C_6H_6 (l)	124.7	49.0	172	4
HCOOH (l)	−361.4	−424.7	129.0	4
CH_3COOH (aq)	−396.6	−485.8	159.8	4
CH_3COO^-	−369.3	−486.0	86.6	4
CO (g)	−137.17	−110.53	197.66	1, 2
	−137.2	−110.5	197.7	3, 4
CO_2 (g)	−394.37	−393.51	213.79	1, 2
	−394.4	−393.5	213.8	3
	−394.4	−393.5	213.7	4

Species or compound	ΔG_f^0 kJ mol^{-1}	ΔH_f^0 kJ mol^{-1}	S^0 J mol^{-1} K^{-1}	Source
$CO_{2\,(aq)}$	—	−413.26	119.36	1
	−386.0	−413.8	117.6	4
$H_2CO_{3\,(aq)}$	−623.14	−699.09	189.31	2
	−623.2	−699.6	187.0	3
	−623.1	−699.6	187.4	4
HCO_3^-	−586.8	−689.93	98.4	1, 2
	−586.8	−692.0	91.2	3, 4
CO_3^{2-}	−527.9	−675.23	−50.0	1, 2
	−527.9	−677.1	−56.9	3
	−527.8	−677.1	−56.9	4
Ca (calcium)				
Ca $_{metal}$	—	0	41.59	1
	0	0	41.6	3
	0	0	41.4	4
Ca^{2+}	−552.8	−543.0	−56.2	1, 2
	−553.5	−542.8	−53.1	3
	−553.6	−542.8	−53.1	4
$Ca(OH)^+$	−718.4	—	—	4
$CaCO_{3\,(aq)}$	−1081.4	−1220.0	−110.0	4
$CaSO_{4\,(aq)}$	−1298.1	−1452.1	−33.1	4
$CaO_{\,lime}$	—	−634.92	38.1	1
	−603.5	−635.1	38.2	3
	−604.0	−635.1	39.8	4
$Ca(OH)_{2\,portlandite}$	−897.5	−985.2	83.4	2
	−898.4	−986.1	83.4	3
	−898.5	−986.1	83.4	4
$CaF_{2\,fluorite}$	−1176.9	−1229.3	68.9	3
	−1167.3	−1219.6	68.9	4
$CaCO_{3\,calcite}$	−1129.07	−1207.6	91.7	2
	−1128.8	−1207.4	91.7	3
	−1128.8	−1206.9	92.9	4
$CaCO_{3\,aragonite}$	−1128.3	−1206.4	93.9	2
	−1127.8	−1207.4	88.0	3
	−1127.8	−1207.1	88.7	4
$CaMg(CO_3)_{2\,dolomite}$	−2161.7	−2324.5	155.2	2, 3
	−2163.4	−2326.3	155.2	4
$CaSO_{4\,anhydrite}$	−1321.98	−1435.5	106.5	2
	−1321.7	−1434.1	106.7	3
	−1321.8	−1434.1	106.7	4
$CaSO_4 \cdot 2H_2O_{\,gypsum}$	−1797.36	−2022.92	193.9	2
	−1797.2	−2022.6	194.1	3
	−1797.3	−2022.6	194.1	4
$Ca_5(PO_4)_3OH_{\,hydroxyapatite}$	−6338.3	−6721.6	390.4	2
	−6338.4	−6721.6	390.4	3
	−6338.5	−6738.5	390.4	4
$Ca_5(PO_4)_3F_{\,fluroroapatite}$	−6508.1	−6872.2	387.9	3
	−6491.5	−6872	387.9	4
$CaAl_2Si_2O_{8\,anorthite}$	−4002.2	−4227.8	199.3	2
	−4017.3	−4243.0	199.3	3
	−4002.3	−4227.9	199.3	4
$CaAl_2Si_4O_{12} \cdot 4H_2O_{\,laumontite}$	−6682.0	−7233.6	485.8	2
$Ca_{0.167}Al_{2.33}Si_{3.67}O_{10}(OH)_{2\,Ca\text{-}beidellite}$	−5346	—	—	2
$CaMgSi_2O_{6\,diopside}$	−3036.6	−3210.8	143.1	3
	−3032.0	−3206.2	142.9	4
$Ca_2Mg_5Si_8O_{22}(OH)_{2\,tremolite}$	−11592.6	−12319.7	548.9	2
	−11627.9	−12355.1	548.9	3
	−11631	−12360	548.9	4

Species or compound	ΔG_f^0 kJ mol^{-1}	ΔH_f^0 kJ mol^{-1}	S^0 J mol^{-1} K^{-1}	Source
Cd (cadmium)				
Cd $_{metal}$	0	0	51.80	1, 3, 4
Cd^{2+}	—	−75.92	−72.8	1
	−77.6	−75.9	−73.2	3, 4
CdO$_2^{2-}$	−284.4	—	—	4
Cd(OH)$^+$	−261.1	—	—	4
Cd(OH)$_{2 (aq)}$	−442.6	—	—	4
Cd(OH)$_3^-$	−600.7	—	—	4
Cd(OH)$_4^{2-}$	−758.4	—	—	4
CdCl$^+$	−224.4	−240.6	43.5	4
CdCl$_3^-$	−487.0	−561.1	202.9	4
Cd(NH$_3$)$_4^{2+}$	−226.1	−450.2	336.4	4
CdO $_{monteponite}$	—	−258.35	54.8	1
	−228.5	−258.2	54.8	3
	−228.4	−258.2	54.8	4
Cd(OH)$_{2 (cr)}$	−473.6	−560.7	96	4
CdCl$_{2 (cr)}$	−343.9	−391.5	115.3	4
CdF$_{2 (cr)}$	−647.7	−400.4	77.4	4
CdS $_{greenockite}$	−145.6	−149.6	70.3	3
	−156.5	−161.9	64.9	4
CdCO$_3$ $_{otavite}$	−669.4	−750.6	92.5	3, 4
CdSO$_{4 (cr)}$	−822.7	−933.3	123.0	4
CdSO$_4$·2H$_2$O $_{(cr)}$	−1068.7	−1239.6	154.0	4
Ce (cerium)				
Ce $_{metal}$	0	0	69.5	3
	0	0	72.0	4
Ce^{3+}	−672.0	−696.2	−205.0	3, 4
Ce^{4+}	−503.8	−537.2	−301	3, 4
Cl (chlorine)				
Cl$_{2 (g)}$	0	0	223.1	3, 4
Cl$^-$	−131.2	−167.08	56.60	1, 2
	−131.3	−167.1	56.7	3
	−131.2	−167.2	56.5	4
HCl $_{(g)}$	—	−92.31	186.90	1
	−95.3	−92.3	186.9	3, 4
Co (cobalt)				
Co $_{metal}$	0	0	30.0	3, 4
Co^{2+}	−54.4	−58.2	−113.0	3, 4
Co^{3+}	134.0	92.0	−305.0	3, 4
CoO $_{(cr)}$	−214.2	−237.9	53.0	3, 4
Co$_3$O$_4$ $_{cobalt spinel}$	−772.6	−891.2	102.5	3
	−774	−891	102.5	4
Co(OH)$_{2 (cr)}$	−450.1	—	—	4
Cr (chromium)				
Cr $_{metal}$	0	0	23.6	3
	0	0	23.8	4
Cr^{2+}	—	−144.0	—	3
	—	−143.5	—	4
CrO$_4^{2-}$	−727.8	−881.2	50.2	4
HCrO$_4^-$	−764.7	−878.2	184.1	4
Cr$_2$O$_7^{2-}$	−1301.1	−1490.3	261.9	4
Cr$_2$O$_{3 (cr)}$	−1058.1	−1139.7	81.2	4

Species or compound	ΔG_f^0 kJ mol^{-1}	ΔH_f^0 kJ mol^{-1}	S^0 J mol^{-1} K^{-1}	Source
Cu (copper)				
Cu $_{metal}$	—	0	33.15	1
	0	0	33.2	3, 4
Cu$^+$	50.0	71.7	41.0	3
	50.0	71.7	40.6	4
Cu^{2+}	—	64.9	−98	1
	65.5	64.8	−99.6	3, 4
CuCl$^+$	−68.2	—	—	4
CuO$_2^{2-}$	−183.6	—	—	4
Cu(OH)$_2$ $_{(aq)}$	−249.0	−395.2	−120.9	4
CuCl$_2$ $_{(aq)}$	−197.9	—	—	4
CuSO$_4$ $_{(aq)}$	−692.2	—	—	4
CuO $_{tenorite}$	−129.6	−157.3	42.6	3
	−129.7	−157.3	42.6	4
Cu$_2$O $_{cuprite}$	−146.0	−168.6	93.1	3, 4
CuCl $_{nantokite}$	−119.9	−137.2	86.2	4
CuCl$_2$ $_{(cr)}$	−175.7	−220.1	108.1	4
CuS $_{covellite}$	−49.1	−48.6	66.6	3
	−53.6	−53.1	66.5	4
Cu$_2$S $_{chalcoite}$	−86.9	−80.1	120.8	3
	−86.2	−79.5	120.9	4
Cu$_2$CO$_3$(OH)$_2$ $_{malachite}$	—	−1054.0	—	3
	−893.6	−1051.4	186.2	4
Cu$_3$(CO$_3$)$_2$(OH)$_2$ $_{azurite}$	—	−1632.2	—	3
	−1315.5	−1632.2	0	4
CuSO$_4$ $_{chalcocyanite}$	—	−771.4	109.2	1
	−662.3	−771.4	109.5	3
	−661.8	−771.4	109	4
CuSO$_4$·5H$_2$O $_{chalcanthite}$	−1879.8	−2279.6	300.4	3
	−1879.7	−2279.6	300.4	4
Cu$_4$SO$_4$(OH)$_6$ $_{brochantite}$	−1818.0	—	—	3
	−1817.7	—	—	4
F (fluorine)				
F$_2$ $_{(g)}$	0	0	202.8	3, 4
F$^-$	−281.5	−335.35	−13.8	1, 2
	−281.7	−335.4	−13.2	3
	−278.8	−332.6	−13.8	4
HF $_{(aq)}$	−296.8	−320.1	88.7	4
HF $_{(g)}$	—	−273.30	173.78	1
	−275.4	−273.3	173.8	3
	−273.2	−271.1	173.8	4
HF $_{(l)}$	—	−299.8	—	4
Fe (iron)				
Fe $_{metal}$	0	0	27.3	2, 3, 4
Fe^{2+}	−82.88	−89.0	—	2
	−78.9	−89.1	−138.0	3
	−78.9	−89.1	−137.7	4
Fe^{3+}	−8.56	−48.85	—	2
	−4.6	−48.5	−316.0	3
	−4.7	−48.5	−315.9	4
FeO$_2^{2-}$	−295.3	—	—	
Fe(OH)$^+$	−277.4	−324.7	−29	4
Fe(OH)$^{2+}$	−233.20	−291.2	—	2
	−229.4	−290.8	−142	4
Fe(OH)$_2^+$	−450.5	−548.9	—	2
	−438.0	—	—	4

Species or compound	ΔG_f^0 kJ mol^{-1}	ΔH_f^0 kJ mol^{-1}	S^0 J mol^{-1} K^{-1}	Source
$Fe(OH)_{3\,(aq)}$	−648.3	−802.5	—	2
	−659.3	—	—	4
$Fe(OH)_4^-$	−833.83	−1058.7	—	2
$FeCl^{2+}$	−143.9	−180.3	−113	4
$FeSO_{4\,(aq)}$	−823.4	−998.3	−117.6	4
$FeO_{\,(cr)}$	−251.2	−272.0	59.8	3
	−251.4	−272.0	60.8	4
Fe_2O_3 hematite	−742.8	−824.7	87.7	2
	−742.7	−824.6	87.4	3
	−742.2	−824.2	87.4	4
Fe_3O_4 magnetite	−1012.9	−1116.1	146.1	2
	−1012.6	−1115.7	146.1	3
	−1015.4	−1118.4	146.4	4
$Fe(OH)_{2\,(s)}$	−486.5	−569.0	88	2, 4
$FeOOH$ goethite	−488.55	−559.3	60.4	2
	−488.6	−559.3	60.4	3
	—	−559.0	—	4
$Fe(OH)_3$ ferrihydrite	−692.07	—	—	2
	−696.5	−823.0	106.7	4
$FeCl_2$ lawrencite	−302.2	−341.6	118.0	3
	−302.3	−341.8	118.0	4
$FeCl_3$ molysite	−333.8	−399.2	142.3	3
	−334.0	−399.5	142.3	4
$FeCO_3$ siderite	−673.05	−753.8	—	2
	−666.7	−737.0	105.0	3
	−666.7	−740.6	92.9	4
$FePO_4 \cdot 2H_2O$ strengite	−1662.9	−1888.2	171.1	2
Fe_2SiO_4 fayalite	−1379.0	−1479.9	145.2	2
	−1379.4	−1479.4	148.3	3
	−1379.0	−1479.9	145.2	4
FeS_2 pyrite	−166.9	−178.2	52.9	2
	−160.2	−171.5	52.9	3
	−166.9	−178.2	52.9	4
FeS_2 marcasite	−158.4	−169.4	53.9	3
	—	−154.8	—	4
FeS pyrrhotite	−100.4	−100.0	60.3	2, 4
FeS mackinawite	−93.0	—	—	2
Fe_3S_4 greigite	−290	—	—	2
$FeAsS$ arsenopyrite	−50	−42	121	4
$FeSO_4 \cdot 7H_2O$ melanterite	−2509.6	−3014.4	409.2	3
	−2509.9	−3014.6	409.2	4
H (hydrogen)				
$H_{2\,(g)}$	0	0	130.57	2
	0	0	130.7	3, 4
$H_2O_{\,(g)}$	−228.58	−241.83	188.84	1, 2
	−228.6	−241.8	188.7	3, 4
$H_2O_{\,(l)}$	−237.14	−285.83	69.95	1, 2
	−237.1	−285.8	70.0	3
	−237.1	−285.8	69.9	4
H^+	0	0	0	1, 2, 3, 4
Hg (mercury)				
Hg metal	0	0	75.9	3
	0	0	76.0	4
$Hg_{\,(l)}$	—	0	75.90	1
Hg^{2+}	—	170.21	−36.19	1
	164.4	171.0	−32.0	3
	164.4	171.0	−32.2	4

Species or compound	ΔG_f^0 kJ mol^{-1}	ΔH_f^0 kJ mol^{-1}	S^0 J mol^{-1} K^{-1}	Source
Hg_2^{2+}	—	166.87	65.74	1
	153.6	172.0	84.5	3
	153.5	172.4	84.5	4
$HHgO_2^-$	−190.3	—	—	4
$Hg(OH)^+$	−52.3	−84.5	71	4
$HgCl^+$	−5.4	−18.8	75	4
$HgCl_4^{2-}$	−446.8	−554.0	293	4
HgS_2^{2-}	41.9	—	—	4
HgO montroydite	—	−90.79	70.25	1
	−58.5	−90.8	70.3	3, 4
Hg_2Cl_2 calomel	—	−265.37	191.6	1
	−210.8	−265.2	192.5	3
	−210.7	−265.2	192.5	4
HgS cinnabar	−50.6	−58.2	82.5	3
	−50.6	−58.2	82.4	4
Hg_2SO_4 (cr)	—	−743.09	200.70	1
	−625.8	−743.1	200.7	4
K (potassium)				
K metal	0	0	64.7	3
	0	0	64.2	4
K^+	−282.5	−252.14	101.2	1, 2
	−282.5	−252.2	101.0	3
	−283.3	−252.4	102.5	4
KCl sylvite	−408.6	−436.5	82.6	2
	−408.6	−436.5	82.6	3
	−409.1	−436.7	82.6	4
K_2SO_4 arcanite	−1319.7	−1437.7	175.6	3
	−1321.4	−1437.8	175.6	4
KNO_3 niter	−394.5	−494.5	133.1	3
	−394.9	−494.6	133.0	4
$KAlSi_3O_8$ microcline	−3742.9	−3681.1	214.2	2
	−3742.3	−3967.7	214.2	3
	−3742.9	−3968.1	214.2	4
$KAl_3Si_3O_{10}(OH)_2$ muscovite	−5608.4	−5984.4	305.3	2
	−5600.7	−5976.7	306.4	3
	−5608.4	−5984.4	306.3	4
$KMg_3AlSi_3O_{10}(OH)_2$ phlogopite	—	—	319.7	3
	−5831.8	—	—	4
La (lanthanum)				
La metal	0	0	56.9	3, 4
La^{3+}	−683.7	−707.1	−217.6	4
Mg (magnesium)				
Mg metal	0	0	32.7	3, 4
Mg^{2+}	−455.4	−467.0	−137	1, 2
	−454.8	−466.8	−138.0	3
	−454.8	−466.8	−138.1	4
$Mg(OH)^+$	−626.7	—	—	4
$MgSO_4$ (aq)	−1212.2	−1356.0	−7.1	4
MgO periclase	−569.3	−601.6	26.95	1, 2
	−269.2	−601.5	26.9	3
	−569.4	−601.7	26.9	4
$Mg(OH)_2$ brucite	−833.51	−924.54	63.18	2
	−833.5	−924.5	63.2	3, 4
$MgCO_3$ magnesite	−1012.1	−1095.8	65.7	2
	−1029.5	−1113.3	65.1	3
	−1012.1	−1095.8	65.7	4

Species or compound	ΔG_f^0 kJ mol^{-1}	ΔH_f^0 kJ mol^{-1}	S^0 J mol^{-1} K^{-1}	Source
MgCO$_3$·3H$_2$O nesquehonite	−1723.7	−1977.3	195.6	3
	−1726.1	—	—	4
MgSO$_4$·7H$_2$O epsomite	−2871.2	−3388.7	372.0	3
	−2871.5	−3388.7	372	4
Mg$_2$SiO$_4$ forsterite	−2056.7	−2175.7	95.2	2
	−2051.3	−2170.4	95.2	3
	−2055.1	−2174.0	95.1	4
MgSiO$_3$ clinoenstatite	−1460.9	−1547.8	67.9	3
	−1462.1	−1549.0	67.7	4
MgSiO$_3$ enstatite	−1459.9	−1546.8	67.8	2
Mg$_3$Si$_2$O$_5$(OH)$_4$ chrysotile	−4035.35	—	—	2
	−4034.0	−4361.7	221.3	3
	−4037.8	−4365.6	221.3	4
Mg$_3$Si$_4$O$_{10}$(OH)$_2$ talc	−5527.1	−5893	260.7	2
	−5536.0	−5915.9	260.8	3
	−5523.7	−5903.3	260.8	4
Mg$_5$Al$_2$Si$_3$O$_{10}$(OH)$_8$ chlorite	−8207.8	−8857.4	465.3	2
Mg$_4$Si$_6$O$_{15}$(OH)$_2$·6H$_2$O sepiolite	−9251.6	−10116.9	613.4	2
Mn (manganese)				
Mn metal	0	0	32.0	2, 3
Mn^{2+}	−228.1	−220.79	−73.6	2, 4
	−228.0	−220.7	−73.6	3
MnOH$^+$	−405.0	−450.6	−17	4
Mn(OH)$_3^-$	−744.2	—	—	4
MnO manganosite	−362.9	−385.2	59.7	3, 4
Mn(OH)$_2$ pyrochroite	−616.5	—	—	2
MnOOH manganite	−133.3	—	—	2
Mn$_3$O$_4$ hausmannite	−1283.2	−1387.8	155.6	2
	−1282.8	−1387.8	154.0	3
	−1283.2	−1387.8	155.6	4
Mn$_2$O$_3$ bixbyite	−881.1	−959.0	110.5	2, 3, 4
MnO$_2$ pyrolusite	−465.14	−520.3	53.1	2
	−465.1	−520.0	53.0	3, 4
MnO$_2$ birnessite	−453.1	—	—	2
MnCO$_3$ rhodochrosite	−816.7	−894.1	85.8	2
	−816.0	−889.3	100.0	3
	−816.7	−894.1	85.8	4
MnS alabandite	−218.0	−213.8	78.2	2
	−218.2	−213.9	78.2	3
	−218.4	−214.2	78.2	4
MnSiO$_3$ rhodonite	−1243.1	−1319.2	102.5	2
	−1243.1	−1319.4	102.5	3
	−1240.5	−1320.9	89.1	4
Mo (molybdenum)				
Mo metal	0	0	28.7	3, 4
MoO$_4$ (aq)	—	−661.1	—	4
MoO$_4^{2-}$	−836.3	−997.9	27.2	4
MoO$_2$ (cr)	−533.0	−587.8	50.0	3
	−533.0	−588.9	46.3	4
MoO$_3$ molybdite	−668.0	−745.2	77.7	3
	−668.0	−745.1	77.7	4
MoS$_2$ molybdenite	−297.4	−306.3	62.6	3
	−225.9	−235.1	62.6	4
N (nitrogen)				
N$_2$ (g)	0	0	191.61	1, 2
	0	0	191.6	3, 4

Species or compound	ΔG_f^0 kJ mol^{-1}	ΔH_f^0 kJ mol^{-1}	S^0 J mol^{-1} K^{-1}	Source
NO$_2^-$	-32.2	-104.6	123.0	4
NO$_3^-$	-108.74	-206.85	146.70	1, 2
	-111.5	-207.4	146.9	3
	-108.7	-205.0	146.4	4
NH$_{3\,(g)}$	-16.45	-45.94	192.77	1, 2
	-16.4	-45.9	192.8	3
	-16.4	-46.1	192.4	4
NH$_{3\,(aq)}$	-26.5	-80.29	111.3	2
	-26.6	-80.3	111.0	3
	-26.5	-80.3	111.3	4
NH$_4^+$	-79.31	-133.26	111.17	1, 2
	-79.4	-133.3	111.2	3
	-79.3	-132.5	113.4	4
NO$_{(g)}$	86.6	90.2	210.8	4
NO$_{2\,(g)}$	51.2	33.1	240.1	3
	51.3	33.2	240.1	4
N$_2$O$_{(g)}$	104.2	82.0	219.8	4
NH$_{3\,(g)}$	-16.4	-45.9	192.8	3
	-16.4	-46.1	192.4	4
(NH$_4$)$_2$SO$_4$ mascagnite	-901.7	-1180.8	220.1	3, 4
NH$_4$NO$_{3\,(cr)}$	-183.8	-365.6	151.1	3
	-183.9	-365.6	151.1	4
Na (sodium)				
Na$_{\text{metal}}$	0	0	51.3	3
	0	0	51.2	4
Na$^+$	-262.0	-240.34	58.45	1, 2
	-261.9	-240.3	58.4	3
	-261.9	-240.1	59.0	4
NaCO$_3^-$	-792.8	-935.9	-49.8	4
NaHCO$_{3\,(aq)}$	-849.7	-943.9	113.8	4
NaSO$_4^-$	-1010.6	-1144.7	108.8	4
NaCl $_{\text{halite}}$	-384.14	-411.15	72.1	2
	-384.2	-411.3	72.1	3
	-384.1	-411.1	72.1	4
NaF $_{\text{villiaumite}}$	-546.3	-576.6	51.3	3
	-543.5	-573.6	51.5	4
Na$_2$CO$_3 \cdot 10$H$_2$O $_{\text{natron}}$	-3427.7	-4081.3	562.7	4
NaHCO$_3$ $_{\text{nahcolite}}$	-851.9	-947.7	102.1	2
	-851.0	-950.8	101.7	4
NaHCO$_3 \cdot$Na$_2$CO$_3 \cdot 2$H$_2$O $_{\text{trona}}$	-2386.6	—	—	2
	-2383.4	-2684.9	301.2	4
Na$_2$SO$_4$ $_{\text{thenardite}}$	-1269.8	-1387.8	149.6	2
	-1270.0	-1387.8	149.6	3
	-1270.2	-1387.1	149.6	4
Na$_2$SO$_4 \cdot 10$H$_2$O $_{\text{mirabilite}}$	-3646.4	-4327.1	592.0	2
	-3646.5	-4327.2	591.9	4
	-3646.8	-4327.3	592.0	4
NaSi$_7$O$_{13}$(OH)$_3$ $_{\text{magadiite}}$	-6651.9	-241.83	188.73	2
NaAlSi$_3$O$_8$ $_{\text{albite}}$	-3711.5	-3935.1	207.4	2
	-3711.7	-3935.1	207.4	3
	-3711.5	-3935.1	207.4	4
NaAlSi$_2$O$_6 \cdot$H$_2$O $_{\text{analcite}}$	-3082.6	-3300.8	234.3	2
	-3091.7	-3309.8	234.4	3
	-3082.6	-3300.8	234.3	4
Na$_{0.33}$Al$_{2.33}$Si$_{3.67}$O$_{10}$(OH)$_2$ $_{\text{Na-beidellite}}$	-5343	—	—	2

Species or compound	ΔG_f^0 kJ mol^{-1}	ΔH_f^0 kJ mol^{-1}	S^0 J mol^{-1} K^{-1}	Source
Ni (nickel)				
Ni $_{metal}$	0	0	29.9	3, 4
Ni^{2+}	−45.6	−54.0	−129.0	3
	−45.6	−54.0	−128.9	4
Ni(OH)$^+$	−230.1	—	—	4
NiO $_{busenite}$	−211.6	−239.7	38.0	3
	−211.7	−239.7	38.0	4
NiCO$_3$ $_{(cr)}$	−612.5	—	—	4
NiS $_{millerite}$	−86.2	−84.9	66.1	3
	−79.5	−82.0	53.0	4
Ni$_3$S$_2$ $_{heazlewoodite}$	−197.1	−202.9	133.9	3, 4
NiSO$_4$ $_{(cr)}$	−759.9	−872.9	92	4
NiSO$_4\cdot$7H$_2$O $_{morenosite}$	−2461.7	−2976.3	378.9	3
	−2461.8	−2976.3	378.9	4
Np (neptunium)				
Np $_{metal}$	0	0	0	5
Np^{3+}	−517.1	—	—	5
Np^{4+}	−502.9	—	—	5
NpO$_2^+$	−915.0	—	—	5
NpO$_2^{2+}$	−795.8	—	—	5
Np(OH)$_5^-$	−1456.4	—	—	5
NpO$_2$(OH)$_2$CO$_3^{2-}$	−1774.2	—	—	5
NpO$_2$ $_{(cr)}$	−1021.8	—	—	5
Np$_2$O$_3$ $_{(cr)}$	−1448.0	—	—	5
Np$_2$O$_5$ $_{(cr)}$	−2013.0	—	—	5
NpO$_2$(OH)$_2$ $_{(cr)}$	−1232.5	—	—	5
Np(CO$_3$)$_2$ $_{(cr)}$	−1630.2	—	—	5
O (oxygen)				
O$_2$ $_{(g)}$	—	0	205.15	1
	0	0	205.2	3
	0	0	205.1	4
O$_2$ $_{(aq)}$	16.4	−11.7	110.9	4
OH$^-$	−157.3	−230.0	−10.7	3
	−157.2	−230.0	−10.8	4
P (phosphorus)				
P $_{(g)}$	—	316.5	163.2	1
P$_2$ $_{(g)}$	—	144.0	218.12	1
	103.7	144.3	218.1	4
P$_4$ $_{(g)}$	—	58.9	280.01	1
	24.4	58.9	280.0	4
PO$_4^{3-}$	−1019.0	−1277.0	−222.0	3
	−1018.7	−1277.4	−222	4
HPO$_4^{2-}$	—	−1299.0	−33.5	1
	−1089.1	−1292.1	−33.5	4
H$_2$PO$_4^-$	—	−1302.6	92.5	1
	−1130.3	−1296.3	90.4	4
H$_3$PO$_4$ $_{(aq)}$	−1142.5	−1288.3	158.2	4
H$_3$PO$_4$ $_{(cr)}$	−1112.3	−1266.9	110.5	3
	−1119.1	−1279.0	110.5	4
Pb (lead)				
Pb $_{metal}$	—	0	64.80	1
	0	0	65.1	3
	0	0	64.8	4
Pb^{2+}	—	0.92	18.5	1
	−24.4	−1.7	10.0	3
	−24.4	−1.7	10.5	4

Species or compound	ΔG_f^0 kJ mol^{-1}	ΔH_f^0 kJ mol^{-1}	S^0 J mol^{-1} K^{-1}	Source
$HPbO_2^-$	−338.4	—	—	4
$Pb(OH)^+$	−226.3	—	—	4
$PbCl^+$	−164.8	—	—	4
$PbCl_{2\,(aq)}$	−297.2	—	—	4
$PbCl_3^-$	−426.3	—	—	4
$PbO_{(cr,\ red)}$	−188.9	−219.0	66.5	4
$PbO_{2\,(cr)}$	−217.3	−277.4	68.6	4
$Pb_3O_{4\,(cr)}$	−601.2	−718.4	211.3	4
$PbS_{\ galena}$	−96.1	−97.7	91.4	3
	−98.7	−100.4	91.2	4
$PbCO_{3\ cerussite}$	−625.3	−699.2	131.0	3
	−625.5	−699.1	131.0	4
$PbSO_{4\ anglesite}$	—	−919.97	148.50	1
	−813.0	−919.9	148.6	3
	−813.1	−919.9	148.6	4
Pd (palladium)				
$Pd_{\ metal}$	0	0	37.8	3
	0	0	37.6	4
Pd^{2+}	176.5	149.0	−184	4
$PdO_{(cr)}$	—	−85.4	—	4
$PdS_{(cr)}$	−67	−75	46	4
$PdS_{2\,(cr)}$	−74.5	−81.2	79	4
Pt (platinum)				
$Pt_{\ metal}$	0	0	41.6	3, 4
Pt^{2+}	254.8	—	—	4
$PtS_{\ cooperite}$	−76.9	−82.4	55.1	3
	−76.1	−81.6	55.1	4
Po (polonium)				
$Po_{\ metal}$	0	0	0	4
Po^{2+}	71.0	—	—	4
Po^{4+}	293.0	—	—	4
$Po(OH)_2^{4+}$	−473.0	—	—	4
$PoO_{2\,(cr)}$	−195.0	—	—	5
$Po(OH)_{4\,(cr)}$	−544.0	—	—	4
$PoS_{(cr)}$	−4.0	—	—	4
Pu (plutonium)				
$Pu_{\ metal}$	0	0	0	5
Pu^{3+}	−578.1	—	—	5
Pu^{4+}	−481.0	—	—	5
PuO_2^+	−849.8	—	—	5
$Pu(OH)_5^-$	−1582.0	—	—	5
$PuO_2(OH)_2CO_3^{2-}$	−1732.1	—	—	5
$PuO_{2\,(cr)}$	−998.0	—	—	5
$Pu(OH)_{4\,(cr)}$	−1426.0	—	—	5
$Pu_2(CO_3)_{3\,(cr)}$	−2918.1	—	—	5
$Pu_2S_{3\,(cr)}$	−979.0	—	—	5
Ra (radium)				
$Ra_{(cr)}$	0	0	71	4
Ra^{2+}	−561.5	−527.6	54	4
$RaSO_{4\,(cr)}$	−1365.6	−1471.1	138	4
Rb (rubidium)				
$Rb_{\ metal}$	—	0	79.78	1
	0	0	76.8	3, 4

Species or compound	ΔG_f^0 kJ mol^{-1}	ΔH_f^0 kJ mol^{-1}	S^0 J mol^{-1} K^{-1}	Source
Rb^+	—	−251.12	121.75	1
	−291.7	−251.1	120.5	3
	−284.0	−251.2	121.5	4
S (sulfur)				
S $_{(cr, rhombic)}$	0	0	32.05	1, 2
	0	0	31.8	3, 4
H_2S $_{(g)}$	−33.4	−20.6	205.81	1, 2
	−33.5	−20.6	205.8	3
	−33.6	−20.6	205.8	4
H_2S $_{(aq)}$	−27.7	−38.6	126	1, 2
	−27.8	−39.7	121	4
HS^-	12.2	−16.3	67	1, 2
	12.1	−17.0	62.8	3
	12.1	−17.6	62.8	4
S^{2-}	85.9	34	—	2
	85.8	33.0	−15.0	3
	85.8	33.1	−14.6	4
SO_2 $_{(g)}$	−300.1	−296.81	248.22	1, 2
HSO_4^-	−755.3	−886.9	131.7	1, 2
SO_3^{2-}	−486.6	−635.6	29.0	3
	−486.5	−635.5	−29	4
SO_4^{2-}	−744.0	−909.34	18.50	1, 2
	−744.6	−909.3	20.0	3
	−744.5	−909.3	20.1	4
Sb (antimony)				
Sb $_{metal}$	0	0	45.5	3
	0	0	45.7	4
SbO^+	−177.1	—	—	4
SbO_2^-	−340.2	—	—	4
$Sb_2S_4^{2-}$	−49.8	−109.6	−26.2	4
$HSbO_2$ $_{(aq)}$	−407.5	−487.9	46.4	4
Sb_2O_4 $_{(cr)}$	−795.7	−907.5	127.2	4
Sb_2O_5 $_{(cr)}$	−829.2	−971.9	125.1	4
Sb_4O_6 $_{valentinite}$	−1252.7	−1417.1	246.0	3
	−1253.0	−1417.1	246.0	4
Sb_2O_5 $_{(cr)}$	−829.2	−971.9	125.1	4
$SbCl_3$ $_{(cr)}$	−323.7	−382.2	184.1	4
SbF_3 $_{(cr)}$	—	−915.5	—	4
Sb_2S_3 $_{stibnite}$	−173.5	−174.9	182.0	3
	−173.6	−174.9	182.0	4
Sc (scandium)				
Sc $_{metal}$	0	0	34.6	3, 4
Sc^{3+}	−586.6	−614.2	−255	4
$Sc(OH)^{2+}$	−801.2	−861.5	−134	4
Sc_2O_3 $_{(cr)}$	−1819.4	−1908.8	77.0	3, 4
Se (selenium)				
Se $_{metal}$	0	0	42.3	3
	0	0	42.4	4
Se^{2-}	129.0	—	—	3
	129.3	—	—	4
SeO_3^{2-}	−369.8	−509.2	13.0	4
SeO_4^{2-}	−441.3	−599.1	54.0	4
HSe^-	44.0	15.9	79	4
$HSeO_3^-$	−411.5	−514.6	135.1	4
$HSeO_4^-$	−452.2	−581.6	149.4	4
H_2Se $_{(aq)}$	22.2	19.2	163.6	4

Species or compound	ΔG_f^0 kJ mol^{-1}	ΔH_f^0 kJ mol^{-1}	S^0 J mol^{-1} K^{-1}	Source
$H_2SeO_{3\,(aq)}$	−426.1	−507.5	207.9	4
$SeO_{2\,(cr)}$	—	−225.4	—	4
Si (silicon)				
$Si_{\,metal}$	0	0	18.8	3, 4
$H_4SiO_{4\,(aq)}$	−1307.9	−1457.3	180	2
	−1308.0	−1460.0	180.0	3
	−1316.6	−1468.6	180	4
$H_3SiO_4^-$	−1251.8	−1431.7	—	2
$H_2SiO_4^{2-}$	−1176.6	−1383.7	—	2
$SiO_{2\,alpha\,quartz}$	−856.3	−910.7	41.46	1, 2
	−856.3	−910.7	41.5	3
	−856.6	−910.9	41.8	4
$SiO_{2\,amorph}$	−849.1	−889.7	—	2
$SiO_{2\,cristobalite}$	−854.5	−908.3	43.4	3
	−855.4	−909.5	42.7	4
$SiO_{2\,tridymite}$	−853.8	−907.5	43.9	3
	−855.3	−909.1	43.5	4
Sn (tin)				
$Sn_{\,metal}$	0	0	51.2	3
	0	0	51.6	4
Sn^{2+}	—	−8.9	−16.7	1
	−27.2	−8.8	−17	4
Sn^{4+}	2.5	30.5	−117	4
$SnOH^+$	−254.8	−286.2	50	4
$SnO\cdot OH^+$	−474.0	—	—	4
$SnO_{\,romarchite}$	—	−280.71	57.17	1
	−256.9	−285.8	56.5	4
$SnO_{2\,cassiterite}$	—	−577.63	49.04	1
	−519.9	−580.7	52.3	3
	−580.7	−519.6	52.3	4
Sr (strontium)				
$Sr_{\,metal}$	0	0	55.4	3
	0	0	52.3	4
Sr^{2+}	−563.83	−550.90	−35.1	2
	−559.4	−545.8	−33.0	3
	−559.5	−545.8	−32.6	4
$Sr(OH)^+$	−721.3	—	—	4
$SrO_{\,(cr)}$	−561.9	−592.0	54.4	4
$SrCO_{3\,strontianite}$	−1144.73	−1225.8	97.2	2
	−1137.6	−1218.7	97.1	3
	−1140.1	−1220.1	97.1	4
$SrSO_{4\,celestite}$	−1345.7	−1456.9	—	2
	−1341.0	−1453.2	118.0	3
	−1340.9	−1453.1	117	4
Ta (tantalum)				
$Ta_{\,metal}$	0	0	41.5	3, 4
TaO_2^+	−842.6	—	—	4
$Ta_2O_{5\,(cr)}$	−1911.0	−2046.0	143.1	3
	−1911.2	−2046.0	143.1	4
Th (thorium)				
$Th_{\,metal}$	—	0	51.8	1
	0	0	53.4	3, 4
Th^{4+}	−705.1	−769.0	−422.6	4
$Th(OH)^{3+}$	−920.5	−1030.1	−343	4
$Th(OH)_2^{2+}$	−1140.9	−1282.4	−218	4
$Th(SO_4)^{2+}$	−1480.7	−1658.5	−230	4

Species or compound	ΔG_f^0 kJ mol^{-1}	ΔH_f^0 kJ mol^{-1}	S^0 J mol^{-1} K^{-1}	Source
ThO$_2$ thorianite	—	−1226.4	65.23	1
	−1168.8	−1226.4	65.2	3, 4
Ti (titanium)				
Ti metal	—	0	30.72	1
	0	0	30.6	3, 4
TiO$_2$ rutile	—	−944.0	50.62	1
	−889.4	−944.8	50.3	3
	−889.5	−944.7	50.3	4
U (uranium)				
U metal	—	0	50.20	1
	0	0	50.3	3
	0	0	50.2	4
U^{3+}	−520.5	−514.6	−126.0	3
	−475.4	−489.1	−192	4
U^{4+}	−579.1	−613.8	−326.0	3
	−531.0	−591.2	−410	4
UO$_2^+$	−962.7	—	—	4
UO$_2^{2+}$	—	−1019.0	−98.2	1
	−953.5	−1019.6	−97.5	4
	−952.5	—	—	5
U(OH)$_5^-$	−1630.8	—	—	5
UO$_2$CO$_3$ (aq)	−1537.9	—	—	5
UO$_2$(CO$_3$)$_2^{2-}$	−2104.4	—	—	5
UO$_2$(CO$_3$)$_3^{4-}$	−2659.2	—	—	5
UO$_2$ uraninite	—	−1085.0	77.03	1
	−1031.8	−1084.9	77.0	3
	−1031.7	−1084.9	77.0	4
UO$_3$ (cr)	—	−1223.8	96.11	1
	−1146.5	−1223.8	98.6	3
	−1145.9	−1223.8	96.1	4
	−1013.9	—	—	5
U$_3$O$_8$ (cr)	—	−3574.8	282.55	1
	−3369.6	—	—	5
UO$_2$CO$_3$ (cr)	−1562.7	—	—	5
USiO$_4$ (cr)	−1861.9	—	—	5
V (vanadium)				
V metal	0	0	28.9	3, 4
VO^{2+}	−446.4	−486.6	−133.9	4
VO$_2^+$	−587.0	−649.8	−42.3	4
VO$_3^-$	−783.6	−888.3	50	4
VO$_4^{3-}$	−899.0	—	—	4
V$_2$O$_7^{4-}$	−1719	—	—	4
HVO$_4^{2-}$	−974.8	−1159.0	17	4
HV$_{10}$O$_{28}^{5-}$	−7702	−8694	222	4
H$_2$VO$_4^-$	−1020.8	−1174.0	121	4
H$_2$V$_{10}$O$_{28}^{4-}$	−7723	—	—	4
VO (cr)	−404.2	−431.8	39.0	3, 4
V$_2$O$_3$ karelianite	−1139.0	−1218.8	98.1	3
	−1139.3	−1218.8	98.3	4
V$_2$O$_4$ (cr)	−1318.4	−1427.4	103.5	3
	−1318.3	−1427.2	102.5	4
V$_2$O$_5$ (cr)	−1419.4	−1550.6	130.5	3
	−1419.5	−1550.6	131.0	4
W (tungsten)				
W metal	0	0	32.6	3, 4
WO$_4^{2-}$	—	−1075.7	—	4

Species or compound	ΔG_f^0 kJ mol^{-1}	ΔH_f^0 kJ mol^{-1}	S^0 J mol^{-1} K^{-1}	Source
WO$_2$ (cr)	−533.8	−589.7	50.5	3
	−533.9	−589.7	50.5	4
WO$_3$ (cr)	−764.1	−842.9	75.9	3
	−764.0	−842.9	75.9	4
WS$_2$ tungstenite	−297.9	95.0	298.3	3
	—	−209	—	4
Y (yttrium)				
Y metal	0	0	44.4	3, 4
Y^{3+}	−693.8	−723.4	−251	4
Y$_2$O$_3$ (cr)	−1816.6	−1905.3	99.1	3, 4
Yb (ytterbium)				
Yb metal	0	0	59.8	3
	0	0	59.9	4
Yb^{2+}	−527	—	—	4
Yb^{3+}	−644.0	−674.5	−238	4
Yb$_2$O$_3$ (cr)	−1726.8	−1814.5	133.0	3
	−1726.8	−1814.6	133.0	4
Zn (zinc)				
Zn metal	—	0	41.63	1
	0	0	41.6	3, 4
Zn^{2+}	—	−153.39	−109.8	1
	−147.3	−153.4	−109.6	3
	−147.1	−153.9	−112.1	4
ZnO$_2^{2-}$	−384.2	—	—	4
ZnOH$^+$	−330.1	—	—	4
HZnO$_2^-$	−457.1	—	—	4
ZnCl$^+$	−275.3	—	—	4
ZnO zincite	—	−350.46	43.65	1
	−320.5	−350.5	43.6	3
	−318.3	−348.3	43.6	4
ZnS sphalerite	−202.5	−206.9	58.7	3
	−201.3	−206.0	57.7	4
ZnSe stilleite	−163	−163	84	4
ZnCO$_3$ smithsonite	−731.5	−812.8	82.4	3
	−731.5	−812.8	82.4	4
ZnSO$_4$ zinkosite	−871.5	−982.8	110.5	3, 4
Zn$_2$SiO$_4$ willemite	−1522.9	−1636.5	131.4	3
	−1523.2	−1636.7	131.4	4
Zr (zirconium)				
Zr metal	0	0	39.0	3, 4
ZrO$_2$ baddeleyite	−1042.8	−1100.6	50.4	3, 4
ZrSiO$_4$ zircon	−1918.9	−2033.4	84.0	3
	−1919.1	−2033.4	84.1	4

Data sources: (1) Cox et al. (1989); (2) Drever (1997); (3) Robie et al. (1978); (4) Wagman et al. (1982); (5) Brookins (1988).

Ionic Radii in Å for Different Charges and Coordination Numbers*

Z	Element	Valence	II	III	IV	V	VI	VII	VIII	IX	X	XII
							Coordination numbers					
1	H	+1					Extremely small					
3	Li	+1			0.68		0.82					
4	Be	+2		0.25	0.35							
5	B	+3		0.10	0.20							
6	C	−4					2.60					
		+4					0.16					
7	N	+3					0.16					
		+5					0.13					
8	O	−2	1.27	1.28	1.30		1.32		1.34			
9	F	−1	1.21	1.22	1.23		1.25					
11	Na	+1			1.07	1.08	1.10	1.21	1.24	1.40		
12	Mg	+2			0.66	0.75	0.80		0.97			
13	Al	+3			0.47	0.56	0.61					
14	Si	+4			0.34		0.48					
15	P	+5			0.25							
16	S	−2			1.56		1.72		1.78			
		+6			0.20							
17	Cl	−1			1.67		1.72		1.65			
		+5		0.20								
		+7			0.28							
19	K	+1					1.46	1.54	1.59	1.63	1.67	1.68
20	Ca	+2					1.08	1.15	1.20	1.26	1.36	1.43
21	Sc	+3					0.83		0.95			
22	Ti	+2					0.94					
		+3					0.75					
		+4				0.61	0.69					
23	V	+2					0.87					
		+3					0.72					
		+4					0.67					
		+5			0.44	0.54	0.62					
24	Cr	+2					0.81(L)					
							0.90(H)					
		+3					0.70					
		+4			0.52		0.63					
		+5			0.43							
		+6			0.38							

Z	Element	Valence	Coordination numbers									
			II	III	IV	V	VI	VII	VIII	IX	X	XII
25	Mn	+2					0.75(L)		1.01			
							0.91(H)					
		+3				0.66	0.66(L)					
							0.73(H)					
		+4					0.62					
		+6			0.35							
		+7			0.34							
26	Fe	+2			0.71(H)		0.69(L)					
							0.86(H)					
		+3			0.57(H)		0.63(L)					
							0.73(H)					
27	Co	+2			0.65(H)		0.73(L)					
							0.83(H)					
		+3					0.61(L)					
							0.69(H)					
28	Ni	+2					0.77					
		+3					0.64(L)					
							0.68(H)					
29	Cu	+1	0.54									
		+2			0.70	0.73	0.81					
30	Zn	+2			0.68	0.76	0.83		0.98			
31	Ga	+3			0.55	0.63	0.70					
32	Ge	+4			0.48		0.62					
33	As	+5			0.42		0.58					
34	Se	−2					1.88		1.90			
		+6			0.37							
35	Br	−1					1.88		1.84			
		+7			0.34							
37	Rb	+1					1.57	1.64	1.68		1.74	1.81
38	Sr	+2					1.21	1.29	1.33		1.40	1.48
39	Y	+3					0.98		1.10	1.18		
40	Zr	+4					0.80	0.86	0.92			
41	Nb	+2					0.79					
		+3					0.78					
		+4					0.77					
		+5			0.40		0.72	0.74				
42	Mo	+3					0.75					
		+4					0.73					
		+5					0.71					
		+6			0.50	0.58	0.68	0.79				
43	Tc	+4					0.72					
44	Ru	+3					0.76					
		+4					0.70					
45	Rh	+3					0.75					
		+4					0.71					
46	Pd	+1	0.67									
		+2			0.72		0.94					
		+3					0.84					
		+4					0.70					
47	Ag	+1	0.75		1.10	1.20	1.23	1.32	1.38			
		+3			0.73							
48	Cd	+2			0.88	0.95	1.03	1.08	1.15			1.39
49	In	+3					0.88		1.00			
50	Sn	+2							1.30			
		+4					0.77					
51	Sb	+3			0.85	0.88						
		+5					0.69					

Z	Element	Valence	II	III	IV	V	VI	VII	VIII	IX	X	XII
							Coordination numbers					
52	Te	+4		0.60								
53	I	−1					2.13		1.97			
		+5					1.03					
55	Cs	+1					1.78		1.82	1.86	1.89	1.96
56	Ba	+2					1.44	1.47	1.50	1.55	1.60	1.68
57	La	+3					1.13	1.18	1.26	1.28	1.36	1.40
58	Ce	+3					1.09		1.22	1.23		1.37
		+4					0.88		1.05			
59	Pr	+3					1.08		1.22			
		+4					0.86		1.07			
60	Nd	+3					1.06		1.20	1.17		
61	Pm	+3					1.04					
62	Sm	+3					1.04		1.17			
63	Eu	+2					1.25		1.33			
		+3					1.03	1.11	1.15			
64	Gd	+3					1.02	1.12	1.14			
65	Tb	+3					1.00	1.10	1.12			
		+4					0.84		0.96			
66	Dy	+3					0.99		1.11			
67	Ho	+3					0.98		1.10			
68	Er	+3					0.97		1.08			
69	Tm	+3					0.96		1.07			
70	Yb	+3					0.95		1.06			
71	Lu	+3					0.94		1.05			
72	Hf	+4					0.79		0.91			
73	Ta	+3					0.75					
		+4					0.74					
		+5					0.72		0.77			
74	W	+4					0.73					
		+6			0.50		0.68					
75	Re	+4					0.71					
		+5					0.60					
		+6					0.60					
		+7			0.48		0.65					
76	Os	+4					0.71					
77	Ir	+3					0.81					
		+4					0.71					
78	Pt	+2			0.68							
		+4					0.71					
79	Au	+3			0.78							
80	Hg	+1		1.05								
		+2	0.77		1.04		1.10		1.22			
81	Tl	+1					1.58		1.68			1.84
		+3					0.97		1.08			
82	Pb	+2			1.02		1.26		1.37	1.41		1.57
		+4					0.86		1.02			
83	Bi	+3				1.07	1.10		1.19			
84	Po	+4							1.16			
87	Ra	+2							1.56			1.72
90	Th	+4					1.08		1.12	1.17		
91	Pa	+4							1.09			
		+5							0.99	1.03		
92	U	+3					1.12					
		+4						1.06		1.08	1.13	
		+5					0.84	1.04				
		+6	0.53		0.56		0.81	0.96				

Z	Element	Valence	II	III	IV	V	VI	VII	VIII	IX	X	XII
							Coordination numbers					
93	Np	+2					1.18					
		+3					1.10					
		+4							1.06			
94	Pu	+3					1.09					
		+4					0.88		1.04			
95	Am	+3					1.08					
		+4							1.03			
96	Cm	+3					1.06					
		+4							1.08			
97	Bk	+3					1.04					
		+4							1.01			
98	Cf	+3					1.03					

*Ionic radii from Whittaker and Muntus, 1970. L = low spin (electrons paired), H = high spin (single electrons).

Piper Diagram for Water Chemistry Problems

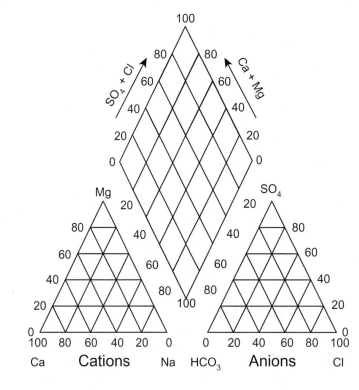

Answers to Selected Numerical Problems

CHAPTER 1

25. $\Delta E = 4.0852 \times 10^{-19}$ J $\lambda = 4.866 \times 10^{-7}$ m $= 487$ nm
26. $\lambda = 9.98 \times 10^{-10}$ m
27. Total number of electrons $= 18$ (2 $3s$, 6 $3p$, 10 $3d$)
29. $1s^2\, 2s^2\, 2p^6\, 3s^2\, 3p^4$
31. 26 protons, 31 neutrons
33. a. 20.18005 amu **b.** 108.97 g
   c. 3.252×10^{24} atoms **d.** 3.35098×10^{-23} g
35. 50.9436 amu
36. 258.16 g
37. a. 0.0214 mol **b.** 0.0214 mol L^{-1}
   c. 1.289×10^{22} atoms (ions)
40. Mole fraction $MgCO_3 = 0.12$
43. a. Empirical formula: $AlSiO_{4.5}H_2$
   b. Molecular formula: $Al_2Si_2O_5(OH)_4$
46. 2.452×10^6 mol
47. 3.97 mol
48. 584.8 cal g^{-1}
49. a. 3549 y **b.** 108 d
50. 0.143×10^{17} kg

CHAPTER 2

23. a. 295 or 823 **d.** 9.3 or 26.3
24. a. 9.0×10^{-6} **b.** 4.978×10^{-3} mol L^{-1} (1.66X)
   c. 8.5×10^{-3} mol L^{-1} (2.7X)
25. b. Positive (0.000165), negative (0.000161)
   c. $Ca^{2+} = 0.93$, $SO_4^{2-} = 0.93$ **d.** $10^{-4.02}$
28. a. TCE: $\log K_{eq} = -1.977$ **b.** TCE: 0.01 g kg^{-1}
   d. TCE: $\log K_t = -2.005$
29. a. $10^{30.4}$ **b.** AP $= 1482$ **c.** AP $= 411$
30. a. 6.94×10^{-7} atm
31. Microcline: dissolution time $= 933$ ky
32. 272 d
33. 3970 m
34. a. 3.63×10^{-10} mol m^{-2} s^{-1} **b.** 36.38 y
   c. $k = 8.15 \times 10^{-10}$ mol m^{-2} s^{-1}
35. a. 14.5% **b.** 4.1%
36. b. $A = 0.13$, $E_a = 49.8$ kJ mol^{-1}
38. c. $\log A = 12.81$ **d.** $k = 3.39 \times 10^{-8}$ s^{-1}
   e. At 25°C: $t_{1/2} = 10.3$ d

CHAPTER 3

21. $10^{-7.934}$ mol L^{-1}
23. 1.40×10^{-3} mol L^{-1}
24. 5.13
27. $H_2CO_{3\,(aq)} = 1.81 \times 10^{-3}$ mol L^{-1}
   $HCO_3^- = 1.19 \times 10^{-3}$ mol L^{-1}
   $CO_3^{2-} = 7.88 \times 10^{-8}$ mol L^{-1}

29. 43 times more abundant
31. 8.3
32. $\log P_{CO_2} = -3.40$ (HCO_3^-) and -3.38 (Ca^{2+})
34. a. $10^{-6.9}$ mol L^{-1} **b.** $10^{-12.9}$ mol L^{-1}
35. $C_A = 4.81 \times 10^{-3}$ eq L^{-1}
36. 5.36 meq L^{-1} pH^{-1}
38. 3.44×10^{-2} meq L^{-1} pH^{-1}
40. 0.657 meq L^{-1} pH^{-1}
42. 120.7 meq L^{-1} pH^{-1}
45. 115.56 kg $CaCO_3$-equivalent ton^{-1}

CHAPTER 4

35. a. -93.9 kJ mol^{-1} **b.** 0.32 V **c.** $10^{16.45}$
37. a. 0.21 V **b.** $10^{10.64}$ **c.** $10^{4.64}$ mol L^{-1}
38. a. 0.83 V
39. b. $10^{-20.9}$ mol L^{-1}
42. Acid mine drainage, $Eh = 0.98$ V
43. a. 0.55 V **b.** 6.31×10^{-32} mol L^{-1} **c.** 730 m V
46. b. -15.18
47. a. 0.69 V **b.** 0.41 V
49. Initial $Eh = 1.28$ V Final $Eh = 1.13$ V
50. 5.5×10^{-3} mol electrons L^{-1}

CHAPTER 5

51. a. $H/C = 1.59$ $O/C = 0.77$
55. 1.13
56. 71
57. Aerobic—2.7 d Anaerobic—1151 d
58. 44.6 ppm
59. 22.9 ppb
60. 4.12×10^{-4} mol L^{-1}
61. 698 d
62. 113 h

CHAPTER 6

37. 2 mrem
38. b. 9050 years before present
   c. 1471 years before present
40. c. 542 m y^{-1}
41. 22,715 years before present
42. a. 1.1 cm y^{-1} **b.** Deposited in 1969
43. a. 0.14 cm/1000 y **b.** 0.5 dpm g^{-1}
44. c. Nainital Lake: 1.27 cm y^{-1}
45. 22%
46. a. At 0°C, $\delta^{18}O_v = -35.2‰$
   b. At -30°C, $\delta^{18}O = -29.9‰$
47. $-7.29‰$

48. b. 40%
50. 20%
51. 42%
52. 39%
54. 51% feedlot runoff
55. b. 10.2°C
57. a. 18.6‰

CHAPTER 7

36. a. 14%   **d.** 32%
38. Eightfold
40. Sixfold
44. 2.32 Å
48. 1.1 sites nm^{-2}
49. 10 meq/100 g
50. 120.4 $m^2 g^{-1}$
52. b. $K = 0.176, n = 1$
54. d. $K = 1.3$
55. 2.93 $cm^3 g^{-1}$
57. Quartz solubility = 9.7 ppm

CHAPTER 8

62. a. Sun: $E = 6.28 \times 10^7$ W m^{-2} $\lambda_M = 0.50$ μm
    b. 61.6×10^4 J m^{-2} min^{-1}
63. 4.31 μm
65. Late Eocene otolith: $T_{min} = 12.49°C$ $T_{max} = 17.05°C$
66. $P_{CO_2} = 10^{-3.20}$ atm
67. $\lambda = 309.7$ nm
68. For first reaction:
    rate = 5.26×10^6 molecules $cm^{-3} s^{-1}$
69. Summer: 1.4×10^{-2} Bq L^{-1}
72. 126 Bq L^{-1}
75. pH = 5.10
77. a. pH = 3.49   **c.** 3.02×10^{-5} mol L^{-1}
79. a. Dust particle: 0.063 m s^{-1} **b.** 43.4 μm
80. a. Cr: $EF_{crust} = 1.1$ **b.** Cr: $X_{noncrustal} = 0.1$ μg m^{-3}
81. a. Fe: $X_{noncrustal} = 97$ ppm
    b. Sample 1: Mn/Fe = 0.009
84. 50% Boston, 30% New York
85. $^{206}Pb/^{207}Pb = 1.192$
86. 22% coal, 71% gasoline, 7% soil
87. Oslo air filters—59% gasoline-derived lead
89. 89% biogenic
90. b. CPI = 1.3

CHAPTER 9

65. Quartz at pH = 5.7, 6.06 ppm
66. 9611 ppm
69. a. 3.80×10^{-9} mol L^{-1}
72. a. 1.40×10^{-6} mol L^{-1}
82. a. Monoun, $p = 9.28$ atm   **d.** $K_{sp} = 10^{-10.49}$
    f. 181.7 mg L^{-1} **g.** 11% modern C
83. 145,000 mg s^{-1}
85. a. 26.4 μg L^{-1}
86. a. For Al, $\tau_{rel} = 0.17$

88. a. 8.91×10^{-7} mol g^{-1}
    b. $C_{soln} = 4.77 \times 10^{-5}$ mol L^{-1}
90. a. 1.0×10^{-7} mol g^{-1}
91. 0.61 meq
93. 0.01 meq L^{-1}
96. d. 212.6 mg L^{-1}
97. 6.45 m
101. 4 m, 1000 m

CHAPTER 10

54. d. 88% deep water
55. c. 62% deep water
56. 34.51‰
57. For N: $f = 0.030$, $g = 0.917$, $\tau = 57,600$ y
58. 0.64 mol charge kg^{-1}
61. a. $\log K_{si} = -9.49$ **b.** 6.06 μmol kg^{-1}
62. a. $\log K_{HSO_4} = 0.44$ **b.** 1.03×10^{-10} mol kg^{-1}
    e. 1.01×10^{-10} mol kg^{-1}
64. a. $\log K_{NH_4^+} = -9.42$ **b.** 1.47×10^{-6} mol kg^{-1}
    e. 4.44×10^{-7} mol kg^{-1}
69. IC = 24.5%
70. 9.4×10^{-4} mol kg^{-1}
71. 7.6%
74. Plankton: 1.20×10^{-6} g kg^{-1}
    Brown algae: 1.17×10^{-6} g kg^{-1}
81. As: 2583 y
82. Chemical half-life: 22.3 y
83. Scavenging turnover time: 0.31 y
84. Pyrene: 40.1 pmol m^{-2} d^{-1}
85. 5.8
87. 2.92×10^6 m^3 d^{-1}
89. 12 d
94. 0.16 y
97. c. $\ln K_D = 7.64$
99. c. 105,830
103. a. -9.69 nmol cm^{-2} y^{-1}
106. 0.0858 g cm^{-2} y^{-1}
107. 0.124 g cm^{-2} y^{-1}

Index

[ ] brackets, enclosing species in, 32
^{234}Th disequilibria, 417

AABW (Antarctic Bottom Water), 390–391
AADW (Antarctic Deep Water), 390–391
AAIW (Antarctic Intermediate Water),
  390–391, 408
absolute humidity, 249
absolute mass, 10
absolute zero, 16
absorption spectrum, 7–8
accelerator mass spectrometry (AMS),
  171–172
accumulation rate, 434
acetic acid, 63
acid–base equilibria, 59–93
  acid/base titrations and, 77–79
  acidity and alkalinity and, 75–77
  acids, 59–60
  amphoteric hydroxides, 75
  bases, 60–61
  buffers, 80–84
  case studies, 88–90
  dissociation of water and pH, 61–62
  mineral-water reactions, 84–88
  pH of natural waters and, 62
  salts, 71–74
  strong acids, 62–63
  strong bases, 75
  weak acids, 63–71
acid–base reactions, 12–13
acid–base titration curve, 78
acid dissociation constant, 59
acid gases, 90
acid mine drainage, 352–354
  chemical reactions in, 352–353
  defined, 49
  mixed with lake water, 327
  sources of, 353
  sulfides and, 353–354
acid mine wastes, 89
acid rain
  buffers and, 80
  defined, 66
  nitrogen and, 193–194
  rainwater chemistry and, 283–286
  sulfur as major contributor to, 195
acid soil, 355
acidic groundwater, 50, 110
acidity
  defined, 75

determining, 76, 78
  mineral–water reactions and, 84
  strong-acid acidity, 78
  titrations, 77–79
  total acidity, 76–77, 78
acids. *see also* acid–base equilibria
  adsorption process and, 157
  amphoteric hydroxides and, 75
  defined, 12
  overview of, 59–60
actinomycetes
  aerobic degradation, 112–113
  anaerobic degradation, 113
  overview of, 112
activation energy, 29–30, 45
activity
  Henry's law, 32–33
  overview of, 30–31
  of water, 40
activity coefficients, 36–40
  activity or fugacity and, 31
  Deby–Hückel model, 36–37
  ionic strength ranges and, 39
  Pitzer model, 39
  software, 48
  Truesdell–Jones model, 37–39
  uncharged species, 39–40
Activity Product (AP), 41
addition reactions, alkenes and alkynes,
  131
adiabatic lapse rates, 249, 250
adsorption–desorption, 341–349. *see also*
  partition/sorption characteristics
  acids and, 157
  carbon, 364–365
  case study, 344
  clay minerals, 224
  colloids, 343–344
  contaminant transport and,
    347–349
  halogens, 365–366
  ion exchange and, 345–347
  isotherms and partitioning of species
    and, 344–345
  metal cations, 342–343
  nitrogen, 368–369
  nuclear wastes, 362–363
  overview of, 341–342
  phosphorus, 369–371
  sulfur, 371–372
  types of particles in, 343
  water column chemistry, 428–430

adsorption isotherms, 221–223
  Freundlich equation, 221–222
  Langmuir equation, 222
advection–diffusion equation, 432–433
aerobes
  aerobic degradation, 112–113
  defined, 111
  denitrification, 113
  sulfide oxidation, 113–114
aerobic degradation, 112–113
aerosols. *see also* particulates
  anthropogenic vs. natural sources, 252,
    288–289
  cloud condensation nuclei, 289
  enrichment factor and, 291–292
  hygroscopic, 289
  suspended particulate matter and, 430,
    431–432
  as trace metal source, 413, 415
  tracking source of, 294–295
  types and sources, 252
age
  of marginal marine sediment, 434
  of seafloor sediments, 421–422
AIR (air) standard, 184, 193
air mass trajectories, 290
air pollution, 251–253
  aerosols, 252
  defined, 251
  indoor pollutants, 275
  smogs, 253
  types of pollutants, 251–252
alcohols, 136
algae, 112
alkaline lakes, 335
alkalinity, 75–77
  carbonate alkalinity, 79
  caustic alkalinity, 78
  defined, 75
  determining, 76
  mineral–water reactions and, 84
  of seawater, 401–403
  titrations, 77–79
  total alkalinity, 76–77, 79
alkanes, 130, 131–132
alkenes, 130, 132–133
alkynes, 130, 132–133
allochthonous carbon sources, 148
alpha particle emission, 165–166
aluminum, 318–320, 410
ambrosia (ragweed) pollen, 435
AMD. *see* acid mine drainage

amino acids, 136
aminocarboxylic acids, 137
ammonium
    continental sources, 193–194
    nitrification of, 113
amorphous silica
    dissolution of silica minerals, 233, 235
    health effects of exposure to, 234–235
    overview of, 233
amphoteric hydroxides, 75
AMS (accelerator mass spectrometry),
        171–172
anaerobes
    anaerobic degradation, 113
    defined, 111
    sulfide reduction, 114
anions
    coordination numbers and, 210
    crystal chemistry and, 208
    ligands and, 41
    negatively charged ions, 6, 94
anoxic environments, 118, 430, 431
Antarctic Bottom Water (AABW), 390–391
Antarctic Deep Water (AADW), 390–391
Antarctic Intermediate Water (AAIW),
        390–391, 408
anthropogenic carbons, 150–156
    dioxins, PCBs, and DFs, 152
    NAPLs, 153–156
    organic carbon, 150
    PAHs, 152–153
    pesticides, 150–152
    sources of, 140–141
anthropogenic vs. natural sources
    aerosols, 252, 288–289
    arsenic, 115
    atmospheric gases, 281–282
    box models and, 20, 22
    CO_2, 244
    Eh–pH diagrams and, 120
    elemental signatures and, 294
    greenhouse gases, 257–261
    heavy metals, 179, 198
    lead (Pb), 181
    marine sediment and, 177
    metals, 352, 358
    nitrates, 193
    organic acids, 63
    radioactive isotopes, 358–359
    sulfur, 196
    VOCs, 302
AP (Activity Product), 41
aqueous complexes, 40–41
aquifers
    acid gas disposal in, 90
    carbon contaminates, 149
    chlorine and bromine contaminates,
        367–368
    CO_2 disposal in, 256
    effluent in, 121
    Floridian aquifer system, 332

leachates, 48, 50, 192
    nitrates, 194
    petroleum contaminates, 123
aragonite
    authigenesis and, 421
    calcareous sediments, 419–420
archaebacteria, 111
aromatic (Aryl) hydrocarbons, 133, 160
Arrhenius equation
    acids, 59
    bases, 60
    overview of, 45
arsenate reduction, case study, 115–116
arsenic
    anthropogenic vs. natural sources, 115
    case study, 115–116
    as human health hazard, 356–357
Aryl (aromatic) hydrocarbons, 133, 160
asbestos, 227–232
    crystal structure of, 228–229
    defined, 227–228
    health effects of exposure to, 229–232
    types and uses of, 229
asbestosis, 229
atmospheric carbon, 139
atmospheric circulation, 247–248
atmospheric environment, 243–312
    air pollution, 251–253
    atmospheric circulation, 247–248
    greenhouse gases and climate change.
        see greenhouse gases
    layers of, 243
    as metal reservoir, 350–351
    ozone. see ozone
    particulates (aerosols). see particulates
    radon and, 275–278
    rainwater chemistry. see rainwater chem-
        istry
    solar and terrestrial radiation, 245–247
    structure and composition, 243–245
    temperature and water vapor variations,
        248–251
    as trace metal source, 413, 415
atmospheric gases, anthropogenic vs. natu-
        ral sources, 281–282
Atmospheric Test Ban Treaty (1963), 169
atomic and molecular weights
    atomic weight defined, 9
    gram-atomic weight, 9
    gram-equivalent weight of an ion, 10
    gram-molecular weight, 10
atomic mass number, 165
atomic number, 3, 8, 165
atoms, 1–8
    atomic weight, 9
    Bohr atom, 1–3
    combining into compounds, 8
    early theories of matter and, 1
    electronegativity and, 209
    ionization and valences, 6–7
    nucleus, 8

quantum mechanics and, 3–6
    Schrödinger wave model, 4
    spectra and elemental analysis, 7–8
authigenesis, 421
authigenic components, 418
autochthonous carbon sources, 148
autotrophs, 111, 113
Avogadro's law of gases, 16
Avogadro's number, 9

bacteria
    deriving methane from, 191
    overview of, 112
    oxidation–reduction and, 112–114
    reduction of arsenate and sulfate,
        115–116
balancing equations, 97–98
Balmer series, 3, 7
Baltic Sea, 431
bar-built estuaries, 424
basalt magmatism, 398–399
base dissociation constant, 60–61
bases. see also acid–base equilibria
    adsorption process and, 157
    amphoteric hydroxides and, 75
    defined, 12
    overview of, 60–61
basins, semi-isolated. see also marginal
        marine environment
    flushing time for, 426–427
    pore water chemistry, 433
    salinity of, 423
    water circulation in, 424–426
batch method, ion-exchange properties and,
        221
becquerel (Bq), 169
benzene, 158–159
beta particle emission, 165–166
binary isotopic mixing, 196
bioactive metals, 409–410
biochemical carbon cycles, 140
biodegradation of contaminants, in soil, 160
biogenic components, 418
biogenic particles, 413–414
bioindicators, 436
biointermediate elements, 398
biolimiting elements, 398
biological oxygen demand (BOD), 150
biological processes, controlling input of
        organic chemicals into water, 159
biological pump, 259
biomarkers, 436
biomass, 140
biosphere, as metal reservoir, 350–351
biounlimited elements, 398
BOD (biological oxygen demand), 150
Bohr atom, 1–3
bonds
    carbon atoms and, 129
    chemical, 8–9
    types of, 208–210

borate, 406–407
boron
    seawater alkalinity and, 401–402
    speciation reaction for, 404–405
box models, 20–23
    anthropogenic vs. natural sources and, 20, 22
    cause-and-effect feedback models, 22–23
    first-order kinetics, 21
    steady-state assumptions, 20–22
Boyle's law, 15–16
Bq (becquerel), 169
brackets ([ ]), enclosing species in, 32
Bragg's law, 212
brass corrosion, 122
brines, oil field, 348
bromine, 366–368
Brönsted–Lowry model, 59, 60
buffered solutions, 80
buffering capacity, 81, 87, 406–407
buffering index
    calcite-carbonic acid, 85–86
    water, 82
    water-silicate, 86–87
    weak acids, 83–84
buffers, 80–84
    buffering capacity, 81
    buffering index for water, 82
    buffering index for weak acids, 83–84
    defined, 80
    functions of, 80–81
    seawater, 406–407

CA (carbonate alkalinity), 79, 402–403
cadmium (Cd), 354–355
calcareous sediments, 419–421
calcite, 419–420, 421
calcite-carbonate system
    buffering index, 85–86
    pH control, 81
    salts of weak acids, 73–74
calcium carbonate
    isotopic fractionation factors, 190
    Th/ U dating, 177
calomel electrode, 99
Calvin cycle, 189
CAM (Crassulacean acid metabolism) cycle, 189
Canyon Diablo troilite (CDT) standard, 184, 194
capacity, oxidative or reductive, 118–119
carbamates, as organonitrogen compound, 137
carbon. *see also* organic compounds
    anthropogenic inputs to surface and groundwaters, 150–156
    anthropogenic sources of, 140–141
    carbon-14 dating, 171–174
    degradation half-life, 159–161
    DIC in aqueous systems, 189–191
    DOC in aqueous systems, 191

esters, 138–139
    formulas, 129–130
    fossil fuels, 146–148
    functional groups, 134
    humic substances, 141–143
    hydrocarbons, 130–134
    isomers, 130
    marine organic matter, 144
    mobility and persistence in natural environment, 364–365
    natural inputs to surface and ground waters, 148–150
    natural sources of, 139–140
    organohalide compounds, 134–136
    organonitrogen compounds, 136–137
    organooxygen compounds, 136
    organophosphorus compounds, 138
    organosulfur compounds, 137
    physical/biological inputs of organic compounds into water, 156–159
    polymers, 138–139
    as reducing agent, 114
    remineralization, 408
    soil organic matter, 143–144
    as source of methane, 191
    tracers in food chains, 191–192
carbon atoms
    bonds, 129
    prefixes, 131
carbon cycle
    cause-and-effect feedback models and, 23
    greenhouse gases and, 258
carbon dioxide. *see* CO_2 (carbon dioxide)
carbon isotopes, radiocarbon dating, 299
carbon preference index (CPI), 300
carbonaceous particles, 289
carbonate alkalinity (CA), 79, 402–403
carbonate compensation depth (CCD), 420
carbonate minerals
    oxidation–reduction reactions, 107
    reactivity of, 87
    water-carbonate interactions, 340
carbonate shell material, 402–403, 419–420
carbonate system. *see* carbonic acid–carbonate system
carbonic acid, 63–69
    abundance of, 63
    equilibrium relationships, 64–65
    groundwater example, 67–69
    open and closed systems, 65
    rainwater example, 66–67
    relative acidity example, 65–66
carbonic acid–carbonate system
    buffering ocean water, 62
    buffering surface and ground waters, 339–340
    buffers and, 80
    case study, 68
    controlling pH, 81
    equations describing, 70–71

gaseous species in rainwater, 281–282
    seawater buffering system, 406–407
carboxylic acids, 136
CAS (Chemical Abstracts System), 129
catalysts, microorganisms acting as, 111
cation-exchange capacity (CEC)
    defined, 219
    surface charge and, 220
    of zeolites, 226
cations
    adsorption–desorption processes and, 347
    adsorption of metal cations, 342–343
    coordination numbers and, 210
    crystal chemistry and, 208
    defined, 6, 94
    particle charges and, 341
cause-and-effect feedback models, 22–23
caustic alkalinity, 78
CCD (carbonate compensation depth), 420
Cd (cadmium), 354–355
CDT (Canyon Diablo troilite) standard, 184, 194
CEAM (Center of Exposure Assessment Modeling), 47–48
CEC (cation-exchange capacity). *see* cation-exchange capacity (CEC)
Cenozoic extinctions, 266
Center of Exposure Assessment Modeling (CEAM), 47–48
Centigrade, 16
CFCs (chlorofluoroxarbons), impact on ozone, 256–257
CH_4. *see* methane (CH_4)
charge balance, 402
Charles's law, of gases, 15–16
chelates, 41, 350
chelation, 349–350
CHEMFORM, 48
Chemical Abstracts System (CAS), 129
chemical bonds, 8–9
chemical composition of open ocean, 393–400
chemical equations, 13–15
    chemical reactions and, 13
    empirical formula, 13–14
    molecular formula, 14–15
chemical facies, 326–328
chemical half-life, 416
chemical mass transfer models, 47
chemical potential, 30
chemical reactions
    acid–base, 12–13
    chemical equations and, 13
    oxidation–reduction, 13
    precipitation, 12
    types of, 12
chemistry
    carbon. *see* carbon
    crystal, 208–212
    equations, 13–15

chemistry (*cont.*)
 pore water, 432–434
 rainwater, 278–281
 reactions, 12–13
 seawater, 400–416
 sediment, 434–437
 surface and ground waters, 324, 329–330
 water, 47–50, 65–66, 325–326
 water column, 428–432
chemolithotrophs
 defined, 111
 mineral–microorganism interactions, 236
 sulfide oxidation and, 113–114
chemoorganotrophs, 236
chemosynthesizers, 111
Chernobyl, 360
chlorinated phenols, 135
chlorine, 366–368
chlorinity, 392–393
chlorofluorocarbons (CFCs), impact on ozone, 256–257
chrysotile exposure, case study, 231
Ci (curies), 169
circulation
 oceanic, 388–392
 in semi-isolated basins, 424–426
clay minerals, 215–224
 adsorption characteristics of, 224
 as deep-sea sediments, 419
 defined, 215
 ion-exchange properties–determination, 221–223
 ion-exchange properties–origin, 219–221
 in marine sediment, 417
 as paleoclimatic indicator, 323
 principal characteristics of, 216
 structures of, 217–219
 types of, 215–216
clays, contrasted with clay minerals, 215
climate change
 Eocene/Oligocene boundary, 266–267
 geologic record of, 262
 greenhouse gases role in, 253–255
 ice cores and, 262–263
 isotopic measurements on ice floes and, 188–189
 marine sediments/carbonate minerals and, 177
 N_2O (nitrous oxide) and, 264
 Paleocene epoch, 265
 sediment record, 263, 266, 268
 variations/causes in last 1000 years, 254
closed systems, 27, 65
cloud condensation nuclei, 289
CO_2 (carbon dioxide), 257–259
 anthropogenic vs. natural sources, 244, 257–258
 atmospheric concentration of, 257
 carbon cycle and, 258
 chemical cycle for seawater species and, 394–395

fertilization effect, 259
 as greenhouse gas, 269
 of natural waters, 66
 ocean uptake of, 258–259
 relationship of groundwater to atmospheric CO_2, 334
 seawater alkalinity and, 401–403
 seawater composition and, 395–399
 seawater pH and, 403–404
 sequestration of, 387
 terrestrial uptake of, 259
coal
 acid mine drainage, 110
 environmental impact of, 146–147
 types of, 146
coesite, 233
colloidal organic carbon (COC), 409–410
colloidal organic material (COM), 410–412
colloids, 343–344
 defined, 343–344
 oxyhydroxides and, 344
 size ranges of, 343–344
 solubility of, 343–344
 trace metals in marine colloids, 410–412
column test method, ion-exchange, 223
COM (colloidal organic material), 410–412
complete ionic equation, 12
complex formations, 349–350
components, 27
composite diagrams
 defined, 106
 Eh–pH diagrams, 107–108
compounds. *see also* organic compounds
 chemical equations, 13–15
 concentration of, 10–12
 mass/weight of, 9–10
 saturated and unsaturated, 129
computer models, water-chemistry, 47–48
concentration, 10–12
 Le Châtelier's principle, 35
 in solution, 11
 units of measurement, 10
condensation
 homogeneous and heterogeneous reactions, 42
 isotopic composition of water and, 186
condensed structural formulas, organic molecules, 130
conservative behavior, of substances, 336–337
conservative species, 76, 401
conservative-type distributions, 414–415
constancy of composition, 392–393
contamination
 carbon, 149
 chlorine and bromine, 367–368
 flushing time and, 426–427
 heavy metal, 358
 in marginal marine sediment, 436
 petroleum, 123
 of San Francisco Bay, 435

source/time of release of contaminates, 172
 transport, 347, 348
 U-234/U-238 isotopes, 180
 water column chemistry and, 428–430
continental crust, 291
continental environment, 313–386
 adsorption–desorption processes. *see* adsorption–desorption
 complex formation and chelation, 349–350
 crust composition, 291
 hydrologic cycle, 313–314
 metals. *see* metals
 nonmetals. *see* nonmetals
 surface and ground waters. *see* surface and ground waters
 weathering. *see* weathering
continental shelf, 387–388
continental slope, 387–388
continuous view, of matter, 1
coordination numbers, crystal chemistry, 210–211
Coriolis deflections
 atmospheric circulation, 247–248
 Ekman spiral and, 390
 estuaries and, 424
corrosion, in marine environment, 122
counter ions, 341
covalent bonding, 9, 208
CPI (carbon preference index), 300
Craig–Gordon evaporation model, 187
Crassulacean acid metabolism (CAM) cycle, 189
crude oil, sweet and sour, 148
crystal chemistry, 208–212
 asbestos, 228–229
 coordination numbers, 210–211
 investigating crystalline structures, 212–214
 ionic substitutions, 211–212
 overview of, 208
 types of bonding and, 208–210
crystalline silica, 233–235
 abundance of, 317
 dissolution of, 233, 235
 health effects of exposure to, 234–235
 overview of, 233
curies (Ci), 169
cyclic hydrocarbons, 133
cyclical salts, 394–395
cyclosilicates, 214

DDT (dichlorodiphenyltrichloroethane), 136
de Broglie wavelength equation, 4
Debye–Hückel model
 calculating activity coefficients, 36–37
 ranges of ionic strength, 39
 selected values for, 37–38
decomposition reactions, 315

deep ocean. *see* ocean, deep

degradation half-life. *see* half-lives

degree of pyritization (DOP), 434

delta values (δ)
  carbon and, 189–193
  overview of, 184–185
  oxygen and hydrogen isotopes in water, 186–189

dendrochronology, correcting carbon-14 variations, 172

denitrification, 113, 193

density, of water, 19

detrital components, 417, 419

diagenesis, 421, 422

diagenetic environment, 147

DIC (dissolved inorganic carbon), 189–191, 401–403

dichlorodiphenyltrichloroethane (DDT), 136

diffuse double layer, 341–342

diffusion, 432–433

diffusion coefficient, 433

dioxins, 136

discontinuous view, of matter, 1

disequilibrium, measurement of, 41–42

dislocations, dissolution/growth and, 47

dispersion coefficient, 433

dissociation constant
  acids, 59
  bases, 60
  inorganic ligands, 408–409
  pressure and, 405–406

dissociation of water, 61–62

dissolution
  aluminum and iron hydroxides and, 318–320
  compared with suspension, 324
  growth and, 46–47
  rates of, 44
  silica and, 317–318

dissolved inorganic carbon (DIC), 189–191, 401–403

dissolved organic carbon (DOC), 148, 191, 409–410

dissolved organic material (DOM), 429

distribution coefficient model, contaminant transport, 348

Dobson unit (DU), 271

DOC (dissolved organic carbon), 148, 191, 409–410

DOM (dissolved organic materials), 429

DOP (degree of pyritization), 434

double bonds, carbon atoms, 129

downwelling, 389–390

drowned-river-valley estuaries, 424

dry adiabatic lapse rate, 249

dry deposition, removal of particulates, 289

DU (Dobson unit), 271

earth's albedo, 247

eddy diffusivity, 433

Eh
  ladders, 119–121
  overview of, 99–100
  oxidation–reduction classification, 118
  oxidation/reduction processes in natural systems, 114, 116–117
  redox buffering and, 117
  redox interface and, 121–124
  relationship to pe, 101

Eh–pH diagrams
  anthropogenic vs. natural sources and, 120
  environmental application of, 109–110
  iron, 103–108
  nitrogen, 108–109
  redox interface, 122
  stability of water, 101–103

Ekman spiral, 390

electrochemical cells, 94–95

electrodes
  Eh–pH diagrams and, 103–109
  electromotive force for, 96–97
  electromotive series for, 96
  as half-reactions, 95
  Nernst equation, Eh and, 99–100
  pe and, 100–101

electromotive force (emf), 95–97

electromotive series, 95–96

electron acceptors
  bases as acceptors, 12, 60
  oxidation–reduction reactions and, 94

electron donors
  acids as donors, 12, 59
  oxidation–reduction reactions and, 94

electron orbitals, 4–6

electronegativity, 209

electrons
  high spin state, 211
  low spin state, 211
  oxidation–reduction reactions and, 13

elemental analysis, 7–8

elemental signatures, 292–294

elements
  atomic weights, 9
  concentration of, 10–12
  enrichment factor and, 291–292
  metals and nonmetals, 208

EM radiation, 3

emf (electromotive force), 95–97

emission spectrum, 7–8

empirical formula, of compounds, 13–14

endothermic reactions, 28

energy
  activation, 29–30
  first law of thermodynamics and, 28
  free, 30, 33–35

enrichment factor
  for metal concentration, 413–414
  of particulates (aerosols), 291–292
  sediment chemistry, 435
  of suspended particulate matter, 430, 431

enthalpy, 28

entropy, 28–29

environmental applications
  Eh–pH diagrams, 109–110
  isotopic measurement of groundwater sources, 187
  isotopic measurement of ice floes, 188–189
  oxidation–reduction. *see* oxidation-reduction reactions
  radiogenic isotopic tracers, 169, 178–181
  stable isotopes, 181
  water-chemistry, 48–50

environmental lapse rate, 250

environmental mineralogy. *see* mineralogy

environmental parameters
  acidity/alkalinity as, 77
  pH as, 62

environments, oxidation–reduction, 118

enzymes, 111

Eocene/Oligocene boundary, 266–267

epidemiological studies, asbestos exposure, 231

epilimnion layer, lakes, 335

equilibrium constant
  amphoteric hydroxides, 75
  carbonate system, 64
  free energy at temperatures other than 25°C, 33–34
  overview of, 31–32
  stoichiometric, speciation and, 404–406
  water column chemistry, 428

equilibrium reactions, 182

equilibrium thermodynamics, 27–58
  activity and fugacity, 30–31
  activity coefficients, 36–40
  aqueous complexes, 40–41
  case studies, 48–50
  chemical potential, 30
  defined, 27
  equilibrium constant, 31–32
  free energies at temperatures other than 25°C, 33–35
  free energy, 30
  Henry's law, 32–33
  kinetics, 42–47
  Le Châtelier's principle, 35–36
  measurement of disequilibrium, 41–42
  overview of, 29–30
  thermodynamic laws, 27–29
  water-chemistry computer models, 47–48

esters, 138–139

estuaries. *see also* marginal marine environment
  alkalinity of, 423
  flushing time for, 426–427
  types of, 424
  water column chemistry, 428–432

ethers, 136

eucaryotes, 111, 235–236

euphotic zone, 396

eutrophication
  lake ecosystems, 150, 334
  role of phosphorus in, 369–371
evaporation
  basin water circulation and, 425–426
  Craig–Gordon model, 187
  isotopic composition of water, 187
event dating, 170, 434
exothermic reactions, 28
extensive properties, 27

facultative anaerobes, 111
feldspar–kaolinite reaction, 84–85, 87
fermentation, redox buffering process and,
  117
ferric iron, weathering, 319–320
fertilization effect, CO_2 (carbon dioxide),
  259
fertilizers
  denitrification in atmosphere, 113
  nitrates in surface and ground waters,
  193
  nitrogen Eh–pH diagram, 108–109
fibrous minerals. *see also* asbestos
  dissolution rate of asbestos, 229–230
  overview of, 227–228
  surface exposure and charges of
  asbestos, 230
Fick's first law, 433
first-order reactions, 43
fjord-type estuaries, 424
Floridian aquifer system, 332
fluorine, 365–367
flushing time, 426–427
food chains, tracers in, 191–192
fossil fuels, 146–148
  coal, 146–147
  nitrogen from combustion, 193–194
  petroleum, 147–148
  sulfur from combustion, 196
fractionation, 182–183
fractionation factor, 183
free energy
  nucleation and, 46
  overview of, 30
  oxidation–reduction reactions and, 95–97
  at temperatures other than 25°C, 33–35
free radicals, 261
freshwater
  estuary types and, 424
  water column chemistry, 428
Freundlich isotherms, 221–222, 345
frontal inversions, 250
fugacity, 30–33
fulvic acid, 141
functional groups
  defined, 129
  organic compounds and, 134
fungi, 112–113
fusion bombs, 169–172

Gaines–Thomas convention, 346
gamma rays (high-energy electromagnetic
  radiation), 166
Garrels model, groundwaters, 332–334
gases. *see also* fugacity; greenhouse gases
  Avogadro's law, 16
  carbonic acid–carbonate system and,
  281–282
  Charles's law, 15–16
  deviation from ideal behavior, 17–18
  Henry's law for, 33–35
  ideal gas law, 15–17
  landfill gases, 192–193
  NADP, 283
  nitrogen cycle, 282–283
  rainwater chemistry and, 281–283
  solubility in rainwater, 281
  sulfur dioxide and, 282
  types of gases, 281–282
  van der Waals constants for, 18
geochemical cycles
  box models of, 20–23
  carbon, 140
  metals, 358
Geochemical Earth Reference Model
  (GERM), 48
The Geochemical Society, 48
Geochemist's Workbench software, 48
geologic record, of climate change, 262
geomagnetic time scale, 422
geometric isomers, 130
geosphere, as metal reservoir, 350–351
geothermal, as methane source, 191
GERM (Geochemical Earth Reference
  Model), 48
Gibbs free energy, 30, 34–35
Gibbs model, river chemistry, 329
glacial marine sediments, 418–419
glaciers, 313
global warming potential (GWP), 256
gram-atomic weight, 9
gram-molecular weight, 10
gray (Gy), 169
greenhouse gases, 253–268
  anthropogenic vs. natural sources,
  257–261
  CH_4 (methane), 259–262
  CO_2 (carbon dioxide), 257–259, 269
  N_2O (nitrous oxide), 262, 264
  overview of, 255–257
  role in climatic changes, 253–255
greenhouse warming, 255
groundwater, 331–334
  acidic groundwater, 50, 110
  case study evaluating contamination of,
  180
  case study of groundwater chemistry, 72
  case study of hydrocarbon spill, 123–124
  case study of sources, 187
  composition, 324

  effect of septic systems on, 121
  Floridian aquifer system, 332
  Garrels model, 332–334
  nitrates in, 193
  phosphate pollution in, 370
  pitfalls in dating, 172
  silicate minerals and, 332
  simplified system of, 331–332
  tritium dating in, 170–172
Gulf Stream, 389
GWP (global warming potential), 256
Gy (gray), 169
gyres, 388–389

H_2O, 82, 395–396. *see also* natural waters;
  water
H_2SO_4 (sulfuric acid), 62
half-lives
  degradation half-life, 159–161
  of radioactive isotopes, 167–168
  of reactions, 44–45
half-reactions. *see* electrodes
halogens, 365–368
  chlorine and bromine, 366–368
  fluorine, 365–366
halophiles, 111
Hatch–Slack cycle, 189
HCl (hydrochloric acid), 62
health effects
  arsenic, 356–357
  crystalline and amorphous silica minerals, 234–235
  fluorine, 365–366
  heavy metals, 356–357
  mercury, 357–358
  selenium, 356–357
health effects, asbestos exposure,
  229–232
  biological interactions, 230–231
  dissolution rate of fibers, 229–230
  fiber surfaces, 230
  measuring effect of exposure,
  230–231
  symptoms and diseases, 229
heat capacity
  defined, 34
  water, 19
heat, compared with temperature, 243
heat of formation, 28
heavy metals
  adsorption characteristics of clays for,
  224
  anthropogenic vs. natural sources, 179,
  198
  arsenic and selenium, 356–357
  defined, 350
  distribution and concentrations, 358
  flushing time for, 426–427
  in humus, 142
  mercury, 357–358

mobility of, 354
overview of, 354
selective extraction and, 355–356
soil and sediment contamination and, 358
transition metals (Zn, Cd, and Pb) and, 354–355
Heisenberg uncertainty principle, 4
Helmholtz free energy, 30
hematite, 104–107
Henderson–Hasselbalch equation, 80
Henry's law, 32–33, 157
heterogeneous condensation, 289
heterogeneous reactions
    nucleation, 45–46
    overview of, 42
heterotrophs, 111, 113, 396–397
high-level radioactive wastes, 360
high spin state, electrons, 211
HMS *Challenger*, 392
HNO_3 (nitric acid), 62
HOCs (hydrophobic organic contaminants), 417, 429
homogeneous condensation, 289
homogeneous reactions
    nucleation, 45–46
    overview of, 42
homolog, 300
homologous series, 300
humic acid, 141
humic coals, 146
humic substances, 141–143
    case studies, 142–143
    defined, 141
    humic and fulvic acids and, 141
    in various environments, 142
humidity, absolute and relative, 249
humin, 141
hydrates, methane, 422–423
hydration shells, 36
hydrocarbon spill, case study, 123
hydrocarbons, 130–134
    alkanes, 131–132
    alkenes and alkynes, 132–133
    aromatic, 133
    cyclic, 133
    overview of, 130–131
    polycyclic aromatic, 133–134
hydrochloric acid (HCl), 62
hydrofacies, classification of water types, 326–328
hydrogen
    isotopic composition of water, 185–189
    remineralization, 408
    tritium dating, 169–172
hydrogen bonding, 9, 209
hydrogen electrode, 99
hydrogenous components, 418
hydrologic cycle
    box models of, 20–21

defined, 314
water reservoirs and, 313
hydrophobic organic contaminants (HOCs), 417, 429
hydrosphere, as metal reservoir, 350–351
hydrostatic equation, 249
hydrothermal systems, 413
hydroxyl radical, 261
hygroscopic aerosols, 289
hypolimnion layer, lakes, 335

IAP (Ion Activity Product), 41–42
ice caps, 313
ice cores
    as indicator of climatic changes, 262–263
    record of acid deposition in, 287
ideal gas law, 16–17
igneous rocks, 332–333
illites, clays, 217
in vitro testing, asbestos exposure, 231
incongruent dissolution, 47
indoor air pollutants, 275. *see also* radon
inhibitors, dissolution/growth and, 47
inner-sphere complex, particles, 341
inorganic ligands, 407–409
inosilicates, 214
insolation, 247
intensive properties, 27
inversions, 250
Ion Activity Product (IAP), 41–42
ion-exchange, 345–347
    adsorption–desorption processes, 345–347
    chemistry of subsurface brines and, 348
    formula, 347
ion-exchange properties–determination
    adsorption isotherms, 221–223
    batch method, 221
    column test method, 223
ion-exchange properties–origin
    defined, 219–221
    surface area, 220–221
    surface charge, 220
ionic bonding, 8–9, 209
ionic species, rainwater chemistry, 278–279
ionic strength
    activity coefficients and, 36–40
    aqueous complexes, 40–41
    of seawater, 400–401
    water column chemistry and, 428
ionic substitutions, crystal chemistry, 211–212
ionization, 6–7, 208
ionization potential, 6–7, 208, 342
ions
    complete ionic equation, 12
    crystal chemistry and, 208
    defined, 6

gram-equivalent weight, 10
net ionic equation, 12
iron
    as hydrogenous component, 418
    as limiting nutrient, 396
    redox indicators and, 434
iron Eh–pH diagram, 103–108, 110
iron hydroxides, weathering, 318–320
isolated systems, 27
isomers, 130
isotherms
    Freundlich equation, 221–222
    Langmuir equation, 222
    partitioning species and, 344–345
isotopes. *see also* radioactive isotopes; stable isotopes
    atomic weight and, 9
    defined, 8
    overview of, 165
    temperature and climate data from, 263
isotopic fractionation
    carbon-14 dating and, 172
    delta value for lake water, 184–185
    DIC in aqueous systems, 189–191
    overview of, 183
    water composition and, 185–189
isotopic signatures, 296–300

Kelvin, 16
kerogen, 147, 191
ketones, 136
kinetics, 42–58
    Arrhenius equation, 45
    case studies, 48–50
    defined, 27
    dissolution and growth, 46–47
    nucleation, 45–46
    overview of, 42
    reactions, 42–45, 182
    using Eh ladder to predict changes, 120

Labrador Current, 389
lakes, 334–339
    acid mine drainage mixed with, 327
    alkaline lakes, 335
    anthropogenic impact on, 334
    box models, 336–339
    conservative and nonconservative behaviors in, 336–337
    eutrophication, 150, 334
    "killer lakes" of Cameroon volcanic line, 336
    layers of, 335
    overview of, 334–335
    residence time/relative residence time, 338–339
    structure and mixing of waters, 335–336
landfills, identifying gases and leachates, 192–193
Langmuir isotherms, 222, 345

lapse rates
adiabatic, 249–250
environmental, 250
laterites, 143, 144
Le Châtelier's principle, 35–36, 80
leachates
landfills, 192–193
using Eh ladder to predict changes in, 120
water-chemistry applied to environmental problems, 50
lead (Pb)
anthropogenic vs. natural sources, 181
dating, 177–178, 179
isotopic signature of, 296–299
sediment chemistry and, 436
as transition metal, 354–355
lever rule, 197
ligands
anion as, 41
defined, 349–350
inorganic/organic in seawater, 407–410
limnology, 334
liquid state, of water, 19
lithogenic particles, 413, 414
lone pairs, structure of water molecule, 19
low-level radioactive wastes, 359
low spin state, electrons, 211
lung cancer
asbestos and, 229
silica exposure and, 235
lysocline, 420

magnetite, 104–107
main thermocline, 388
manganese Eh-pH diagram, 110
manganese oxyhydroxides, 418
mantle, deriving methane from, 191
marginal marine environment, 423–437
estuaries, 424
flushing time, 426–427
importance of, 387
pore water chemistry, 432–434
sediment chemistry, 434–437
types of, 423
water circulation in semi-isolated basins, 424–426
water column chemistry, 428–432
marine environment, 387–457
chemical composition of open ocean, 393–400
marginal, 423–437
marine sediments, geochemistry of, 416–423
physical aspects of open ocean, 387–393
seawater chemistry, 400–416
marine influence, rainwater chemistry and, 281
marine invertebrates, 266
marine organic matter, 144

marine sediments
anthropogenic vs. natural sources, 177
case studies in dating, 179–180
reasons for dating, 177
sedimentation-rate relationships and activity, 176–177
Th dating of, 174–176
marine sediments, geochemistry of, 416–423
authigenesis in, 421
deep-sea sediments, distribution of, 418–421
methane hydrates, 422–423
seafloor sediments, age of, 421–422
sedimentary component types, 416–418
Mars, arsenate-reducing bacteria and, 116
mass, 9
mass number, 8
mass transfer models, 47
matter, 1
Mediterranean Intermediate Water (MIW), 390–392
Mediterranean Sea, 425
mercury
box models of, 21–22
case study of deep-sea sediments, 123
as human health hazard, 357–358
mesothelioma cancer, 229
metal cations, 342–343
metal cycle, 350–352
human impact on, 352
interactions between metal reservoirs, 351
types of metal reservoirs, 350
metal oxalates, 350
metallic bonding, 9, 209
metallic iron, 104–107
metalloids, 350
metals, 350–364
acid drainage, 352–354
adsorption characteristics of clays for heavy, 224
adsorption–desorption processes, 355
adsorption process of, 429–430
anthropogenic, 352
crystal chemistry and, 208
distribution in coals, 147
distribution types, 414–415
heavy metals, 354–358
ligands and, 407, 408–410
in marginal marine sediment, 434–436
metal cycle and, 350–352
mobility of, 354, 356
natural, 350–351
Oklo fossils fission reactors and, 363–364
pore water chemistry and, 432–434
radioactive isotopes and radioactive waste disposal, 358–363
reservoirs, 350

suspended particulate matter, 430–432
trace metals in marine colloids, 410–412
trace metals in seawater, 412–414
metastable state, 29
methane (CH_4)
anaerobic degradation as major source of, 113
hydrates, 422–423
landfill, 192–193
sinks for, 259–261
sources of, 191, 259, 261
methanogenesis, 192–193
methanogens, 111
microbial decomposition, 192
microorganisms, 110–114
energy pathways and, 111–112
factors effecting reaction rate with minerals, 237–238
oxidation–reduction reactions, 110–114
role in dissolution and precipitation of minerals, 236–237
types of, 235
mid-depth maxima and minima, 415
mineral dust, particulate types, 288
mineral–microorganism interactions, 235–238
dissolution and precipitation of minerals, 236–237
factors affecting reaction rate, 237–238
types of microorganisms, 235
mineral–water reactions, 84–88
acidity/alkalinity and, 84
calcite-carbonic acid buffering index, 85–86
natural waters, buffering capacity, 87
rate of mineral reactivity, 87–88
water-K-feldspar example, 84–85
water-silicate buffering index, 86–87
mineralogy, 207–242
asbestos minerals. *see* asbestos
classes of minerals, 208
clay minerals. *see* clay minerals
crystal chemistry, 208–212
crystalline and amorphous silica minerals, 233–235
defined, 207–208
dissolution and suspension and, 324
dissolution/growth of minerals, 46–47
microorganisms and. *see* mineral–microorganism interactions
reactivity of, 87–88
as salts of weak acids and strong bases, 71
silicate minerals, 214–215
types of minerals, 207–208
water and. *see* mineral-water reactions
X-ray crystallography and, 212–214
zeolites minerals, 225–227
mining wastes, 88, 89
MINTEQA2, 48

MIW (Mediterranean Intermediate Water), 390–392
mixed-anion facies, water, 326
mixed-cation facies, water, 326
mixed-layer clays, 219
mixing
    binary isotopic mixing, 196
    multi-end-member mixing, 196–197
MLS (multilevel sampling systems), 172
molar volume, gases, 17
molarity, 11
mole fraction, 11
mole (mol), 9, 11
molecular formula
    of compounds, 14–15
    of organic compounds, 130
molecular signatures, 294–296
molecules
    electronegativity, 209
    gram-molecular weight, 10
    nonmetallic molecules, 18
    water, 18–19
monoprotic acid, 83
multi-end-member mixing, 196–197
multilevel sampling systems (MLS), 172

N_2O (nitrous oxide), 262, 264
NABW (North Atlantic Bottom Waters), 390–391
NADP (National Atmospheric Deposition Program), 283
NADW (North Atlantic Deep Waters), 390–392
National Atmospheric Deposition Program (NADP), 283
natron, salts of weak acids, 74–75
natural sources. see anthropogenic vs. natural sources
natural waters
    buffering capacity, 87
    buffers to acid/base additions, 80–81
    environmental parameters and, 77
    pH values of, 62
    stong acids, 62–63
    weak acids, 63–71
NBS-19 standard, 184
neritic environment, 388
Nernst equation, 98–100, 101–102
nesosilicates, 214
net ionic equation, 12
neutrons
    alpha particles, 165
    atom and, 8
    defined, 165
nitrate reduction, 117
nitrates
    anthropogenic vs. natural sources, 193
    nitrogen isotopes indicating sources of, 194
    in surface and ground waters, 193

nitric acid (HNO_3), 62
nitrification, 113
nitrogen
    adsorption–desorption processes, 368–369
    concentrations in rivers, 371
    denitrification/nitrification, 113
    Eh–pH diagrams and, 108–109
    isotopes, 193–194
    organonitrogen compounds and, 136
nitrogen cycle, 282–283. see also greenhouse gases
nitrous oxide (N2O), 262, 264
noble gases, 7
nonconservative behavior, 336–337
nonconservative species, 76
nonmetallic molecules, 18
nonmetals, 364–372
    carbon, 364–365
    defined, 208
    halogens, 365–368
    nitrogen, 368–369
    phosphorus, 369–371
    sulfur, 371–372
nonsulfidic environments, 118
normality, 11
North Atlantic Bottom Waters (NABW), 390–391
North Atlantic Deep Waters (NADW), 390–392
Northeastern Atlantic Ocean, 411
nuclear fission, 360–364
    formula for reaction in, 360
    Oklo fossils fission reactors, 363–364
    radioactive wastes due to, 360–362
nuclear fuel cycle, 359–360
nucleation, 42, 45–46
nucleus, of atom, 8, 165
nutrient-type distributions, 414, 415

obligate anaerobes, 111
OC (organic carbon), 150, 411–412
ocean basin, 387–388
ocean, deep
    deep-sea sediments, 418–421
    pH in, 404
    structure of ocean, 388
    thermohaline circulation and, 390–392
ocean margin sediments, 418
ocean, open
    chemical cycle for seawater species, 394–395
    composition and salinity, constancy of, 392–393
    continental shelves/ocean basin, 387–388
    oceanic circulation, 388–392
    residence time of seawater species, 393–394
    seawater composition, processes controlling, 395–400

    vertical structure of, 388
ocean water. see also marine sediments
    case study of corrosion of brass in, 122
    case study of mercury and selenium in, 123
    paleothermometry, 197–198
    redox interface and, 121–122
    source of methane in, 191
    surface water reservoirs, 313
    tritium concentration in, 170
    uptake of CO_2 by, 258–259
oceanic circulation, 388–392
oceanic environment, 388
octahedral clays, 215–216
oil spill, case study, 123
Oklo fossils fission reactors, 363–364
open ocean. see ocean, open
open systems, 27, 65
order of reactions, 42–45
organic acids, 63
organic carbon (OC), 150, 411–412
organic compounds. see also carbon
    half-lives, 161
    overview of, 129
organic contaminants, 436–437
organic ligands
    in seawater, 407–408, 409–410
    trace metals and, 411–412
organic matter, soil and, 143–144
organohalide compounds, 134–136
    chlorinated phenols, 135
    DDT (dichlorodiphenyltrichloroethane), 136
    dioxins, 136
    naming, 134–135
    notorious examples, 134
    polychlorinated biphenyls (PCBs), 135
organonitrogen compounds, 136–137
organooxygen compounds, 136
organophosphorus compounds, 138
organosulfur compounds, 137
outer-sphere complex, particles, 341
OXC (oxidative capacity), 118–119
oxic environments, 118, 123
oxidation, 94
oxidation–reduction ladders, 119–120
oxidation–reduction reactions. see also redox reactions
    balancing equations, 97–98
    brass corrosion case study, 122
    classifying environments, 118
    Eh–pH diagrams and, 103–109
    electrochemical cells, 94–95
    electromotive force, 95–97
    electromotive series, 95–96
    heavy metals and, 354–355
    hydrocarbon spill case study, 123
    mercury and selenium in deep-sea sediments, 123
    in natural systems, 114

oxidation–reduction reactions (*cont.*)
  Nernst equation and Eh and, 98–100
  nuclear wastes and, 362
  overview of, 13
  oxidation–reduction ladders, 119–120
  oxidative and reductive capacity,
    118–119
  pe and, 100–101
  questions and problems, 124–128
  redox buffering, 117
  redox interface, 121–124
  reduction of arsenate and sulfate,
    115–116
  role of microrganisms in, 110–114
  seawater–sediment interactions,
    398–399
  septic effluent case study, 121
  stability of water diagram, 101–103
  water column chemistry and, 428
oxidative capacity (OXC), 118–119
oxidizing agents, 94, 114
oxycline, 430
oxygen
  chemical cycle for seawater species and,
    394–395
  isotopic composition of water, 185–189
  as major oxidizing agent, 114
  redox reactions and, 430
  remineralization, 408
  seawater composition and, 396–397
  using Eh ladder to predict changes in,
    120
oxygenases, 111
oxyhydroxides
  pore water chemistry and, 432, 434
  redox reactions and, 430
  removal of, 414
  removal of zinc, 344
  solubility of, 343–344, 365
ozone, 268–275
  Antarctic ozone hole, 273
  anthropogenic impact on, 271
  formation/destruction processes, 270
  kinetic data for ozone destruction,
    272–273
  measuring in DUs, 271
  ozone-destroying reactions, 271–272
  stratospheric, 268–271
  tropospheric, 273–275
  UV radiation and, 273

Pa dating, 176
PAHs. *see* polycyclic aromatic hydrocarbons (PAHs)
Paleocene epoch, 265
paleoclimatic indicators, 323
paleothermometry, 197–198
palygorskite group minerals, 219
particulate organic carbon (POC), 148
particulates, 288–300. *see also* aerosols;
  partition/sorption characteristics

adsorption–desorption processes and,
  343
air mass trajectories and, 290
charges of, 341–342
elemental signatures, 292–294
enrichment factor, 291–292
isotopic signatures, 296–300
molecular signatures, 294–296
removal from atmosphere, 289–290
sampling, 290
source identification, 291
temporal variations in, 300
trace metal removal with, 413–414
types of, 288–289
partition/sorption characteristics. *see also*
  adsorption-desorption isotherms and,
  344–345
  organic compounds, 156–158
Pauli exclusion principle, 4
Pb. *see* lead (Pb)
PCBs (polychlorinated biphenyls), 135
PDB (PeeDee beleminite) standard, 184
pe
  overview of, 100–101
  relationship to Eh, 101
  stability of water diagram, 102–103
pedalfers, soils, 143
pedocals, soils, 143
PeeDee beleminite (PBD) standard, 184
percent carbon of biogenic origin, 299
percent modern carbon (pmc), 299
perfect radiators, 245
periodic table
  of metals, 352
  quantum mechanics and, 4–6
pesticides, carbamates, 137
petroleum, 147–148
  contamination of aquifer, 123
  deriving methane from, 191
  elemental composition, 148
  hydrocarbons, 436
  overview of, 147
  sulfur derived from, 196
pH. *see also* Eh–pH diagrams
  adsorption and, 157, 343
  dissociation of water and, 61
  as environmental parameter, 62
  of natural waters, 62, 66
  nonconservative and conservative species
    and, 76
  particle charges and, 342
  rainwater chemistry and, 283–286
  of seawater, 403–404
  seawater buffers and, 406–407
  of surface and ground waters, 339–341
  types of reactions controlling, 81
phases, 27, 42
phenols, 136
phosphorus
  in natural waters, 369–371
  in organic compounds, 138

remineralization, 408
photolithotrophs, 111, 236
photons
  EM radiation and, 3
  microorganisms and, 236
photoorganotrophs, 236
photosynthesis, 140, 395–396
photosynthesizers, 111
phyllosilicates, 215
phytoplankton, 411
picocurie, 169
Piper diagrams, 325–326
Pitzer model, 39
Plummer and MacKenzie model, 39
POC (particulate organic carbon), 148
point of zero charge (PZC), 342
point of zero net proton charge (PZNPC),
  342
polar covalent bonds, 9
polychlorinated biphenyls (PCBs), 135
polycyclic aromatic hydrocarbons (PAHs),
  133–134
  anthropogenic carbons, 152–153
  in marginal marine sediment,
    436–437
  molecular signatures, 294
  removal rates for, 417
  sources of, 296
polymers, 138–139
polymorphs, 207, 233
polyprotic acid, 84
pore waters
  authigenesis in, 421
  authigenic components and, 418
  chemistry, 432–434
postoxic environments, 123
precipitates, of solutions, 12
precipitation reactions. *see also* rainwater
  chemistry
  aluminum and iron hydroxides and,
    318–320
  basin water circulation and, 425–426
  hydrologic cycle and, 314
  organic compounds and, 158
  overview of, 12
  silica and, 317–318
pressure
  Le Châtelier's principle, 35
  stoichiometric equilibrium constants and,
    405–406
primary minerals, weathering, 316
primary pollutants, 251
primary production, biomass, 140
primordial carbon, 139
procaryotes, microorganisms, 111, 235–236
properties, system, 27
protons
  alpha particles, 165
  in atom, 8
  defined, 165
  electron acceptors, 12, 60

electron donors, 12, 59
protozoa, 112–113
pulmonary disease
    asbestos and, 229
    quartz dust and, 234
PZC (point of zero charge), 342
PZNPC (point of zero net proton charge),
    342

quantum mechanics
    beyond the Bohr atom, 3–4
    periodic table and, 4–6
    relationship of quantum numbers to
        electron orbitals, 4–6
quartz
    resistance to weathering, 317–318
    respiratory diseases and, 234

radiation
    inversions, 250
    solar and terrestrial, 245–247
    UV, 273
radiative forcing, 255–256
radiators, perfect, 245
radioactive decay
    breakdown of nucleus as cause of, 8
    overview of, 167–168
    process of, 165–166
radioactive isotopes
    basic principles, 165–166
    carbon-14 dating, 171–174
    case studies, 172, 179–180
    defined, 165
    environmental studies and, 169
    low- and high-level wastes, 359–360
    measuring radioactivity, 169
    mobility and sequestering in surface
        environment, 362–363
    natural and anthropogenic sources,
        358–359
    nuclear fission and, 360–362
    nuclear fuel cycle and, 359–360
    Oklo fossils fission reactors, 363–364
    radioactive decay and growth, 167–169
    radiogenic isotopic tracers, 178–181
    tritium dating, 169–172
    U-series dating methods, 174–181
radioactive parents, 167–168
radioactive wastes, 358–363
    low and high levels, 359–360
    mobility and sequestering radioactive
        isotopes, 362–363
    nuclear fission and, 360–362
    nuclear fuel cycle and, 359–360
    Oklo fossils fission reactors, 363–364
    overview of, 358–359
    zeolite, using to contain, 227
radioactivity
    defined, 165
    measuring, 169
    for scavenging rate determination, 416

radiocarbon dating, 299
radiogenic isotopic tracers
    landfill gas and leachate identification,
        192–193
    Rb–Sr isotopic system, 178, 180–181
    Th–U–Pb isotopic system, 181
radiogenic progeny, 167–168
radiometric dating, 434
radius ratio, 209
radon, 275–278
    case study, 277–278
    concentrations of, 276
    index, 276
    as indoor air pollutant, 275
    radioactive nature of, 275–276
rainout, wet deposition, 290
rainwater chemistry, 278–281
    acidity of, 66
    composition and excess ions, 280
    delta value for, 184
    gaseous species in, 281–283
    ionic species in, 278–279
    isotopic composition of, 186
    marine influence on, 281
    nitrates in, 193–194
    pH and acid rain, 283–286
    primary associations, 278
    tritium concentration in, 170–172
rate-controlling step, 47
Rayleigh distillation equation, 185–186
Rb–Sr isotopic system, 178, 180–181
RBE (relative biological effectiveness), 169
RDC (reductive capacity), 118–119
reactions
    Arrhenius equation, 45
    dissolution/growth and, 46–47
    nucleation, 45–46
    order of, 42–45
    overview of, 42
Redfield ratio
    defined, 395–396
    for open-ocean plankton, 413
    suspended particulate matter and,
        430–431
redox buffering, 117
redox front. *see* redox interface
redox indicators, 434
redox interface, 121–124
redox reactions. *see also* oxidation–
        reduction reactions
    defined, 94
    pore water chemistry and, 432
    water column chemistry and, 430, 431
redoxcline, 430
reducing agents, 94, 114
reductases, 111
reduction, 94
reductive capacity (RDC), 118–119
reference electrode, 99
relative biological effectiveness (RBE), 169
relative humidity, 249

relative instantaneous radiative forcing
        (RIRF), 256
relative residence time, lake substances,
        338–339
rem (roentgen equivalent man), 169
remineralization, 403–404, 408
residence time
    defined, 415
    of lake substances, 338
    of seawater species, 393–394
respiratory diseases
    asbestos and, 229
    silica exposure and, 234, 235
RIRF (relative instantaneous radiative
        forcing), 256
rivers
    challenges to Gibbs model, 329
    chemical composition of, 393, 394
    composition, 324
    detrital components from, 417
    Gibbs model, 328–329
    ionic elements in, 330–331
    marginal marine environment and,
        423
    seawater composition and, 397
    Stallard and Edmond model, 329–330
    as trace metal source, 413
rocks
    detrital components from, 417
    isotopic composition of water and, 187
roentgen equivalent man (rem), 169

salinity
    constancy of, 392–393
    of marginal marine environment, 423
    in semi-isolated basins, 425
    thermohaline circulation and, 390–392
    of water, 19
    water column chemistry, 428
salt-wedge estuaries, 424
salting out effect, 429–430
salts. *see also* buffers
    of strong bases, 75
    of weak acids, 71–74
saltwater, 424, 428
San Francisco Bay, CA, 435
sapropelic coals, 146
SAR (sediment accumulation rate), 434
saturated compounds, 129
saturated zone, 331
saturation horizon, 420–421
saturation indices (SI), 49
scavenging
    rates, determining, 415–416
    of trace metals, 414
scavenging turnover time, 416
Schrödinger wave model, 4
sea salt, particulate types, 288
seafloor sediments, 421–422. *see also*
        marine sediments, geochemistry of
        seawater chemistry, 400–416

seafloor sediments (*cont.*)
    alkalinity, dissolved inorganic carbon,
      401–403
    buffers, buffering capacity, 406–407
    chemical composition of open ocean,
      393–400
    ionic strength of, 400–401
    ligands, inorganic and organic, 407–410
    metal distributions, 414–415
    scavenging rates, determination of,
      415–416
    seawater pH, 403–404
    speciation, stoichiometric equilibrium
      constants, 404–406
    trace metals in marine colloids, 410–412
    trace metals in seawater, 412–414
seawater pH scale, 403, 406
seawater species
    chemical cycle for, 394–395
    constancy of composition, salinity,
      392–393
    residence time of, 393–394
second-order reactions, 43
secondary minerals, 316
secondary pollutants, 251
secular equilibrium
    carbon-14 dating and, 172
    radioactive decay and growth, 168
sediment. *see also* marine sediments, geo-
    chemistry of
    contamination by heavy metals, 358
    of marginal marine environment,
      434–437
    seawater interaction with, 398
    traps, 415
sediment accumulation rate (SAR), 434
sediment record, as indicator of climatic
    changes, 263, 266, 268
sedimentary organic matter, recording
    environmental changes, 145
sedimentation rate, 176–177
selective extraction, 355–356
selenium
    in deep-sea sediments, 123
    as human health hazard, 356–357
semi-isolated basins. *see* basins, semi-
    isolated
septic systems, 121
Setchenow equation, 39–40
sewage treatment, 112–113
SHE (standard hydrogen electrode), 96
shelf break, 387
short wavelength radiation, 245
SI (saturation indices), 49
SI (Système international d'unites), 10
sievert (Sy), 169
silica phases, 233
silicate minerals, 214–215
    abundance of, 208
    authigenesis and, 421

dissolution of, 233
    groundwaters and, 332
    properties of, 215
    as siliceous sediment, 419
    types of, 214–215
    water–silicate interactions in surface and
      ground water, 340–341
    weathering and, 316–318
siliceous sediments, 419
silicic acid, 69–70, 86–87
silicosis, 235
silicotuberculois, 235
sill depth, 424–425
single bonds, carbon atoms, 129
SLAP (standard light Antarctic precipita-
    tion), 184
smectites, clays, 217–218
smogs, 253
SMOW (standard mean ocean water), 184
soil
    acid soil, 355
    classification, 144
    composition, 144
    contamination by heavy metals, 358
    formation, 143–144
solar and terrestrial radiation
    earth's albedo, 247
    insolation, 247
    solar constant, 246
solar constant, 246
solar radiation, 245–247
solid state, of water, 19
solutions. *see also* partition/sorption char-
    acteristics
    balance of positive/negative charges in,
      65–66
    buffered, 80
    molarity and, 11
    precipitates and, 12
    solubility of metal oxalates, 350
    solubility of oxyhydroxides, 343–344
sorosilicates, 214
sorption characteristics. *see* partition/sorp-
    tion characteristics
sounding curve, 250
sparging remediation technique, 158
speciation
    models, 47
    stoichiometric equilibrium constants and,
      404–406
    zinc, 411
species. *see also* seawater species
    calculating activity coefficients for un-
      charged, 39–40
    in chemical reaction, 12
    convention of enclosing in brackets, 32
    distribution by adsorption–desorption,
      341
    gaseous species in rainwater, 281–283
    ionic species in rainwater, 278–279

isotherms and partitioning of, 344–345
    nonconservative and conservative, 76
spectator ions, 12
spectra, 7–8
spectrometry, 9
SPM (suspended particulate matter)
    composition of, 414
    water column chemistry, 430–432
stability diagrams
    defined, 320
    thermodynamic data for, 321
    water–rock interactions, 322–323
stable isotopes, 181–198
    carbon, 189–193
    case studies, 187, 192–195, 198
    defined, 165
    delta notation, 184–185
    fractionation, 182–183
    isotopic fractionation factor, 183
    mixing, 196–197
    nitrogen, 193–194
    overview of, 181–182
    oxygen and hydrogen isotopes in water,
      185–189
    paleothermometry, 197–198
    sulfur, 194–196
stable state, 29
Stallard and Edmond model, river chem-
    istry, 329–330
standard heat of formation, 28
standard hydrogen electrode (SHE), 96
standard light Antarctic precipitation
    (SLAP), 184
standard mean ocean water (SMOW), 184
standard state, 28
standard temperature and pressure (STP),
    of gases, 17
standards, stable isotope ratios for, 184–185
states, system, 29
steady-state assumptions, 20
Stiff diagrams, 325–326
stoichiometric equilibrium constants
    speciation and, 404–406
    water column chemistry, 428
stoichiometric ionic strength, 400–401
STP (standard temperature and pressure),
    of gases, 17
stratosphere, tritium concentration in, 170
stratospheric ozone, 268–271
    Antarctic ozone hole, 273
    kinetic data for ozone destruction,
      272–273
    ozone-destroying reactions, 271–272
stream-cut estuaries, 424
strong-acid acidity, 78
strong acids
    acid–base reactions and, 12
    dissociation constant and, 59
    natural water and, 62–63
    titrations for alkalinity, 77–78

strong bases
    dissociation constant and, 60–61
    salts of, 75
    titrations for acidity, 77–78
strong force, of nucleus, 8
structural formulas, organic molecules, 129–130
suboxic environments, 118
substitution reactions, 130
subsurface brines, oil fields, 348
subtropical inversions, 250
sulfate reduction
    case study, 115–116
    overview of, 114
    redox buffering and, 117
    seawater–sediment interactions and, 398
sulfates, 137, 288–289
sulfide oxidation, 113–114, 123
sulfides
    acid mine drainage and, 353–354
    Eh–pH diagram and, 107–108
    organosulfur compounds, 137
sulfidic environments, 118
sulfones, 137
sulfoxides, 137
sulfur
    adsorption–desorption processes, 371–372
    anthropogenic vs. natural sources, 196
    case study identifying sources of, 195
    in coals, 147
    isotopes and, 194–196
    organosulfur compounds, 137
    pore water chemistry and, 433–434
sulfur dioxide, 282
sulfuric acid (H_2SO_4), 62
surface and ground waters, 324–341
    case study, 327
    graphical representation of water chemistry, 325–326
    groundwaters, 331–334
    hydrofacies, 326–328
    inventory of reservoirs, 313
    lakes, 334–339
    pH in, 404
    pH of, 339–341
    rivers, 328–331
    variations in chemistry of, 324
surface currents, 388–390
surface ocean, 388
surface poisons, 47
surface–water enrichments, 415
suspended particulate matter (SPM)
    composition of, 414
    water column chemistry, 430–432
suspension of minerals, 324
swamps, sulfide oxidation and, 113–114
Sy (sievert), 169
systematic names, 129

Système international d'unites (SI), 10
systems, 27

tectonic estuaries, 424
tektosilicates, 215
temperature
    Arrhenius equation relating to, 45
    compared with heat, 243
    fractionation factor and, 183
    isotopic measurements for climate changes, 188–189
    Le Châtelier's principle, 35
    thermohaline circulation and, 390–392
temperature and water vapor, 248–251
    absolute and relative humidity, 249
    adiabatic lapse rates, 249–250
    environmental lapse rate, 250
    inversions, 250
    sounding curve, 250
temporal variations, in particulates (aerosols), 300
terrestrial radiation, 245–247
terrestrial uptake, of CO_2, 259
terrigenous sediments, 418
tetrahedra structure, of silicates, 214
tetrahedral clays, 215–216
Th (thorium) decay chain
    dating of calcium carbonate, 177
    dating of marine sediments, 174–176
    overview of, 166
Th–U–Pb isotopic system, 181
thermal decomposition, 192
thermal energy, 388
thermocline (mesolimnion) layer, lakes, 335
thermodynamics, 27–29, 32. see also equilibrium thermodynamics
thermohaline circulation, 390–392
thermophiles, 111
thiols, 137
thiourea compounds, 137
thorium. see Th (thorium) decay chain
TOC (total carbon content), 148
total acidity, 76, 78
total alkalinity, 76, 79
total carbon content (TOC), 148
total pH scale, 403, 406
trace metals
    in marine colloids, 410–412
    metal distributions, types of, 414–415
    in seawater, 412–414
tracers, in food chains, 191–192
transition metals (Zn, Cd, and Pb), 354–355
triple bonds, carbon atoms, 129
tritium dating, 169–172
tritium–helium-3 dating method, 170–171, 172
tropopause, 170
troposphere, 243

tropospheric ozone, 273–275
Truesdell–Jones model, 37–39
tundra, soils, 144

U-234/U-238 isotopes, 180
U. S. Environmental Protection Agency, 47–48
U. S. Geological Survey (USGS), 47–48
U-series disequilibrium dating, 174–178
    activity and sedimentation-rate relationships, 176–177
    Pb dating, 177–178
    Th dating of calcium carbonate, 177
    Th dating of marine sediments, 174–176
unsaturated compounds, 129
unstable state, 29
upwelling
    deep-sea sediments and, 419
    described, 389–390
    seawater composition and, 397
uranium decay chain, 165–166, 180, 416.
    see also U-series disequilibrium dating
uranium ore, 359
USGS (U. S. Geological Survey), 47–48
UV radiation, 273

V-SMOW (Vienna standard mean ocean water), 184
vadose zone, 331
valence, 6
valence shell electron-pair repulsion (VSEPR) model, 18
van der Waals equation, 17
Vanselow convention, ionic concentration, 346
vapor state, of water, 19
vermiculites, clays, 218
vertical circulation, 390–392
Vienna standard mean ocean water (V-SMOW), 184
volatile organic compounds (VOCs)
    anthropogenic vs. natural sources, 302
    molecular signatures, 294
volatility, of organic compounds, 158
Vostok ice core, 262
VSEPR (valence shell electron-pair repulsion) model, 18

washout, wet deposition, 290
waste. see also radioactive wastes
    acid mine wastes, 89
    animal and human, 193
    radioactive, 358–363
water. see also natural waters; surface and ground waters
    acidity and alkalinity, 75
    activity of, 40
    carbonate interactions, 340

water (*cont.*)
circulation in semi-isolated basins, 424–426
classification of types (hydrofacies), 326–328
condensation process, 42
electrode, 101–103
heat capacity of, 19
input of organic chemicals, 156–159
mineral interactions. *see* mineral–water reactions
oxidation–reduction, 159
oxygen and hydrogen isotopes in, 185–189
precipitation, 158
properties of, 19
rock interactions, 322–323
silicate mineral interactions, 86–87, 340–341
sorption/partition characteristics, 156–158
stability diagrams, 101–103
states of, 19–20
structure of, 18–20
thermohaline circulation and, 390–392
volatilization, 158–159
water chemistry
balance of positive/negative charges, 65–66

case studies, 48–50
computerized, 47–48
graphical representation of, 325–326
water column
chemistry, 428–432
release of metals to, 432
water vapor, 248–251
absolute and relative humidity, 249
adiabatic lapse rates, 249–250
aerosols attracting, 289
environmental lapse rate, 250
weak acids, 63–71
acetic, 63
acid–base reactions and, 12
buffering index for, 83–84
carbonic, 63–69
carbonic acid–carbonate system, 70–71
dissociation constant and, 59
salts of, 71–74
silicic, 69–70
weak bases, 60
weathering, 314–323
case study, 323
defined, 314
dissolution and precipitation of aluminum and iron hydroxides, 318–320
dissolution and precipitation of silica, 316–318

overview of, 314–317
primary and secondary minerals and, 316
silicates and, 316–317
stability diagrams, 320–323
table of decomposition reactions, 315
types of, 314–315
wet adiabatic lapse rates, 250
wet deposition
removal of particulates, 289
washout and rainout, 290
wind, 388–390

X-ray crystallography, 212–214

zeolite minerals, 225–227
containing radioactive wastes, 227
defined, 225
environmental applications, 226–227
properties of, 225
structure of, 225–226
zeroth-order reactions, 43
zeta potential, asbestos fibers, 230
zinc (Zn)
removal by oxyhydroxide colloids, 344
speciation, 411
as transition metal, 354–355

Credits

This page constitutes an extension of the copyright page. We have made every effort to trace the ownership of all copyrighted material and to secure permission from copyright holders. In the event of any question arising as to the use of any material, we will be pleased to make the necessary corrections in future printings. Thanks are due to the following authors, publishers, and agents for permission to use the material indicated.

Chapter 1

Figure 1–1. From GEOCHEMISTRY, 2/e by Arthur Brownlow. Copyright © 1996. Reprinted by permission of Pearson Education, Inc., Upper Saddle River, NJ.

Figure 1–2. From GEOCHEMISTRY, 2/e by Arthur Brownlow. Copyright © 1996. Reprinted by permission of Pearson Education, Inc., Upper Saddle River, NJ.

Figure 1–8. From OCEANOGRAPHY: A VIEW OF THE EARTH, 7/e by M. G. Gross and E. Gross. Copyright © 1996. Reprinted by permission of Pearson Education, Inc., Upper Saddle River, NJ.

Figure 1–9. From THE EARTH AND ITS OCEANS by Duxbury. Copyright © 1971. Upper Saddle River, NJ: Prentice Hall, Inc.

Figure 1–10. From THE EARTH AND ITS OCEANS by Duxbury. Copyright © 1971. Upper Saddle River, NJ: Prentice Hall, Inc.

Figure 1–11. From GEOCHEMISTRY OF NATURAL WATERS, 3/e by James I. Drever. Copyright © 1997. Reprinted by permission of Pearson Education, Inc., Upper Saddle River, NJ.

Figure 1–12. From CHEMICAL CYCLES AND THE GLOBAL ENVIRONMENT: Assessing Human Influences by R. M. Garrels, F. T. Mackenzie and C. Hunt. Los Altos: William Kaufmann, Inc., 1975.

Figure 1–14a. From "A new look at the long-term carbon cycle" by R. A. Berner in GSA TODAY, 9, 11, 1999, pp. 1–6. Copyright © 1999 by Geological Society of America. Reproduced with permission of Geological Society of America via Copyright Clearance Center.

Figure 1–14b. From "A new look at the long-term carbon cycle" by R. A. Berner in GSA TODAY, 9, 11, 1999, pp. 1–6. Copyright © 1999 by Geological Society of America. Reproduced with permission of Geological Society of America via Copyright Clearance Center.

Figure 1–15. From "A new look at the long-term carbon cycle" by R. A. Berner in GSA TODAY, 9, 11, 1999, pp. 1–6. Copyright © 1999 by Geological Society of America. Reproduced with permission of Geological Society of America via Copyright Clearance Center.

Chapter 2

Figure 2–3. Figure from GEOCHEMISTRY, GROUNDWATER AND POLLUTION by C. A. J. Appelo and D. Postma, p. 536. Copyright © 1996. Reprinted by permission of A. A. Balkema Publishers, The Netherlands.

Figure 2–5. From AQUEOUS ENVIRONMENTAL CHEMISTRY by Donald Langmuir. Copyright © 1997. Reprinted by permission of Pearson Education, Inc., Upper Saddle River, NJ.

Figure 2–6. From GEOCHEMISTRY OF NATURAL WATERS, 3/e by James I. Drever. Copyright © 1997. Reprinted by permission of Pearson Education, Inc., Upper Saddle River, NJ.

Figure 2–7. From GEOCHEMISTRY OF NATURAL WATERS, 3/e by James I. Drever. Copyright © 1997. Reprinted by permission of Pearson Education, Inc., Upper Saddle River, NJ.

Chapter 3

Figure 3–1. From AQUEOUS ENVIRONMENTAL GEOCHEMISTRY by Donald Langmuir. Copyright © 1997. Reprinted by permission of Pearson Education, Inc., Upper Saddle River, NJ.

Figure 3–C1–1. From AQUEOUS ENVIRONMENTAL GEOCHEMISTRY by Donald Langmuir. Copyright © 1997. Reprinted by permission of Pearson Education, Inc., Upper Saddle River, NJ.

Figure 3–C2–1. From "Groundwater chemistry within a plateau neighboring Matsumoto City, Japan" by H. Li and S. Misawa in ENVIRONMENTAL GEOLOGY, 1994, #24, pp. 166–175. Reprinted by permission of Springer-Verlag GmbH & Co.

Figure 3–C3–1. From "A process model of natural attenuation in drainage from a historic mining district" by A. C. Berger, C. M. Bethke and J. L. Krumhansl in APPLIED GEOCHEMISTRY, #15, pp. 655–666. Copyright 2000, with permission from Elsevier Science.

Figure 3–C4–1. From "Aquifer disposal of acid gases: modelling of water–rock reactions for trapping of acid wastes" by W. D. Gunter, E. H. Perkins and I. Hutcheon in APPLIED GEOCHEMISTRY, #15, pp. 1085–1095. Copyright 2000, with permission from Elsevier Science.

Chapter 4

Figure 4–1. From PRINCIPLES AND APPLICATIONS OF GEOCHEMISTRY by G. Faure. Copyright © 1998. Reprinted by permission of Pearson Education, Inc., Upper Saddle River, NJ.

Figure 4–2. From SOLUTIONS, MINERALS, AND EQUILIBRIA by Garrels and Christ. Copyright © 1965. Upper Saddle River, NJ: Prentice Hall, Inc.

Figure 4–8. From SOLUTIONS, MINERALS, AND EQUILIBRIA by Garrels and Christ. Copyright © 1965. Upper Saddle River, NJ: Prentice Hall, Inc.

Figure 4–C2–1. From "Bacterial dissimilatory reduction of arsenate and sulfate in meromictic Mono Lake, California" by R. S. Oremland et al. in GEOCHIMICA ET COSMOCHIMICA ACTA, #64, pp. 3073–3084. Copyright 2000, with permission from Elsevier Science.

Figure 7–16. From "Illustrations of the occurrence and diversity of mineral-microbe interactions involved in weathering minerals" by J. Berthelin, C. Leyval and C. Mustin in ENVIRONMENTAL MINERALOGY edited by J. D. Cotter-Howells, L. S. Campbell, E. Valsami-Jones and M. Batchelder, pp. 7–25. This figure is reproduced with permission of the Mineralogical Society of Great Britain and Ireland.

Chapter 8

Figure 8–1. From UNDERSTANDING OUR ATMOSPHERIC ENVIRONMENT, 2/e by M. Neiburger, J. G. Edinger and W. D. Bonner, p. 453. Copyright © 1982. NY: Wm. Freeman.

Figure 8–3. From AN INTRODUCTION TO ATMOSPHERIC PHYSICS by R. G. Fleagle and J. A. Businger, p. 346. Copyright © 1963. Reprinted by permission of Elsevier Science.

Figure 8–5. From ELEMENTS OF METEOROLOGY by Miller/Thompson/Peterson/Harigan. Copyright © 1975. Reprinted by permission of Pearson Education, Inc., Upper Saddle River, NJ.

Figure 8–7. From GLOBAL ENVIRONMENT: WATER, AIR, and GEOCHEMICAL CYCLES by Berner/Berner. Copyright © 1996. Reprinted by permission of Pearson Education, Inc., Upper Saddle River, NJ.

Figure 8–8. From GLOBAL ENVIRONMENT: WATER, AIR, and GEOCHEMICAL CYCLES by Berner/Berner. Copyright © 1996. Reprinted by permission of Pearson Education, Inc., Upper Saddle River, NJ.

Figure 8–9. From GLOBAL ENVIRONMENT: WATER, AIR, and GEOCHEMICAL CYCLES by Berner/Berner. Copyright © 1996. Reprinted by permission of Pearson Education, Inc., Upper Saddle River, NJ.

Figure 8–11. From "The ice core record" in ANTARCTICA AND ENVIRONMENTAL CHANGE by C. Lorius, J. Jouzel and D. Raynaud, pp. 27–32. Copyright © 1993. Oxford, UK: Oxford University Press.

Figure 8–C6–1. From "Cooler winters as a possible cause of mass extinctions at the Eocene/Oligocene boundary" by L. C. Ivany, W. P. Patterson and K. C. Kohmann in NATURE, 2000, No. 407, pp. 887–890. Reprinted by permission of Macmillan Magazines Limited.

Figure 8–C6–2. From "Cooler winters as a possible cause of mass extinctions at the Eocene/Oligocene boundary" by L. C. Ivany, W. P. Patterson and K. C. Kohmann in NATURE, 2000, No. 407, pp. 887–890. Reprinted by permission of Macmillan Magazines Limited.

Figure 8–C7–1. From "Boron isotopic composition and concentration in modern marine carbonates" by N. G. Hemming and G. N. Hanson in GEOCHIMICA ET COSMOCHIMICA ACTA, #56, pp. 537–543. Copyright 1992, with permission of Elsevier Science.

Figure 8–13. From ENVIRONMENTAL CHEMISTRY by G. W. van Loon and S. J. Duffy, p. 492. Copyright © 2000. Oxford, UK: Oxford University Press.

Figure 8–15. From "Progress towards a quantitative understanding of Antarctic ozone depletion" by S. Solomon in NATURE, 1990, #347, pp. 347–354. Reprinted by permission of Macmillan Magazines Ltd.

Figure 8–16. From ENVIRONMENTAL CHEMISTRY by G. W. van Loon and S. J. Duffy, p. 492. Copyright © 2000. Oxford, UK: Oxford University Press.

Figure 8–17. From GLOBAL ENVIRONMENT: WATER, AIR, and GEOCHEMICAL CYCLES by Berner/Berner. Copyright © 1996. Reprinted by permission of Pearson Education, Inc., Upper Saddle River, NJ.

Figure 8–18. From GLOBAL ENVIRONMENT: WATER, AIR, and GEOCHEMICAL CYCLES by Berner/Berner. Copyright © 1996. Reprinted by permission of Pearson Education, Inc., Upper Saddle River, NJ.

Figure 8–19. From GLOBAL ENVIRONMENT: WATER, AIR, and GEOCHEMICAL CYCLES by Berner/Berner. Copyright © 1996. Reprinted by permission of Pearson Education, Inc., Upper Saddle River, NJ.

Figure 8–20. From GLOBAL ENVIRONMENT: WATER, AIR, and GEOCHEMICAL CYCLES by Berner/Berner. Copyright © 1996. Reprinted by permission of Pearson Education, Inc., Upper Saddle River, NJ.

Figure 8–21. From AQUEOUS ENVIRONMENTAL GEOCHEMISTRY by Donald Langmuir. Copyright © 1997. Reprinted by permission of Pearson Education, Inc., Upper Saddle River, NJ.

Figure 8–22. From AQUEOUS ENVIRONMENTAL GEOCHEMISTRY by Donald Langmuir. Copyright © 1997. Reprinted by permission of Pearson Education, Inc., Upper Saddle River, NJ.

Figure 8–23. From AQUEOUS ENVIRONMENTAL GEOCHEMISTRY by Donald Langmuir. Copyright © 1997. Reprinted by permission of Pearson Education, Inc., Upper Saddle River, NJ.

Figure 8–C9–1. From "The record of anthropogenic pollution in snow and ice in Svalbard, Norway" by J. Simões and V. S. Zagorodnov in ATMOSPHERIC ENVIRONMENT, #35, pp. 403–413. Copyright 2001, with permission of Elsevier Science.

Figure 8–25. From "A graphical technique for determining major components in a mixed aerosol" by K. A. Rahn in ATMOSPHERIC ENVIRONMENT, #33, pp. 1441–1455. Copyright 1999, with permission of Elsevier Science.

Figure 8–C13–1. From "The origin of atmospheric lead in Oslo, Norway" by G. Åberg et al. in ATMOSPHERIC ENVIRONMENT, #33, pp. 3335–3344. Copyright 1999, with permission of Elsevier Science.

Chapter 9

Figure 9–1. From GLOBAL ENVIRONMENT: WATER, AIR, and GEOCHEMICAL CYCLES by Berner/Berner. Copyright © 1996. Reprinted by permission of Pearson Education, Inc., Upper Saddle River, NJ.

Figure 9–C2–1. From "Geochemical modeling of coal mine drainage, Summit County, Ohio" by Annabelle Foos in ENVIRONMENTAL GEOLOGY, No. 31, pp. 205–210. Copyright © 1997. Reprinted by permission of Springer-Verlag GmbH & Co.

Figure 9–C3–1. From "Comparison of chemical hydrology of Florida and Yucatan" by W. Back and B. B. Hanshaw in JOURNAL OF HYDROLOGY, 1970, #10, pp. 360–368. Oxford, UK: Elsevier Science.

Figure 9–C8–1. From "The effect of acid soil leaching on trace element abundance in a medium-sized stream" by M. Åström in APPLIED GEOCHEMISTRY, #16, pp. 387–396. Copyright 2001, with permission of Elsevier Science.

Figure 9–7. From "Hydrochemical facies and groundwater flow patterns in northern part of Atlantic coastal plain" by Back from USGS Professional Paper 498-A, 1966.

Figure 9–11. From "Genesis of some ground waters from igneous rocks" by R. K. Garrels in RESEARCHES IN GEOCHEMISTRY, Vol. 2, ed. by P. H. Abelson, pp. 405–420.

Figure 9–12. From "Genesis of some ground waters from igneous rocks" by R. K. Garrels in RESEARCHES IN GEOCHEMISTRY, Vol. 2, ed. by P. H. Abelson, pp. 405–420.

Figure 9–19. From "An overview of trace metals in the environment, from mobilization to remediation" by A. C. L. Larocque and P. E. Rasmussen in ENVIRONMENTAL GEOLOGY, 33, 1998, pp. 85–91. Reprinted with permission from Elsevier Science.

Figure 9–20. From "An overview of trace metals in the environment, from mobilization to remediation" by A. C. L. Larocque and P. E. Rasmussen in ENVIRONMENTAL GEOLOGY, 33, 1998, pp. 85–91. Reprinted with permission from Elsevier Science.

Figure 9–21. From "The biogeochemical cycling of elemental mercury: Anthropogenic influences" by R. P. Mason, W. F. Fitzgerald and F. M. M. Morel in GEOCHIMICA ET COSMOCHIMICA ACTA 58, 1994, pp. 3191–3198. Reprinted with permission from Elsevier Science.

Figure 9–22. From PRINCIPLES AND APPLICATIONS OF GEOCHEMISTRY by G. Faure. Copyright © 1998. Reprinted by permission of Pearson Education, Inc., Upper Saddle River, NJ.

Figure 9–26. From GLOBAL ENVIRONMENT: WATER, AIR, and GEOCHEMICAL CYCLES by Berner/Berner. Copyright © 1996. Reprinted by permission of Pearson Education, Inc., Upper Saddle River, NJ.

Figure 9–27. From GLOBAL ENVIRONMENT: WATER, AIR, and GEOCHEMICAL CYCLES by Berner/Berner. Copyright © 1996. Reprinted by permission of Pearson Education, Inc., Upper Saddle River, NJ.

Chapter 10

Figure 10–C5–1. From "A record of hydrocarbon input to San Francisco Bay as traced by biomarker profiles in surface sediment and sediment cores" by F. D. Hostettler et al. in MARINE GEOLOGY, #64, pp. 115–127. Copyright 1999, with permission of Elsevier Science.

Figure 10–C5–2. From "The impact of human activities on sediments of San Francisco Bay, California: an overview"

by A. Van Geen and S. N. Luoma in MARINE CHEMISTRY, #64, pp. 1–6. Copyright 1999, with permission of Elsevier Science.

Figure 10–3. From OCEANOGRAPHY by Jerome Williams. Copyright © 1962. NY: Little, Brown & Co.

Figure 10–4. From OCEANOGRAPHY by Jerome Williams. Copyright © 1962. NY: Little, Brown & Co.

Figure 10–7. From INTRODUCTION TO PHYSICAL OCEANOGRAPHY by von Arx. Copyright © 1962. Upper Saddle River, NJ: Prentice Hall, Inc.

Figure 10–9. Figure by Ian Worpole from "Hot springs on the ocean floor" article by Edmond and van Damm in SCIENTIFIC AMERICAN, 1983, #248, p. 83. Reprinted by permission of Ian Worpole.

Figure 10–12. From "A kinetic model for the chemical composition of seawater" by W. S. Broecker in QUATERNARY RESEARCH 1, 1971, pp. 188–207. Reprinted with permission from Elsevier Science.

Figure 10–13. Figure from CHEMICAL OCEANOGRAPHY by W. S. Broecker, p. 214. Copyright © 1974. Reprinted with permission from Elsevier Science.

Figure 10–19. From "Modern marine sediments" by A. C. Alpin in ENVIRONMENTAL MINERALOGY, EUROPEAN MINERALOGICAL UNION NOTES IN MINERALOGY, Vol. 2, edited by D. J. Vaughan and R A. Wogelius, pp. 125–172.

Figure 10–20. From "Modern marine sediments" by A. C. Alpin in ENVIRONMENTAL MINERALOGY, EUROPEAN MINERALOGICAL UNION NOTES IN MINERALOGY, Vol. 2, edited by D. J. Vaughan and R. A. Wogelius, pp. 125–172.

Figure 10–21. From "Methane hydrate: a major reservoir of organic carbon in the shallow geosphere?" by K. A. Kvenvolden in CHEMICAL GEOLOGY, #71, pp. 41–51. Copyright 1988, with permission of Elsevier Science.

Figure 10–25. From "A geochemical record of eutrophication and anoxia in Chesapeake Bay sediments" by A. R. Zimmerman and E. A. Canuel in MARINE CHEMISTRY, #69, pp. 117–137. Copyright 2000, with permission of Elsevier Science.

Figure 10–26. From "Aerobic biodegradation of alkylated aromatic hydrocarbons by a bacterial community" by H. Budzinski, N. Raymond, T. Nadalig, M. Gilewicz, P. Garrigues, J. C. Bertrand and P. Caumette in ORGANIC GEOCHEMISTRY 28, 1998, pp. 337–348. Reprinted by permission of Elsevier Science.

Figure 10–P95. From "Antimony cycling in the western Mediterranean" by K. Takayanagi, D. Cossa and J. M. Martin in MARINE CHEMISTRY, 54, 1996, pp. 303–312. Reprinted with permission from Elsevier Science.

SI Base Units

Quantity	Name	Symbol
Length	meter	m
Mass	kilogram	kg
Time	second	s
Electric current	ampere	A
Thermodynamic temperature	kelvin	K
Amount of substance	mole	mol

SI Derived Units

Quantity	Name	Symbol	Equivalent
Frequency	hertz	Hz	s^{-1}
Force	newton	N	$m\,kg\,s^{-2}$
Pressure	pascal	Pa	$N\,m^{-2}$
Energy, work, heat	joule	J	$N\,m$
Power, radiant flux	watt	W	$J\,s^{-1}$
Electric charge	coulomb	C	$s\,A$
Electric potential, emf	volt	V	$J\,C^{-1}$
Conductance	siemens	S	$A\,V^{-1}$
Radioactivity	becquerel	Bq	s^{-1}
Adsorbed dose	gray	Gy	$J\,kg^{-1}$
Dynamic viscosity	pascal second	Pa s	$kg\,m^{-1}\,s^{-1}$
Surface tension	newton per meter	$N\,m^{-1}$	$kg\,s^{-2}$
Heat flux density, irradiance	watt per square meter	$W\,m^{-2}$	$kg\,s^{-3}$
Molar entropy, molar heat capacity	joule per mole kelvin	$J\,mol^{-1}\,K^{-1}$	$m^2\,kg\,s^{-2}\,K^{-1}\,mol^{-1}$

SI Prefixes

Prefix	Symbol	Factor	Prefix	Symbol	Factor
peta	P	10^{15}	deci	d	10^{-1}
tera	T	10^{12}	centi	c	10^{-2}
giga	G	10^{9}	milli	m	10^{-3}
mega	M	10^{6}	micro	μ	10^{-6}
kilo	k	10^{3}	nano	n	10^{-9}
hecto	h	10^{2}	pico	p	10^{-12}
deca	da	10^{1}	femto	f	10^{-15}